Gmelin Handbook of Inorganic Chemistry

8th Edition

Gmelin Handbook of Inorganic Chemistry

8th Edition

Gmelin Handbuch der Anorganischen Chemie

Achte, völlig neu bearbeitete Auflage

Prepared and issued by

Gmelin-Institut für Anorganische Chemie
der Max-Planck-Gesellschaft
zur Förderung der Wissenschaften

Director: Ekkehard Fluck

Founded by — Leopold Gmelin

8th Edition — 8th Edition begun under the auspices of the Deutsche Chemische Gesellschaft by R. J. Meyer

Continued by — E. H. E. Pietsch and A. Kotowski, and by Margot Becke-Goehring

Springer-Verlag Berlin Heidelberg GmbH 1988

STAFF GMELIN HANDBOOK

Dr. G. Bär, D. Barthel, Dr. N. Baumann, Dr. W. Behrendt, D. Benzaid, Dr. L. Berg, Dipl.-Chem. E. Best, K. D. Bonn, P. Born-Heck, Dipl.-Ing. V. A. Chavizon, E. Cloos, Dipl.-Phys. G. Czack, I. Deim, Dipl.-Chem. H. Demmer, Dipl.-Geol. R. Ditz, R. Dowideit, Dipl.-Chem. M. Drößmar, Dr. D. Düster, U. Duggen, Dr. P. Eigen, M. Engels, Dr. H.-J. Fachmann, Dr. J. Faust, V. Frick, G. Funk, Dipl.-Ing. N. Gagel, Dr. U. W. Gerwarth, Dipl.-Phys. D. Gras, Dipl.-Bibl. W. Grieser, Dr. I. Haas, Dr. R. Haubold, B. Heibel, Dipl.-Min. H. Hein, G. Heinrich-Sterzel, A. Heise-Schuster, H.-P. Hente, H. W. Herold, U. Hettwer, Dr. G. Hönes, Dr. W. Hoffmann, Dr. W. Huisl, R. Jonuschat, Dr. R. Jotter, Dr. J. von Jouanne, Dr. B. Kalbskopf, Dipl.-Chem. W. Karl, H.-G. Karrenberg, Dipl.-Phys. H. Keller-Rudek, B. Kirchner, A. Klusch, C. Koeppel, Dipl.-Chem. H. Köttelwesch, R. Kolb, E. Kranz, Dipl.-Chem. I. Kreuzbichler, Dr. A. Kubny, Dr. M. Kunz, Dr. W. Kurtz, M. Langer, Dr. U. Lanzendörfer, Dr. B. Ledüc, Dr. A. Leonard, Dipl.-Chem. H. List, H. Mathis, E. Meinhard, C. Metz, K. Meyer, Dr. M. Mirbach, Dipl.-Chem. B. Mohsin, Dr. U. Neu-Becker, V. Neumann, K. Nöring, Dipl.-Chem. R. Nohl, Dipl.-Min. U. Nohl, Dr. U. Ohms-Bredemann, Dr. W. Petz, I. Rangnow, Dipl.-Phys. H.-J. Richter-Ditten, Dipl.-Chem. H. Rieger, E. Rieth, A. Rosenberger, E. Rudolph, G. Rudolph, Dipl.-Chem. S. Ruprecht, Dr. B. Sarbas, Dr. D. Schiöberg, V. Schlicht, B. Schmitt, Dipl.-Chem. D. Schneider, Dipl.-Min. P. Schubert, Dr. K. Schücke, A. Schwärzel, Dr. B. Schwager, Dipl.-Ing. H. M. Somer, Dr. C. Strametz, Dr. G. Swoboda, Dr. U. Taubald, M. Teichmann, Dr. D. Tille, Dipl.-Ing. H. Vanecek, Dipl.-Chem. P. Velić, Dipl.-Ing. U. Vetter, H.-M. Wagner, Dipl.-Phys. J. Wagner, R. Wagner, Dr. E. Warkentin, Dr. C. Weber, Dr. A. Wietelmann, Dr. M. Winter, Dr. B. Wöbke, K. Wolff

STAFF GMELIN ONLINE DATA SYSTEM

Dr. R. Froböse, Dr. P. Kuhn, Dr. G. Olbrich

Organometallic Compounds in the Gmelin Handbook

The following listing indicates in which volumes these compounds are discussed or are referred to:

Ag	Silber B5 (1975)
Au	Organogold Compounds (1980)
Be	Organoberyllium Compounds 1 (1987)
Bi	Bismut-Organische Verbindungen (1977)
Co	Kobalt-Organische Verbindungen 1 (1973), 2 (1973), Kobalt Erg.-Bd. A (1961), B1 (1963), B2 (1964)
Cr	Chrom-Organische Verbindungen (1971)
Cu	Organocopper Compounds 1 (1985), 2 (1983), 3 (1986), 4 (1987), Index (1987)
Fe	Eisen-Organische Verbindungen A1 (1974), A2 (1977), A3 (1978), A4 (1980), A5 (1981), A6 (1977), A7 (1980), A8 (1985), B1 (partly in English; 1976), Organoiron Compounds B2 (1978), Eisen-Organische Verbindungen B3 (partly in English; 1979), B4 (1978), B5 (1978), Organoiron Compounds B6 (1981), B7 (1981), B8 to B10 (1985), B11 (1983), B12 (1984), Eisen-Organische Verbindungen C1 (1979), C2 (1979), Organoiron Compounds C3 (1980), C4 (1981), C5 (1981), C7 (1985), and Eisen B (1929–1932)
Ga	Organogallium Compounds 1 (1986)
Hf	Organohafnium Compounds (1973)
Nb	Niob B4 (1973)
Ni	Nickel-Organische Verbindungen 1 (1975), 2 (1974), Register (1975), Nickel B3 (1966), and C1 (1968), C2 (1969)
Np, Pu	Transurane C (partly in English; 1972)
Pb	Organolead Compounds 1 (1987)
Pt	Platin C (1939) and D (1957)
Ru	Ruthenium Erg.-Bd. (1970)
Sb	Organoantimony Compounds 1 (1981), 2 (1981), 3 (1982), 4 (1986)
Sc, Y, La to Lu	D6 (1983)
Sn	Zinn-Organische Verbindungen 1 (1975), 2 (1975), 3 (1976), 4 (1976), 5 (1978), 6 (1979), Organotin Compounds 7 (1980), 8 (1981), 9 (1982), 10 (1983), 11 (1984), 12 (1985), 13 (1986), 14 (1987), 15 (1988) **present volume**
Ta	Tantal B2 (1971)
Ti	Titan-Organische Verbindungen 1 (1977), 2 (1980), Organotitanium Compounds 3 (1984), 4 and Register (1984)
U	Uranium Suppl. Vol. E2 (1980)
V	Vanadium-Organische Verbindungen (1971), Vanadium B (1967)
Zr	Organozirconium Compounds (1973)

Gmelin Handbook of Inorganic Chemistry

8th Edition

Sn

Organotin Compounds

Part 15

Dibutyltin-Oxygen Compounds

With 5 illustrations

by **Herbert Schumann** and **Ingeborg Schumann**

AUTHORS Herbert Schumann, Ingeborg Schumann
Technische Universität Berlin

FORMULA INDEX Edgar Rudolph, Gmelin-Institut, Frankfurt am Main

EDITOR Ulrich Krüerke, Gmelin-Institut, Frankfurt am Main

Springer-Verlag Berlin Heidelberg GmbH 1988

LITERATURE CLOSING DATE: 1985

Library of Congress Catalog Card Number: Agr 25-1383

ISBN 978-3-662-06614-0 ISBN 978-3-662-06612-6 (eBook)
DOI 10.1007/978-3-662-06612-6

Originally published by Springer-Verlag, Berlin · Heidelberg · New York · Tokyo in 1988
Softcover reprint of the hardcover 8th edition 1988

Typesetting

Preface

The present volume continues the series on "Organotin Compounds" which first appeared in 1975 and now comprises a collection of fifteen volumes. The overall plan of the series has been given in the preface of Volume 1. The present Volume 15 continues the description of the mononuclear organotin compounds with tin-oxygen bonds (Chapter 1.4.1). Treatment of organotin compounds with tin-sulfur, -selenium, and -tellurium bonds appeared earlier in Chapters 1.4.2 to 1.4.4 of Volumes 9, 1982, and 10, 1983.

Volume 15 contains R_2Sn-oxygen compounds with R = butyl, iso-butyl, sec-butyl, and tert-butyl, covering the literature completely to the end of 1985. Additional volumes describing tin-oxygen compounds with other R_2Sn and RSn groups will follow in the near future.

Abbreviations and symbols are explained on pp. X/XI along with other remarks.

We thank Prof. Dr. Dr. h.c. Ekkehard Fluck and his coworkers at the Gmelin Institute for their excellent cooperation. In particular, we thank Dr. Ulrich Krüerke for his accurate and sympathetic editing, Mr. Edgar Rudolph for preparing the index, Dr. Grant for reading the English text, and Mr. Hans-Georg Karrenberg for drawing the numerous formulas and molecular structures. We thank Mrs. Ellen Redlinger, Miss Susanne Schumann, and Miss Stefanie Schumann for their meticulous handling of the literature index and the members of the chemical department of the library at the Technische Universität Berlin for their assistance in searching and acquiring the references from the literature.

Berlin-Lichtenrade
November 1987

Herbert Schumann
Ingeborg Schumann

Explanations, Abbreviations, and Units

Many compounds in this volume are presented in tables in which numerous abbreviations are used and the units are omitted for the sake of conciseness. This necessitates the following clarification.

The term "special" in the second column of the tables indicates preparative methods which are described under the further information section following the table.

Temperatures are given in °C, otherwise K stands for Kelvin. Abbreviations used with temperatures are m.p. for melting point, b.p. for boiling point, dec. for decomposition, and subl. for sublimation. Terms like 80°/0.1 mean the boiling or sublimation point at a pressure of 0.1 Torr. **Densities** D are given in g/cm^3. D_c and D_m distinguish calculated and measured values, respectively.

NMR represents **nuclear magnetic resonance**. Chemical shifts are given as δ values in ppm and positive to low field from the following reference substances: $Si(CH_3)_4$ for ^1H and ^{13}C, $BF_3 \cdot O(C_2H_5)_2$ for ^{11}B, $CFCl_3$ for ^{19}F, H_3PO_4 for ^{31}P, and $Sn(CH_3)_4$ for ^{119}Sn. Multiplicities of the signals are abbreviated as s, d, t, q (singlet to quartet), quint, sext, sept (quintet to septet), and m (multiplet); terms like dd (double doublet) and t's (triplets) are also used. Assignments referring to labeled structural formulas are given in the form C-4, H-3,5. Carbon and hydrogen atoms of the C-bonded ligands (C_4H_9 in this volume) are labeled α, β, γ, and δ. Coupling constants J in Hz appear usually in parentheses behind the δ value, along with the multiplicity and the assignment, and refer to the respective nucleus. If a more precise designation is necessary, they are given as, e.g., $^nJ(C,H)$ or J(1,3) referring to labeled formulas. The coupling to the two nuclei ^{117}Sn and ^{119}Sn is written as J(Sn,X) = A/B. A single value, J(Sn,X) = A, always refers to the ^{119}Sn nucleus.

Nuclear quadrupole resonance is abbreviated NQR, with the transitions in MHz.

Mössbauer spectra are represented by ^{119}Sn-γ; both the isomer shift δ (vs. $BaSnO_3$ or SnO_2 at room temperature) and the quadrupole splitting Δ are given in mm/s; the experimental error has generally been omitted. Other reference substances for δ are indicated after the numerical value, e.g., δ = −0.31 (α-Sn).

Optical spectra are labeled as IR (infrared), R (Raman), and UV (electronic spectrum including the visible region). IR bands and Raman lines are given in cm^{-1}; the assigned bands are usually labeled with the symbols ν for stretching vibration and δ for deformation vibration. Intensities occur in parentheses either in the common qualitative terms (s, m, w, vs, etc.) or as numerical relative intensities. The UV absorption maxima, λ_{max}, are given in nm followed by the extinction coefficient ε ($L \cdot cm^{-1} \cdot mol^{-1}$) or log ε in parentheses; sh means shoulder.

Photoelectron spectra are abbreviated PE, e.g., PE/He(I), with the ionization energies in eV.

Solvents or the **physical state** of the sample and the temperature (in °C or K) are given in parentheses immediately after the spectral symbol, e.g., R (solid), ^{13}C NMR (C_6D_6, 50°C), or at the end of the data if spectra for various media are reported. Common solvents are given by their formula (C_6H_{12} = cyclohexane) except THF, DMF, and HMPT, which represent tetrahydrofuran, dimethylformamide, and hexamethylphosphoric triamide, respectively.

The data of **mass spectra**, abbreviated MS, are given as m/e, relative intensity in parentheses, and fragment ions in brackets; $[M]^+$ is the molecular ion.

References, quoted in the last column, are occasionally also placed in the first and second column if statements from different sources must be distinguished.

Figures give only selected parameters. Barred bond lengths (in Å) or angles are mean values for parameters of the same type.

Table of Contents

Organotin Compounds

1.4.1.2 Diorganotin-Oxygen Compounds

1.4.1.2.1 Diorganotin-Oxygen Compounds of the $R_2Sn(OR')_2$ Type

Dimethyltin-oxygen compounds (1.4.1.2.1.1), diethyltin-oxygen compounds (1.4.1.2.1.2), dipropyltin-oxygen compounds (1.4.1.2.1.3), and diisopropyltin-oxygen compounds (1.4.1.2.1.4) have been described in "Organotin Compounds" 14, 1987, pp. 24, 147, 198, and 211, respectively.

1.4.1.2.1.5 Dibutyltin-Oxygen Compounds, $(C_4H_9)_2Sn(OR)_2$

1.4.1.2.1.5.1 Dibutyltin Dihydroxide, $(C_4H_9)_2Sn(OH)_2$

When the oily dibutyltin residues, gathered from several reduction reactions between $(C_4H_9)_2SnH_2$ and aldehydes or ketones, are covered with CH_3OH, warmed to 45 to 50°C, and brought in contact with air by bubbling it through the solution for 1 d, colorless crystals can be isolated from the solvent. This material, melting at 113 to 116°C, shows the typical hydroxyl absorption in its infrared spectrum and the results of analysis correspond, as it is said "surprisingly", to $(C_4H_9)_2Sn(OH)_2$, the structure of which "remains to be examined in more detail" [3]. The compound is also cited in the patent literature concerning its use for the in situ production of $(-(C_4H_9)_2SnOOCCH_2CH(SCOC_6H_5)COO-)_n$ as a stabilizer for poly(vinyl)chloride [1, 2]. The effect of heat on the dissipation factors of silicone rubber in the presence of the title compound and lauric acid [5] and the usefulness of $(C_4H_9)_2Sn(OH)_2$ as a mosquito larvicide [4] have been studied.

References:

[1] Société anonyme des manufactures des glaces et produits chimiques de Saint-Gobain, Chauny et Cirey (Fr. 1105652 [1955]; C.A. **1959** 6082; Belg. 538646 [1954/59]).

[2] Société anonyme des manufactures des glaces et produits chimiques de Saint-Gobain, Chauny et Cirey (Brit. 775242 [1957]; C.A. **1958** 6398).

[3] Kuivila, H. G., Beumel, O. F. (J. Am. Chem. Soc. **83** [1961] 1246/50).

[4] Cardarelli, N. F. (Mosq. News **38** [1978] 328/33).

[5] Yokoyama, T., Suzuki, H., Mukai, J. (IEEE Trans. Electr. Insul. EI-15 [1980] 373/81; C.A. **94** [1981] No. 31852).

1.4.1.2.1.5.2 Dibutyltin Bis(Organyl Oxides), $(C_4H_9)_2Sn(OR)_2$

1.4.1.2.1.5.2.1 Dibutyltin Bis(Alkoxides), $(C_4H_9)_2Sn(OR)_2$

The compounds belonging to this section are listed in Table 1. They are prepared by the following general methods.

Method I: Reaction of $(C_4H_9)_2SnCl_2$ with RONa or ROH (1:2 mole ratio).
The reactions of $(C_4H_9)_2SnCl_2$ with the appropriate RONa compounds, which, except otherwise noted, are usually conducted in ROH as the solvent and in an inert N_2 atmosphere at temperatures between 0 and 20°C, yield No. 1 [10, 20, 26, 45, 78, 103, 124], No. 2 [26, 78] (in C_2H_5OH-C_6H_6, reflux/8 h [33]), No. 3 [26, 78], No. 4 [23, 78, 118] (in C_6H_6, reflux/3 h [72]), No. 5 [10, 21, 26, 78] (in $C_6H_5CH_3$, 0 to 5°C [5]), No. 8 [78], No. 11 [78], No. 12 [5], and Nos. 27, 28, 31, 32, and 33 in $C_6H_5CH_3$ (reflux) [17].
Nos. 23 and 29 are prepared by the reaction of $(C_4H_9)_2SnCl_2$ with the appropriate ROH in the presence of $N(C_2H_5)_3$ [120] and C_5H_5N [13], respectively.

Method II: Condensation of $(-(C_4H_9)_2SnO-)_n$ with ROH (1:2 mole ratio).
The dehydration of mixtures of $(-(C_4H_9)_2SnO-)_n$ with the appropriate ROH has been used for the synthesis of Nos. 1 and 2 (in the presence of $Mg(OR)_2$ as the dehydrating agent) [64], No. 5 [55] in C_6H_{12} [3] or $C_6H_{11}CH_3$ [4], No. 13 in C_6H_{12}, and No. 44 in C_6H_{12} [3] (with azeotropic removal of the water formed), and No. 46 at 160 to 190°C for 30 to 40 min [151].

Method III: Transamination reactions.
The transamination of $(C_4H_9)_2Sn(N(C_2H_5)_2)_2$ with CH_3OH and C_2H_5OH in refluxing C_5H_{12} [28], or with $C_2H_5CH(CH_3)OH$ and $(CH_3)_2CHCH_2OH$ in C_6H_6 at 20°C [87] yields Nos. 1, 2, 6, and 7, respectively.

Method IV: The transalkoxylation of $(C_4H_9)_2Sn(OCH_3)_2$ with the appropriate ROH derivative leads to No. 9 [40], No. 14 (90°C/1.5 h) [7, 9], and No. 45 [9]. The action of the appropriate ROH compound on $(C_4H_9)_2Sn(OC_2H_5)_2$ in refluxing C_6H_6 for 2 h affords Nos. 34, 35, and 38 to 41 [89]. The reaction between $(C_4H_9)_2Sn(OC_3H_7\text{-}i)_2$ and the appropriate ROH compound, carried out in C_6H_6, gives Nos. 16 [153], 36, and 37 [72].

Method V: Insertion reactions.
The reaction of $(C_4H_9)_2Sn(OCH_3)_2$ with the twofold molar amount of CH_3CHO or CCl_3CHO is rapid and exothermic, leading to the Sn-O bond insertion products No. 19 and 20, respectively [46]. The likewise exothermic and rapid reaction between $(C_4H_9)_2Sn(ON{=}C(CH_3)_2)_2$ and CCl_3CHO yields No. 21 [138], whereas the insertion of $\overline{O\text{-}CH_2\text{-}C}HCH_2OC_6H_5$ into the Sn-O bonds of $(C_4H_9)_2Sn(OOCC_6H_5)_2$ needs 165 to 180°C for 5 to 8 h to afford No. 23 [120].

References on p. 22

Table 1
Dibutyltin Bis(Alkoxides), $(C_4H_9)_2Sn(OR)_2$.
Further information on compounds preceded by an asterisk is given at the end of the table.
Explanations, abbreviations, and units on p. X.

No.	OR group method of preparation (yield in %)	properties and remarks	Ref.
*1	OCH_3 I [45,78,124] (46 [26], 67 [20], 93 [10,103]) II (39 [64]) III (80 [28])	b.p. 110°/0.1 [28], 124°/0.3 [45], 126 to 128°/0.05 [20], 126 to 130°/0.05 [26], 131 to 132°/0.3 [78], 136 to 139°/1 to 1.5 [10,28,77,103], 138 to 140°/22 [64], 150 to 152°/3 [47]	[10,20,26, 28,45,47, 64,77,78, 103]
		D^{20} = 1.2862 [10], 1.291 [47]	[10,47]
		n_D = 1.4752 [64], 1.4831 [10,77], 1.4845 [139], 1.4876 [47], 1.4880 [26] at 20°, 1.4852 [20] at 25°	[10,20,26, 47,64,77, 139]
		μ = 1.33 D (in C_6H_6, 25°)	[32,53]
		^{1}H NMR (CCl_4): 3.52 [28] or 3.58 (OCH_3) [45] at 32 to 33°	[28, 45]
		^{119}Sn NMR (neat): −165	[87,94,142]
		^{13}C NMR (neat): 13.8 (C-δ), 19.5 (C-α, J(Sn,C) = 642), 27.2 (C-γ, J(Sn,C) = 85), 27.8 (C-β, J(Sn,C) = 30), 52.0 (OCH_3)	[93]
		^{119}Sn-γ: δ = 1.10, Δ = 2.32 at 77 K [87] at 78 K [74]; 0.97, 2.3 at 83 K [65]	[65,74,87]
		IR in Table 2 on p. 8	
*2	OC_2H_5 I [78] (60 [26], 88 [33]) II (42 [64]) III (80 [28]) special [19,51]	extremely moisture-sensitive liquid	[19,26]
		b.p. 95°/0.15 [19], 96 to 97°/0.2 [26], 107°/0.1 [28], 115 to 120°/0.2 [78], 117°/1 [51], 128 to 129°/12 [64]	[19,26,28, 51,64,78]
		n_D^{20} = 1.4762 [64], 1.4790 [26]	[26,64]
		^{1}H NMR (CCl_4): 1.13 (CH_3 of C_2H_5), 3.74 (OCH_2)	[28]
		^{119}Sn NMR: −154 (50% in CCl_4) [94], −161 (neat) [87,94]	[87,94]
		^{119}Sn-γ (77 K): δ = 1.30, Δ = 2.00	[24,87]
*3	OC_3H_7 I [78] (54 [26])	viscous, moisture-sensitive liquid	[26]
		b.p. 99 to 100°/0.1 [26], 119 to 121°/0.25 [78]	[26,78]
		n_D^{20} = 1.4782	[26]
		^{119}Sn NMR (neat): −159	[87]
*4	OC_3H_7-i I [23,78,118] (95 [72])	b.p. 81 to 85°/0.15 [78], 87°/1.5 [72], 131 to 132°/10 [23,118]	[23,72,78, 118]
		n_D^{30} = 1.4657	[23]
		^{119}Sn NMR (neat, 25°): −90	[87,94]
		^{13}C NMR: 27.8 (CH_3 of C_3H_7), 66.5 (OCH)	[93]

Table 1 (continued)

No.	OR group method of preparation (yield in %)	properties and remarks	Ref.
*5	OC_4H_9	m.p. 110 to 115°?	[3,4]
	I [5,10,78] (41 [26], 68 to 75 [21])	viscous, moisture-sensitive liquid	[26]
	II [3,4,55] special [75]	b.p. 125 to 127°/0.1 [78], 130 to 135°/4 [44], 136 to 138°/0.05 [26], 142°/0.1 [55], 178 to 180°/0.3 [75]	[21,26,44, 55,75,78]
		D^{20} = 1.2056	[21]
		n_D^{20} = 1.4721	[26]
		^1H NMR: 3.75 (OCH_2)	[75]
		^{119}Sn NMR (neat): −161	[87,94,119]
6	OC_4H_9-s	b.p. 92 to 94°/0.07	[87]
	III [87]	^{119}Sn NMR (neat): −34	
		with CS_2 → (s-$C_4H_9O)_2CO$ (36%), s-$C_4H_9SC(S)OC_4H_9$-s (59%), and (-($C_4H_9)_2SnX$-$)_n$ (X = O, S)	[109]
		with $\overline{OCH_2CH_2OCO}$ → (s-$C_4H_9O)_2CO$ (27%) and $(C_4H_9)_2\overline{SnOCH_2CH_2O}$ (31%)	[110]
		with BCl_3 → $(C_4H_9)_2SnCl_2$ and s-$C_4H_9OBCl_2$	[27]
*7	OC_4H_9-i	b.p. 100 to 104°/0.2	[87]
	III [87]	^{119}Sn NMR (neat, 22°): −154 [119], −150 [87]	[87,119]
*8	OC_4H_9-t	b.p. 152 to 154°/27	[78]
	I [78]	^{119}Sn NMR (neat or CCl_4): −34	[87,94]
		^{13}C NMR: 13.6 (C-δ), 21.4 (C-α, J(Sn,C) = 496), 34.1 (CH_3 of t-C_4H_9), 70.3 (OC)	[93]
		IR in Table 7 on p. 20	
*9	OC_5H_{11} IV [40]	IR in Table 8 on p. 21	[35,40]
10	$OCH(CH_3)C_3H_7$	no preparation reported	
		with CS_2 (65°/48 h) → $(RO)_2CO$ (95%)	[109]
*11	OC_8H_{17}	b.p. 173 to 175°/0.4	[78]
	I [78]	IR in Table 9 on p. 21	[35,40,41]
		with CS_2 (120°/20 h) → $C(OR)_4$ (dec. on distillation into $(RO)_2CO$)	[78]
		with $P(O)(OC_6H_{13})_2OH$ → $(C_4H_9)_2Sn(OC_8H_{17})OP(O)(OC_6H_{13})_2$	[91,92]
		stabilizer for PVC yielding translucent to transparent material	[2]

References on p. 22

Table 1 (continued)

No.	OR group method of preparation (yield in %)	properties and remarks	Ref.
12	$OCH_2CH(C_2H_5)C_4H_9$ I [5]	hardening agent for epoxide resins	[15]
13	$OC_{10}H_{21}$ II	no properties reported	[3]
14	$OC_{12}H_{25}$ IV [7,9]	D = 1.025 at 20° [9], n_D^{20} = 1.4730 [7,9] stabilizer for PVC improves dyeability of polypropylene fibers improves workability of styrene resins	[7,9] [7] [62,63] [129]
*15	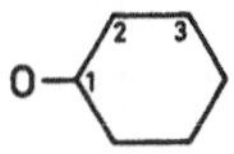 special [98]	m.p. 37 to 40° b.p. 145 to 149°/10^{-3} ^{13}C NMR (neat): 13.6 (C-δ), 19.4 (C-α, J(Sn,C) = 515), 38.2 (C-3), 72.8 (C-2) IR in Table 10 on p. 22 with $CS_2 \rightarrow (RO)_2CO$ (65°/72 h, 44%; 40°/72 h, 88%)	[98] [93] [35,40] [109]
16	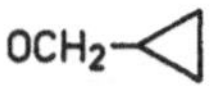 IV (50)	b.p. 138 to 140°/0.5 ^1H NMR ($CDCl_3$ or CCl_4): 3.48 (CH_2O) IR (neat or Nujol): ν(CO) 1130 (m), 1060 (vs), and 1040 (vs), ν(SnO) 655 (s), 605 (s), and 525 (s)	[153]
*17	OCH_2CF_3 special	b.p. 220 to 222°/0.1 ^1H NMR: 4.15 (CH_2O)	[75]
*18	$OCH(CF_3)_2$ special	b.p. 75°/10^{-3}	[31]
19	$OCH(CH_3)OCH_3$ V	dec. at 0.1 Torr into the starting components $(C_4H_9)_2Sn(OCH_3)_2$ and CH_3CHO ^1H NMR: 3.38 (OCH_3), 5.13 (q, CH, J(H,H) = 4.2)	[46]
20	$OCH(CCl_3)OCH_3$ V	very viscous oil ^1H NMR: 3.82 (OCH_3), 5.19 (CH) IR: 1155, ν(CO?) 1085, ν(CCl_3) 830, 805	[46]
21	$OCH(CCl_3)ON{=}C(CH_3)_2$ V	yellow-red, viscous liquid IR: ν(C=N) 1648 (m), ν(NO) 925 (s)	[138]
22	$O(CH_2)_3OOCCH{=}CHCOO(CH_2)_3OH$	no preparation reported stabilizer for PVC	 [58]

References on p. 22

Table 1 (continued)

No.	OR group method of preparation (yield in %)	properties and remarks	Ref.
23	$OCH(CH_2OC_6H_5)CH_2$- $OOCC_6H_5$ I, V [120]	viscous, crystallizing resin influence of donor solvents on the formation by Method V	[120] [152]
24	$O(CH_2)_4OOCC_9H_{19}$	no preparation reported stabilizer for PVC	 [71]
25	$OCH_2CH_2(OCH_2CH_2)_{24}$- $OOCC_{17}H_{35}$	no preparation reported antistatic agent for thermoplastics	 [52]
26	$OCH_2COOCH_2CH(C_2H_5)C_4H_9$	no preparation reported heat and light stabilizer for PVC TLC investigations	 [76] [131]
27	$OCH(CH_3)COOC_2H_5$ I	stabilizer for PVC	[17]
28	$OCH(C_2H_5)COOC_2H_5$ I	stabilizer for PVC	[17]
29	$OC(CH_3)_2COOC_2H_5$ I	stabilizer for PVC	[13]
30	$OCH_2COCH_2COCH_3$	no preparation reported used in the recovery of polyols from polyurethane wastes	 [113]
31	$OCH(COOC_4H_9)CH(OH)$- $COOC_4H_9$ I	stabilizer for PVC	[17]
32	$OC(COOC_2H_5)$- $(CH_2COOC_2H_5)_2$ I	stabilizer for PVC	[13, 17]
33	$OC(COOC_4H_9)$- $(CH_2COOC_4H_9)_2$ I	stabilizer for PVC	[13, 17]
34	$OCH_2CH_2NHCH_3$ IV (52)	yellow liquid b.p. 130 to 135°/0.2 IR: 1060 (vs), 685 (m), 650 (m, br), 590 (m), 505 (w, br)	[89]
35	$OCH_2CH_2NHC_2H_5$ IV (53)	yellow liquid b.p. 135 to 140°/0.2 IR: 1070 (vs), 1020 (m), 670 to 655 (w, br), 595 (m, br), 505 (w)	[89]
36	$OCH_2CH_2N(CH_3)_2$ IV	b.p. 145°/2.0	[72]

References on p. 22

Table 1 (continued)

No.	OR group method of preparation (yield in %)	properties and remarks	Ref.
37	$OCH_2CH_2N(C_2H_5)_2$ IV	b.p. 174 to 178°/4.5	[72]
38	$OCH(CH_3)CH_2NH_2$ IV (36)	colorless liquid b.p. 155 to 165°/0.1 IR: 1065 (vs, br), 1015 (s, br), 680 (s, br), 590 (m), 500 (w)	[89]
39	$OCH(CH_3)CH_2N(CH_3)_2$ IV (59)	yellow liquid b.p. 120 to 125°/0.2 IR: 1075 (vs), 1038 (s), 675 to 660 (w, br), 590 (m, br), 510 (w, br)	[89]
40	$OCH_2CH_2CH_2NH_2$ IV (45)	yellow liquid b.p. 155 to 160°/0.1 IR: 1070 (vs, br), 1020 (m), 685 (m), 660 (m), 590 (m), 555 (w), 510 (w)	[89]
41	$OCH_2CH_2CH_2N(C_2H_5)_2$ IV (64)	yellow liquid b.p. 155 to 160°/0.3 IR: 1065 (vs), 1015 (m), 670 (m, br), 600 (m, br), 500 (w, br)	[89]
*42	$OCH_2C_6H_5$ special [75]	b.p. 210°/0.4 ^{1}H NMR: 4.59 (CH_2O) stabilizer for PVC catalyst in peptide synthesis	[75] [1] [137]
43	$OCH(CH_3)C_6H_5$	no preparation reported with CS_2 (65°/72 h)→$(RO)_2CO$ (traces)	 [109]
44	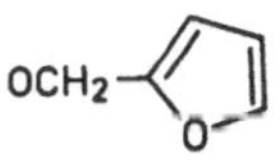 II [3]	stabilizer for PVC	[1]
45	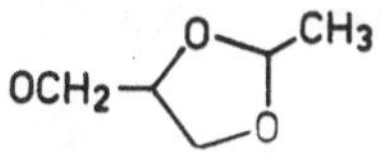 IV	stabilizer for PVC	[9]
46	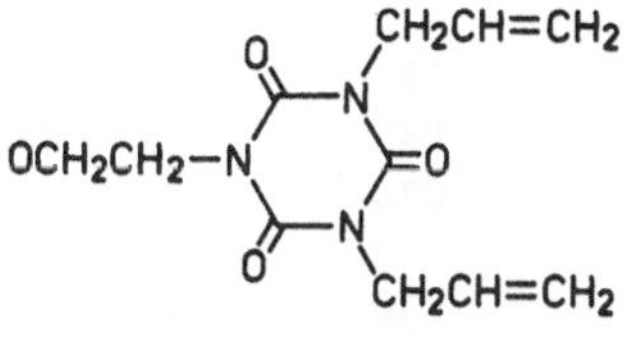 II (95)	n_D^{20} = 1.5300 catalyst in polyurethane synthesis	[151]

References on p. 22

* Further information:

$(C_4H_9)_2Sn(OCH_3)_2$ (Table **1**, No. **1**) is also formed in numerous methanolysis reactions of functionally substituted dibutyltin compounds.

The concentration and temperature dependence of the ^{119}Sn chemical shift is demonstrated by the following values [94]:

CCl_4 solution mol %	δ (ppm)	neat liquid t (°C)	δ (ppm)
18.9	−159	70	−165
27.3	−161	120	−159
45.2	−163	160	−147

The IR spectrum of $(C_4H_9)_2Sn(OCH_3)_2$ in the region 1500 to 400, 1300 to 400, and 1090 to 1025 cm^{-1} is depicted and qualitatively discussed in [32], [40], or [41], respectively. Assigned frequencies are summarized in Table 2. The following frequencies are obtained from solution

Table 2
IR Spectra of Neat $(C_4H_9)_2Sn(OCH_3)_2$.
Wave numbers in cm^{-1}.

[40]	[54]	assignment	[25]	assignment
2965 (vs, sh)		$\nu_{as}(CH_3)$		
2954 (vs)		$\nu'_s(CH_3)$		
2930 (vs, sh)		$\nu_s(CH_3)$		
2924 (vs)		$\nu_{as}(CH_2)$/C ?		
2898 (vs, sh)		$\nu_s(CH_2)$/C		
2871 (s)		$\nu_{as}(CH_2)$/Sn ?		
2849 (s)		$\nu_s(CH_2)$/Sn		
1464 (ms)	1462 (ms)	$\delta_{as}(CH_3)$		
1456 (ms, sh)		$\delta(CH_2)$/C ?		
1447 (ms)		$\delta'_s(CH_3)$	1445	$\delta_{as}(CH_3)$
1435 (mw, sh)		$\delta(CH_2)$/C ?		
1416 (mw)	1416 (mw)	$\delta(CH_2)$/Sn	1417 (w)	$\delta_{as}(CH_2)$/Sn
1377 (ms)	1376 (m)	$\delta_s(CH_3)$	1377	$\delta_s(CH_3)$
			1155 (w)	$\delta_s(CH_2)$/Sn
1075 (ms)		$\gamma(CH_3)$		
1063 (vs)	1063 (vs)	$\nu(CO)$	1064	$\nu_{as}(CO)$
			1036	$\nu_s(CO)$
960 (w)		$\nu(CC)$	960	$\nu(CC)$
748 (vw)	744 (w)	$\gamma(CH_2C)$		
689 (w, sh)	676 (m)	$\gamma(CH_2Sn)$	694	$\varrho(CH_2Sn)$
			673	$\nu_{as}(SnO)$
616 (w)	610 (m)	$\nu_{as}(SnO) + \nu_{as}(SnC)$	615	$\nu_s(SnO)$
603 (w, sh)	587 (m)			
515 (mw)	512 (s)	$\nu_s(SnC)$	516	$\nu_{as}(SnC)$
471 (mw)	475 (s)	$\nu_s(SnO)$	470	$\nu_s(SnC)$

References on p. 22

spectra: ν(CO) 1063 (vs) for 0.04 to 0.6 mol/L CCl_4, or 1053 (vs) for 0.01 to 0.2 mol/L $CHCl_3$, ν_{as}(SnO) + ν_{as}(SnC) 616 (w) and 603 (w), ν_s(SnC) 515 (mw), and ν_s(SnO) 471 (mw) for 0.3 to 0.6 mol/L C_6H_{14} [35]. The basicity of the oxygen atoms in $(C_4H_9)_2Sn(OCH_3)_2$ and in the analogous Si and Ge compound increases in the series Si < Ge < Sn as concluded from the ν(OH) shift of CH_3OH when it is added to solutions of these compounds in CCl_4 [61, 100].

From the UV spectrum, λ_{max} = 215 nm (log ε = 3.18), it is concluded that there is no dπ-pπ interaction in the Sn-O bond, in contrast to the Sn-S bond [47, 48].

Cryoscopic [32] and spectroscopic evidence [41, 87, 93, 94, 142] suggest a dimeric structure with five-coordinate Sn (Formula I) or a higher associated structure with six-coordinate Sn (Formula II).

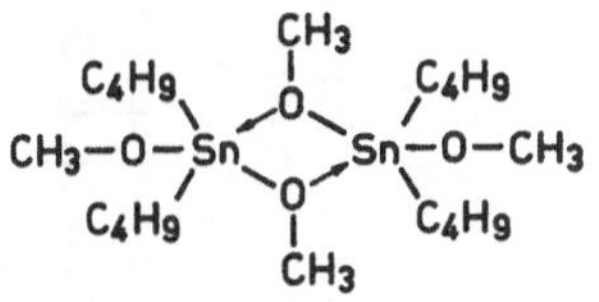

I

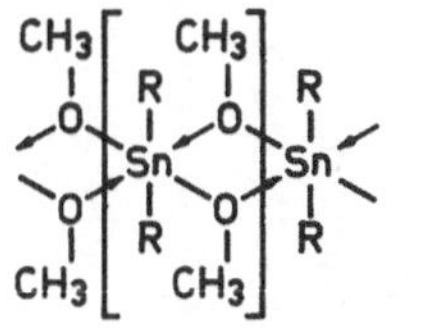

II R = C_4H_9

The determination of the title compound by atomic absorption spectroscopy is possible with a sensitivity of 2.3 or 3.0 µg/mL using the two wavelengths λ = 2246 or 2863 Å of an air-acetylene flame. The results with other organotin compounds indicate that the sensitivity decreases with decreasing energy of the Sn-C bonds [102].

Selected chemical reactions of $(C_4H_9)_2Sn(OCH_3)_2$ are listed in Table 3.

$(C_4H_9)_2Sn(OCH_3)_2$ shows anthelmintic activity towards the cestode Raillietina cesticillus when administered in the feed or by capsule as a single oral dose to infested chickens (50 to 100 mg/kg chicken cause an 80% removement), whereas it shows no effect against the nematode Ascaridia galli from chickens [12]. The usefulness of the title compound as a fungicide and bactericide, especially against Alternaria species, Botrytis cinerea, and Sclerotinia fructicola on seeds and growing plants, is claimed in [6].

$(C_4H_9)_2Sn(OCH_3)_2$ catalyzes the reactions between isocyanates and alcohols [135, 154, 160], the polymerization and copolymerization of diisocyanates [122, 134], the preparation of isocyanates from carbamates [155], or from N- and O-substituted mono- or bis-urethanes [147]. It is used as a catalyst in the preparation of poly(ethylene terephthalate) [79, 99, 126], polylactones [158, 166, 167], carbonates or polycarbonates from urea and alcohols [157, 162] and for carbonate ester interchange [133], in the synthesis of acrylic and methacrylic acid esters by ester exchange [128], the synthesis of dialkylaminoethylacrylates and methacrylates [123] and of N-substituted methacrylamides [144, 145, 148]. $(C_4H_9)_2Sn(OCH_3)_2$ is also useful as a catalyst for the vulcanization of polyalkoxy-terminated organopolysiloxanes [140, 156, 159], for room-temperature vulcanizable fluid silicone compositions [95, 111], for the preparation of crosslinkable silane-modified polyethylene sheets [163], for the polymerization of N-carboxyanhydrides [97], for the polymerization of the C≡N groups in polyacrylonitrile, polymethacrylonitrile, and poly(α-cyanoethyl acrylate) [107], and for the preparation of polyacetal resins [59].

References on p. 22

Table 3
Reactions of $(C_4H_9)_2Sn(OCH_3)_2$.
The C_4H_9 groups on the Sn atom are abbreviated R.

No.	reactant	conditions	product (yield) and remarks	Ref.
1	H_2O	slowly in air, 3 s in 95% C_2H_5OH	$(\text{-}R_2SnO\text{-})_n$ (100)	[20]
2	H_2O_2 (98%)	in ether	$(\text{-}R_2Sn\text{-}O\text{-}O\text{-})_n$	[81]
3	CO_2	passing at 20° in C_6H_6, 100°/10 h, CO_2 under pressure	$R_2Sn(OCOOCH_3)_2$ R_2SnCO_3 and $(CH_3O)_2CO$	[46] [109]
4	COS	C_6H_6 under reflux, 2 h	$(\text{-}R_2SnS\text{-})_3$ and $(CH_3O)_2CO$	[109]
5	CS_2	20°/5 min	$R_2Sn(OCH_3)SCSOCH_3$ and $R_2Sn(SCSOCH_3)_2$	[46]
		65°/72 h [109], 100°/5 to 10 h, sealed tube [78, 83, 103]	$(\text{-}R_2SnS\text{-})_3$ and $C(OCH_3)_4$ (95)	[78, 83, 103, 109]
6	SO_2	in C_6H_6, 15 min bubbling, slightly exothermic	$R_2Sn(OCH_3)OSO_2CH_3$	[46]
7	R′NCO $R' = C_2H_5$, C_6H_5	1:1, exothermic, rapid 1:2, exothermic, rapid	$R_2Sn(OCH_3)NR'COOCH_3$ $R_2Sn(NR'COOCH_3)_2$ (82 for $R' = C_6H_5$)	[46]
8	R′NCS $R' = C_6H_5$, $CH_2{=}CHCH_2$	1:1, exothermic, rapid 1:2, exothermic ($R' = C_6H_5$), rapid 1:1, $(CH_2Cl)_2$, 70°/16 h	$R_2Sn(OCH_3)SC(=NR')OCH_3$ $R_2Sn(SC(=NR')OCH_3)_2$ $(\text{-}R_2SnS\text{-})_3$, $R'N{=}C(OCH_3)_2$ ($R' = C_6H_5$)	[46, 73, 121] [121]
9	R′CHO $R' = CH_3$, CCl_3	1:1, exothermic 1:2, exothermic	$R_2Sn(OCH_3)OCHR'OCH_3$ $R_2Sn(OCHR'OCH_3)_2$	[46]
10	CCl_3CN	1:1 or 1:2, 20°/12 h	$R_2Sn(OCH_3)N{=}C(CCl_3)OCH_3$	[46]
11	$C_{10}H_7N{=}C{=}NC_{10}H_7$ ($C_{10}H_7$ = 1-naphthyl)	1:1 1:2	$R_2Sn(OCH_3)N(C_{10}H_7)C(=NC_{10}H_7)OCH_3$ $R_2Sn(N(C_{10}H_7)C(=NC_{10}H_7)OCH_3)_2$	[46]
12	$BrCH_2CH_2Br$	—	$R_2Sn(Br)OCH_3$, R_2SnBr_2, $BrCH_2CH_2OCH_3$, and $CH_3OCH_2CH_2OCH_3$	[36]

References on p. 22

13	$BrCH_2CH{=}CH_2$	—	$R_2Sn(Br)OCH_3$, R_2SnBr_2, and $CH_3OCH_2CH{=}CH_2$	[36]
14	$C_{12}H_{25}OH$	90°/1.5 h	$R_2Sn(OC_{12}H_{25})_2$	[9]
15	$\overline{CH_2OCH(CH_3)OC}HCH_2OH$	similar to No. 14	$R_2Sn(OCH_2\overline{CHOCH(CH_3)OC}H_2)_2$	[9]
16	$2\text{-}CH_3OOCC_6H_4OH$	similar to No. 14	$R_2Sn(OC_6H_4COOCH_3\text{-}2)_2$	[9]
17	$CH_2{=}CHCH_2OH$	—	$R_2Sn(OCH_2CH{=}CH_2)_2$	[101]
18	$CH_2{=}CHCH_2C(CH_3)(OH)C_4H_9\text{-}t$	140°/1 h, then 170°/4 h	$R_2Sn(CH_2CH{=}CH_2)_2$, CH_3OH, and $CH_3COC_4H_9\text{-}t$	[125]
19	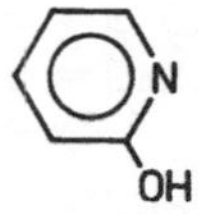	90° with stirring	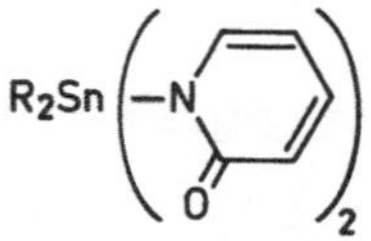	[39]
20	$CH_3N(CH_2CH_2OH)_2$	xylene, reflux, 20 min	$R_2Sn(OCH_2CH_2)_2NCH_3$ (87)	[104]
21	$C_3H_7CH{=}NOH$	similar to No. 14	$R_2Sn(ON{=}CHC_3H_7)_2$	[9]
22	$(CH_3)_2N(CH_2)_3C(CH_3){=}NOH$	110°	$R_2Sn(ON{=}C(CH_3)(CH_2)_3N(CH_3)_2)_2$	[22]
23	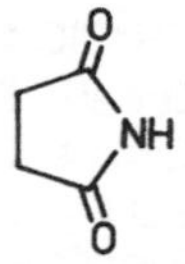	similar to No. 14	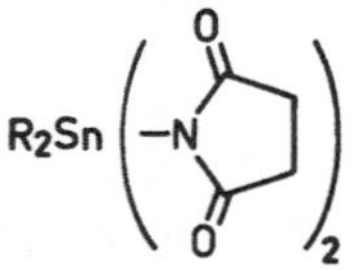	[9]
24	$CH_2(COOC_4H_9)_2$	similar to No. 14	$R_2Sn(CH(COOC_4H_9)_2)_2$	[9]
25	$HOOCCH{=}CHCH{=}CHCH_3$	similar to No. 14	$R_2Sn(OCH_3)OOCCH{=}CHCH{=}CHCH_3$	[9]
26	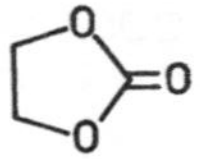	20°/2 h	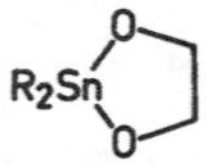 (83), $(CH_3O)_2CO$ (70)	[110]
27	$R'R''CO + H_2O_2$ (98%) $R' = H$, $R'' = CH_3$, C_2H_5, C_6H_5, $4\text{-}CH_3C_6H_4$, $4\text{-}ClC_6H_4$, $4\text{-}NO_2C_6H_4$, $4\text{-}CH_3OC_6H_4$, $2\text{-}C_5H_4N$, $4\text{-}HOOCC_6H_4$; $R' = CH_3$, $R'' = C_2H_5$; $R',R'' = (CH_2)_5$	ether, 20°	$(\text{-}R_2Sn\text{-}O\text{-}O\text{-}CR'R''\text{-}O\text{-})_n$	[60,66,81]

References on p. 22

Table 3 (continued)

No.	reactant	conditions	product (yield) and remarks	Ref.
28	i-$C_8H_{17}OOCCH_2SH$	–	$R_2Sn(SCH_2COOC_8H_{17}$-i$)_2$, stabilizer for PVC	[14]
29	$R'CH(SCOCH_3)CH_2SCOCH_3$ R' = t-C_4H_9, $(CH_3)_3Si$, $F(CH_3)_2Si$, $CH_3O(CH_3)_2Si$	1:1, 70°/2 h	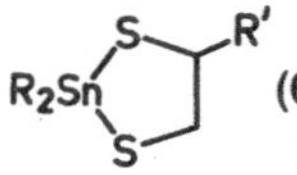 (65 to 80)	[150]
30	$R'C(S)SCH_3$ R' = i-C_3H_7, C_6H_5, 2-, or 4-$CH_3C_6H_4$, 3-, or 4-$CH_3OC_6H_4$, 4-ClC_6H_4	3:2, 60 to 65°/4 to 16 h	$(-R_2SnS-)_3$, $R_2Sn(SCH_3)_2$ and $R'C(OCH_3)_3$ (60 to 90)	[117]
31	$P(O)(OC_2H_5)_2OH$	1:1, C_6H_6 under reflux, 2.5 h	$R_2Sn(OCH_3)OP(O)(OC_2H_5)_2$, catalyst for hardening organopolysiloxanes	[91, 92]
32	$R'(C_5H_5)GeCl_2$ $R' = CH_3$, C_2H_5	heating	R_2SnCl_2, $R'(C_5H_5)Ge(OCH_3)_2$ (50 to 70)	[114]
33	$CH_3(C_6H_5)Si(H)Cl$	1:1, 20°	$R_2Sn(H)Cl$ and $CH_3(C_6H_5)Si(OCH_3)_2$	[43]
34	R'_nSiX_{4-n} $R' = CH_3$, C_2H_5 X = F, Cl, Br, I n = 2 or 3	exothermic, rapid	R_2SnX_2 and $R'_nSi(OCH_3)_{4-n}$	[88]
35	$(C_6H_5)_2Si(OCH_3)_2$	dioxane, reflux	polymer	[11]
36	$(CH_3)_3SiSC_2H_5$	heating	$R_2Sn(SC_2H_5)_2$ and $(CH_3)_3SiOCH_3$	[105]
37	R_2SnH_2	1:1, 50°/24 h 20°	$(-R_2Sn-)_n$ $R_2Sn(H)OCH_3$ (80)	[30, 42] [101]
38	SnI_4	CCl_4, reflux, 5 h	R_2SnI_2 (88) and $Sn(OCH_3)_4$ (95)	[84]

References on p. 22

39	R_2SnX_2 X = Cl [45, 53, 132], Br [45, 53], F, I, SCN [45]	1:1, mostly mixed without a solvent	$R_2Sn(X)OCH_3$	[45, 53, 132]
40	$(\text{-}R_2SnO\text{-})_n$	C_6H_6, reflux	$R_2(CH_3O)SnOSn(OCH_3)R_2$	[75]
41	$R_2Sn(OC(CH_3){=}CHCOCH_3)_2$	in C_6H_{14} [98], 1:1, cooling to −60° [93, 98]	$R_2Sn(OCH_3)OC(CH_3){=}CHCOCH_3$	[93, 98]
42	$R_2Sn(OOCR')_2$ R′ = CH_3, $C_{11}H_{23}$ [45, 53], CH_2Cl [53]	1:1, in CCl_4 or C_5H_{12} [45], C_6H_6 or C_6H_{14} [53], at 0 to 20°	$R_2Sn(OCH_3)OOCR'$ (60 to 90)	[45, 53]
	R′ = CH_3, CH_2Cl	1:1, sealed tube, −70 to +140°, 2 to 3 h	$R'COO(R_2)SnOSn(R_2)OCH_3$ and $R'COOCH_3$	[38]
43	$R_2Sn(OSO_2R')_2$ OSO_2R' = camphorsulfonate	—	$R_2Sn(OCH_3)OSO_2R'$	[45]
44	$(\text{-}R_2SnS\text{-})_3$	CCl_4, exothermic	$CH_3O(R_2)SnSSn(R_2)OCH_3$ (61)	[67]
45	$TiCl_4$	—	1:1 complex	[53]
46	$Hg(Si(CH_3)_3)_2$	25°/0.1 h	$(\text{-}R_2Sn\text{-})_n$ (89), $(CH_3)_3SiOCH_3$ (90), and Hg (95)	[108]
47	$Hg(Si(CH_3)_3)_2$ $+ CH_3I$	1:1:5, C_6H_{14}, −15 to +20°	$R_2(CH_3)SnI$ (64), $(CH_3)_3SiOCH_3$, and Hg	[108]
48	$(CO)_4OsH_2$	1:1, in CH_3CN, cooling	$(CO)_4Os(\mu\text{-}SnR_2)_2Os(CO)_4$	[96]

References on p. 22

$(C_4H_9)_2Sn(OCH_3)_2$ stabilizes PVC [2, 18, 77] as well as those liquid stabilizers which consist partly of dialkyltin sulfide and organotin maleate (used for Cl-containing resins) [110]. For the use of the title compound as a stabilizer for polyacrylo- and polymethacrylonitrile, see [82, 112, 168]; for polyacetals, see [69]; and for the copolymer of ethylene glycol and 1,2-bis(4-carbomethoxyphenoxy)ethane, see [70]. Combinations of $(C_4H_9)_2Sn(OCH_3)_2$ with epoxides, like 5,6-epoxyoctanol [29], or benzophenone derivatives also stabilize resin compositions [86]. Addition of $(C_4H_9)_2Sn(OCH_3)_2$ to polyester molding compositions causes increased flame resistance [56]. The addition to PVC compositions containing azodicarbonamide causes an activation of this foaming agent [115, 116].

$(C_4H_9)_2Sn(OC_2H_5)_2$ (Table **1**, No. **2**) is formed in the reaction of excess C_2H_5OH with $(C_4H_9)_2Sn(CF{=}CF_2)_2$ (reflux, 20 h, 95% yield) [19] or with $(C_4H_9)_2Sn(SC_3H_7)_2$ (very slow reaction even with $4\text{-}CH_3C_6H_4SO_3H$ as a catalyst; azeotropic removal of C_3H_7SH at 120 to 125°C during 200 h, 90% yield) [51].

IR spectrum (neat, assigned bands): $\delta(CH_2)$/O 1459, $\delta_{as}(CH_3)$ 1450, $\delta_{as}(CH_2)$/Sn 1412 (w), $\delta_s(CH_3)$ 1372, $\delta_s(CH_2)$/Sn 1151 (w), $\nu_{as}(CO)$ 1055, $\nu_s(CO)$ 1047 (sh), $\varrho(CH_2)$ 751 (w), $\varrho(CH_2Sn)$ 697 (sh), $\nu_{as}(SnO)$ 679, $\nu_s(SnO)$ 605, $\nu_{as}(SnC)$ 500, $\nu_s(SnC)$ 453 [25]. A comparison of the spectroscopic results with those of $(C_4H_9)_2Sn(OCH_3)_2$ suggests association in the pure liquid state (cf. Formulas I and II on p. 9) [87].

The chemical reactions of $(C_4H_9)_2Sn(OC_2H_5)_2$ are listed in Table 4.

$(C_4H_9)_2Sn(OC_2H_5)_2$ is used as a stabilizer for PVC [2] and for the manufacture of liquid crystal displays [141]. It catalyzes the transesterification of $CH_2{=}C(CH_3)COOCH_3$ with $(CH_3)_2NCH_2CH_2OH$ and increases the yield of $CH_2{=}C(CH_3)COOCH_2CH_2N(CH_3)_2$ by preventing side reactions [130, 136]. $(C_4H_9)_2Sn(OC_2H_5)_2$ is an effective catalyst in the preparation of alkylsilicate binders used in the refractory technology [143]. It also catalyzes the polymerization of ethylene carbonate which is accompanied by partial decarboxylation yielding $(\text{-}CH_2CH_2OCOOCH_2CH_2O\text{-})_n$ [161] but is ineffective as a catalyst for the macrocyclization of lactone, $\overline{OCH_2CH_2C}{=}O$ [146].

$(C_4H_9)_2Sn(OC_3H_7)_2$ (Table **1**, No. **3**). The IR spectra of the neat compound and of its solutions in CCl_4 (1, 0.1, and 0.01 mol/L) in the 1090 to 1025 cm^{-1} region are depicted [41] and discussed in terms of the existence of only dimeric species in the pure liquid state [41, 87]. The following bands have been assigned for solution spectra: $\nu(CO)$ 1068 (ms) in CCl_4, or 1068 (ms) and 1058 (sh) in $CHCl_3$, $\nu_{as}(SnO)$ and $\nu_{as}(SnC)$ 603 (w) and 581 (w, sh), $\nu_s(SnC)$ 518 (w), $\nu_s(SnO)$ 485 (w) in C_6H_{14} [35]. Other assigned frequencies are listed in Table 5.

$(C_4H_9)_2Sn(OC_3H_7)_2$ reacts with COS in C_6H_6 under reflux for 2 h to yield $(C_3H_7O)_2CO$ (69% yield) and with CS_2 at 65°C for 72 h [109] or in a sealed tube at 120°C for 20 h [78] to give $C(OC_3H_7)_4$ (94% yield [109]) each along with $(\text{-}(C_4H_9)_2SnS\text{-})_3$. The reaction with $\overline{OCH_2CH_2OCO}$ at 100°C for 8 h affords $(C_4H_9)_2\overline{SnOCH_2CH_2O}$ (45% yield) and $(C_3H_7O)_2CO$ (51% yield) [110].

Table 4
Reactions of $(C_4H_9)_2Sn(OC_2H_5)_2$.
The C_4H_9 groups on the Sn atom are abbreviated R.

No.	reactant	conditions	product (yield) and remarks	Ref.
1	CO_2	in C_6H_6, 100°/10 h, CO_2 under pressure	R_2SnCO_3 and $(C_2H_5O)_2CO$	[109]
2	COS	C_6H_6 under reflux, 2 h	$(\text{-}R_2SnS\text{-})_3$ and $(C_2H_5O)_2CO$	[109]
3	CS_2	excess CS_2, 20°	$R_2Sn(OC_2H_5)SC(S)OC_2H_5$ and $R_2Sn(SC(S)OC_2H_5)_2$	[78]
		65°/72 h [109], 96°/5 h, sealed tube [78]	$(\text{-}R_2SnS\text{-})_3$ and $C(OC_2H_5)_4$ (80 [78], 97 [109])	[78, 109]
4	$(CH_3)_2C(CH_2OH)_2$	?	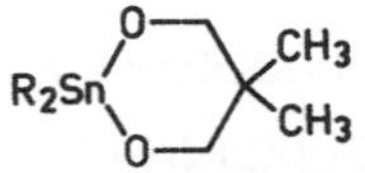 intermediate in the preparation of lipophilic lithium ionophores delivering lithium into rat brain	[164]
5	$CH_3COCH_2COCH_3$	1:2*)	$R_2Sn(OC(CH_3)\text{=}CHCOCH_3)_2$ (75)	[34]
6	CH_3COCH_2COOR' $R' = CH_3, C_2H_5$	1:2*)	$R_2Sn(OC(CH_3)\text{=}CHCOOR')_2$ (83, 62)	[34]
7	$NH_2CH_2CH_2OH$	1:1*)	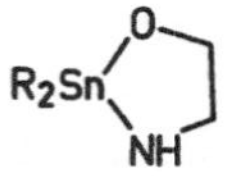 (35)	[50]
8	$NH(CH_2CH_2OH)_2$	1:1*)	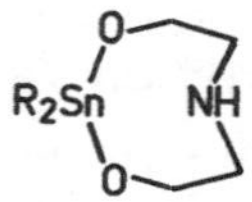 (98)	[50]
9	$N(CH_2CH_2OH)_3$	1:1*)	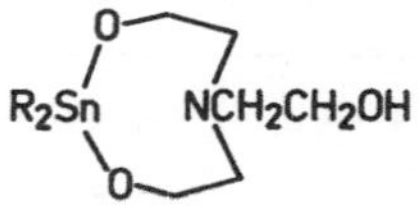 (98)	[50]
10	$NR'_2CH_2CH_2CH_2OH$ $R' = H, C_2H_5$	1:1*)	$R_2Sn(OC_2H_5)OCH_2CH_2CH_2NR'_2$ (43, 62)	[89]
		1:2*)	$R_2Sn(OCH_2CH_2CH_2NR'_2)_2$ (45, 64)	

References on p. 22

Table 4 (continued)

No.	reactant	conditions	product (yield) and remarks	Ref.
11	$NR'_2CH_2CH(CH_3)OH$ R′ = H, CH_3	1:1*) 1:2*)	$R_2Sn(OC_2H_5)OCH(CH_3)CH_2NR'_2$ (37, 65) $R_2Sn(OCH(CH_3)CH_2NR'_2)_2$ (36, 59)	[89]
12	$NH(R')CH_2CH_2OH$ R′ = CH_3, C_2H_5	1:1*) 1:2*)	$R_2Sn(OC_2H_5)OCH_2CH_2NHR'$ (54, 67) $R_2Sn(OCH_2CH_2NHR')_2$ (52, 53)	[89]
13	R′SH R′ = C_3H_7, C_4H_9, i-C_4H_9, t-C_4H_9, $C_{12}H_{25}$, $C_6H_5CH_2$, C_6H_5	1:2*)	$R_2Sn(SR')_2$	[51]
14	$HSCH_2CH_2OH$	1:1*)	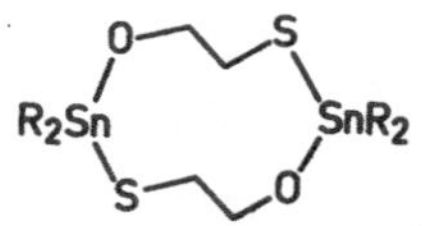 (96)	[57]
15	$HSCH_2CH(OH)CH_2OH$	1:1*)	$(-R_2SnSCH_2CH(OH)CH_2O-)_3$ (100)	[57]
16	$HSCH_2CH_2COOH$	1:1*)	$(-R_2SnSCH_2CH_2O-)_n$ (98)	[57]
17	$HSCH(CH_3)COOH$	1:1*)	$(-R_2SnSCH(CH_3)COO-)_n$ (98)	[57]
18	CH_3CONH_2	1:1*)	$(-R_2SnN(COCH_3)-)_n$ (97)	[90]
19	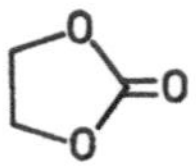	N_2, 40°/4 h, or in C_6H_6 under reflux, 1 d	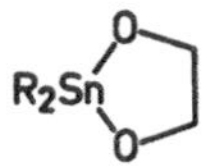 (82) and $(C_2H_5O)_2CO$ (73 to 93)	[110]
20	$R'C(S)SCH_3$ R′ = CH_3, i-C_3H_7, C_6H_5, 4-ClC_6H_4, 2-, 4-$CH_3C_6H_4$, 3-, 4-$CH_3OC_6H_4$	excess dithiocarboxylate, N_2, 60 to 95°/3 to 36 h	$(-R_2SnS-)_3$, $R_2Sn(SCH_3)_2$, and $R'C(OC_2H_5)_3$ (50 to 100)	[117]
21	$(-CH_3(H)SiO-)_n$	1:1, cooling 1:2, cooling	$R_2Sn(H)OC_2H_5$ (76) R_2SnH_2 (66)	[49]

*) In C_6H_6, with removal of the binary azeotrope C_2H_5OH-C_6H_6.

References on p. 22

Table 5
IR Spectra of Neat $(C_4H_9)_2Sn(OC_3H_7)_2$.
Wave numbers in cm^{-1}.

[40]	assignment	[25]	assignment
2962 (vs, sh)	$\nu_{as}(CH_3)$		
2953 (vs)	$\nu'_s(CH_3)$		
2931 (vs, sh)	$\nu_s(CH_3)$		
2920 (vs)	$\nu_{as}(CH_2)/C$?		
2902 (s, sh)	$\nu_s(CH_2)/C$		
2867 (s)	$\nu_{as}(CH_2)/Sn$		
2848 (s)	$\nu_s(CH_2)/Sn$		
1464 (ms)	$\delta_{as}(CH_3)$	1461	$\delta(CH_2)/O$
1456 (ms, sh)	$\delta(CH_2)/C$?	1455	$\delta_{as}(CH_3)$
1449 (m, sh)	$\delta'_s(CH_3)$		
1436 (mw, sh)	$\delta(CH_2)/C$?		
1416 (mw)	$\delta(CH_2)/Sn$	1415 (w)	$\delta_{as}(CH_2)/Sn$
1379 (ms, sh)	$\delta_s(CH_3)$	1379	$\delta_s(CH_3)$
		1152 (w)	$\delta_s(CH_2)/Sn$
1072 (ms)	$\gamma(CH_3)$		
1068 (ms)	$\nu(CO)$	1066	$\nu_{as}(CO)$
		1048	$\nu_s(CO)$
963 (mw)	$\nu(CC)$		
746 (vw, sh)	$\gamma(CH_2C)$	753 (w)	$\varrho(CH_2)$
689 (w)	$\gamma(CH_2Sn)$	697 (sh)	$\varrho(CH_2Sn)$
		679	$\nu_{as}(SnO)$
603 (mw) }	$\nu_{as}(SnC) + \nu_{as}(SnO)$	605	$\nu_s(SnO)$
581 (w, sh) }			
		532	$\nu_{as}(SnC)$
518 (w)	$\nu_s(SnC)$		
485 (w)	$\nu_s(SnO)$	482	$\nu_s(SnC)$

$(C_4H_9)_2Sn(OC_3H_7\text{-}i)_2$ (Table 1, No. 4). The dependence of the ^{119}Sn chemical shift on the concentration of the compound in CCl_4 and of the neat compound on the temperature shows the following values:

CCl_4 mol%	δ (ppm) ±5 ppm	Ref.	neat t (°C)	δ (ppm) ±5 ppm	Ref.
10.2	−31	[94]	22	−100 (±2)	[119]
12.2	−29		25	− 90	[87, 94]
15.4	−31		41	− 65	[94]
21.0	−29		50	− 38	
30.0	−40		59	− 33	
52.0	−71		80	− 27	
62.0	−76		93	− 27	

References on p. 22

The nonlinear increase in the δ values demonstrate an increase in dimerization with increasing concentration and decreasing temperature, even being incomplete in the neat liquid at ambient temperature [87, 119]:

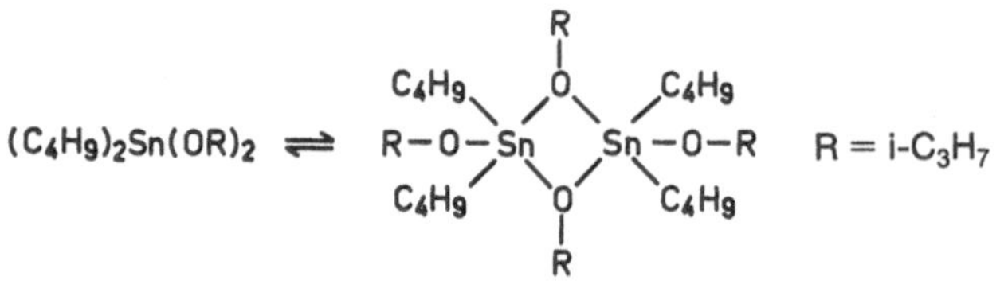

III IV

The observation of only one ^{119}Sn resonance signal shows that the equilibration between III and IV is rapid. The changes in enthalpy and entropy calculated for the dissociation of IV are: $\Delta H = 80 \pm 6$ [119] or 100 ± 17 kJ/mol [87] and $\Delta S = 255 \pm 17\ J \cdot mol^{-1} \cdot K^{-1}$. The extrapolated δ values for the monomer and the dimer are −22.5 and ~140 ppm, respectively [119].

^{13}C NMR data for the $(C_4H_9)_2Sn$ group (δ in ppm, $^1J(Sn,C)$ in Hz): 13.6 (C-δ), 19.2 (C-α, J = 520), 27.0 (C-γ, J = 88), 27.5 (C-β, J = 38). The $^1J(Sn,C)$ value of 520 Hz is intermediate between the corresponding values of the exclusively dimeric $(C_4H_9)_2Sn(OCH_3)_2$ (642 Hz) and the exclusively monomeric $(C_4H_9)_2Sn(OC_4H_9\text{-}t)_2$ (496 Hz), confirming the equilibrium III ⇌ IV [93].

The reaction of $(C_4H_9)_2Sn(OC_3H_7\text{-}i)_2$ with COS yields $(i\text{-}C_3H_7O)_2CO$ and $(\text{-}(C_4H_9)_2SnS\text{-})_3$, the reaction with CS_2 affords $(i\text{-}C_3H_7O)_2CO$, $(i\text{-}C_3H_7O)(i\text{-}C_3H_7S)CO$, and $(\text{-}(C_4H_9)_2SnS\text{-})_3$ in yields dependent on the reaction temperature [78, 109]. A 61% yield of $(i\text{-}C_3H_7O)_2CO$ is obtained when $(C_4H_9)_2Sn(OC_3H_7\text{-}i)_2$ is reacted with ethylene carbonate, $\overline{OCH_2CH_2OCO}$, at 100°C for 8 h, a 99% yield when the reaction is conducted for 26 h in refluxing $C_6H_5CH_3$. The Sn-containing product is $(C_4H_9)_2\overline{SnOCH_2CH_2O}$ [110]. In the reactions of the title compound with $CH_3CH{=}CHCH(CH_3)OH$ (1:2 mole ratio, in refluxing C_6H_6) or with $C_6H_5CH_2OH$ (heating in vacuum), the exchange products $(C_4H_9)_2Sn(OCH(CH_3)CH{=}CHCH_3)_2$ (70% yield) [127] and $(C_4H_9)_2Sn(OCH_2C_6H_5)_2$ [23] are formed. The reaction with $N(C_2H_5)_2CH_2CH_2OH$ gives $(C_4H_9)_2Sn(OC_3H_7\text{-}i)OCH_2CH_2N(C_2H_5)_2$ (1:1 mole ratio) or $(C_4H_9)_2Sn(OCH_2CH_2N(C_2H_5)_2)_2$ (1:2 mole ratio) [72]. $(C_4H_9)_2\overline{SnSCH_2COO}$ is formed in a quantitative yield by heating 1:1 but also 1:2 mixtures of $(C_4H_9)_2Sn(OC_3H_7\text{-}i)_2$ and $HSCH_2COOH$ for 3 to 4 h in C_6H_6, with azeotropic removal of $i\text{-}C_3H_7OH$ [85]. The reactions with CH_3COX (X = CN, NCS) in refluxing C_6H_{12} for 2 h yield, dependent on the mole ratio, $(C_4H_9)_2Sn(X)OC_3H_7\text{-}i$ or $(C_4H_9)_2SnX_2$ along with $CH_3COOC_3H_7\text{-}i$ [118].

$(C_4H_9)_2Sn(OC_4H_9)_2$ (Table **1**, No. **5**) is formed by heating $(C_4H_9)_2(C_4H_9O)SnOSn(OC_4H_9)\text{-}(C_4H_9)_2$ (the isolable intermediate in the reaction between $(\text{-}(C_4H_9)_2SnO\text{-})_n$ and C_4H_9OH according to Method II) under reduced pressure at 180 to 220°C, along with $(\text{-}(C_4H_9)_2SnO\text{-})_n$ as the second disproportionation product [75].

On the basis of the temperature dependence of ^{119}Sn chemical shift, it is assumed that only dimeric species are present in the neat liquid at ambient temperature. With increasing temperature, the dimers dissociate forming an equilibrium with the monomeric species (neat, δ ± 5 ppm/°C): −161/25 [87, 94, 119], −146/80, −126/99, and −115/118 (cf. No. 4) [94]. The change in the enthalpy is $\Delta H = 59 \pm 13$ kJ/mol [119].

References on p. 22

The IR spectra of $(C_4H_9)_2Sn(OC_4H_9)_2$ in the ν(CO) region are depicted for the neat liquid and for 1 to 0.01 M solutions in CCl_4 at room temperature [41] and are discussed in terms of the monomer-dimer equilibrium. The following bands of solution spectra are assigned: ν(CO) 1074 (s) in CCl_4 or 1068 (s) in $CHCl_3$, $\nu_{as}(SnO)$ and $\nu_{as}(SnC)$ 611 (w, sh) and 598 (w), $\nu_s(SnC)$ 509 (w), and $\nu_s(SnO)$ 487 (w, sh) cm^{-1} in C_6H_{14} [35]. The assigned bands in the region 3000 to 400 cm^{-1} for the liquid are listed in Table 6 [25, 40].

Table 6
IR Spectra of Neat $(C_4H_9)_2Sn(OC_4H_9)_2$.
Wave numbers in cm^{-1}.

[40]	assignment	[25]	assignment
2962 (vs, sh)	$\nu_{as}(CH_3)$		
2954 (vs)	$\nu'_s(CH_3)$		
2929 (vs, sh)	$\nu_s(CH_3)$		
2921 (vs)	$\nu_{as}(CH_2)/C$?		
2903 (s, sh)	$\nu_s(CH_2)/C$		
2869 (s)	$\nu_{as}(CH_2)/Sn$?		
2853 (s)	$\nu_s(CH_2)/Sn$		
1464 (ms)	$\delta_{as}(CH_3)$	1462	$\delta(CH_2)/O$
1456 (ms, sh)	$\delta(CH_2)/C$?	1456	$\delta_{as}(CH_3)$
1448 (m, sh)	$\delta'_s(CH_3)$		
1438 (mw, sh)	$\delta(CH_2)/C$		
1416 (mw)	$\delta(CH_2)/Sn$	1414 (w)	$\delta_{as}(CH_2)/Sn$
1376 (ms)	$\delta_s(CH_3)$	1376	$\delta_s(CH_3)$
		1152 (w)	$\delta_s(CH_2)/Sn$
1074 (s)	ν(C-O)	1057	$\nu_{as}(CO)$
		1027	$\nu_s(CO)$
963 (m)	ν(CC)		
745 (vw)	$\gamma(CH_2C)$	749 (w)	$\varrho(CH_2)$
688 (w, sh)	$\gamma(CH_2Sn)$	697 (sh)	$\varrho(CH_2Sn)$
		678	$\nu_{as}(SnO)$
611 (w, sh) 598 (w)	$\nu_{as}(SnC) + \nu_{as}(SnO)$	611	$\nu_s(SnO)$
		535	$\nu_{as}(SnC)$
509 (w)	$\nu_s(SnC)$		
487 (w)	$\nu_s(SnO)$	489	$\nu_s(SnC)$

The reaction of $(C_4H_9)_2Sn(OC_4H_9)_2$ with COS for 2 h in refluxing C_6H_6 leads to $(C_4H_9O)_2CO$ (70% yield) [109], that with CS_2 at 65°C for 2 h [109] or at 120°C for 20 h [78] to $C(OC_4H_9)_4$ (85 to 95% yield), each along with $(\text{-}(C_4H_9)_2SnS\text{-})_3$ [78, 109]. Heating the title compound with $\overline{OCH_2CH_2OCO}$ to 120°C for 16 h affords $(C_4H_9O)_2CO$ and $(C_4H_9)_2\overline{SnOCH_2CH_2O}$ [110], with $C_6H_5C(S)SCH_3$ to 95°C for 18 h yields $C_6H_5C(OC_4H_9)_3$ (45% yield), $(C_4H_9)_2Sn(SCH_3)_2$, and $(\text{-}(C_4H_9)_2SnS\text{-})_3$ [117]. The treatment of $(C_4H_9)_2Sn(OC_4H_9)_2$ with equimolar amounts of $P(O)(OC_2H_5)_2OH$ in refluxing $C_6H_5CH_3$ for 3 h affords $(C_4H_9)_2Sn(OC_4H_9)OP(O)(OC_2H_5)_2$ [91, 92]. $(C_4H_9)_2Sn(OC_4H_9)_2$ and NH_4X (X = Cl, I, NCS) react in refluxing $C_6H_5CH_3$ forming $(C_4H_9)_2SnX_2$ compounds in yields of 70 to 90%, along with C_4H_9OH and NH_3 [55]. The products obtained

References on p. 22

from mixing $(C_4H_9)_2Sn(OC_4H_9)_2$ with equimolar amounts of $(C_4H_9)_2SnCl_2$, $(C_4H_9)_2Sn(OOCCH_3)_2$, or $SnCl_4$ without a solvent, are formulated as 1:1 complexes in [53], whereas the exothermally formed products of the reactions of $(C_4H_9)_2Sn(OC_4H_9)_2$ with $(C_4H_9)_2SnX_2$ (X = Cl, Br) are described as $(C_4H_9)_2Sn(X)OC_4H_9$ compounds [44].

The anthelmintic activity of $(C_4H_9)_2Sn(OC_4H_9)_2$ has been tested against Raillietina cesticillus and Ascaridia galli from chicken; activity is very low when administered in the food, whereas, when administered by capsule as a single oral dose, 100 or 200 mg/kg body weight, the compound is at least 80 or 60% active, respectively [12].

The title compound is used as a stabilizer for PVC [2], as a catalyst in the synthesis of polyesters [8, 80], polyurethane coating compositions [37], diarylcarbonates [165], or polycarbonates [133], and for the recovery of polyols from polyurethane wastes [113]. It is also used as a nonstaining, nondiscoloring antioxidant for rubber [16], as an additive to dental impression material [149], and in compositions which improve the release properties and abrasion resistance of paper [68].

$(C_4H_9)_2Sn(OC_4H_9\text{-i})_2$ (Table **1**, No. **7**). The ^{119}Sn chemical shift shows the following dependence on concentration of the compound in solution at ambient temperature ($\delta \pm 2$ ppm/mol% in CCl_4): −105/7%, −114/9%, and −132/17% [94]; this is caused by the rapid monomer-dimer equilibrium (cf. No. 4). Together with the value for the neat compound, $\delta = -154$ ppm, the limiting chemical shifts $\delta = -24$ and −160 ppm for the monomer and dimer, respectively, have been extrapolated. The change in enthalpy and entropy associated with the dissociation of the dimer have been calculated to be $\Delta H = 74 \pm 2$ kJ/mol and $\Delta S = 226 \pm 17\ J \cdot mol^{-1} \cdot K^{-1}$ [119].

$(C_4H_9)_2Sn(OC_4H_9\text{-t})_2$ (Table **1**, No. **8**). Since the ^{119}Sn chemical shift value is independent of the concentration of the compound in CCl_4, $\delta = -34$ ppm (4 to 100 mol%) [94], No. 8 is suggested to be exclusively monomeric both in solution and as neat liquid [94, 87].

The monomeric state is also deduced from the depicted IR spectra in the ν(CO) region of the neat compound and its solutions in CCl_4 (1, 0.1, and 0.01 mol/L) [41]. Assigned frequencies of solution spectra: ν(CO) 959 (s) in CCl_4 or 959 (s) and 945 (ms, sh) in $CHCl_3$; $\nu_{as}(SnO)$ and $\nu_{as}(SnC)$ 611 (w, sh) and 602 (w, sh); $\nu_s(SnC)$ 513 (vw), $\nu_s(SnO)$ 476 (vw, sh) in C_6H_{14} [35]. The assigned frequencies of the spectrum of the neat compound are listed in Table 7 [40].

Table 7
IR Spectrum of Neat $(C_4H_9)_2Sn(OC_4H_9\text{-t})_2$ [40].
Wave numbers in cm^{-1}.

wave number	assignment	wave number	assignment
2964 (vs, sh)	$\nu_{as}(CH_3)$	1413 (w)	$\delta(CH_2)/Sn$
2955 (vs)	$\nu'_s(CH_3)$	1378 (m, sh)	$\delta_s(CH_3)$
2924 (vs, sh)	$\nu_s(CH_3)$	1074 (w)	$\gamma(CH_3)$
2918 (vs)	$\nu_{as}(CH_2)/C$?	959 (s)	ν(CO)
2896 (s, sh)	$\nu_s(CH_2)/C$	742 (vw)	$\gamma(CH_2C)$
2870 (s)	$\nu_{as}(CH_2)/Sn$?	692 (w)	$\gamma(CH_2Sn)$
2849 (s, sh)	$\nu_s(CH_2)/Sn$	611 (w, sh)	$\nu_{as}(SnC) + \nu_{as}(SnO)$
1464 (ms)	$\delta_{as}(CH_3)$	602 (w, sh)	$\nu_{as}(SnC) + \nu_{as}(SnO)$
1456 (ms, sh)	$\delta(CH_2)/C$?	513 (vw)	$\nu_s(SnC)$
1447 (mw, sh)	$\delta'_s(CH_3)$	476 (vw, sh)	$\nu_s(SnO)$
1435 (w, sh)	$\delta(CH_2)/C$?		

References on p. 22

The reaction of No. 8 with COS in refluxing C_6H_6 for 2 h yields only traces of $(t\text{-}C_4H_9O)_2CO$ and No. 8 is largely recovered [109]. The product isolated from the reaction of No. 8 with CS_2 at 120°C for 20 h is $CH_2{=}C(CH_3)_2$ (80%) [78]. The title compound and $\overline{OCH_2CH_2OCO}$ react at 120°C within 16 h forming $(t\text{-}C_4H_9O)_2CO$ (62%) and $(C_4H_9)_2\overline{SnOCH_2CH_2O}$ (57%) [110].

$(C_4H_9)_2Sn(OC_5H_{11})_2$ (Table **1**, No. **9**). The assigned IR frequencies of the compound in solution are: ν(CO) 1076 (s) and 1057 (ms) in CCl_4 or 1073 (ms) and 1054 (s) in $CHCl_3$; ν_{as}(SnO) and ν_{as}(SnC) 611 (w) and 584 (w); ν_s(SnC) 506 (vw), ν_s(SnO) 475 (vw) in C_6H_{14} [35]. The assigned frequencies of the neat compound are listed in Table 8 [40].

Table 8
IR Spectrum of Neat $(C_4H_9)_2Sn(OC_5H_{11})_2$ [40].
Wave numbers in cm^{-1}.

wave number	assignment	wave number	assignment
2962 (vs, sh)	$\nu_{as}(CH_3)$	1415 (mw)	$\delta(CH_2)$/Sn
2955 (vs)	$\nu'_s(CH_3)$	1377 (m)	$\delta_s(CH_3)$
2930 (vs)	$\nu_s(CH_3)$	1076 (s)	ν(CO)
2923 (vs)	$\nu_{as}(CH_2)$/C ?	1057 (ms)	
2901 (s, sh)	$\nu_s(CH_2)$/C	960 (w)	γ(CC)
2869 (s)	$\nu_{as}(CH_2)$/Sn ?	745 (vw)	$\gamma(CH_2C)$
2849 (s)	$\nu_s(CH_2)$/Sn	693 (vw)	$\gamma(CH_2Sn)$
1465 (ms)	$\delta_{as}(CH_3)$	611 (w)	ν_{as}(SnC) + ν_{as}(SnO)
1453 (m, sh)	$\delta(CH_2)$/C ?	584 (w, sh)	
1450 (m, sh)	$\delta'_s(CH_3)$	506 (w)	ν_s(SnC)
1437 (mw, sh)	$\delta(CH_2)$/C ?	475 (vw)	ν_s(SnO)

$(C_4H_9)_2Sn(OC_8H_{17})_2$ (Table **1**, No. **11**). The ν(CO) range of the IR spectra of the compound as liquid film and in CCl_4 solutions (1 or 0.01 M) is depicted; it indicates no association in both states [41]. Assigned frequencies of solution spectra are: ν(CO) 1072 (ms) in CCl_4 or 1071 (ms, sh) and 1054 (ms, sh) in $CHCl_3$; ν_{as}(SnO) and ν_{as}(SnC) 603 (w, sh) and 591 (w); ν_s(SnC) 504 (w), and ν_s(SnO) 478 (w, sh) cm^{-1} in C_6H_{14} [35]. Assigned frequencies of the IR spectrum are summarized in Table 9 [40].

Table 9
IR Spectrum of Neat $(C_4H_9)_2Sn(OC_8H_{17})_2$ [40].
Wave numbers in cm^{-1}.

wave number	assignment	wave number	assignment
2962 (s, sh)	$\nu_{as}(CH_3)$	1414 (w)	$\delta(CH_2)$/Sn
2953 (s)	$\nu'_s(CH_3)$	1377 (m)	$\delta_s(CH_3)$
2927 (vs)	$\nu_s(CH_3)$	1072 (ms)	ν(CO)
2921 (vs)	$\nu_{as}(CH_2)$/C ?	959 (w)	γ(CC)
2901 (s, sh)	$\nu_s(CH_2)$/C	744 (vw)	$\gamma(CH_2C)$
2868 (s)	$\nu_{as}(CH_2)$/Sn ?	692 (w, sh)	$\gamma(CH_2Sn)$
2852 (vs)	$\nu_s(CH_2)$/Sn	603 (w, sh)	ν_{as}(SnC) + ν_{as}(SnO)
1465 (ms)	$\delta_{as}(CH_3)$	591 (w)	
1455 (m, sh)	$\delta(CH_2)$/C ?	504 (w)	ν_s(SnC)
1450 (m, sh)	$\delta'_s(CH_3)$	478 (w)	ν_s(SnO)
1436 (mw, sh)	$\delta(CH_2)$/C ?		

References on p. 22

$(C_4H_9)_2Sn(OC_6H_{11}\text{-c})_2$ (Table **1**, No. **15**) is formed in a 70% yield by the reaction of $(C_4H_9)_2SnH_2$ with cyclohexanone in the presence of $(C_4H_9)_2Sn(acac)_2$ and $(C_4H_9)_2Sn(OR')_2$ or traces of No. 15 itself as the catalysts (1:2:0.01:0.02 mole ratio; R' = CH_3, C_2H_5) in C_6H_{12} at 30°C for 40 min. The mechanism of catalysis and the influence of the solvent were studied thoroughly [98].

The assigned IR frequencies are listed in Table 10 [40]. Solution spectra show ν(CO) 1066 (vs) in CCl_4, or 1066 (vs) and 1056 (vs, sh) in $CHCl_3$; $\nu_{as}(SnO)$ and $\nu_{as}(SnC)$ 604 (w) and 586 (w, sh); $\nu_s(SnC)$ 516 (w), and $\nu_s(SnO)$ 474 (vw, sh) cm^{-1} in C_6H_{12} [35].

Table 10
IR Spectrum of Neat $(C_4H_9)_2Sn(OC_6H_{11}\text{-c})$ [40].
Wave numbers in cm^{-1}.

wave number	assignment	wave number	assignment
2962 (s, sh)	$\nu_{as}(CH_3)$	1414 (mw)	$\delta(CH_2)/Sn$
2954 (s, sh)	$\nu'_s(CH_3)$	1377 (m)	$\delta_s(CH_3)$
2929 (vs)	$\nu_s(CH_3)$	1076 (s, sh)	$\gamma(CH_3)$
2921 (vs, sh)	$\nu_{as}(CH_2)/C$?	1066 (vs)	$\nu(CO)$
2902 (vs, sh)	$\nu_s(CH_2)/C$	960 (mw, sh)	$\gamma(CC)$
2869 (s, sh)	$\nu_{as}(CH_2)/Sn$?	742 (vw)	$\gamma(CH_2C)$
2853 (s)	$\nu_s(CH_2)/Sn$	691 (mw)	$\gamma(CH_2Sn)$
1463 (m)	$\delta_{as}(CH_3)$	604 (w)	$\nu_{as}(SnC) + \nu_{as}(SnO)$
1453 (ms, sh)	$\delta(CH_2)/C$?	586 (w, sh)	
1449 (ms)	$\delta'_s(CH_3)$	516 (w)	$\nu_s(SnC)$
1436 (m, sh)	$\delta(CH_2)/C$?	474 (vw, sh)	$\nu_s(SnO)$

$(C_4H_9)_2Sn(OCH_2CF_3)_2$ (Table **1**, No. **17**) is formed along with $(\text{-}(C_4H_9)_2SnO\text{-})_n$ when $CF_3CH_2O(C_4H_9)_2SnOSn(C_4H_9)_2OCH_2CF_3$ is heated to 180 to 220°C under reduced pressure [75].

$(C_4H_9)_2Sn(OCH(CF_3)_2)_2$ (Table **1**, No. **18**) is the product of the exothermic reaction between $(C_4H_9)_2SnH_2$ and $(CF_3)_2CO$ (100% yield) [31].

$(C_4H_9)_2Sn(OCH_2C_6H_5)_2$ (Table **1**, No. **42**) is formed along with $(\text{-}(C_4H_9)_2SnO\text{-})_n$ in the thermolysis of $C_6H_5CH_2O(C_4H_9)_2SnOSn(C_4H_9)_2OCH_2C_6H_5$ under reduced pressure [75].

References:

[1] Burt, S. L., Bakelite Corp. (U.S. 2489518 [1949]; C.A. **1950** 5639).
[2] Cleverdon, D., Staudinger, J. J. P., Distillers Co., Ltd. (U.S. 2481086 [1949]; C.A. **1950** 4718).
[3] Bakelite Corp. (Brit. 664133 [1952]; C.A. **1952** 11230).
[4] Burt, S. L., Union Carbide and Carbon Corp. (U.S. 2583048 [1952]; C.A. **1953** 146).
[5] Mack, G. P., Parker, E., Advanced Solvents and Chemical Corp. (U.S. 2592926 [1952]; C.A. **1952** 11767; U.S. 2626953 [1953]; C.A. **1953** 11224; Brit. 694944 [1953]; C.A. **1954** 7625).
[6] Farbwerke Hoechst A.-G. (Brit. 797073 [1953]; C.A. **1959** 22714).

[7] Mack, G. P., Parker, E., Advanced Solvents and Chemical Corp. (Fr. 1078569 [1953/54]; C. **1950/54** 1864).
[8] Caldwell, J. R., Eastman Kodak Co. (U.S. 2720507 [1955]; C.A. **1956** 2205).
[9] Mack, G. P., Parker, E., Advanced Solvents and Chemical Corp. (U.S. 2727917 [1955]; C.A. **1959** 5197; Ger. 953079 [1956]; Brit. 766857 [1957]; C.A. **1957** 8788).
[10] Mack, G. P., Parker, E., Advanced Solvents and Chemical Corp. (U.S. 2700657 [1955]; C.A. **1956** 397).

[11] Foster, W. E., Koenig, P. E., Ethyl Corp. (U.S. 2998407 [1956]; C.A. **56** [1962] 6170).
[12] Kerr, K. B., Walde, A. W. (Exptl. Parasitol. **5** [1956] 560/70).
[13] Weinberg, E. L., Metal and Thermit Corp. (Brit. 753998 [1956]; C.A. **1957** 6683).
[14] Carlisle Chemical Works, Inc. (Brit. 781452 [1957]; C.A. **1958** 3864).
[15] Chemische Werke Albert (Brit. 783764 [1957]; C.A. **1958** 5883).
[16] Tomka, L. A., Weinberg, E. L., Metal and Thermit Corp. (U.S. 2798862 [1957]; C.A. **1957** 15166).
[17] Weinberg, E. L., Metal and Thermit Corp. (U.S. 2796412 [1957]; C.A. **1957** 16537).
[18] Longman, S. H., Carlisle Chemical Works, Inc. (U.S. 2921917 [1960]; C.A. **1960** 8152).
[19] Seyferth, D., Raab, G., Brändle, K. A. (J. Org. Chem. **26** [1961] 2934/7).
[20] Alleston, D. L., Davies, A. G. (J. Chem. Soc. **1962** 2050/4).

[21] Zhivukhin, S. M., Dudikova, E. D., Ter-Sarkisyan, E. M. (Zh. Obshch. Khim. **32** [1962] 3059/61; J. Gen. Chem. [USSR] **32** [1962] 3010/1).
[22] Farbenfabriken Bayer A.-G. (Brit. 945068 [1960/63]; C.A. **60** [1964] 12051).
[23] Katsumura, T. (Nippon Kagaku Zasshi **83** [1963] 729/31; C.A. **59** [1963] 5185).
[24] Aleksandrov, A. Yu., Ohklobystin, O. Yu., Polak, L. S., Shpinel', V. S. (Dokl. Akad. Nauk SSSR **157** [1964] 934/7; Dokl. Phys. Chem. Proc. Acad. Sci. USSR **154/159** [1964] 768/71).
[25] Butcher, F. K., Gerrard, W., Mooney, E. F., Rees, R. G., Willis, H. A. (Spectrochim. Acta **20** [1964] 51/61).
[26] Gerrard, W., Mooney, E. F., Rees, R. G. (J. Chem. Soc. **1964** 740/5).
[27] Gerrard, W., Rees, R. G. (J. Chem. Soc. **1964** 3510/1).
[28] Lorberth, J., Kula, M. R. (Chem. Ber. **97** [1964] 3444/51).
[29] Mack, G. P., M and T Chemicals, Inc. (U.S. 3147285 [1956/64]; C.A. **62** [1965] 11973).
[30] Neumann, W. P., Schneider, B. (Angew. Chem. **76** [1964] 891).

[31] Cullen, W. R., Styan, G. E. (Inorg. Chem. **4** [1965] 1437/40).
[32] Goldshtein, I. P., Zemlyanskii, N. N., Shamagina, O. P., Guryanova, E. N., Panov, E. M., Slovokhotova, N. A., Kocheshkov, K. A. (Dokl. Akad. Nauk SSSR **163** [1965] 880/3; Dokl. Chem. Proc. Acad. Sci. USSR **160/165** [1965] 715/8).
[33] Mehrotra, R. C., Gupta, V. D. (J. Organometal. Chem. **4** [1965] 145/50).
[34] Mehrotra, R. C., Gupta, V. D. (J. Organometal. Chem. **4** [1965] 237/40).
[35] Mendelsohn, J., Marchand, A., Valade, J. (Compt. Rend. **261** [1965] 135/8).
[36] Pommier, J. C., Valade, J. (Compt. Rend. **260** [1965] 4549/52).
[37] Wyandotte Chemicals Corp. (Brit. 994348 [1961/65]; C.A. **63** [1965] 5904).
[38] Zemlyanskii, N. N., Panov, E. M., Shamagina, O. P., Kocheshkov, K. A. (Zh. Obshch. Khim. **35** [1965] 1029/31; J. Gen. Chem. [USSR] **35** [1965] 1034/5).
[39] Ismail, R. M. (J. Organometal. Chem. **6** [1966] 663/4).
[40] Mendelsohn, J., Marchand, A., Valade, J. (J. Organometal. Chem. **6** [1966] 25/44).

[41] Mendelsohn, J., Pommier, J. C., Valade, J. (Compt. Rend. C **263** [1966] 921/4).
[42] Neumann, W. P., Pedain, J., Sommer, R. (Liebigs Ann. Chem. **694** [1966] 9/18).

[43] Bellegarde, B., Pereyre, M., Valade, J. (Bull. Soc. Chim. France **1967** 3082/3).
[44] Chadha, R. N., Pande, K. C., Stauffer Chemical Co. (Neth. Appl. 6612421 [1965/67]; C.A. **67** [1967] No. 44618; Ger. Offen. 1542495 [1966/70]).
[45] Davies, A. G., Harrison, P. G. (J. Chem. Soc. C **1967** 298/300).
[46] Davies, A. G., Harrison, P. G. (J. Chem. Soc. C **1967** 1313/7).
[47] Goldshtein, I. P., Guryanova, E. N., Zemlyanskii, N. N., Syutkina, O. P., Panov, E. M., Kocheshkov, K. A. (Izv. Akad. Nauk SSSR Ser. Khim. **1967** 2201/7; Bull. Acad. Sci. USSR Div. Chem. Sci. **1967** 2115/9).
[48] Goldshtein, I. P., Guryanova, E. N., Zemlyanskii, N. N., Syutkina, O. P., Panov, E. M., Kocheshkov, K. A. (Dokl. Akad. Nauk SSSR **175** [1967] 836/9; Dokl. Chem. Proc. Acad. Sci. USSR **172/177** [1967] 688/91).
[49] Hayashi, K., Iyoda, J., Shiihara, I. (J. Organometal. Chem. **10** [1967] 81/94).
[50] Mehrotra, R. C., Gupta, V. D. (Indian J. Chem. **5** [1967] 643/5).

[51] Sukhani, D., Gupta, V. D., Mehrotra, R. C. (J. Organometal. Chem. **7** [1967] 85/90).
[52] Takeda, T., Iino, K., Ando, M., Kawakami, Y., Seki, T., Japan Telegram and Telephone Corp. and Nitto Chemical Industrial Co., Ltd. (Japan. 67-24044 [1964/67]; C.A. **68** [1968] No. 87885).
[53] Zemlyanskii, N. N., Goldshtein, I. P., Guryanova, E. N., Syutkina, O. P., Panov, E. M., Slovokhotova, N. A., Kocheshkov, K. A. (Izv. Akad. Nauk SSSR Ser. Khim. **1967** 728/35; Bull. Acad. Sci. USSR Div. Chem. Sci. **1967** 707/12).
[54] Maire, J. C., Ouaki, R. (Helv. Chim. Acta **51** [1968] 1150/4).
[55] Pande, K. C. (J. Organometal. Chem. **13** [1968] 187/94).
[56] Raichle, K., Alfes, F., Schnell, H., Prater, K., Farbenfabriken Bayer A.-G. (Ger. 1266497 [1965/68]; C.A. **69** [1968] No. 3448).
[57] Sukhani, D., Gupta, V. D., Mehrotra, R. C. (Australian J. Chem. **21** [1968] 1175/9).
[58] Argus Chemical Corp. (Brit. Amended 988955 [1962/69]; C.A. **76** [1972] No. 46928).
[59] Asahi Kasei Kogyo Kabunshiki Kaisha (Brit. 1151927 [1965/69]; C.A. **71** [1969] No. 13755).
[60] Dannley, R. L., Aue, W. A., Farrant, G. C. (U.S. 3458546 [1966/69]; C.A. **71** [1969] No. 102016).

[61] Marchand, A., Mendelsohn, J., Lebedeff, M., Valade, J. (J. Organometal. Chem. **17** [1969] 379/88).
[62] Senda, K., Ichikawa, A., Ozeki, T., Nakajima, E., Sakai, M., Nishikawa, T., Mitsubishi Rayon Co., Ltd. (Japan. 69-13589 [1966/69]; C.A. **72** [1970] No. 13783).
[63] Senda, K., Ichikawa, A., Oseki, T., Nakajima, E., Sakai, M., Yasui, A., Hirose, M., Mitsubishi Rayon Co., Ltd. (Japan. 69-28872 [1966/69]; C.A. **72** [1970] No. 122813, **73** [1970] No. 131931, **73** [1970] No. 57071).
[64] Voronkov, M. G., Romadan, Yu. P. (Zh. Obshch. Khim. **39** [1969] 2785/6; J. Gen. Chem. [USSR] **39** [1969] 2721).
[65] Chapman, A. C., Davies, A. G., Harrison, P. G., McFarlane, W. (J. Chem. Soc. C **1970** 821/4).
[66] Dannley, R. L. (Ger. Offen. 1929077 [1969/70]; C.A. **75** [1971] No. 6089).
[67] Davies, A. G., Harrison, P. G. (J. Chem. Soc. C **1970** 2035/8).
[68] Gibbon, R. M., Pierpoint, E. K., Imperial Chemical Industries Ltd. (U.S. 3527728 [1968/70]).
[69] Ishida, S., Sato, K., Fujita, M., Fukuda, H., Mori, K., Asahi Chemical Industry Co., Ltd. (Japan. 70-20914 [1966/70]; C.A. **73** [1970] No. 88600).
[70] Kobayashi, H., Komoto, H., Yoshino, M., Asahi Chemical Industry Co., Ltd. (Ger. 1816678 [1968/70]; C.A. **74** [1971] No. 112613).

[71] Kresta, J., Jadrnicek, B. (Chem. Prumysl **20** [1970] 222/3; C.A. **73** [1970] No. 99299).
[72] Mehrotra, R. C., Bachlas, B. P. (J. Organometal. Chem. **22** [1970] 121/8).
[73] Sakai, S., Uchida, A., Ishii, Y. (Kogyo Kagaku Zasshi **73** [1970] 2320/4).
[74] Smith, P. J. (Organometal. Chem. Rev. A **5** [1970] 373/402).
[75] Davies, A. G., Kleinschmidt, D. C., Palan, P. R., Vasishtha, S. C. (J. Chem. Soc. C **1971** 3972/6).
[76] Klimsch, P., Kühnert, P. (Ger. [East] 71359 [1968/71]; C.A. **73** [1970] No. 78136).
[77] Minsker, K. S., Fedoseyeva, G. T., Zavarova, T. B., Krats, E. O. (Vysokomol. Soedin. A **13** [1971] 2265/78; J. Polym. Sci. [USSR] A **13** [1971] 2544/60).
[78] Sakai, S., Kobayashi, Y., Ishii, Y. (J. Org. Chem. **36** [1971] 1176/80).
[79] Teraaski, I., Okamoto, T., Sasaki, M., Asahi Chemical Industry Co., Ltd. (Japan. 71-19944 [1968/71]; C.A. **75** [1971] No. 152378).
[80] Vizurraga, L. R., Fiber Industries, Inc. (Ger. Offen. 2057614 [1969/71]; C.A. **75** [1971] No. 110731).

[81] Dannley, R. L., Aue, W. A., Shubber, A. K. (J. Organometal. Chem. **38** [1972] 281/6).
[82] Freireich, S., Gertner, D., Zilkha, A. (J. Polym. Sci. Polym. Chem. Ed. **10** [1972] 3109/10).
[83] Ishii, Y., Sakai, S., Mitsubishi Chemical Industries Co., Ltd. (Japan. 72-40780 [1969/72]; C.A. **77** [1972] No. 164053).
[84] Melnichenko, L. S., Zemlyanskii, N. N., Kocheshkov, K. A. (Izv. Akad. Nauk SSSR Ser. Khim. **1972** 2055/8; Bull. Acad. Sci. USSR Div. Chem. Sci. **1972** 1993/6).
[85] Mehrotra, R. C., Gupta, V. D., Sharma, C. K. (Indian J. Chem. **10** [1972] 645/8).
[86] Seki, T., Suzuki, K., Matsuzaki, T., Nitto Kasei Co., Ltd. (U.S. 3647746 [1965/72]; C.A. **77** [1972] No. 20626).
[87] Smith, P. J., White, R. F. M., Smith, L. (J. Organometal. Chem. **40** [1972] 341/53).
[88] Armitage, D. A., Tarassoli, A. (Inorg. Nucl. Chem. Letters **9** [1973] 1225/7).
[89] Gaur, D. P., Srivastava, G., Mehrotra, R. C. (J. Organometal. Chem. **63** [1973] 213/9).
[90] Gaur, D. P., Srivastava, G., Mehrotra, R. C. (J. Organometal. Chem. **63** [1973] 221/31).

[91] Lengnick, G. F., Stauffer-Wacker Silicone Corp. (Ger. 2028320 [1970/73]).
[92] Lengnick, G. F., Stauffer-Wacker Silicone Corp. (Brit. 1326075 [1970/73]; C.A. **80** [1974] No. 38098).
[93] Mitchell, T. N. (J. Organometal. Chem. **59** [1973] 189/97).
[94] Smith, P. J., Smith, L. (Inorg. Chim. Acta Rev. **7** [1973] 11/33).
[95] Beers, M. D., General Electric Co. (Can. 960393 [1972/74]; C.A. **82** [1975] No. 172780).
[96] Collman, J. P., Murphy, D. W., Fleischer, E. B., Swift, D. (Inorg. Chem. **13** [1974] 1/6).
[97] Freireich, S., Gertner, D., Zilkha, A. (Eur. Polym. J. **10** [1974] 439/43).
[98] Knocke, R., Neumann, W. P. (Liebigs Ann. Chem. **1974** 1486/95).
[99] Komoto, H., Toyomoto, K., Matsumoto, Y., Asahi Chemical Industry Co., Ltd. (Japan. Kokai 74-99794 [1973/74]; C.A. **82** [1975] No. 98848).
[100] Marchand, A., Gerval, P., Soulard, M. H., Massol, M., Barran, J. (J. Organometal. Chem. **74** [1974] 227/38).

[101] Massol, M., Barrau, J., Satge, J., Bouyssieres, B. (J. Organometal. Chem. **80** [1974] 47/69).
[102] Peetre, I. B., Smith, B. E. F. (Mikrochim. Acta **1974** 301/10).
[103] de Wolfe, R. H. (Synthesis **1974** 153/72).
[104] Tzschach, A., Pönicke, K. (Z. Anorg. Allgem. Chem. **404** [1974] 121/8).
[105] Armitage, D. A., Sinden, A. W. (J. Organometal. Chem. **90** [1975] 285/90).

[106] Itsukaichi, Y., Kondo, T., Sankyo Organic Chemicals Co., Ltd. (Japan. Kokai 75-33243 [1973/75]; C.A. **83** [1975] No. 98347).
[107] Minke, R., Freireich, S., Zilkha, A. (Israel J. Chem. **13** [1975] 212/20).
[108] Mitchell, T. N. (J. Organometal. Chem. **92** [1975] 311/9).
[109] Sakai, S., Fujinami, T., Yamada, T., Furusawa, S. (Nippon Kagaku Kaishi **1975** 1789/94; C.A. **84** [1976] No. 5089).
[110] Sakai, S., Furusawa, S., Matsunaga, H., Fujinami, T. (J. Chem. Soc. Chem. Commun. **1975** 265/6).

[111] Beers, M. D., Smith, A. H., General Electric Co. (Can. 997891 [1972/76]; C.A. **86** [1977] No. 156825).
[112] Cohen, D., Marom, G., Zilkha, A. (Eur. Polym. J. **12** [1976] 795/800).
[113] Kayama, I., Kawamura, K., Kawakami, H., Iwasaki, K., Sawachika, Y., Kokoku Chemical Industry Co., Ltd. (Japan. Kokai 76-06909 [1974/76]; C.A. **84** [1976] No. 165849).
[114] Kocheshkov, K. A., Zemlyanskii, N. N., Shriro, V. S., Strelenko, Yu. A., Ustynyuk, Yu. (Izv. Akad. Nauk SSSR Ser. Khim. **1976** 950; Bull. Acad. Sci. USSR Div. Chem. Sci. **1976** 933).
[115] Minagawa, M., Abe, M., Kurita, N., Adeka Argus Chemical Co., Ltd. (Japan. Kokai 76-138758 [1974/76]; C.A. **86** [1977] No. 107488).
[116] Müller, H., Sander, H. J., Büssing, J., Ciba-Geigy Marienberg GmbH (Ger. Offen. 2444991 [1974/76]; C.A. **85** [1976] No. 22370).
[117] Sakai, S., Fujinami, T., Kosugi, K., Matsunaga, K. (Chem. Letters **1976** 891/2).
[118] Gorsi, B. L., Mehrotra, R. C. (Indian J. Chem. A **15** [1977] 1099/101).
[119] Kennedy, J. D. (J. Chem. Soc. Perkin Trans. II **1977** 242/8).
[120] Klebanov, M. S., Shogolon, I. M., Novikova, T. V. (Zh. Obshch. Khim. **47** [1977] 1078/81; J. Gen. Chem. [USSR] **47** [1977] 987/9).

[121] Sakai, S., Niimi, H., Kobayashi, Y., Ishii, Y. (Bull. Chem. Soc. Japan **50** [1977] 3271/5).
[122] Freireich, S., Gertner, D., Zilkha, A. (Israeli 36312 [1971/78]; C.A. **89** [1978] No. 130370).
[123] Kametani, M., Lio, Y., Nitto Chemical Industry Co., Ltd. (Japan. Kokai Tokkyo Koho 78-144522 [1977/78]; C.A. **90** [1979] No. 169290).
[124] Mehrotra, R. C., Rai, A. K., Jain, N. C. (Proc. Indian Acad. Sci. A **87** [1978] 61/7).
[125] Perozzo, V., Tagliavini, G. (J. Organometal. Chem. **162** [1978] 37/44).
[126] Tokiyama, Y., Matsunaga, N., Japan Ester Co., Ltd. (Japan. Kokai 78-52595 [1976/78]; C.A. **89** [1978] No. 130172).
[127] Goel, S. C., Singh, V. K., Mehrotra, R. C. (Syn. React. Inorg. Metal-Org. Chem. **9** [1979] 459/70).
[128] Kametani, M., Io, Y., Nitto Chemical Industry Co., Ltd. (Japan. Kokai Tokkyo Koho 79-41814 [1977/79]; C.A. **91** [1979] No. 57794).
[129] Minagawa, G., Ito, M., Sekiguchi, T., Adeka Argus Chemical Co., Ltd. (Japan. Tokkyo Koho 79-37014 [1974/79]; C.A. **92** [1980] No. 147852).
[130] Nitto Chemical Industry Co., Ltd. (Fr. Demande 2400006 [1977/79]; C.A. **91** [1979] No. 108462).

[131] Novitskaya, L. P., Dregval, G. F., Brodskaya, N. M. (Gig. Sanit. **1979** No. 6, pp. 48/51; C.A. **91** [1979] No. 158466).
[132] Peruzzo, V., Tagliavini, G., Gambaro, A. (Inorg. Chim. Acta **34** [1979] L263/L265).
[133] Yamazaki, N., Nakahama, S., Endo, K., Mitsubishi Chemical Industries Co., Ltd. (Japan. Kokai Tokkyo Koho 79-63023 [1977/79]; C.A. **91** [1979] No. 92272).
[134] Cinnamon, S., Freireich, S., Zilkha, A. (Eur. Polym. J. **16** [1980] 147/8).
[135] Dabi, S., Zilkha, A. (Eur. Polym. J. **16** [1980] 475/8).

[136] Kamentani, Y., Iino, Y., Nitto Chemical Industry Co., Ltd. (Brit. 1572438 [1977/80]; C.A. **94** [1981] No. 16327).
[137] Oleinik, N. M., Garkusha-Bozhko, I. P., Litvinenko, L. M. (Zh. Org. Khim. **16** [1980] 1094/5; C.A. **93** [1980] No. 132792).
[138] Rupani, P., Singh, A., Rai, A. K., Mehrotra, R. C. (J. Organometal. Chem. **185** [1980] 209/17).
[139] Thiele, L. (Z. Chem. [Leipzig] **20** [1980] 315/6).
[140] van der Weij, F. W. (Makromol. Chem. **181** [1980] 2541/8).

[141] Bernhard, J., Siemens A.-G. (Ger. Offen. 3001125 [1980/81]).
[142] Blunden, S. J., Smith, P. J., Beynon, P. J., Gillies, D. G. (Carbohydr. Res. **88** [1981] 9/18).
[143] Jones, K., Biddle, K. D., Das, A. K., Emblem, H. G. (Silicates Ind. **46** [1981] 107/11; C.A. **96** [1982] No. 73389).
[144] Nitto Chemical Industry Co., Ltd. (Japan. Kokai Tokkyo Koho 81-131555 [1980/81]; C.A. **96** [1982] No. 85038).
[145] Nitto Riken Industries, Inc., Nitto Chemical Industry Co., Ltd. (Japan. Kokai Tokkyo Koho 81-100749 [1980/81]; C.A. **95** [1981] No. 204660).
[146] Shanzer, A., Libman, J., Frolow, F. (J. Am. Chem. Soc. **103** [1981] 7339/40).
[147] Krimm, H., Buysch, H. J., Bayer A.-G. (Eur. Appl. 48368 [1980/82]; C.A. **97** [1982] No. 55324).
[148] Nitto Chemical Industry Co., Ltd., Nitto Riken Industries, Inc. (Japan. Kokai Tokkyo Koho 57-193436 [1981/82]; C.A. **98** [1983] No. 180045).
[149] Pavlenko, V. M., Zinovev, G. I., Klemin, V. A., Donetsk State Medical Institute (U.S.S.R. 978859 [1980/82]; C.A. **98** [1983] No. 113760).
[150] Voronkov, M. G., Mirskov, R. G., Kuznetsova, G. V., Yarosh, N. K., Yarosh, O. G., Stankevich, O. S., Albanov, A. I., Vitkovskii, V. Yu. (Zh. Obshch. Khim. **52** [1982] 1820/4; J. Gen. Chem. [USSR] **52** [1982] 1612/6).

[151] Gordetsov, A. S., Noskov, N. M., Tasalova, M. E., Pavlova, L. A., Karlik, V. M., Kuzina, V. I., Dergunov, Yu. I. (Zh. Prikl. Khim. **56** [1983] 2635/8; C.A. **100** [1984] No. 210021).
[152] Klebanov, M. S., Shoshina, L. V., Shologon, I. M. (Zh. Obshch. Khim. **53** [1983] 1131/7; J. Gen. Chem. [USSR] **53** [1983] 1004/9).
[153] Mehrotra, R. C., Gupta, K. K. (J. Indian Chem. Soc. **60** [1983] 1172/4).
[154] Robins, J., Edwards, B. H., Tokach, S. K. (Polym. Mater. Sci. Eng. **49** [1983] 331/5; C.A. **101** [1984] No. 6275).
[155] Spohn, R. J., Exxon Research and Engineering Co. (Ger. Offen. 3204973 [1982/83]; C.A. **99** [1983] No. 213033).
[156] White, M. A., Smith, R. A., Beers, M. D., Swiger, R. T., Lucas, G. M., General Electric Co. (Eur. Appl. 69256 [1981/83]; C.A. **100** [1984] No. 122568).
[157] Ball, P., Füllmann, H., Schwalm, R., Heitz, W. (C_1 Mol. Chem. **1** [1984] 95/108).
[158] Kriecheldorf, H. R., Mang, T., Jonte, J. M. (Macromolecules **17** [1984] 2173/81).
[159] Lucas, G. M., General Electric Co. (Eur. Appl. 110251 [1982/84]; C.A. **101** [1984] No. 92599).
[160] Robins, J., Edwards, B. H., Tokach, S. K. (Advan. Urethane Sci. Technol. **9** [1984] 65/76; C.A. **102** [1985] No. 7130).

[161] Sakai, S., Suzuki, M., Aono, T., Hasebe, K., Kanbe, H., Hiroe, M., Kakei, T., Fujinami, T., Takemura, H. (Kobunshi Ronbunshu **41** [1984] 151/8; C.A. **101** [1984] No. 24005).
[162] Schwalm, R., Ball, P., Pullmann, H., Heitz, W. (Polym. Prepr. Am. Chem. Soc. Div. Polym. Chem. **25** [1984] 272/3).

[163] Sumitomo Bakelite Co., Ltd. (Japan. Kokai Tokkyo Koho 59-221324 [1983/84]; C.A. **102** [1985] No. 150610).
[164] Yeda Research and Development Co., Ltd. (Israeli 59148 [1980/84]; C.A. **102** [1985] No. 95269).
[165] Bolon, D. A., Gorczyca, T. B., Hallgren, J. E., General Electric Co. (U.S. 4533504 [1982/85]; C.A. **104** [1986] No. 50677).
[166] Kricheldorf, H. R., Jonte, J. M., Berl, M. (Makromol. Chem. **1985** Suppl. 12, pp. 25/38; C.A. **104** [1986] No. 130384).
[167] Kricheldorf, H. R., Mang, T., Jonte, J. M. (Makromol. Chem. **186** [1985] 955/76).
[168] Shiedlin, A., Marom, G., Zilkha, A. (Polymer **26** [1985] 447/51).

1.4.1.2.1.5.2.2 Dibutyltin Bis(Alkenyl Oxides), $(C_4H_9)_2Sn(OR)_2$

The compounds belonging to this section are summarized in Table 11 and are prepared by the following general methods.

Method I: Reaction of $(C_4H_9)_2SnCl_2$ with ROH, RONa, or ROAg in a 1:2 mole ratio.
$(C_4H_9)_2SnCl_2$ reacts with methyl ricinoleate or diethyleneglycol diricinoleate in refluxing ether and in the presence of C_5H_5N with formation of No. 13 [4, 6] and 14 [6], respectively.
The reaction of $(C_4H_9)_2SnCl_2$ with CH_2=$CHCH_2ONa$ in CH_2=$CHCH_2OH$ at low temperature [3] and with sodium tropolonate, $C_7H_5O_2Na$, in C_2H_5OH at 20°C/0.5 h [8] or refluxing CH_3OH [10], yields Nos. 1 and 16, respectively.
Addition of $(C_4H_9)_2SnCl_2$ to a refluxing suspension of $(CN)_2C$=$C(C_6H_5)OAg$ in CH_3COCH_3 leads to the formation of No. 12 within 5 to 30 h [15].

Method II: Condensation between $(-(C_4H_9)_2SnO-)_n$ and ROH in a 1:2 mole ratio.
Azeotropic dehydration of $(-(C_4H_9)_2SnO-)_n$ with CH_2=$CHCH_2OH$ (C_6H_6, reflux/2 h) [7], (C_6H_{14}, reflux) [2], with $CH_3COCH(CN)_2$ (C_6H_6, reflux/0.5 h) in the presence of activated charcoal [15], and with $CH_3COCH_2C(CH_3)$=NCH_2CH_2N=$C(CH_3)CH_2COCH_3$ (C_6H_6-C_2H_5OH, reflux/2 h) [22] gives Nos. 1, 11, and 15, respectively.

Method III: Transalkoxylation of $(C_4H_9)_2Sn(OR')_2$ with ROH in a 1:2 mole ratio or with an excess of ROH.
The action of CH_2=$CHCH_2OH$ on $(C_4H_9)_2Sn(OCH_3)_2$ leads to No. 1 with liberation of CH_3OH [13]. $(C_4H_9)_2Sn(OC_3H_7\text{-}i)_2$ has been used as the starting material for the preparation of Nos. 2 [21], 3 [17], 5 [24], 6 [20], 7 [19], 8 [18], 9 [23], and 10 [25]. All reactions are conducted in refluxing C_6H_6 and forced to completion by simultaneous removal of the i-C_3H_7OH-C_6H_6 azeotrope formed.

Table 11
Dibutyltin Bis(Alkenyl Oxides), $(C_4H_9)_2Sn(OR)_2$.
Further information on compounds preceded by an asterisk is given at the end of the table.
Explanations, abbreviations, and units on p. X.

No.	OR group method of preparation (yield in %)	properties and remarks	Ref.
*1	$OCH_2CH{=}CH_2$ I [3] II [2,7] III [13] special [9]	liquid [3], air-sensitive oil [9] b.p. 220°/0.3 ^1H NMR (CCl_4): 4.08 (CH_2O) IR: 1470, 1380, 1130, 920 with $(C_4H_9)_2SnH_2 \rightarrow \{(C_4H_9)_2Sn(H)OR\} \rightarrow$ 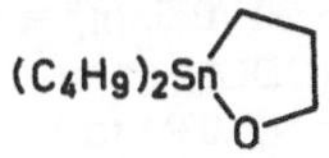 (30%) and $(-(C_4H_9)_2Sn-)_n$ stabilizer for PVC	[3,9] [9] [12,13] [1]
2	$OCH_2CH_2CH{=}CH_2$ III (75 to 90)	b.p. 126°/0.5, n_D^{39} = 1.479 ^1H NMR ($CDCl_3$): 2.32 (H-2), 3.72 (H-1), 5.05 and 5.16 (H-4), 5.80 (H-3) IR spectrum discussed	[21]
3	$OCH_2CH{=}CHCH_3$ III (71)	light yellow liquid b.p. 140 to 142°/0.6, n_D^{25} = 1.4905 ^1H NMR ($CDCl_3$): 0.75 to 1.83 (H-4, masked by C_4H_9 multiplet), 4.1 (H-1), 5.58 to 5.95 (H-2,3) IR ($CHCl_3$): ν(C=C) 1665 (vw), ν(CO) 1075 (s), ν(=CH) 960 (vs), ν(SnO) 550 (s) and unassigned frequencies	[17]
4	$OCH_2C(CH_3){=}CH_2$	no preparation reported, probably Method I or/and II with $(C_4H_9)_2SnH_2 \rightarrow \{(C_4H_9)_2Sn(H)OR\} \rightarrow$ 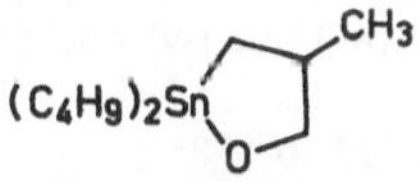 and $(-(C_4H_9)_2Sn-)_n$	 [12]
5	$OCH_2CH_2CH_2CH{=}CH_2$ III (83)	b.p. 137 to 140°/0.4, n_D^{32} = 1.484 ^1H NMR: 1.65 (H-2), 2.08 (H-3), 3.74 (H-1), 4.87 and 5.02 (H-5), 5.81 (H-4) IR spectrum discussed	[24]
6	$OCH(CH_3)CH_2CH{=}CH_2$ III (85)	b.p. 104°/0.5, n_D^{34} = 1.464 ^1H NMR ($CDCl_3$): 2.18 (H-2), 3.83 (H-1), 5.02 and 5.15 (H-4), 5.75 (H-3); CH_3 signal masked by C_4H_9 multiplet IR (neat): ν(C=C) 1645 (s), ν(CH) 1830 (s), 1000 (w), 960 (s), and 915 (vs), ν(CO) 1125 (vs), 1080 (vs), and 1030 (s), ν(SnO) 675 (m), 610 (s), and 560 (m, br)	[20]

References on p. 32

Table 11 (continued)

No.	OR group method of preparation (yield in %)	properties and remarks	Ref.
7	$OCH(CH_3)CH{=}CHCH_3$ III (71)	b.p. 130 to 132°/0.5 ^{1}H NMR ($CDCl_3$): 4.25 (H-1), 5.60 (H-2,3); CH_3 signals masked by C_4H_9 multiplet IR (neat): ν(C=C) 1670 (w), ν(CO) 1070 (vs) and 1025 (s), ν(=CH) 970 (vs), ν(SnO) 550 (m) and 515 (m)	[19]
8	$OC(CH_3)_2CH{=}CH_2$ III (61)	b.p. 88 to 90°/0.5, n_D^{25} = 1.464 ^{1}H NMR ($CDCl_3$; some confusion between δ and τ values!): 0.91 to 1.65 (CH_3, masked by C_4H_9 multiplet), 5.04 and 5.27 (H-3), 6.27 (H-2) IR ($CHCl_3$): ν(C=C) 1630 (w), ν(CH) 1840 to 1820 (w), 945 (vs), and 910 (vs), ν(CO) 1140 (vs) and 1025 (w), ν(SnO) 575 (vs) and 550 (s, sh)	[18]
9	$OCH(C_2H_5)CH{=}CHCH_3$ III (84)	b.p. 133°/0.4, n_D^{30} = 1.465 ^{1}H NMR ($CDCl_3$): 0.88 (CH_3/C_2H_5), 1.65 (H-4), 3.92 (H-1), 5.34 to 5.61 (H-2,3) IR spectrum discussed	[23]
10	$OC(CH_3)(C_2H_5)CH{=}CH_2$ III (85)	b.p. 150°/0.6, n_D^{32} = 1.462 ^{1}H NMR ($CDCl_3$): 1.23 (CH_3-1), 4.96 and 5.12 (H-3, 5.97 (H-2); C_2H_5 signals masked by C_4H_9 multiplet IR (neat): ν(C=C) 1640 (vw), ν(CO) 1175 (m), 1120 (m), 1075 (w), and 1015 (s, sh), ν($CH{=}CH_2$) 995 (vs) and 910 (vs), ν(SnO) 605 (w) and 520 (vw)	[25]
11	$OC(CH_3){=}C(CN)_2$ II (47)	m.p. 171 to 173° coordination polymer bridged by O and/or N atoms yielding five- or six-coordinate Sn	[15]
*12	$OC(C_6H_5){=}C(CN)_2$ I (61)	m.p. 125 to 131° ^{119}Sn-γ (78 K): δ = 1.87, Δ = 4.40 IR (KBr): ν(CN) 2235 (vs), 2225 (vs), 2210 (vs), and 2178 (m), ν(CO) 1495 (vs)	[15]
13	$OCH(C_6H_{13})CH_2CH{=}CH$-$(CH_2)_7COOCH_3$ derivative of methyl ricinoleate I [17, 18]	stabilizer for PVC nonstaining, nondiscoloring antioxidant for rubber	[4, 6] [5]

References on p. 32

Table 11 (continued)

No.	OR group method of preparation (yield in %)	properties and remarks	Ref.
14	$OCH(C_6H_{13})CH_2CH{=}CH$- $(CH_2)_7COOCH_2CH_2O$- $CH_2CH_2OOC(CH_2)_7$- $CH{=}CHCH_2CH(C_6H_{13})OH$ derivative of diethylene glycol diricinoleate I	stabilizer for PVC	[6]
15	$OC(CH_3){=}CHC(CH_3){=}N$- $CH_2CH_2N{=}C(CH_3)$- CH_2COCH_3 II	dirty white solid m.p. 175° IR (Nujol): ν(C=N) 1610 (vs), ν(C-N) 1280 (vs), ν(SnC) 570 (m), ν(SnN) 390 (m), ν(SnO) 360 (sh) six-coordinate Sn atom	[22]
16	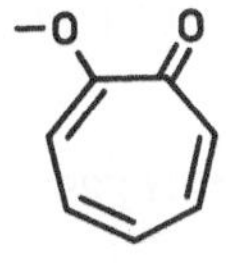 I [8,10]	m.p. 110 to 111° [10], 110 to 111.5° [8] ^{119}Sn-γ (80 K): $\delta = 1.30$ [10,16], $\Delta = 3.67$ [16], 3.68 [10]; temperature dependence calculations IR (Nujol): ν(C=C) 1597 (vs), ν(C=O) 1513 (vs), ν(SnO) 513 (vs) [8], or 510 [10]; unassigned bands (errors in the table heading!) [16] R: unassigned bands (errors also) Raman lattice modes and their relationship to the Mössbauer-Debye-Waller factor of the ^{119}Sn-γ resonance monomeric, distorted *trans* octahedral structure with anisobidentate tropolone ligands, bonded via both O atoms to Sn	[8,10] [10,16] [14,16] [8,10,16] [16] [11] [10,11,14,16]

* Further information:

$(C_4H_9)_2Sn(OCH_2CH{=}CH_2)_2$ (Table **11**, No. **1**) is formed on distillation of $CH_2{=}CHCH_2O(C_4H_9)_2$-$SnOSn(C_4H_9)_2OCH_2CH{=}CH_2$ at 220°C/0.3 Torr along with $(-(C_4H_9)_2SnO-)_n$ [9].

$(C_4H_9)_2Sn(OC(C_6H_5){=}C(CN)_2)_2$ (Table **11**, No. **12**). On the basis of the IR and Mössbauer spectra, along with the monomeric species, an associated structure with one nonbridging and one N-bridging CN group is discussed, causing five-coordination at the Sn atom [15].

References on p. 32

References:

[1] Burt, S. L., Bakelite Corp. (U.S. 2489518 [1949]; C.A. **1950** 5639).
[2] Bakelite Corp. (Brit. 664133 [1952]; C.A. **1952** 11230).
[3] Mach, G. P., Parker, E., Advance Solvents and Chemicals Corp. (U.S. 2700675 [1955]; C.A. **1956** 397).
[4] Weinberg, E. L., Metal and Thermit Corp. (Brit. 753998 [1956]; C.A. **1957** 6683).
[5] Tomka, L. A., Weinberg, E. L., Metal and Thermit Corp. (U.S. 2798862 [1957]; C.A. **1957** 15166).
[6] Weinberg, E. L., Metal and Thermit Corp. (U.S. 2796412 [1957]; C.A. **1957** 16537).
[7] Laliberte, B. R., Davidsohn, W. E., Henry, M. C. (J. Organometal. Chem. **5** [1966] 526/31).
[8] Komura, M., Tanaka, T., Okawara, R. (Inorg. Chim. Acta **2** [1968] 321/4).
[9] Davies, A. G., Kleinschmidt, D. C., Palan, P. R., Vashishtha, S. C. (J. Chem. Soc. C **1971** 3972/6).
[10] Naik, D. V., May, J. C., Curran, C. (J. Coord. Chem. **2** [1973] 309/15).

[11] Hazony, Y., Herber, R. H. (J. Phys. Colloq. [Paris] **35** [1974] C6-131/C6-137).
[12] Marchand, A., Gerval, P., Soulard, M. H. (J. Organometal. Chem. **74** [1974] 209/25).
[13] Massol, M., Barrau, J., Satge, J., Bouyssieres, B. (J. Organometal. Chem. **80** [1974] 47/69).
[14] Rein, A. J. (Diss. Rutgers State Univ. 1974, pp. 1/233; Diss. Abstr. Intern. B **35** [1975] 4859).
[15] Köhler, H., Neef, L., Korecz, L., Burger, K. (J. Organometal. Chem. **90** [1975] 159/71).
[16] Rein, A. J., Herber, R. H. (J. Chem. Phys. **63** [1975] 1021/9).
[17] Goel, S. C., Mehrotra, R. C. (Z. Anorg. Allgem. Chem. **440** [1978] 281/5).
[18] Goel, S. C., Singh, V. K., Mehrotra, R. C. (Z. Anorg. Allgem. Chem. **447** [1978] 253/6).
[19] Goel, S. C., Singh, V. K., Mehrotra, R. C. (Syn. React. Inorg. Metal-Org. Chem. **9** [1979] 459/70).
[20] Goel, S. C., Mehrotra, R. C. (Syn. React. Inorg. Metal-Org. Chem. **11** [1981] 35/45).

[21] Goel, S. C., Mehrotra, R. C. (Indian J. Chem. A **20** [1981] 1054/6).
[22] Sandhu, G. K., Sandhu, S. S. (Syn. React. Inorg. Metal-Org. Chem. **12** [1982] 215/28).
[23] Goel, S. C. (Syn. React. Inorg. Metal-Org. Chem. **13** [1983] 725/33).
[24] Goel, S. C. (Indian J. Chem. A **24** [1985] 880/2).
[25] Goel, S. C. (Syn. React. Inorg. Metal-Org. Chem. **15** [1985] 533/44).

1.4.1.2.1.5.2.3 Dibutyltin Bis(Aryl Oxides), $(C_4H_9)_2Sn(OR)_2$

The compounds belonging to this class are listed in Table 12 and are prepared by the following general methods.

Method I: Reaction of $(C_4H_9)_2SnX_2$ (X = Cl or I) with RONa (1:2 mole ratio).
The reaction between $(C_4H_9)_2SnCl_2$ and the appropriate RONa compound, usually prepared from CH_3ONa and ROH, has been used for the synthesis of No. 1 in refluxing C_7H_{16} [8], or Nos. 17, 22 [26], 18, 19 [36], 25, and 26 [38], each in refluxing C_6H_6. No. 22 has also been prepared in refluxing $C_6H_5CH_3$ [5]. Instead of the chloro derivative, $(C_4H_9)_2SnI_2$ has been used as the starting material in the synthesis of Nos. 12 (xylene, reflux, 48 h) and 17 (C_6H_6, reflux, 4 h) [21].

Method II: Reaction of $(-(C_4H_9)_2SnO-)_n$ with ROH (1:2 mole ratio).
The dehydration between $(-(C_4H_9)_2SnO-)_n$ and the appropriate ROH compound, usually forced to go to completion by distillative removal of the water or the solvent-water azeotrope formed, leads to No. 1 [31] in C_6H_{12} [1], $C_6H_5CH_3$ [17], or tetraline [14], Nos. 2 and 6 in tetraline [14], No. 4 in $C_6H_5CH_3$ [23], Nos. 5, 14, and 15 in C_6H_6 [10], Nos. 7 to 11 (solvent?) [31], No. 12 (no solvent, 60 to 90°C under vacuum for 2 h) [21], No. 13 in $C_6H_5CH_3$ [9], and Nos. 21 [20], 24, 31 [42], 27, 28 [40], each in C_6H_6.

Method III: Transalkoxylation of $(C_4H_9)_2Sn(OR')_2$ with ROH (1:2 mole ratio or excess of ROH).
Heating of mixtures of $(C_4H_9)_2Sn(OCH_3)_2$ and the appropriate ROH compound leads to the liberation of CH_3OH and the formation of No. 1 (in C_6H_6) [24], Nos. 19 and 20 (90°C/0.5 h) [11, 18], No. 22 [3], or No. 29 (80 to 100°C) [12, 13, 27]. The reaction between $(C_4H_9)_2Sn(OC_3H_7\text{-}i)_2$ and 1-nitroso-2-naphthol in dry C_6H_6 yields No. 30 [19].

Table 12
Dibutyltin Bis(Aryl Oxides), $(C_4H_9)_2Sn(OR)_2$.
Further information on compounds preceded by an asterisk is given at the end of the table.
Explanations, abbreviations, and units on p. X.

No.	OR group method of preparation (yield in %)	properties and remarks	Ref.
*1	OC_6H_5 I (90 [8]) II [1,31], (84 [14], 100 [17]) III [24]	pale yellow oil m.p. 45 to 48° [8], 51 to 53° [14], 79 to 82° [31] b.p. 146 to 148°/0.1 [14], 161°/0.35 [8] ^{119}Sn NMR: −120 (neat liquid at m.p.), −138 (sat. sol. in CCl_4 or C_6H_6 at room temp.) [24], −142 (supercooled neat liquid at 23°) [32]	[14] [8,14,31] [8,14] [24,32]
2	OC_6H_4Cl-4 II (75)	m.p. 80 to 85°, b.p. 178 to 180°/0.07	[14]
3	OC_6Cl_5	no preparation reported claimed as a fungicide and bactericide stabilizer for PVC against degradation by UV	 [2,6] [7]
4	$OC_6H_4NH_2$-2 II	brown solid IR (neat or Nujol): ν(SnO) 665 (m) and 585 (m), ν(SnC) 558 (m), ν(SnN) 483 (vw); other unassigned bands in the 4000 to 400 range reported monomeric; six-coordinate Sn with two N→Sn bonds	[23]
5	$OC_6H_2(Br_2\text{-}4,6)NO_2$-2 II	yellow oil, stable against moisture active as a fungicide, insecticide, and herbicide	[10]

References on p. 37

Table 12 (continued)

No.	OR group method of preparation (yield in %)	properties and remarks	Ref.
6	$OC_6H_4CH_3$-4 II (73)	m.p. 65 to 71° b.p. 166.5 to 168.5°/0.05	[14]
7	$OC_6H_4C_4H_9$-t-4 II	m.p. 88 to 91° additive to lubricants	[31]
8	$OC_6H_4C_8H_{17}$-i-4 II [31]	prepared from technical 4-i-$C_8H_{17}C_6H_4OH$ m.p. −9° [31], D^{20} = 1.0989, n_D^{20} = 1.5314 [31, 33] viscosity reported [31] additive to lubricants [31, 33]	[31] [31, 33]
9	$OC_6H_4C_8H_{17}$-t-4 II	m.p. −9°, D^{20} = 1.0732, n_D^{20} = 1.5262 viscosity reported additive to lubricants	[31]
10	$OC_6H_4C_9H_{19}$-i-4 II	m.p. −40°, D^{20} = 1.1035, n_D^{20} = 1.5238 viscosity reported additive to lubricants	[31]
11	$OC_6H_4C_{10}H_{21}$ II [31]	m.p. −48° [31], D^{20} = 1.0754, n_D^{20} = 1.5204 [31, 33] viscosity reported [31] additive to lubricants [31, 33]	[31, 33]
12	$OC_6H_4C_{15}H_{31}$-3 I, II	semisolid tested as a stabilizer for PVC	[21]
13	$OC_6H_2(CH_2N(CH_3)_2)_3$-2,4,6 II	catalyst in the production of polyurethanes	[9]
14	$OC_6H_3(NO_2$-2)CH_3-4 II	m.p. 46° tested as a fungicide, insecticide, and herbicide	[10]
15	$OC_6H_2((NO_2)_2$-2,4)C_4H_9-t-6 II	yellow, nondistillable oil tested as a fungicide, insecticide, and herbicide	[10]
16	$OC_6H_4(C_6H_4C_3H_7$-i-2)-4	no preparation reported anthelmintic activity tested against Raillietina cesticillus and Ascaridia galli	 [4]
17	OC_6H_4CHO-2 I [21, 26]	m.p. 143° with $Br_2 \rightarrow C_4H_9SnBr(OR)_2$ with $SO_2 \rightarrow$ pale yellow mono-insertion product (m.p. 190° with dec.) stabilizer for PVC	[21, 26] [26] [28] [21]

References on p. 37

Table 12 (continued)

No.	OR group method of preparation (yield in %)	properties and remarks	Ref.
18	$OC_6H_4COC_6H_5$-2 I	yellow to orange, low melting solid IR: ν(C=O) 1590, ν(C-O) 1260 six-coordinate Sn with two C=O→Sn bonds	[36]
19	$OC_6H_3(OCH_3$-5)COC_6H_5-2 I [36] III [11, 18]	yellow crystals m.p. 53 to 55° [11, 18], 120° [36] b.p. 155 to 160°/0.2 [11, 18] IR: ν(C=O) 1590, ν(C-O) 1280	[11, 18] [11, 18, 36] [36]
20	$OC_6H_3(OCH_3$-5)$(COC_6H_4OH$-2)-2 III	yellow-brown solid UV absorber (200 to 400 nm)	[11]
21	$OC_6H_3(OC_8H_{17}$-5)COC_6H_5-2 II	stabilizer for PVC against light and heat	[20]
22	$OC_6H_4COOCH_3$-2 I [5, 26] III [3]	slightly pink, low melting solid [5] m.p. 45° [26], completely molten at 100° [5] with SO_2→white mono-insertion product (m.p. 105°) stabilizer for PVC	[5, 26] [28] [3, 5]
*23	$OC_6H_4COOCH_2CH(OH)CH_2Cl$-2 special	—	[34]
24	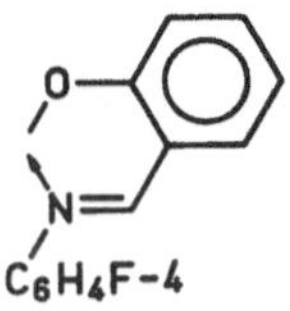 II (95)	cream yellow solid, m.p. 234° ^{1}H NMR: 0.70 to 1.90 (C_4H_9), 6.70 to 8.15 (C_6H_4), 9.23 (CH=N) ^{13}C NMR: 13.8 (C-δ), 26.1 (C-γ), 26.9 (C-α), 27.5 (C-β); 114.8 (C-2), 115.6 (C-6), 116.6 (C-9), 119.1 (C-4), 119.4 (C-10), 123.3 (C-5), 132.5 (C-3), 133.2 (C-8), 160.3 (C-7), 163.3 (C-11), 167.2 (C-1) ^{119}Sn-γ (77 K): δ = 1.01, Δ = 2.13 IR discussion effective bactericide against Bacillus subtilis and Staphylococcus aureus	[42]
25	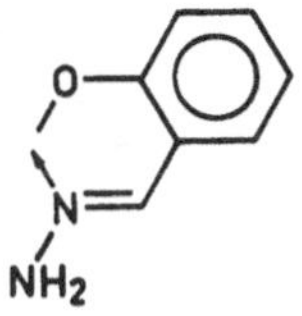 I (75)	m.p. 200° (dec.) IR: ν(C=N) 1615, ν(C-O) 1275, ν(N-N) 970, 940, and 930; only one ν(SnC) in the 570 to 560 region; misinterpreted in terms of a *cis*- instead of *trans*-octahedral structure	[38]

References on p. 37

Table 12 (continued)

No.	OR group method of preparation (yield in %)	properties and remarks	Ref.
26	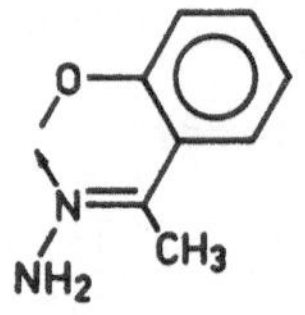 I (70)	m.p. 200° (dec.) IR: ν(C=N) 1625, ν(C-O) 1295, ν(N-N) 960 and 930, ν(SnC); interpretation like No. 27	[38]
27	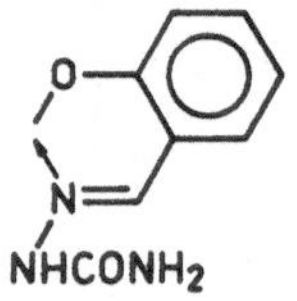 II (90)	cream yellow solid; m.p. 192° ^{119}Sn-γ (77 K): δ = 1.08, Δ = 2.36 IR discussion: distorted *cis*-octahedral geometry inferred	[40]
28	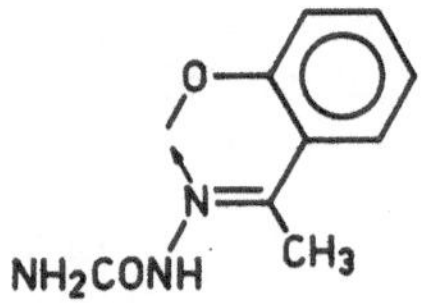 II (94)	brown-yellow solid, m.p. 185° ^{119}Sn-γ (77 K): δ = 0.98, Δ = 2.23 IR discussion: structure like No. 27	[40]
29	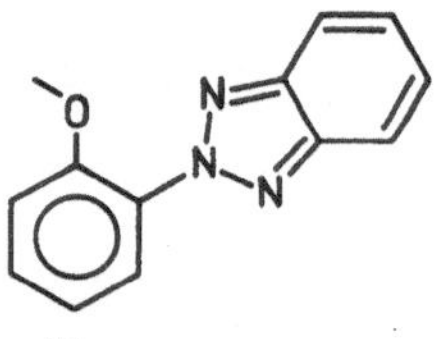 III	m.p. 115 to 120° stabilizer for synthetic resins	[12, 13, 27]
30	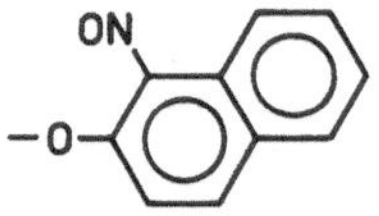 III	greenish yellow substance; m.p. 104 to 105° IR (Nujol, 3000 to 430 range, selected bands): ν(N=O) 1620 (s), ν(C-O) 1580 (m), ν(NCO) + $\delta_{as}(CH_3)$ + $\delta_{as}(CH_2)$/Sn 1375 (vs), 1340 (w), and 1300 (w), ν(C-O) 1140 (m), ν_{as}(SnO) 675 (m), ν_s(SnO) 635 (s), ν(SnC) 575 (m)	[19]
31	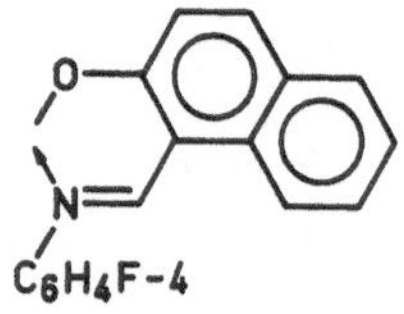 II (85)	orange-red solid; m.p. 118° ^{1}H NMR: 0.55 to 1.82 (C_4H_9), 7.0 to 8.05 (C_6H_4), 9.6 (CH=N) ^{119}Sn-γ (77 K): δ = 0.87, Δ = 2.26 effective bactericide against Bacillus subtilis and Staphylococcus aureus	[42]

* Further information:

$(C_4H_9)_2Sn(OC_6H_5)_2$ (Table **12**, No. **1**). The ^{119}Sn NMR shift of the neat compound is temperature dependent and has the limiting values $\delta = -48$ ppm at high temperature and $\delta = -148$ ppm at low temperature attributed to the monomer and the five-coordinate dimer, respectively (cf. Formulas I and II on p. 18, $R = C_6H_5$). The enthalpy and entropy changes associated with the dissociation of the dimer are $\Delta H = 78 \pm 2$ kJ/mol and $\Delta S = 238 \pm 7\ J \cdot mol^{-1} \cdot K^{-1}$. These values demonstrate that there is no significant difference in the O→Sn bond strength of the title compound and other $(C_4H_9)_2Sn(OR)_2$ dimers with $R = i\text{-}C_3H_7$ or $i\text{-}C_4H_9$ (No. 4 or 7 in Table 1 on p. 3 or 4). The differences in ΔS among the three compounds reflect the different steric effects of the respective OR group [32].

$(C_4H_9)_2Sn(OC_6H_5)_2$ is hydrolyzed by H_2O to give dimeric $(C_6H_5O(C_4H_9)_2SnOSn(C_4H_9)_2\text{-}OC_6H_5)_2$ [8]. It reacts with $\overline{OCH_2CH_2OCO}$ at 180°C for 30 to 60 h to give $(C_6H_5O)_2CO$ as an intermediate (detected by ^1H NMR and IR) which is further decarboxylated [29]. The action of $(C_2H_5)_2NLi$ on No. 1 (ether-HMPT, 0°C) leads to a 50% formation of $(C_4H_9)_2Sn(N(C_2H_5)_2)_2$ [15, 22]. The redistribution between No. 1 and $(C_4H_9)_2SnCl_2$ in xylene at 100°C yields $(C_4H_9)_2Sn(Cl)OC_6H_5$ [16].

$(C_4H_9)_2Sn(OC_6H_5)_2$ has been tested with respect to its anthelmintic activity against Raillietina cesticillus and Ascaridia galli [4].

$(C_4H_9)_2Sn(OC_6H_5)_2$ can be used as a transesterification catalyst [39] in the preparation of polycarbonates [35, 37] or alkylene oxide polymers [30] or as a catalyst in the regioselective reaction of C_6H_5OH with HCHO [41]. It has been used as an additive to lubricants [31], in polysiloxane compositions suitable for the treatment of paper to improve its release properties and abrasion resistance [17] and as well as a stabilizer for PVC [25].

$(C_4H_9)_2Sn(OC_6H_4COOCH_2CH(OH)CH_2Cl\text{-}2)_2$ (Table **12**, No. **23**) is formed by the reaction of $(C_4H_9)_2Sn(OC_6H_4COOH\text{-}2)_2$ with epichlorohydrin, $ClCH_2\overline{CHCH_2O}$, at 80 to 90°C, along with its isomer $(C_4H_9)_2Sn(OOCC_6H_4OCH_2CH(OH)CH_2Cl\text{-}2)_2$ (Table 27, No. 13). Graphs demonstrating the kinetics of the autocatalytic reaction are depicted [34].

References:

[1] Bakelite Corp. (Brit. 664133 [1952]; C.A. **1952** 11230).
[2] Farbwerke Hoechst A.-G. (Brit. 797073 [1953]; C.A. **1959** 22714).
[3] Mack, G. P., Parker, E., Advan. Solvents and Chemical Corp. (U.S. 2727917 [1955]; C.A. **1959** 5197; Ger. 953079 [1956]; Brit. 766875 [1957]; C.A. **1957** 8788).
[4] Kerr, K. B., Walde, A. W. (Exptl. Parasitol. **5** [1956] 560/70).
[5] Weinberg, E. L., Metal and Thermit Corp. (U.S. 2796412 [1957]; C.A. **1957** 16537).
[6] Brückner, H., Härtel, K., Farbwerke Hoechst A.-G. (Ger. 1025198 [1958]; C.A. **1960** 12468).
[7] Harrington, R. C., Smith, J. L., Eastman Kodak Co. (U.S. 3038877 [1960/62]; C.A. **57** [1962] 7465).
[8] Considine, W. J., Ventura, J. J. (J. Org. Chem. **28** [1963] 221/2).
[9] Hulse, R., Twitchett, H. J., Imperial Chemical Industries Ltd. (Brit. 957841 [1961/64]; C.A. **61** [1964] 9526).
[10] Stamm, W. A., Stauffer Chemical Co. (Fr. 1405428 [1963/65]; C.A. **63** [1965] 14904).

[11] Ismail, R. M., Dynamit Nobel A.-G. (Fr. 1454605 [1965/66]; C.A. **67** [1967] No. 32782).
[12] Ismail, R. M., Dynamit Nobel A.-G. (Fr. 1457966 [1965/66]; C.A. **67** [1967] No. 54265).

[13] Ismail, M. R., Raalf, H., Dynamit Nobel A.-G. (Fr. 1531509 [1966/68]; C.A. **71** [1969] No. 13832).
[14] Rees, R. G., Webb, A. F. (J. Organometal. Chem. **12** [1968] 239/40).
[15] Cuvigny, T., Normant, H. (Compt. Rend. C **269** [1969] 1389/402; C.A. **72** [1970] No. 67059).
[16] Moedritzer, K., van Wazer, J. R., Monsanto Co. (U.S. 3470220 [1965/69]; C.A. **72** [1970] No. 12883).
[17] Gibbon, R. M., Pierpoint, E. K., Imperial Chemical Industries, Ltd. (U.S. 35277728 [1968/70]; C.A. **70** [1969] No. 116355).
[18] Ismail, R. M. (Z. Naturforsch. **25b** [1970] 14/8).
[19] Mehrotra, R. C., Bachlas, B. P. (J. Organometal. Chem. **22** [1970] 129/37).
[20] Seki, T., Suzuki, K., Matsuzaki, T., Nitto Chemical Industrial Co., Ltd. (U.S. 3498947 [1966/70]; C.A. **72** [1970] No. 101439).

[21] Ghatge, N. D., Vernekar, S. P. (Ind. Eng. Chem. Prod. Res. Develop. **10** [1971] 214/6).
[22] Cuvigny, T., Normant, H. (J. Organometal. Chem. **38** [1972] 217/28).
[23] Mehrotra, R. C., Bachlas, B. P. (J. Organometal. Chem. **40** [1972] 129/33).
[24] Smith, P. J., White, R. F. M., Smith, L. (J. Organometal. Chem. **40** [1972] 341/53).
[25] Kubota, N., Adeka-Argus Chemical Co., Ltd. (Japan. 73-03826 [1969/73]; C.A. **79** [1973] No. 79724).
[26] Gopinathan, S., Gopinathan, C., Gupta, J. (Indian J. Chem. **12** [1974] 626/8).
[27] Seki, T., Suzuki, K., Matsuzaki, T., Nitto Kasei Co., Ltd. (U.S. 3856727 [1965/74]; C.A. **83** [1975] No. 207045).
[28] Gopinathan, S., Gopinathan, C., Jose, C. I., Gupta, J. (Indian J. Chem. **13** [1975] 78/80).
[29] Sakai, S., Furusawa, S., Matsunaga, H., Fujinami, T. (J. Chem. Soc. Chem. Commun. **1975** 265/6).
[30] Nakata, T., Kawamata, K., Osaka Soda Co., Ltd. (Japan. 76-40920 [1969/76]; C.A. **86** [1977] No. 122042).

[31] Tsvetkov, O. N., Komarova, N. N., Belov, P. O., Ermolov, F. N., Kulagin, V. V., Korenev, K. D., Vipper, A. B. (Tr. Mosk. Inst. Neftekhim. Gazov Prom. No. 126 [1976] 87/93; C.A. **88** [1978] No. 76227).
[32] Kennedy, J. D. (J. Chem. Soc. Perkin Trans. II **1977** 242/8).
[33] Lashki, V. L., Tsvetkov, O. N., Ermolov, F. N., Komarova, N. N., Vipper, A. B., Belov, P. S., Kaidala, E. V., Korenev, K. D., Kulagin, V. V., Markov, A. A. (Khim. Tekhnol. Topl. Masel **1977** No. 4, pp. 54/7; Chem. Technol. Fuels Oils **13** [1977] 289/92).
[34] Klebanov, M. S., Shoshina, L. V., Shologon, I. M., Nikonova, L. P. (Zh. Obshch. Khim. **48** [1978] 138/41; J. Gen. Chem. [USSR] **48** [1978] 117/20).
[35] Krimm, H., Buysch, H. J., Rudolph, H., Bayer A.-G. (Ger. Offen. 2736062 [1977/79]; C.A. **90** [1979] No. 186578).
[36] Awasarkar, P. A., Gopinathan, S., Gopinathan, C. (Indian J. Chem. A **19** [1980] 127/9).
[37] Yamazaki, N., Higashi, F., Nakahama, S., Yamaguchi, K. (Kenkyu Hokoku Asahi Garasu Kogyo Gijutsu Shoreikai **37** [1980] 33/48; C.A. **95** [1981] No. 180053).
[38] Pardhy, S. A., Gopinathan, S., Gopinathan, C. (Syn. React. Inorg. Metal.-Org. Chem. **13** [1983] 385/95).
[39] Pilati, F., Munari, A., Manaresi, P. (Polym. Commun. **25** [1984] 187/9; C.A. **101** [1984] No. 91494).

[40] Saxena, A., Tandon, J. P. (Polyhedron **3** [1984] 681/8).
[41] Ninagawa, A., Ohnishi, Y., Takeuchi, H., Matsuda, H. (Makromol. Chem. Rapid Commun. **6** [1985] 793/6).
[42] Saxena, A., Tandon, J. P., Crowe, A. J. (Polyhedron **4** [1985] 1085/9).

1.4.1.2.1.5.2.4 Dibutyltin Bis(Organyl Oxides), $(C_4H_9)_2Sn(OR)_2$, with R = Heterocycle

The compounds belonging to this class are listed in Table 13. They have been prepared by the following methods.

Method I: Reaction of $(C_4H_9)_2SnCl_2$ with ROH or RONa (1:2 mole ratio).
No. 2 has been prepared from $(C_4H_9)_2SnCl_2$ and kojic acid in refluxing C_2H_5OH for 1 h, followed by addition of NaOH to precipitate NaCl [16]. The reaction between $(C_4H_9)_2SnCl_2$ and 8-hydroxyquinoline [12] in C_2H_5OH [4] yields No. 7 after addition of CH_3COONa in aqueous C_2H_5OH and a little aqueous NH_3 [1]. No. 8 is obtained by mixing solutions of $(C_4H_9)_2SnCl_2$ in C_2H_5OH and 5,7-dichloro-8-quinolinol in hot C_2H_5OH, containing a few drops of CH_3COOH, and heating the mixture for 5 to 10 min on a hot plate [18]. Addition of aqueous CH_3COONa to an ethanolic mixture of $(C_4H_9)_2SnCl_2$ and 5-nitro-8-quinolinol in C_2H_5OH and refluxing for 2 h affords No. 12 [18]. Nos. 9, 11, and 13 to 15 were prepared by mixing ethanolic solutions of $(C_4H_9)_2SnCl_2$ and the ligand, refluxing the mixtures for 4 to 5 h. After cooling, the addition of CH_3COONa causes the precipitation of the appropriate compound [20]. Nos. 16, 20, 21 [24], and 19 [23], or Nos. 16 to 18 [28] have been synthesized in a similar manner, adjusting the pH of the reaction mixture to ~6.5 either with dilute NH_3 (1:4) or with a saturated CH_3COONa solution [23,24], or by a small amount of C_5H_5N [28].
The metathetic reaction of $(C_4H_9)_2SnCl_2$ with RONa in C_2H_5OH has been used for the synthesis of Nos. 5 [17] and 7 [10].

Method II: Reaction of $(-(C_4H_9)_2SnO-)_n$ with ROH (1:2 mole ratio).
The dehydration reaction between $(-(C_4H_9)_2SnO-)_n$ and the appropriate ROH affords Nos. 4 and 7 (in C_6H_6, $C_6H_5CH_3$ or C_6H_{12}, reflux/30 min) [21], No. 3 (in $C_6H_5CH_3$, azeotropic removal of the H_2O formed) [26], No. 6 (no solvent, heating under vacuum [27] or heating to 160 to 190°C for 30 to 40 min [29]), and No. 19 (in C_6H_6 or C_6H_{12}, reflux/2 to 12 h) [31].

Method III: The transalkoxylation method has been used for the synthesis of No. 4 from $(C_4H_9)_2Sn(OCH_3)_2$ and 2-hydroxypyridine in refluxing C_6H_6 [7].

Table 13
Dibutyltin Bis(Organyl Oxides), $(C_4H_9)_2Sn(OR)_2$ with R = Heterocycle.
Further information on compounds preceded by an asterisk is given at the end of the table.
Explanations, abbreviations, and units on p. X.

No.	OR group method of preparation (yield in %)	properties and remarks	Ref.
1	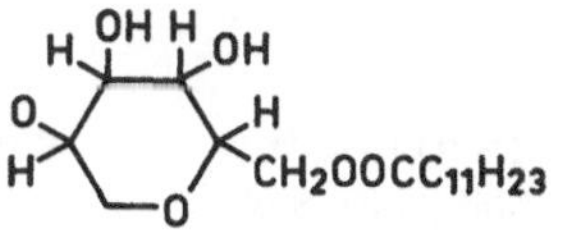	no preparation reported antistatic agent for thermoplastic resins	[8]
2	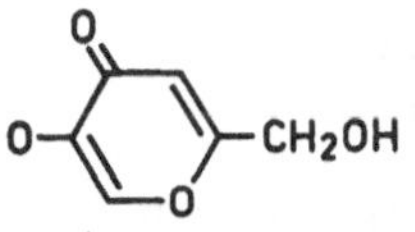 $C_6H_6O_4$ = kojic acid I	m.p. 157 to 159° ^{119}Sn-γ (80 K): δ = 1.35, Δ = 3.70 IR (KBr): ν(SnO) 550 octahedral structure with bidentate OR ligands and *trans* C-Sn-C bonds	[16]

References on p. 44

Table 13 (continued)

No.	OR group method of preparation (yield in %)	properties and remarks	Ref.
3	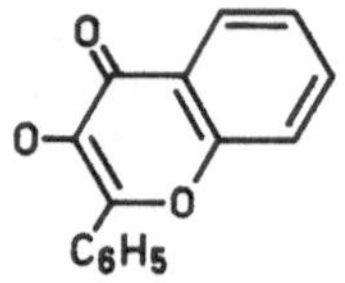 $C_{15}H_{10}O_3$ = 3-hydroxyflavone II	m.p. 178 to 180° ^{119}Sn-γ (80 K): δ = 1.25, Δ = 3.27 fluorescence emission (397 nm excitation): 465 (w) nm IR (solid): ν_{as}(CO) 1550 distorted *trans*-octahedral structure suggested	[26]
4	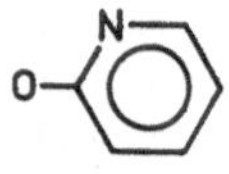 II [21] III [7]	m.p. 105° [21], clear yellow oil which crystallizes at ca. 80° [7]	[7,21]
		D = 1.5663	[7]
		IR: 620, 570, 530, 450, 435, 315, 400; N→Sn coordination indicated by bands at 420 and 385	[21]
5	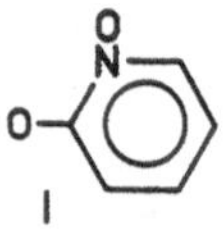 I	bactericide and fungicide LD_{50} estimations	[17]
6	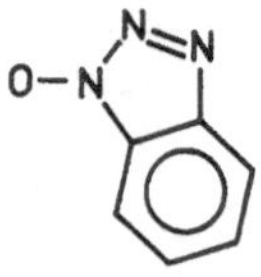 II (96 [27,29])	m.p. 59 to 60° [27,29], explodes on heating over an open flame! [27]	[27,29]
		IR (Nujol): ν(CH/C_4H_9) 2940, 2910, and 2850; unassigned bands in the 3060 to 510 range	[27]
		tested as a catalyst in polyurethane synthesis	[29]
*7	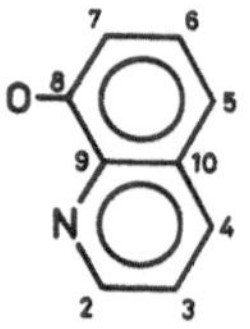 I [1,4,10,12] II [21]	yellow crystals [1,10]	[1,10]
		m.p. 150 to 152° [4], 152 to 153.5° [31], 154.5 to 155.5° [1], 155° [2,21], 155 to 156° [3]	[1 to 4, 21,31]
		μ = 4.63 D at 25°	[10]
		spectra given on pp. 43/4	
		with SO_2→yellow colored monoinsertion product (m.p. 200° with dec.)	[19]
		tested as a fungicide	[5]
8	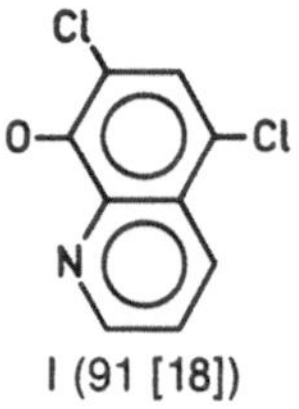 I (91 [18])	m.p. 144 to 145°	[18]
		^{1}H NMR: 0.84 (H-δ), 1.26 (H-α) in CCl_4; 7.73 (H-3), 7.93 (H-6), 8.53 (H-2), 9.50 (H-4) in CD_3SOCD_3	[20]
		^{119}Sn-γ (77 K): δ = 1.01, Δ = 2.26; octahedral structure with bidentate OR and *cis*-C_4H_9 groups	[23]
		MS (70 eV): $[M-2\ C_4H_9\text{-}OC_9H_4NCl_2]^+$ (100), $[M-OC_9H_4NCl_2]^+$ (28.2), $[M-C_4H_9]^+$ (14.7), $[M-2\ C_4H_9]^+$ (4.9), $[M]^+$ (<0.5)	[18]

References on p. 44

Table 13 (continued)

No.	OR group method of preparation (yield in %)	properties and remarks	Ref.
9	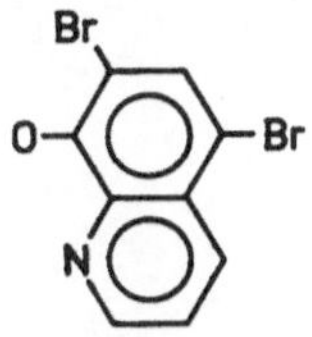 I [20]	m.p. 208 to 210°	[20]
		^{1}H NMR: 0.83 (H-δ), 1.23 (H-α) in CCl_4; 7.75 (H-3), 7.93 (H-6), 8.47 (H-2), 9.50 (H-4) in CH_3SOCH_3	
		structure like No. 8 predicted	[23]
10	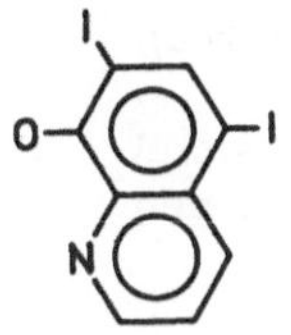	no preparation reported	
		^{1}H NMR (CH_3SOCH_3): 7.73 (H-3), 7.93 (H-6), 8.40 (H-2), 9.33 (H-4)	[20]
11	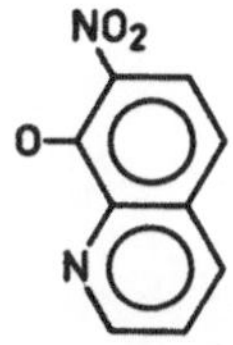 I (62)	m.p. 149 to 150°	[20]
12	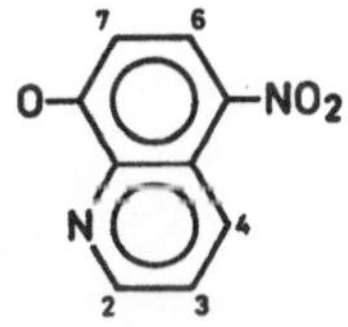 I (50 [18])	m.p. 229 to 230°	[18]
		^{1}H NMR: 0.83 (H-δ), 1.21 (H-α) in CS_2; 7.00 (H-7), 8.01 (H-3), 8.68 (H-6), 9.47 (H-4), 9.63 (H-2) in CH_3SOCH_3	[20]
		^{119}Sn-γ (77 K): $\delta = 1.38$, $\Delta = 3.05$; structure like No. 8 predicted	[23]
		MS (70 eV): $[M-2\ C_4H_9\text{-}OC_9H_5NNO_2]^+$ (100), $[M-C_4H_9]^+$ (24.1), $[M-OC_9H_5NNO_2]^+$ (14.3), $[M-2\ OC_9H_5NNO_2]^+$ (2.1); $[M]^+$ absent	[18]
13	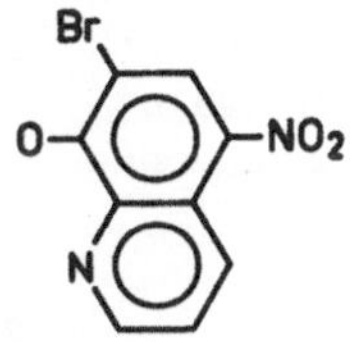 I	m.p. 276° (dec.)	[20]

References on p. 44

Table 13 (continued)

No.	OR group method of preparation (yield in %)	properties and remarks	Ref.
14	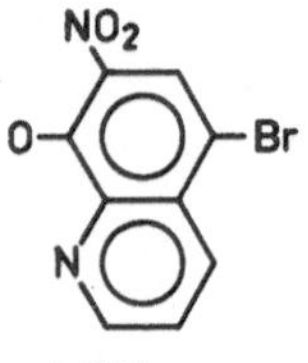 I (80)	m.p. 234 to 235° (dec.)	[20]
15	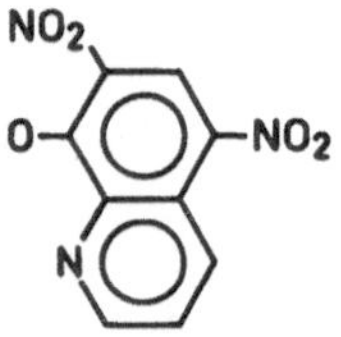 I (91 [20])	m.p. 253 to 255° ^{1}H NMR (CF_3COOH): 8.53 (H-3), 9.01 (H-6), 9.27 (H-4), 9.90 (H-2) ^{119}Sn-γ (77 K): δ = 1.4, Δ = 4.3; octahedral structure with bidentate OR ligands and *trans*-C_4H_9 groups suggested	[20] [23]
16	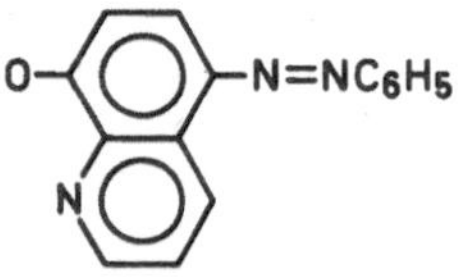 I [24], (65 [28])	m.p. 70° (dec.) [24], 118° [28] ^{119}Sn-γ (77 K): δ = 1.12, Δ = 2.38; *cis*-octahedral structure suggested UV (C_6H_6 or CH_3OH): λ_{max} = 430	[24, 28] [24] [28]
17	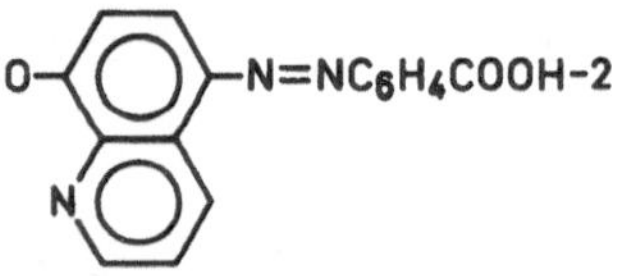 I (25)	m.p. 163° IR: ν_{as}(OCO) 1725 UV: λ_{max} = 420 in CH_3OH; 420 (sh) and 480 in C_6H_6	[28]
18	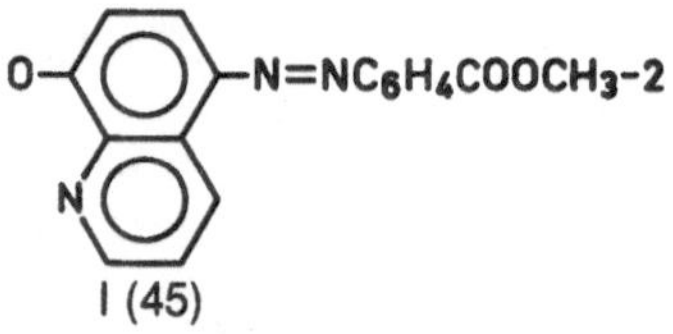 I (45)	m.p. 174° IR: ν_{as}(OCO) 1720 UV: (C_6H_6 or CH_3OH): λ_{max} = 420	[28]
19	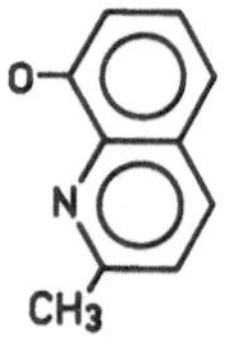 I (85 [23]) II [31]	ivory solid, m.p. 231 to 233° [31], 255° (dec.) [23] ^{119}Sn NMR ($CDCl_3$): −248.7 (J(Sn, H-α) = 72.6, J(Sn, C-α) = 805, J(Sn, C-β) = 45.3, J(Sn, C-γ) = 122.1); correlations between the C-Sn-C angle and ^{2}J(Sn, H) and ^{1}J(Sn, C) discussed ^{119}Sn-γ (77 K): δ = 1.49, Δ = 3.04; structure like No. 8 predicted	[23, 31] [31] [23]

References on p. 44

Table 13 (continued)

No.	OR group method of preparation (yield in %)	properties and remarks	Ref.
20	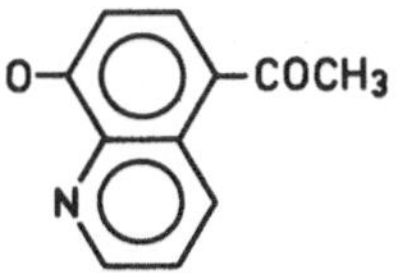 I (77)	m.p. 117° ^{119}Sn-γ (77 K): δ = 1.01, Δ = 2.20; *cis*-octahedral structure suggested	[24]
21	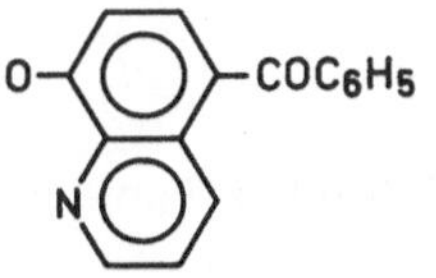 I (75)	m.p. 80° ^{119}Sn-γ (77 K): δ = 0.99, Δ = 2.06; *cis*-octahedral structure suggested	[24]

* Further information:

$(C_4H_9)_2Sn(OC_9H_6N)_2$ (Table **13**, No. **7**) is also formed in the reaction of $(C_4H_9)_2ClSnOSn(OC_2H_5)(C_4H_9)_2$ with 8-hydroxyquinoline in the presence of $N(C_2H_5)_3$ in a good yield [2]. In the following spectroscopic part, the carbon and hydrogen atoms of the C_4H_9 groups are labeled α, β, γ, and δ; the formula in Table 13 on p. 40 shows the labeling of the OC_9H_6N ligand; δ values are given in ppm and the J values in Hz.

^{1}H NMR spectrum: 0.5 (H-δ), 1.36 (H-α) in CS_2 [20]; 1.25 (C_4H_9) [21], 7.06 (H-3, J(3,4) = 8.3), 8.06 (q, H-4, J(2,4) = 1.7), 8.44 (q, H-2, J(2,3) = 4.4) in CH_2Cl_2 [13]; 6.91, 7.32, 7.86, 8.3 (OC_9H_6N) in CCl_4 [21].

^{13}C NMR spectrum: 13.7 (C-δ), 25.6 (C-α, J(Sn,C) = 614), 26.8 (C-γ, J(Sn,C) = 102), 27.8 (C-β, J(Sn,C) = 26), 112.6 (C-7), 113.4 (C-5), 121.0 (C-3), 129.7 (C-10), 130.1 (C-6), 136.6 (C-9), 138.2 (C-4), 142.4 (C-2), and 158.1 (C-8) [32].

^{15}N NMR spectrum (in $CDCl_3$, nearly saturated): 123.7 to low frequency from neat $CH_3{}^{15}NO_2$ [32].

^{119}Sn NMR spectrum: −260.7 (J(Sn,H-α) = 72.6, J(Sn,C-α) = 612, J(Sn,C-β) = 23.6, J(Sn,C-γ) = 102.4) in $CDCl_3$ [31]; −262 in $CHCl_3$ [25, 32] (J(Sn,C-α) = 614) [32].

The following ^{119}Sn Mössbauer data were measured:

temperature	δ (mm/s)	Δ (mm/s)	Ref.
78 K	0.86	2.06	[15]
	1.00	2.00	[9,15]
	1.02	2.04	[12,14,15]
80 K	0.93	2.05	[10,15]
	1.10	2.21	[11,15,23]

References on p. 44

The following IR data are reported (wave numbers in cm^{-1}):

Nujol [12]	assignment [12]	Nujol [6]	[21]
645 (s)	ν_{as}(SnO)		
613 (s)			620
586 (m)	ν(SnC)		
567 (m)			570
515 (s)			530
445 (w)	ν_s(SnO)	444 (m)	450
435 (w)		434 (s)	435
		416 (m)	415
394 (m)	ν(SnN)	393 (vs)	400
386 (m)		383 (s)	

Other bands in the far IR region (Nujol spectrum) were found at 366 (m), 346 (m), 301 (m), 279 (s, br), 231 (vs), 194 (s), 165 (m, br), and 117 (m) cm^{-1} [6].

UV spectrum: λ_{max} (log ε) = 372(3.66) nm in $CHCl_3$ [32] or 375(3.89) nm in CH_3OH [22]. The formation of the title compound from $(C_4H_9)_2SnCl_2$ and 8-quinolinol in CH_3OH, $CHCl_3$, and C_6H_{14} has been studied by electronic spectra using the molar ratio method [30].

In the 70 eV mass spectrum of the title compound, the molecular ion $[M]^+$ is not observed. The following Sn-containing ions are reported: $[M-2\ C_4H_9\text{-}OC_9H_6N]^+$ (100), $[M-C_4H_9]^+$ (76.1), $[M-2\ C_4H_9]^+$ (14.1), and $[M-OC_9H_6N]^+$ (9.0) [18].

From spectroscopic evidence, the monomeric compound possesses a distorted octahedral structure with *cis*-C_4H_9 groups (C-Sn-C angle between 109° and 120°, calculated from ^2J(Sn, H-α), or between 123° and 127°, calculated from ^1J(Sn, C-α) [31]) and bidentate 8-quinolinolate ligands [10, 11, 12, 21, 23, 25, 31].

References:

[1] Blake, D., Coates, G. E., Tate, J. M. (J. Chem. Soc. **1961** 756).
[2] Alleston, D. L., Davies, A. G., Hancock, M. (J. Chem. Soc. **1964** 5744/8).
[3] Gerrard, W., Mooney, E. F., Rees, R. G. (J. Chem. Soc. **1964** 740/5).
[4] Tanaka, T., Komura, M., Kawasaki, Y., Okawara, R. (J. Organometal. Chem. **1** [1963/64] 484/9).
[5] Földesi, I., Stranger, G. (Acta Chim. Acad. Sci. Hung. **45** [1965] 313/22).
[6] Douek, I., Frazer, M. J., Goffer, Z., Goldstein, M., Rimmer, B., Willis, H. A. (Spectrochim. Acta A **23** [1967] 373/81).
[7] Ismail, R. M., Dynamit Nobel A.-G. (Fr. 1475896 [1966/67]; C.A. **68** [1968] No. 13178).
[8] Takeda, T., Iino, K., Ando, M., Kawakami, Y., Seki, T., Japan. Telegram and Telephone Corp. and Nitto Chemical Industrial Co., Ltd. (Japan. 67-24047 [1964/67]; C.A. **68** [1968] No. 87887).
[9] Gol'danskii, V. I., Khrapov, V. V., Okhlobystin, O. Yu., Rochev, V. Ya. (in: Gol'danskii, V. I., Herber, R. H., Chemical Application of Mössbauer Spectroscopy, New York 1968, pp. 336/769).
[10] Mullins, M. A., Curran, C. (Inorg. Chem. **7** [1968] 2581/8).

[11] Fitzsimmons, B. W., Seeley, N. J., Smith, A. W. (J. Chem. Soc. A **1969** 143/6).
[12] Poller, R. C., Ruddick, J. N. R. (J. Chem. Soc. A **1969** 2273/6).
[13] Kawasaki, Y. (Org. Magn. Resonance **2** [1970] 165/72).
[14] Poller, R. C., Ruddick, J. N. R., Taylor, B., Toley, D. L. B. (J. Organometal. Chem. **24** [1970] 341/6).
[15] Smith, P. J. (Organometal. Chem. Rev. A **5** [1970] 373/402).
[16] Naik, D. V., May, J. C., Curran, C. (J. Coord. Chem. **2** [1972/73] 309/15).
[17] Olin Corp. (Fr. Demande 2199538 [1972/74]; C.A. **82** [1975] No. 140307).
[18] Barsode, C. D., Umapathy, P., Sen, D. N. (J. Indian Chem. Soc. **52** [1975] 942/6).
[19] Gopinathan, S., Gopinathan, C., Jose, C. I., Gupta, J. (Indian J. Chem. **13** [1975] 78/80).
[20] Barsode, C. D., Umapathy, P., Sen, D. N. (J. Indian Chem. Soc. **53** [1976] 761/3).

[21] Sandhu, S. S., Pushkarna, S. K. (J. Indian Chem. Soc. **53** [1976] 1214/20).
[22] Barsode, C. D., Umapathy, P., Sen, D. N. (J. Indian Chem. Soc. **54** [1977] 1172/7).
[23] Bhide, S. N., Umapathy, P., Gupta, M. P., Sen, D. N. (J. Inorg. Nucl. Chem. **40** [1978] 1003/7).
[24] Ghuge, K. D., Umapathy, P., Gupta, M. P., Sen, D. N. (J. Inorg. Nucl. Chem. **43** [1981] 653/8).
[25] Otera, J. (J. Organometal. Chem. **221** [1981] 57/61).
[26] Blunden, S. J., Smith, P. J. (J. Organometal. Chem. **226** [1982] 157/63).
[27] Gordetsov, A. S., Pereshein, V. V., Skobeleva, S. E., Pavlova, L. A., Tyutina, T. P., Karlik, V. M., Dergunov, Yu. I. (Zh. Obshch. Khim. **52** [1982] 2762/7; J. Gen. Chem. [USSR] **52** [1982] 2435/9).
[28] Basu Baul, T. S., Chattopadhyay, T. K., Majee, B. (Polyhedron **2** [1983] 635/40).
[29] Gordetsov, A. S., Noskov, N. M., Tasalova, M. E., Pavlova, L. A., Karlik, V. M., Kuzina, V. I., Dergunov, Yu. I. (Zh. Prikl. Khim. **56** [1983] 2635/8; J. Appl. Chem. [USSR] **56** [1983] 2453/7; C.A. **100** [1984] No. 210021).
[30] Langseth, W. (Inorg. Chim. Acta **87** [1984] 47/51).

[31] Howard, W. F., Crecely, R. W., Nelson, W. H. (Inorg. Chem. **24** [1985] 2204/8).
[32] Jain, V. K., Mason, J., Saraswat, B. S., Mehrotra, R. C. (Polyhedron **4** [1985] 2089/96).

1.4.1.2.1.5.2.5 Dibutyltin Alkoxides of the $(C_4H_9)_2\overline{Sn\text{-}O\text{-}R\text{-}O}$ Type

The compounds belonging to this section are summarized in Table 14 and they are prepared by the following general methods.

Method I: a. Reaction between $(C_4H_9)_2SnCl_2$ and HO-R-OH (1:1 mole ratio).
No. 1 (no conditions given) [6], (nonaqueous interfacial technique with excess diol containing $N(C_2H_5)_3$ in one phase and a solution of $(C_4H_9)_2SnCl_2$ in C_6H_{14} in the other phase) [22]; No. 2 (in $C_6H_5CH_3$, in the presence of Na_2CO_3) [6]; Nos. 12, 16, and 17 (in C_6H_6, in the presence of Na_2CO_3) [77]; Nos. 20, 39, 49, and 103 (nonaqueous interfacial technique, see No. 1) [22]; No. 21 (1:6 mole ratio, 100°C, in the presence of two equivalent of NaOH and traces of $C_6H_5CH_3$) [6]; No. 48 (in ether, in the presence of $NaNH_2$ or C_5H_5N) [9]; and No. 104 (in ether, in the presence of C_5H_5N) [5] were synthesized.

b. Reaction of $(C_4H_9)_2SnCl_2$ with NaO-R-ONa (1:1 mole ratio).
This method has been used for the synthesis of the oligomers No. 20 and 54 (in C_6H_{14}-$C_6H_4(CH_3)_2$-1,2) [27]; No. 46 (in C_6H_6, reflux/2 h) [98]; No. 47 (in $C_6H_5CH_3$, 0°C) [3]; Nos. 51 [31], 117 [68], 126 [64], 127, and 128 [67] (in CH_3OH with stirring

for 2 h); No. 53 (in C_6H_6) [93]; Nos. 121 and 123 (in C_6H_6, reflux/3 h) [106]; and No. 108 (in THF, 20°C/2 h, then reflux for 10 min) [44].

Method II: Condensation of (-($(C_4H_9)_2$SnO-)$_n$ with HO-R-OH (1:1 mole ratio).

This type of reaction, used for the synthesis of most of the compounds belonging to this section, is usually conducted in solvents like C_6H_6, $C_6H_5CH_3$, $C_6H_4(CH_3)_2$-1,2, or CH_3OH. The reaction comes to completion by precipitation of the organotin diolate or by removal of the water formed either by azeotropic distillation or by means of a Dean-Stark water separator. In most of the cases in which the organotin derivative is used for further reactions, "in situ" solutions are directly used, or the appropriate solvent is evaporated under normal or reduced pressure and the remaining product is triturated and used without further purification. Characterized compounds were purified by distillation or recrystallization from a suitable solvent or solvent mixture.

Compounds prepared in C_6H_6:

No.	Ref.	No.	Ref.	No.	Ref.
1	[30,101]	33	[35,41,42,49]	89	[60]
2	[30]	34	[41]	90	[60,62]
5	[107]	35	[35,41,42,43,49]	91	[62]
7	[62]	36	[42,49]	94	[89]
8	[12,30,99,107]	37,38	[85]	95	[40,62]
10	[91]	42	[100]	96	[40,123]
13	[133]	43	[87,100,101]	97	[40]
14	[107]	55	[63]	98	[62]
15,16	[39]	73	[40]	116	[69]
17	[12,39]	74,80	[40]	118,119	[122]
18	[39]	77	[127]	120,123	[63]
27	[132]	81	[117]		
28 to 32	[41]	88	[50]		

Compounds prepared in $C_6H_5CH_3$:

No.	Ref.	No.	Ref.	No.	Ref.
2	[52]	50	[3]	105,106	[126]
6	[52,134]	56,75,76,81	[129]	123	[24]
13,19	[118]	85	[119]		
48	[10,30]	92	[129]		

Compounds prepared in $C_6H_4(CH_3)_2$:

No.	Ref.	No.	Ref.	No.	Ref.
103	[13]	109	[8,44]	114,115	[8]
107	[36,44]	110,112	[44]		
108	[44]	113	[36,44]		

References on p. 92

Compounds prepared in CH_3OH:

No.	Ref.	No.	Ref.	No.	Ref.
57	[95]	66	[109]	79	[78]
58	[45, 59, 121]	67	[121]	82	[53, 111]
59	[55]	68	[112]	83	[53]
60	[45]	69	[79]	84	[88]
61	[121]	70, 71	[105]	86	[81]
62	[45, 121]	72	[125]	87	[54, 57, 92]
63	[72]	75	[110]	94, 99	[54]
64, 65	[86]	78	[74]		

Nos. 123, 124, and 125 were synthesized in C_6H_6 or $C_6H_5CH_3$ [65]; Nos. 39, 40, and 41 in C_6H_6, $C_6H_5CH_3$, or $C_6H_4(CH_3)_2$ [13]; and No. 95 in a CH_3OH-C_6H_6 mixture [53]. C_6H_{12} has been used in the preparation of No. 1 [1], $C_6H_{11}CH_3$ for No. 1 [2], and No. 103 [1, 2]. No solvent is reported in [97] for No. 11, and no conditions at all are given for the synthesis of Nos. 8 [75], 45 [131], 93 [115], 100 [128], and 101 [130].

Method III: Transalkoxylation of $(C_4H_9)_2Sn(OR')_2$ ($R' = CH_3$, C_2H_5, or i-C_3H_7) with diols (1:1 mole ratio, azeotropic removal of R'OH).

Nos. 1, 2, 8, 9, 20, 21, 22, 24, 25, and 39 have been prepared by mixing $(C_4H_9)_2Sn(OCH_3)_2$ with the appropriate diol in ether or THF and allowing the mixture to stand at room temperature until precipitation of the product occurs [15]. The reaction of $(C_4H_9)_2Sn(OCH_3)_2$ with $C_6H_5COCH(OH)C_6H_5$ in C_6H_6 at 40°C for 45 min affords No. 43 [100]; the reaction with $CH_3N(CH_2CH_2OH)_2$ in refluxing $C_6H_4(CH_3)_2$ for 20 min yields No. 108 [44].

Transalkoxylation of $(C_4H_9)_2Sn(OC_2H_5)_2$ with the appropriate diol has been used for the synthesis of Nos. 2, 3, and 14 (in refluxing C_6H_6 for 12 h) [94]; for Nos. 20, 23, 26, and 39 (in refluxing C_6H_6 for 15 min) [30]; for Nos. 1, 2, 4, 8, 9, 20, 21, 24, 39, 40, and 41 (in C_6H_6 at 110 to 115°C) [11]; for Nos. 107 and 111 (in refluxing C_6H_6) [14]; and for No. 23 [108].

$(C_4H_9)_2Sn(OC_3H_7\text{-i})_2$ has been transalkoxylated by $C_6H_4(OH)_2$-1,2 using C_6H_6 as the solvent [29].

Method IV: Reaction of $(C_4H_9)_2Sn$ with carbonyl compounds.

Irradiation of benzene solutions containing a 1:1 molar mixture of cyclic $(-(C_4H_9)_2Sn-)_n$ and a carbonyl compound, with day-light lamps at 20 to 24°C for 2 to 76 h, leads to formation and subsequent addition of intermediate $(C_4H_9)_2Sn$ to the C=O group of the substrate. Thus, the reaction of $(C_4H_9)_2Sn$ with CH_3CHO, CH_3COCH_3, $CH_3COCOCH_3$, C_6H_5CHO, or $C_6H_5COCOC_6H_5$ yields Nos. 8, 9, 42, 12, or 43, respectively [46].

General Remarks:

Dibutylstannylene derivatives of diols play an important role in the organic chemistry and biochemistry because of their high regioselective and stereospecific reactions with alkylating or acylating agents, without requiring a blocking of hydroxyl groups which are not directly involved in the reaction. Thus, the so-called "dibutyltin oxide method" became very important not only for synthetic purposes, e.g., for the synthesis of macrocyclic crown esters, but also

References on p. 92

as characterization and isolation technique in the carbohydrate chemistry, especially of nucleosides and glucosides.

Both dibutylstannylene nucleosides and glucosides listed in Table 14 are only those for which the position of the O-$(C_4H_9)_2$Sn-O unit within the compound is clearly defined in the literature.

Table 14
Dibutyltin Alkoxides of the $(C_4H_9)_2$Sn-O-R-O Type.
Further information on compounds preceded by an asterisk is given at the end of the table.
Explanations, abbreviations, and units on p. X.

No.	O-R-O group method of preparation (yield in %)	properties and remarks	Ref.
*1	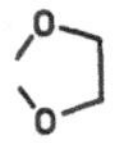 Ia [6], (90 [22]) II [1,2,30], (90 [101]) III [15], (98 [11]) special [17,48,101]	m.p. 195 to 200° [1,2], 198 to 205° (softening range) [22], 215 to 220° [15], 222 to 223° [17], 224 to 225° [101], 224 to 226° [25], 225 to 228° [30]	[1,2,15,17, 22,25,30, 101]
		sublimation at 220 to 240°/0.3	[11]
		^{1}H NMR: 0.80 to 1.40 (m, C_4H_9) in $CHCl_3$ [17]; 0.907 (t, H-δ, J(H,H) = 7.5), 1.287 (t(m), H-α, J(H,H) = 8.2), 1.362 (sept, H-γ, J(H,H) = 7.5), 1.624 (m, H-β) in $CDCl_3$ [135]; 3.33 (s, CH_2O) in CCl_4 or C_6H_{12} [15]; 3.49 in $CHCl_3$ [17]; 3.611 (J(Sn,H) = 31) for associated species (dimer predominant) and 3.674 (s(br)) for the monomer [135]; 3.68 (J(Sn,H) = 32) [101] in $CDCl_3$	[15,17,101, 135]
		^{13}C NMR (33°, 0.85 M in $CDCl_3$): 13.34 (s, C-δ, J(Sn,C) = 38), 22.27 (s, C-α, J(Sn,C) = 643), 26.76 (s, C-γ, J(Sn,C) = 101), 27.27 (s, C-β, J(Sn,C) = 31), 62.89 (s, CH_2O, J(Sn,C) = 38) for associated species (dimer predominant) and 64.06 for the monomer	[135]
		^{119}Sn NMR: −181 (33°, 0.85 M in $CDCl_3$) [135], −189 ± 5 (saturated in $CDCl_3$) [30]	[30,135]
		^{119}Sn-γ: δ = 1.10, Δ = 2.80 at 77 K [20, 30, 101]; 1.212 and 2.931, or 1.223 and 3.002 (frozen solution in $C_6H_5C_4H_9$) at 78 K [116]	[20,30,101, 116]
		IR (KBr): 1120, 1060, 895, and 710	[17]

Table 14 (continued)

No.	O-R-O group method of preparation (yield in %)	properties and remarks	Ref.
*2	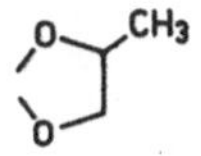 I a (94 [6]) II [30,52] III [15,94], (92 [11]) special [17,48]	m.p. 181 to 183° [17], 182° [6], 182 to 185° [52], 183 to 185° [18,25], 185° [15] sublimation at 210 to 230°/0.1 ^{1}H NMR on p. 81 ^{119}Sn NMR (saturated in $CDCl_3$): -164 ± 5 ^{119}Sn-γ: $\delta = 1.13$, $\Delta = 2.72$ (77 K) [20,30], $\delta = 1.144$, $\Delta = 2.747$ (racemic), or 1.166 and 2.829 (chiral isomer) (78 K) [116]; dimeric structure with five-coordinate Sn IR (KBr): 1140, 1050, 930, 855, and 680	[6,15,17, 18,25,52] [11] [52,107] [30] [20,30,116] [17]
*3	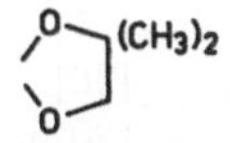 III [94]	^{119}Sn-γ (78 K): $\delta = 1.144$, $\Delta = 2.627$; dimeric species with five-coordinate Sn	[116]
4	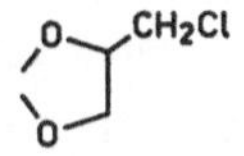 III	white solid decomposes on distillation	[11]
*5	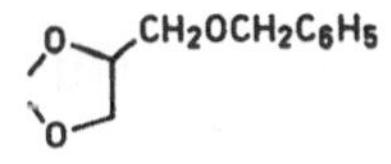 II (40)	m.p. 125 to 128°	[107]
6	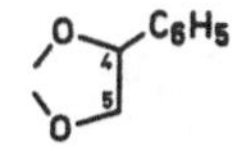 II [52], (100 [134])	m.p. 186 to 188° (dec.) ^{1}H NMR (half-saturated in $CDCl_3$): 3.21 (H-5′, $^3J_{cis}(H,H) = 4.46$, $^2J_{gem}(H,H) = -9.20$), 3.80 (H-5, $^3J_{trans}(H,H) = 9.50$), 4.53 (H-4) with $SbCl_3 \rightarrow (C_4H_9)_2SnCl_2$ and $\overline{OCH(C_6H_5)CH_2OSbCl}$ (85%) with C_6H_5COCl, $C_6H_5(CH_3)_2SiCl$, and aqueous HCl (subsequently added) $\rightarrow$ $C_6H_5COOCH(C_6H_5)CH_2OH$ (95%, regio-selective)	[52] [84] [134]

References on p. 92

Table 14 (continued)

No.	O-R-O group method of preparation (yield in %)	properties and remarks	Ref.
7	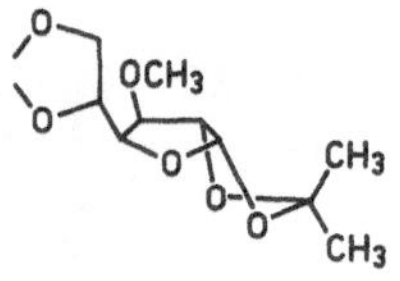 II	with C_6H_5COCl (in C_5H_5N) → 6-C_6H_5COO-R-OH (75%) with Br_2 (4 Å molecular sieves) → 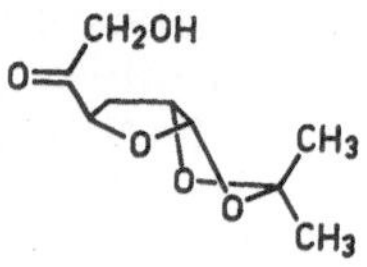 (48%)	[62]
*8	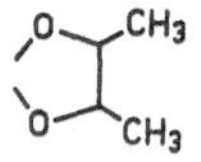 II [30,75], (98 [12], 50 [107], 77 [99]) III [15], (93 [11]) IV [46]	m.p. 119 to 121° [18,25], 120° [15], 121 to 124° [30], or 124 to 126° [12] for the meso-diol derivative; 134 to 136° [99], 138 to 141° [46], or 142 to 145° [107] for the derivative of optically pure diol	[12,15,18, 25,30,46, 99,107]
		b.p. 171 to 173°/0.2	[11]
		^{119}Sn NMR ($CDCl_3$): -155 ± 5	[30]
		^{119}Sn-γ (77 or 78 K): $\delta = 1.25$, $\Delta = 2.85$ [20,30], or 1.257 and 3.084 (racemic), 1.251 and 3.058 (R,R) [116], respectively; dimeric structure with five-coordinate Sn suggested	[20,30,116]
*9	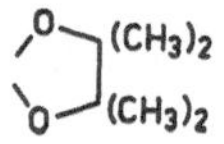 III [15], (92 [11]) IV [46]	m.p. 45° [15], 46° [46]	[15,46]
		b.p. 188 to 189°/0.1 [18,25], 195°/0.4 [15], 204 to 205°/1.6 [11]	[11,15,18, 25]
10	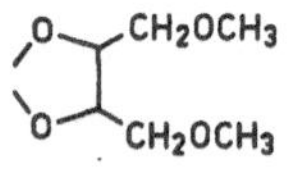 II (>95)	low melting solid with $C_6H_3(OCH_3\text{-}1)(CH_2Br)_2\text{-}2,6$ (2:1 mole ratio, hot DMF, aqueous workup) → 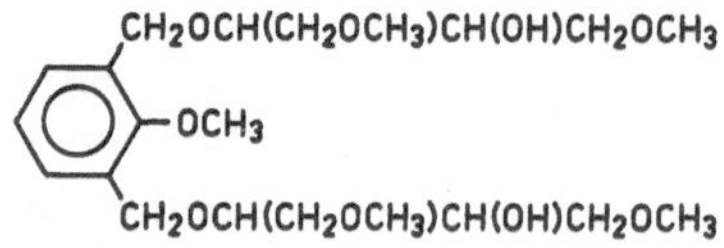	[91]
11	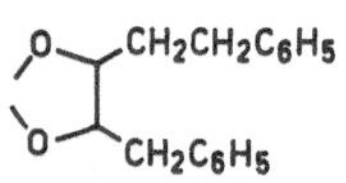 II	dimeric with five-coordinate Sn with oxidants like $[N(C_4H_9)_4]IO_4$, $Pb(OOCCH_3)_4$, $C_6H_5I(OOCCH_3)_2$, or $(C_6H_5)_3Bi(OOCCH_3)_2$ (CH_2Cl_2, 20 to 40°, 5 min to 2 h) → $C_6H_5CH_2CHO$ and $C_6H_5CH_2CH_2CHO$ (60 to 100%)	[97]

References on p. 92

Table 14 (continued)

No.	O-R-O group method of preparation (yield in %)	properties and remarks	Ref.
12	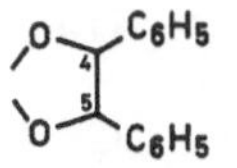 Ia (45 meso, 47 (±) [77]) IV [46]	m.p. 113 to 115° (±) [77], 195 to 197° [46] (meso) [77]	[46,77]
		^{1}H NMR: 0.66 to 1.90 (m, C_4H_9), 5.16 (m, H-4,5), 7.01 to 7.40 (m, C_6H_5) for the meso-form in CF_3COOH; 0.73 to 1.80 (m, C_4H_9), 4.31 (m, H-4,5), 6.83 to 7.43 (m, C_6H_5) for the (±)-form in $CDCl_3$ IR (KBr): unassigned bands in the 4000 to 600 range for each diastereomer MS (70 eV): 457 $[M]^+$ with R'COCl (R' = CH_3, C_6H_5, 1:1 or 1:2, in C_6H_6, reflux) → $(C_4H_9)_2SnCl_2$ and 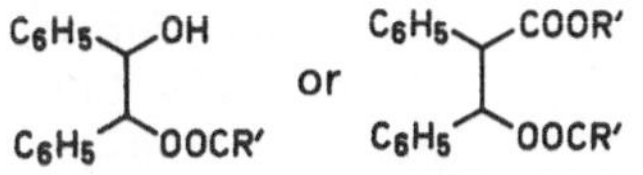 for each diastereomer	[77]
		with $\overset{*}{R}$'COCl → $(C_4H_9)_2SnCl_2$ and 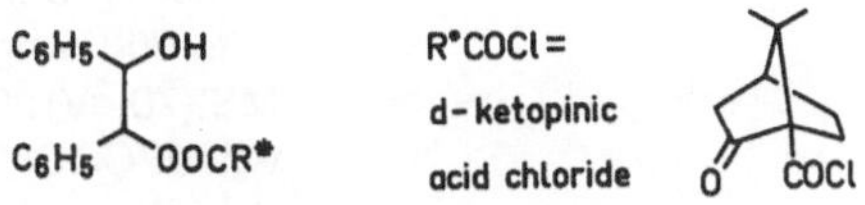 in a high diastereomeric selectivity	[118]
13	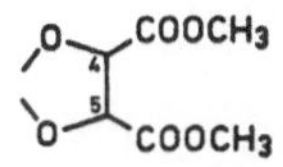 II [118], (86 [133])	L(+)-HO-R-OH derivative: m.p. 180 to 185° ^{1}H NMR ($CDCl_3$): 0.86 to 1.83 (m, C_4H_9), 3.58 (s, CH_3O), 4.63 (s, H-4,5) IR (KBr): ν(C=O) 1760 with $ClOC(CH_2)_nCOCl$ (1:1 mole ratio, C_6H_6, reflux/12 h) → $(C_4H_9)_2SnCl_2$ and 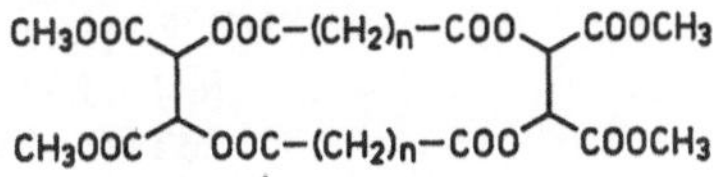 (n = 3 (35%), 5 (26%), 7 (25%), 8 (24%))	[133]
		with d-ketopinic acid chloride like No. 12	[118]

References on p. 92

Table 14 (continued)

No.	O-R-O group method of preparation (yield in %)	properties and remarks	Ref.
*14	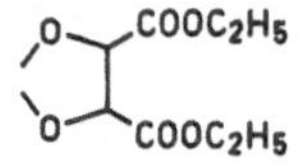 II (75 to 85 [107]) III [94]	m.p. 140 to 145° (meso and chiral) ^{1}H NMR on p. 84	[107]
		^{119}Sn-γ (78 K): $\delta = 1.20$, $\Delta = 3.20$ [20]; 1.302 and 3.200 (racemic), 1.290 and 3.143 (racemic, frozen solution in $C_6H_5C_4H_9$), 1.308 and 3.182 (S,S); dimeric species with five-coordinate Sn [114]	[20,114]
15	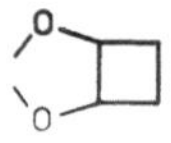 II	amorphous powder with $Br_2 \rightarrow (C_4H_9)_2SnBr_2$ and 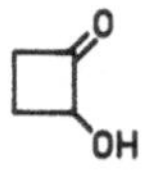	[39]
16	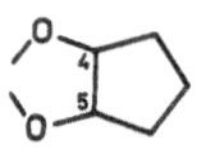 Ia (48 (meso), 44 (±) [77]) II [39]	m.p. 143.5° (±), 192.3° (meso) [77], 228 to 233° (impure) [39]	[39,77]
		^{1}H NMR: 0.80 to 2.26 (m, C_4H_9, $(CH_2)_3$), 4.33 (m, H-4,5) for the (±)-form in CF_3COOH; 0.80 to 1.96 (m, C_4H_9, $(CH_2)_3$), 3.92 (m, H-4,5) for the meso-form in $CDCl_3$ IR (KBr): unassigned bands in the 3000 to 600 range for each diastereomer MS (70 eV): 333 $[M]^+$.	[77]
		with R′COCl (R′ = CH_3, C_6H_5) and d-ketopinic acid chloride like No. 12	[118]
		with $Br_2 \rightarrow (C_4H_9)_2SnBr_2$ and 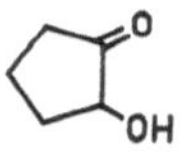	[39]
17	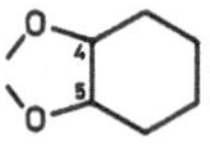 Ia (56 (±), 25 (meso) [77]) II (100 [39], 96 (trans), 96 (cis) [12])	m.p. 164 to 165° (cis) [12], 168 to 170° (meso), 230 to 232° (±) [77], 234 to 236° (trans) [12]	[12,77]
		^{1}H NMR ($CDCl_3$): 0.85 to 1.90 (m, C_4H_9, $(CH_2)_4$), 3.05 (m, H-4,5) for the (±)-form; 0.85 to 2.04 (m, C_4H_9, $(CH_2)_4$), 3.50 (m, H-4,5) for the meso-form IR (KBr): unassigned bands in the 3000 to 600 range for each diastereomer MS (70 eV, m/e (% for (±), % for meso)): $[M]^+$, 347 (<1, <1), 291 (22, 5), 235 (74, 13), 232 (30, 8), 191 (24, 11), 175 (20, 5), 135 (42, 10), 57 (46, 66)	[77]
		with $Br_2 \rightarrow (C_4H_9)_2SnBr_2$ and 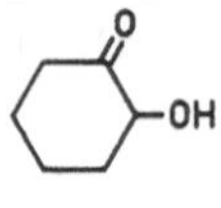	[39]

References on p. 92

Table 14 (continued)

No.	O-R-O group method of preparation (yield in %)	properties and remarks	Ref.
		with R′COCl (R′ = CH_3, C_6H_5) like No. 12	[77]
		with ClOC-X-COCl→$(C_4H_9)_2SnCl_2$ and	[113,120]
		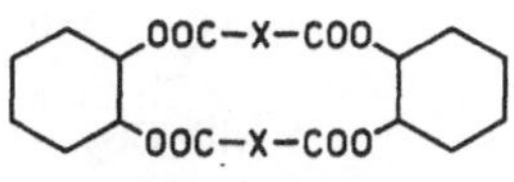	
		X = CH_2-O-CH_2 [120] X = $(CH_2)_n$, n = 1, 3, 5, 7, 8 [113]	
		with $C_6H_3((CH_2Br)_2$-2,6)OCH_3 → $(C_4H_9)_2SnBr_2$ and	[91]
		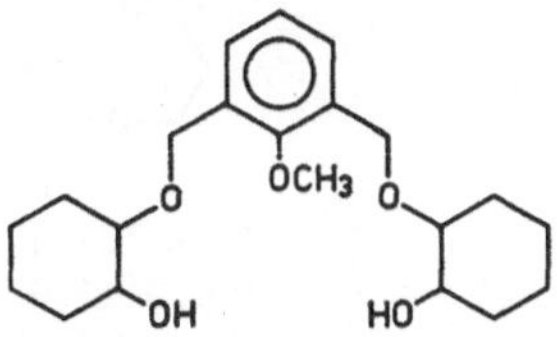	
		with $SbCl_3$→$(C_4H_9)_2SnCl_2$ and	[113]
		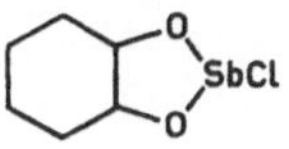	
18	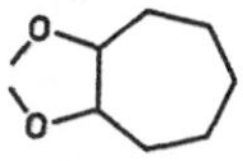 II	m.p. 162 to 164° with Br_2→$(C_4H_9)_2SnBr_2$ and 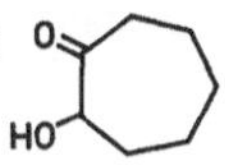	 [39]
19	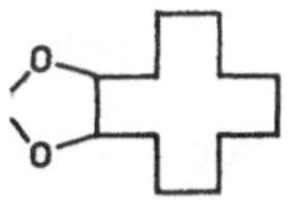 II	no properties reported with d-ketopinic acid chloride like No. 12	[118]
20	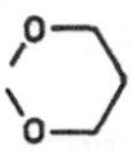	m.p. 86 to 88° [25], 88° [15]	[15,25]
		softening point for $(-(C_4H_9)_2SnO(CH_2)_3O-)_n$: 105 to 109°	[22,27]
	Ia (1 [22]) Ib (3 [27])	(3 < n < 120) [22], > 290° (78 < n < 1400) [27]	
	III [15,30], (78 [11])	b.p. 182 to 186°/0.3 [30], 184 to 185°/0.3 [11], 200°/1.5 [15]	[11,15,30]
		^{119}Sn NMR: −228 ± 10 (saturated in $CDCl_3$), −288 ± 2 (neat liquid, at 96°); oligomeric species in solution, containing six-coordinate Sn, suggested	[30]

References on p. 92

Table 14 (continued)

No.	O-R-O group method of preparation (yield in %)	properties and remarks	Ref.
20 (continued)		with COS→(-$(C_4H_9)_2$SnS-)$_3$ and 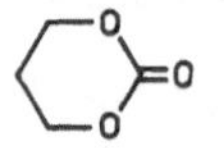 (38%)	[47]
		with CS_2→(-$(C_4H_9)_2$SnS-)$_3$ and 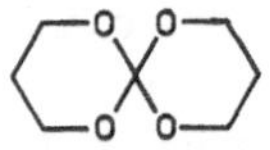 (92%)	[18, 25]
		with C_6H_5NCS→(-$(C_4H_9)_2$SnS-)$_3$ and 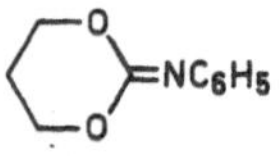 (60%)	[58]
21	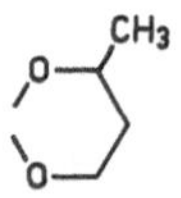 Ia [6] III [15], (83 [11])	m.p. 134° [15], liquid at 130° [6]	[6, 15]
		b.p. 165°/0.25 [11], 185°/1.5 [15]	[11, 15]
		dimeric in C_6H_6	[11]
		stabilizer for PVC	[6]
22	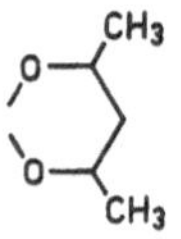 III	m.p. 155° b.p. 159°/0.7 monomer ⇆ dimer equilibrium	[15]
23	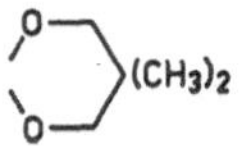 III [30, 108]	b.p. 188 to 190°/0.3 ^{119}Sn NMR (neat liquid): -213 ± 5; oligomeric species with six-coordinate Sn suggested	[30]
		with $CH_3C_6H_4SO_2Cl$ → $(CH_3)_2C(CH_2OH)CH_2OSO_2C_6H_4CH_3$	[75]
		with $BrCH_2CONHC_7H_{15}$ → $(CH_3)_2C(CH_2OCH_2CONHC_7H_{15})_2$	[108]
		with $ClOC(CH_2)_5COCl$ → $(CH_3)_2C(CH_2OOC(CH_2)_5COOCH_2)_2$-$C(CH_3)_2$	[80]
24	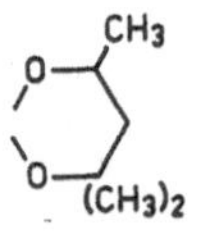 III [15], (77 [11])	m.p. 80°	[15]
		b.p. 143°/1.5 [15], 163 to 165°/0.7 to 0.8 [11]	[11, 15]
		monomer ⇆ dimer equilibrium	

References on p. 92

Table 14 (continued)

No.	O-R-O group method of preparation (yield in %)	properties and remarks	Ref.
		with C_6H_5NCO (1:1 or 1:2, C_6H_6, reflux) → 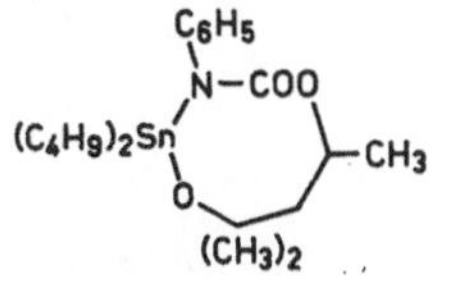 (100%) or 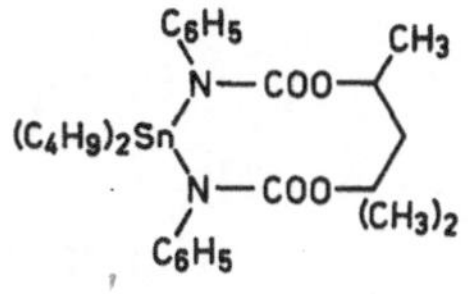 (100%)	[37]
25	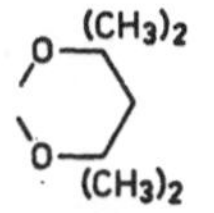 III	m.p. 70° b.p. 119°/0.8 monomer ⇆ dimer equilibrium	[15]
26	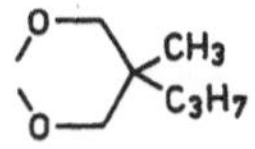 III	b.p. 200°/0.03 ^{119}Sn NMR (saturated in CCl_4): -233 ± 5; oligomeric species with six-coordinate Sn suggested	[30]
27	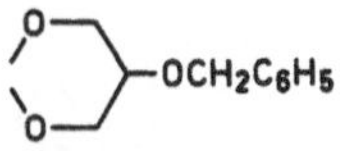 II (100)	not isolated in a pure state with C_6H_5COCl ($CHCl_3$, reflux/3 h) → $C_6H_5CH_2OCH(CH_2OH)CH_2OOCC_6H_5$	[132]
28	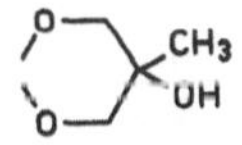 II	catalyst in polyurethane synthesis	[41]
29	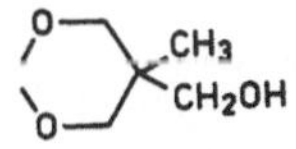 II	catalyst in polyurethane synthesis	[41]
30	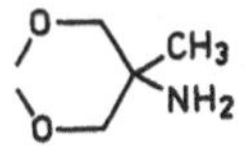 II	catalyst in polyurethane synthesis	[41]

References on p. 92

Table 14 (continued)

No.	O-R-O group method of preparation (yield in %)	properties and remarks	Ref.
31	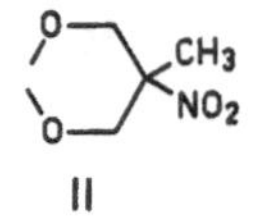 II	catalyst in polyurethane synthesis	[41]
32	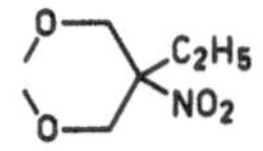 II	catalyst in polyurethane synthesis	[41]
33	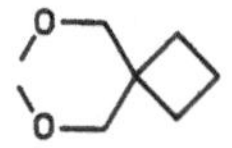 II [41], (64 [35,42,49])	m.p. 137 to 139° monomeric stabilizer for halogen-containing resins catalyst in polyurethane synthesis	[42,49] [35,42,49] [41]
34	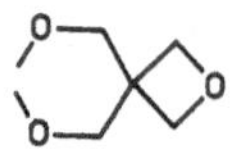 II	catalyst in polyurethane synthesis	[41]
35	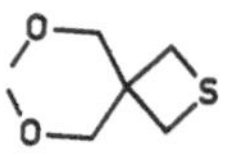 II [35,41,43] (95 [42,49])	m.p. 185 to 187° monomeric ^{1}H NMR: 2.83 (s, CH_2SCH_2) IR (TMS mull): γ(ring) 1176 catalyst in polyurethane synthesis stabilizer for halogen-containing resins	[49] [42,49] [41,43] [35,42,49]
36	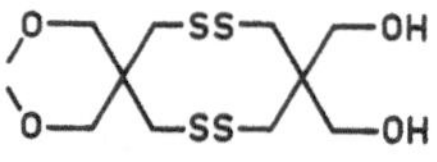 II	yellow crystals m.p. 135 to 142° UV ($CHCl_3$): λ_{max} 326 stabilizer for halogen-containing resins	[42,49]
37	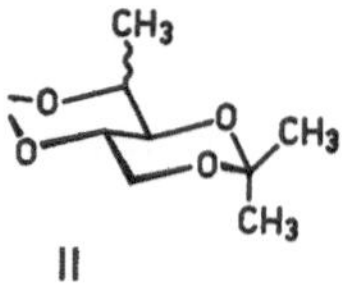 II	with $Br_2 \rightarrow (C_4H_9)_2SnBr_2$ and 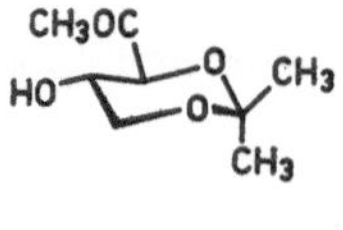 (91)	[85]
38	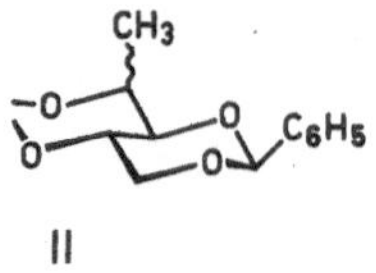 II	not isolated with $Br_2 \rightarrow (C_4H_2)_2SnBr_2$ and 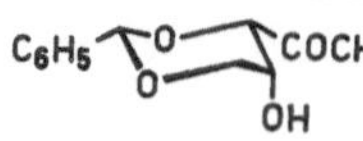 (33)	[85]

References on p. 92

Table 14 (continued)

No.	O-R-O group method of preparation (yield in %)	properties and remarks	Ref.
39	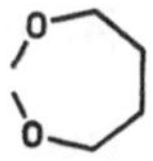	softening point: 190 to 198° (polymer, $3 < n < 120$)	[22]
		m.p. 195° (dec., probably polymer)	[13]
	Ia (52 [22])	viscous liquid	[11, 15, 25, 30]
	II (92 [13])	b.p. 160°/0.2 [15], 162 to 164°/0.2 [30], 175 to 178°/0.4 [11], 177 to 178° 0.6 [25]	
	III [15, 30], (78 [11])	^{119}Sn NMR: -154 ± 5 (in CCl_4) [30, 34]; -161 ± 2 (neat liquid) [30]	[30, 34]
		with $CS_2 \rightarrow (\text{-}(C_4H_9)_2SnS\text{-})_3$, 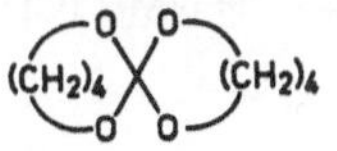 (37), and $(\text{-}O(CH_2)_4OCO\text{-})_n$ (62)	[25]
		with $C_6H_5NCS \rightarrow (\text{-}(C_4H_9)_2SnS\text{-})_3$ and polyurethane	[58]
		with $ClOC(CH_2)_4COCl \rightarrow (C_4H_9)_2SnCl_2$ and 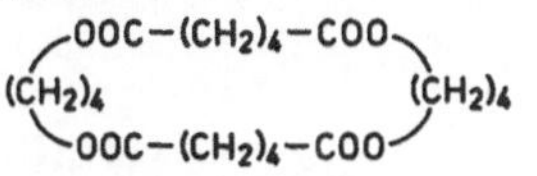 (21)	[80]
40	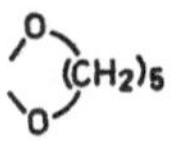	white solid	[11]
		m.p. 174° (dec., probably polymer)	[13]
	II (89 [13])	b.p. 165 to 175°/0.5 [25], 185 to 187°/0.15 [11]	[11, 25]
	III (72 [11])	with $CS_2 \rightarrow (\text{-}(C_4H_9)_2SnS\text{-})_3$, $(\text{-}O(CH_2)_5OCO\text{-})_n$, and 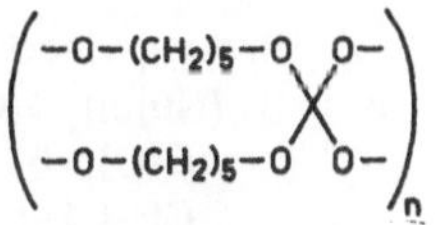	[25]
41	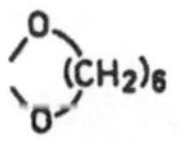	white solid	[11]
		m.p. 148° (dec., probably polymer)	[11, 13]
	II (93 [13])	dec. on distillation	[11]
	III [11]		
42	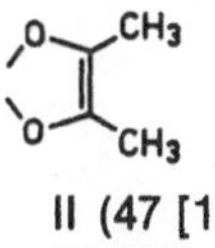	m.p. 80°	[100]
		dec. 112°	[46]
	II (47 [100])	IR: ν(SnOC) 670	[100]
	IV [46]		

References on p. 92

Table 14 (continued)

No.	O-R-O group method of preparation (yield in %)	properties and remarks	Ref.
*43	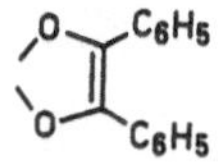 II [87, 101], (81 [100]) III [100] IV [46] special [100]	m.p. 89 to 90° (from C_5H_{12} at −50°) [100], dec. 161° [46]	[46, 100]
		^1H NMR ($CDCl_3$): 0.7 to 1.9 (m, C_4H_9), 7.2 to 7.7 and 7.9 to 8.1 (m, C_6H_5) IR: ν(C=C) 1650, ν(SnOC) 670	[100]
*44	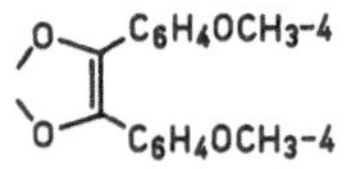 special [100]	^1H NMR ($CDCl_3$): 3.80 (CH_3O)	[100]
45	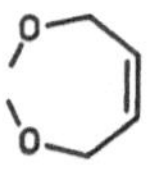 II	with cis-$CH_3COOCH_2CH{=}CHCH_2OOCCH_3$ and $Pd(P(C_6H_5)_3)_4$ → 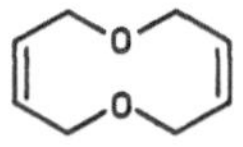 with $C_6H_4(CH(OOCCH_3)CH{=}CH_2)_2$-1,2 and Pd(0) → 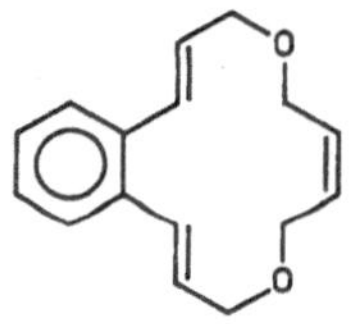	[131]
46	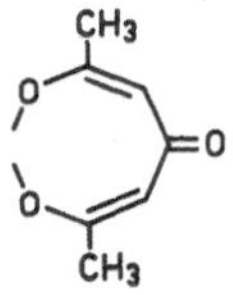 Ib	brown, viscous liquid IR (Nujol): ν(CH) 2960 to 2920 (s), ν(C=O) 1600 and 1510 (w), δ(CH_2) 1400 (w), δ(CH_3) 1300 (w), $ν_{as}$(SnO) 680 (s), $ν_s$(SnO) 610 (vw), ? 590 (w), $ν_{as}$(SnC) 550 (vw), $ν_s$(SnC) 520 (m)	[98]
47	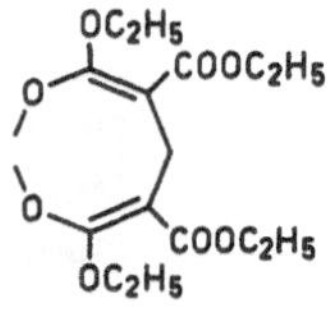 Ib	D^{20} = 1.1844, n_D^{20} = 1.4710 stabilizer for halogen-containing resins additive to lubricants	[3]

References on p. 92

Table 14 (continued)

No.	O-R-O group method of preparation (yield in %)	properties and remarks	Ref.
*48	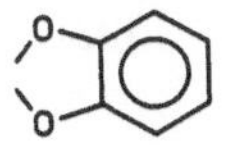 I a [9] II [10,30] III [29] special [63,82,100,101]	m.p. 269 to 271° [30], 272 to 274° [100] sublimation at 210°/0.2 ^{119}Sn-γ (77 K): δ = 1.29, Δ = 3.40 [21,28], or 1.52 and 3.62 [16,20,30]; linear polymeric structure with six-coordinate Sn deduced IR in Table 18 on p. 87 with $ClOCCH_2OCH_2COCl \rightarrow (C_4H_9)_2SnCl_2$ and 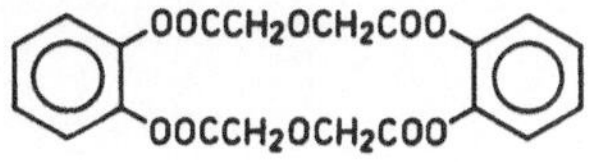 with $SnCl_2 \rightarrow (C_4H_9)_2SnCl_2$ and 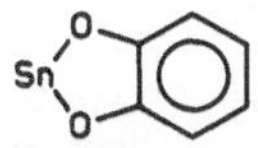	[30,100] [30] [16,20,21, 28,30] [16,20,30] [29] [120] [63]
49	 I a	polymeric hydrolytically unstable	[22]
*50	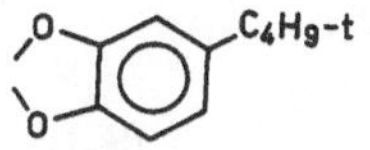 II [3] special [82]	dec. 245° m.p. 262 to 264° catalyst in polyurethane production	[82] [3]
51	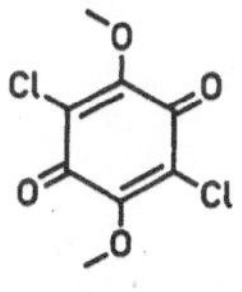 I b	dark blue solid dec. >200° ^{119}Sn-γ (80 K): δ = 1.49, Δ = 3.88; polymeric trans-octahedral structure IR (KBr): ν(SnO) 488 to 445	[31]
52	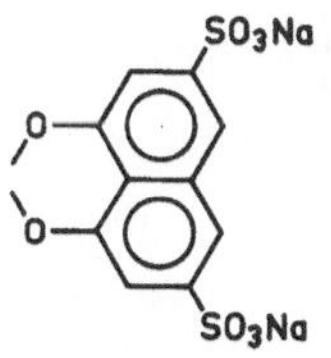	indicated by potentiometric titration of $(C_4H_9)_2SnCl_2$ with NaOH in the presence of chromotropic acid in water-dioxane	[33]

References on p. 92

Table 14 (continued)

No.	O-R-O group method of preparation (yield in %)	properties and remarks	Ref.
53	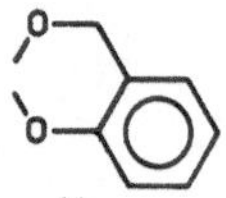 Ib	m.p. 57° IR discussion	[93]
54	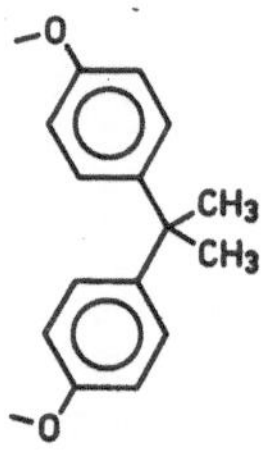 Ib (39)	softening ≧290° polymer with $78 < n < 1400$	[27]
55	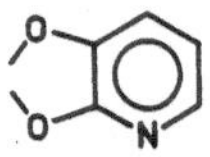 II (90)	m.p. 199 to 201° ^{119}Sn-γ (77 K): $\delta = 1.26$, $\Delta = 2.82$ IR (Nujol): unassigned bands in the 1609 to 476 region	[63]
56	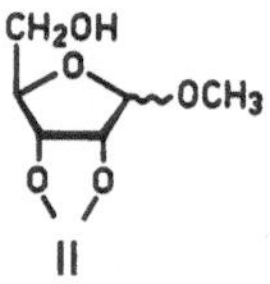 II	with $SOCl_2$ → 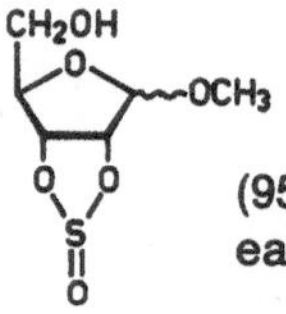 (95%; α and β, each 100% exo)	[129]
57	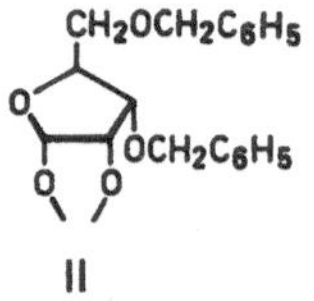 II	with R'X (CH_3I or $C_6H_5CH_2Br$) → 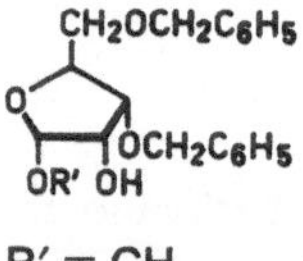 and 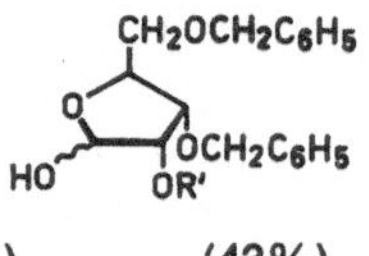 R' = CH_3 (49%) (43%) R' = $CH_2C_6H_5$ (83%) (13%)	[95]
*58	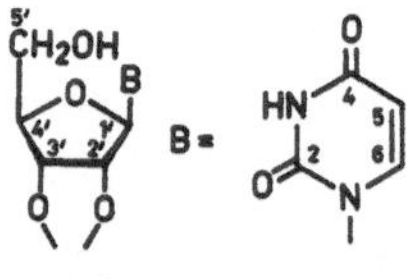 HO-R-OH = uridine II [121], (90 [59], 96 [45])	m.p. 232 to 233° [59], 232 to 234° [45], 233° [121]	[45, 59, 121]
		^{119}Sn-γ (78 K): $\delta = 1.29$, $\Delta = 3.02$	[121]
		IR (Nujol): ν(NH, OH) 3470 (s), ν(C=O) 1680 (s, br), ν(ring) 1470 (s) and 1445 (s), ν(SnC) 585 (m) and 510 (w), ν(SnO) 420 (w)	
		UV (CH_3OH): λ_{max} (ε) = 261 (9400)	[45]
		MS (70 eV): $[M]^+$, $[M-C_4H_9]^+$, $[M-uracil]^+$, $(M-C_4H_9-uracil]^+$	

References on p. 92

Table 14 (continued)

No.	O-R-O group method of preparation (yield in %)	properties and remarks	Ref.
59	$C_5H_7O_4B$, cf. No. 58 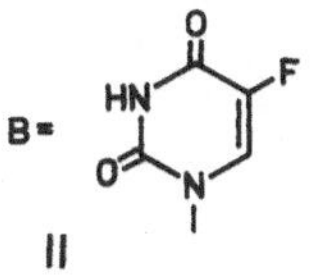 II	no physical properties reported anticarcinogenic activity	[55]
*60	$C_5H_7O_4B$, cf. No. 58 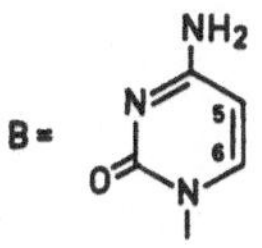 HO-R-OH = cytidine II (91)	m.p. 217 to 218° ^{1}H NMR (CD_3OD): 0.90 (t, H-δ), 1.2 to 1.8 (m, H-α, β, γ), 3.88 (m, H-5′), 4.3 (m, H-2′, 3′), 5.88 (s, H-1′), 5.92 (d, H-5, J(H-5, H-6) = 7.5), 7.97 (d, H-6) UV: λ_{max} (ε) = 275(8200) in CH_3OH/OH^-, or 282(13300) in CH_3OH/H^+	[45]
*61	$C_5H_7O_4B$, cf. No. 58 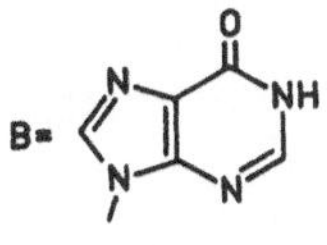 HO-R-OH = inosine II	m.p. 202 to 204° ^{119}Sn-γ (78 K): δ = 1.29, Δ = 3.08 IR (Nujol): ν(NH, OH) 3530 (m), 3300 (s), 3120 (w), 3060 (w), and 3040 (w), ν(C=O) 1710 (s), ν(ring) 1590 (s) and 1565 (s), ν(CO)/Sn 1115 (s) and 1095 (s), ν(SnC) 595 (s) and 530 (m), ν(SnO) 460 (m)	[121]
*62	$C_5H_7O_4B$, cf. No. 58 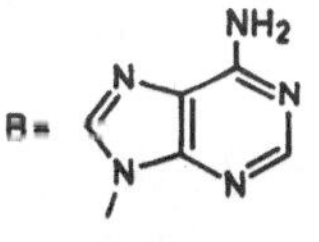 HO-R-OH = adenosine II [121], (70 [45])	m.p. 154 to 156° [45], 156 to 158° [121]	[45, 121]
		^{119}Sn-γ (78 K): δ = 1.28, Δ = 3.06 IR (Nujol): ν(NH_2, OH) 3320 (s, br) and 3140 (s, br), ν(ring) 1605 (s) and 1575 (s), ν(CO)/Sn 1105 (s, sh) and 1080 (s, br), ν(SnC) 570 (m), ν(SnO) 475 (m) and 450 (m, sh)	[121]
		UV (CH_3OH): λ_{max} (ε) = 259(14700) MS (70 eV): $[M]^+$, $[M-C_4H_9]^+$, $[M-adenine]^+$	[45]
63	$C_5H_7O_4B$, cf. No. 58 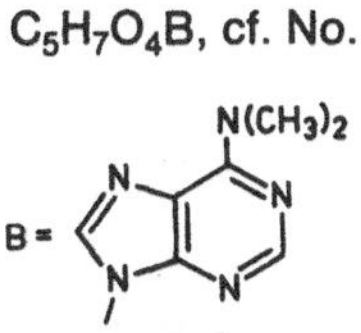 II (73)	m.p. 223 to 225.5° MS: m/e = 531 $[M]^+$ and 523 to 529 (0.4 to 0.9), 192 (7.2), 163, 148, 134, 120, 93, 57 (18.6) $[C_4H_9]^+$	[72]

References on p. 92

Table 14 (continued)

No.	O-R-O group method of preparation (yield in %)	properties and remarks	Ref.
64	$C_5H_7O_4B$, cf. No. 58 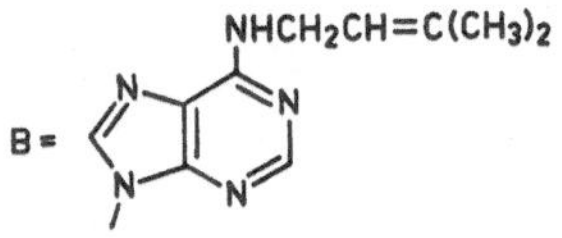 II	with $\overline{CH_2CO\text{-}O\text{-}OCCH_2}$ (only in C_5H_5N in the presence of $[N(C_4H_9)_4]Br$) → 3′-$HOOCCH_2CH_2COO$-R-OH(100%)	[86]
65	$C_5H_7O_4B$, cf. No. 58 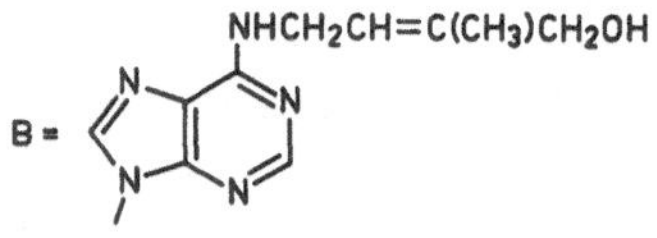 *cis* or *trans* II	with $\overline{CH_2CO\text{-}O\text{-}OCCH_2}$ like No. 64	[86]
66	$C_5H_7O_4B$, cf. No. 58 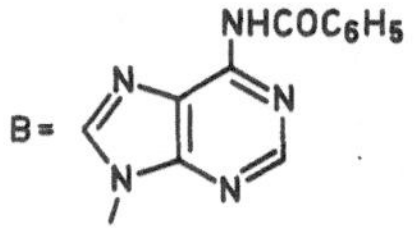 II (56 [109])	m.p. 167 to 169° UV (CH_3OH): $\lambda_{max} = 278$, $\lambda_{min} = 247$ with C_6H_5COCl → 3′-C_6H_5CO-O-R-OH (48%) with 2-$CHBr_2C_6H_4COCl$ → 3′-(2-$CHBr_2C_6H_4CO$)O-R-OH (83%)	[109] [70]
*67	$C_5H_7O_4B$, cf. No. 58 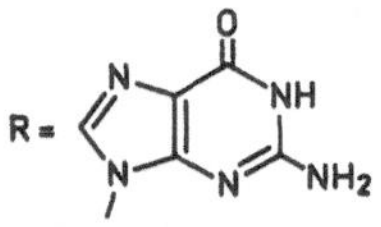 HO-R-OH = guanosine II	m.p. 235° (dec.) ^{119}Sn-γ (78 K): $\delta = 1.27$, $\Delta = 3.12$ IR (Nujol): ν(NH, OH) 3480 (s), 3310 (s) and 3160 (s), ν(C=O) 1695 (s), ν(ring) 1485 (s) and 1410 (m), ν(CO)/Sn 1120 (s) and 1095 (s), ν(SnC) 590 (m) and 555 (w), ν(SnO) 465 (m) and 440 (m)	[121]
68	$C_5H_7O_4B$, cf. No. 58 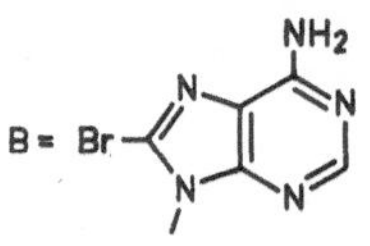 II	not isolated in a pure state with 4-$CH_3C_6H_4SO_2Cl$ → 2′-(4-$CH_3C_6H_4SO_2$)O-R-OH (94%)	[112]

References on p. 92

Table 14 (continued)

No.	O-R-O group method of preparation (yield in %)	properties and remarks	Ref.
69	$C_5H_7O_4B$, cf. No. 58 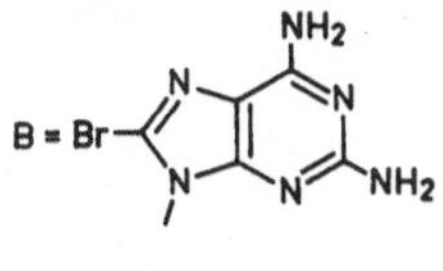 II (25)	0.5-H_2O solvate m.p. 221 to 223° UV: λ_{max} = 285, 263, 218 (50% C_2H_5OH), 290, 260 (50% C_2H_5OH/H^+), or 280, 263 (50% C_2H_5OH/OH^-) MS: m/e = 229, 231, 331 to 339 with 4-$CH_3C_6H_4SO_2Cl$ → 2′-(4-$CH_3C_6H_4SO_2$)O-R-OH (56%)	[79]
70	$C_5H_7O_4B$, cf. No. 58 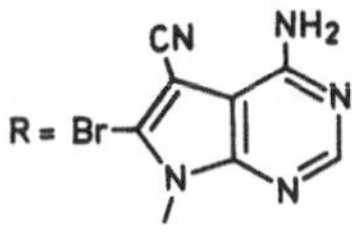 II (92)	m.p. 206 to 208° with 4-$CH_3C_6H_4SO_2Cl$ → 2′-(4-$CH_3C_6H_4SO_2$)O-R-OH (64%)	[105]
71	$C_5H_7O_4B$, cf. No. 58 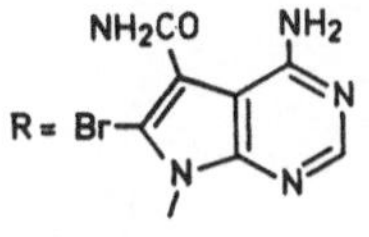 II (91)	m.p. 215 to 217° with 4-$CH_3C_6H_4SO_2Cl$ → 2′-(4-$CH_3C_6H_4SO_2$)O-R-OH (66%)	[105]
72	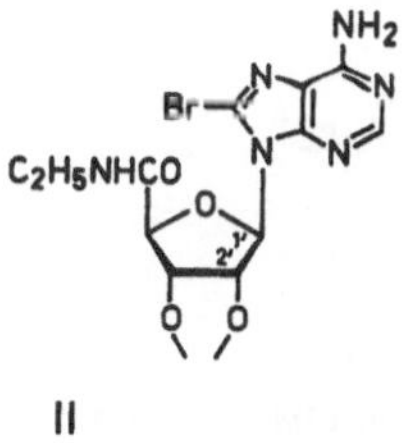 II	with 4-$CH_3C_6H_4SO_2Cl$ → 2′-(4-$CH_3C_6H_4SO_2$)O-R-OH (90%)	[125]
73	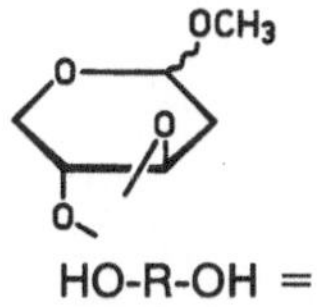 HO-R-OH = methyl-2-desoxy-α,β-D-*threo*-pentopyranoside II (small)	m.p. 207° (α-anomer), 212° (β-anomer) (each not analytically pure) with 4-$NO_2C_6H_4COOH$ → 4-$NO_2C_6H_4$COO-R-OOC$C_6H_4NO_2$-4	[40]

References on p. 92

Table 14 (continued)

No.	O-R-O group method of preparation (yield in %)	properties and remarks	Ref.
74	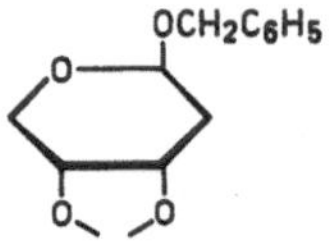 HO-R-OH = benzyl-2-desoxy-β-D-*erythro*-pentopyranoside II [40]	m.p. 127 to 132° [39], 132 to 135° [40] $[\alpha]_D^{23} = -78.3$ ($CHCl_3$, c ~ 1)	[39,40] [40]
75	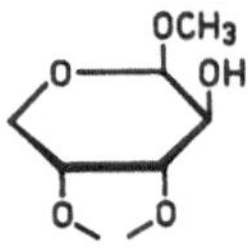 HO-R-OH = methyl-β-L-arabinopyranoside II [110,129]	with Br_2 → 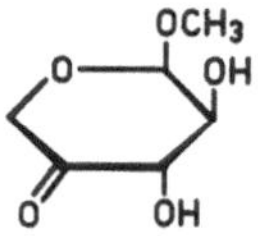 (60 to 70%) with $SOCl_2$ → 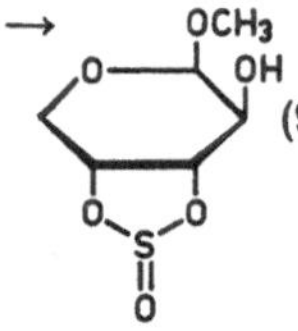 (90% (100% exo))	[110] [129]
76	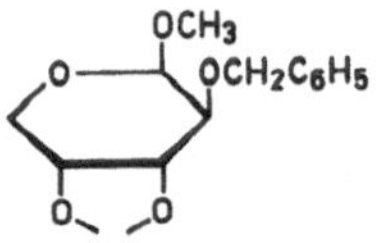 HO-R-OH = methyl-2-O-benzyl-β-L-arabino-pyranoside II	with $SOCl_2$ → 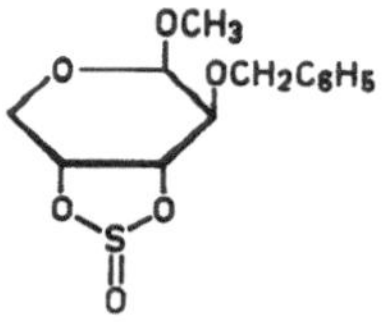 (96%; exo/endo = 75/25)	[129]
77	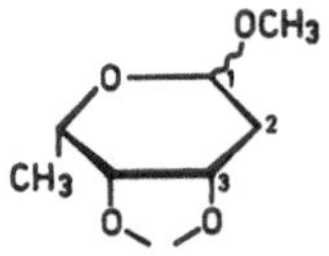 HO-R-OH = methyl-2,6-di-desoxy-α,β-L-lyxo-hexopyranoside II α- [127] or β-anomer [104]	with $C_6H_5CH_2Br$→3-$C_6H_5CH_2O$-R-OH	[104, 127]

References on p. 92

Table 14 (continued)

No.	O-R-O group method of preparation (yield in %)	properties and remarks	Ref.
78	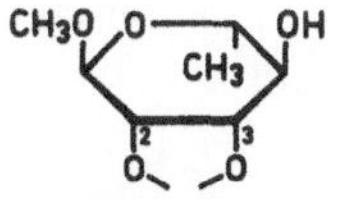 HO-R-OH = methyl α-L-rhamnopyranoside II	with $C_6H_5CH_2Br$→ 3-$C_6H_5CH_2O$-R-OH (50%)	[74]
79	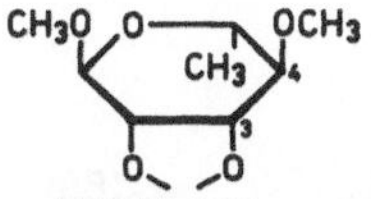 HO-R-OH = methyl 4-O-methyl-α-L-rhamno-pyranoside II	with $C_6H_5CH_2Br$→ 3-$C_6H_5CH_2O$-R-OH (62%)	[78]
80	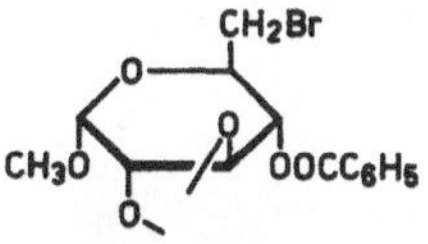 HO-R-OH = methyl 4-O-benzoyl-6-bromo-6-desoxy-α-D-glucopyranoside II (41)	m.p. 166° $[\alpha]_D^{23}$ = 33.6 ($CHCl_3$, c ~ 1)	[40]
81	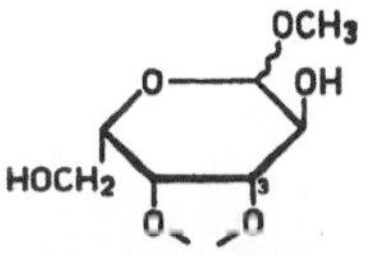 HO-R-OH = α- or β-methyl D-galactopyranoside II [117, 129]	with CH_2=$CHCH_2Cl$→ 3-CH_2=$CHCH_2O$-R-OH (79%; β-anomer) with $SOCl_2$ → 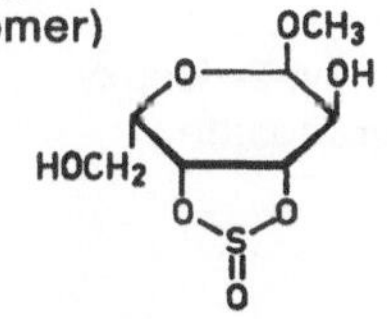 (70%; α-anomer, exo/endo = 82/18)	[117] [129]
82	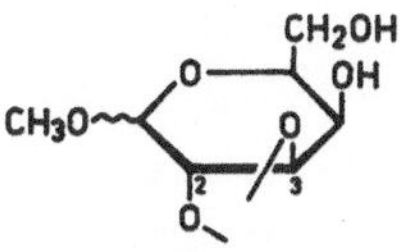 HO-R-OH = α- or β-methyl D-galactopyranoside II [53, 111]	with 4-$CH_3C_6H_4SO_2Cl$ → 2-(4-$CH_3C_6H_4SO_2$)O-R-OH [53], or, in contrast, predominantly (>95%) the 3-O-tosylate [111] (α-anomer) with C_6H_5COCl → 3-C_6H_5COO-R-OH (β-anomer)	[53, 111] [111]

References on p. 92

Table 14 (continued)

No.	O-R-O group method of preparation (yield in %)	properties and remarks	Ref.
83	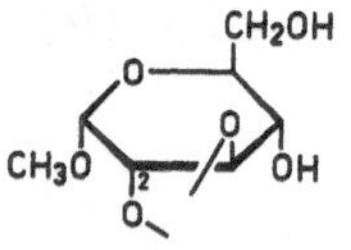 HO-R-OH = methyl α-D-glucopyranoside II	m.p. 105 to 115° with $C_{13}H_{27}COCl$ (tetradecanoic acid chloride) → 2-$C_{13}H_{27}COO$-R-OH (73%)	[53]
84	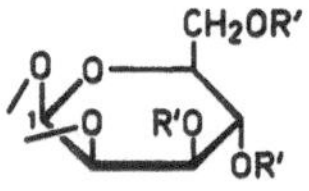 R′ = -$CH_2CH{=}CH_2$ HO-R-OH = 3,4,6-tri-O-allyl-β-D-mannopyranose II	with CH_3I → 1-CH_3O-R-OH (88%)	[88]
85	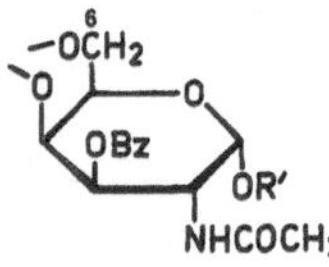 R′ = -$CH_2CH{=}CH_2$ Bz = -$CH_2C_6H_5$ HO-R-OH = allyl 2-acetamido-3-O-benzyl-2-desoxy-α-D-galactopyranoside II	with $C_6H_5CH_2Br$ → 6-$C_6H_5CH_2O$-R-OH (75%)	[119]
86	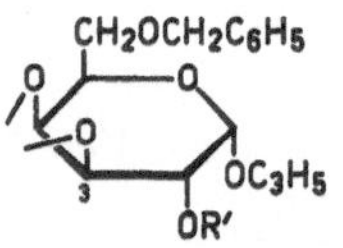 R′ = -COC_6H_5 or -$CH_2CH{=}CHCH_3$ HO-R-OH = allyl 2-O-benzoyl- or allyl 2-O-crotyl-6-O-benzyl-α-D-galactopyranoside II	with $C_6H_5CH_2Br$ → 3-$C_6H_5CH_2O$-R-OH	[81]

References on p. 92

Table 14 (continued)

No.	O-R-O group method of preparation (yield in %)	properties and remarks	Ref.
87	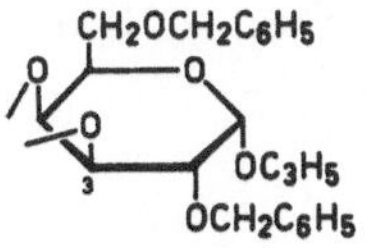 HO-R-OH = allyl 2,6-di-O-benzyl-α-D-galactopyranoside II	with R'X → 3-R'O-R-OH (R'X = C_6H_5COCl (81%) [54, 57, 111], CH_2=$CHCH_2I$ (79%), CH_3I (77%) [54, 57], $C_6H_5CH_2Br$ (72%) [54], $CH_3OCH_2CH_2OCH_2Cl$ (81%) [92])	[54, 57, 92, 111]
88	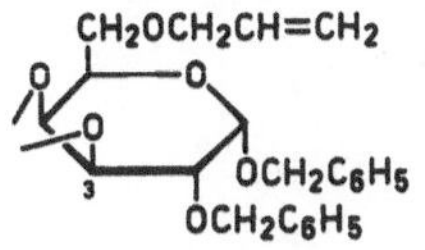 HO-R-OH = benzyl 6-O-allyl-2-O-benzyl-α-D-galactopyranoside II	with $C_6H_5CH_2Br$ → 3-$C_6H_5CH_2O$-R-OH (66%)	[50]
*89	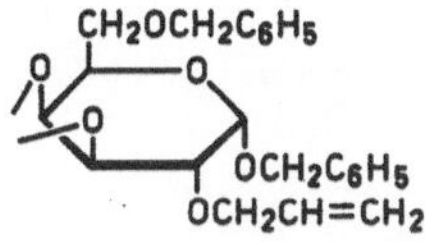 HO-R-OH = benzyl 2-O-allyl-6-O-benzyl-α-D-galactopyranoside II	for reactions, see further information	[60]
*90	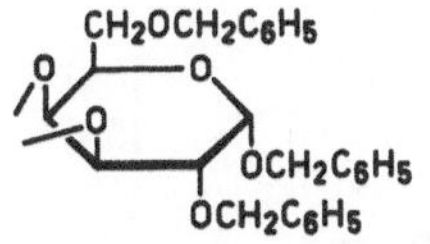 HO-R-OH = benzyl 2,6-di-O-benzyl-α-D-galactopyranoside II [60, 62]	for reactions, see further information	[60, 62]

References on p. 92

Table 14 (continued)

No.	O-R-O group method of preparation (yield in %)	properties and remarks	Ref.
91	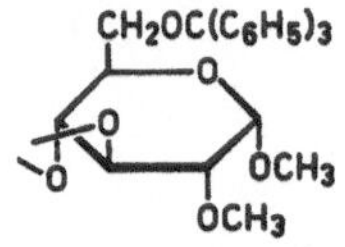 HO-R-OH = methyl 2-O-methyl-6-O-trityl-α-D-glucopyranoside II	with Br_2 → 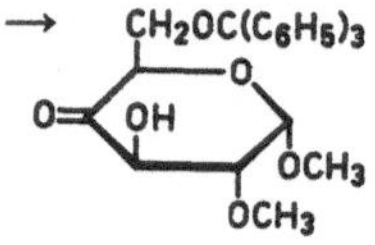 (72%)	[62]
92	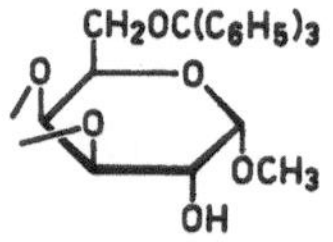 HO-R-OH = methyl 6-O-trityl-α-D-galactopyranoside II	with $SOCl_2$ → 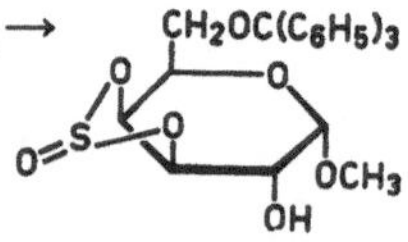 (65%; exo/endo = 53/47)	[129]
93	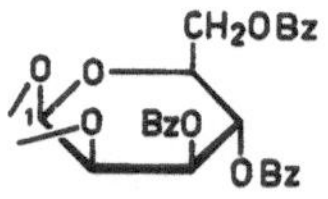 Bz = $-CH_2C_6H_5$ HO-R-OH = 3,4,6-tri-O-benzyl-β-D-mannose II	with $C_6H_5CH_2Br$ → 1-$C_6H_5CH_2O$-R-OH (70%)	[115]
94	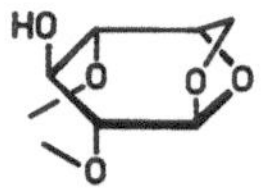 HO-R-OH = 1,6-anhydro-D-glucopyranose II	benzylation	[130]
*95	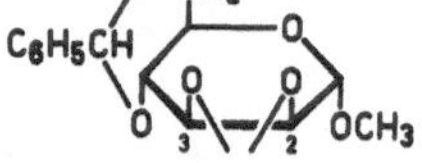 HO-R-OH = methyl 4,6-di-O-benzylidene-α-D-mannopyranoside II [54,89]	m.p. 105 to 107° with $C_6H_5CH_2Br$ → 3-$C_6H_5CH_2O$-R-OH (85%) with C_6H_5COCl → mixture of 2- and 3-C_6H_5COO-R-OH	[89] [54]

References on p. 92

Table 14 (continued)

No.	O-R-O group method of preparation (yield in %)	properties and remarks	Ref.
*96	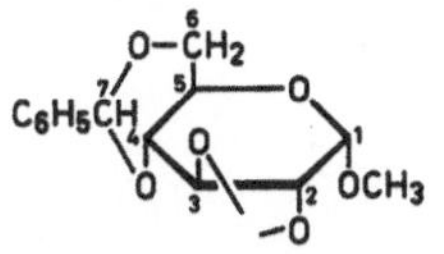 HO-R-OH = methyl 4,6-di-O-benzyliden-α-D-glucopyranoside II [62], (63 [40], 91 [53])	m.p. 194 to 195° [53], 203 to 206° [40]	[40, 53]
		$[\alpha_D^{23}]$ = 18.1 ($CHCl_3$; c ~1)	[40]
		^1H NMR ($CDCl_3$): 3.45 (s, OCH_3), 4.85 (d, H-1, J(H-1,2) = 3.0), 5.45 (s, H-7)	[53]
		^{119}Sn NMR ($C_6H_5CH_3$, 0.6 M): −131.6 (J(Sn, C-2/C-3) = 12/<10, J(Sn, C-1) = 20, J(Sn, C-4) = 34) [76], −132 [61]	[61, 76]
		with 4-$CH_3C_6H_4SO_2Cl$ → 2-(4-$CH_3C_6H_4SO_2$)O-R-OH (70%) with C_6H_5COCl → 2-C_6H_5COO-R-OH (86%)	[53]
		with $C_6H_5CH{=}CHCH_2OOCCH_3$ in the presence of $Pd(P(C_6H_5)_3)_4$ → 2- and 3-$C_6H_5CH{=}CHCH_2O$-R-OH (47 and 34%)	[131]
		with Br_2 → 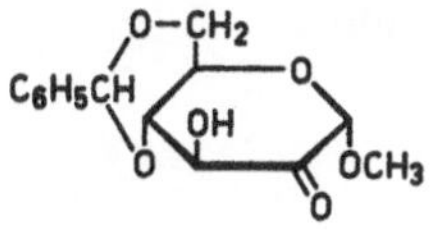 (60%)	[62]
97	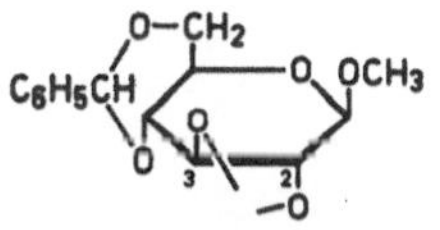 HO-R-OH = methyl 4,6-di-O-benzyliden-β-D-glucopyranoside II [123], (8 [40])	m.p. 236 to 238°	[40]
		with $C_6H_5CH_2Br$ → 2-$C_6H_5CH_2O$-R-OH (29%) and 3-$C_6H_5CH_2O$-R-OH (61%) with $CH_2{=}CHCH_2Br$ → 2-$CH_2{=}CHCH_2O$-R-OH (33%) and 3-$CH_2{=}CHCH_2O$-R-OH (60%) with CH_3I → 2-CH_3O-R-OH (22%) and 3-CH_3O-R-OH (66%) with C_6H_5COCl → 2-C_6H_5COO-R-$OOCC_6H_5$-3 (4%), 2-C_6H_5COO-R-OH (49%), and 3-C_6H_5COO-R-OH (39%) with 4-$CH_3C_6H_4SO_2Cl$ → 2-(4-$CH_3C_6H_5SO_2$)O-R-O- ($O_2SC_6H_4CH_3$-4)-3 (5%), 2-(4-$CH_3C_6H_4SO_2$)O-R-OH (21%), and 3-(4-$CH_3C_6H_4SO_2$)O-R-OH (63%)	[123]

References on p. 92

Table 14 (continued)

No.	O-R-O group method of preparation (yield in %)	properties and remarks	Ref.
98	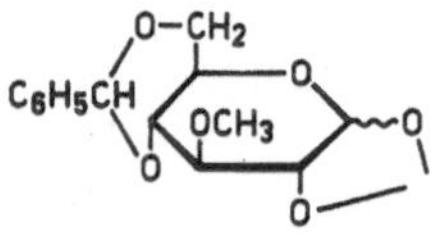 HO-R-OH = 3-O-methyl-4,6-di-O-benzyliden-α,β-D-glucopyranoside II (30)	m.p. 104 to 107° $[\alpha_D^{23}]$ = 1.9 ($CHCl_3$; c ~1)	[40]
99	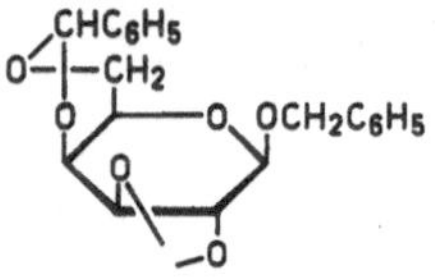 HO-R-OH = benzyl 4,6-di-O-benzyliden-β-D-galactopyranoside II	with Br_2 → 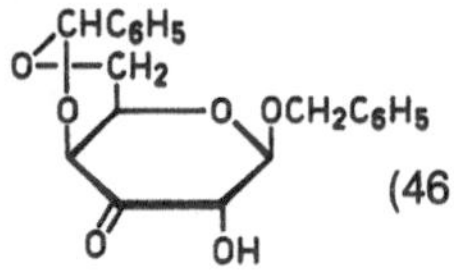 (46 to 72%)	[62]
100	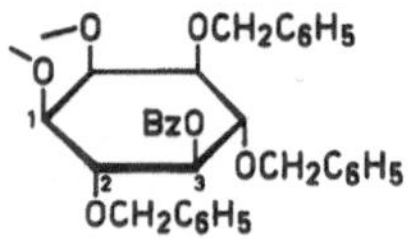 Bz = $-CH_2C_6H_5$ HO-R-OH = 2,3,4,5-tetra-O-benzyl-myo-inositol II	with R′X → 1-R′O-R-OH (R′X = $C_6H_5CH_2Br$ (60%), CH_2=$CHCH_2I$ (73%), CH_3I (66%))	[54]
101	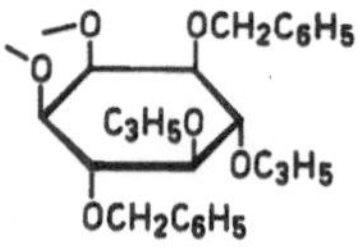 HO-R-OH = 2,5-O-benzyl-3,4-O-allyl-myo-inositol II	with CH_2=$CHCH_2Br$ → 1-CH_2=$CHCH_2O$-R-OH	[128]
102	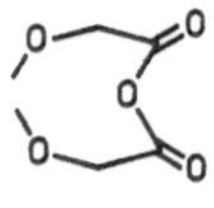	no preparation reported anthelmintic activity tested against Raillietina cesticillus and Ascaridia galli from chickens	[4]

References on p. 92

Table 14 (continued)

No.	O-R-O group method of preparation (yield in %)	properties and remarks	Ref.
103	 Ia (2 [22]) II [1,2], (76 [13])	m.p. 35 to 47° [13], 120 to 130° [1,2]; softening point for $(-(C_4H_9)_2SnOCH_2CH_2OCH_2CH_2O-)_n$: 220 to 227° ($3 < n < 120$) [22] heat stabilizer for PVC	[1,2,13,22] [2]
104	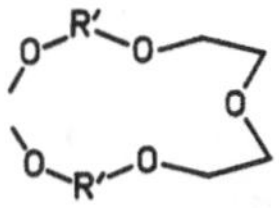 R′ = $-CH(C_6H_{13})CH_2CH=$ $CH(CH_2)_7CO-$ Ia	mobile liquid stabilizer for resin compositions, especially for PVC	[5]
105	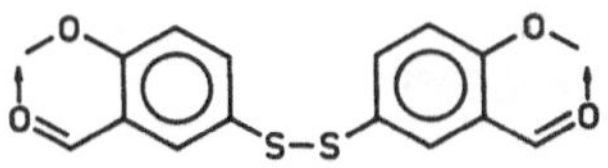 II (80)	yellow, low-melting solid, associated in $C_6H_5CH_3$ ^{1}H NMR: 1.00 to 1.80 (C_4H_9), 6.93 and 7.53 (C_6H_3), 9.83 (CHO) IR: ν(C=O) 1630, ν(C-O) 1285, ν(SnC) 590, ν(SnO) 430 polymeric, six-coordinate structure with bidentate ligands and *trans*-C_4H_9 groups suggested	[126]
106	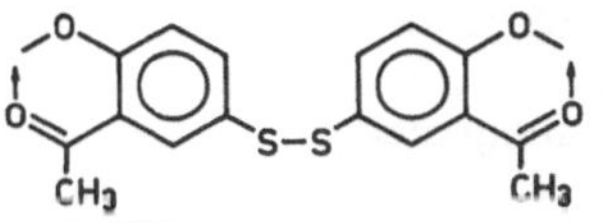 II (85)	red, low-melting solid ^{1}H NMR: 1.00 to 1.17 (C_4H_9), 2.60 (CH_3), 6.86 to 7.30 (C_6H_3) IR: ν(C=O) 1600, ν(C-O) 1290, ν(SnO) 430 structure like No. 105 suggested	[126]
	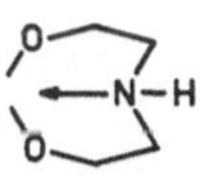 II (90 [36,44]) III (98 [14])	white solid m.p. 170 to 210°, b.p. 174°/1.0 $\mu = 4.06$ D (C_6H_6, 20°) ^{119}Sn-γ (78 K): $\delta = -1.74$ (β-Sn) [36]; $\delta = 0.91$, $\Delta = 2.20$ [36,51] trigonal-bipyramidal structure with one C_4H_9 group and the coordinating nitrogen in apical positions suggested extremely active against Bacillus sub- tilis, Bacillus mesentericus, and Chaetomium globosum in vitro	[14] [44] [36,51] [36,44] [56]

References on p. 92

Table 14 (continued)

No.	O-R-O group method of preparation (yield in %)	properties and remarks	Ref.
108	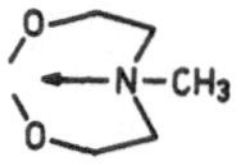 Ib (84) II (93) III (87)	b.p. 125°/0.04 $\mu = 3.97$ D (C_6H_6, 20°) structure like No. 107	[44]
109	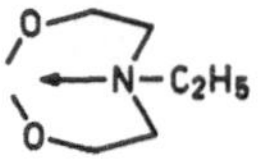 II [8], (96 [44])	b.p. 132°/0.1 $\mu = 3.98$ D (C_6H_6, 20°) ^{119}Sn-γ (77 K): $\delta = 0.91$, $\Delta = 2.20$; in contrast to [44, 51] a trigonal-bipyramidal structure, with both C_4H_9 groups and the coordinating nitrogen in the equatorial plane, deduced catalyst for polyurethane formation	[44] [90] [8]
110	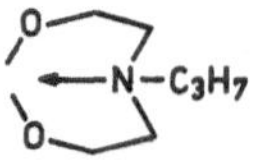 II (91)	b.p. 134°/0.02 structure like No. 107 suggested	[44]
111	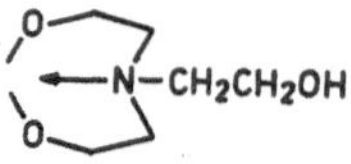 III (98)	colorless, viscous liquid state of coordination uncertain	[14]
112	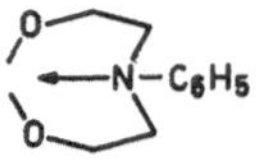 II (88)	m.p. 175 to 200°, b.p. 168°/0.06 $\mu = 2.76$ D (C_6H_6, 20°) structure like No. 107 suggested	[44]
113	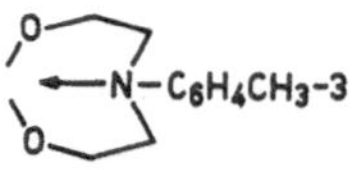 II [36], (88 [44])	m.p. 120 to 180°, b.p. 178°/0.1 $\mu = 2.56$ D (C_6H_6, 20°) ^{119}Sn-γ (78 K): $\delta = -1.80$ (β-Sn) [36]; $\delta = 0.85$, $\Delta = 2.23$ [36, 51]; structure like No. 107 suggested	[44] [36, 51]
114	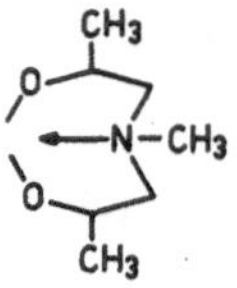 II	catalyst for polyurethane formation	[8]

Table 14 (continued)

No.	O-R-O group method of preparation (yield in %)	properties and remarks	Ref.
115	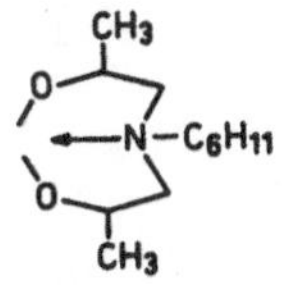 II	catalyst for polyurethane formation	[8]
116	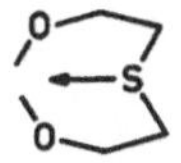 II (80 [69])	m.p. 98 to 100° ^{119}Sn-γ (77 K): δ = 1.16, Δ = 2.08; structure deduction like No. 109	[69] [90]
117	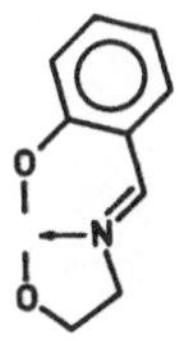 Ib (55)	m.p. 155 to 157° (dec.) IR: ν(C=N) 1620 (m), ν(CO) 1310 (w)	[68]
118	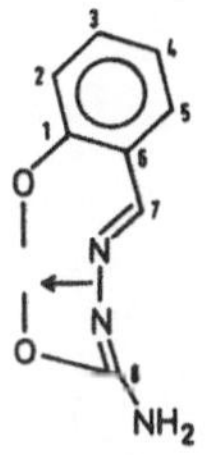 II (90)	yellow semi-solid ^{1}H NMR: 0.7 to 1.8 (C_4H_9), 4.95 (NH_2), 7.01 (m, C_6H_4), 8.35 (CH=N) ^{13}C NMR: 13.5 (C-δ), 21.9 (C-α, J(Sn,H) = 570.2/607.7); 26.5 (C-γ, J(Sn,H) = 83.5), 27.0 (C-β, J(Sn,H) = 33.3), 113.3 (C-3), 116.9 (C-2), 117.7 (C-6), 121.4 (C-4), 132.9 (C-5), 153.6 (C-8), 165.4 (C-1), 167.0 (C-7) IR discussion	[122]
119	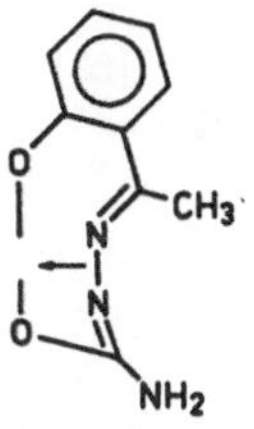 II	green-yellow solid m.p. 91° IR discussion	[122]

References on p. 92

Table 14 (continued)

No.	O-R-O group method of preparation (yield in %)	properties and remarks	Ref.
120	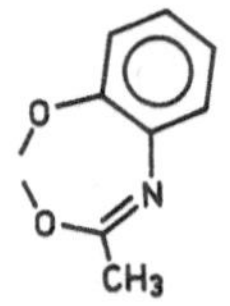 II (95)	m.p. 208 to 210° (dec.) ^{119}Sn-γ (77 K): δ = 1.39, Δ = 3.30 IR: ν(C=N) 1555, unassigned bands in the 1600 to 500 region; the intracyclic ν(C=N) together with the stability toward air and H_2O excludes the isomeric structure 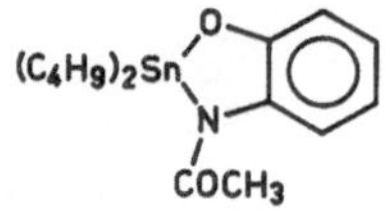	[63]
121	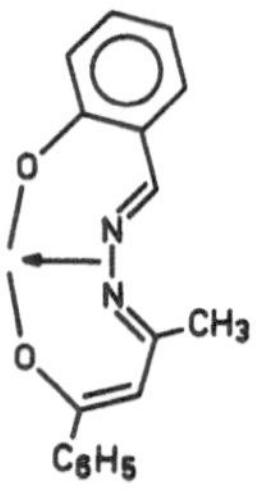 Ib	m.p. 138° IR: ν_{as}(SnC) 555, ν_s(SnC) 515, ν(SnN) 460; trigonal-bipyramidal structure with *cis*-C_4H_9 groups suggested	[106]
122	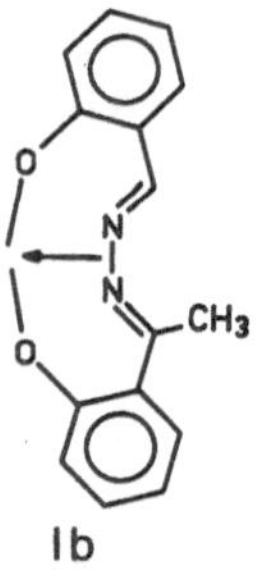 Ib	m.p. 200° (dec.) IR like No. 121	[106]
123	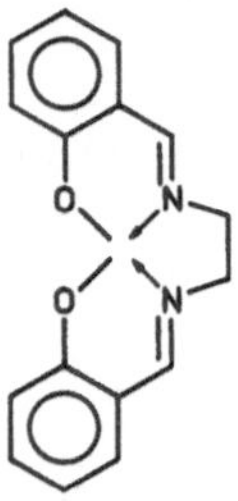 "salen" II [24, 65], (96 [63])	yellow crystals m.p. 151 to 153° [63], 152 to 154° [24], 215° [65] ^{119}Sn-γ (77 K): δ = 1.27, Δ = 3.50 IR: unassigned bands in the 1624 to 462 range distorted octahedral structure with *trans*-C_4H_9 groups suggested [63]; from ^{1}H NMR studies an equilibrium between *cis* and *trans* forms in solution deduced [65]	[24] [24, 63, 65] [63] [63, 65]

Table 14 (continued)

No.	O-R-O group method of preparation (yield in %)	properties and remarks	Ref.
124	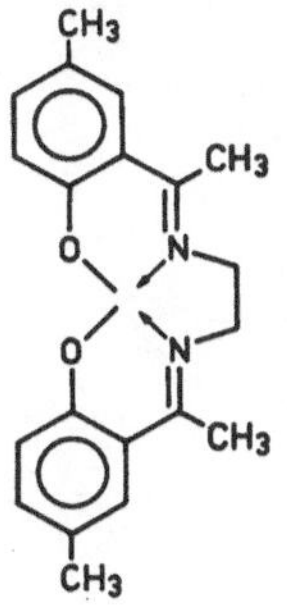 II	m.p. 235° distorted octahedral structure with *trans*-C_4H_9 groups suggested, both in the solid and in solution	[65]
125	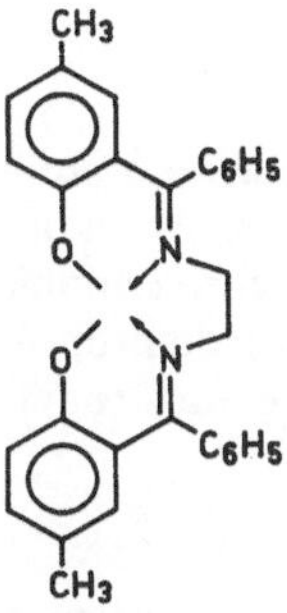 II	m.p. 221° structure suggestion like No. 124	[65]
126	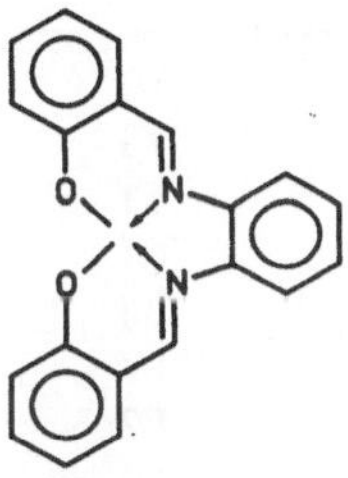 Ib	orange solid m.p. 62 to 64° ^{119}Sn-γ (80 K): δ = 1.35, Δ = 3.78; octahedral structure with *trans*-C_4H_9 groups suggested	[64]
127	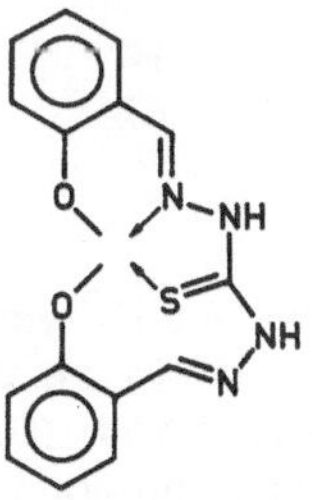 Ib (70 to 90)	yellow crystals m.p. 135° IR discussion; spectroscopic data suggest a structure with six-coordinate Sn, involving two C_4H_9 groups, two oxygen, one nitrogen, and the sulfur atom	[67]

References on p. 92

Table 14 (continued)

No.	O-R-O group method of preparation (yield in %)	properties and remarks	Ref.
128	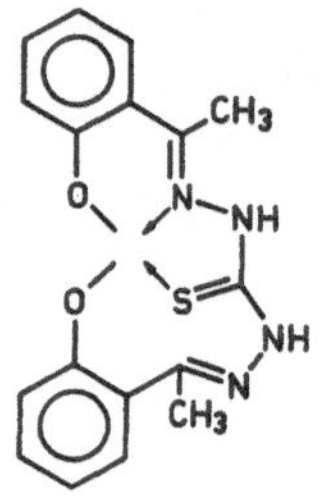 Ib (70 to 90)	yellow crystals m.p. 113° ^{1}H NMR ($CDCl_3$): 0.1 to 1.7 (m, C_4H_9), 2.30 to 2.50 (d, $CH_3C{=}N$); structure suggestion like No. 127	[67]

* Further information:

$(C_4H_9)_2\overline{SnOCH_2CH_2O}$ (Table **14**, No. **1**) is also obtained by the reaction between equimolar amounts of $(C_4H_9)_2Sn(OR')_2$ compounds and $\overline{OCH_2CH_2OCO}$ in dry C_6H_6 or $C_6H_5CH_3$, along with $(R'O)_2CO$. $(C_4H_9)_3SnOR'$ compounds react with $\overline{OCH_2CH_2OCO}$ only under severe conditions to give No. 1, $(R'O)_2CO$, and $Sn(C_4H_9)_4$ because of the energy-consuming first step disproportionation of $(C_4H_9)_3SnOR'$ into $(C_4H_9)_2Sn(OR')_2$ and $Sn(C_4H_9)_4$, followed by the fast reaction of $(C_4H_9)_2Sn(OR')_2$ with the carbonate [48]:

starting alkoxide	temperature in °C	time in hours	No. 1 (yield in %)	$(R'O)_2CO$ (yield in %)
$(C_4H_9)_2Sn(OCH_3)_2$	20	2	83	70
$(C_4H_9)_2Sn(OC_2H_5)_2$	40	4	82	73
$(C_4H_9)_2Sn(OC_3H_7)_2$	100	8	45	51
$(C_4H_9)_2Sn(OC_3H_7\text{-}i)_2$	100	8	48	61
$(C_4H_9)_2Sn(OC_4H_9)_2$	120	16	42	35
$(C_4H_9)_2Sn(OC_4H_9\text{-}s)_2$	120	16	31	27
$(C_4H_9)_2Sn(OC_4H_9\text{-}t)_2$	120	16	57	62
$(C_4H_9)_3SnOCH_3$	185	2	68	70
$(C_4H_9)_3SnOC_2H_5$	185	24	24	28
$(C_4H_9)_3SnOC_3H_7\text{-}i$	185	24	—	20
$(C_4H_9)_3SnOC_4H_9\text{-}t$	185	48	45	47

$(C_4H_9)_3SnOSn(C_4H_9)_3$ reacts with $\overline{OCH_2CH_2OCO}$ at 180°C for 6 h to give a 90% yield of No. 1; whereas at 100 or 120°C $(C_4H_9)_3SnOCH_2CH_2OSn(C_4H_9)_3$ is obtained in an 88 or 86% yield, respectively [17]. No. 1 is reformed in the reaction of catechol with $(C_4H_9)_2\overline{SnOCH_2CH_2OSn(C_4H_9)_2O}$ which is the product of the reaction between the title compound and $(\text{-}(C_4H_9)_2SnO\text{-})_n$ (reaction No. 15, Table 15, p. 79) [101]:

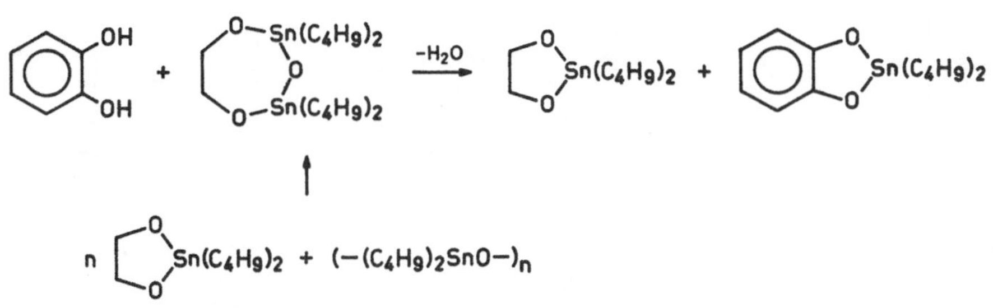

The structure of the solid has been determined by X-ray analysis using 1011 reflections and was refined to $R = 0.069$. The compound crystallizes orthorhombic, space group $P2_12_12_1$-D_2^4 with $a = 8.103(4)$, $b = 7.409(3)$, and $c = 21.563(4)$ Å; $Z = 4$ gives $D_c = 1.39$ g/cm^3. The molecular structure, see **Fig. 1**, consists of infinite chains, formed from units of $(C_4H_9)_2\overline{SnOCH_2CH_2O}$ which are associated along the 2_1 screw axis by O→Sn coordination bonds [114]. The Sn atom is six-coordinated [96, 114] and is in the center of a highly distorted octahedron [114].

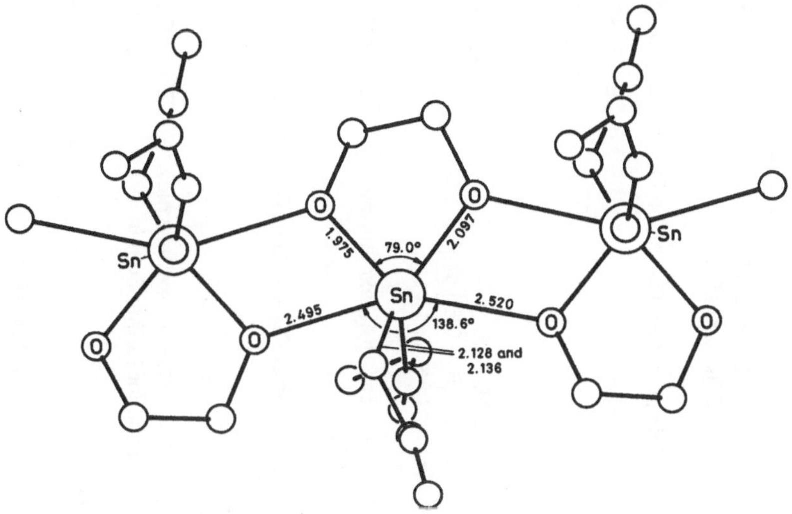

Fig. 1. Molecular structure of $(C_4H_9)_2\overline{SnOCH_2CH_2O}$ [114].

In solution there is an equilibrium between the cyclic monomer, dimeric molecules, and higher associated species with interconversion of four-, five-, and six-coordinate Sn, respectively. Under normal conditions and concentrations, the dimer is to be considered the predominant species in the fast exchange. For the dimeric species, different types of structures have been assigned, as shown in Formulas I to VI. All experimental data, obtained from solution molecular weight determinations [11, 15, 17], from ^{119}Sn Mössbauer investigations [30, 116], and especially from concerted NMR spectroscopic measurements [15, 17, 30] at variable temperatures [135] are best accounted for by the "fluxional type" structure (Formulas III to VI), in which a fast intramolecular shift with inversion at Sn is responsible for the apparent high symmetry features of the molecule [135].

The chemical reactions of $(C_4H_9)_2\overline{SnOCH_2CH_2O}$ are listed in Table 15, p. 78.

Table 15
Reactions of $(C_4H_9)_2\overline{SnOCH_2CH_2O}$.
The C_4H_9 groups on the Sn atom are abbreviated Bu, the CH_3CO group is abbreviated Ac.

No.	reactant	conditions	product (yield in %) and remarks	Ref.
1	CS_2	excess CS_2, $ClCH_2CH_2Cl$, 100 to 110°/10 h, sealed tube	$(\text{-}Bu_2SnS\text{-})_3$ and  (82)	[18,25]
2	C_6H_5NCO	1:1, C_6H_6, exothermic, reflux/1 h	$Bu_2\overline{SnN(C_6H_5)COOCH_2CH_2O}$ (100)	[37]
		1:2, as above	$Bu_2\overline{SnN(C_6H_5)COOCH_2CH_2OOCN}(C_6H_5)$ (100)	
		heating	$(\text{-}Bu_2SnO\text{-})_n$ and 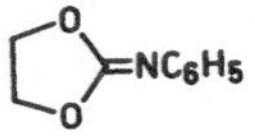	[23]
3	RNCS R = CH_3, C_6H_5	1:1, $ClCH_2CH_2Cl$, 70°/16 h (R = CH_3) or 55°/2 h (R = C_6H_5)	$(\text{-}Bu_2SnS\text{-})_3$ and 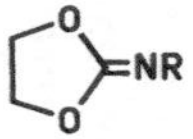 (70 for R = CH_3) (98 for R = C_6H_5)	[19,58]
4	Cl_2CO	1:1, CCl_4 or $CHCl_3$, exothermic, stirring	Bu_2SnCl_2 and 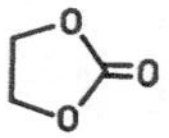 (78)	[102]
5	ClOC-COCl	1:1, CCl_4 or $CHCl_3$, exothermic at 0°	Bu_2SnCl_2 and 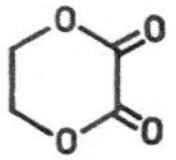 (85)	[102]
6	$ClOCCH_2COCl$	1:1, CCl_4 or $CHCl_3$, exothermic, stirring for further 30 min	Bu_2SnCl_2 and $(\text{-}OCH_2CH_2OOCCH_2CO\text{-})_n$ (19 for n = 2), (77 for the polymer)	[102]
7	$ClOC(CH_2)_2COCl$	like No. 6	Bu_2SnCl_2 and $(\text{-}OCH_2CH_2OOCCH_2CH_2CO\text{-})_2$	[102]
8	$ClOCCH_2OCH_2COCl$	1:1, in C_6H_6, reflux/12 h	Bu_2SnCl_2 and $(\text{-}OCH_2CH_2OOCCH_2OCH_2CO\text{-})_2$ (20)	[120]
9	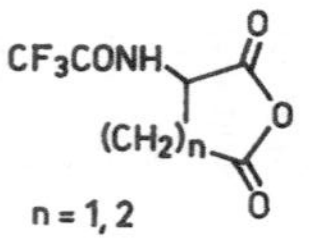 n = 1, 2	1:2, in $CHCl_3$, reflux/2 h	$(\text{-}Bu_2SnO\text{-})_n$ and $(\text{-}OCH_2CH_2OOC(CH_2)_nCH(NHCOCF_3)CO\text{-})_2$	[94]

References on p. 92

10	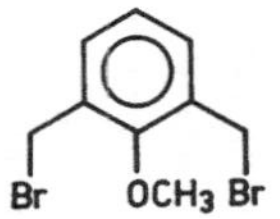	2:1, in hot DMFA, aqueous workup	Bu_2SnBr_2 and an excellent yield of 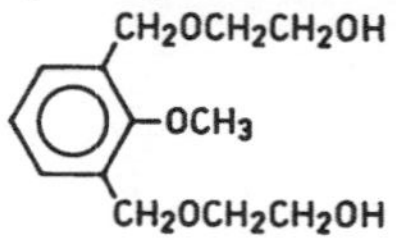	[91]
11	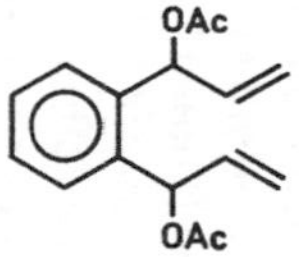	1:1, THF, in the presence of [Pd(0)], stirring at 20°/30 min	$Bu_2Sn(OAc)_2$ and 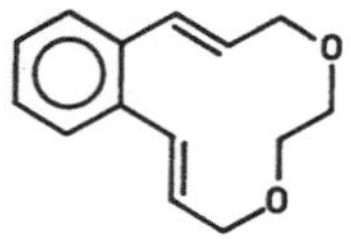 (94)	[131]
12	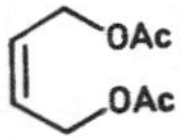	in the presence of $[Pd(P(C_6H_5)_3)_4]$	$Bu_2Sn(OAc)_2$ and 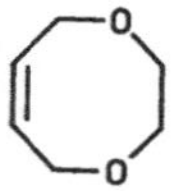 (66)	[131]
13	Lewis bases D D = C_5H_5N, THF, $HCON(CH_3)_2$, CH_3SOCH_3, $(CH_2)_4SO_2$	in excess D, or in $CHCl_3$ for D = $(CH_2)_4SO_2$	1:1 complexes	[103]
14	MCl_3 M = Sb, As	in C_6H_6, reflux/4 h	Bu_2SnCl_2 and 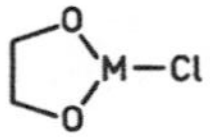 (96 for M = Sb)	[84]
15	$(-Bu_2SnO-)_n$	appropriate amount of $(-Bu_2SnO-)_n$, in $C_6H_5CH_3$, reflux/10 min	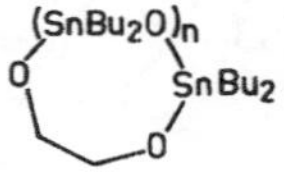 (92 for n = 1) (93 for n = 2) (90 for n = 10)	[101]

References on p. 92

The title compound can serve as nonstaining, nondiscoloring antioxidant for rubber [7] and as a stabilizer for PVC [2, 6]. It catalyzes the formation of poly-ε-caprolactones of high molecular weight [26] or the esterification of dibasic acids, e.g., terephthalic acid [38].

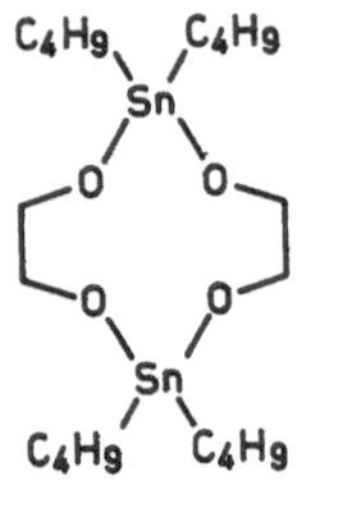

I

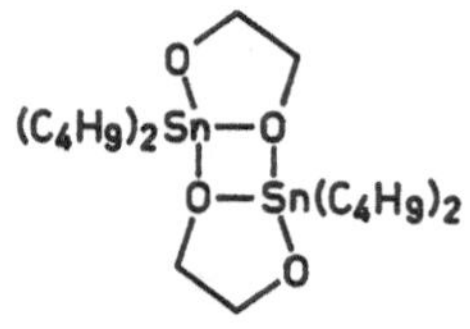

II

III

VI

IV

V

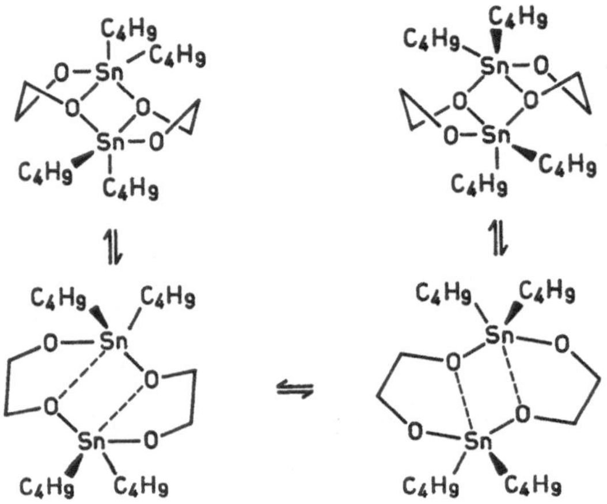

$(C_4H_9)_2\overline{SnOCH(CH_3)CH_2O}$ (Table **14**, No. **2**) is obtained in a 30% yield by the reaction of $(C_4H_9)_3SnOCH_3$ with $\overline{OCH(CH_3)CH_2OCO}$ without a solvent at 185°C during 13 h, along with 59% $(CH_3O)_2CO$ [48]. Only 3% of the title compound is formed during the high temperature distillation (up to 240°C) of $(C_4H_9)_3SnOCH(CH_3)CH_2OSn(C_4H_9)_3$, which is the product of the reaction between $(C_4H_9)_3SnOSn(C_4H_9)_3$ and $\overline{OCH(CH_3)CH_2OCO}$ (75%) [17].

From ^1H NMR spectroscopic investigations it is concluded that, under normal solution conditions, the compound exists predominantly as the dimer which equilibrates slowly with the monomer and rapidly with oligomeric species. As a derivative from an asymmetric diol, both the monomeric and the dimeric species exist as a racemic ("meso"-type complex VII, composed of two molecules of opposite configuration) or as an optically active sample (chiral complex VIII, composed of two molecules of identical configuration) distinguishable by ^1H NMR (Table 16) [107]. In addition, the following ^1H NMR data are reported (δ in ppm, J in Hz, half-saturated solution in $CDCl_3$, atom-numbering according to Formula IX): $\delta = 3.00$ (H-5′, $^3J_{cis}(H,H) = 3.98$, $^2J_{gem}(H,H) = -9.23$), and 3.71 (H-4,5, $^3J_{trans}(H,H) = 8.95$) [52].

References on p. 92

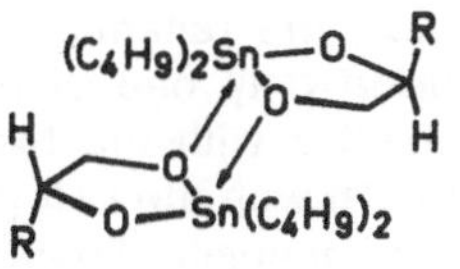

VII

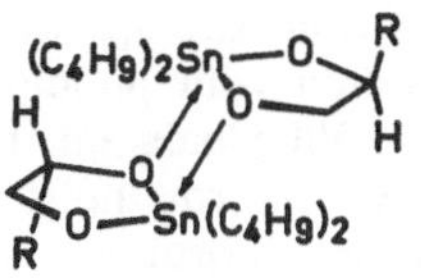

VIII

R = CH_3

Table 16
^1H NMR Spectrum of $(C_4H_9)_2SnOCH(CH_3)CH_2O$ [107].
δ in ppm, J in Hz, 0.2 M solution in $CDCl_3$.

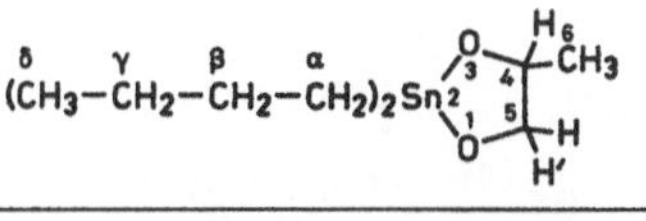

proton (multiplicity)	J	δ racemic monomer	VII	δ chiral species monomer	VIII
H-5, H-4 (m's)*)			3.706		3.719
H-5′ (t)	9.5		3.000		3.014
H-6 (d)	6	1.134	1.118	1.154	1.138
(0.05 M)		1.159	1.123	1.158	1.123
(0.01 M)		1.172	1.127	1.171	1.127
H-α (m)			1.637		1.628
H-β (m)			1.297		1.300
H-γ (m)			1.386		1.387
H-δ (t)	7	0.946	0.918	0.948	0.919
(0.05 M)		0.948	0.921	0.947	0.920
(0.01 M)		0.947	0.921	0.948	0.921

*) The line shapes of the two multiplets are different.

Recrystallization of No. 2 from C_6H_6, with separate treatment of the precipitate and the concentrated mother liquid with HOOC-COOH in CH_3CN, leads to a regeneration and optical enrichment of the diol [107].

$(C_4H_9)_2SnOCH(CH_3)CH_2O$ reacts with CS_2 in a sealed tube at 105°C for 10 h to yield $OCH_2CH(CH_3)O$-C-$OCH(CH_3)CH_2O$ (87%) and $(-(C_4H_9)_2SnS-)_3$ [18, 25]. The action of C_6H_5NCS on the title compound in $ClCH_2CH_2Cl$ at 55°C for 2 h affords 95% of $OCH(CH_3)CH_2OC$=NC_6H_5 and $(-(C_4H_9)_2SnS-)_3$ [58]. The subsequent treatment of No. 2 with C_6H_5COCl and $C_6H_5(CH_3)_2SiCl$ (1 : 1 : 1) in concentrated $CHCl_3$ solution at 0 to 5°C affords, after mild hydrolysis with cold HCl, the secondary ester, $HOCH_2CH(CH_3)OOCC_6H_5$, in a yield of 85%; whereas by the standard acylation method in C_6H_6 and in the presence of stoichiometric amounts of C_5H_5N, the primary ester, $HOC(CH_3)CH_2OOCC_6H_5$, is formed as the regioselective product (91%) [134]. A 79% yield of $OCH(CH_3)CH_2OCO$ is obtained from the exothermic reaction of No. 2 with Cl_2C=O in CCl_4 or $CHCl_3$. The analogous reaction with oxalyl chloride leads to the formation of $OCH(CH_3)CH_2OOCCO$ (66%), but also the carbonate, $OCH(CH_3)CH_2OCO$, together with CO (19% each). The use of malonyl dichloride or succinyl dichloride affords the oligomers $(-OCH(CH_3)CH_2OOC(CH_2)_nCO-)_x$ with n = 1 and x = 4.47 or n = 2 and x = 5.54, respectively. In all these reactions, $(C_4H_9)_2SnCl_2$ is the organotin product [102]. The high degree of orientation in the dimeric complexes VII and VIII is demonstrated by their regiospecific and stereospecific

References on p. 92

reaction with pimeloyl dichloride, $ClOC(CH_2)_5COCl$. Thus, the racemic VII as well as the chiral VIII form on condensation with pimeloyl dichloride only one of the four possible diastereomeric tetralactones: VII yields the tetralactone IXa with the two CH_3 groups in diagonal positions, whereas VIII affords the tetralactone IXb with the CH_3 groups in parallel positions. In both cases, the methyl groups are pointing to opposite directions in relation to the ring surface [94, 107]. The title compound, unlike acyclic $(C_4H_9)_2Sn(OR)_2$ and cyclic, six-membered $(C_4H_9)\overline{SnOCR_2CR_2CR_2O}$ compounds, reacts with excess C_5H_5N, $(CH_3)_2SO$, and $(CH_3)_2NCHO$ on warming and with $\overline{CH_2(CH_2)_3SO_2}$ in $CHCl_3$ with formation of 1:1-complexes [103].

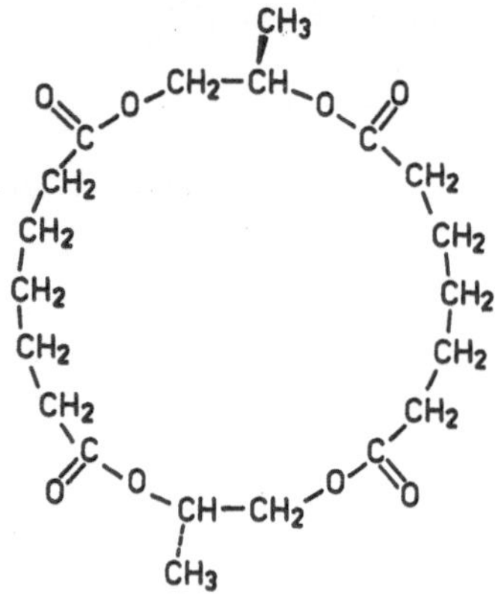

IXa

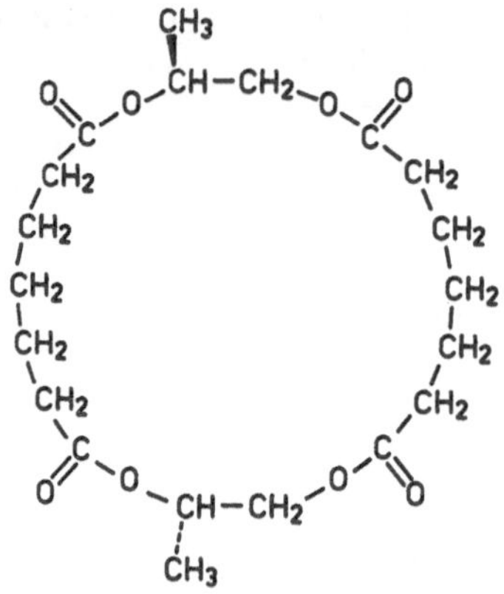

IXb

$(C_4H_9)_2\overline{SnOC(CH_3)_2CH_2O}$ (Table **14**, No. **3**) reacts with pimeloyl dichloride, $ClOC(CH_2)_5COCl$, to give the tetralactone X as the sole macrocyclic product with antiparallel arrangement of the diol units, demonstrating the regiospecifity of the tin template [94].

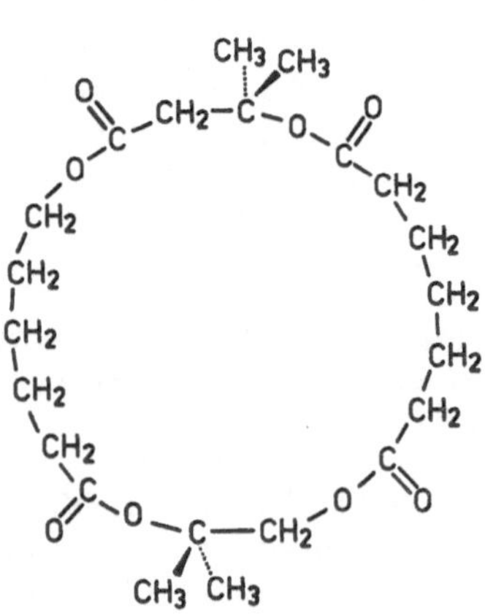

X

References on p. 92

$(C_4H_9)_2SnOCH(CH_2OCH_2C_6H_5)CH_2O$ (Table **14**, No. **5**) is shown to form dimers which are composed either of two molecules of opposite configuration yielding a racemic, "meso"-type complex (cf. No. 2, Formula VII) or of two molecules of identical configuration producing a chiral complex (cf. No. 2, Formula VIII) [107].

Recrystallization of No. 5 from CH_2Cl_2-C_6H_{14} and subsequent regeneration of the free diol by treatment of the precipitate and the concentrated mother liquid with HOOC-COOH in CH_3CN, can be used for the optical enrichment of the diol [107].

$(C_4H_9)_2SnOCH(CH_3)CH(CH_3)O$ (Table **14**, No. **8**). The compound, derived from the asymmetric diol $HOCH(CH_3)CH(CH_3)OH$, forms dimeric complexes which are composed of either two molecules of opposite configuration (meso complex XI) or of identical configuration (chiral complex XII). The diastereomeric relationship of the complexes can be used for the optical enrichment of the diol by fractional crystallization of No. 8 from C_6H_6, and regeneration of the diol by reaction with HOOC-COOH in CH_3CN [107].

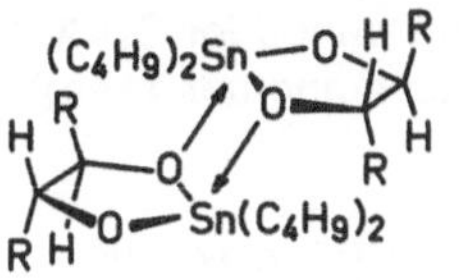

XI

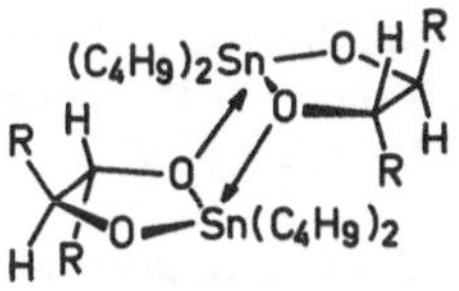

XII

R = CH_3

$OCH(CH_3)CH(CH_3)O$-C-$OCH(CH_3)CH(CH_3)O$ (90%) and $(-(C_4H_9)_2SnS-)_3$ are formed when No. 8 reacts with CS_2 in $ClCH_2CH_2Cl$ at 110°C for 10 h [18, 25]. Heating of the title compound with C_6H_5NCO [23] or C_6H_5NCS [58] affords $OCH(CH_3)CH(CH_3)OC{=}NC_6H_5$ and $(-(C_4H_9)_2SnO-)_n$ or $(-(C_4H_9)_2SnS-)_3$, respectively. The treatment of No. 8 with equivalent amounts of R′Cl (R′ = C_6H_5CO or $CH_3C_6H_4SO_2$) in $CHCl_3$ under reflux leads to the formation of $R'OCH(CH_3)CH(CH_3)O$-$SnCl(C_4H_9)_2$ which can react with, e.g., R″Cl to give the nonsymmetric derivatization product $R'OCH(CH_3)CH(CH_3)OR''$ (R″ = CH_3CO) [66, 75]. The title compound reacts exothermally with Cl_2CO to give $OCH(CH_3)CH(CH_3)OCO$ in a 93% yield. However, the exothermal reaction with ClOC-COCl leads to only 36% of the expected cyclic oxalate, $OCH(CH_3)CH(CH_3)OOCCO$, along with the cyclic carbonate (36%) and CO (55%) [102]. Macrocyclic tetralactones of Formula XIII are formed from equimolar amounts of No. 8 and the diacyl dichlorides, ClOC-X-COCl, of malonic acid (X = CH_2) [113], diglycolic acid (X = CH_2OCH_2) [120], glutaric acid (X = $(CH_2)_3$), pimelic acid (X = $(CH_2)_5$) [113], suberic acid (X = $(CH_2)_6$) [80], azelaic acid (X = $(CH_2)_7$), and sebacic acid (X = $(CH_2)_8$) [113]. In [102], instead of the dimeric structure XIII, the formation of an octamer or of a tetramer for the malonic (X = CH_2) or succinic acid derivative (X = $(CH_2)_2$), is reported. Addition of $SbCl_3$ to a refluxing C_6H_6 solution of No. 8 yields $OCH(CH_3)CH(CH_3)OSbCl$ [84]. Unlike $(C_4H_9)_2Sn(OR')_2$ and the cyclic $(C_4H_9)_2Sn(OCR'_2)_2CR'_2$ compounds, the title compound forms 1:1 complexes with C_5H_5N, CH_3SOCH_3, $HCON(CH_3)_2$, and $CH_2(CH_2)_3SO_2$ [103].

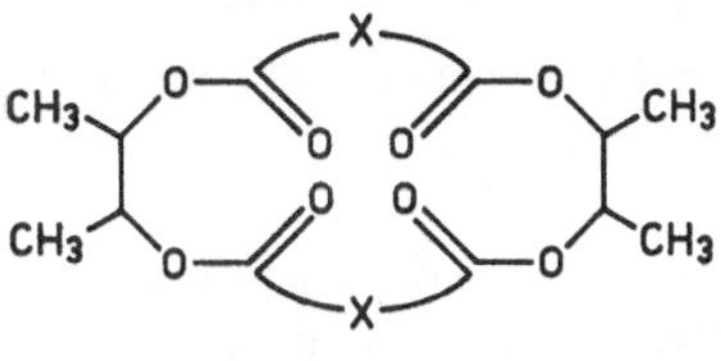

XIII

References on p. 92

$(C_4H_9)_2\overline{SnOC(CH_3)_2C(CH_3)_2O}$ (Table **14**, No. **9**) reacts with C_6H_5NCO [23] or with C_6H_5NCS in $ClCH_2CH_2Cl$ at 85°C for 40 h [18, 58] yielding $\overline{OC(CH_3)_2C(CH_3)_2OC}{=}NC_6H_5$ along with $(\text{-}(C_4H_9)_2SnO\text{-})_n$ or $(\text{-}(C_4H_9)_2SnS\text{-})_3$, respectively. Treatment of No. 9 with Cl_2CO affords $\overline{OC(CH_3)_2C(CH_3)_2OCO}$ in a 79% yield. Unlike Nos. 1, 2, and 8, the title compound does not form a cyclic oxalate in the reaction with ClOC-COCl, but the above carbonate (72% yield) along with the equivalent amount of CO. In the reaction of No. 9 with malonyl dichloride or succinyl dichloride, the oligolactones of the type $(\text{-}OC(CH_3)_2C(CH_3)_2OOC(CH_2)_nCO\text{-})_x$ (n = 1, x = 4; n = 2, x = 3) are formed. In all these reactions of No. 9 with acyl chlorides, the organotin product formed is $(C_4H_9)_2SnCl_2$ [102]. The title compound forms a 1:1 complex with CH_3SOCH_3 [103].

No. 9 can serve as a catalyst in the esterification of dibasic acids, e.g., terephthalic acid with glycol [38].

$(C_4H_9)_2\overline{SnOCH(COOC_2H_5)CH(COOC_2H_5)O}$ (Table **14**, No. **14**) forms dimeric molecular complexes XI and XII (R = $COOC_2H_5$) in solution which are in a slow equilibrium with the monomer and in a fast equilibrium with higher associated structures. The dimer to monomer ratio depends on the concentration and has been determined for $CDCl_3$ solutions by ^1H NMR measurements (concentration, XI to monomer, XII to monomer ratio): 0.200 M, 10.0, 11.3; 0.050 M, 5.3, 4.8; 0.010 M, 1.8, 2.4 [107].

The ^1H NMR data for the title compound are listed in Table 17 [107].

From the temperature dependence of the ^{119}Sn Mössbauer recoil free fraction, it can be concluded that in the solid the intermolecular forces between the racemic or chiral dimeric units are much weaker than the intramolecular forces causing the formation of the dimers [116].

Table 17
^1H NMR Spectrum of
$(C_4H_9)_2\overline{SnOCH(COOC_2H_5)CH(COOC_2H_5)O}$ [107].
δ in ppm, J in Hz, 0.2 M solution in $CDCl_3$.

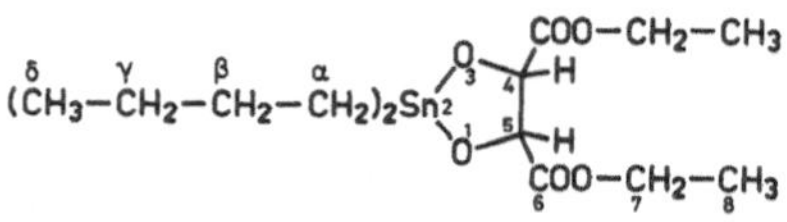

proton (multiplicity)	J	δ racemic monomer	XI	δ chiral monomer	XII
H-5, 4 (s)			4.552		4.562
H-7 (q) *)	7	4.296	4.207 *)	4.297	4.209 *)
					4.201 *)
H-8 (t)	7		1.306		1.307
H-α, β, γ (m)			0.9 to 1.6		0.9 to 1.6
H-δ (t)	7		0.900		0.902

*) Two quartets are observed for the optical pure XII, one broad quartet in the racemic XI.

Racemic No. 14 (structure XI) reacts with pimeloyl dichloride, $ClOC(CH_2)_5COCl$, yielding exclusively the tetralactone XIV, whereas chiral No. 14 (structure XII) solely provides, in the analogous reaction, the diasteromeric tetralactone XV, thus proving the high degree of

References on p. 92

orientation in the molecular complexes as well as their regio- and stereospecific behavior [94, 107]. The fractional recrystallization of No. 14 (derived from the (S, S)-diol followed by separate treatment of the precipitate and of the saturated mother liquid (in C_6H_6) with HOOC-COOH) can be used for the optical enrichment of the diol. In this case, the precipitate contains the pure optical material [107].

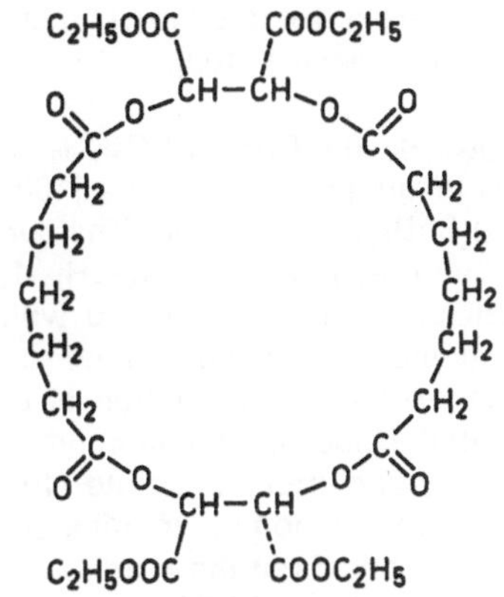

XIV

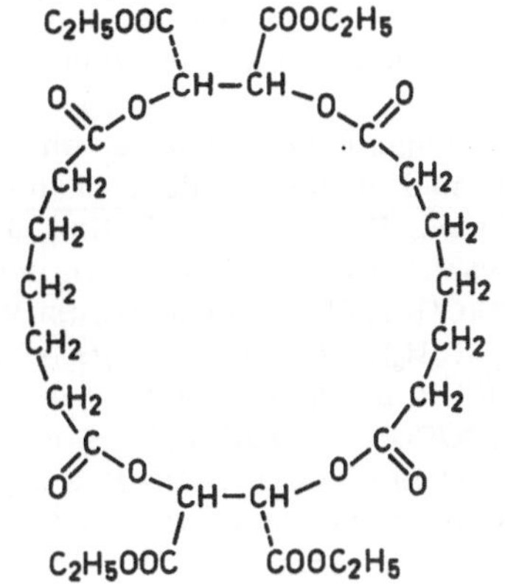

XV

$(C_4H_9)_2\overline{SnOCH_2CH_2CH_2O}$ (Table **14**, No. **20**) has been investigated by X-ray analysis. The preliminary results show that in the solid state this compound forms chains by strong O→Sn coordination, keeping the Sn atom in a distorted octahedral environment, formed by four O atoms and the two C_4H_9 groups in trans positions (**Fig. 2**). In basic solvents, like C_5H_5N or bipyridine, the compound is monomeric because of the coordination of solvent molecules [32]. In noncoordinating solvents like C_6H_6 [11] or $CDCl_3$ [30], the compound seems to be at least dimeric.

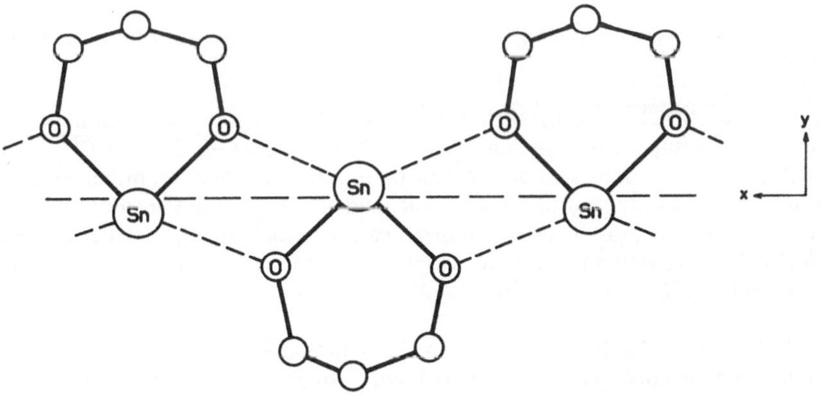

Fig. 2. Molecular structure of $(C_4H_9)_2\overline{SnOCH_2CH_2CH_2O}$ [32]; C_4H_9 groups omitted since unclear in the original.

References on p. 92

$(C_4H_9)_2\overline{SnOC(C_6H_5)=C(C_6H_5)O}$ (Table **14**, No. **43**) is also obtained by the reaction of diphenylvinylene carbonate, $\overline{OC(C_6H_5)=C(C_6H_5)OCO}$, with $(-(C_4H_9)_2SnO-)_n$ (in $C_6H_5CH_3$, 1 h reflux, CO_2 evolution starts at 90°C), along with $\overline{C(C_6H_5)=C(C_6H_5)(-O(C_4H_9)_2Sn-)_2O}$; or with $(C_4H_9)_2Sn(OCH_3)_2$ (no solvent, 120 to 130°C/1.5 h) and $(CH_3O)_2CO$ as pale yellow crystals in a yield of 86% [100].

In the crystal, the compound is suggested to be associated so that the coordination number of Sn is five or six. In concentrated solution, the molecules are associated to at least dimers; but in dilute solution, dissociation occurs to yield monomeric species [100].

The compound decomposes on heating into $(-(C_4H_9)_2Sn-)_n$ and $C_6H_5COCOC_6H_5$. Benzil is also formed in the reactions of the title compound with air [87, 100] or Br_2 [100], along with $\overline{C(C_6H_5)=C(C_6H_5)(-O(C_4H_9)_2Sn-)_nO}$ (n = 2, 3) or $(C_4H_9)_2SnBr_2$, respectively. The compound hydrolyses to $C_6H_5CH(OH)COC_6H_5$ and $(-(C_4H_9)_2SnO-)_n$. The reaction with $C_6H_4(OH)_2$-1,2 gives $C_6H_5CH(OH)COC_6H_5$ and $(C_4H_9)_2\overline{SnOC_6H_4O}$-2. Treatment of the title compound with $COCl_2$ affords $(C_4H_9)_2SnCl_2$ and $\overline{OC(C_6H_5)=C(C_6H_5)OCO}$ [100]. Addition of a twofold molar amount of CH_3COCl to the title compound in $C_6H_5CH_3$ causes the exothermic formation of only *cis*-$C_6H_5C(OOCCH_3)=C(OOCCH_3)C_6H_5$ and $(C_4H_9)_2SnCl_2$ [87, 100]. Under the same conditions, the less reactive $(CH_3CO)_2O$ yields a mixture of the *cis* (90%) and *trans* diacetate (10%) as a consequence of competitive acylation and metallotropic isomerisation of the initial monoacyl derivative, *cis*-$C_6H_5C(OOCCH_3)=C(OSn(C_4H_9)_2Cl)C_6H_5$. The products of the reaction of No. 43 with phthalic anhydride are $C_6H_5COCOC_6H_5$ and compounds XVI and XVII.

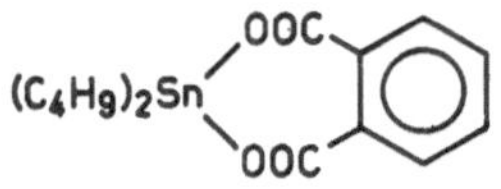

XVI

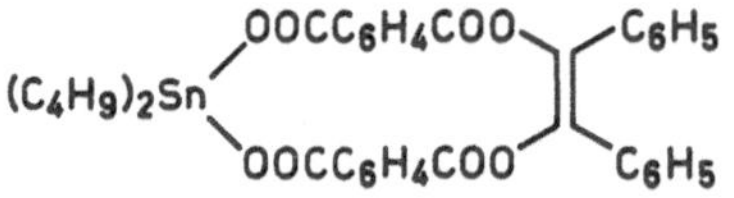

XVII

The reaction of $SOCl_2$ with the title compound causes the formation of $C_6H_5COCOC_6H_5$ and, presumably, SO. No. 43 reacts exothermally with $(CH_3)_2SiCl_2$ leading to $(CH_3)_2\overline{SiOC(C_6H_5)=C(C_6H_5)O}$ as indicated by NMR analysis [100]. The compound undergoes a telomerization reaction with $(-(C_4H_9)_2SnO-)_n$ in refluxing C_6H_6 or $C_6H_5CH_3$ to give a series of oligomers, $\overline{C(C_6H_5)=C(C_6H_5)(-O(C_4H_9)_2Sn-)_nO}$, with n = 1 to 5, 7, and 10, which can be separated [87, 100, 101].

$(C_4H_9)_2\overline{SnOC(C_6H_4OCH_3\text{-}4)=C(C_6H_4OCH_3\text{-}4)O}$ (Table **14**, No. **44**). The formation of the title compound from dianisylvinylene carbonate, $\overline{OC(C_6H_4OCH_3\text{-}4)=C(C_6H_4OCH_3\text{-}4)OCO}$, and $(C_4H_9)_2Sn(OCH_3)_2$ can be achieved only at low temperatures. Thus, as monitored by NMR, in $CDCl_3$ at 37°C for 3 weeks, the title compound and $(CH_3O)_2CO$ are formed in a 63% yield, contaminated with only traces of the thermolysis product anisil, 4-$CH_3OC_6H_4COCOC_6H_4$-OCH_3-4; at 78°C, the reaction leads to an enediolate/dione ratio of 1:1, and without a solvent at 130°C for 1.5 h to a 79% yield of anisil [100].

$(C_4H_9)_2\overline{SnOC_6H_4O}$-2 (Table **14**, No. **48**). Besides Methods I to III, the compound is also formed in the reactions of $(C_4H_9)_3SnOSn(C_4H_9)_3$ with catechol (1:1, in CH_3OH, C_2H_5OH, or C_6H_6, in the presence of air, via a paramagnetic complex, precipitation after standing for 1 to 2 d, 65% yield) [82]; of $\overline{CH_2CH_2(O(C_4H_9)_2Sn)_2O}$ as a soluble and reactive source of dibutyltin oxide with equivalent amounts of catechol along with No. 1 (in refluxing $C_6H_5CH_3$) [101]; of No. 43

References on p. 92

with catechol along with $C_6H_5CH(OH)COC_6H_5$ (1:1, in refluxing C_6H_6, 2 h, 59% yield) [100]; or of $(C_4H_9)_2\overline{SnOC_6H_4COO}$-2 with catechol (1:1, in refluxing C_6H_6, 1 h) along with $C_6H_4(OH)COOH$-2 [63].

The IR spectrum of the compound is listed in Table 18 [29].

The potentiometric titration of $(C_4H_9)_2SnCl_2$ with NaOH in the presence of catechol in dioxane-water proves the formation of the hydroxy anion $[(C_4H_9)_2SnOC_6H_4O\text{-}2(OH)_2]^{2-}$ [29].

Table 18
IR Spectrum of $(C_4H_9)_2\overline{SnOC_6H_4O}$-2 [29].
Wave numbers in cm^{-1}.

wave number	assignment	wave number	assignment
3068 (vw)		1104 (s)	
3015 (m)	$\nu(CH)/C_6H_4$	1082 (w)	
2956 (vs)		1030 (m)	C_6H_5 vibration
2915 (vs)	$\nu(CH)$	968 (vw)	
2858 (vs)		900 (vw)	
1583 (m)	$\nu(C{=}C)$	876 (vw)	
1483 (vs)		856 (vs)	$\pi(CH)$?
1455 (s)	$\delta(CH) + \nu(C{=}C)$	787 (vs)	$\varrho(CH_3)$
1394 (vw)		733 (vs)	$\varrho(CH_3) + C_6H_4 + \nu(SnO)$
1370 (s)		680 (m)	
1318 (m)	$\delta(CH)$	625 (s)	$\nu_{as}(SnO)$
1270 (w)		616 (vs)	$\nu_s(SnO)$
1248 (vs)	$\gamma(C\text{-}O)$	542 (w)	$\nu_{as}(SnC)$
1198 (s)		526 (w)	$\nu_s(SnC)$?
1176 (w)			

$(C_4H_9)_2\overline{SnOC_6H_3(C_4H_9\text{-}t\text{-}4)O}$-2 (Table **14**, No. **50**) is formed in a 66% yield from $(C_4H_9)_3SnOSn(C_4H_9)_3$ and $C_6H_3(C_4H_9\text{-}t\text{-}4)(OH)_2$-1,2 in $C_6H_5CH_3$ in the presence of air via a paramagnetic intermediate [82].

$(C_4H_9)_2\overline{SnO\text{-}C_9H_{10}N_2O_4\text{-}O}$ (Table **14**, No. **58**). As inferred on the basis of ^{119}Sn Mössbauer and IR spectroscopic data, the compound exists in the solid state as a dimer with one bridging ribose oxygen causing a distorted trigonal bipyramidal environment of Sn with the C_4H_9 groups in equatorial positions. The idealized regular trigonal bipyramidal environment of Sn in dibutyltin nucleosidates is shown below [121]:

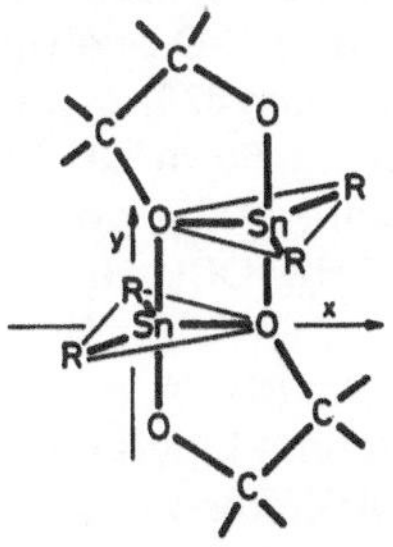

References on p. 92

The dibutylstannylene function of the title compound serves not as much as a protecting group, but rather as an activating group for the 2'- and 3'-oxygen functions. Thus the compound reacts with CH_3COCl in the presence of $N(C_2H_5)_3$ (in CH_3OH, 20°C/10 min) and subsequent hydrolysis with formation of a 30:70 mixture of 2'- and 3'-O-acetyluridine (Formulas XVIII and XIX, R' = $-OCCH_3$) which was crystallized twice from C_2H_5OH, giving 69% of pure XIX. The reaction with CH_3CO-O-$OCCH_3$ in DMF (0°C/4 h) or in CH_3OH in the presence of $N(C_2H_5)_3$ (20°C/3 min) also yields pure XIX (54% or 34%, respectively). 3'-O-benzoyluridine (XIX, R' = $-OCC_6H_5$) is obtained in a 78% yield by treatment of a CH_3OH solution of the title compound with C_6H_5COCl and $N(C_2H_5)_3$ (20°C/10 min), followed by partition between water and ether. A roughly equal mixture of 2'- and 3'-O-benzyluridine (65% yield) is obtained by the reaction of $C_6H_5CH_2Br$ with No. 58 in DMF (100°C/1 h), which can be separated into the isomers XVIII and XIX (R' = $-CH_2C_6H_5$, 31% or 26%, respectively) by crystallization. CH_3I and the title compound react in DMF at 37°C to give a 70% yield of a 55:45-mixture of XVIII and XIX (R' = CH_3) for which no resolution was achieved. The reaction between No. 58 and 4-$CH_3C_6H_4SO_2Cl$ in CH_3OH-CH_3COCH_3 (20°C/10 min) affords 2'-O-toluenesulfonyluridine (XVIII, R' = $-SO_2C_6H_4CH_3$-4) in a 62% yield. A 2:3-mixture of 2'- and 3'-uridine phosphate (XVIII and XIX, R' = $-PO(OH)_2$) is the product of the reaction of No. 58 with $(CH_3)_3SnOSn(CH_3)_3$ and $POCl_3$ in CH_3OH (20°C/20 min; 87% yield) [45]. Analogously, the reaction of No. 58 with $PSCl_3$ in CH_3OH in the presence of $N(C_4H_9)_3$ (0 to 20°C/1 h) affords a mixture of XVIII and XIX (R' = $-PS(OH)_2$), isolated as lithium salts in a 40% yield [71]. Heating a solution of No. 58 and 2-$NO_2C_6H_4CH_2Br$ in DMF (110°C/3 to 4 h) leads to 24% of pure XVIII (R' = $-CH_2C_6H_4NO_2$-2) [59]. Treatment of No. 58 with 4-$CH_3OC_6H_4CH_2Br$ in DMF (100°C/2 h) causes the formation of a 1:1-mixture of XVIII and XIX (R' = $-CH_2C_6H_4OCH_3$-4) from which the isomer XVIII was separated by crystallization (21%) [83, 124]. From the reaction with 4-nitro-2-picolyl-1-oxide chloride in DMF (110°C/1 h), pure XVIII can be obtained (12%; R' = $C_5H_3NO(NO_2$-4)CH_2-2) [73]. The reaction of Br_2 with No. 58 in CH_3OH results in bromination of the uracil ring system in the position C-5 (Formula XX) [136].

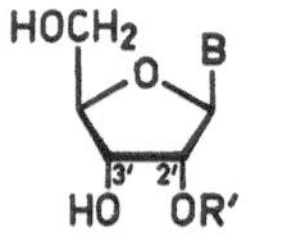

XVIII

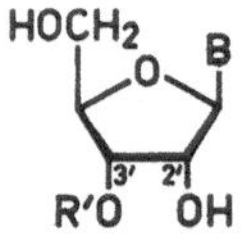

XIX

B = uracil-1-yl

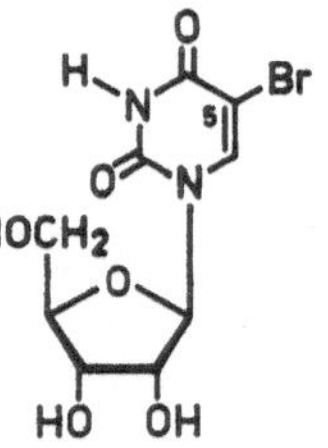

XX

$(C_4H_9)_2\overline{SnO\text{-}C_9H_{11}N_3O_3\text{-}O}$ (Table **14**, No. **60**). Addition of CH_3CO-O-$OCCH_3$ and $N(C_2H_5)_3$ to a solution of No. 60 in CH_3OH prepared in situ according to Method II (20°C/5 min) causes the formation of 96% of a 1:3 mixture of 2'- and 3'-O-acetylcytidine. Under the same conditions, the title compound reacts with C_6H_5COCl giving 87% of pure 3'-O-bencoylcytidine. The reaction of $(C_4H_9)_3SnOSn(C_4H_9)_3$ and $POCl_3$ with No. 60 in CH_3OH affords 73% of a mixture of cytidine 2'-and 3'-phosphate (for the structure of the products, see No. 58, Formulas XVIII and XIX with B = cytosine-1-yl and R' = $-OCCH_3$, $-OCC_6H_5$, or $-PO(OH)_2$), respectively [45].

$(C_4H_9)_2\overline{SnO\text{-}C_9H_{10}N_4O_3\text{-}O}$ (Table **14**, No. **61**). A dimeric structure with one coordinating ribose O atom and the Sn atom in a trigonal bipyramidal environment with equatorial C_4H_9 groups (cf. No. 58) is deduced from spectroscopic results [121]. The compound reacts with $PSCl_3$ in CH_3OH in the presence of $N(C_2H_5)_3$ to give a mixture of 2′- and 3′-inosine phosphorothioate (see No. 58, Formulas XVIII and/or XIX with B = hypoxanthine-9-yl and R′ = $-PS(OH)_2$), isolated as lithium salts [71].

$(C_4H_9)_2\overline{SnO\text{-}C_{10}H_{11}N_5O_2\text{-}O}$ (Table **14**, No. **62**). Based on spectroscopic data, the solid compound is suggested to possess a dimeric structure, with Sn in a trigonal bipyramidal environment [121], cf. No. 58. The reaction of the title compound with C_6H_5COCl gives a 70% yield of pure 3′-O-benzoyladenosine; with 2-$CHBr_2C_6H_4COCl$, a 72% yield of 3′-O-(2-dibromomethylbenzoyl)adenosine [70]; with 4-$CH_3C_6H_4SO_2Cl$ a 70% yield of pure 2′-O-toluenesulfonyladenosine; with $POCl_3$ and $(C_4H_9)_3SnOSn(C_4H_9)_3$, a 78% yield of a ca. 2:3 mixture of adenosine-2′- and -3′-phosphate (isolated as barium salts) [45]; and with $P(S)Cl_3$ a 59% yield of a mixture of adenosine-2′- and -3′-phosphorothioate [71]. The reaction with succinic anhydride, $\overline{CH_2CO\text{-}O\text{-}OCCH_2}$, which only proceeds in dry C_5H_5N in the presence of $[N(C_4H_9)_4]Br$, affords a 93% yield of the half-ester 3′-O-succinoyladenosine [86]; with 4-methoxy-2-picolyl-N-oxide chloride, a separable mixture of 2′- and 3′-O-(4-methoxy-2-picolyl-N-oxide)adenosine (for the structure of the products, see No. 58, Formulas XVIII or/and XIX, B = adenine-9-yl, R = $-OCC_6H_5$, $-OCC_6H_4CHBr_2$-2, $-SO_2C_6H_4CH_3$-4, $-PO(OH)_2$, $-PS(OH)_2$, $-OCCH_2CH_2COOH$, and $C_5H_3NO(OCH_3\text{-}4)CH_2\text{-}2$) is obtained [73].

$(C_4H_9)_2\overline{SnO\text{-}C_{10}H_{11}N_5O_3\text{-}O}$ (Table **14**, No. **67**) is suggested to possess a dimeric structure in the solid state with a trigonal bipyramidal environment of Sn, cf. No. 58 [121].

$(C_4H_9)_2\overline{SnO\text{-}C_{23}H_{26}O_4\text{-}O}$ (Table **14**, No. **89**). The addition of a solution of the title compound in HMPT to a solution of 2,3,4,6-tetra-O-benzyl-α-D-galactopyranosyl chloride in the same solvent containing anhydrous LiI (1:2 mole ratio) and stirring of the mixture for 2 d in the dark at room temperature yields 74% of α-linked (XXI) and 8% of β-linked (XXII) disaccharide [60]:

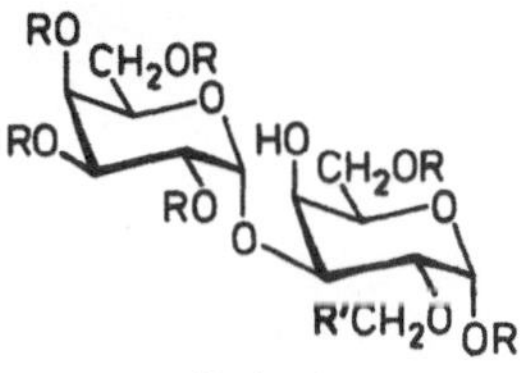

α-linked

XXI

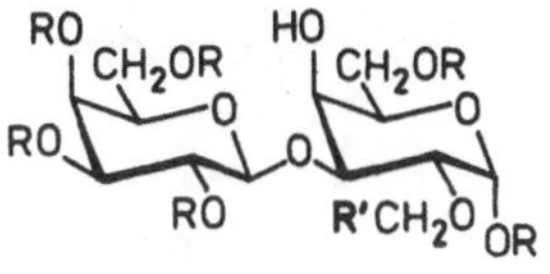

β-linked

XXII

R = $CH_2C_6H_5$
R′ = $CH{=}CH_2$

$(C_4H_9)_2\overline{SnO\text{-}C_{27}H_{28}O_4\text{-}O}$ (Table **14**, No. **90**) reacts with the twofold amount of 2,3,4,6-tetra-O-acetyl-α-D-glucopyranosyl bromide in the presence of $SnCl_4$ in $ClCH_2CH_2Cl$ (0 to 20°C/1.5 d) to give 25% of the trisaccharide XXIII. Equivalent amounts of the title compound and 2,3,4,6-tetra-O-acetyl-β-D-glucopyranosyl chloride react in HMPT (60°C/3 h) with formation of 49% of the ortho-ester XXIV; in $ClCH_2CH_2Cl$ this reaction (45°C/12 h) affords the orthoester XXIV along with the di-orthoester XXV in an approximate 2:1 ratio. The condensation of 2,3,4,6-tetra-O-benzyl-α-D-galactopyranosyl bromide with No. 90 in HMPT in the dark at room temperature for 2 d leads to the pure α- and β-linked disaccharides XXI (29%) and XXII (19%) (cf.

References on p. 92

No. 89, $R^1 = R^2 =$ -$CH_2C_6H_5$, $R^3 = H$) [60]. The reaction of Br_2 in CH_2Cl_2 with No. 90 in C_6H_6 in the presence of 4 Å molecular sieves, or in CH_2Cl_2 in the presence of $(C_4H_9)_3SnOCH_3$, causes the formation of the keto-derivative XXVI in yields of 69 or 74%, respectively [62].

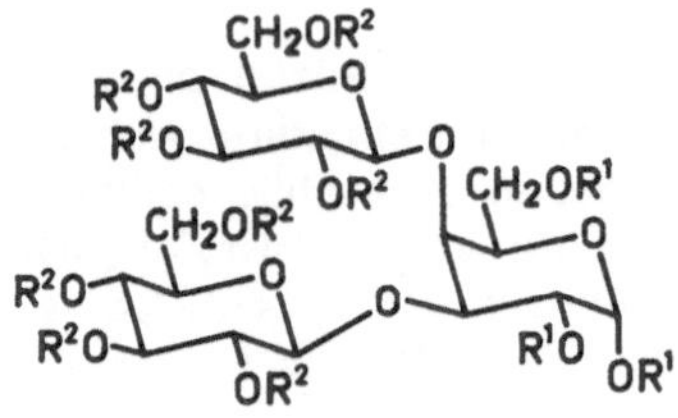

$R^1 = -CH_2C_6H_5$
$R^2 = -COCH_3$

XXIII

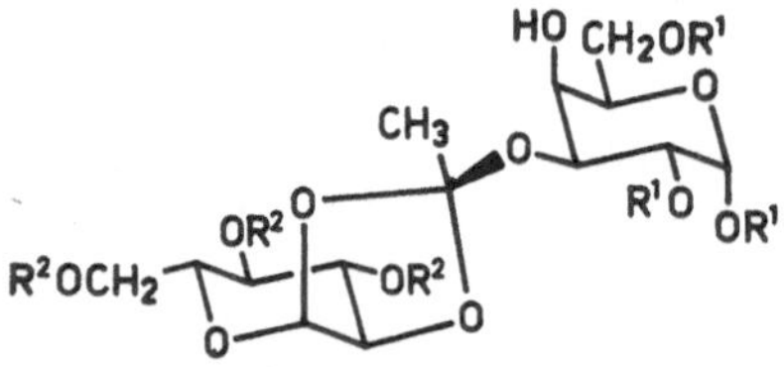

$R^1 = -CH_2C_6H_5$, $R^2 = -COCH_3$

XXIV

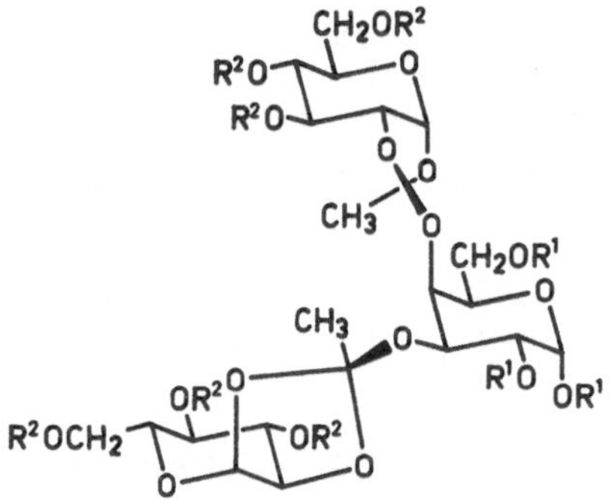

XXV

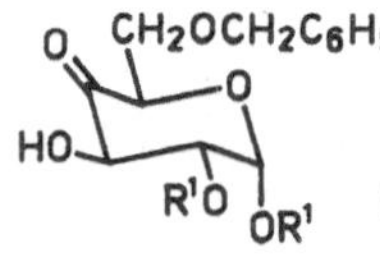

$R^1 = -CH_2C_6H_5$

XXVI

$(C_4H_9)_2\overline{SnO\text{-}C_{14}H_{16}O_4\text{-}O}$ (Table **14**, No. **95**). The crystal and molecular structure has been determined by X-ray analysis using 6180 unique reflections ($R = 0.067$). The crystals are monoclinic, space group $P2_1-C_2^2$ with $a = 22.604(4)$, $b = 21.027(4)$, $c = 14.140(3)$ Å, and $\beta = 105.4(2)°$; $Z = 2$ and $D_c = 1.31$ g/cm^3. They are composed of pentamers, the five-coordinated molecules being arranged in a linear chain. Because there are no strong polar groups sticking out from this pentameric building block, there is no hydrogen bonding or other intermolecular association between the individual blocks. The mean plane of the pentamer is tilted away from a plane normal to the twofold screw axis causing a nonparallel stacking of the blocks. Within the pentamers, coordination exists between each Sn atom and one oxygen atom of each neighboring molecule resulting in a octahedral environment for the three central Sn atoms and a pentagonal bipyramidal geometry around the terminal Sn atoms, both highly

References on p. 92

distorted; see **Fig. 3**. The five individual sugar moieties have similar configurations; there are only slight differences in the conformation of the C_4H_9 ligands [89].

^{119}Sn NMR spectrum (0.5 M in C_6D_6, 40°C): δ (relative intensity) = −132.5 (2.8), −143.7 (2.2), −145.9 (2.5), and −156.6(1) ppm. It is suggested that each signal can be assigned to a different oligomeric species existing in solution, their interconversion being slow on the NMR time scale as indicated by the sharpness of the resonances [89].

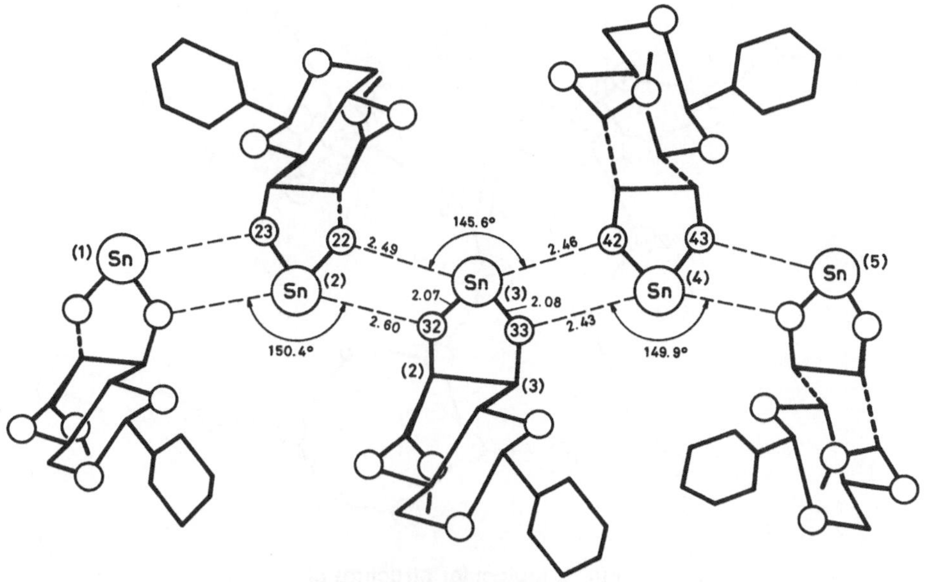

Fig. 3. Schematic molecular structure of methyl 4,6-di-O-benzyliden-2,3-dibutylstannylene-α-D-mannopyranoside [89]. C_4H_9 groups omitted.

Other bond angles (°) at the Sn atoms (Bu = C_4H_9):

O(32)-Sn(2)-O(22)	66.6	O(22)-Sn(3)-O(32)	69.2
-O(23)	143.9	-O(33)	146.8
-C(Bu)	81.5 and 80.8	-C(Bu)	85.6 and 81.4
O(33)-Sn(4)-O(42)	67.5	O(42)-Sn(3)-O(32)	145.2
-O(43)	144.4	-O(33)	68.0
-C(Bu)	82.2 and 87.7	-C(Bu)	85.6 and 82.0

$(C_4H_9)_2\overline{SnO\text{-}C_{14}H_{16}O_4\text{-}O}$ (Table **14**, No. **96**). The crystal structure of the compound has been determined by X-ray analysis from 1767 independent reflections and has been refined to R = 0.109. The crystals are orthorhombic, space group $P2_12_12_1-D_2^4$ with a = 19.696(3), b = 19.390(3), and c = 12.517(2) Å, Z = 4, D_c = 1.42 and D_m = 1.40 g/cm³ (flotation). The structure consists of discrete dimeric units, in which every Sn atom is in a distorted trigonal bipyramidal environment, with the two C_4H_9 groups and O(3) in the equatorial plane and with O(2) and the coordinating O(3′) of the second molecule in the apical positions, see **Fig. 4**, p. 92 [61].

References on p. 92

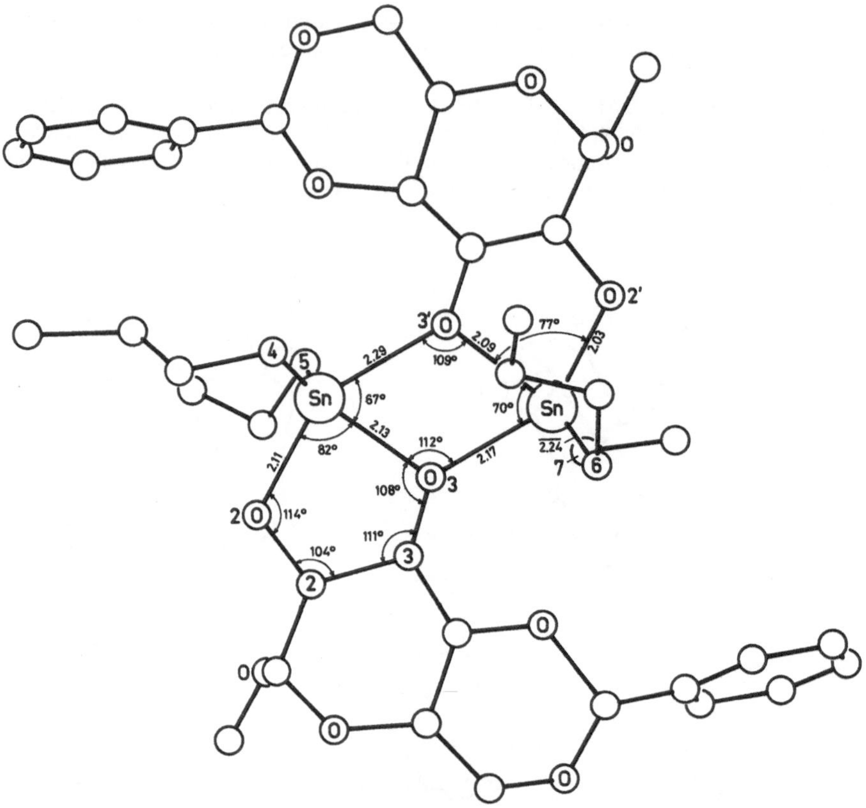

Fig. 4. Molecular structure of
methyl 4,6-di-O-benzyliden-2,3-O-dibutylstannylene-α-D-glucopyranoside [61].

Other bond angles (°):

O(2)-Sn-O(3′)	149	O(2′)-Sn-O(3)	147
-C(4)	97	-C(6)	105
-C(5)	101	-C(7)	100
O(3)-Sn-C(4)	113	O(3′)-Sn-C(6)	116
-C(5)	106	-C(7)	116
O(3′)-Sn-C(4)	92	O(3)-Sn-C(6)	90
-C(5)	91	-C(7)	95
C(4)-Sn-C(5)	139	C(6)-SN-C(7)	126

References:

[1] Bakelite Corp. (Brit. 664133 [1952]; C.A. **1952** 11230).
[2] Burt, S. L., Union Carbide and Carbon Corp. (U.S. 2583084 [1952]; C.A. **1953** 146).
[3] Mack, G. P., Parker, E., Advance Solvents and Chemical Corp. (U.S. 2604483 [1952]; C.A. **1953** 4358; Ger. 939028 [1956]; C.A. **1958** 11119).
[4] Kerr, K. B., Walde, A. W. (Experim. Parasitol. **5** [1956] 560/70).
[5] Weinberg, E. L., Metal and Thermit Corp. (Brit. 753998 [1956]; C.A. **1957** 6683).

[6] Ramsden, H. E., Banks, C. K., Metal and Thermit Corp. (U.S. 2789994 [1957]; C.A. **1957** 14786).
[7] Tomka, L. A., Weinberg, E. L., Metal and Thermit Corp. (U.S. 2798862 [1957]; C.A. **1957** 15166).
[8] Hulse, R., Twitchett, H. J., Imperial Chemical Industries Ltd. (Brit. 899948 [1959/62]; C.A. **58** [1963] 9137).
[9] Emeleus, H. J., Zuckerman, J. J. (J. Organometal. Chem. **1** [1964] 328/35).
[10] Hulse, R., Twitchett, H. J., Imperial Chemical Industries Ltd. (Brit. 957841 [1961/64]; C.A. **61** [1964] 9526).

[11] Mehrotra, R. C., Gupta, V. D. (J. Organometal. Chem. **4** [1965] 145/50).
[12] Considine, W. J. (J. Organometal. Chem. **5** [1966] 263/6).
[13] Voronkov, M. G., Romadane, I. (Khim. Geterotsikl. Soedin. **1966** 892/6; Chem. Heterocycl. Compounds [USSR] **1966** 682/5).
[14] Mehrotra, R. C., Gupta, V. D. (Indian J. Chem. **5** [1967] 643/5).
[15] Pommier, J. C., Valade, J. (J. Organometal. Chem. **12** [1968] 433/42).
[16] Davies, A. G., Milledge, H. J., Puxley, D. C., Smith, P. J. (J. Chem. Soc. A **1970** 2862/6).
[17] Sakai, S., Fujimura, Y., Ishii, Y. (J. Org. Chem. **35** [1970] 2344/7).
[18] Sakai, S., Kobayashi, Y., Ishii, Y. (J. Chem. Soc. Chem. Commun. **1970** 235/6).
[19] Sakai, S., Uchida, A., Ishii, Y. (Kogyo Kagaku Zasshi **73** [1970] 2320/4).
[20] Smith, P. J. (Organometal. Chem. Rev. A **5** [1970] 373/402).

[21] Zuckermann, J. J. (Advan. Organometal. Chem. **9** [1970] 21/134).
[22] Carraher, C. E., Scherubel, G. A. (J. Polym. Sci. Polym. Chem. Ed. **9** [1971] 983/9).
[23] Ishii, Y., Sakai, S., Mitsubishi Chemical Industries Co., Ltd. (Japan. 71-28002 [1969/71]; C.A. **75** [1971] No. 129791).
[24] Kawakami, K., Miya-Uchi, M., Tanaka, T. (J. Inorg. Nucl. Chem. **33** [1971] 3773/80).
[25] Sakai, S., Kobayashi, Y., Ishii, Y. (J. Org. Chem. **36** [1971] 1176/80).
[26] Amann, H., Rauch, H., Deutsche Gold- und Silber-Scheideanstalt, vorm. Roessler (Ger. Offen. 2056729 [1970/72]; C.A. **77** [1972] No. 89125).
[27] Carraher, C. E., Scherubel, G. A. (Makromol. Chem. **160** [1972] 259/61).
[28] Debye, N. W. G., Fenton, D. E., Zuckerman, J. J. (J. Inorg. Nucl. Chem. **34** [1972] 352/7).
[29] Mehrotra, R. C., Gupta, V. D., Sharma, C. K. (Z. Naturforsch. **27b** [1972] 386/91).
[30] Smith, P. J., White, R. F. M., Smith, L. (J. Organometal. Chem. **40** [1972] 341/53).

[31] Naik, D. V., May, J. C., Curran, C. (J. Coord. Chem. **2** [1972/73] 309/15).
[32] Pommier, J. C., Mendes, E., Valade, J. (J. Organometal. Chem. **55** [1973] C19/C23).
[33] Sharma, C. K., Gupta, V. D., Mehrotra, R. C. (J. Indian. Chem. Soc. **50** [1973] 207/8).
[34] Smith, P. J., Smith, L. (Inorg. Chim. Acta Rev. **7** [1973] 11/33).
[35] Throckmorton, P. E., McKillip, W. J., Richards, H. J., Ashland Oil, Inc. (U.S. 3767615 [1972/73]; C.A. **80** [1974] No. 96772).
[36] Tzschach, A., Pönicke, K., Korecz, L., Burger, K. (J. Organometal. Chem. **59** [1973] 199/206).
[37] Agur, D. P., Srivastava, G., Mehrotra, R. C. (Indian J. Chem. **12** [1974] 1193/6).
[38] Chimura, K., Takashima, S., Kawashima, M., Mitsubishi Rayon Co., Ltd. (Japan. 74-14741 [1970/74]; C.A. **81** [1974] No. 136302).
[39] David, S. (Compt. Rend. C **278** [1974] 1051/3).
[40] David, S., Thieffry, A. (Compt. Rend. C **279** [1974] 1045/7).

[41] Throckmorton, P. E., McKillip, W. J., Ashland Oil, Inc. (U.S. 3786067 [1972/74]; C.A. **80** [1974] No. 83246).
[42] Throckmorton, P. E., McKillip, W. J., Richards, H. J., Sitz, G. E. (Can. Sulphur Symp. Papers, Calgary, Alta, Can., 1974, R, pp 1/12).

[43] Throckmorton, P. E., McKillip, W. J., Slagel, R. C., Ashland Oil Inc. (U.S. 3828007 [1972/74]; C.A. **82** [1975] No. 58523).
[44] Tzschach, A., Pönicke, K. (Z. Anorg. Allgem. Chem. **404** [1974] 121/8).
[45] Wagner, D., Verheyden, P. H., Moffatt, J. G. (J. Org. Chem. **39** [1974] 24/30).
[46] Neumann, W. P., Schwarz, A. (Angew. Chem. **87** [1975] 844/5; Angew. Chem. Intern. Ed. Engl. **14** [1975] 812).
[47] Sakai, S., Fujinami, T., Yamada, T., Furusawa, S. (Nippon Kagaku Kaishi **1975** 1789/94; C.A. **84** [1976] No. 5090).
[48] Sakai, S., Furusawa, S., Matsunaga, H., Fujinami, T. (J. Chem. Soc. Chem. Commun. **1975** 265/6).
[49] Throckmorton, P. E., Richards, H. J., Sitz, G. E., McKillip, W. J. (J. Elastomers Plast. **7** [1975] 276/84).
[50] Auge, C., David, S., Veyrieres, A. (J. Chem. Soc. Chem. Commun. **1976** 375/6).

[51] Bancroft, G. M., Kumar Das, V. G., Sham, T. K., Clark, M. G. (J. Chem. Soc. Dalton Trans. **1976** 643/54).
[52] Mathiasch, B. (Z. Anorg. Allgem. Chem. **425** [1976] 249/56).
[53] Munavu, R. M., Szmant, H. H. (J. Org. Chem. **41** [1976] 1832/6).
[54] Nashed, M. A., Anderson, L. (Tetrahedron Letters **1976** 3503/6).
[55] Ozaki, S., Sakai, H., Mitsui Seiyaku Kogyo Co., Ltd. (Japan. Kokai 76-100089 [1975/76]; C.A. **86** [1977] No. 140408).
[56] Plum, H., Buschhoff, M., Cejka, A., Schering A.-G. (Ger. Offen. 2526711 [1975/76]; C.A. **86** [1977] No. 116058; Swiss 620573 [1975/80]; C.A. **94** [1981] No. 134151).
[57] Nashed, M. A., Anderson, L. (Carbohydr. Res. **56** [1977] 419/22).
[58] Sakai, S., Niimi, H., Kobayashi, Y., Ishii, Y. (Bull. Chem. Soc. Japan **50** [1977] 3271/5).
[59] Ohtsuka, E., Tanaka, S., Ikehara, M. (Nucleic Acid Chem. **1** [1978] 401/4).
[60] Auge, C., Veyrieres, A. (J. Chem. Soc. Perkin Trans. I **1979** 1825/32).

[61] David, S., Pascard, C., Cesario, M. (Nouv. J. Chim. **3** [1979] 63/8).
[62] David, S., Thieffry, A. (J. Chem. Soc. Perkin Trans. I **1979** 1568/73).
[63] Honnick, W. D., Zuckerman, J. J. (Inorg. Chem. **18** [1979] 1437/43).
[64] Petridis, D., Lockwood, T., O'Rourke, M., Naik, D. V., Mullins, F. P., Curran, C. (Inorg. Chim. Acta **33** [1979] 107/11).
[65] Sandhu, S. S., Sandhu, G. K., Pushkarna, S. K. (Indian J. Chem. A **17** [1979] 425/7).
[66] Shanzer, A. (Israel J. Chem. **18** [1979] 354/8).
[67] Srivastava, T. N., Agarwal, M. (Indian J. Chem. A **17** [1979] 613/5).
[68] Srivastava, T. N., Chauhan, A. K. S., Agarwal, M. (J. Inorg. Nucl. Chem. **41** [1979] 896/7).
[69] Verma, R., Gupta, V. D., Mehrotra, R. C. (Natl. Acad. Sci. Letters [India] **2** [1979] 130/2).
[70] Chattopadhyaya, J. B. (Tetrahedron Letters **21** [1980] 4113/6).

[71] Holy, A., Kois, P. (Collection Czech. Chem. Commun. **45** [1980] 2817/29).
[72] Kato, T., Zemlicka, J. (J. Org. Chem. **45** [1980] 4006/10).
[73] Mizuno, Y., Endo, T., Takahashi, A., Inaki, A. (Chem. Pharm. Bull. [Tokyo] **28** [1980] 3041/8).
[74] Rana, S. S., Barlow, J. J., Matta, K. L. (Carbohydr. Res. **85** [1980] 313/7).
[75] Shanzer, A. (Tetrahedron Letters **21** [1980] 221/2).
[76] Blunden, S. J., Smith, P. J., Beynon, P. J., Gillies, D. G. (Carbohydr. Res. **88** [1981] 9/18).
[77] David, S., Thieffry, A. (Tetrahedron Letters **22** [1981] 2885/8).
[78] Hong, N., Funabashi, M., Yoshimura, J. (Carbohydr. Res. **96** [1981] 21/8).

[79] Muraoka, M. (Chem. Pharm. Bull. [Tokyo] **29** [1981] 3449/54).
[80] Shanzer, A., Mayer-Shochet, N., Frolow, F., Rabinovich, D. (J. Org. Chem. **46** [1981] 4662/5).
[81] Slife, C. W., Nashed, M. A., Anderson, L. (Carbohydr. Res. **93** [1981] 219/30).
[82] Stegmann, H. B., Schrade, R., Saur, H., Schuler, P., Scheffler, K. (J. Organometal. Chem. **214** [1981] 197/213).
[83] Takaku, H., Kamaike, K., Tsuchiya, H. (Nucleic Acid Symp. Ser. No. 10 [1981] 171/4).
[84] Anchisi, C., Corda, L., Maccioni, A., Podda, G. (J. Heterocycl. Chem. **19** [1982] 141/4).
[85] David, S., Estramareix, B., Fischer, J. C., Therisod, M. (J. Chem. Soc. Perkin Trans. I **1982** 2131/7).
[86] David, S., de Sennyey, G., Sotta, B. (Tetrahedron Letters **23** [1982] 1817/20).
[87] Davies, A. G., Hawari, J. A. A. (J. Organometal. Chem. **224** [1982] C37/C39).
[88] El Ashry, E. S. H., Schuerch, C. (Carbohydr. Res. **105** [1982] 33/43).
[89] Holzapfel, C. W., Koekemoer, J. M., Marais, C. F., Kruger, G. J., Pretorius, J. A. (S. African J. Chem. **35** [1982] 80/8).
[90] Korecz, L., Saghier, A. A., Burger, K., Tzschach, A., Jurkschat, K. (Inorg. Chim. Acta **58** [1982] 243/9).
[91] McKervey, M. A., O'Connorm, T. (J. Chem. Soc. Chem. Commun. **1982** 655/7).
[92] Nashed, M. A., Chowdhary, M. S., Anderson, L. (Carbohydr. Res. **102** [1982] 99/110).
[93] Pandit, S. K., Gopinathan, S., Gopinathan, C. (Indian J. Chem. A **21** [1982] 726/7).
[94] Shanzer, A., Libman, J., Gottlieb, H., Frolow, F. (J. Am. Chem. Soc. **104** [1982] 4220/5).
[95] Su, T. L., Klein, R. S., Fox, J. J. (J. Org. Chem. **47** [1982] 1506/9).
[96] Swisher, R. G. (Diss. Univ. Massachusetts 1982, pp. 1/285; Diss. Abstr. Intern. B **42** [1982] 4786).
[97] Anchisi, C., Maccioni, A., Maccioni, A. M., Podda, G. (Gazz. Chim. Ital. **113** [1983] 73/6).
[98] Bachlas, B. P., Kumar, A., Sharma, H., Maire, J. C. (Bull. Soc. Chim. France **1983** 46/8).
[99] Crout, D. H. G., Morrey, S. M. (J. Chem. Soc. Perkin Trans. I **1983** 2435/40).
[100] Davies, A. G., Hawari, J. A. A. (J. Chem. Soc. Perkin Trans. I **1983** 875/82).
[101] Davies, A. G., Hawari, J. A. A., Hua-De, P. (J. Organometal. Chem. **251** [1983] 203/8).
[102] Davies, A. G., Hua-De, P., Hawari, J. A. A. (J. Organometal. Chem. **256** [1983] 251/60).
[103] Davies, A. G., Price, A. J. (J. Organometal. Chem. **258** [1983] 7/13).
[104] Martin, A., Pais, M., Monneret, C. (Carbohydr. Res. **113** [1983] 21/9).
[105] Maruyama, T., Wotring, L. L., Townsend, L. B. (J. Med. Chem. **26** [1983] 25/9).
[106] Pardhy, S. A., Gopinathan, S., Gopinathan, C. (Indian J. Chem. A **22** [1983] 73/5).
[107] Shanzer, A., Libman, J., Gottlieb, H. E. (J. Org. Chem. **48** [1983] 4612/7).
[108] Shanzer, A., Samuel, D., Korenstein, R. (J. Am. Chem. Soc. **105** [1983] 3815/8).
[109] Takaku, H., Ueda, S. (Bull. Chem. Soc. Japan **56** [1983] 1424/7).
[110] Tsuda, Y., Hanajima, M., Yoshimoto, K. (Chem. Pharm. Bull. [Tokyo] **31** [1983] 3778/80).
[111] Tsuda, Y., Haque, M. E., Yoshimoto, K. (Chem. Pharm. Bull. [Tokyo] **31** [1983] 1612/24).
[112] Uesugi, S., Kaneyasu, T., Matsugi, J., Ikehara, M. (Nucleosides & Nucleotides **2** [1983] 373/85).
[113] Anchisi, C., Corda, L., Fadda, A. M., Maccioni, A., Podda, G. (J. Heterocycl. Chem. **21** [1984] 577/81).
[114] Davies, A. G., Price, A. J., Dawes, H. M., Hursthouse, M. B. (J. Organometal. Chem. **270** [1984] C1/C3).
[115] Dessinges, A., Olesker, A., Likacs, G., Thang, T. T. (Carbohydr. Res. **126** [1984] C6/C8).
[116] Herber, R. H., Shanzer, A., Libman, J. (Organometallics **3** [1984] 586/91).
[117] Kohata, K., Abbas, S. A., Matta, K. L. (Carbohydr. Res. **132** [1984] 127/35).
[118] Mukaiyama, T., Tomioka, I., Shimizu, M. (Chemi. Letters **1984** 49/52).

[119] Nashed, M. A., El-Sokkary, R. I., Rateb, L. (Carbohydr. Res. **131** [1984] 47/52).
[120] Podda, G., Anchisi, C., Corda, L., Fadda, A. M., Maccioni, A. (J. Heterocycl. Chem. **21** [1984] 1789/91).
[121] Ruisi, G., Lo Giudice, M. T., Pellerito, L. (Inorg. Chim. Acta **93** [1984] 161/5).
[122] Saxena, A., Tandon, J. P. (Polyhedron **3** [1984] 681/8).
[123] Takeo, K., Shibata, K. (Carbohydr. Res. **133** [1984] 147/51).
[124] Takaku, H., Kamaike, K., Tsuchiya, H. (J. Org. Chem. **49** [1984] 51/6).
[125] Akhrem, A. A., Ermolenko, T. M., Timoshchuk, V. A. (Zh. Org. Khim. **21** [1985] 2190/5; J. Org. Chem. [USSR] **21** [1985] 2004/8).
[126] Awarsarkar, P. A., Gopinathan, S., Gopinathan, C. (Syn. React. Inorg. Metal-Org. Chem. **15** [1985] 133/47).
[127] Brimacombe, J. S., Rahman, K. M. M. (J. Chem. Soc. Perkin Trans. I **1985** 1067/72).
[128] Gigg, J., Gigg, R., Payne, S., Conant, R. (Carbohydr. Res. **140** [1985] C1/C3).
[129] Guiller, A., Gagnieu, C. H., Pacheco, H. (Tetrahedron Letters **26** [1985] 6067/70).
[130] Haque, M. E., Kikuchi, T., Yoshimoto, K., Tsuda, Y. (Chem. Pharm. Bull. **33** [1985] 2243/55).
[131] Keinan, E., Sahai, M., Roth, Z., Nudelman, A., Herzig, J. (J. Org. Chem. **50** [1985] 3558/66).
[132] Marx, M. H., Wiley, R. A. (Tetrahedron Letters **26** [1985] 1379/80).
[133] Podda, G., Traldi, P. (J. Heterocycl. Chem. **22** [1985] 1129/33).
[134] Ricci, A., Roelens, S., Vannucchi, A. (J. Chem. Soc. Chem. Commun. **1985** 1457/8).
[135] Roelens, S., Taddei, M. (J. Chem. Soc. Perkin Trans. II **1985** 799/804).
[136] Tronchet, J. M. J., Mekhael, K., Graf-Poncet, J., Benhamza, R., Geoffroy, M. (Helv. Chim. Acta **68** [1985] 1893/6).

1.4.1.2.1.5.3 Dibutyltin Derivatives of β-Diketones, -Ketoacidesters or -amides, and -Ketimines

The compounds of this section possess a chelate structure represented by Formulas I and II. They are listed in Table 19 according to the substituents R, R′, and R″.

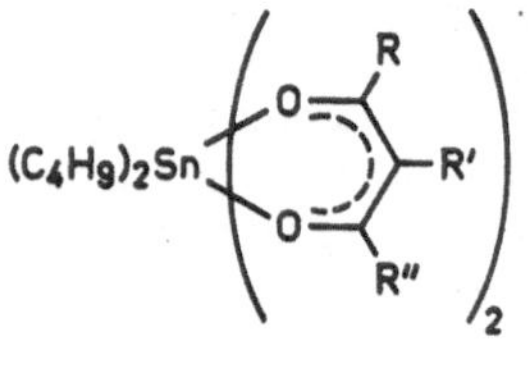

I

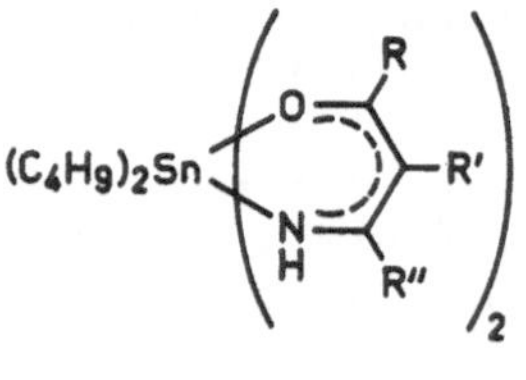

II

The compounds have been prepared by the following general methods.

Method I: a. $(C_4H_9)_2SnCl_2$ and diketones (1:2 mole ratio).
No. 1 has been prepared by the reaction of $(C_4H_9)_2SnCl_2$ with $CH_3COCH_2COCH_3$ in dilute NH_4OH solution at 20°C [3], or by the reaction of $(C_4H_9)_2SnCl_2 \cdot 2NH_3$ with acetylacetone in $C_6H_4(CH_3)_2$ [7] or C_6H_6 [4, 5]. Nos. 25 to 28 were obtained by addition of the appropriate 3-methyl-1-phenyl-4-acyl-5-pyrazolinone derivative to a C_6H_6 solution of $(C_4H_9)_2SnCl_2$ in the presence of $N(C_2H_5)_3$ [30].

b. $(C_4H_9)_2SnCl_2$ and Tl or Na enolates (1:2 mole ratio).
Addition of solutions of $(C_4H_9)_2SnCl_2$ in $C_6H_5CH_3$ to a dispersion of CH_3ONa and the appropriate β-diketo-derivative in $C_6H_5CH_3$ (prepared at 0°C) leads to Nos. 1 to 4, 13 to 16, and 21 to 23 [1, 2]. Nos. 1, 4, 11, 20, and 24 were obtained by mixing $(C_4H_9)_2SnCl_2$ with the appropriate Na enolate in C_6H_6 and subsequent refluxing of the mixture for 2 h [14]. No. 17 [29] and No. 29 [31] have been synthesized under analogous conditions. The preparation of No.1 from $(C_4H_9)_2SnCl_2$ and $TlOC(CH_3)$=$CHCOCH_3$ in hot petroleum ether (b.p. 65 to 70°C) with precipitation of TlCl is reported in [9].

Method II: $(-(C_4H_9)_2SnO-)_n$ and β-diketones (1:2 mole ratio).
The condensation reaction between $(-(C_4H_9)_2SnO-)_n$ and the appropriate β-diketone in an organic solvent, usually with azeotropic removal of the water formed in the reaction, affords No. 1 [6], (in refluxing C_6H_6) [21], (solvent ?, 0 to 3°C; products prepared at higher temperatures are difficult to purify) [24], No. 4 [21, 33], No. 7 [17], No. 10 [11], and No. 11 (all in refluxing C_6H_6) [9], Nos. 8 and 12 (in refluxing $C_6H_5CH_3$) [32], and No. 9 (in petroleum ether, 20°C, rapid) [9].

Method III: $(C_4H_9)_2Sn(OC_2H_5)_2$ or $(C_4H_9)_2Sn(OC_3H_7\text{-}i)_2$ and β-diketones (1:2 mole ratio).
The transalkoxylation of $(C_4H_9)_2Sn(OC_2H_5)_2$ by the appropriate β-diketo-derivative in refluxing C_6H_6 has been used for the synthesis of Nos. 1, 18, 19 [8], 5, and 6 [22]; the transalkoxylation of $(C_4H_9)_2Sn(OC_3H_7\text{-}i)_2$ under the same conditions for the preparation of Nos. 25 to 28 is reported in [30].

Table 19
Dibutyltin Derivatives of β-Diketones, -Ketoacidesters or -amides, and -Ketimines.
Further information on compounds preceded by an asterisk is given at the end of the table.
Explanations, abbreviations, and units on p. X.

No.	R, R′, and R″ in Formulas I and II method of preparation (yield in %)	properties and remarks	Ref.
compounds of Formula I			
*1	CH_3 H CH_3 Ia [3,4,5], (99 [7]) Ib [1,2,9,14] II [21,24], (80 to 90 [6]) III (75 [8])	30° [6,14,24], 31 to 33° (dec. to a yellow oil upon melting) [9]	[6,9, 14,24]
		b.p. 130 to 132°/1.6 [1,2], 132°/0.3 [21], 132°/0.4 [8], 144°/0.015 [6]	[1,2,6, 8,21]
		n_D^{20} = 1.5243	[4,5,7]
		μ (25°) = 2.58 (C_6H_6) or 2.73 D (C_6H_{12})	[9]
		^{1}H NMR ($CHCl_3$): 1.97 ($OCCH_3$), 5.30 (CH)	[9]
		^{13}C NMR ($CDCl_3$ [21], neat [10, 20]): 13.9 (C-δ) [10,20,21], 26.4 [20,21] or 26.5 [10] (C-γ, J(Sn,C) = 122 [20,21], 130 [10]), 27.3 [20,21] or 27.7 [10] (C-α, J(Sn,C) = 910 [20], 914 [10], 925 [21]), 27.4 [10] or 27.7 [20,21] (C-β, J(Sn,C) = 41 [10,21], 42 [20]); 27.7 (CH_3) [21], 99.8 (CH) [10,21], 190.7 [21] or 191 (CO) [10]	[10,20, 21]
		IR (KBr): ν(CO) 1582 (vs), ν(SnC) 550 (s), ν(SnO) 410 (s)	[21]

References on p. 102

Table 19 (continued)

No.	R, R′, and R″ in Formulas I and II, p. 96 method of preparation (yield in %)			properties and remarks	Ref.
				UV (C_2H_5OH): $\lambda_{max}(\varepsilon)$ = 273 (264)	[9]
				PE (He(I), 35 to 90°): 8.31 and 8.69 (π^3), 9.24 and 9.49 eV (n^-)	[19]
2	CH_3 Ib	H	C_2H_5	stabilizer for halogen-containing resins additive to lubricants	[1,2]
3	CH_3 Ib	H	C_4H_9	uses like No. 2	[1,2]
*4	CH_3 Ib [1,2,14] II [21,33]	H	C_6H_5	sticky solid [14], yellow-orange crystals; m.p. 47.5 to 49.5 [33]	[14,33]
				^1H NMR ($CDCl_3$): 2.18 ($OCCH_3$), 6.0 (CH), 7.4 to 7.9 (C_6H_5)	[21]
				^{119}Sn NMR ($CDCl_3$): −390.4 ($^2J(Sn,H)$ = 96.9, $^1J(Sn,C\text{-}\alpha)$ = 902, $^2J(Sn,C\text{-}\beta)$ = 41.2, $^3J(Sn,C\text{-}\gamma)$ = 129.9)	[33]
				^{13}C NMR (neat): 13.8 (C-δ), 26.4 (C-γ), 27.3 (C-β), 27.9 (C-α, J(Sn,C) = 901) [20,21], 28.5 (CH_3O), 96.8 (CH), 182.7 and 192.9 (CO) [21]	[20,21]
				IR (KBr): ν(CO) 1590 (vs), ν(SnC) 541 (m), ν(SnO) 416 (m)	[21]
				from spectroscopic data octahedral structure with *trans*-C_4H_9 groups suggested	[21,33]
				uses like No. 2	[1,2]
5	CH_3 III	H	C_6H_4F-4	yellow-red liquid, b.p. 170°/2 IR discussion	[22]
6	CH_3 III	H	C_6H_4Cl-4	yellow-orange liquid, b.p. 175°/1.5 IR discussion	[22]
7	CH_3 II	H	C_6H_4Br-4	creamy solid, m.p. 83° IR discussion	[17]
*8	CH_3 II (70) (Formula III, p. 101)	H	C_6H_4S-	red, thermally stable, low melting polymeric solid	[32]
				^1H NMR: 1.00 to 1.80 (C_4H_9), 2.13 ($COCH_3$), 7.50 (C_6H_4)	
				IR: ν(CO) 1600, ν(SnO) 440	
9	CF_3 II	H	CF_3	m.p. −2°	[9]
				μ = 1.96 D in C_6H_6 at 25°; octahedral structure with *cis*-C_4H_9 groups	
				^1H NMR ($CHCl_3$): 6.02 (CH)	

References on p. 102

Table 19 (continued)

No.	R, R′, and R″ in Formulas I and II, p. 96 method of preparation (yield in %)			properties and remarks	Ref.
10	CF_3 II	H	c-C_4H_3S-2 (2-thienyl)	orange liquid, b.p. 148°/0.5 IR: ν_{as}(SnC) 590 (m) and ν_s(SnC) 520 (s); IR suggests a *cis*-octahedral structure	[11]
*11	C_6H_5 Ib [14] II [9]	H	C_6H_5	yellow solid [9], m.p. 90 to 91.5° [33], 92 to 94° [9], 93° [14] $\mu = 2.07$ (at 25°) in C_6H_{12}, 2.22 (at 16°), 2.15 (at 25° and 40°), 2.18 D (at 60°) in C_6H_6 molar Kerr constant (C_6H_{12}, 632.8 nm laser source): $K_m = +666 \cdot 10^{-12}$ esu/mol ^{1}H NMR ($CHCl_3$): 6.70 (CH/aliphatic) ^{119}Sn NMR ($CDCl_3$): −385.1 (^{2}J(Sn,H) = 98.1, J(Sn,C-α) = 880, J(Sn,C-β) = 41.3, J(Sn,C-γ) = 126.0) UV: λ_{max} (log ε) = 337(4.49) and 252(4.31) in C_6H_{12}; 341.5(4.65) and 252(4.39) in $CHCl_3$ [26]; 343(2.42) in C_2H_5OH [9] with $SO_2 \rightarrow$ mono-insertion product melting at 110° with $Br_2 \rightarrow C_4H_9SnBr(C_{15}H_{11}O_2)_2$	[9, 14, 33] [9] [25] [9] [33] [9, 26] [16] [14]
*12	C_6H_5 II (73) (Formula III, p. 101)	H	C_6H_4S-	red, thermally stable, low melting polymeric substance IR: ν(CO) 1620, ν(SnO) 435	[32]
13	CH_3 Ib	C_4H_9	C_2H_5	uses like No. 2	[1, 2]
14	CH_3 Ib	OC_2H_5	CH_3	uses like No. 2	[1, 2]
15	CH_3 Ib	$CH_2C_6H_5$	CH_3	uses like No. 2	[1, 2]
16	CH_3 Ib	C_8H_{17}-i	CH_3	uses like No. 2	[1, 2]
17	CH_3 Ib	H	CH_2COCH_3	brown liquid IR: ν(CH) 2950 and 2910 (s), ν(CO) 1710 (s) and 1635 to 1610 (w), δ(CH) 1420 to 1380 (s), δ(CH_3) 1300 (m), ν_{as}(SnO) 680 (vs), ν_s(SnO) 590 (s), ν_{as}(SnC) 550 (w), ν_s(SnC) 515 (s)	[29]
18	CH_3 III (83 [8])	H	OCH_3	b.p. 136 to 138°/0.2 to 0.3 catalyst for the esterification of dibasic acids	[8] [13]

References on p. 102

Table 19 (continued)

No.	R, R′, and R″ in Formulas I and II, p. 96 method of preparation (yield in %)			properties and remarks	Ref.
19	CH_3 III (62 [8])	H	OC_2H_5	b.p. 154°/0.3 catalyst for the formation of polyurethane	[8] [18]
20	C_6H_5 Ib	H	OC_2H_5	viscous liquid	[14]
21	OC_2H_5 Ib	H	OC_2H_5	D = 1.161 at 20°, n_D^{20} = 1.4665 uses like No. 2	[1,2]
22	OC_4H_9 Ib	H	OC_4H_9	D = 1.1055 at 20°, n_D^{20} = 1.4544 uses like No. 2	[1,2]
23	OC_2H_5 Ib	$CH_2C_6H_5$	OC_2H_5	uses like No. 2	[1,2]
24	CH_3 Ib	H	NHC_6H_5	m.p. 140°	[14]
*25	CH_3 Ia, III	$-C(CH_3)=N-N(C_6H_5)-$		yellow crystals, m.p. 135 to 140°	[30]
*26	C_2H_5 Ia, III	$-C(CH_3)=N-N(C_6H_5)-$		pink crystals, m.p. 135 to 140° ^1H NMR ($CDCl_3$): 0.6 to 1.6 (C_4H_9), 1.3 (t, CH_3(R)), 2.3 (s, CH_3C=N), 2.7 (q, CH_2), 6.9 to 7.0 (m, C_6H_5)	[30]
*27	C_6H_5 Ia, III	$-C(CH_3)=N-N(C_6H_5)-$		yellow crystals, m.p. 125 to 130° ^1H NMR ($CDCl_3$): 0.5 to 1.7 (C_4H_9), 1.9 (s, CH_3), 6.8 to 8.0 (m, C_6H_5) active as insecticide against Tribolium castaneum and Trogoderma granarium (LD_{50} = 0.6112 or 0.6432%, respectively)	[30] [35]
*28	4-ClC_6H_4 Ia, III	$-C(CH_3)=N-N(C_6H_5)-$		light yellow crystals, m.p. 140 to 145° ^1H NMR ($CDCl_3$): 0.6 to 1.6 (C_4H_9), 1.7 (s, CH_3), 6.8 to 7.9 (m, C_6H_5)	[30]
compound of Formula II					
29	CH_3 Ib	H	CH_3	μ = 0.885 D six-coordinate *trans* structure with $I_2 \rightarrow$ 1:1 complex	[31]

* Further information:

$(C_4H_9)_2Sn(OC(CH_3)=CHCOCH_3)_2$ (Table **19**, No. **1**) is suggested to possess an octahedral molecular structure with the Sn coordinated by the two bidentate ligands [9, 10, 14] and the two butyl groups in *cis*- [9], or, more likely, in *trans* positions [20, 21].

References on p. 102

The title compound reacts with an equivalent amount of $(C_4H_9)_2Sn(OCH_3)_2$ at −60°C [10] or at +20°C in C_6H_{14} [15] with formation of dimeric $(C_4H_9)_2Sn(OCH_3)OC(CH_3)$=$CHCOCH_3$.

Dibutyltin bis(acetylacetonate) in combination with organoindium compounds is used in coating solutions for the preparation of transparent, electrically conducting films on glass in the production of transparent electrodes [37 to 41]. It catalyzes the formation of polyurethane [18, 23, 24, 27], the esterification of dibasic acids [12], the hydrostannylation of carbonyl compounds [15], and the crosslinking in polyorganosiloxanes [34]. The compound is also useful as a stabilizer for halogen-containing resins and as an additive to lubricants [1, 2].

The title compound has been tested as a fungicide [3].

$(C_4H_9)_2Sn(OC(CH_3)$=$CHCOC_6H_5)_2$ (Table **19**, No. **4**). The stepwise formation constants of the complex have been determined potentiometrically at constant ionic strength (0.1 M, NaCl) at 20, 30, and 40°C in dioxane-water (75/25 v/v) starting from $(C_4H_9)_2SnCl_2$ and the chelating ligand (°C, log K_1, log K_2): 20°, 11.29, 7.40; 30°, 11.34, 6.73; 40°, 10.98, 6.29. The thermodynamic parameters for the overall reaction are $\Delta H° = -32.71 \pm 0.45$ kcal/mol and $\Delta S° = -26.07 \pm 1.37$ cal · mol^{-1} · K^{-1} [36].

$(C_4H_9)_2Sn(OC(R)$=$CHCOC_6H_4S\text{-})_2$ (Table **19**, No. **8** with R = CH_3 and No. **12** with R = C_6H_5). The structure of the polymeric compounds prepared from $(\text{-}(C_4H_9)_2SnO\text{-})_n$ and $RCOCH_2C_6H_4$-(-S-S-$C_6H_4COCH_2COR$-4)-4 is illustrated by Formula III [32].

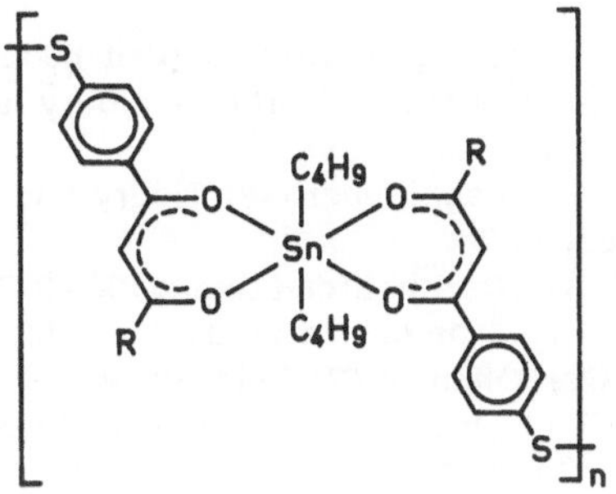

III

$(C_4H_9)_2Sn(OC(C_6H_5)$=$CHCOC_6H_5)_2$ (Table **19**, No. **11**). The depolarized Rayleigh light scattering of C_6H_{12} solutions of the title compound has been studied interferometrically at the wavelengths 457.9, 488.0, and 514.5 nm. The data clearly indicate a significant preresonance enhancement of the central Rayleigh line. The intensities and line shapes indicate a monomeric *trans* or slightly distorted *trans* structure with only moderate distortion from ligand-ligand coplanarity and excludes a *cis* arrangement [28] which has been deduced solely from dipole measurements [9]. The sign and the value of the molar Kerr constant confirm this structure which can best be characterized as a skew-trapezoidal-bipyramidal structure (Formula IV, p. 102) [25].

The stepwise formation constants and the overall changes in the enthalpy, ΔH°, and entropy, ΔS°, accompanying the formation of the title compound were estimated under the

References on p. 102

conditions reported for No. 4 (°C, log K_1, log K_2): 20°, 11.58, 7.78; 30°, 11.32, 7.28; 40°, 11.34, 6.64; $\Delta H° = -31.57 \pm 0.46$ kcal/mol, $\Delta S° = -19.18 \pm 1.01$ cal · $mol^{-1} \cdot K^{-1}$ [36].

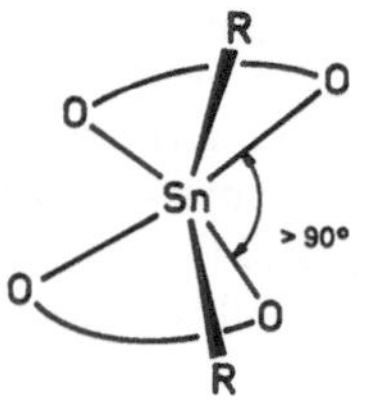

IV

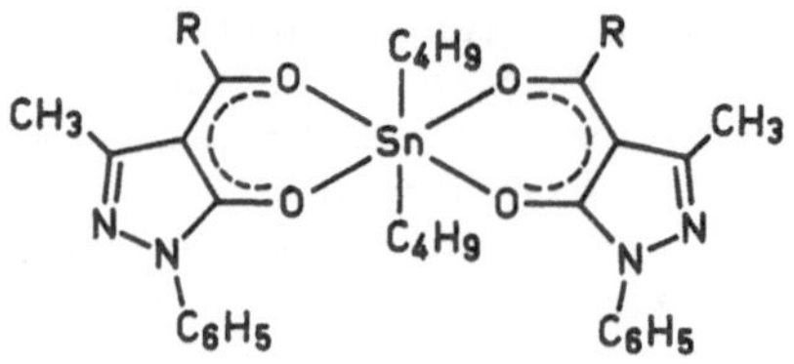

V

$(C_4H_9)_2Sn(OCR{=}C\text{-}C(CH_3){=}N\text{-}N(C_6H_5)CO)_2$ (Table **19**, Nos. **25** to **28** with R = CH_3, C_2H_5, C_6H_5, or 4-ClC_6H_4). The derivatives of 3-methyl-1-phenyl-4-acyl-5-pyrazolinone possess the structure of Formula V [30].

References:

[1] Mack, G. P., Parker, E., Advance Solvents and Chemical Corp. (U.S. 2604483 [1952]; C.A. **1953** 4358).

[2] Mack, G. P., Parker, E., Advance Solvents and Chemical Corp. (Ger. 939028 [1956]; C.A. **1958** 11119).

[3] Ethyl Co. and Esso Research and Engineering Co. (Brit. 808535 [1958]; C.A. **1959** 13496).

[4] Koninklijke Industrieele Maatschappij voorheen Noury and van der Lande, N.V. (Fr. 1320473 [1961/63]; C.A. **59** [1963] 8788).

[5] Koninklijke Industrieele Maatschappij voorheen Noury and van der Lande, N.V. (Neth. 109491 [1962]; C.A. **62** [1965] 9173).

[6] Neumann, W. P., Kleiner, F. G. (Tetrahedron Letters **1964** 3779/82).

[7] Vivieen, W. J. C., Schröder, A., Koninklijke Industrieele Matschaappij voorheen Noury and van der Lande, N. V. (Ger. Offen. 1167836 [1962/64]).

[8] Mehrotra, R. C., Gupta, V. D. (J. Organometal. Chem. **4** [1965] 237/40).

[9] Moore, C. Z., Nelson, W. H. (Inorg. Chem. **8** [1969] 138/43).

[10] Mitchell, T. N. (J. Organometal. Chem. **59** [1973] 189/97).

[11] Bachlas, B. P., Jain, R. R. (J. Organometal. Chem. **82** [1974] 359/65).

[12] Chimura, K., Takashima, S., Kawashima, M., Mitsubishi Rayon Co., Ltd. (Japan. 74-16853 [1970/74]; C.A. **82** [1975] No. 97856).

[13] Chimura, K., Takashima, S., Kawashima, M., Mitsubishi Rayon Co., Ltd. (Japan. 74-17252 [1970/74]; C.A. **82** [1975] No. 98718).

[14] Gopinathan, S., Gopinathan, C., Gupta, J. (Indian J. Chem. **12** [1974] 626/8).

[15] Knocke, R., Neumann, W. P. (Liebigs Ann. Chem. **1974** 1486/95).

[16] Gopinathan, S., Gopinathan, C., Jose, C. I., Gupta, J. (Indian J. Chem. **13** [1975] 78/80).

[17] Bachlas, B. P., Jain, R. R. (Indian J. Chem. A **14** [1976] 359).

[18] Thiele, L., Becker, R., Frommelt, H. (Plaste Kautschuk **24** [1977] 549/50).

[19] Cauletti, C., Furlani, C., Piancastelli, M. N. (J. Organometal. Chem. **149** [1978] 289/95).

[20] Domazetis, G., Magee, R. J., James, B. D. (J. Organometal. Chem. **148** [1978] 339/54).

[21] Domazetis, G., Magee, R. J., James, B. D. (J. Inorg. Nucl. Chem. **41** [1979] 1547/53).

[22] Jain, R. R., Maire, J. C., Maire-Limouzin, Y. Y., Baldy, A., Bachlas, B. P. (Indian J. Chem. A **19** [1980] 482/4).

[23] Meyborg, H., Mormann, W., Groegler, G., Schwindt, J., Bayer A.-G. (Ger. Offen. 2832253 [1978/80]; C.A. **92** [1980] No. 164764).
[24] Thiele, L. (Z. Chem. [Leipzig] **20** [1980] 315/6).
[25] Brahma, S. K., Nelson, W. H. (Inorg. Chem. **21** [1982] 4076/9).
[26] Howard, W. F., Nelson, W. H. (Inorg. Chem. **21** [1982] 2283/5).
[27] Leiner, H. H., Bossert, E. C., M and T Chemicals Inc. (Eur. Appl. 59632 [1981/82]; C.A. **98** [1983] No. 18184).
[28] Nelson, W. H., Howard, W. F., Pecora, R. (Inorg. Chem. **21** [1982] 1483/6).
[29] Bachlas, B. P., Kumar, A., Sharma, H., Maire, J. C. (Bull. Soc. Chim. France **1983** II 46/8).
[30] Saxena, S., Singh, Y. P., Rai, A. K. (Indian J. Chem. A **23** [1984] 878/80).
[31] Agrawal, M., Tandon, J. P. (J. Indian. Chem. Soc. **62** [1985] 61/3).
[32] Awarsarkar, P. A., Gopinathan, S., Gopinathan, C. (Syn. React. Inorg. Metal-Org. Chem. **15** [1985] 133/47).
[33] Howard, W. F., Crecely, R. W., Nelson, W. H. (Inorg. Chem. **24** [1985] 2204/8).
[34] Letoffe, M., Rhone-Poulenc Spécialités Chimiques (Eur. Appl. 147323 [1983/85]; C.A. **104** [1986] No. 6999).
[35] Saxena, P. N., Saxena, S., Rai, A. K., Saxena, S. C. (Indian. Biol. **17** [1985] 23/4).
[36] Singh, G., Gupta, V. D. (Indian J. Chem. A **24** [1985] 440/3).
[37] Tanaka, T., Yamanashi, F., Alps Electric Co., Ltd. (Japan. Kokai Tokkyo Koho 60243278 [1984/85]; C.A. **104** [1986] No. 217591).
[38] Tanaka, T., Yamanashi, F., Alps Electric Co., Ltd. (Japan. Kokai Tokkyo Koho 60243279 [1984/85]; C.A. **104** [1986] No. 212093).
[39] Tanaka, T., Yamanashi, F., Alps Electric Co., Ltd. (Japan. Kokai Tokkyo Koho 60243280 [1984/85]; C.A. **104** [1986] No. 212094).
[40] Tanaka, T., Yamanashi, F., Alps Electric Co., Ltd. (Japan. Kokai Tokkyo Koho 60245779 [1984/85]; C.A. **105** [1986] No. 8069).
[41] Tanaka, T., Yamanashi, F., Alps Electric Co., Ltd. (Japan. Kokai Tokkyo Koho 60245780 [1984/85]; C.A. **105** [1986] No. 62345).

1.4.1.2.1.5.4 Dibutyltin Biscarboxylates, $(C_4H_9)_2Sn(OOCR)_2$

1.4.1.2.1.5.4.1 Dibutyltin Biscarboxylates, $(C_4H_9)_2Sn(OOCR)_2$ with R = Unsubstituted Alkyl

1.4.1.2.1.5.4.1.1 Dibutyltin Diformiate, $(C_4H_9)_2Sn(OOCH)_2$

The synthesis as well as the physical or chemical properties of this compound are not mentioned in the literature. The compound can be determined in nanogram amounts by use of molecular emission cavity analysis [2]. The compound has been tested as an anthelmintic for the removal of Raillietina cesticillus and Ascaridia galli from chickens [1].

References:

[1] Kerr, K. W., Walde, A. W. (Exptl. Parasitol. **5** [1956] 560/70).
[2] Akpofure, C. O., Belcher, R., Bogdanski, S. L. (Anal. Letters **8** [1975] 921/9).

1.4.1.2.1.5.4.1.2 Dibutyltin Diacetate, $(C_4H_9)_2Sn(OOCCH_3)_2$

1.4.1.2.1.5.4.1.2.1 Synthesis, Properties, and Reactions

$(C_4H_9)_2Sn(OOCCH_3)_2$ is prepared in 95 to 98% yield from $(-(C_4H_9)_2SnO-)_n$ and CH_3COOH at 60 to 70°C [20], or from $(-(C_4H_9)_2SnO-)_n$ and acetic anhydride, with heating to drive off the water formed [38]. The compound is also obtained from $(C_4H_9)_2SnCl_2$ and $NaOOCCH_3$ in an 88% [30] or 90% yield [75], or using $KOOCCH_3$ in ethanol at 75°C [1]. $Sn(C_4H_9)_4$ reacts with $Sn(OOCCH_3)_4$ at 200 to 210°C within 10 h with formation of $(C_4H_9)_2Sn(OOCCH_3)_2$ (73% yield) [53, 55]. $(C_4H_9)_2SnBr_2$ reacts with $Pb(OOCCH_3)_2$ in aqueous CH_3COOH to give $(C_4H_9)_2Sn(OOCCH_3)_2$ in an 85% yield [21], and $(C_4H_9)_2Sn(C_6H_5)_2$ reacts with $Pb(OOCCH_3)_2$ in the presence of $Hg(OOCCH_3)_2$ in CH_3COOH within 12 h with formation of dibutyltin diacetate in a 70% yield [59]. It is also formed in yields up to 90% from $(C_4H_9)_2SnH_2$ and CH_3COOH without solvent [12, 18, 19].

Dibutyltin diacetate is formed after the oxidation of $Sn(C_4H_9)_4$ with CrO_3 [27], in the decomposition of $(C_4H_9)_2Sn(SC_2H_4OOCCH_3)_2$ at 240 to 260°C [54], and in the redistribution reaction between $(C_4H_9)_3SnOOCCH_3$ and $C_4H_9Sn(OOCCH_3)_3$ at 180°C [31].

1-^{14}C labelled dibutyltin diacetate has been obtained from 1-^{14}C tetrabutyltin by cleavage with Br_2 and further reaction with CH_3COOH [65]. 1-^{14}C labelled $(C_4H_9)_2Sn(OOCCH_3)CH_2CH(OH)C_2H_5$ reacts with CH_3COOH to form the labelled compound [62].

$(C_4H_9)_2Sn(OOCCH_3)_2$ can be separated from other organotin compounds or from residues, food, and plant materials by different methods, especially thin-layer chromatography [28, 64, 68, 69, 72], paper chromatography [17], polarography [6, 41], and amperometric precipitating titration [46]. The use of $NaIO_3$ for the volumetric determination of tin in organic and inorganic compounds was critically studied including dibutyltin diacetate as an example [7].

Melting points for the compound are 8°C [38] and 8 to 10°C [18]. The following boiling points are given: 82 to 86°C/0.2 Torr [71], 110 to 112°C/0.5 Torr [18], 112 to 114°C/1 Torr [18], 114 to 115°C/1 Torr [53], 114 to 117°C/2 Torr [54], 122 to 123°C/1 Torr [55], 125 to 126°C/1 Torr [59], 130°C/5 Torr [38], 135 to 138°C/3 to 4 Torr [20], 142 to 145°C/10 Torr [1, 31], 144.5 to 155.5°C/10 Torr [21], 145°C/10 Torr [30], and 146 to 147°C/10 Torr [26].

Refractivity index n_D^{20} = 1.4703 [55], 1.4706 [21], 14702 [59], 14775 [15], 1.5142 [5], and n_D^{26} = 1.4688 [18], 1.4692 [18], 1.475 [15]. The molar refraction is 76.50 [2, 3].

Dipole moment in benzene at 25°C: μ_D = 1.39 D [29, 39, 61].

^{1}H NMR spectrum (in $CDCl_3$): δ = 2.04 (CH_3CO) ppm [71].

^{13}C NMR spectrum (in $CDCl_3$, ppm): δ = 13.6 (C-δ), 20.4 (CH_3), 25.0 (C-α, J(Sn, C) = 601 Hz), 26.4 (C-γ), 26.8 (C-β), 181.3 (CO) ppm [70, 71]; δ = 13.6 (C-δ), 20.0 (CH_3), 25.0 (C-α, J(Sn, C) = 630 Hz), 26.5 (C-γ, J(Sn, C) = 92 Hz), 27.0 (C-β, J(Sn, C) = 35 Hz), 181 (CO) ppm [58].

^{119}Sn NMR spectrum (neat): $\delta = -195$ ppm [13]. Dipole-dipole relaxations through nonbonded protons cause a negative nuclear Overhauser effect. The spin-lattice relaxation time T_1 is found to be 0.84 s and the nuclear Overhauser enhancement factor η is 0.29 at 300 K for the mixture of the α and β anomer [74].

^{17}O NMR spectrum (neat): δ = 291 ppm relative to D_2O [76].

^{119}Sn Mössbauer spectrum (in mm/s): δ = 1.35, Δ = 3.56 (80 K) [52], δ = 1.36, Δ = 3.50 (78 K) [23], δ = 1.40, Δ = 3.45 (78 K) [22], δ = 1.45, Δ = 3.51 (84 K) [51]. These data indicate a monomeric molecule with a trans-octahedral configuration and bidentate acetate ligands [51].

References on p. 107

Table 20
Infrared Spectrum of $(C_4H_9)_2Sn(OOCCH_3)_2$.
Wave numbers in cm^{-1}.

neat solid [38]	[63]	Nujol [25]	C_6H_{12} (5%) [38]	CCl_4 [63]	assignment
				1634 (sh)	$\nu_{as}(OCO)$
1605 (s)	1610 (s)	1600 (vs)	1609 (s)	1610 (s)	$\nu_{as}(OCO)$
1570 (s)	1592 (sh)			1592 (sh)	$\nu_{as}(OCO), \delta(CH_2)$
1425 (s)	1433	1425 (s)	1425 (sh)	1433	$\nu_s(OCO), \delta(CH_3)$
			1400 (sh)		$\nu_s(OCO)$
1380 (s)	1385 (s)	1384 (vs)	1377 (s)	1385 (s)	$\nu_s(OCO), \delta(CH_3)$
1333 (m)	1333 (s)	1335 (s)	1333 (s)	1333 (s)	$\delta(CH_3)$
		1254 (m)			
		1182 (m)			
		1160 (m)			
		1090 (m)			
		1018 (s)			
		950 (m)			
		892 (m)			
		852 (w)			
		772 (w)			
		750 (w)			
690 (s)	697 (s)	690 (vs)	694 (s)	694 (s)	$\delta(CH_2), \delta(OCO)$
623 (m)	625 (s)	640 (s)	622 (m)	625 (s)	$\gamma(OCO), \nu_{as}(SnC)$
		540 (m)			
491 (w)	494 (s)	490 (m)	490 (w)	495 (s)	$\varrho(OCO), \nu_s(SnC)$
302 (s)			303 (s)		$\nu(SnO)$

The IR spectrum of $(C_4H_9)_2Sn(OOCCH_3)_2$ has been investigated for the neat liquid, and in solutions of CCl_4 and C_6H_{12}. The assignments [25, 38, 63] are shown in Table 20. In other discussions of the IR spectrum of dibutyltin diacetate, in comparison with spectra of other compounds, the following bands are assigned: ν(OCO) 1605, 1585, 1382 cm^{-1} [61], ν(OCO) 1580 cm^{-1} [4], ν(OCO) 1582, ν(SnC) 550, ν(SnO) 410 cm^{-1} [71].

UV spectrum (in cyclohexane): $\lambda_{max} = 210$ nm ($\log \varepsilon = 3.08$) [36, 37].

In the chemical ionization mass spectrum of $(C_4H_9)_2Sn(OOCCH_3)_2$, using CH_4 as the reagent gas, the only fragmentation ion is $[(C_4H_9)_2SnOOCCH_3]^+$ (100% with respect to the normalized relative abundance of the ^{120}Sn isotope) [60].

Photolysis of dibutyltin diacetate in C_6H_6 in the presence of $C_6H_5CH{=}N(C_4H_9\text{-t})O$ gave in the ESR spectrum a weak signal due to the n-butyl spin adduct $C_6H_5CH(C_4H_9)\text{-}N(C_4H_9\text{-t})O^{\bullet}$ [42].

$(C_4H_9)_2Sn(OOCCH_3)_2$ decomposes above 400°C with formation of SnO [47]. The thermal decomposition of this compound in a sealed tube at 400°C gave tin(II) in a brown residue (containing 15% C, 2.3% H, 20.3% O, and 62% Sn) besides 70% C_4H_{10}, 18% C_4H_8, and <0.01% C_8H_{18} [44].

References on p. 107

Chemical reactions of $(C_4H_9)_2Sn(OOCCH_3)_2$ are summarized in Table 21 and are arranged by reactions with inorganic compounds (Nos. 1 to 10), organic compounds (Nos. 11 to 19), and organometallic compounds (Nos. 20 to 40).

Table 21
Reactions of $(C_4H_9)_2Sn(OOCCH_3)_2$.
In column 4, the C_4H_9 groups on the Sn atom are abbreviated Bu.

No.	reactant	conditions	products (yield in %)	Ref.
1	H_2O		$(Bu_2SnOOCCH_3)_2O$	[38]
2	H_2O	in ether, acetone	$(Bu_2SnOOCCH_3)_2$, $CH_3COOSnBu_2O$-(-Bu_2SnO-)$_2SnBu_2OOCCH_3$	[26]
3	NaOH	in CH_3OH	(-Bu_2SnO-)$_n$	[31]
4	H_2SO_4	conc. H_2SO_4	$[Bu_2Sn(OSO_3H)_2H]^+$, HSO_4^-, $CH_3COOH_2^+$	[33]
5	$GeBr_4$	in ether	Bu_2SnBr_2, $Ge(OOCCH_3)_4$	[49]
6	$SnCl_4$	in ether	Bu_2SnCl_2, $Sn(OOCCH_3)_4$, $Cl_2Sn(OOCCH_3)_2$	[49]
7	$SnCl_4$	reflux in C_6H_{14}	Bu_2SnCl_2, $Sn(OOCCH_3)_4$	[55]
8	$SnCl_4$	reflux in C_6H_6	Bu_2SnCl_2, $Cl_2Sn(OOCCH_3)_2$	[55]
9	$SnBr_4$	in ether	Bu_2SnBr_2, $Sn(OOCCH_3)_4$	[49, 55]
10	SnI_4	reflux in C_6H_{14}	Bu_2SnI_2, $Sn(OOCCH_3)_4$	[55]
11	$ClC(CH_3)_2C_2H_5$	45 min	Bu_2SnCl_2, $(CH_3)_3COOCCH_3$	[45]
12	$CH_3CH_2ClCH{=}CH_2$	180 min	$Bu_2Sn(Cl)OOCCH_3$, $CH_3CH{=}CHCH_2$-$OOCCH_3$, $CH_2{=}CHCH(CH_3)OOCCH_3$	[45]
13	$ClCH_2CH{=}CHCH_3$	180 min	$Bu_2Sn(Cl)OOCCH_3$, $CH_3CH{=}CHCH_2$-$OOCCH_3$, $CH_2{=}CHCH(CH_3)OOCCH_3$	[45]
14	CH_3OH		1:1 complex	[73]
15	CH_2-CH_2 \ O /	180°C	polymers	[8]
16	t-$C_4H_9OOC_4H_9$-t	in $C_6H_5CH_3$, −80 to −40°C, UV	C_4H_9-radicals	[57]
17	$KOOC(CH_2)_4COOK$		(-$Bu_2SnOOC(CH_2)_4COO$-)$_n$	[30]
18	$ClCH_2COOH$	in C_6H_6	$Bu_2Sn(OOCCH_2Cl)_2$	[32]
19	$HSCH_2COOCH_3$		$Bu_2Sn(SCH_2COOCH_3)_2$	[50]
20	LiC_4H_9	in ether, 0°C	$SnBu_4$	[40]
21	LiC_6H_5	ether-C_6H_6, −70°C	$Bu_3SnC_6H_5$, $CH_3COC_6H_5$	[48]
22	$LiN(C_2H_5)_2$	C_6H_6, 24 h reflux	$Bu_2Sn(N(C_2H_5)_2)_2$	[43]

Table 21 (continued)

No.	reactant	conditions	products (yield in %)	Ref.
23	C_4H_9MgBr		$SnBu_4$, $(C_4H_9)_2C(OH)CH_3C_4H_9COCH_3$	[34]
24	$Al(C_8H_{17}\text{-i})_3$		$Bu_2Sn(C_8H_{17}\text{-i})_2$	[24]
25	$Si(OC_2H_5)_4$	1 week, 25°C	$(\text{-}Bu_2SnOSi(OC_2H_5)_2O\text{-})_2$	[35]
26	$Si(OC_2H_5)_4$	130°C	$Si(OSnBu_2OOCCH_3)_4$	[67]
27	$Si(OC_2H_5)_4$	130°C, 1.5 h	polyorganostannosiloxane	[9]
28	$(CH_3)_3SiCl$	1:2	Bu_2SnCl_2, $(CH_3)_3SiOOCCH_3$	[56]
29	$(CH_3)_3SiX$	1:1	$Bu_2Sn(X)OOCCH_3$; X = Cl, Br, I	[56]
30	$(CH_3)_2SiCl_2$	1:1	Bu_2SnCl_2, $(CH_3)_2Si(OOCCH_3)_2$	[56]
31	$(\text{-}(CH_3)_2SiO\text{-})_n$	100°C, 50 min	polyorganostannosiloxane	[10]
32	$(C_4H_9)_2SnH_2$	−70°C	$Bu_2Sn(H)OOCCH_3$	[19]
33	$(C_4H_9)_2SnH_2$	18 h, 25°C	$Bu_2(CH_3COO)SnSn(OOCCH_3)Bu_2$	[12, 18]
34	$(C_4H_9)_2SnCl_2$	100°C, 5 min	$Bu_2Sn(Cl)OOCCH_3$	[14, 39]
35	$(C_4H_9)_2SnBr_2$	pentane	$Bu_2Sn(Br)OOCCH_3$	[16]
36	$Sn(OC_2H_5)_4$		$CH_3COO(\text{-}SnBu_2OSn(C_2H_5)_2O\text{-})_n\text{-}C_2H_5$	[11]
37	$(C_4H_9)_2Sn(OCH_3)_2$	2 h, 130°C	$Bu_2Sn(OOCCH_3)OSn(OCH_3)Bu_2$, CH_3COOCH_3	[39]
38	$(C_4H_9)_2Sn(OR)_2$		equilibrium with an 1:1 complex; R = CH_3, C_4H_9	[39]
39	$(C_4H_9)_2Sn(CH_3)_2$	9 h, 200°C	$Bu_2(CH_3)SnOOCCH_3$	[55]
40	$Hg(Si(CH_3)_3)_2$	C_6H_6, 2.5 h, 25°C	$(\text{-}Bu_2Sn\text{-})_n$, $(CH_3)_3SiOOCCH_3$, Hg	[66]

References:

[1] Eberly, K. C., Firestone Tire and Rubber Co. (U.S. 2560034 [1951]; C.A. **1952** 7583).

[2] West, R., Rochow, E. G. (J. Am. Chem. Soc. **74** [1952] 2490/1).

[3] Vogel, A. I., Cresswell, W. T., Leicester, J. (J. Phys. Chem. **58** [1954] 174/7).

[4] Freeman, J. P. (J. Am. Chem. Soc. **80** [1958] 5954/6).

[5] Merten, R., Bayer, O., Simmler, W., Loew, G., Farbenfabriken Bayer A.-G. (Ger. 1111378 [1958]; C.A. **56** [1962] 3664).

[6] Baker, R. A. (Diss. Univ. New Hampshire 1959; Diss. Abstr. **20** [1959] 897).

[7] Farnsworth, M., Pekola, J. (Anal. Chem. **31** [1959] 410/4).

[8] Dörfelt, C., Reindl, E., Härtel, K., Farbwerke Hoechst A.-G. (Ger. 1079329 [1960]; C.A. **1961** 14328).

[9] Ishizuka, S., Momoi, M., Fujita, T., Shin-Etsu Chemical Industry Co. (Japan. 60-12945 [1961]; C.A. **1961** 10319).

[10] Ishizuka, S., Momoi, M., Fujita, T., Shin-Etsu Chemical Industry Co. (Japan. 60-12946 [1961]; C.A. **1961** 10319).

[11] Koton, M. M., Kiseleva, T. M. (Dokl. Akad. Nauk SSSR **130** [1960] 86/7; C.A. **1960** 10839).
[12] Sawyer, A. K., Kuivila, H. G. (J. Am. Chem. Soc. **82** [1960] 5958/9).
[13] Burke, J. J., Lauterbur, P. C. (J. Am. Chem. Soc. **83** [1961] 326/31).
[14] Sawyer, A. K., Kuivila, H. G. (Chem. Ind. [London] **1961** 260).
[15] Sayre, R. (J. Chem. Eng. Data **6** [1961] 560/4).
[16] Alleston, D. L., Davies, A. G. (J. Chem. Soc. **1962** 2050/4).
[17] Gasparic, J., Cee, A. (J. Chromatog. **8** [1962] 393/8).
[18] Sawyer, A. K., Kuivila, H. G. (J. Org. Chem. **27** [1962] 610/4).
[19] Sawyer, A. K., Kuivila, H. G. (J. Org. Chem. **27** [1962] 837/41).
[20] Sheverdina, N. I., Abramova, L. V., Paleeva, I. E., Kocheshkov, K. A. (Khim. Prom. **1962** No. 10, pp. 707/8; C.A. **59** [1963] 8776).

[21] Zemlyanskii, N. N., Panov, E. M., Kocheshkov, K. A. (Zh. Obshch. Khim. **32** [1962] 291/3; J. Gen. Chem. [USSR] **32** [1962] 284/6).
[22] Aleksandrov, A. Yu., Delyagin, N. N., Mitrofanov, K. P., Polak, L. S., Shpinel, V. S. (Dokl. Akad. Nauk SSSR **148** [1963] 126/8; Dokl. Chem. Proc. Acad. Sci. USSR **148/153** [1963] 1/3).
[23] Gol'danskii, V. I., Makarov, E. F., Stukan, R. A., Trukhtanov, V. A., Khrapov, V. V. (Dokl. Akad. Nauk SSSR **151** [1963] 357/60; Dokl. Phys. Chem. Proc. Acad. Sci. USSR **148/153** [1963] 598/601).
[24] Mangham, J. R., Ethyl Corp. (U.S. 3095433 [1960/63]; C.A. **59** [1963] 6440).
[25] Slobokhotova, N. A., Faizi, N. A., Zemlynskii, N. N., Panov, E. M., Kocheshkov, K. A. (Zh. Obshch. Khim. **33** [1963] 2610/3; J. Gen. Chem. [USSR] **33** [1963] 2544/6).
[26] Zemlyanski, N. M., Panov, E. M., Slovokhotova, N. A., Shamagina, O. P., Kocheshkov, K. A. (Dokl. Akad. Nauk SSSR **149** [1963] 312/4; Dokl. Chem. Proc. Acad. Sci. USSR **148/153** [1963] 205/8).
[27] Deblandre, C., Gielen, M., Nasielski, J. (Bull. Soc. Chim. Belges **73** [1964] 214/25).
[28] Neubert, G. (Z. Anal. Chem. **203** [1964] 265/72).
[29] Zemlyanski, N. N., Gol'dshtein, I. P., Gur'yanova, E. N., Panov, E. M., Slovokhotova, K. A., Kocheshkov, N. A. (Dokl. Akad. Nauk SSSR **156** [1964] 131/4; Dokl. Phys. Chem. Proc. Acad. Sci. USSR **154/159** [1964] 452/5).
[30] Frankel, M., Gertner, D., Wagner, D., Zilkha, A. (J. Appl. Polym. Sci. **9** [1965] 3383/8).

[31] Suenobu, Y., Yoshitomi Pharmaceutical Industries, Ltd. (Japan. 65-2336 [1962/65]; C.A. **62** [1965] 14726).
[32] Zemlyanskii, N. N., Panov, E. M., Shamagina, O. P., Kocheshkov, K. A. (Zh. Obshch. Khim. **35** [1965] 1029/31; J. Gen. Chem. [USSR] **35** [1965] 1034/5).
[33] Gillespie, R. J., Kapoor, R., Robinson, E. A. (Can. J. Chem. **44** [1966] 1197/202).
[34] Moore, M. D., Lanning, F. C. (Quart. J. Florida Acad. Sci. **29** [1966] 243/7).
[35] Nagy, J., Borbely-Kuszmann, A. (Period. Polytech. Chem. Eng. **10** [1966] 139/45).
[36] Gol'dshtein, I. P., Gur'yanova, E. N., Zemlyanskii, N. N., Syutkina, O. P., Panov, E. M., Kocheshkov, K. A. (Dokl. Akad. Nauk SSSR **175** [1967] 836/9; Dokl. Chem. Proc. Acad. Sci. USSR **172/177** [1967] 688/91).
[37] Gol'dshtein, I. P., Gur'yanova, E. N., Zemlyanskii, N. N., Syutkina, O. P., Panov, E. M., Kocheshkov, K. A. (Izv. Akad. Nauk SSSR Ser. Khim. **1967** 2201/7; Bull. Acad. Sci. [USSR] Div. Chem. Sci. **1967** 2115/9).
[38] Maeda, Y., Okawara, R. (J. Organometal. Chem. **10** [1967] 247/56).
[39] Zemlyanskii, N. N., Gol'dshtein, I. P., Gur'yanova, E. N., Syutkina, O. P., Panov, E. M., Shovokhotova, N. A., Kocheshkov, K. A. (Izv. Akad. Nauk SSSR Ser. Khim. **1967** 728/35; Bull. Acad. Sci. [USSR] Div. Chem. Sci. **1967** 707/12).
[40] Reiff, H. F., La Liberte, B. R., Davidsohn, W. E., Henry, M. C. (J. Organometal. Chem. **15** [1968] 247/50).

[41] Geyer, T., Rotermund, U. (Acta Chim. [Budapest] **59** [1969] 201/10).
[42] Janzen, E. G., Blackburn, B. J. (J. Am. Chem. Soc. **91** [1969] 4481/90).
[43] Lorberth, J. (J. Organometal. Chem. **19** [1969] 435/8).
[44] Razuvaev, G. A., Domrachev, G. A., Kochetikhina, K. G. (Zh. Obshch. Khim. **39** [1969] 1106/8; J. Gen. Chem. [USSR] **39** [1969] 1076/7).
[45] Ayrey, G., Poller, R. C., Siddiqui, I. H. (J. Polym. Sci. B **8** [1970] 1/6).
[46] Haasova, L., Pribyl, M. (Z. Anal. Chem. **249** [1970] 35/8).
[47] Kochetikhina, G. H., Domrachev, G. A., Razuvaev, G. A. (Metody Poluch. Anal. Veshchestv Osoboi Chist. **1970** 125/8; C.A. **76** [1972] No. 140977).
[48] La Liberte, B. R., Reiff, H. F., Davidsohn, W. E. (Org. Prep. Proced. Intern. **2** [1970] 325/8).
[49] Melnichenko, L. S., Zemlyanskii, N. N., Samurskaya, K. A., Kocheshkov, K. A. (Dokl. Akad. Nauk SSSR **190** [1970] 351/3; Dokl. Chem. Proc. Acad. Sci. USSR **190/195** [1970] 48/50).
[50] Riethmayer, S., Chemische Werke München Otto Bärlocher GmbH (Ger. Offen. 1905898 [1969/70]; C.A. **73** [1970] No. 109904).

[51] Herber, R. H. (J. Chem. Phys. **54** [1971] 3755/60).
[52] Maddock, A. G., Platt, R. H. (J. Chem. Soc. A **1971** 1191/5).
[53] Melnichenko, L. S., Zemlyanskii, N. N., Kolosova, N. D., Kocheshkov, K. A. (Dokl. Akad. Nauk SSSR **198** [1971] 1348/9; Dokl. Chem. Proc. Acad. Sci. USSR **196/201** [1971] 534/5).
[54] Yamaji, Y., Nakagawa, Y., Matsuda, H., Matsuda, S. (Kogyo Kagaku Zasshi **74** [1971] 735/8; C.A. **75** [1971] No. 49257).
[55] Melnichenko, L. S., Zemlyanskii, N. N., Kocheshkov, K. A. (Izv. Akad. Nauk SSSR Ser. Khim. **1972** 2055/8; Bull. Acad. Sci. [USSR] Div. Chem. Sci. **1972** 1993/6).
[56] Armitage, D. A., Tarassoli, A. (Inorg. Nucl. Chem. Letters **9** [1973] 1225/7).
[57] Davies, A. G., Scaiano, J. C. (J. Chem. Soc. Perkin Trans. **II 1973** 1777/80).
[58] Mitchell, T. N. (J. Organometal. Chem. **59** [1973] 189/97).
[59] Syutkina, O. P., Panov, E. M., Kocheshkov, K. A. (Zh. Obshch. Khim. **43** [1973] 1322/4; J. Gen. Chem. [USSR] **43** [1973] 1313/5).
[60] Fish, R. H., Holmstead, R. L., Casida, J. E. (Tetrahedron Letters **1974** 1303/6).

[61] Guryanova, E. M., Syutkina, O. P., Panov, E. M., Kocheshkov, K. A. (Zh. Obshch. Khim. **44** [1974] 1974/8; J. Gen. Chem. [USSR] **44** [1974] 1937/40).
[62] Fish, R. H., Kimmel, E. C., Casida, J. E. (J. Organometal. Chem. **93** [1975] C1/C4).
[63] Morikawa, T. (Kagaku To Kogyo [Osaka] **49** [1975] 98/113; C.A. **84** [1976] No. 5795).
[64] Moshkovskaya, M. V., Nefedov, V. D., Molchanova, N. G., Zhuravlev, V. E. (Tr. Estestvennonauch. Inst. Permsk. Gos. Univ. **13** No. 3 [1975] 210/3; C.A. **86** [1977] No. 121468).
[65] Fish, R. H., Kimmel, E. C., Casida, J. E. (J. Organometal. Chem. **118** [1976] 41/54).
[66] Mitchell, T. N. (J. Organometal. Chem. **92** [1975] 311/9).
[67] Wohlfarth, E., Hechtl, W., Hittmair, P., Wacker-Chemie GmbH (Ger. Offen. 2524000 [1975/1976]; C.A. **86** [1977] No. 91490).
[68] Woidich, H., Pfannhauser, W. (Z. Lebensm. Unters. Forsch. **162** [1976] 49/54).
[69] Woidich, H., Pfannhauser, W., Blaicher, G. (Deut. Lebensm.-Rundsch. **72** [1976] 421/2).
[70] Domazetis, G., Magee, R. J., James, B. D. (J. Organometal. Chem. **148** [1978] 339/54).

[71] Domazetis, G., Magee, R. J., James, B. D. (J. Inorg. Nucl. Chem. **41** [1979] 1547/53).
[72] Vasundhara, T. S., Parihar, D. B. (Z. Anal. Chem. **294** [1979] 408).
[73] Varentsova, N. V., Gol'dshtein, I. P., Paleeva, I. E., Tarakanov, O. G., Gur'yanova, E. N. (Zh. Obshch. Khim. **50** [1980] 2085/93; J. Gen. Chem. [USSR] **50** [1980] 1688/94).

[74] Blunden, S. J., Frangou, A., Gillies, D. G. (Org. Magn. Resonance **20** [1982] 170/4).
[75] Kizlink, J., Rattay, V. (Czech. 218093 [1981/84]; C.A. **103** [1985] No. 105155).
[76] Lycka, A., Holecek, J. (J. Organometal. Chem. **294** [1985] 179/82).

1.4.1.2.1.5.4.1.2.2 Toxicity and Biocidal Properties

Dibutyltin acetate is less effective as a biocide than triorganotin compounds. Its bactericidal, insecticidal, molluscicidal, and especially fungicidal properties, as well as its toxicity and phytotoxicity, have been studied.

Tests as a potato blight fungicide (Phytophthora infestans) showed the compound as effective as tributyltin acetate, but only about one-tenth as effective as triphenyltin acetate, but less phytotoxic [7, 9]. For the use as a bactericide and fungicide, the compound has been tested against Alternaria tenius [3, 4, 15], Aspergillus flaver [15], Aspergillus niger [15], Aspergillus terreus [10], Botrytis cinerea [3, 4], Memmiella eschinata [15], Myrotecium verrucaria [15], Rhizopus nigricans [10], Sclerotinia fructicola [3, 4], and Trichophyton mentagrophytes [10].

The anthelmintic activity of $(C_4H_9)_2Sn(OOCCH_3)_2$ has been tested against the nematode, Ascaridia galli, and the cestode, Raillietina cesticillus, in chickens [1, 2].

Dibutyltin diacetate inhibits the population growth and cell survival of marine unicellular algae Skeletonema costatum and Thalassiosira pseudonana less than tributyl- and triphenyltin compounds [17].

The compound is used as part of coatings for wood, glass, and metals against flies, bugs, cockroaches and mosquitoes [6, 8].

The LD_{50} value of $(C_4H_9)_2Sn(OOCCH_3)_2$ for mice is 109.7 mg/kg bodyweight [5, 14]. A more recent investigation showed a LD_{50} value of 1070 mg/kg bodyweight for deer mice Peromyscus maniculatus [16]. An intensive study of the toxicity of dibutyltin diacetate showed after intubation of mice and rats with daily one-fourth LD_{50} amounts of this compound early mitochondrial damage with cristae swelling and subsequent hydropic degeneration. For this and some other effects, a mechanism was proposed via a dithiol inhibition of ATP production in the mitochondria with consequent interference of ATP-dependent membrane activity in other parts of the cell [5]. Other investigations also showed a swelling of rat liver mitochondria [11] and of the glutathione-S-aryltransferase from rat liver [12]. The metabolism of dibutyltin diacetate in the biological oxidation process proceeds via a monooxygenase or nonenzymatic cleavage to butyltin derivatives [13].

References:

[1] Kerr, K. B., Walde, A. W., Dr. Salsburys Laboratories Iowa (U.S. 2702775 [1953/55]; C.A. **1955** 7816).
[2] Kerr, K. B., Walde, A. W. (Exptl. Parasitol. **5** [1956] 560/70).
[3] Brückner, H., Härtel, K., Farbwerke Hoechst A.-G. (Ger. 1025198 [1958]; C.A. **1960** 12468).
[4] Farbwerke Hoechst A.-G. (Brit. 797073 [1953]; C.A. **1959** 22714).
[5] Calley, D., Guess, W. L., Austian, J. (J. Pharm. Sci. **56** [1967] 1267/72).
[6] Kochkin, D. A., Azerbaev, I. N. (Vestn. Akad. Nauk Kaz. SSR **22** No. 12 [1966] 53/61).
[7] McIntosh, A. H. (Ann. Appl. Biol. **66** [1970] 115/8).

[8] Kochkin, D. A., Novoderzhkina, I. S., Voronkov, N. A., Zubov, P. I., Azerbaev, I. N. (Fiziol. Opt. Aktiv. Polim. Veshchestva Tr. 2nd Vses. Simp., Riga 1969 [1971], pp. 89/102).
[9] McIntosh, A. H. (Ann. Appl. Biol. **69** [1971] 43/6).
[10] Burk, G. A., Dow Chemical Co. (U.S. 3649673 [1970/72]; C.A. **76** [1972] No. 126430).

[11] Wulf, R. G., Byington, K. H. (Arch. Biochem. Biophys. **167** [1975] 176/85).
[12] Henry, R. A., Byington, K. H. (Biochem. Pharmacol. **25** [1976] 2291/5).
[13] Kimmel, E. C., Fish, R. H., Casida, J. E. (J. Agric. Food Chem. **25** [1977] 1/9; C.A. **86** [1977] No. 38306).
[14] Smith, P. J., Luijten, J. G. A., Klimmer, O. R. (Intern. Tin Res. Inst. Publ. No. 538 [1978]).
[15] Smith, A. H., General Electric Co. (Eur. Appl. 34877 [1980/81]; C.A. **96** [1982] No. 8319).
[16] Schafer, E. W., Bowles, W. A. (Arch. Environ. Contam. Toxicol. **14** [1985] 111/29).
[17] Walsh, G. E., McLaughlan, L. L., Lores, E. M., Louie, M. K., Deans, C. H. (Chemosphere **14** [1985] 383/92).

1.4.1.2.1.5.4.1.2.3 Uses

Dibutyltin diacetate is a catalyst for the formation of urethane [21, 30, 31] and of polyurethanes for rigid foam, elastomers and coatings [3, 4, 5, 7, 9, 13, 20, 24, 25, 27, 34, 35], for the nucleophilic substitution of urethanes with hydroxyl-containing compounds [12], for the reaction of substituted ureas with isocyanates to form biuret [28], for the formation of carbamic esters from isocyanates and alcohols [32], for transesterification reactions [23], for the hydrostannation of aldehydes and ketones [15], and for the polymerization of nitriles [18], ethylene oxide [8], and the copolymerization of glycolide with lactones [40]. It has been found that the reaction steps in the formation of urethanes are first the complexation of CH_3OH to dibutyltin diacetate, dissociation of this complex into a proton and an anion of the type $[(C_4H_9)_2Sn(OOCCH_3)_2(OCH_3)]^-$, insertion of the isocyanate into the Sn-OCH_3 bond, and methanolysis of the thus-formed urethane precursor with simultaneous regeneration of the anion [31].

$(C_4H_9)_2Sn(OOCCH_3)_2$ is a stabilizer for poly(vinyl chloride) [1, 6, 14] and for polyester resins [11].

Another important use for dibutyltin diacetate is based on the action of this compound as a catalyst for condensation reactions of silanols. Linear polydiorganosiloxanediols and other polysiloxanes can be vulcanized with hydrogen evolution, by using polyfunctional silanes in the presence of $(C_4H_9)_2Sn(OOCCH_3)_2$ [2, 10, 16, 22, 26]. The mechanism of this silane-silanole condensation reactions involves complexation of the silanol with $(C_4H_9)_2Sn(OOCCH_3)_2$, followed by the attack of a silane or silanole molecule [22]. Dibutyltin diacetate catalyzes the crosslinking of butyl rubber with silane coupling agents [41], and it catalyzes the reaction of silanole groups on the surface of glass fibers with diisocyanates [37]. The compound is used to catalyze the hydrolysis and gelation of technical ethylsilicate to prepare alkyl silicate binders for refractory technology [29, 36], and the formation of mullite via gels prepared from technical ethyl silicate and aluminium chloride [38].

Transparent, electrically conducting layers of tin oxyde on glass can be made by chemical vapor deposition using dibutyltin diacetate, O_2, and H_2O in N_2 as the carrier gas at a substrate temperature of 400 to 550°C. $SbCl_5$ or CCl_3CF_3 are used as dopants [17, 19, 33, 39, 42].

Patents in which the use of dibutyltin diacetate for several purposes is claimed are listed at the end of the reference list, pp. 113/26.

References:

[1] Kenyon, A. S. (NBS-C-525 [1953] 81/94).
[2] Baranovskaya, N. B., Berlin, A. A., Zakharova, M. Z., Mizikin, A. I. (Khim. Prakt. Primen. Kremneorg. Soedin. Tr. 2nd Konf., Leningrad 1958 [1961], pp. 88/95; C.A. **1960** 7206).
[3] Hostettler, F., Cox, E. F. (Ind. Eng. Chem. **52** [1960] 609/10).
[4] Farkas, A., Mills, G. A. (Advan. Catal. **13** [1962] 363/446).
[5] Entelis, S. G., Nesterov, O. V., Tiger, R. P. (J. Cell. Plast. **3** [1967] 360/3).
[6] Geddes, W. C. (Rubber Chem. Technol. **40** [1967] 177/216).
[7] Nesterov, O. V., Chirkov, Yu. N., Entelis, S. G. (Kinetika Kataliz **8** [1967] 137/3; Kinet. Catal. **8** [1967] 1161/3).
[8] Matsuda, S., Matsuda, H., Ninagawa, A., Iwamoto, N. (Kogyo Kagaku Zasshi **71** [1968] 2054/9; C.A. **70** [1969] No. 47920).
[9] Dunn, D. (J. Cell. Plast **5** [1969] 341/7).
[10] Nagy, J., Borbely-Kuszmann, A. (Kinet. Mech. Polyreactions Intern. Symp. Macromol. Chem. Prepr., Budapest 1969, Vol. 4, pp. 297/301).

[11] Arkdzhovskii, V. N., Belotserkovskaya, K. B., Leikina, M., Vasileva, A. N. (Steklyannye Volokna Stekloplast. **1970** 194/9; C.A. **78** [1973] No. 4913).
[12] Volodarskaya, Yu. I., Sal'nikova, G. A., Shmidt, Ya. A. (Dokl. Akad. Nauk SSSR **195** [1970] 841/4, Dokl. Chem. Proc. Acad. Sci. USSR **193/195** [1970] 865/67).
[13] Zabrodin, V. B., Nesterov, O. V., Entelis, S. G. (Kinetika Kataliz **11** [1970] 114/9; Kinet. Catal. **11** [1970] 91/5).
[14] Ayrey, G., Poller, R. C., Siddiqui, I. H. (Polymer **13** [1972] 299).
[15] Knocke, R., Neumann, W. P. (Liebigs Ann. Chem. **1974** 1486/95).
[16] Martyakova, N. I., Dolgoplsk, S. B., Kagan, E. G., Kostikyan, T. S., Petrukhno, L. A. (Zh. Obshch. Khim. **44** [1974] 298/303; J. Gen. Chem. [USSR] **44** [1974] 283/7).
[17] Kane, J., Schweizer, H. P., Kern, W. (J. Electrochem. Soc. **123** [1976] 270/7).
[18] Minke, R., Freireich, S., Zilkha, A. (Israel J. Chem. **13** [1975] 212/20).
[19] Kane, J., Schweizer, H. P., Kern, W. (J. Electrochem. Soc. **122** [1975] 1144/9).
[20] Dearlove, T. J., Campbell, G. A. (J. Appl. Polym. Sci. **21** [1977] 1499/509).

[21] Thiele, L., Becker, R., Frommelt, H. (Faserforsch. Textiltech. **28** [1977] 343/8).
[22] Fierens, P., Vandendunghen, G., Segers, W., van Elsuwe, R. (React. Kinet. Catal. Letters **8** [1978] 179/87).
[23] Poller, R. C., Retout, S. P. (J. Organometal. Chem. **173** [1979] C7/C8).
[24] Anonymous communication (Res. Discl. No. 195 [1980] 272).
[25] Karpel, S. (Tin Its Uses No. 125 [1980] 1/6).
[26] Van der Weij, F. W. (Makromol. Chem. **181** [1980] 2541/8).
[27] Anonymous communication (Res. Discl. No. 208 [1981] 309/10).
[28] Joel, D., Thiele, L., Kühn, G. (Acta Polym. **32** [1981] 27/30).
[29] Jones, K., Biddle, K. D., Das, A. K., Emblem, H. G. (Silicates Ind. **46** [1981] 107/11).
[30] Van der Weij, F. W. (J. Polym. Sci. Polym. Chem. Ed. **19** [1981] 381/8).

[31] Van der Weij, F. W. (J. Polym. Sci. Polym. Chem. Ed. **19** [1981] 3063/8).
[32] Varentsova, N. V., Gol'dshtein, I. P., Paleeva, I. E., Tarakanov, O. G., Gur'yanova, E. N. (Zh. Obshch. Khim. **52** [1982] 1612/20; J. Gen. Chem. [USSR] **52** [1982] 1424/31).
[33] Belanger, D., Bartkowski, M., Dodelet, J. P., Lombos, B. A., Dickson, I., Dao, L. (J. Can. Ceram. Soc. **52** [1983] 28/32).
[34] Robins, J., Edwards, B. H., Tokach, S. K. (Polym. Mater. Sci. Eng. **49** [1983] 331/5).
[35] Bats, J. P., Ducasse, Y., Lalande, R., Tauzia, M. (Eur. Polym. J. **20** [1984] 997/1001).

[36] Jones, K., Emblem, H. G., Hafez, H. M. (J. Non-Cryst. Solids **63** [1984] 201/8).
[37] Yosomiya, R., Morimoto, K. (Polym. Bull. [Berlin] **12** [1984] 41/8).
[38] Al-Jarsha, Y. M. M., Biddle, K. D., Das, A. K., Davies, T. J., Emblem, H. G., Jones, K., McCullough, J. M., Rhaman, M. A. M. A., Sharf El Deen, A. N. A. El. M., Wakefield, R. (J. Mater. Sci. **20** [1985] 1773/81).
[39] Belanger, D., Dodelet, J. P., Lombos, B. A., Dickson, J. I. (J. Electrochem. Soc. **132** [1985] 1398/405).
[40] Kricheldorf, H. R., Jonte, J. M., Berl, M. (Makromol. Chem. Suppl. **12** [1985] 25/38).
[41] Matsuo, T., Kakei, M., Ikeda, S., Matsumoto, T., Yamashita, S. (Nippon Gomu Kyokaishi **58** [1985] 94/100; C.A. **102** [1985] No. 221945).
[42] Thomas, S., Amouroux, J. (Vide Couches Minces **40** [1985] 381/3).

Patented Uses for $(C_4H_9)_2Sn(OOCCH_3)_2$

Catalyst for the Formation of Urethane, Polyurethanes and Other Monomers or Polymers

Anspon, H. D., Pschorr, F. E., General Aniline & Film Corp., Polymerization of α-Chloroacrylates, U.S. 2683705 [1954]; C.A. **1954** 13275; Brit. 725215 [1954]; C.A. **1955** 9326.

Caldwell, J. R., Eastman Kodak Co., Organometallic Tin Compounds as Catalysts in the Preparation of Polyesters, U.S. 2720507 [1955]; C.A. **1956** 2205.

Caldwell, J. R., Eastman Kodak Co., Linear Polyurethanes from Alkyl Diurethanes of Aromatic Diamines, U.S. 2801231 [1957]; C.A. **1957** 18704.

Merten, R., Bayer, O., Simmler, W., Loew, G., Farbenfabriken Bayer A.-G., Isocyanate-based Foams, Ger. 1111378 [1958]; C.A. **56** [1962] 3664.

Asahi Chemical Industry Co., Ltd., Improving the Workability of High-Molecular Weight Polypropylene, Fr. 1341537 [1962/63]; C.A. **60** [1964] 10880.

Mangham, J. R., Ethyl Corp., Alkyltin Compounds, U.S. 3095433 [1960/63]; C.A. **59** [1963] 6440.

Asahi Kasei Kogyo Kabushiki Kaisha, Brit. 1151927 [1965/69]; C.A. **71** [1969] No. 13755.

Yamamoto, K., Nakatsuka, K., Mitsubishi Rayon Co., Ltd., Thermoplastic Polyacetal Resins, Japan. 69-15185 [1965/69]; C.A. **72** [1970] No. 13232.

Imperial Chemical Industries, Ltd., Procedure for the Synthesis of Homomolecular Linear Polyesters, Neth. 6915283 [1969/70].

Nakata, T., Kawamata, K., Osaka Soda Co., Ltd., Organotin-Phosphorus Ester Catalysts for Alkylene Oxide Polymerization, Ger. Offen. 1941690 [1968/70]; C.A. **72** [1970] No. 133372.

Vizurraga, L. R., Fiber Industries, Inc., Polymerization Catalysts for Producing Polyesters, Ger. Offen. 2057614 [1969/71]; C.A. **75** [1971] No. 110731.

Nawada, K., Tsunawaki, K., Tanaka, I., Watanabe, K., Teijin Ltd., Polycondensation Catalysts of Poly(ethylene terephthalate) Preparation, Japan. 71-28788 [1967/71]; C.A. **76** [1972] No. 46712.

Klug, E. D., Spurlin, H. M., Young III, W. L., Hercules Inc., Water-Desalinating Reverse Osmosis Membranes from Hydroxyalkyl Derivatives of Cellulose, U.S. 3620970 [1969/71]; C.A. **76** [1972] No. 73988.

Doerge, H. P., Wismer, M., PPG Industries, Inc., Polyurethane Foams with Reduced Smoke Levels, U.S. 3637542 [1969/72]; C.A. **76** [1972] No. 141739.

Watanabe, T., Yamamoto, A., Yamashita, Y., Kansai Paint Co., Ltd., Modified Urethane Oil, Japan. 74-06559 [1970/74]; C.A. **82** [1975] No. 45268.

Yasuda, N., Yamashita, T., Ariyoshi, Y., Ajinomoto Co., Inc., N-Protected Amino Acid Esters, Japan. Kokai 74-133301 [1973/74]; C.A. **82** [1975] No. 171443.

Reymore, H. E., Zane, J. K., Upjohn Co., Catalyst System for Trimerizing a Polyisocyanate, Ger. Offen. 2502260 [1974/75]; C.A. **83** [1975] No. 207100.

Kimura, K., Toa Gosei Chemical Industry Co., Ltd., Self-Extinguishing Resins, Japan. Kokai 75-75286 [1973/75]; C.A. **84** [1976] No. 5838.

Carlos, D. D., Hicks, D. D., Thermoset Coatings from Nonreactive Polymers, U.S. Appl. B 415847 [1973/75]; C.A. **82** [1975] No. 141773.

Narayan, T. L., Cenker, M., BASF Wyandotte Corp., Carbodiimide-Isocyanurate Foams Containing Urethane Linkages, U.S. 3922238 [1974/75]; C.A. **84** [1976] No. 90954.

Sukhoveeva, I. B., Pushina, M. Ya., Shvetsova-Shitovskaya, K. D., Melmikov, N. N., N-Carbamoylated Phosphorylacetamides, U.S.S.R. 461629 [1973/75]; C.A. **84** [1976] No. 90298.

Naka, R., Narahara, T., Mukai, J., Hitachi, Ltd., Polyurethane Foam Composition, Ger. Offen. 2600183 [1975/76]; C.A. **85** [1976] No. 125079.

Kayama, I., Kawamura, K., Kawakami, H., Iwasaki, K., Sawachika, Y., Kokuko Chemical Industry Co., Recovery of Polyols from Polyurethane Wastes, Japan. Kokai 76-06909 [1974/76]; C.A. **84** [1976] No. 165849.

Kume, S., Kuribayashi, A., Motohashi, S., Dainippon Ink and Chemicals, Inc., Oxazolines and Oxazines, Japan. Kokai 76-34155 [1974/76]; C.A. **85** [1976] No. 177394.

Seki, M., Narahara, T., Hitachi, Ltd., Catalysts for Foam Production, Japan. Kokai 76-54695 [1974/76]; C.A. **85** [1976] No. 178504.

Seki, M., Narahara, T., Hitachi, Ltd., Catalysts for Foam Production, Japan. Kokai 76-54698 [1974/76]; C.A. **85** [1976] No. 193596.

Groves, J. D., Minnesota Mining and Mfg. Co., Organotin Catalyst System for Isocyanate Reactions to Manufacture Dielectric Urethane Material, Ger. Offen. 2721492 [1976/77]; C.A. **88** [1978] No. 38827.

Suzuki, K., Saiki, N., Teijin, Ltd., High-Molecular-Weight Polyester Elastomers with Good Whitness, Japan. Kokai 77-06796 [1975/77]; C.A. **86** [1977] No. 156806.

Maeda, S., Kondo, A., Nishimura, S., Abe, Y., Hodogaya Chemical Co., Ltd., Catalysts for Polymerization of Dimethyl Terephthalate and p-Xylylene Glycol, Japan. Kokai 77-23196 [1975/77]; C.A. **86** [1977] No. 172161.

Vogt, H. C., Parekh, M., Patton, J. T., BASF Wyandotte Corp., Tin-Titanium Complexes as Esterification/Transesterification Catalysts, U.S. 4081708 [1975/77]; C.A. **86** [1977] No. 178187.

Narayan, L., Cenker, M., BASF Wyandotte Corp., Carbodiimide-Isocyanurate Foams Containing Urethane Linkages, U.S. 4029610 [1974/77]; C.A. **87** [1977] No. 85754.

Nishikaji, T., Watanabe, K., Sawayama, S., Mitsubishi Chemical Industries Co., Ltd., Dialkylaminoethyl Acrylates and Methacrylates, Japan. Kokai 78-34714 [1976/78]; C.A. **89** [1978] No. 44410.

Tokiyama, Y., Nishinohara, M., Matsunaga, N., Japan Ester Co., Ltd., Manufacture of Polyesters, Japan. Kokai Tokkyo Koho 79-163996 [1978/79]; C.A. **92** [1980] No. 147532.

Schuermann, H., Van Aalten, H. A. A., Bung, J., AKZO N. V., Sizing Agent for Paper, Austrian 356510 [1974/80]; C.A. **93** [1980] No. 74253.

Cenker, M., Narayan, T., BASF Wyandotte Corp., Alumina Trihydrate as Flame Retardant Agent for Urethane-Modified Carbodiimide-Isocyanurate Foams, Eur. Appl. 8589 [1978/80]; C.A. **93** [1980] No. 47776.

Patton, T. L., Exxon Research and Engineering Co., Cyanoformamidyl Isocyanate, Ger. 2066059 [1969/80]; C.A. **94** [1981] No. 30396.

Huebel, W., Kriebel, G., Chemische Werke Huels A.-G., Polyurethane Cellular Plastic, Ger. Offen. 2831026 [1978/80]; C.A. **93** [1980] No. 48062.

Tokiyama, Y., Nishinohara, M., Matsunaga, N., Japan Ester Co., Ltd., Continuous Esterification of Terephthalic Acid, Japan. Kokai Tokkyo Koho 80-15423 [1978/80]; C.A. **93** [1980] No. 150009.

Lion Akzo K. K., Surface Sizing of Paper, Japan. Kokai Tokkyo Koho 80-90699 [1978/80]; C.A. **94** [1981] No. 5118.

Schimmel, K. F., Sturni, L. C., Robles, M. J., PPG Industries, Inc., Pigment Grinding Vehicle, U.S. 4186124 [1975/80]; C.A. **93** [1980] No. 27821.

Narayan, T., BASF Wyandotte Corp., Liquid Carbodiimide-Modified Organic Polyisocyanates Employing Organotin Catalysts, U.S. 4228095 [1979/80]; C.A. **94** [1981] No. 16335.

Woo, J. T., SCM Corp., Polyurethane Electrocoating Composition, U.S. 4241200 [1979/80]; C.A. **94** [1981] No. 193779.

Dainippon Ink and Chemicals, Inc., Blocked Isocyanate Compositions, Japan. Kokai Tokkyo Koho 82-42718 [1980/82]; C.A. **97** [1982] No. 57246.

Dainippon Ink and Chemicals, Inc., Thermosetting Resin Compositions, Japan. Kokai Tokkyo Koho 82-42719 [1980/82]; C.A. **97** [1982] No. 74036.

Toray Industries, Inc., Presensitized Plates for Waterless Lithographic Plates, Japan. Kokai Tokkyo Koho 57129442 [1981/82]; C.A. **99** [1983] No. 149588.

Dainippon Ink and Chemicals, Inc., Blocked Polyisocyanates, Japan. Kokai Tokkyo Koho 57139054 [1981/82]; C.A. **98** [1983] No. 108950.

Dainippon Ink and Chemicals, Inc., Water-Thinned Thermosetting Coating Compositions, Japan. Kokai Tokkyo Koho 57 143366 [1981/82]; C.A. **98** [1983] No. 55744.

Dainippon Ink and Chemicals, Inc., Thermosetting Polyurethane Coating Compositions, Japan. Kokai Tokkyo Koho 57145161 [1981/82]; C.A. **98** [1983] No. 127816.

Dainippon Ink and Chemicals, Inc., Polyurethane Emulsions for Coatings, Japan. Kokai Tokkyo Koho 57 159858 [1981/82]; C.A. **98** [1983] No. 91172.

Singh, H., Hutt, J. E., Williams, M. E., Products Research and Chemical Corp., Linear Liquid Polythioethers, PCT Intern. Appl. WO 82-01879 [1980/82]; C.A. **97** [1982] No. 164711.

Huemmer, T. F., Lakshmanan, P. R., Gulf Oil Corp., Photopolymerizable Acrylate Monomers, U.S. 4309561 [1980/82]; C.A. **96** [1982] No. 86129.

Rizk, S. D., Hsieh, H. W. S., Prendergast, J. J., Essex Chemical Corp., Silicon-Terminated Polyurethane Polymer, U.S. 4345053 [1981/82]; C.A. **97** [1982] No. 184132.

Dainippon Ink and Chemicals, Inc., Scratch-Resistant Glossy Coatings, Japan. Kokai Tokkyo Koho 58101121 [1981/83]; C.A. **100** [1984] No. 8601.

Dainippon Ink and Chemicals, Inc., Emulsifiers, Japan. Kokai Tokkyo Koho 58128135 [1982/83]; C.A. **99** [1984] No. 214303.

Dainippon Ink and Chemicals, Inc., Thermosetting Polyurethane Emulsion Coatings, Japan. Kokai Tokkyo Koho 58132051 [1982/83]; C.A. **100** [1984] No. 35914.

Russell, G. C., McBridge, P., Kodak Ltd.; Eastman Kodak Co., Solid Blocked Crosslinking Agents Based on 1,4-Cyclohexanebis(methyl isocyanate), PCT Intern. Appl. WO 83-00328 [1981/83]; C.A. **98** [1983] No. 162558.

Calundann, G. W., East, A. J., Celanese Corp., Anisotropic Melt Phase Forming Poly(ester-carbonate) Derived from 6-Hydroxy-2-napthoic Acid, Aromatic Diol, Organic Compound Capable of Forming a Carbonate Linkage, and Optionally, Other Aromatic Hydroxy-acid and Carbocycle Dicarboxylic Acid, U.S. 4371660 [1981/83]; C.A. **98** [1983] No. 144967.

Oriel, S. L., Flowers, J. D., Dow Chemical Co., Moisture Curing of Polymer Coating Compositions Containing Isocyanatoalkyl Esters of Unsaturated Carboxylic Acids, U.S. 4401794 [1980/83]; C.A. **100** [1984] No. 8572.

Ueno, R., Tsuchiya, H., Itoh, S., Yamamoto, I., Kabushiki Kaisha Ueno Seiyaku Oyo Kenkyujo, Benzyl Esters of Aromatic Hydroxy Carboxylic Acids, Eur. Appl. 117502 [1983/84]; C.A. **102** [1985] No. 45629.

Burchner, I., Szabo, L., Szathmari, F., Budalakk Festek es Mugyantagyar, Polyester Resins, Hung. 31759 [1982/84]; C.A. **101** [1984] No. 132610.

Dainippon Ink and Chemicals, Inc., Polyester Decorative Laminates with Improved Hardness, Japan. Kokai Tokkyo Koho 59124851 [1983]; C.A. **101** [1984] No. 212833.

Blem, A. R., McGuinness, J. A., Uniroyal, Inc., Defoliating Composition Containing Substituted Thiadiazole Ureas, Eur. Appl. 145329 [1983/85]; C.A. **103** [1985] No. 155826.

Wick, G., AKZO G.m.b.H., Packing Material and Its Use, Ger. Offen. 3411361 [1984/85]; C.A. **104** [1986] No. 69724.

Lange, A., Parge, A., Wuerzer, B., BASF A.-G., Thiazolyureas and Their Use in Combatting Unwanted Vegetation, Ger. Offen. 3413755 [1984/85]; C.A. **104** [1986] No. 68849.

Toyobo Co., Ltd., Discoloration-Resistant Polyurethane Fibers, Japan. Kokai Tokkyo Koho 6036560 [1983/85]; C.A. **103** [1985] No. 23751.

Saito, M., Arimatsu, G., Tani, K., Mitamura, H., Katsuo, K., Toyobo Co., Ltd., Discoloration-Resistant Spandex Fibers, Japan. Kokai Tokkyo Koho 60155262 [1984]; C.A. **104** [1986] No. 7173.

Johnson, M. R., Dow Chemical Co., Polymerizable UV Light Stabilizer from Isocyanatoalkyl Esters of Unsaturated Carboxylic Acids, U.S. 4504628 [1982/85]; C.A. **102** [1985] No. 221738.

Stabilizer for PVC or Other Polymers

Quattlebaum, W. M., Young, D. M., Carbide and Carbon Chemicals Corp., Vinyl Resin Compositions Suitable for Coating Paper, etc., U.S. 2307075 [1943]; C.A. **1943** 3553.

Yngve, V., Carbide and Carbon Chemicals Corp., Vinyl Resin Compositions, U.S. 2307092 [1943]; C.A. **1943** 3532.

Rugeley, E. W., Quattlebaum, W. M., Carbide and Carbon Chemicals Corp., Heat Stabilizers for Vinyl Resin Filaments, U.S. 2344002 [1944]; C.A. **1944** 3392.

Eberly, K. C., Firestone Tire and Rubber Co., Alkyltin Carboxylates, U.S. 2560034 [1951]; C.A. **1952** 7538.

Harding, J., Union Carbide & Carbon Corp., Heat and Light-Stabilized Vinyl Resins, U.S. 2597987 [1952]; C.A. **1952** 8893.

Weinberg, E. L., Tomka, L. A., Metal & Thermit Corp., Antioxydant for Rubber, U.S. 2789107 [1957]; C.A. **1957** 10939.

Lutz, W. D., Metallgesellschaft A.-G., Stabilization of Poly(vinyl chloride), Brit. 874574 [1960/61]; C.A. **57** [1962] 1080.

Cohen, S., Brecker, L. R., Thee, A., Argus Chemical N. V., Stabilizers for Poly(vinyl chloride) Resins, Ger. Offen. 1934188 [1968/70]; C.A. **72** [1970] No. 79853.

Nakanome, I., Takeya, K., Suzuki, H., Japan Exlan Co., Ltd., Color-Stabilized Compositions from Halogen-Containing Acrylonitrile Polymers, Ger. Offen. 1946330 [1968/70]; C.A. **73** [1970] No. 121473.

Enoki, K., Yoshihara, Y., Kirai, Y., Japan Soda Co., Ltd., Stabilization of Ethylene Terephthalate Polyesters, Ger. Offen. 1954959 [1968/70]; C.A. **73** [1970] No. 26275.

Enoki, Y., Kiki, Y., Japan Soda Co., Ltd., Stabilization of Polyesters, Japan. 71-32067 [1968/71]; C.A. **77** [1972] No. 49557.

Kunststoffwerk Karl Egger, Polymer Foams, Neth. Appl. 8101923 [1980/81]; C.A. **96** [1982] No. 53490.

Asahi-Dow Ltd., Stabilized Methylchloroform Compositions, Japan. Kokai Tokkyo Koho 82-116021 [1981/82]; C.A. **97** [1982] No. 21515.

Catalysts for the Condensation of Silicones and Other Organosilicon Materials

Braley, O. A., Dow Corning Corp., Organopolysiloxane Compositions, U.S. 2504388 [1950]; C.A. **1950** 6196.

Rossiter, W. T., Currie, C. C., Dow Corning Ltd., Organosilicon Compositions, Brit. 713233 [1952/54]; C. **1955** 9922.

Rossiter, W. T., Currie, C. C., Dow Corning Corp., Stabilized, Corrosion-Inhibiting Polymeric Organosilicon Compositions, U.S. 2742368 [1956]; C.A. **1956** 11713.

Polmanteer, K. E., Dow Corning Corp., Siloxane Elastomers Curing at Room Temperature, Ger. 1019462 [1957]; C.A. **1960** 10378.

Simmler, W., Merten, R., Farbenfabriken Bayer A.-G., Stannous Siloxanes for Use as Lubricants, Ger. 1099743 [1961]; C.A. **1961** 27875.

Dathe, C., Stable Silicone Adhesives, Ger. [East] 36060 [1962/65]; C.A. **63** [1965] 15066.

Nelson, M. E., Dow Corning Corp., Treatment of Glass Fibers, Fr. 1577541 [1967/69]; C.A. **72** [1970] No. 91445.

Wessel, J. K., Dow Corning Corp., Room Temperature Hardenable Organopolysiloxane Elastomer, Ger. Offen. 1907980 [1968/69]; C.A. **72** [1970] No. 13621.

Kusuhara, M., Oki, D., Tokyo Shibaura Electric Co., Water-Repellent Treatment of Fibers and Paper, Japan. 69-32035 [1967/69]; C.A. **72** [1970] No. 112749.

Smith, R. L., General Electric Co., Room-Temperature-Vulcanizing Silicone Rubber Foam, U.S. 3516951 [1967/70]; C.A. **73** [1970] No. 36312.

Gibbon, R. M., Kilbride, W., Pierpoint, E. K., Imperial Chemical Industries, Ltd., Paper Release Coating Comprising Polysiloxanes, Organotin Compounds, and Organic Amines, U.S. 3527728 [1968/70].

Dyck, M., Cordis Corp., Nonthrombogenic Plastics for Blood-Handling Apparatus, U.S. 3549409 [1967/70]; C.A. **74** [1971] No. 54813.

Ross, M. E., Dow Corning Corp., Polysiloxane Compositions Hardenable at Room Temperature, Ger. Offen. 2106766 [1970/71]; C.A. **76** [1972] No. 47153.

McKellar, R. L., Dow Corning Corp., Organosilicon Compounds Containing Two Isocyanate Groups, Ger. 1568393 [1965/72]; C.A. **77** [1972] No. 140274.

Patterson, W. J., Morris, D. E., United States National Aeronautics and Space Administration, Polymerizable Disilanols Having Perfluoroalkylene Groups, U.S. 3763204 [1970/73]; C.A. **80** [1974] No. 28170.

Gibbon, R., Imperial Chemical Industries Ltd., Surface Treating Compositions, Brit. 1372061 [1971/74]; C.A. **82** [1975] No. 113430.

Nitzsche, S., Hittmair, P., Hechtl, W., Mittermeier, M., Wohlfarth, E., Wacker-Chemie GmbH, Rebound-Plastic Siloxanes, Ger. Offen. 2259802 [1972/74]; C.A. **81** [1974] No. 154803.

Creasey, N. G., Pike, L. C., Imperial Chemical Industries Ltd., Organopolysiloxane Coating Compositions, Ger. Offen. 2404399 [1973/74]; C.A. **82** [1975] No. 45267.

Schulz, J. R., Dow Corning Corp., Silicone Rubber Compositions, Ger. Offen. 2407290 [1973/74]; C.A. **82** [1975] No. 74210.

Kim, Y. K., Dow Corning Corp., Imidoorganosilicon Compounds, U.S. 3901913 [1974/75]; C.A. **84** [1976] No. 61417.

Knox, W. B., Imperial Chemical Industries Ltd., Elastomer-Forming Compositions, Brit. 1401804 [1972/75]; C.A. **83** [1975] No. 207378.

Ichikawa, K., Mie, T., Ichihara, C., Ona, I., Toray Silicone Co., Ltd., Silicone Composition, Ger. Offen. 2720457 [1976/77]; C.A. **88** [1978] No. 38825.

Fukuoka, M., Kokan, S., Sumitomo Bakelite Co., Ltd., Vulcanization of Ethylene-Propene-Diene Rubber, Japan. Kokai 77-09054 [1975/77]; C.A. **86** [1977] No. 141425.

Mine, K., Toray Silicone Co., Ltd., Vulcanization of Silicone Rubber at Room Temperature, Japan. Kokai 78-102956 [1977/78]; C.A. **89** [1978] No. 216631.

Beers, M. D., General Electric Co., Silicone Composition Settable by Humidity, Belg. 877267 [1978/79]; C.A. **92** [1980] No. 60173.

Okada, F., Kishita, H., Sato, N., Shin-Etsu Chemical Co., Ltd., Curable Organopolysiloxane Compositions, Brit. Appl. 2005288 [1977/79]; C.A. **92** [1980] No. 7943.

Toray Silicone Co., Ltd., Silicone Resin Product, Neth. Appl. 79-08004 [1979/81]; C.A. **95** [1981] No. 134528.

Franz, H., Vanek, J. C., PPG Industries, Inc., Coupling Agent for Bonding an Organic Polymer to an Inorganic Surface, U.S. 4154638 [1977/79]; C.A. **91** [1979] No. 58380.

Inmont Corp., Vulcanizable Composition of a Polyurethane with Silicon-Containing End Groups Having a Higher Setting Speed, Belg. 883685 [1980]; C.A. **94** [1981] No. 67031.

West, G. C., Polfus, W. F., Reeves Brothers, Inc., Waterproofing Textile Materials, Fr. Demande 2436799 [1978/80]; C.A. **93** [1980] No. 169627.

Hagen, P., Jonas, R., Wienecke, M., Perennatorwerk Alfred Hagen GmbH, Flame-Protected Room Temperature-Hardenable Silicone Compositions, Ger. Offen. 2909462 [1979/80]; C.A. **93** [1980] No. 205958.

Innertsberger, E., Sommer, O., Bosch, E., Mueller, J., Schiller, A., Wacker Chemie GmbH, Molding Compositions Crosslinkable to Elastomers at Room Temperature, Ger. 2911301 [1979/80]; C.A. **93** [1980] No. 96552.

Fujita, T., Iwamoto, M., Toray Industries, Inc., Presensitized Negative Lithographic Printing Plate, Ger. Offen. 2943379 [1978/80]; C.A. **95** [1981] No. 15986.

Ito, H., Noro, J., Ito, K., Imai, S., Toyota Motor Co., Ltd., Coating of Wiper Blades for Automobiles, Japan. Kokai Tokkyo Koho 80-15873 [1978/80]; C.A. **93** [1980] No. 9364.

Mikami, R., Hanada, T., Toray Silicone Co., Ltd., Silicone Resin Compositions, Japan. Kokai Tokkyo Koho 80-31806 [1978/80]; C.A. **93** [1980] No. 27900.

Mikami, R., Toray Silicone Co., Ltd., Room-Temperature-Curable Siloxane Release Coating Materials, Japan. Kokai Tokkyo Koho 80-48245 [1978/80]; C.A. **93** [1980] No. 74028.

Toray Industries, Inc., Waterless Lithographic Plates, Japan. Kokai Tokkyo Koho 80-70847 [1978/80]; C.A. **93** [1980] No. 195517.

Toray Industries, Inc., Correction of Defective Waterless Lithographic Plates, Japan. Kokai Tokkyo Koho 80-89842 [1978/80]; C.A. **94** [1981] No. 93653.

Toray Industries, Inc., Waterless Lithographic Plates, Japan. Kokai Tokkyo Koho 80-124149 [1979/80]; C.A. **94** [1981] No. 165684.

Toshiba Silicone Co., Ltd., Moisture-Vulcanizable Silicone Rubber Sealing Compositions, Japan. Kokai Tokkyo Koho 80-133452 [1979/80]; C.A. **94** [1981] No. 85455.

Toray Silicone Co., Ltd., Finishing Compositions for Textiles, Japan. Kokai Tokkyo Koho 80-137275 [1979/80]; C.A. **94** [1981] No. 67250.

Hitachi, Ltd., Surface Stabilization of Semiconductor Devices, Japan. Kokai Tokkyo Koho 80-156330 [1979/80]; C.A. **94** [1981] No. 184422.

Toray Silicone Co., Ltd., Polymer Composition Containing Silicone Resin, Neth. Appl. 80-00303 [1980/81]; C.A. **95** [1981] No. 188784.

Bryant, E. R., Knittel, G. H., Dircks, L. E., Inmont Corp., Vulcanizable Silicon-Terminated Polyurethane Polymer Compositions Having Improved Cure Speed, U.S. 4222925 [1978/80]; C.A. **93** [1980] No. 241371.

Scott, M. A., Brown, A. E., Columbia Ribbon and Carbon Mfg. Co., Inc., Transfer Ribbons and Process for Producing Same, Brit. Appl. 2053305 [1979/81]; C.A. **95** [1981] No. 26307.

Kinashi, T., Fujita, T., Kawabe, N., Toray Industries, Inc., Dry Planographic Printing Plate and Preparation Thereof, Brit. Appl. 2064803 [1979/81]; C.A. **96** [1982] No. 113495.

Mikami, R., Toray Silicone Co., Ltd., Room Temperature Curing Silicone Resin Compositions, Brit. Appl. 2067212 [1980/81]; C.A. **96** [1982] No. 70108.

Sattlegger, H., Schnurrbusch, K., Degen, B., Achtenberg, T., Bayer A.-G., Polysiloxane Molding Compositions and Process for Their Preparation, Eur. Appl. 22976 [1979/81]; C.A. **94** [1981] No. 140980.

Alberts, H., Friemann, H., Sattlegger, H., Moretto, H., Bayer A.-G., Stable Graft Copolymer Dispersions, Eur. Appl. 29948 [1979/81]; C.A. **95** [1981] No. 117248.

Mikami, R., Toray Silicone Co., Ltd., Silicone Resin Compositions Hardenable at Room Temperature and Articles Coated with a Hardened Film of These Compositions, Fr. Demande 2474518 [1980/81]; C.A. **96** [1982] No. 8280.

Alberts, H., Moretto, H. H., Sattlegger, H., Bayer A.-G., Graft Polymer Dispersions, Ger. Offen. 2947963 [1979/81]; C.A. **95** [1981] No. 44926.

Friemann, H., Moretto, H. H., Alberts, H., De Montigny, A., Toepsch, H., Bayer A.-G., Siloxane Release Coating, Ger. Offen. 2947965 [1979/81]; C.A. **95** [1981] No. 82558.

Smith, A. H., Dziark, J. J., General Electric Co., Stable Catalyst Composition for Preparing Room Temperature Vulcanizable Silicone Rubber Compositions, Ger. Offen. 3022978 [1979/81]; C.A. **94** [1981] No. 158160.

Beers, M. D., General Electric Co., Hardenable Preparations and Their Use, Ger. Offen. 2925443 [1979/81]; C.A. **94** [1981] No. 193498.

Toray Industries, Inc., Presensitized Waterless Lithographic Plates, Japan. Tokkyo Koho 81-23150 [1973/81]; C.A. **96** [1982] No. 133209.

Toshiba Silicone Co., Ltd., Fungicides for Moisture-Curable Silicone Rubber Sealants, Japan. Kokai Tokkyo Koho 81-38348 [1979/81]; C.A. **95** [1981] No. 63899.

Mitsui Real Estate Development Co., Ltd., Yushiro Chemical Industry Co., Ltd.; Toray Silicone Co., Ltd., Anticorrosive Coating Materials, Japan. Kokai Tokkyo Koho 81-65055 [1979/81]; C.A. **95** [1981] No. 134513.

Toray Industries, Inc., Waterless Lithographic Plates, Japan. Kokai Tokkyo Koho 81-109788 [1980/81]; C.A. **96** [1982] No. 133211.

Beers, M. D., General Electric Co., Curable Compositions, U.S. 4257932 [1978/81]; C.A. **95** [1981] No. 8998.

Reusser, R. E., Phillips Petroleum Co., Polysilicones-Polybutadienediol Blends as Release Coatings, U.S. 4261876 [1978/81]; C.A. **95** [1981] No. 8977.

Homan, G. R., Lee, Chi-Long, Dow Corning Corp., Compositions Including Mercaptoorganopolysiloxanes and Stannic Salts of Carboxylic Acids, U.S. 4265792 [1979/81]; C.A. **95** [1981] No. 170753.

Hanada, T., Mikami, R., Toray Silicone Co., Ltd., Room-Temperature-Curable Silicon Resins, U.S. 4267297 [1979/81]; C.A. **95** [1981] No. 8993.

Suezawa, M., Asano, M., Toray Industries, Inc., Developer for Dry Planographic Printing Plates, Eur. Appl. 43132 [1980/82]; C.A. **96** [1982] No. 133232.

Makino, F., Shimokawa, Y., Toray Industries, Inc., Planographic Printing Plate for Direct Printing, Eur. Appl. 44220 [1980/82]; C.A. **96** [1982] No. 172182.

Alberts, H., Friemann, H., Moretto, H. H., Sattlegger, H., Mietzsch, F., Bayer A.-G., Room Temperature Crosslinkable Graft Polymer Dispersion, Eur. Appl. [1980/82]; C.A. **96** [1982] No. 201396.

Sattlegger, H., Schnurrbusch, K., Achtenberg, T., Bayer A.-G., Polysiloxane Molding Compositions, Ger. Offen. 3104645 [1981/82]; C.A. **97** [1982] No. 128926.

Alberts, H., Friemann, H., Moretto, H. H., Sattlegger, H., Bayer A.-G., Stable Organic Multicomponent Dispersions, Ger. Offen. 3105499 [1981/82]; C.A. **97** [1982] No. 184131.

Sattlegger, H., Schnurrbusch, K., Metzinger, H. G., Bayer A.-G., Organopolysiloxane Molding Composition Storable in the Absence of Water, Ger. Offen. 3114773 [1981/82]; C.A. **98** [1983] No. 74022.

Toray Industries, Inc., Photosensitive Silicones for Printing Plates, Japan. Kokai Tokkyo Koho 82-08243 [1980/82]; C.A. **97** [1982] No. 118228.

Toray Industries, Inc., Correction of Waterless Lithographic Plates, Japan. Kokai Tokkyo Koho 82-13447 [1980/82]; C.A. **97** [1982] No. 101709.

Toray Industries, Inc., Waterless Lithographic Plates, Japan. Kokai Tokkyo Koho 82-31596 [1981/82]; C.A. **97** [1982] No. 118231.

Toshiba Silicone Co., Ltd., Self-Bonding Room-Temperature-Curable Siloxane, Japan. Kokai Tokkyo Koho 57108880 [1980/82]; C.A. **98** [1983] No. 5209.

Toray Industries, Inc., Electrophotographic Fabrication of High-Resolution Lithographic Plates, Japan. Kokai Tokkyo Koho 57178893 [1981/82]; C.A. **100** [1984] No. 77403.

Toray Industries, Inc., Fixing of Negative Type Waterless Lithographic Plates, Japan. Kokai Tokkyo Koho 57192956 [1981/82]; C.A. **100** [1984] No. 112270.

Toray Industries, Inc., Lithographic Plates, Japan. Kokai Tokkyo Koho 57192957 [1981/82]; C.A. **100** [1984] No. 13555.

Toray Industries, Inc., Selective Permeable Membranes, Japan. Kokai Tokkyo Koho 57194004 [1981/82]; C.A. **98** [1983] No. 162070.

Shin-Etsu Chemical Industry Co., Ltd., Siloxane-Based Release Coating Materials, Japan. Kokai Tokkyo Koho 57207646 [1981/82]; C.A. **98** [1983] No. 217318.

Asano, M., Mori, Y., Takayama, Y., Toray Industries, Inc., Dry Planographic Printing with Organic Acid Additive, U.S. 4347303 [1978/82]; C.A. **97** [1982] No. 191262.

Koda, Y., Ona, I., Takeda, A., Toray Silicone Co., Inc., Cyclofluorsilicone-Containing Compositions for the Treatment of Fibers, U.S. 4355149 [1981/82]; C.A. **98** [1983] No. 5431.

Counsell, P. J. C., Evode, Ltd., Metallic Cladding of Concrete and Other Structures, Brit. Appl. 2107211 [1981/83]; C.A. **99** [1983] No. 57976.

Bosch, E., Craubner, I., Dorsch, N., Schiller, A., Sommer, O., Wacker-Chemie GmbH, Silicone Rubber with Delayed Skin Formation, Ger. Offen. 3133564 [1981/83]; C.A. **98** [1983] No. 180895.

Schiller, A., Dorsch, N., Bosch, E., Sommer, O., Wacker-Chemie GmbH, Moisture-Curable Rubber Stable in the Absence of Moisture, Ger. Offen. 3212008 [1982/83]; C.A. **100** [1984] No. 23447.

Beers, M. D., General Electric Co., Cold-Curing Single-Component Silicone Rubber Compositions of a Low Modulus, Ger. Offen. 3305356 [1982/83]; C.A. **99** [1983] No. 213958.

Dainippon Ink and Chemicals, Inc., Room-Temperature-Curable Coating Materials, Japan. Kokai Tokkyo Koho 5838766 [1981/83]; C.A. **99** [1983] No. 124166.

Yokohama Rubber Co., Ltd., Mold Releasing Agents, Japan. Kokai Tokkyo Koho 5845298 [1981/83]; C.A. **99** [1983] No. 142881.

Toray Industries, Inc., Photosensitive Printing Plate, Japan. Kokai Tokkyo Koho 58102234 [1981/83]; C.A. **99** [1983] No. 96867.

Toray Industries, Inc., Retouching Solution for a Waterless Lithographic Plate, Japan. Kokai Tokkyo Koho 58102235 [1981/83]; C.A. **99** [1983] No. 80083.

Dainippon Ink and Chemicals, Inc., Resin Compositions Curable at Room Temperature, Japan. Kokai Tokkyo Koho 58173158 [1982/83]; C.A. **101** [1984] No. 8799.

Seiren Co., Ltd., Waterproofing of Synthetic Fabrics, Japan. Kokai Tokkyo Koho 58208473 [1982/83]; C.A. **100** [1984] No. 140786.

Dziark, J. J., General Electric Co., Scavenger for One-Component Alkoxyfunctional RTV Compositions and Processes, PCT Intern. Appl. WO 83-02948 [1982/83]; C.A. **100** [1984] No. 8241.

Lucas, G. M., Dziark, J. J., General Electric Co., Adhesion Promoters for One-Component RTV Silicone Compositions, S. African 83-02242 [1983]; C.A. **101** [1984] No. 39994.

Chung, R. H., Beers, M. D., Swiger, R. T., General Electric Co., Scavenger for One-Component RTV Compositions, S. African 83-07984 [1983/84]; C.A. **101** [1984] No. 232060.

Dziark, J. J., Shinohara, K., General Electric Co., Silicone Products Division, Self-Bonding One-Component Dimedone RTV Silicone Rubber Compositions, U.S. 4395507 [1981/83]; C.A. **99** [1983] No. 141391.

Penn, H. I., Morton Thiokol, Inc., Adhesives Containing Reaction Products of γ-Isocyanatopropyltriethoxysilane and Thermoplastic Polyesters, U.S. 4408021 [1981/83]; C.A. **99** [1983] No. 213831.

Kitamura, K., Suizawa, M., Toray Industries, Inc., Coating Compositions and Its Use, Eur. Appl. 126192 [1984]; C.A. **102** [1985] No. 133693.

Lucas, G. M., Dziark, J. J., General Electric Co., Stable Single Component Organopolysiloxane Preparation Vulcanizable at Room Temperature, Ger. Offen. 3313451 [1983/84]; C.A. **102** [1985] No. 80083.

Hong-Son, R., General Electric Co., Silicone-Polyimide Copolymers and Condensation Vulcanizable Compositions Containing Them, Ger. Offen. 3341700 [1982/84]; C.A. **101** [1984] No. 132223.

Lucas, G. M., General Electric Co., Single-Component Silicone Rubber Compositions Vulcanizable at Room Temperature, Ger. Offen. 3411691 [1983/84]; C.A. **102** [1985] No. 47184.

White, M. A., Hallgren, J. E., Lockhart, T. P., General Electric, Stable, Essentially Water-Free Single-Component Organopolysiloxane Composition Vulcanizable at Room Temperature, Ger. Offen. 3411703 [1983/84]; C.A. **102** [1985] No. 47185.

Lockhart, T. P., General Electric Co., Stable, Moisture-Hardenable, Single Component, Alkoxy End Group-Containing Organopolysiloxane Compositions, Ger. Offen. 3411716 [1983/84]; C.A. **102** [1985] No. 63441.

Chung, R. H., General Electric Co., Crosslinking Agent for Room Temperature Vulcanizable Compositions, Ger. Offen. 3416700 [1983/84]; C.A. **102** [1985] No. 96781.

Chung, R. H., Chung, B., Lucas, G. M., General Electric Co., Transparent and Colorless Compositions Vulcanizable at Room Temperature, Ger. Offen. 3416851 [1983/84]; C.A. **102** [1985] No. 96790.

Toray Industries, Inc., Waterless Lithographic Plates, Japan. Kokai Tokkyo Koho 5915248 [1982/84]; C.A. **101** [1984] No. 101249.

Cemedine Co., Ltd., Room-Temperature-Curable Sealent Compositions, Japan. Kokai Tokkyo Koho 5924771 [1982/84]; C.A. **101** [1984] No. 112573.

Toray Industries, Inc., Dampening-Free Lithographic Plate, Japan. Kokai Tokkyo Koho 5948768 [1982/84]; C.A. **101** [1984] No. 238195.

General Electric Co., Manufacture of Room-Temperature-Curable Silicone Rubbers in a Devolatilizing Extruder, Japan. Kokai Tokkyo Koho 5996163 [1982/84]; C.A. **101** [1984] No. 212495.

General Electric Co., Ore-Package Silicone Rubber Compositions Curable at Room Temperature, Japan. Kokai Tokkyo Koho 59109557 [1982/84]; C.A. **101** [1984] No. 212496.

Toshiba Silicone Co., Ltd., Room-Temperature-Curable Rubbers, Japan. Kokai Tokkyo Koho 59126459 [1983/84]; C.A. **101** [1984] No. 193486.

Toray Industries, Inc., Waterless Lithographic Printing Plates, Japan. Kokai Tokkyo Koho 59135471 [1983/84]; C.A. **102** [1985] No. 36791.

Institute of Physical and Chemical Research, Biocompatible Plastic Film, Japan. Kokai Tokkyo Koho 59155432 [1983/84]; C.A. **102** [1985] No. 67427.

General Electric Co., Silicone Rubber Compositions Curable at Room Temperature, Japan. Kokai Tokkyo Koho 59155463 [1983/84]; C.A. **102** [1985] No. 47186.

Sasaki, S., Kaiya, N., Toray Silicone Co., Inc., Composition for Forming Releasable Film, PCT Intern. Appl. WO 84-00034 [1982/84]; C.A. **100** [1984] No. 158320.

Penn, H. J., Morton Thiokol, Inc., Adhesive for Solar Control Film, U.S. 4429005 [1981/84]; C.A. **100** [1984] No. 157940.

Ashby, B. A., General Electric Co., Composition for Promoting Adhesion of Curable Silicones to Substrates, U.S. 4460739 [1983/84]; C.A. **101** [1984] No. 172860.

Lockhart, T. P., General Electric Co., One-Package, Stable, Moisture-Curable, Alkoxy-Terminated Organopolysiloxane Rubber Compositions, U.S. 4467063 [1983/84]; C.A. **101** [1984] No. 153326.

Ryang, H. S., General Electric Co., Vulcanizable Silicone-Polyimides, U.S. 4472565 [1982/84]; C.A. **102** [1985] No. 8055.

Lucas, G. M., Dziark, J. J., General Electric Co., Adhesion Promotors for One-Component Room Temperature Vulcanizable Silicone Compositions, U.S. 4483973 [1982/84]; C.A. **102** [1985] No. 80080.

Chung, R. H., General Electric Co., Silane Scavengers for Hydroxy Radicals Containing Silicon-Hydrogen Bonds, U.S. 4489191 [1983/84]; C.A. **102** [1985] No. 114956.

Dziark, J. J., General Electric Co., One-Component RTV Silicone Rubber Compositions with Good Self-Bonding Properties to Acrylate Plastics, U.S. 4489200 [1983/84]; C.A. **102** [1985] No. 96784.

Bryant, E. R., Knittel, G. H., Dircks, L. E., Inmont Corp., Vulcanizable Sealant Compositions, Austrian Specif. 542856 [1980/85]; C.A. **103** [1985] No. 125207.

Beers, M. D., General Electric Co., Low Modulus One Component Room Temperature Vulcanizable Compositions, Can. 1195033 [1983/85]; C.A. **104** [1986] No. 52192.

Blizzard, J. D., Vanwert, B., Dow Corning Corp., Bonding Surfaces With Permanent-Bond Adhesive, Eur. Appl. 131854 [1983/85]; C.A. **102** [1985] No. 96620.

Gutek, B. I., Vanwert, B., Dow Corning Corp., Flexible Silicone Resin-Coated Fabric, Eur. Appl. 131870 [1983/85]; C.A. **103** [1985] No. 55375.

Redman, R. P., Imperial Chemical Industries PLC; Dulux Australia Ltd., Coating Compositions Useful in an Electrodeposition Process, Eur. Appl. 137603 [1983/85]; C.A. **103** [1985] No. 38753.

Gaa, P. C., Dana, D. E., PPG Industries, Inc., Aqueous Dispersion, Internally Silylated and Dispersed Polyurethane Resins, and Surfaces Containing Same, Eur. Appl. 163214 [1984/85]; C.A. **104** [1986] No. 187509.

Lucas, G. M., Dziark, J. J., General Electric Co., Composition of Organopolysiloxane Having Alkoxy Groups Partially Vulcanizable at Room Temperature, Fr. Demande 2557122 [1983/85]; C.A. **104** [1986] No. 131670.

Wuerminghausen, T., Saykowski, F., Sattlegger, H., Bayer A.-G., Polyorganosiloxane Compositions Hardenable to Elastomers at Room Temperature, Ger. Offen. 3323911 [1983/85]; C.A. **102** [1985] No. 133369.

Pyrlik, M., Remmers, G., Schill und Seilacher GmbH und Co., Single Component Polysiloxane Molding Compositions Storable in the Absence of Moisture and Cold-Curable in the Presence of Moisture, Ger. Offen. 3342026 [1983/85]; C.A. **103** [1985] No. 16175.

Pyrlik, M., Schamar, L., Huetker, H., Schill und Seilacher GmbH und Co., Cold-Curing Single-Component Siloxane Compositions and Their Use, Ger. Offen. 3342027 [1983/85]; C.A. **103** [1985] No. 143521.

Sproesser, U., Plass, H., Ivancic, J., Chemie Dr. Buechtemann GmbH und Co., Cold-Hardenable Permanently Elastic Single-Component Composition, Ger. Offen. 3439745 [1983/85]; C.A. **103** [1985] No. 143220.

Toray Industries, Inc., Waterless Printing Plates, Japan. Kokai Tokkyo Koho 6011845 [1983/85]; C.A. **102** [1985] No. 212731.

Toray Industries, Inc., Dampening Water-Free Lithographic Plates, Japan. Kokai Tokkyo Koho 6021050 [1983/85]; C.A. **103** [1985] No. 150994.

Toray Industries, Inc., Processing of Photoimaging Laminate Having Silicone Rubber Surface Layer, Japan. Kokai Tokkyo Koho 6028656 [1983/85]; C.A. **102** [1985] No. 212735.

Toray Industries, Inc., Preparation of Dampening-Free Lithographic Plate, Japan. Kokai Tokkyo Koho 6029750 [1983/85]; C.A. **103** [1985] No. 113377.

Dainippon Ink and Chemicals, Inc., Resin Compositions for Room-Temperature-Curable Coating Materials, Japan. Kokai Tokkyo Koho 6047076 [1983/85]; C.A. **103** [1985] No. 7820.

Dainippon Ink and Chemicals, Inc., Clear Topcoatings for Metallic Undercoatings, Japan. Kokai Tokkyo Koho 6048172 [1983/85]; C.A. **103** [1985] No. 7827.

Toray Industries, Inc., Photoimaging Laminates with Direct Print-Out Type Protective Coatings, Japan. Kokai Tokkyo Koho 6049933 [1983/85]; C.A. **102** [1985] No. 229504.

Dainippon Ink and Chemicals, Inc., Resin Compositions for Room-Temperature-Curable Coatings Having Good Appearance, Japan. Kokai Tokkyo Koho 6067573 [1983/85]; C.A. **103** [1985] No. 72719.

Toray Industries, Inc., Dampening-Free Lithographic Printing Plates, Japan. Kokai Tokkyo Koho 6078452 [1983/85]; C.A. **103** [1985] No. 186930.

Toray Industries, Inc., Preparation of Waterless Lithographic Plates, Japan. Kokai Tokkyo Koho 6080854 [1983/85]; C.A. **103** [1985] No. 79535.

Dainippon Ink and Chemicals, Inc., Mastic Coating Materials, Japan. Kokai Tokkyo Koho 60129168 [1983/85]; C.A. **104** [1986] No. 35644.

Toray Industries, Inc., Waterless Presensitized Lithographic Plates, Japan. Kokai Tokkyo Koho 60131536 [1983/85]; C.A. **104** [1986] No. 79258.

Kawabe, N., Tsuda, M., Asano, M., Toray Industries, Inc., Waterless Lithographic Plates, Japan. Kokai Tokkyo Koho 60153048 [1984/85]; C.A. **104** [1986] No. 59439.

Suezawa, M., Ikeda, N., Kinashi, T., Toray Industries, Inc., Dry Planographic Printing Inks, Japan. Kokai Tokkyo Koho 60158276 [1984/85]; C.A. **104** [1986] No. 90683.

Kuwamura, S., Ooka, M., Murakami, Y., Dainippon Ink and Chemicals, Inc., Room-Temperature-Curable Coating Compositions, Japan. Kokai Tokkyo Koho 60168769 [1984/85]; C.A. **104** [1986] No. 131613.

Ooka, M., Kuwamura, S., Murakami, Y., Dainippon Ink and Chemicals, Inc., Formation of Weathering- and Cracking-Resistant Coating on Roof Tiles, Japan. Kokai Tokkyo Koho 60171280 [1984/85]; C.A. **105** [1986] No. 10796.

Yamada, M., Sakai, K., Toray Industries, Inc., Wetting-Free Lithographic Plates, Japan. Kokai Tokkyo Koho 60188947 [1984/85]; C.A. **104** [1986] No. 234326.

Kitamura, K., Ito, T., Toray Industries, Inc., Inks for Preparation of Waterless Lithographic Plates, Japan. Kokai Tokkyo Koho 60196346 [1984/85]; C.A. **104** [1986] No. 139349.

Kuwamura, S., Oka, M., Danippon Ink and Chemicals, Inc., Hydrolyzable Silane-Containing, Room-Temperature-Curable Coatings, Japan. Kokai Tokkyo Koho 60206812 [1984/85]; C.A. **104** [1986] No. 188307.

Chung, R. H., General Electric Co., Scavengers for RTV Silicone Rubber Compositions, U.S. 4495331 [1983/85]; C.A. **102** [1985] No. 114957.

Lockhart, T. P., General Electric Co., One-Package, Stable, Moisture Curable, Alkoxy-Terminated Organopolysiloxane Compositions, U.S. 4499229 [1983/85]; C.A. **102** [1985] No. 133378.

Lockhart, T. P., General Electric Co., One-Package, Stable, Moisture-Curable Siloxanes Containing Zeolites, U.S. 4499230 [1983/85]; C.A. **102** [1985] No. 133380.

Ashby, B. A., Lucas, G. M., General Electric Co., Self-Bonding Room-Temperature-Vulcanizable Silicone Compositions, U.S. 4506058 [1984/85]; C.A. **102** [1985] No. 222005.

Chung, R. H., General Electric Co., End-Capping Catalysts for Forming Alkoxyfunctional One-Component RTV Compositions, U.S. 4515932 [1982/85]; C.A. **103** [1985] No. 143215.

Lockhart, T. P., Wengrovius, J. H., General Electric Co., Room-Temperature-Vulcanizable Organopolysiloxane Compositions, U.S. 4517337 [1984/85]; C.A. **103** [1985] No. 38553.

Swiger, R. T., Hallgren, J. E., General Electric Co., Scavengers for One-Component Alkoxy-Functional RTV Compositions, U.S. 4523001 [1983/85]; C.A. **103** [1985] No. 88912.

Ryang, H. S., General Electric Co., Silane Derivatives of Norbonanedicarboximides, U.S. 4533737 [1984/85]; C.A. **104** [1986] No. 111159.

Elias, R. C., PPG Industries, Inc., Coating for Reducing the Temperature Rise of Heat Sensitive Substrates, U.S. 4546045 [1984/85]; C.A. **104** [1986] No. 70367.

Kitamura, K., Suezawa, M., Toray Industries, Inc., Coating Composition and Its Use, U.S. 4551516 [1983/85]; C.A. **104** [1986] No. 90656.

Lucas, G. M., General Electric Co., Curable Silicone Compositions, U.S. 4555420 [1984/85]; C.A. **104** [1986] No. 150585.

Nakayama, N., Isoda, T., Watanabe, Y., Aoki, M., Ricoh Co., Ltd., Carrier Particles for a Two-Component Dry Developer, Ger. Offen. 3511171 [1984/85]; C.A. **104** [1986] No. 119923.

Kamaishi, T., Kitamura, K., Ito, T., Toray Industries, Inc., Imaging Film for Printing Plate Preparation, Japan. Kokai Tokkyo Koho 60192693 [1984/85]; C.A. **105** [1986] No. 88704.

Chemical Vapor Deposition of SnO_2 Layers

Mukaiyama, T., Furuuchi, S., Asahi Glass Co., Ltd., Forming Conductive Layers on Glass Plates, Japan. Kokai 75-61415 [1973/75]; C.A. **84** [1976] No. 78600.

Mukaiyama, T., Furuuchi, S., Kanai, E., Asahi Glass Co., Ltd., Electrically Conductive Glass, Japan. Kokai 75-61416 [1973/75]; C.A. **84** [1976] No. 78601.

Mukaiyama, T., Furuuchi, S., Asahi Glass Co., Ltd., Forming Conductive Layers on Glass Plates, Japan. Kokai Tokkyo Koho 75-61695 [1973/75]; C.A. **84** [1976] No. 78598.

Henery, V. A., Wagner, W. A. E., PPG Industries, Inc., High Resistivity Electroconductive Tin Oxide Films, U.S. 4235945 [1978/80]; C.A. **94** [1981] No. 94688.

Wagner, W. E., Davis, J. A., PPG Industries, Inc., Airless Spray Method for Depositing Electroconductive Tin Oxide Coatings, U.S. 4263335 [1978/81]; C.A. **95** [1981] No. 107281.

Sanyo Electric Co., Ltd., Transparent Conducting Film, Japan. Kokai Tokkyo Koho 58-30007 [1981/83]; C.A. **98** [1983] No. 226429.

Middleton, D. J., Grenier, J. I., Ford Motor Co., Infrared Reflective Glass Sheet, U.S. 4547400 [1985]; C.A. **104** [1986] No. 94126.

Middleton, D. J., Grenier, J. I., Ford Motor Co., Infrared Reflective Glass Sheet, U.S. 4548836 [1985]; C.A. **104** [1986] No. 94127.

1.4.1.2.1.5.4.1.3 Dibutyltin Biscarboxylates, $(C_4H_9)_2Sn(OOCR)_2$ with R = C_2H_5, C_3H_7, C_4H_9, t-C_4H_9, C_5H_{11}, C_6H_{13}, and i-C_6H_{13}

$(C_4H_9)_2Sn(OOCC_2H_5)_2$

Dibutyltin dipropionate was obtained in an 89% yield by dissolving $(\text{-}(C_4H_9)_2SnO\text{-})_n$ in propionic acid and subsequent vacuum distillation [14].

The compound boils at 113 to 118°C/2 Torr [9] or 131 to 132°C/3 Torr [14]. It has a density of $D_4^{20} = 1.2576$ g/cm³ and a refractive index $n_D^{20} = 1.4707$ [14].

^{119}Sn Mössbauer spectrum (at 77 K): $\delta = 1.21$, $\Delta = 3.70$ mm/s [6].

The compound decomposes on heating to form $(C_4H_9)_3SnOOCC_2H_5$ and organic compounds free of tin [13]. Above 250°C, $C_2H_5COC_2H_5$, $C_2H_5COC_3H_7$, and $C_3H_7COC_3H_7$ could be isolated as decomposition products [9]. The compound reacts with CCl_4 under UV irradiation with formation of $(C_4H_9)_2SnCl_2$, $CHCl_3$, $C_4H_9CH{=}CH_2$, C_2H_5COOH, and Cl_3CCCl_3 [20], with $SnCl_4$ and $SnBr_4$ in ether or hexane with formation of $Sn(OOCC_2H_5)_4$ and $(C_4H_9)_2SnCl_2$ or $(C_4H_9)_2SnBr_2$, respectively [11, 14], and with $GeBr_4$ with formation of $Ge(OOCC_2H_5)_4$ and $(C_4H_9)_2SnBr_2$ [11]. The reaction with Lewis bases in C_6H_6 at 30°C gives 1:1 adducts. For the following Lewis bases the thermodynamic parameters for the formation of adducts have been determined by colorimetric titration (K in L/mol, $-\Delta H°$ and $-\Delta G°$ in kJ/mol, $-\Delta S°$ in $J \cdot mol^{-1} \cdot K^{-1}$, respectively): $N(C_4H_9)_3$ (10.3, 15.8, 5.9, 33), $N(CH_3)_2CH_2CH_2N(CH_3)_2$ (46.0, 26.6, 9.7, 56), C_5H_5N (2.6, 21.2, 2.4, 62), 4-$CH_3C_5H_4N$ (3.5, 25.5, 3.2, 74), 1,10-phenanthroline (13.9, 37.0, 6.6, 100), C_5H_5NO (23.4, 15.3, 8.0, 24), and $P(C_4H_9)_3$ (7.3, 26.2, 5.0, 70) [19].

Dibutyltin dipropionate is an anthelminticum. It has been tested for the removal of Raillietina cesticillus and Ascaridia galli from chickens [1]. The compound is only about one-tenth as effective as tributyltin acetate in controlling potato haulm blight, Phytophthora infestans in laboratory tests on detached leaflets, but it was less phytotoxic [10].

A mixture of $(C_4H_9)_2Sn(OOCC_2H_5)_2$ and compounds like $HSCH_2COOC_8H_{17}$-i or $HSCH_2$-$CH_2OOCC_7H_{15}$-i is used as a stabilizer for poly(vinyl chloride) [12]. Dibutyltin dipropionate is also used as a thermooxidative stabilizer for ester type synthetic lubricants [23] or as an accelerator for the hardening of poly(dimethylsiloxane) rubber [15].

$(C_4H_9)_2Sn(OOCC_3H_7)_2$

A method for the synthesis of this compound is not described in the literature. The compound has a boiling point of 148 to 150°C/3 Torr. It decomposes above 250°C forming $C_2H_5COC_2H_5$, $C_2H_5COC_3H_7$, $C_3H_7COC_3H_7$, and $(C_4H_9)_3SnOOCC_3H_7$ [9].

A method for the separation of $(C_4H_9)_2Sn(OOCC_3H_7)_2$ from food by paper chromatography is described [17].

The compound is tested as an anthelminticum for the removal of Raillietina cesticillus and Ascaridia galli from chickens [1]. It is used, but is only about one-tenth as effective as tributyltin acetate in controlling potato haulm blight, but it was less phytotoxic [10]. Dibutyltin dibutyrate is also used as an accelerator for the hardening of poly(dimethylsiloxane) rubber [15].

$(C_4H_9)_2Sn(OOCC_4H_9)_2$

Dibutyltin divalerate is prepared by the reaction of $(\text{-}(C_4H_9)_2SnO\text{-})_n$ with C_4H_9COOH in C_6H_6 [6].

^{119}Sn Mössbauer spectrum (at 77 K): $\delta = 1.23$, $\Delta = 3.23$ mm/s [6].

The enthalpies of mixing $(C_4H_9)_2Sn(OOCC_4H_9)_2$ with C_5H_5N or $CHCl_3$ at 298 K were measured as −23.0 and −12.0 kJ/mol, respectively [21].

Dibutyltin divalerate has been tested as an anthelminthicum for the removal of Raillietina cesticillus and Ascaridia galli from chickens [1].

$(C_4H_9)_2Sn(OOCC(CH_3)_3)_2$

A method for the preparation of this compound is not given in the literature. The compound is used as a stabilizer for poly(vinyl chloride) [7] and as a catalyst for the synthesis of polyurethanes [22].

$(C_4H_9)_2Sn(OOCC_5H_{11})_2$

$(C_4H_9)_2Sn(OOCC_5H_{11})_2$ has been prepared in a 93% yield by the reaction of $(\text{-}(C_4H_9)_2SnO\text{-})_n$ with $C_5H_{11}COOH$ in C_6H_6 or ligroine [8]. It is also formed by the thermal decomposition of $(C_4H_9)_2Sn(OO\text{-}OCC_5H_{11})_2$ or $(C_4H_9)_2Sn(OOCC_5H_{11})(OO\text{-}OCC_5H_{11})$ in C_6H_6 [16].

Boiling point: 173 to 185°C/3 Torr [9]. Refractive index $n_D^{20} = 1.4695$ [8].

The separation and identification of $(C_4H_9)_2Sn(OOCC_5H_{11})_2$ from poly(vinyl chloride) residues and other material is possible using paper chromatography in 1-$BrC_{10}H_7$-CH_3COOH (1:1) as a solvent [5].

$(C_4H_9)_2Sn(OOCC_5H_{11})_2$ decomposes above 250°C with formation of $(C_4H_9)_3SnOOCC_5H_{11}$ and $C_5H_{11}COC_5H_{11}$ [9].

The compound is tested as an anthelminticum for the removal of Raillietina cesticillus and Ascaridia galli from chickens [1]. It is used as a stabilizer for poly(vinyl chloride) [2, 4], as a catalyst for the synthesis of polyurethanes [22], and as a catalyst for the reaction between polydimethylsiloxanes and polyalkoxysilanes [3].

$(C_4H_9)_2Sn(OOCC_6H_{13})_2$

The compound has been prepared by the reaction of $(\text{-}(C_4H_9)_2SnO\text{-})_n$ with $C_6H_{13}COOH$ in C_6H_6 [6].

^{119}Sn Mössbauer spectrum (at 77 K): $\delta = 1.24$, $\Delta = 3.64$ mm/s [6].

A method for the separation of this compound from food by paper chromatography is described [17].

The compound is tested as an anthelminticum for the removal of Raillietina cesticillus and Ascaridia galli from chickens [1]. It is also used as a stabilizer for poly(vinyl chloride) [2].

$(C_4H_9)_2Sn(OOCCH_2CH(CH_3)CH_2CH_2CH_3)_2$

This compound is obtained quantitatively by the reaction of $(\text{-}(C_4H_9)_2SnO\text{-})_n$ with $CH_3(CH_2)_3CH(CH_3)CH_2COOH$ at 70 to 80°C for 1.5 h [18].

References:

[1] Kerr, K. B., Walde, A. W. (Exptl. Parasitol. **5** [1956] 560/70).

[2] Berlin, A. A., Popova, Z. V., Yanovskii, D. M. (Zh. Prikl. Khim. **33** [1960] 871/7; J. Appl. Chem. [USSR] **33** [1960] 870/5).

[3] Novikov, A. S., Nudelmann, Z. N. (Kauch. Rezina **19** No. 12 [1960] 3/7; Soviet Rubber Technol. **19** No. 12 [1960] 5/7).

[4] Luz, W. D., Metallgesellschaft A.-G. (Brit. 874574 [1960/61]; C.A. **57** [1962] No. 1080).

[5] Gasparic, J., Cee, A. (J. Chromatog. **8** [1962] 393/8).

[6] Krizhanskii, L. M., Okhlobystin, O. Yu., Popov, A. V., Rogozev, B. I. (Dokl. Akad. Nauk SSSR **160** [1965] 1121/3; Dokl. Chem. Proc. Acad. Sci. USSR **160/165** [1965] 142/4).

[7] Hoch, S., Tenneco Chemicals, Inc. (Ger. 1801274 [1968/69]; C.A. **71** [1969] No. 13811).

[8] Mikhailov, G. D., Chegolya, A. S. (Sin. Volokna **1969** 18/21; C.A. **74** [1971] No. 3711).

[9] Matsuda, S., Matsuda, H., Yamaji, Y., Ninimiya, K. (Kogyo Kagaku Zasshi **73** [1970] 1007/9).

[10] McIntosh, A. H. (Ann. Appl. Biol. **66** [1970] 115/8).

[11] Melnichenko, L. S., Zemlyanskii, N. N., Samurskaya, K. A., Kocheshkov, K. A. (Dokl. Akad. Nauk SSSR **190** [1970] 351/3; Dokl. Chem. Proc. Acad. Sci. USSR **190/195** [1970] 48/50).

[12] Riethmayer, S., Chemische Werke München Otto Bärlocher GmbH (Ger. Offen. 1905898 [1969/70]; C.A. **73** [1970] No. 109904).

[13] Yamaji, Y., Moriguchi, J., Matsuda, H., Matsuda, S. (Kogyo Kagaku Zasshi **73** [1970] 1013/7; C.A. **73** [1970] No. 87985).

[14] Melnichenko, L. S., Zemlyanskii, N. N., Kocheshkov, K. A. (Izv. Akad. Nauk SSSR Ser. Khim. **1972** 2055/8; Bull. Acad. Sci. [USSR] Div. Chem. Sci. **1972** 1993/6).

[15] Kim, S. K., Kim, B. K., Moon, M. S., Kim, K. B., Won, K. D. (Hwahak Kwa Hwahak Kongop No. 17 [1974] 89/93 from C.A. **82** [1975] No. 99661).

[16] Malkov, V. D., Maslennikov, V. P., Vyshinskii, N. N., Aleksandrov, Yu. A. (Zh. Obshch. Khim. **44** [1974] 2708/12; J. Gen. Chem. [USSR] **44** [1974] 2661/5).

[17] Moshkovskaya, M. V., Nefedov, V. D., Molchanova, N. G., Zhuravlev, V. E. (Tr. Estestvennonauch. Inst. Permsk. Gos. Univ. **13** No. 3 [1975] 210/3; C.A. **86** [1977] No. 121468).

[18] Boboli, E., Malasnicki, W. L., Lato, C., Instytut Przemyslu Organicznego (Pol. 93206 [1975/77]; C.A. **89** [1978] No. 129714).

[19] Graddon, D. P., Rana, B. A. (J. Organometal. Chem. **136** [1977] 19/24).

[20] Ayrey, G., Man, F. P., Poller, R. C. (J. Organometal. Chem. **173** [1979] 171/4).

[21] Tsvetkov, V. G., Zabotin, K. P., Shmeleva, A. N., Bryukhanov, A. N., Suldin, B. V., Aleksandrov, Yu. A. (Zh. Obshch. Khim. **53** [1983] 388/92; J. Gen. Chem. [USSR] **53** [1983] 339/42).

[22] Bats, J. P., Ducasse, Y., Lalande, R., Tauzia, M. (Eur. Polym. J. **20** [1984] 997/1001).

[23] Hronec, M., Cvengrosova, Z., Kizlink, J., Stolcova, M., Ilavsky, J., Sitek, J. (Oxid. Commun. **8** [1985/86] 51/64).

1.4.1.2.1.5.4.1.4 Dibutyltin Dicaprylate, $(C_4H_9)_2Sn(OOCC_7H_{15})_2$, and Dibutyltin Bis(2-ethylhexanoate), $(C_4H_9)_2Sn(OOCCH(C_2H_5)C_4H_9)_2$

$(C_4H_9)_2Sn(OOCC_7H_{15})_2$ is prepared from $(C_4H_9)_2SnCl_2$ and $C_7H_{15}COONa$ in water at 80°C. Heating of this mixture for 2 h gives 90 to 100% yield [21, 25]. When $(C_4H_9)_2SnCl_2$ was dissolved in CH_3OH and combined with a solution of $C_7H_{15}COONa$ in CH_3OH-H_2O, a yield of 88% was possible, which could be increased to nearly 100% using a 20% excess of sodium caprylate [17]. In C_2H_5OH as a solvent, and after addition of dry C_6H_6 (to precipitate the NaCl), a

93% yield of a viscous, slightly cloudy liquid was formed [6]. Dibutyltin dichloride reacts with $C_7H_{15}COOH$ in the presence of $N(C_2H_5)_3$ in C_6H_6. After an exothermic reaction $(C_4H_9)_2Sn(OOCC_7H_{15})_2$ was obtained in a 78% yield [6]. The compound is also obtained from $(-(C_4H_9)_2SnO-)_n$ and $C_7H_{15}COOH$ and can be isolated in yields up to 98% by heating under vacuum to remove the H_2O formed [8, 23].

The yellow liquid has a boiling point of 177 to 178°C/0.01 Torr [6] or 215 to 220°C/3 Torr [8]. A melting point of −6°C is reported in [36]. The refractive index is $n_D^{20} = 1.4668$ [23], or $n_D^{24} = 1.4673$ [6]; a range between 1.4653 and 1.4681 is given in [8].

^{119}Sn Mössbauer spectrum (at 74 K): $\delta = 1.35$, $\Delta = 3.45$ mm/s [9, 11]. The IR spectrum shows the ν(OCO) at 1600 cm^{-1} [6].

$(C_4H_9)_2Sn(OOCC_7H_{15})_2$ decomposes at 250°C to give $C_7H_{15}COC_7H_{15}$ and $(C_4H_9)_3SnOOCC_7H_{15}$ [29]. The compound reacts with $(-(C_4H_9)_2SnS-)_3$ in C_6H_6 after warming with formation of $C_7H_{15}COO(C_4H_9)_2SnSSn(C_4H_9)_2OOCC_7H_{15}$ [26]. Mössbauer spectroscopic studies on poly(vinyl chloride), stabilized with dibutyltin dicaprylate, showed that $(C_4H_9)_2SnCl_2$ is formed by capture of HCl during heating of poly(vinyl chloride), but complexes between the stabilizer and the polymer are also formed [32].

Dibutyltin dicaprylate is effective as a fungicide [5, 17]. Its anthelmintic activity has been tested against Raillietina cesticillus and Ascaridia galli [1]. The compound is therefore used in wood preservation [5].

Other uses for dibutyltin dicaprylate are the stabilization of poly(vinyl chloride) against dehydrochlorination [18, 32, 35, 37] and the catalysis of the reaction of sulfohydrazides with isocyanates [43], the polymerization of ethylene carbonate [89], and the reaction of acidic alcohols and phenols with diisocyanates [20]. The compound is a very effective crosslinking agent for the vulcanization of silicones [2, 17, 50].

$(C_4H_9)_2Sn(OOCCH(C_2H_5)C_4H_9)_2$ was prepared in three radioactively tagged forms, dibutyl-1-^{14}C-tin bis(2-ethylhexanoate) from $(C_3H_7{}^{14}CH_2)_2SnCl_2$ and dibutyl-^{113}Sn bis(2-ethylhexanoate) from $(C_4H_9)_2{}^{113}SnCl_2$, both with $C_4H_9(C_2H_5)CHCOOH$ in CH_3COCH_3 in the presence of $NaHCO_3$ [14, 15], and dibutyltin bis(2-ethylhexanoate-1-^{14}C) from $(-(C_4H_9)_2SnO-)_n$ and C_4H_9-$(C_2H_5)^{14}CHCOOH$ in nearly quantitative yields [16]. The compounds have been isolated in form of waxy cakes and have been used for investigating the mechanism of PVC stabilization effected by this compound [15, 16].

A method for the determination of Sn in inorganic and organic material using $NaIO_3$ was tested also using dibutyltin bis(2-ethylhexanoate) [3].

The IR spectrum of $(C_4H_9)_2Sn(OOCCH(C_2H_5)C_4H_9)_2$ was measured and assigned in [46], and it is listed below (wave numbers in cm^{-1}):

in Nujol	in CCl_4	assignment
1587 (s)	1595 (s)	$\nu_{as}(OCO)$
1464 (s)	1464 (s)	$\delta_s(CH_2)$
1400 (s)	1400 (s)	$\delta_s(CH_2 \leftarrow CH_2CO)$
1385 (s)	1385 (s)	$\nu_s(OCO)$, $\delta_s(CH_3)$
1332		
1299	1299	$\delta_w(CH_2)$
1282	1282	
735		$\delta_r(CH_2)$
728		

in Nujol	in CCl_4	assignment
685 (s)		$\delta_s(CH_2 \leftarrow CH_2Sn)$
671 (s)	671 (s)	
658 (s)	658 (s)	
602	613	$\nu_{as}(SnC)$
571 (s)	571 (s)	$\nu_s(SnC)$
470 (w)	470 (w)	
460 (w)	460 (w)	

$(C_4H_9)_2Sn(OOCCH(C_2H_5)C_4H_9)_2$ is tested as an insecticide. The LD_{50} against the house fly, Musca domestica is 130×10^{-10} moles per fly [4]. The compound is also effective as a wood preservative. It was tested in xylene solution against Limnoria [13], and as a potato blight fungicide. It is only about one-tenth as effective as $(C_4H_9)_3SnOOCCH_3$ in controlling potato haulm blight (Phytophthora infestans), but it was less phytotoxic [30]. The acute oral toxicity of dibutyltin bis(2-ethylhexanoate) to deer mice was found to be $LD_{50} = 1075$ mg/kg [90].

The compound is a stabilizer for poly(vinyl chloride) [10, 22, 44], polyesters [33], and ethylene terephthalate polyesters [27]. Investigations of the mechanism of PVC stabilization, using radioactively tagged $(C_4H_9)_2Sn(OOCCH(C_2H_5)C_4H_9)_2$, showed that, under conditions of accelerated thermal degradation, the extent to which the polymer becomes butylated is extremely slight, whereas a relatively large proportion of the 2-ethylhexanoate groups become firmly bonded to the polymer [10, 14, 15, 16].

Other uses for dibutyltin bis(2-ethylhexanoate) are as a catalyst for silane-silanole condensations in silicone materials [24, 31, 45, 48, 53, 54, 59, 67, 85], as a crosslinking catalyst for the synthesis of urethane prepolymer compositions and polymers [7, 42, 49, 52, 55, 57, 61, 68, 73, 83], and as an additive for lubricants, avoiding the annealing of aluminium [19].

A mixture of dibutyltin dicarboxylates, made from a technical mixture of fatty acids $C_7H_{15}COOH$ by the methods for the synthesis of either dibutyltin dicaprylate or dibutyltin bis(2-ethylhexanoate) is referred to as dibutyltin dioctoate or dibutyltin dioctanoate and serves as an additive to improve the workability of ethylene-propylene block copolymers [38], as a crosslinking catalyst for acrylate-cyanoacetate-ureadiacetoacetamide-diacetoacetamide copolymers [63], as a catalyst for the synthesis of various silicone materials [28, 34, 40, 41, 47, 51, 56, 58, 60, 62, 64, 65, 66, 69, 70, 72, 74, 75, 77, 79, 81, 82, 84, 86, 87, 88, 91, 92, 93], and for the synthesis of polyurethanes [12, 39, 71, 76, 78, 80].

References:

[1] Kerr, K. B., Walde, A. W. (Exptl. Parasitol. **5** [1956] 560/70).

[2] Baranovskaya, N. B., Berlin, A. A., Zakharova, M. Z., Mizikin, A. I. (Khim. Prakt. Primen. Kremneorg. Soedin., Tr. 2nd Konf., Leningrad 1958 [1961], pp. 88/95; C.A. **1960** 7206).

[3] Farnsworth, M., Pekola, J. (Anal. Chem. **31** [1959] 410/4).

[4] Blum, M. S., Pratt, J. J. (J. Econ. Entomol. **53** [1960] 445/8).

[5] Fuse, G., Nishimoto, K. (Mokuzai Kenkyu No. 26 [1961] 34/48; C.A. **56** [1962] No. 13137).

[6] Alleston, D. L., Davies, A. G. (J. Chem. Soc. **1962** 2050/4).

[7] Farkas, A., Mills, G. A. (Advan. Catal. **13** [1962] 363/446).

[8] Sheverdina, N. I., Abramova, L. V., Paleeva, I. E., Kocheshkov, K. A. (Khim. Prom. **1962** No. 10, pp. 707/8; C.A. **59** [1963] No. 8776).

[9] Aleksandrov, A. Yu., Delyagin, N. N., Mitrofanov, K. P., Polak, L. S., Shpinel, V. S. (Dokl. Akad. Nauk SSSR **148** [1963] 126/8; Dokl. Chem. Proc. Acad. Sci. USSR **148/153** [1963] 1/3).

[10] Frye, A. H., Horst, R. W., Paliobagis, M. A. (Am. Chem. Soc. Div. Polym. Chem. Preprints **4** [1963] 260/82).

[11] Gol'danskii, V. I., Makarov, E. F., Stukan, R. A., Trukhtanov, V. A., Khrapov, V. V. (Dokl. Akad. Nauk SSSR **151** [1963] 357/60; Dokl. Phys. Chem. Proc. Acad. Sci. USSR **148/153** [1963] 598/601).

[12] Strickland, A., Minnesota Mining & Manufacturing Co. (Ger. Offen. 1159633 [1960/63]; C.A. **60** [1964] 9458).

[13] Vind, H. P., Hochmann, H. (Tin Its Uses No. 57 [1963] 10/2).

[14] Frye, A. H., Horst, R. W. (Intern. J. Appl. Radiation Isotopes **15** [1964] 169/74).
[15] Frye, A. H., Horst, R. W., Paliobagis, M. A. (J. Polym. Sci. A **2** [1964] 1765/84).
[16] Frye, A. H., Horst, R. W., Paliobagis, M. A. (J. Polym. Sci. A **2** [1964] 1801/4).
[17] Vita-Zahnfabrik H. Rauter K.-G. (Fr. 1413601 [1963/65]; C.A. **62** [1965] 5138).
[18] Gelfman, Ya. A., Lauris, I. V., Kusova, V. P. (Sb. Tr. Vses. Nauchn. Issled. Inst. Nov. Stroit. Mater. No. 14 [1966] 54/8; C.A. **68** [1968] No. 79010).
[19] Sun Oil Company (Brit. 1182800 [1967/70]).
[20] Zhitinkina, A. K., Shoshtaeva, M. V. (Sin. Fiz. Khim. Polim. No. 5 [1968] 129/32; C.A. **70** [1969] No. 2976).

[21] Boboli, E., Rajewski, M., Kowalski, M., Pazgan, A., Lato, W., Institut Przemyslu Organicznego (Fr. 1580291 [1968/69]).
[22] Hoch, S., Tenneco Chemicals Inc. (Ger. 1801274 [1968/69]; C.A. **71** [1969] No. 13811).
[23] Mikhailov, G. D., Chegolya, A. S. (Sin. Volokna **1969** 18/21; C.A. **74** [1971] No. 3711).
[24] Wessel, J. K., Dow Corning Corp. (Ger. 1907980 [1968/69]; C.A. **72** [1970] No. 13621).
[25] Boboli, E., Rajewski, M., Kowalski, M., Pazgan, A., Lato, C., Institut Przemyslu Organicznego (Pol. 59754 [1967/70]; C.A. **74** [1971] No. 31850).
[26] Davies, A. G., Harrison, P. G. (J. Chem. Soc. C **1970** 2035/8).
[27] Enoki, K., Yoshihara, Y., Kirai, Y., Japan Soda Co., Ltd. (Ger. 1954959 [1968/70]; C.A. **73** [1970] No. 26275).
[28] Imperial Chemical Industries Ltd. (Fr. Demande 2011554 [1968/70]; C.A. **74** [1971] No. 4348).
[29] Matsuda, S., Matsuda, H., Yamaji, Y., Ninimiya, K. (Kogyo Kagaku Zasshi **73** [1970] 1007/9).
[30] McIntosh, A. H. (Ann. Appl. Biol. **66** [1970] 115/8).

[31] Miller, R. A. L., Imperial Chemical Industries Ltd. (Brit. 1179040 [1967/70]; C.A. **72** [1970] No. 79814).
[32] Aleksandrov, A. Yu., Gol'danskii, V. I., Zavarova, T. B., Korytko, L. A. (Vysokomol. Soedin. B **13** [1971] 784/6; C.A. **76** [1972] No. 60429).
[33] Enoki, Y., Kiki, Y., Japan Soda Co., Ltd. (Japan. 71-32067 [1968/71]; C.A. **77** [1972] No. 49557).
[34] Johnson and Johnson (Brit. Amended 1127625 [1964/71]; C.A. **76** [1972] No. 26035).
[35] Krats, E. O., Zavarova, T. B. (Vysokomol. Soedin. A **13** [1971] 899/905; C.A. **75** [1971] No. 21634).
[36] Minsker, K. S., Fedoseeva, G. T., Zavarova, T. B., Krats, E. O. (Vysokomol. Soedin. A **13** [1971] 2265/78; J. Polym. Sci. [USSR] A **13** [1971] 2455/60).
[37] Zavarova, T. B., Okladnov, N. A., Fedoseeva, G. T., Minsker, K. S. (Vysokomol. Soedin. A **13** [1971] 1003/8; C.A. **75** [1971] No. 37171).
[38] Koga, S., Chisso Corp. (Japan. Kokai 72-34934 [1971/72]; C.A. **81** [1974] No. 26526).
[39] Murakami, T., Ishii, H., Ishii, H., Matsushita Electric Works, Ltd. (Japan. 72-00750 [1967/72]; C.A. **77** [1972] No. 102738).
[40] Kaburaki, K., Ishizawa, A., Tokyo Shibaura Electric Co., Ltd. (Japan. Kokai 73-16939 [1971/73]; C.A. **79** [1973] No. 67345).

[41] Mestetsky, T. S., GAF Corp. (U.S. 3770687 [1971/73]; C.A. **80** [1974] No. 61137).
[42] Gemeinhardt, P. G., Britain, J. W., Baychem Corp. (U.S. 3822223 [1958/74]; C.A. **81** [1974] No. 121781).
[43] Grekov, A. P., Otroshko, G. V. (Kratk. Tezisy Vses. Soveshch. Probl. Mekh. Geteroliticheshkikh Reakts. **1974** 19/20 from C.A. **85** [1976] No. 77131).
[44] Kim, S. K., Kim, B. K., Moon, M. S., Kim, K. B., Won, K. D. (Hwahak Kwa Hwahak Kongop **17** [1974] 89/93 from C.A. **82** [1975] No. 99661).

[45] Kim, S. G., Won, G. D., So, J. N. (Hwahak Kwa Hwahak Kongop **18** [1975] 84/7 from C.A. **84** [1976] No. 5800).
[46] Morikawa, T. (Kagaku To Kogyo [Osaka] **49** [1975] 98/113; C.A. **84** [1976] No. 5795).
[47] Tsukada, T., Kabuki, K., Tokyo Shibaura Electric Co., Ltd. (Japan. 75-15261 [1969/75]; C.A. **83** [1975] No. 180508).
[48] Keiser, LeRoy H., Phillips Petroleum Co. (U.S. 3936582 [1974/76]; C.A. **84** [1976] No. 137384).
[49] Groves, J. D., Minnesota Mining and Mfg. Co. (Ger. Offen. 2721492 [1976/77]; C.A. **88** [1978] No. 38827).
[50] Severnyi, V. V., Minasyan, R. M., Minasyan, O. I. (Vysokomol. Soedin. A **19** [1977] 1549/55; J. Polym. Sci. [USSR] A **19** [1977] 1775/83).

[51] Takamizawa, M., Yamamoto, Y., Inoue, Y., Noshiro, A., Fujii, H., Shin-Etsu Chemical Industry Co., Ltd. (Japan. Kokai 77-106864 [1976/77]; C.A. **88** [1978] No. 52114).
[52] Vogt, H. C., Parekh, M., Patton, J. T., BASF Wyandotte Corp. (U.S. 4018708 [1975/77]; C.A. **86** [1977] No. 178187).
[53] Clark, W. H., Skinner, C. E., Dow Corning Corp. (U.S. 4105617 [1975/78]; C.A. **90** [1979] No. 24528).
[54] Fierens, P., Vandendunghen, G., Sgers, W., Van Elsuwe, R. (React. Kinet. Catal. Letters **8** [1978] 179/87).
[55] Groves, J. D., D'Zuro, D. S. A., Minnesota Mining and Mfg. Co. (U.S. 4102716 [1976/78]; C.A. **90** [1979] No. 73000).
[56] Ishikawa, T. (Japan. Kokai 78-77265 [1976/78]; C.A. **89** [1978] No. 147638).
[57] Kobayashi, Y., Nakata, Y., Iizuka, H., Mitsubishi Chemical Industries Co., Ltd. (Japan. Kokai Tokkyo Koho 78-146797 [1977/78]; C.A. **90** [1979] No. 139190).
[58] Kokan, S., Fukuoka, M., Sumitomo Bakelite Co., Ltd. (Japan. Kokai 78-88050 [1977/78]; C.A. **89** [1978] No. 216345).
[59] Clark, W. H., Skinner, C. E., Dow Corning Corp. (U.S 4144216 [1975/79]; C.A. **91** [1979] No. 22213).
[60] Arai, M., Maruyama, M., Shin-Etsu Chemical Industry Co., Ltd. (Japan. Kokai Tokkyo Koho 80-36268 [1978/80]; C.A. **93** [1980] No. 27069).

[61] Baughman, R. H., Preziosi, A. F., Yee, K. C., Allied Chemical Corp. (U.S 4220747 [1975/80]; C.A. **93** [1980] No. 240258).
[62] Dow Corning K. K. (Japan. Tokkyo Koho 80-19950 [1971/80]; C.A. **93** [1980] No. 205967).
[63] Heckles, J. S. (U.S. 4218515 [1979/80]; C.A. **94** [1981] No. 4577).
[64] Shin-Etsu Chemical Industry Co., Ltd., Dainippon Printing Co., Ltd. (Japan. Tokkyo Koho 80-49624 [1977/80]; C.A. **94** [1981] No. 193821).
[65] Toshiba Silicone Co., Ltd. (Japan. Kokai Tokkyo Koho 80-90556 [1978/80]; C.A. **93** [1980] No. 205819).
[66] Watanabe, T., Fujii, S., Sato, T., Yasuda, H., Chisso Corp. (Brit. Appl. 2039454 [1979/80]; C.A. **94** [1981] No. 86635).
[67] Grenoble, M. E., Goossens, J. C., General Electric Co. (Can. 1114247 [1976/81]; C.A. **96** [1982] No. 87184).
[68] Preziosi, A. F., Patel, G. N., Denkewalter, R. G., Baughman, R. H., Allied Corp. (Eur. Appl. 42069 [1980/81]; C.A. **96** [1982] No. 163352).
[69] Shin-Etsu Chemical Industry Co., Ltd. (Japan. Kokai Tokkyo Koho 81-05850 [1979/81]; C.A. **94** [1981] No. 210077).
[70] Toshiba Silicone Co., Ltd. (Japan. Kokai Tokkyo Koho 81-78960 [1979/81]; C.A. **95** [1981] No. 171175).

[71] Asahi Chemical Industry Co., Ltd. (Japan. Kokai Tokkyo Koho 57195777 [1981/82]; C.A. **98** [1983] No. 216992).
[72] Chisso Corp. (Japan. Kokai Tokkyo Koho 8217668 [1980/82]; C.A. **97** [1982] No. 40999).
[73] Pokorny, R. J., Minnesota Mining and Mfg. Co. (U.S 4329442 [1981/82]; C.A. **97** [1982] No. 39901).
[74] Shin-Etsu Chemical Industry Co., Ltd. (Japan. Kokai Tokkyo Koho 57162753 [1981/82]; C.A. **98** [1983] No. 127852).
[75] Shin-Etsu Chemical Industry Co., Ltd. (Japan. Kokai Tokkyo Koho 57174347 [1981/82]; C.A. **98** [1983] No. 144850).
[76] Mitsubishi Monsanto Chemical Co. (Japan. Kokai Tokkyo Koho 5891705 [1981/83]; C.A. **100** [1984] No. 7386).
[77] Shin-Etsu Chemical Industry Co., Ltd. (Japan. Kokai Tokkyo Koho 58167647 [1982/83]; C.A. **100** [1984] No. 122946).
[78] Yokohama Rubber Co., Ltd. (Japan. Kokai Tokkyo Koho 58101119 [1981/83]; C.A. **99** [1983] No. 196063).
[79] Chisso Corp. (Japan. Kokai Tokkyo Koho 59191746 [1983/84]; C.A. **102** [1985] No. 79805).
[80] Hitachi Cable, Ltd. (Japan. Kokai Tokkyo Koho 59100127 [1982/84]; C.A. **101** [1984] No. 172855).

[81] McVie, J., Dow Corning Corp. (Eur. Appl. 114512 [1983/84]; C.A. **101** [1984] No. 212827).
[82] Niemi, R. G., Dow Corning Corp. (U.S. 4476155 [1983/84]; C.A. **101** [1985] No. 232059).
[83] Pokorny, R. J., Minnesota Mining and Mfg. Co. (U.S. 4444975 [1983/84]; C.A. **101** [1984] No. 39980).
[84] Shin-Etsu Chemical Industry Co., Ltd. (Japan. Kokai Tokkyo Koho 5980463 [1982/84]; C.A. **101** [1984] No. 17223).
[85] Yosomiya, R., Morimoto, K. (Polym. Bull. [Berlin] **12** [1984] 41/8).
[86] Chisso Corp. (Japan. Kokai Tokkyo Koho 6006740 [1983/85]; C.A. **102** [1985] No. 221670).
[87] Inoue, Y., Arai, M., Inoe, T., Shin-Etsu Chemical Industry Co., Ltd. (Japan. Kokai Tokkyo Koho 60258260 [1984/85]; C.A. **105** [1986] No. 80389).
[88] Inoue, Y., Okami, T., Shin-Etsu Chemical Industry Co., Ltd. (Japan. Kokai Tokkyo Koho 60231761 [1984/85]; C.A. **104** [1986] No. 208683).
[89] Otsu, T., Endo, K., Hozawa, K., Komatsu, M. (Mem. Fac. Eng. Osaka City Univ. **26** [1985] 101/7).
[90] Schafer, E. W., Bowles, W. A. (Arch. Environ. Contam. Toxicol. **14** [1985] 111/29).

[91] Shin-Etsu Chemical Industry Co., Ltd. (Japan. Kokai Tokkyo Koho 6008361 [1983/85]; C.A. **102** [1985] No. 186496).
[92] Shin-Etsu Chemical Industry Co., Ltd. (Japan. Kokai Tokkyo Koho 6060160 [1983/85]; C.A. **102** [1985] No. 72451).
[93] Takaai, T., Arai, M., Shin-Etsu Chemical Industry Co., Ltd. (Japan. Kokai Tokkyo Koho 60199057 [1984/85]; C.A. **104** [1986] No. 131242).

1.4.1.2.1.5.4.1.5 Dibutyltin Biscarboxylates, $(C_4H_9)_2Sn(OOCR)_2$ with R = C_8H_{17}, i-C_8H_{17}, C_9H_{19}, and neo-C_9H_{19}

$(C_4H_9)_2Sn(OOCC_8H_{17})_2$

The synthesis of dibutyltin dipelargonate is not described in the literature. It is used as an anthelminticum against Raillietina cesticillus and Ascaridia galli from chickens [1], and as a stabilizer for poly(vinyl chloride) [6].

$(C_4H_9)_2Sn(OOC\text{-}i\text{-}C_8H_{17})_2$

A synthesis of dibutyltin bis(isopelargonate) is not described in the literature. The LD_{50} value is 199.9 mg/kg bodyweight for mice [5].

$(C_4H_9)_2Sn(OOCC_9H_{19})_2$

Dibutyltin dicaprate is prepared in 95 to 98% yield by the reaction of $(\text{-}(C_4H_9)_2SnO\text{-})_n$ with capric acid between 60 and 70°C. The yellow liquid has a refractive index $n_D = 1.4675$ to 1.4701 [2].

^{119}Sn Mössbauer spectrum (at 77 K): $\delta = 1.21$, $\Delta = 3.61$ mm/s [3].

The compound is used as an anthelminticum for the removal of Raillietina cesticillus and Ascaridia galli from chickens [1], as a stabilizer for poly(vinyl chloride) [4], and a plasticizer for polyesters [7].

$(C_4H_9)_2Sn(OOCCH_2CH_2C(C_2H_5)_3)_2$

Dibutyltin bis(neodecanoate) is mentioned in the patent literature as a stabilizer for poly(vinyl chloride) [6].

References:

[1] Kerr, K. B., Walde, A. W. (Exptl. Parasitol. **5** [1956] 560/70).
[2] Sheverdina, N. I., Abramova, L. V., Paleeva, I. E., Kocheshkov, K. A. (Khim. Prom. **1962** No. 10, pp. 707/8; C.A. **59** [1963] No. 8776).
[3] Krizhanskii, L. M., Okhlobystin, O. Yu., Popov, A. V., Rogozev, B. I. (Dokl. Akad. Nauk SSSR **160** [1965] 1121/3; Dokl. Chem. Proc. Acad. Sci. USSR **160/165** [1965] 142/4).
[4] Gelfman, Ya. A., Lauris, I. V., Kuskova, V. P. (Sb. Tr. Vses. Nauchn. Issled. Inst. Nov. Stroit. Mater. No. 14 [1966] 54/8; C.A. **68** [1968] No. 79010).
[5] Calley, D. J., Guess, W. L., Autian, J. (J. Pharm. Sci. **56** [1967] 1/240), according to Smith, P. J. (Intern. Tin Res. Inst. Publ. No. 538 [1978]).
[6] Hoch, S., Tenneco Chemicals Inc. (Ger. Offen. 1801274 [1968/69]; C.A. **71** [1969] No. 13811).
[7] Barshtein, R. S., Li, P. Z., Gorbunova, V. G., Gazin, V. A., Martynov, S. F., Markina, E. E., Pertsov, L. D., Kalinkin, S. F., Shcherbakov, V. S. (U.S.S.R. 311930 [1967/71]; C.A. **76** [1972] No. 26012).

1.4.1.2.1.5.4.1.6 Dibutyltin Dilaurate, $(C_4H_9)_2Sn(OOCC_{11}H_{23})_2$

1.4.1.2.1.5.4.1.6.1 Synthesis, Properties, and Reactions

Dibutyltin dilaurate is prepared from $(\text{-}(C_4H_9)_2SnO\text{-})_n$ and $C_{11}H_{23}COOH$ at 60 to 70°C in 95 to 98% yield [10], or from $(C_4H_9)_2SnCl_2$ and $C_{11}H_{23}COONa$ in water at 80 to 90°C in quantitative yield after 2 [30] or 5 h [25]. Another method, which allows the synthesis of pure dibutyltin dilaurate free of tributyltin and monobutyltin compounds on a technical scale uses $(C_4H_9)_2SnCl_2 \cdot 2\,NH_3$ in the reaction with $C_{11}H_{23}COOH$ in benzene with 90% yield after 2 h at room temperature [14] or 99% yield after 2 h reflux [15, 17].

1-^{14}C-labeled dibutyltin dilaurate, $(C_4H_9)_2Sn(OO^{14}CC_{11}H_{23})_2$ has been obtained from $(-(C_4H_9)_2SnO-)_n$ and $C_{11}H_{23}{}^{14}COOH$ as a pale yellow oil, which was purified by molecular distillation [51, 52].

Melting points for the compound are 22 to 24°C [10], 23°C [34], 26.8 to 27°C [52], and 37°C [32]. The refractive index is $n_D^{20} = 1.4683$ [14, 15, 17], 1.4724 [7], and $n_D^{26} = 1.470$ [7], and the molar refraction is 168.29 [1, 2, 7]. Dipole moment $\mu = 1.45$ D in benzene at 25° [18].

^{119}Sn Mössbauer spectrum (in mm/s): $\delta = 1.36$, $\Delta = 3.43$ in the pure compound, but $\delta =$ 1.42 to 1.49, and $\Delta = 2.86$ to 3.24 in PVC (80 K) [53], $\delta = 1.45$, $\Delta = 3.45$ [11], $\delta = -1.36$ (β-Sn), $\Delta = 3.35$ (78 K) [13], $\delta = -1.22$ (β-Sn), $\Delta = 3.4$ (77 K) [20].

The IR spectrum is depicted between 4000 and 500 cm^{-1} [12]. The ν(OCO) is assigned at 1600 cm^{-1} [53].

Analytical methods have been developed for the determination of Sn and other elements in organometallic compounds including dibutyltin dilaurate. The quantitative analysis of Sn is possible by X-ray fluorescence [26], or after wet oxidation of the organometallic compound using nitric and sulfuric acid followed by colorimetric titration [4, 5]. Oxygen is determined in dibutyltin dilaurate by pyrolysis using a carrier gas containing hydrogen [38], and the determination of the $C_{11}H_{23}COO$ group is possible by potentiometric titration using Sb electrodes [24].

The separation and determination of dibutyltin dilaurate from other organotin compounds, from residues or from poly(vinyl chloride), and other plastic material can be achieved by different methods, like paper chromatography [6, 9], thin layer chromatography [8, 16, 35, 40, 42, 43, 45, 47, 48], gel chromatography [41, 55], potentiometry [31], polarography [3, 23, 27, 39, 56], atomic absorption spectrometry [44], and even ^{13}C NMR spectrometry [49], and ESR spectroscopy [46]. An analysis of the content of dibutyltin dilaurate in benzene is also possible via the determination of the dielectric data of the solution, especially for the use in automatic control of the concentration of such solutions [37], and a rapid determination of the concentration of $(C_4H_9)_2Sn(OOCC_{11}H_{23})_2$ in solution is possible by β-ray reflection. The concentration of dibutyltin dilaurate in solution is proportional to the intensity of back-scattered β-radiation of a ^{90}Sr source. An analysis takes 10 to 15 min, the relative accuracy is 0.05 to 3% and the relative sensitivity is 0.2% [19].

Dibutyltin dilaurate decomposes above 250°C with formation of $C_{11}H_{23}COC_{11}H_{23}$ [32]. The compound reacts with H_2O [28], CH_3OH [28, 50], C_4H_9OH [54], dioxane [28], $C_4H_9OOCNH(CH_2)_6NHCOOC_4H_9$ [33], $CH_3OCH_2CH_2OOCNH(CH_2)_6NHCOOCH_2CH_2OCH_3$ [33], and with $(C_4H_9)_2Sn(OCH_3)_2$ [21] with formation of complexes. The stability constants of the 1:2 complexes (K at 25°C) with H_2O (3.0), CH_3OH (5.3), and dioxane (3.1) were determined by UV spectroscopy in heptane [28]. The enthalpy of formation of the 1:1 complex is $\Delta H_f =$ -1.2 kcal/mol for CH_3OH in hexane [50], and -9.1 kcal/mol for $(C_4H_9)_2Sn(OCH_3)_2$ [21].

The reactions of dibutyltin dilaurate with alkyl and alkenyl halides have been studied in connection with investigations of the mechanism of PVC stabilization. The compound reacts at 180°C with t-C_4H_9Cl with formation of $C_{11}H_{23}COOC_4H_9$-t, but with $CH_3CHClCH{=}CH_2$ with formation of $(C_4H_9)_2SnCl_2$ [29, 36], as well as with $(CH_3)_2CClCH_2CH_3$ [29]. Investigations of the reaction of $(C_4H_9)_2Sn(OO^{14}CC_{11}H_{23})_2$ with PVC under controlled conditions showed that exchange of the laureate ligand with labile chlorine atoms in the polymer is the important step besides the reaction of the stabilizer dibutyltin dilaurate with HCl formed during degradation of the poly(vinyl chloride) [51, 52]. Mössbauer spectroscopy is used to investigate the mechanism of the thermostabilizing action of dibutyltin dilaurate on polyethylene. Upon irradiation of stabilized polyethylene, it could be shown that addition of butyl radicals, formed from the radiolysis of dibutyltin dilaurate, to the polymer chain blocked the free valencies arising in the polymer [22].

References:

[1] West, R., Rochow, E. G. (J. Am. Chem. Soc. **74** [1952] 2490/1).

[2] Vogel, A. I., Cresswell, W. T., Leicester, J. (J. Phys. Chem. **58** [1954] 174/7).

[3] Baker, R. A. (Diss. Univ. New Hampshire 1958, pp. 1/70; Diss. Abstr. **20** [1959] 897).

[4] Chapman, A. H., Duckworth, M. W., Price, J. W. (Brit. Plast. **32** [1959] 78).

[5] Farnsworth, M., Pekola, J. (Anal. Chem. **31** [1959] 410/4).

[6] Williams, D. J., Price, J. W. (Analyst [London] **85** [1960] 579/82).

[7] Sayre, R. (J. Chem. Eng. Data **6** [1961] 560/4).

[8] Türler, M., Högl, O. (Mitt. Gebiete Lebensmittelunters. Hyg. **52** [1961] 123/30).

[9] Gasparic, J., Cee, A. (J. Chromatog. **8** [1962] 393/8).

[10] Sheverdina, N. I., Abramova, L. V., Paleeva, I. E., Kocheshkov, K. A. (Khim. Prom. [Kiev] **1962** No. 10, pp. 707/8; C.A. **59** [1963] No. 8776).

[11] Aleksandrov, A. Yu., Delyagin, N. N., Mitrofanov, K. P., Polak, L. S., Shpinel, V. S. (Dokl. Akad. Nauk SSSR **148** [1963] 126/8; Dokl. Phys. Chem. Proc. Acad. Sci. USSR **148/153** [1963] 1/3).

[12] Cummins, R. A., Dunn, P. (Rept. Defence Stand. Lab. Australia No. 266 [1963] 1/106).

[13] Gol'danskii, V. I., Makarov, E. F., Stukan, R. A., Trukhtanov, V. A., Khrapov, V. V. (Dokl. Akad. Nauk SSSR **151** [1963] 357/60; Dokl. Phys. Chem. Proc. Acad. Sci. USSR **148/153** [1963] 598/601).

[14] Koninklijke Industrieele Maatschappij voorheen Noury and van der Lande N. V. (Fr. 1320473 [1961/63]; C.A. **59** [1963] 8788/9).

[15] Koninklijke Industrieele Maatschappij voorheen Noury and van der Lande N. V. (Neth. 109491 [1962/64]; C.A. **62** [1965] No. 9173).

[16] Neubert, G. (Z. Anal. Chem. **203** [1964] 265/72).

[17] Viveen, W. J. C., Schröder, A., Koninklijke Industrieele Maatschappij vorheen Noury and van der Lande N. V. (Ger. Offen. 1167836 [1962/64]).

[18] Zemlyanskii, N. N., Gol'dshtein, I. P., Gur'yanova, E. N., Panov, E. M., Slovokhotova, N. A., Kocheshkov, K. A. (Dokl. Akad. Nauk SSSR **156** [1964] 131/4; Dokl. Phys. Chem. Proc. Acad. Sci. USSR **154/159** [1964] 452/5).

[19] Ashbel, F. B., Parshina, A. M., Goizman, M. S., Zhizhina, L. I., Kuptsova, K. M. (Zavodsk. Lab. **31** [1965] 1062/3; Ind. Lab. [USSR] **31** [1965] 1316/7).

[20] Krizhanskii, L. M., Okhlobystin, O. Yu., Popov, A. V., Rogozev, B. I. (Dokl. Akad. Nauk SSSR **160** [1965] 1121/3; Dokl. Phys. Chem. Proc. Acad. Sci. USSR **160/165** [1965] 142/4).

[21] Zemlyanskii, N. N., Gol'dshtein, I. P., Gur'yanova, E. N., Sytkina, O. P., Panov, E. M., Shovokhotova, N. A., Kocheshkov, K. A. (Izv. Akad. Nauk SSSR Ser. Khim. **1967** 728/35; Bull. Acad. Sci. [USSR] Div. Chem. Sci. **1967** 707/12).

[22] Aleksandrov, A. Yu., Baldokhin, Yu. V., Braginskii, R. P., Gol'danskii, V. I., Korytko, L. A., Leshchenko, S. S., Finkel, G. E. (Khim. Vysokikh Energ. **2** [1968] 331/7; High Energy Chem. [USSR] **2** [1968] 285/90).

[23] Bork, V. A., Selivokhin, P. I. (Plasticheskie Massy **1968** No. 4, pp. 56/7; C.A. **69** [1968] No. 10519).

[24] Groagova, A., Pribyl, M. (Z. Anal. Chem. **234** [1968] 423/8).

[25] Boboli, E., Rajewski, M., Kowalski, M., Pazgan, A., Lato, W., Instituut Przemyslu Organicznego (Fr. 1580291 [1968/69]).
[26] Guenther, F., Geyer, R., Stevenz, D. (Neue Hütte **14** [1969] 563/6).
[27] Shkorbatova, T. L., Pegusova, L. D. (Ochistka Proizvod. Stochnykh Vod **1969** 224/31; C.A. **73** [1970] No. 91030).
[28] Zabrodin, V. B., Bekhli, L. S., Tiger, R. P., Entelis, S. G. (Zh. Fiz. Khim. **43** [1969] 2371/3; Russ. J. Phys. Chem. **43** [1969] 1329/31).
[29] Ayrey, G., Poller, R. C., Siddiqui, I. H. (J. Polym. Sci. Polym. Letters Ed. **8** [1970] 1/6).
[30] Boboli, E., Rajewski, M., Kowalski, M., Pazgan, A., Lato, C., Instituut Przemyslu Organicznego (Pol. 59754 [1967/70]; C.A. **74** [1971] No. 31850).

[31] Kapisinska, V., Caplovic, J. (Chem. Prumysl **20** [1970] 487/8; C.A. **74** [1971] No. 60697).
[32] Matsuda, S., Matsuda, H., Yamai, Y., Ninimiya, K. (Kogyo Kagaku Zasshi **73** [1970] 1007/9).
[33] Lipatova, T. E., Bakalo, L. A., Sirotinskaya, A. L. (Sin. Fiz. Khim. Polim. No. 8 [1971] 68/70; C.A. **77** [1972] No. 20132).
[34] Minsker, K. S., Fedoseeva, G. T., Zavarova, T. B., Krats, E. O. (Vysokomol. Soedin. A **13** [1971] 2265/78; Polym. Sci. [USSR] A **13** [1971] 2544/60).
[35] Simpson, D., Curell, B. R. (Analyst **96** [1971] 515/21).
[36] Ayrea, G., Poller, R. C., Siddiqui, I. H. (J. Polym. Sci. Polym. Chem. Ed. **10** [1972] 725/35).
[37] Volodin, V. A., Frangulyan, L. A., Parshina, A. M. (Khim. Prom. [Moscow] **49** [1973] 832/3; C.A. **80** [1974] No. 70904).
[38] Imaeda, K., Kuriki, T. (Bunseki Kagaku **23** [1974] 47/52; C.A. **80** [1974] No. 103500).
[39] Fleet, B., Fouzder, N. B. (J. Electroanal. Chem. Interfacial Electrochem. **63** [1975] 69/78; C.A. **84** [1976] No. 841210).
[40] Moshkovskaya, M. V., Nefedov, V. D., Molchanova, N. G., Zhuravlev, V. E. (Tr. Estestvennonauch. Inst. Permsk. Gos. Univ. **13** No. 3 [1975] 210/3; C.A. **86** [1977] No. 121468).

[41] Kazarinova, N., Kozitskaya, L. (Ukr. Khim. Zh. **42** [1976] 526/8; C.A. **85** [1976] No. 95066).
[42] Woidich, H., Pfannhauser, W. (Z. Lebensm. Untersuch. Forsch. **162** [1976] 49/54).
[43] Woidich, H., Pfannhauser, W., Blaicher, G. (Deut. Lebensm. Rundschau **72** [1976] 421/2).
[44] George, G. M., Frahm, L. J., McDonnell, J. P. (J. Assoc. Off. Anal. Chem. **60** [1977] 1054/8; C.A. **87** [1977] 1054/8; C.A. **87** [1977] No. 182711).
[45] Oki, Y., Mori, F., Koyama, M. (Kobunshi Ronbunshu **34** [1977] 43/7; C.A. **86** [1977] No. 122184).
[46] Stegmann, H. B., Über, W., Scheffler, K. (Z. Anal. Chem. **286** [1977] 59/64).
[47] Novitskaya, L. P., Dregval, G. F., Brodskaya, N. M. (Gig. Saint. **1979** No. 6, pp. 48/51).
[48] Vasundhara, T. S., Parihar, D. B. (Z. Anal. Chem. **294** [1979] 408).
[49] Sebenik, A., Osredkar, U., Zigon, M., Vizovisek, I. (Hem. Ind. **34** [1980] 316/8).
[50] Varentsova, N. V., Goldshtein, I. P., Paleeva, I. E., Tarakanov, O. G., Guryanova, E. N. (Zh. Obshch. Khim. **50** [1980] 2085/93; J. Gen. Chem. [USSR] **50** [1980] 1688/94).

[51] Ayrey, G., Hsu, S. Y., Poller, R. C. (Polym. Sci. Technol. **26** [1984] 171/87).
[52] Ayrey, G., Hsu, S. Y., Poller, R. C. (J. Polym. Sci. Polym. Chem. Ed. **22** [1984] 2871/86).
[53] Allen, D. W., Brooks, J. S., Clarkson, R. W., Richard, W., Unwin, J., Smith, P. J. (Polym. Degrad. Stab. **13** [1985] 191/200).
[54] Berlin, P. A., Tiger, R. P., Entelis, S. G., Zaporozhskaya, S. V. (Zh. Fiz. Khim. **59** [1985] 262/3; Russ. J. Phys. Chem. **59** [1985] 159/63).
[55] Jirackova-Audouin, L., Raceze, D., Verdu, J. (Analusis **13** [1985] 59/64).
[56] Kitamura, H., Sugimae, A., Nakamoto, M. (Bull. Chem. Soc. Japan **58** [1985] 2641/7).

1.4.1.2.1.5.4.1.6.2 Toxicity and Biocidal Properties

Because of the use of dibutyltin dilaurate as a stabilizer for poly(vinyl chloride), the toxicity of this compound was tested extensively in comparison to the toxicity of other organotin compounds used for the same purpose [8, 15, 37, 38, 44]. Toxicity in aqueous media was especially investigated [43] with Phytobacterium phosphoreum, Pseudomonas putida, fish, algae, and daphnids [49].

Tests as a potato blight fungicide (Phytophthora infestans) showed the compound as effective as tributyltin acetate, but only about one-tenth as effective as triphenyltin acetate, but less phytotoxic [28]. Dibutyltin dilaurate was used as a fungicide in rubber based antifouling coating materials [39], in room temperature-curable silicone rubber sealing compositions [46], in agglomerating agents for mortar, concrete and coating materials [29, 30], as an antibactericidal agent in textile dry cleaning compositions [27], and in antiseptic treatments of leather substitutes [34]. $(C_4H_9)_2Sn(OOCC_{11}H_{23})_2$ was also tested against Pseudomonas aeruginos and Mycobacterium tuberculosis for the application in hard surface disinfectants for hospitals [26].

The insecticidal action of $(C_4H_9)_2Sn(OOCC_{11}H_{23})_2$ on rice stem borer, Chilo suppressalis and Callosobruchus chinensis was 9.4-fold less effective than that of $(C_4H_9)_3SnCl$, but it showed significant sustained insecticidal action on Chilo suppressalis. The germination and initial growth of the rice plant were strongly inhibited by using dibutyltin dilaurate [11]. The LD_{50} value for larvae of Culex pipiens pipiens is 0.71 ppm [20].

The anthelmintic activity of $(C_4H_9)_2Sn(OOCC_{11}H_{23})_2$ has been tested against tapeworms from chickens [1, 2, 3, 5, 6, 7, 9, 18, 19, 21]. The compound is very effective in doses of 280 mg/kg live weight [14]. Tapeworms of the species Hymenolepis fraterna were killed with doses of 2 ppm of $(C_4H_9)_2Sn(OOCC_{11}H_{23})_2$ [17]. Another test with the standard therapeutic dose of 125 mg/kg $(C_4H_9)_2Sn(OOCC_{11}H_{23})_2$ in gelatine capsules resulted in a total elimination of the adult forms of Choanotaenia infundibulum, Raillietina tetragona, Raillietina echinobothrida, Raillietia cesticillus, Cotugnia digonopora, and Hymenolepis carioca [13]. $(C_4H_9)_2Sn(OOCC_{11}H_{23})_2$ was also used as an anthelminticum for livestock [24], angus calves [10], sheep [12], and cats [22].

The LD_{50} value of dibutyltin dilaurate for rats is found to be 175 mg/kg oral [25, 31, 32, 33, 35, 40], and the lethal dose intraperitoneal is found to be 85 mg/kg for rats [4], and 710 mg/kg for mice [48]. Chronic feeding experiments with some organotin compounds used as stabilizers in mice showed $(C_4H_9)_2Sn(OOCC_{11}H_{23})_2$ to have a negative influence on the health condition of this animals [36]. Another toxicity test, the injection of $(C_4H_9)_2Sn(OOCC_{11}H_{23})_2$ into the yolk sac of fertile eggs prior to incubation resulted in a high order of toxicity and in teratogenic effects for this compound [16].

A poisoning caused by $(C_4H_9)_2Sn(OOCC_{11}H_{23})_2$ was diagnosed in 1975 in cattle, mink, and palm doves. The accidental addition of dibutyltin dilaurate to calf concentrates at levels up to 25000 ppm caused poisoning in 1000 cattle, of which 171 died. Palm doves ingesting concentrates containing 12500 ppm $(C_4H_9)_2Sn(OOCC_{11}H_{23})_2$ were also poisoned and had high concentrations of tin in tissues. Mink were inadvertently fed a vitamin-mineral supplement containing about 1700 ppm $(C_4H_9)_2Sn(OOCC_{11}H_{23})_2$ and showed a high susceptibility to this compound [41].

The tissue toxicity of $(C_4H_9)_2Sn(OOCC_{11}H_{23})_2$ was investigated by implantation experiments with plastic material that contained dibutyltin dilaurate as a stabilizer. Tissue necrosis could be found when such a material was implanted into the paravertebral muscle of a rabbit and death of cells when evaluated by the cell culture technique [23]. A significant decrease in body weight gain of rats was observed after 15 days oral exposure. No effect was observed in the activities of brain enzymes, succinic dehydrogenase, adenosine triphosphatase, acetylcholine esterase and monoamine oxidase. $(C_4H_9)_2Sn(OOCC_{11}H_{23})_2$ treatment resulted in a significant decrease in the activities of microsomal enzymes glucose-6-phosphatase, aminopyrine-N-demethylase, benzphetamines-N-demethylase, aniline hydroxylase, benzopyrene hydroxylase and also on cytochrome P-450 content, whereas no difference in the activities of mitochondrial enzymes, succinic dehydrogenase, Mg^{2+}-adenosine triphosphatase as well as in the activity of lysomal enzyme acid phosphatase was observed. The treatment with dibutyltin dilaurate produced an induction in heme oxygenase activity whereas the activity of aminolevulinic acid synthetase remained unaltered. Dibutyltin dilaurate affects the biotransformation mechanism and heme metabolism of hepatocytes [45]. Oral doses between 15 and 40 mg $(C_4H_9)_2Sn(OOCC_{11}H_{23})_2$ per kg body weight of rabbits for 6 weeks caused an increase of serum glutamate-oxalacetate transaminase, glutamate-pyruvate transaminase and lactate dehydrogenase, indicating liver damage [42].

Dibutyltin dilaurate is exempted from the requirement of a tolerance when used as ingredient in plastic ear tags and tail devices for ruminants and swine in pesticide formulations [47].

References:

[1] Kerr, K. B. (Poultry Sci. **31** [1952] 328/36).
[2] Hedges, E. S. (Metall **9** [1955] 23/6).
[3] Kerr, K. B., Walde, A. W., Dr. Salzburys Laboratories Iowa (U.S. 2702775 [1953/55]; C.A. **1955** 7816).
[4] Stoner, H. B., Barnes, J. M., Duff, J. I. (Brit. J. Pharmacol. Chemother. **10** [1955] 16/25).
[5] Abdou, A. H. (J. Helminthol. **30** [1956] 121/8; C.A. **1957** 6018).
[6] Edgar, S. A. (Poultry Sci. **35** [1956] 64/73; C.A. **1956** 13272).
[7] Edgar, S. A., Teer, P. A. (Poultry Sci. **36** [1957] 329/34).
[8] Kubota, S. (Yuki Gosei Kagaku Kyokaishi **16** [1958] 94/5).
[9] Enigk, K., Duewel, D. (Deut. Tierärztl. Wochenschr. **1** [1959] 10/6).
[10] Kohler, P. H., Rogoff, W. M. (J. Econ. Entomol. **52** [1959] 1223/4; C.A.**1961** 26349).

[11] Koike, H. (Bochu-Kagaku **26** [1961] 51/6; C.A. **63** [1965] 12255).
[12] Dorsman, W. (Versl. Landbouwk. Onderz. No. 68.14 [1962] 1/176 from C.A. **58** [1963] 14639).
[13] Gras, G., Graber, M., Vidal, A. (Trav. Soc. Pharm. Montpellier **22** [1962] 151/65; C.A. **59** [1963] 12067).
[14] Willomitzer, J. (Ved. Prace Vyzkum. Ustavu Vet. Lekar. Brno **1962** 265/76; C.A. **61** [1964] 16675).
[15] Kinoshita, Y., Muraoka, K. (Shokuhin Eiseigaku Zasshi **4** [1963] 78/85; C.A. **59** [1963] 9232).
[16] McLaughlin, J., Marliac, J. P., Verrett, M. J., Mutchler, M. K., Fitzhugh, O. G. (Toxicol. Appl. Pharmacol. **5** [1963] 760/71; C.A. **60** [1964] 6084).
[17] Gras, G., Castel, J. (Trav. Soc. Pharm. Montpellier **24** [1964] 116/9).
[18] Peardon, D. L., Haberman, W. O., Marr, J. E., Garland, F. W., Wilcke, H. L. (Poultry Sci. **44** [1964] 413/24; C.A. **62** [1965] 16674).
[19] Fry, J. L., Wilson, H. R. (Sunshine State Agr. Res. Rept. **10** [1965] 18/9; C.A. **62** [1965] 16675).
[20] Gras, G., Rioux, J. A. (Arch. Inst. Pasteur Tunis **42** [1965] 9/22).

[21] Graber, M., Gras, G. (Rev. Elevage Med. Vet. Pays. Trop. **19** [1966] 7/14; C.A. **65** [1966] 4327).

[22] Tareeva, A. I., Borodina, G. M. (Farmakol. Toksikol. **30** [1967] 207/9; C.A. **67** [1967] No. 20351).
[23] Stetson, J. B. (Inorg. Congr. Ser. Excerpta Med. No. 200 [1968/70] 455/7; C.A. **75** [1971] No. 33280).
[24] Thorburn, J. A., Bosman, C. J., Cooper, McDougall and Robertson Ltd. (S. African 67-00061 [1967/68]; C.A. **70** [1969] No. 60807).
[25] Klimmer, O. R. (Arzneim.-Forsch. Drug. Res. **19** [1969] 934/9).
[26] Verdiechio, R. J. (Proc. Ann. Meet. Chem. Spec. Manuf. Assoc. **56** [1969/70] 114/7).
[27] Wallhäuser, K. H., Heiss, L., Farbwerke Hoechst A.-G. (Ger. 1289937 [1966/69]; C.A. **70** [1969] No. 107730).
[28] McIntosh, A. H. (Ann. Appl. Biol. **66** [1970] 115/8).
[29] Vyzkumny a Vyvojovy Ustav Maltovin a Osikocementu Radotin (Fr. 2034075 [1970/70]; C.A. **75** [1971] No. 67149).
[30] Bartl, M., Vyskumny a Vyvojovy Ustav Maltovin a Osinkocementu (Ger. Offen. 2007514 [1970/71]; C.A. **75** [1971] No. 143678).

[31] Klimsch, P., Heibel, F. (Inform. Chim. No. 96 [1971] 125/6).
[32] Foakes, E. H. (Develop. PVC Technol. Proc. Symp., London 1973, pp. 78/89; C.A. **83** [1975] No. 11234).
[33] Luijten, J. G. A., Klimmer, O. R. (Tin Res. Inst. Publ. No. 501 [1973]).
[34] Petrova, V. I. (Kozh. Obuv. Prom. **1973** No. 12, pp. 22/4; C.A. **81** [1974] No. 38976).
[35] McCollister, D. D., Schober, E. A. (Environ. Qual. Saf. **4** [1975] 80/95).
[36] Majlathova (Cesk. Hyg. **21** [1976] 60/6; C.A. **85** [1976] No. 138133).
[37] Maksimova, N. S., Krynskaya, I. L., Evsyukov, V. I. (Plasticheskie Massy **1976** No. 12, pp. 33/4; C.A. **86** [1977] No. 101549).
[38] Statsek, N. K., Kucherenko, T. V. (Gig. Primen. Polim. Mater. **1976** 61/2; C.A. **86** [1977] No. 184164).
[39] Kimura, M., Fujikura Rubber Works, Ltd. (Japan. Kokai 77-80343 [1975/77]; C.A. **87** [1977] No. 186206).
[40] Smith, P. J., Luijten, J. G. A., Klimmer, O. R. (Intern. Tin Res. Inst. Publ. No. 538 [1978]).

[41] Shlosberg, A., Egyed, M. N. (Vet. Hum. Toxicol. **21** [1979] 1/3).
[42] Tanaka, S. (Tokyo Ika Daigaku Zasshi **38** [1980] 607/24; C.A. **95** [1981] No. 55779).
[43] Mazaev, V. T., Medvedev, Yu. M. (Gig. Sanit. **1981** No. 6, pp. 75/6; C.A. **95** [1981] No. 144846).
[44] Mishchenko, V. F., Zubov, V. A. (Obrastanie Biokorroz. Vodn. Srede **1981** 185/8).
[45] Mushtaq, M. N., Mukhtar, H., Datta, K. K., Tandon, S. G., Seth, P. K. (Drug Chem. Toxicol. **4** [1981] 75/88).
[46] Toshiba Silicone Co., Ltd. (Japan. Kokai Tokkyo Koho 82-96044 [1982]; C.A. **97** [1982] No. 184129).
[47] United States Environmental Protection Agency (Fed. Regist. **49** [1984] 31694/5).
[48] Schafer, E. W., Bowles, W. A. (Arch. Environ. Contam. Toxicol. **14** [1985] 111/29).
[49] Steinhaeuser, K. G., Amann, W., Spaeth, A., Polenz, A. (Vom Wasser **65** [1985] 203/14).

1.4.1.2.1.5.4.1.6.3 Uses

Dibutyltin dilaurate is made commercially in large amounts for several uses, but mainly in producing crystal clear extrusions and moulding stock, plastisols, and organosols, where it prevents accumulation of HCl although it is not suitable for use at high temperatures [89]. It is manufactured under different trade names like Mark 1038, Clear 1, Advastab DBTL, Thermolite 12, or Niax D-22 [62].

The compound is a stabilizer for poly(vinyl chloride) [2, 3, 14, 19, 54, 58, 62, 70, 78, 82, 94, 151, 172, 197]. The mechanism of the stabilization reaction was studied in numerous papers using several methods in different solvents and at various temperatures [7, 8, 11, 12, 33, 36, 37, 45, 48, 69, 79, 81, 90, 91, 102, 158, 170, 179, 193]. It could be demonstrated that the complexing of labile chlorine atoms by the tin atom is the most important step, followed by substitution of the complexed chlorine atom by a carboxylate ligand bound to the tin. This stabilization mechanism requires that the new allylic substituent is more thermally stable than the allylic chlorine [20]. The reaction of 2,4,6-trichloro-n-heptane as a PVC model compound with $(C_4H_9)_2Sn(OOCC_{11}H_{23})_2$ at 200 °C suggests that the stabilizer reacts with poly(vinyl chloride) only through structural defects rather than the normal structures [39]. Other investigations give indication for a radical mechanism via thermal or photochemical generated butyl radicals [71, 85].

$(C_4H_9)_2Sn(OOCC_{11}H_{23})_2$ is also a stabilizer for polyethylene [18], chloroparaffin oil [46], chlorinated rubber paints [121] and -enemals [1], ester type synthetic lubricants [192], and epoxyisocyanurate oligomers [96, 97].

Dibutyltin dilaurate is the most important catalyst for the reactions between NCO- and hydroxy groups, which is used for the synthesis of polyurethanes and for a lot of reactions of urethanes. The catalytic effect of $(C_4H_9)_2Sn(OOCC_{11}H_{23})_2$ on reactions between isocyanates and alcohols, on the formation of urethanes and polyurethanes and on the reactions of urethanes are studied in numerous papers [6, 25, 30, 38, 43, 49, 50, 52, 53, 57, 59, 61, 68, 77, 80, 83, 84, 88, 93, 95, 106 to 110, 112, 114, 115, 116, 118, 119, 120, 122 to 127, 129, 130, 131, 134, 139, 142, 143, 144, 147 to 150, 152, 153, 154, 157, 159, 160, 161, 164 to 167, 173, 177, 178, 181, 182, 185, 186, 194, 195, 201, 202, 204, 206, 208, 211]. Some of them are more oriented towards the influence of the catalyst on the use of the polyurethane material made [4, 5, 23, 35, 44, 65, 67, 75, 87, 92, 105, 136, 137, 138, 140, 141, 145, 146, 156, 176, 189, 190, 198]. The kinetics and the mechanism of the urethane formation, catalyzed by $(C_4H_9)_2Sn(OOCC_{11}H_{23})_2$ is investigated by several methods [9, 10, 17, 21, 22, 26, 27, 28, 31, 32, 40, 41, 51, 56, 74, 76, 98, 104, 117, 132, 155, 163, 175, 180, 187, 203, 207]. A complex formation between $(C_4H_9)_2Sn(OOCC_{11}H_{23})_2$ and the alcohols, participating in the reaction is postulated [15, 16, 29, 42, 72, 162].

$(C_4H_9)_2Sn(OOCC_{11}H_{23})_2$ is also used as a catalyst for the synthesis of polybutadiene ioneses [128] or copolymers containing glucose derivatives [86], and for the polymerization of formaldehyde [111], methyl methacrylate [55, 63, 64], vinyl acetate [63, 64], and styrene [63, 64].

Dibutyltin dilaurate is a silane-silanol condensation catalyst [73], and a catalyst for the crosslinking of silicone rubbers [34, 47, 99, 100, 103, 135, 168, 183, 188, 191], of halogenated butyl rubbers with silane coupling agents [199, 200, 210], of silicone-containing polyurethanes [60, 169], for the reaction of isocyanates with silanol groups on the surface of glass fibers [184], and for the condensation of chlorosulfonated polyethylene with 3-aminopropyltriethoxysilane [209]. A study of the reaction mechanism of the vulcanization of silicone rubber showed that well-defined tin-siloxane compounds are formed during this process. The rate of the vulcanization is dependent on the rate of formation of these compounds [13, 24].

$(C_4H_9)_2Sn(OOCC_{11}H_{23})_2$ is useful as crosslinking agent for polyethers [101], polyesters [171, 196], and epoxy resins [205]. The compound is a catalyst for the synthesis of adamantyl-substituted organosilicon carbamic esters [66], and for various transesterification reactions [174]. It can be used as a curing catalyst for polyacrylonitrile on cotton [133], and as a catalyst for alkyl silicate binders in refractory technology [113].

Patents in which the use of dibutyltin dilaurate for numerous uses is claimed are listed at the end of the reference list (p. 149) arranged by the year of the issue of the patent.

References:

[1] Gay, P. J. (Tin Its Uses No. 29 [1953] 6).
[2] Luijten, J. G. A., Pezarro, S. (British Plast. **30** [1957] 183/6).
[3] Winkler, D. E. (Ind. Eng. Chem. **50** [1958] 863/4).
[4] Anonymous (Chem. Eng. News **1959** No. 16, p. 43).
[5] Hostettler, F., Cox, E. F. (Ind. Eng. Chem. **52** [1960] 609/10).
[6] Farkas, A., Mills, G. A. (Advan. Catal. **13** [1962] 363/446).
[7] Shteding, M. M., Karpov, V. L. (Vysokomol. Soedin. **4** [1962] 1806/11; C.A. **59** [1963] 2989).
[8] Morikawa, T., Yoshida, K. (Kagaku To Kogyo [Osaka] **38** [1964] 667/71 from C.A. **62** [1965] 14891).
[9] Rand, L., Thir, B., Reegen, S. L., Frisch, K. C. (J. Appl. Polym. Sci. **9** [1965] 1787/95).
[10] Entelis, S. G., Nesterov, O. V., Zabrodin, V. B. (Kinetika Kataliz **7** [1966] 627/31; Kinet. Catal. **7** [1966] 552/5).

[11] Gel'fman, Ya. A., Lauris, I. V., Kuskova, V. P. (Sb. Tr. Vses. Nauchno Issled. Inst. Nov. Stroit. Mater. **1966** No. 14, pp. 54/8; C.A. **68** [1968] No. 79010).
[12] Gel'fman, Ya. A., Zemlyanskii, N. N., Lauris, I. V., Syutkina, O. P., Kuskova, V. P., Panov, E. M. (Sb. Tr. Vses. Nauchno Issled. Inst. Nov. Stroit. Mater. **1966** No. 14, pp. 58/61; C.A. **68** [1968] No. 79011).
[13] Nagy, J., Borbely-Kuszmann, A. (Period. Polytech. [Budapest] **10** [1966] 139/45).
[14] Oakes, V., Hughes, B. (Plastics [London] **31** [1966] 1132/4; C.A. **66** [1967] No. 47003).
[15] Entelis, S. G., Nesterov, O. V., Tiger, R. P. (J. Cell. Plast. **3** [1967] 360/3).
[16] Frisch, K. C., Reegen, S. L., Floutz, W. V., Oliver, J. P. (J. Polym. Sci. Polym. Chem. Ed. **5** [1967] 35/42).
[17] Nesterov, O. V., Chirkov, Yu. N., Entelis, S. G. (Kinetika Kataliz **8** [1967] 1371/3; Kinet. Catal. [USSR] **8** [1967] 1161/3).
[18] Aleksandrov, A. Yu., Baldokhin, Yu. V., Braginskii, R. P., Gol'danskii, V. I., Korytko, L. A., Leshchenko, S. S., Finkel, G. E. (Khim. Vysokikh Energ. **2** [1968] 331/7; High Energy Chem. [USSR] **2** [1968] 285/90).
[19] Fuchsman, C. H. (Advan. Chem. Ser. **85** [1968] 18/37).
[20] Klemchuk, P. P. (Advan. Chem. Ser. **85** [1968] 1/17).

[21] Lipatova, T. E., Bakalo, L. A., Loktionova, R. A. (Vysokomol. Soedin. A **10** [1968] 1554/60; Polym. Sci. [USSR] A **10** [1968] 1799/807).
[22] Zhitinkina, A. K., Shoshtaeva, M. V. (Sin. Fiz. Khim. Polim. **1968** No. 5, pp. 129/32; C.A. **70** [1969] No. 2976).
[23] Dunn, D. (J. Cell. Plast. **5** [1969] 341/7).
[24] Nagy, J., Borbely-Kuszmann, A. (Kinet. Mech. Polyreactions Intern. Symp. Macromol. Chem. Prepr., Budapest 1969, Vol. 4, pp. 297/301).
[25] Pogosov, Yu. L., Ratnikov, E. N. (Stogi Nauki Vysokomol. Soedin. **1969** 48/53 from C.A. **77** [1972] No. 90252).
[26] Lipatova, T. E., Bakalo, L. A., Ishchenko, S. S., Gongalo, Ř. F. (Sin. Fiz. Khim. Polim. **1970** No. 7, pp. 35/40; C.A. **75** [1971] No. 36747).
[27] Lipatova, T. E., Bakalo, L. A., Nizelski, Yu. N. (J. Macromol. Sci. Chem. **4** [1970] 1743/58).
[28] Lipatova, T. E., Bakalo, L. A., Sirotinskaya, A. L., Lopatina, V. S. (Vysokomol. Soedin. A **12** [1970] 911/6; Polym. Sci. [USSR] A **12** [1970] 1036/42).
[29] Reegen, S. L., Frisch, K. C. (J. Polym. Sci. A I **8** [1970] 2883/91).
[30] Volodarskaya, Yu. I., Sal'nikova, G. A., Shmidt, Ya. A. (Dokl. Akad. Nauk SSSR **195** [1970] 841/4; Dokl. Chem. Proc. Acad. Sci. USSR **190/195** [1970] 865/7).

[31] Zabrodin, V. B., Nestrov, O. V., Entelis, S. G. (Kinetika Kataliz **11** [1970] 114/9; Kinet. Catal. [USSR] **11** [1970] 91/5).
[32] Zabrodin, V. B., Nesterov, O. V., Entelis, S. G. (Kinetika Kataliz **11** [1970] 1060/1; Kinet. Catal. [USSR] **11** [1970] 877/8).
[33] Andersson, K. B., Sorvik, E. M. (J. Polym. Sci. Polym. Symp. No. 33 [1971] 247/67; C.A. **76** [1972] No. 34791).
[34] Evans, C. J. (Tin Its Uses No. 89 [1971] 5/8).
[35] Evans, C. J. (Tin Its Uses No. 90 [1971] 6/8).
[36] Horun, S. (Mater. Plast. [Bucharest] **8** [1971] 526/9; C.A. **76** [1972] No. 127967).
[37] Minsker, K. S., Fedoseeva, G. T., Zavarova, T. B., Krats, E. O. (Vysokomol. Soedin. A **13** [1971] 2265/78; Polym. Sci. [USSR] A **13** [1971] 2544/60).
[38] Mozzhukhina, L. V., Melamed, V. I., Evdokimova, V. A., Antipova, V. F., Apukhtina, N. P. (Uretanovye Elastomery **1971** 45/9).
[39] Suzuki, T., Takakura, I., Yoda, M. (Eur. Polym. J. **7** [1971] 1105/10; C.A. **76** [1972] No. 15265).
[40] Chirkov, Yu. N., Zabrodin, V. B., Nesterov, O. V., Entelis, S. G. (Kinetika Kataliz **13** [1972] 228/30; Kinet. Catal. [USSR] **13** [1972] 200/2).

[41] Grigoreva, V. A., Baturin, S. M., Entelis, S. G. (Vysokomol. Soedin. A **14** [1972] 1345/9; Polym. Sci. [USSR] A **14** [1972] 1507/12).
[42] Imoto, M. (Setchaku **16** [1972] 110/2; C.A. **77** [1972] No. 52580).
[43] Matsunaga, K., Tosaka, A., Sato, H., Yamashita, T. (Nippon Kagaku Kaishi **1972** 95/9; C.A. **76** [1972] No. 85077).
[44] Otey, F. H., Westhoff, R. P., Mehltretter, C. L. (J. Cell. Plast. **8** [1972] 156/9; C.A. **77** [1972] No. 89318).
[45] Suzuki, M., Tsuge, S., Takeuchi, T. (J. Polym. Sci. A I **10** [1972] 1051/60).
[46] Viska, J., Sulc, J., Galle, A. (Chem. Prumysl **22** [1972] 341/4; C.A. **78** [1973] No. 6235).
[47] Nagy, J., Lipovetz, I., Becker-Palossy, K., Borbely-Kuszmann, A. (Muanyag Gumi **10** [1973] 221/4; C.A. **80** [1974] No. 48918).
[48] Oki, Y., Mori, F. (Kobunshi Kagaku **30** [1973] 737/41).
[49] Olkhov, Yu. A., Baturin, S. M., Entelis, S. G. (Vysokomol. Soedin. A **15** [1973] 2058/62; Polym. Sci. [USSR] A **15** [1973] 2330/5).
[50] Dieter, J. A., Frisch, K. C., Wolgemuth, L. G. (Am. Chem. Soc. Div. Org. Coatings Plastics Chem. Papers **34** [1974] 703/8; C.A. **84** [1976] No. 75775).

[51] Grekov, A. P., Otroshko, G. V. (Zh. Org. Khim. **10** [1974] 1905/8; J. Org. Chem. [USSR] **10** [1974] 1916/8).
[52] Grekov, A. P., Otroshko, G. V. (Kratk. Tezisy Vses. Soveshch. Probl. Mekh. Geteroliticheskikh Reakts. **1974** 19/20; C.A. **85** [1976] No. 77131).
[53] Grigoreva, V. A., Belonogova, O. V., Baturin, S. M., Entelis, S. G. (Izv. Akad. Nauk SSSR Ser. Khim. **1974** 812/5; Bull. Acad. Sci. USSR Div. Chem. Sci. **1974** 777/9).
[54] Kulas, F. R., Thorshaug, N. P. (Polym. Eng. Sci. **14** [1974] 366/70).
[55] Nakamura, Y., Ouchi, T., Imoto, M. (Kobunshi Ronbunshu **31** [1974] 676/81; C.A. **82** [1975] No. 86676).
[56] Nesterov, O. V., Zabrodin, V. B., Chirkov, Y. N., Entelis, S. G. (Kinetika Kataliz **15** [1974] 1341/2; Kinet. Catal. [USSR] **15** [1974] 1183/4).
[57] Bakalo, L. A., Sirotinskaya, A. L., Dergunov, Yu. I., Gerega, V. F. (Sint. Fiz. Khim. Polim. **16** [1975] 6/10; C.A. **84** [1976] No. 123490).
[58] Malyshev, L. N., Zavarova, T. B. (Plasticheskie Massy **1975** No. 8, pp. 42/3; C.A. **83** [1975] No. 180196).
[59] Potter, R. M., Knutson, G. J., Zimmer, M. F. (Compat. Propellants Explos. Pyrotech. Plast. Addit. Conf., Dover, N. J., 1974 [1975], II-D, pp. 1/12 from C.A. **86** [1977] No. 142411).

[60] Kuznetsova, V. P., Zapunnaya, K. V., Omel'chenko, S. I., Bakalo, L. A., Soboleva, A. P. (Sint. Fiz. Khim. Polim. **18** [1976] 3/9; C.A. **86** [1977] No. 17501).

[61] Lipatova, T. E., Bakalo, L. A., Sirotinskaya, A. L., Blagonravova, A. A., Pronina, I. A. (Sint. Fiz. Khim. Polim. **18** [1976] 31/3; C.A. **86** [1977] No. 6414).
[62] Nass, L. I. (Encycl. PVC **1976** 295/384; C.A. **85** [1976] No. 6425).
[63] Nikolaeva, Yu. V., Antonova, G. M., Ryazanov, A. S., Duvakina, N. I., Nikolaev, A. F. (Sb. Tr. Leningr. Tekhnol. Inst. **1976** No. 2, pp. 42/5; C.A. **86** [1977] No. 5888).
[64] Nikolaeva, Y. V., Zhuk, E. G., Duvakina, N. I., Nikolaev, A. F. (Sb. Tr. Leningr. Tekhnol. Inst. **1976** No. 2, pp. 45/6; C.A. **86** [1977] No. 5882).
[65] Patten, W., Seefried Jr., C. G., Whitman, R. D. (4th SPI Intern. Cell. Plast. Conf. Proc., Montreal 1976, pp. 88/94; C.A. **86** [1977] No. 156395).
[66] Ushchenko, V. P., Kim, A. D., Khardin, A. P., Brel, V. K. (Zh. Obshch. Khim. **46** [1976] 2157/8; J. Gen. Chem. [USSR] **46** [1976] 2077).
[67] Brecker, L. R. (Plast. Eng. **33** [1977] 39/42; C.A. **86** [1977] No. 172221).
[68] Gromova, M. F., Bakalo, L. A., Chirkova, L. I., Mozheiko, L. N. (Khim. Drev. **1977** No. 4, pp. 71/4; C.A. **87** [1977] No. 186275).
[69] Kim, B. H., Kang, D. Y. (Chongi Hakhoechi **26** [1977] 185/90; C.A. **87** [1977] No. 24039).
[70] Plitz, I. M., Willingham, R. A., Starnes, W. H. (Macromolecules **10** [1977] 499/500).

[71] Rabek, J. F., Canback, G., Ranby, B. (J. Appl. Polym. Sci. **21** [1977] 2211/23).
[72] Thiele, L., Becker, R., Frommelt, H. (Faserforsch. Textiltech. **28** [1977] 343/8).
[73] Fierens, P., Vandendunghen, G., Segers, W., van Elsuwe, R. (React. Kinet. Catal. Letters **8** [1978] 179/87; C.A. **89** [1978] No. 163633).
[74] Mamutova, N. N., Rakhimova, T. F. (Funkts. Org. Soedin. Polim. **1978** 5/10; C.A. **92** [1980] No. 75479).
[75] Sarakuz, O. N., Sinaiskii, A. G., Sarakuz, V. N., Romanovskii, G. K. (Prom. Sint. Kauch. **1978** 15/9; C.A. **90** [1979] No. 24468).
[76] Volkova, N. N., Olkhov, Yu. A., Baturin, S. M., Smirnov, L. P. (Vysokomol. Soedin. A **20** [1978] 199/206; Polym. Sci. [USSR] A **20** [1978] 230/8).
[77] Volkova, N. N., Olkhov, Yu. A., Baturin, S. M., Smirnov, L. P. (Vysokomol. Soedin. B **20** [1978] 827/30).
[78] Miyazawa, T., Massaya, T. (Jigyo Gaiyo-Nagano-ken Seimitsu Kogyo Shikenjo **1979/80** 103/6; C.A. **95** [1981] No. 63042).
[79] Mori, F., Koyama, M., Oki, Y. (Angew. Makromol. Chem. **75** [1979] 123/35).
[80] Olczyk, W. (Organika **1979** 81/9).

[81] Park, G. S., Tran Van Hoang (Eur. Polym. J. **15** [1979] 817/22).
[82] Cooray, B. B., Scott, G. (Polym. Degrad. Stab. **2** [1980] 35/51).
[83] Erae, V. A., Jaeaeskelaeinen, P., Ukkonen, K. (Angew. Makromol. Chem. **88** [1980] 79/88).
[84] Erae, V. A., Lehtinen, A. (Finn. Chem. Letters **1980** 59/61).
[85] Gupta, V. P., Pierre, L. E. S. (J. Polym. Sci. Polym. Chem. Ed. **18** [1980] 1483/8).
[86] Hayakawa, T., Yamada, T., Hidaka, S., Yamagishi, M., Takeda, K., Toda, F. (Polym. Prepr. Am. Chem. Soc. Div. Polym. Chem. **20** [1980] 530/1).
[87] Karpel, S. (Tin Its Uses No. 125 [1980] 1/6).
[88] Kutyanina, L. G., Baramboim, N. K. (Izv. Vysshikh Uchebn. Zavedenii Tekhnol. Legk. Prom. **1980** No. 23, pp. 60/6).
[89] Lower, E. S. (Pigm. Resin. Technol. **9** [1980] 10/1).
[90] Minsker, K. S., Kolesov, S. V., Kotsenko, L. M. (Vysokomol. Soedin. A **22** [1980] 2253/8; Polym. Sci. [USSR] A **22** [1980] 2471/7).

[91] Mitani, K., Ogata, T., Nakatsukasa, M., Mitzutani, Y. (Polymer **21** [1980] 1463/8; C.A. **94** [1981] No. 122461).
[92] Morozov, Yu. L., Alter, Yu. M., Knyazhanskii, S. L., Belov, I. B. (Kauch. Rezina **1980** 30/3; C.A. **93** [1980] No. 73340).
[93] Olczyk, W., Szpykowska, H. (Organika **1980** 103/12; C.A. **95** [1981] No. 43769).
[94] Park, G. S., Tran Van Hoang (Eur. Polym. J. **16** [1980] 779/83).
[95] Richter, E. B., Macosko, C. W. (Polym. Eng. Sci. **20** [1980] 921/4).
[96] Sorokin, M. F., Shode, L. G., Onosova, L. A. (Tr. Mosk. Khim. Tekhnol. Inst. No. 110 [1980] 102/14).
[97] Sorokin, M. F., Shode, L. G., Onosova, L. A., Murzakhanova, L. A. (Tr. Mosk. Khim. Tekhnol. Inst. No. 110 [1980] 1002).
[98] Steinle, E. C., Critchfield, F. E., Castro, J. M., Macosko, C. W. (J. Appl. Polym. Sci. **25** [1980] 2317/29).
[99] Van der Weij, F. W. (Makromol. Chem. **181** [1980] 2541/8).
[100] Yokoyama, T., Suzuki, H., Mukai, J. (IEEE Trans. Electr. Insul. **EI-15** [1980] 373/81).

[101] Andre, D., Le Nest, J. F., Cheradame, H. (Eur. Polym. J. **17** [1981] 57/61).
[102] Ayrey, G., Hsu, S. Y., Poller, R. C. (Org. Coat. Appl. Polym. Sci. Proc. **46** [1981] 639/4).
[103] Bajaj, P., Varshney, S. K. (Polymer **22** [1981] 372/6; C.A. **95** [1981] No. 26369).
[104] Bechara, I. S. (ACS Symp. Ser. No. 172 [1981] 393/402).
[105] Brecker, L. R. (Poliplasti Plast. Rinf. **29** [1981] 63/6; C.A. **95** [1981] No. 25931).
[106] Dmitrieva, T. S., Baraban, O. P., Valuev, V. I., Belov, I. B. (Prom. Sint. Kauch. **1981** 14/6; C.A. **96** [1982] No. 53535).
[107] Duraiaj, B., Anbazhagan, K., Venkatarao, K. (Angew. Makromol. Chem. **96** [1981] 157/65; C.A. **95** [1981] No. 62789).
[108] Galla, E. A., Mascioli, R. L., Bechara, I. S. (J. Elastomers Plast. **13** [1981] 205/23).
[109] Galla, E. A., Mascioli, R. L., Bechara, I. S. (Proc. S. P. I. 26th Ann. Urethane Div. Tech. Conf., 1981, pp. 24/31; C.A. **96** [1982] No. 53341).
[110] Huynh-Ba, G., Jerome, R. (ACS Symp. Ser. No. 172 [1981] 205/17).

[111] Ishida, S. (J. Appl. Polym. Sci. **26** [1981] 2743/50).
[112] Joel, J., Thiele, L., Kühn, G. (Acta Polym. **32** [1981] 27/30).
[113] Jones, K., Biddle, K. D., Das, A. K., Emblem, H. G. (Silicates Ind. **46** [1981] 107/11).
[114] Klempner, D., Frisch, K. C. (ACS Symp. Ser. No. 172 [1981] 533/51).
[115] Korzyuk, E. L., Zharkov, V. V. (Kinetika Kataliz **22** [1981] 522/5; Kinet. Catal. [USSR] **22** [1981] 399/400).
[116] Kresta, J. E., Garcia, A., Frisch, K. C., Linden, G. (ACS Symp. Ser. No. 172 [1981] 403/17).
[117] Nadler, M. P., Reed, R., Doyle, J., Chan, M. (CPIA Publ. **340** [1981] 51/62).
[118] Olczyk, W. (Chem. Stosow. **25** [1981] 41/51; C.A. **95** [1981] No. 187972).
[119] Olkhov, Yu. A., Kalmykov, Yu. B., Baturin, S. M. (Vysokomol. Soedin. A **23** [1981] 677/81; Polymer. Sci. [USSR] A **23** [1981] 763/8).
[120] Prajsnar, B., Osterheld, K., Schütz, W. (Chemiker Ztg. **105** [1981] 359/63).

[121] Quian, H., Chung, M. Y., Sun, C. K. (Tuliao Gongye No. 61 [1981] 1/3 from C.A. **95** [1981] No. 188701).
[122] Rusch, T. E., Raden, D. S. (Rev. Gen. Caoutch. Plast. No. 610 [1981] 75/82; C.A. **95** [1981] No. 25655).
[123] Sorokin, M. F., Shode, L. G., Mirenskii, R. B., Kabanova, N. S. (Lakokras. Mater. Ikh Primen **1981** No. 1, pp. 7/9; C.A. **94** [1981] No. 176788).
[124] Ulrich, H. (ACS Symp. Ser. No. 172 [1981]; C.A. **96** [1982] No. 53885).

[125] Valuev, V. I., Gordeeva, S. B., Romanovskii, G. K., Sarakuz, O. N. (Kauch. Rezina **1981** 8/9; C.A. **95** [1981] No. 170731).
[126] Volkova, N. N., Olkhov, Yu. A., Baturin, S. M., Smirnov, L. P. (Sint. Poliuretanov **1981** 90/5).
[127] Wong, S. W., Frisch, K. C. (Advan. Urethane Sci. Technol. **8** [1981] 75/92; C.A. **96** [1982] No. 105544).
[128] Yamashita, S., Itoi, M., Kohjiya, S. (Kobunshi Ronbunshu **38** [1981] 189/94; C.A. **95** [1981] No. 170690).
[129] Zaplatin, A. A., Kafengauz, I. M., Khudyak, E. P., Frolov, Yu. M., Samigullin, F. K. (Plasticheskie Massy **1981** 36/7).
[130] Zhigotskii, A. G., Dyaminov, M. S. (Khim. Tekhnol. [Kiev] **1981** 44/5).

[131] Chujo, Y., Tatsuda, T., Yamashita, Y. (Polym. Bull. [Berlin] **8** [1982] 239/44).
[132] Dorozhkin, V. P., Kimelblat, V. I. (J. Polym. Sci. Polym. Chem. Ed. **20** [1982] 2863/78).
[133] Guise, G. B., Freeland, G. N. (Textile Res. J. **52** [1982] 182/5).
[134] Ilavsky, M., Dusek, K. (Polym. Bull. [Berlin] **8** [1982] 359/66).
[135] Jiang, S., Zhou, S., Chen, S. (Taiyangneng Xuebao **3** [1982] 464/6).
[136] Karpel, S. (Tin Its Uses No. 132 [1982] 14/6).
[137] Marciano, J. H., Rojas, A. J., Williams, R. J. J. (Polymer **23** [1982] 1489/92).
[138] McBride, P. (J. Oil Colour Chem. Assoc. **65** [1982] 257/62).
[139] Merrill, E. W., Salzman, E. W., Wan, S., Mahmud, N., Kushner, L., Lindon, J. N., Curme, J. (Trans. Am. Soc. Artif. Intern. Organs **28** [1982] 482/7).
[140] Movsesyan, L. G., Andreasyan, G. A. (Prom. Arm. **1982** No. 12, pp. 28/31; C.A. **98** [1983] No. 108621).

[141] Rojas, A. J., Marciano, J. H., Williams, R. J. (Polym. Eng. Sci. **22** [1982] 840/4).
[142] Smirnova, T. F., Naimark, N. I., Kuzmin, V. N., Tepteleva, L. A., Egorov, S. F. (Plasticheskie Massy **1982** 22/4).
[143] Varentsova, N. V., Gol'dshtein, I. P., Paleeva, I. E., Tarakanov, O. G., Gur'yanova, E. N. (Zh. Obshch. Khim. **52** [1982] 1612/20; J. Gen. Chem. [USSR] **52** [1982] 1424/31).
[144] Dang Thi, A. T., Camberlin, Y. L., Pascault, J. P. (Angew. Makromol. Chem. **111** [1983] 29/51).
[145] Halasa, E. (Polimery [Warsaw] **28** [1983] 126/9).
[146] Hourston, D. J., Zia, Y. (J. Appl. Polym. Sci. **28** [1983] 2139/49).
[147] Idage, B. B., Vernekar, S. P., Ghatge, N. D. (J. Appl. Polym. Sci. **28** [1983] 3559/63).
[148] Ihms, D., Stoffer, J. O., Schneider, D. F., McClain, C. (10th Proc. Water-Borne Higher-Solids Coat. Symp., Hattiesburg, Miss., 1983, 233/42).
[149] Ilavsky, M., Dusek, K. (Polymer **24** [1983] 981/90).
[150] Kim, J. K., Kim, S. C. (Pollimo **7** [1983] 28/38).

[151] Lisitskii, V. V., Gataullin, R. F., Ryazantseva, V. G., Kislitsin, V. K., Davidenko, N. V., Minsker, K. S. (Plasticheskie Massy **1983** 18/9).
[152] Mikheev, V. V., Svetlakov, N. V., Semenova, L. V. (Lakokras. Mater. Ikh Primen **1983** No. 6, pp. 5/6; C.A. **100** [1984] No. 105162).
[153] Minatono, S., Takamatsu, H., Yamashita, S., Mutsumoto, T. (Nippon Setchaku Kyokaishi **19** [1983] 132/8).
[154] Rakhmatullina, G. M., Averko-Antonovich, L. A., Kirpichnikov, P. A. (Izv. Vysshikh Uchebn. Zavedenii Khim. Khim. Tekhnol. **26** [1983] 735/9).
[155] Robins, J., Edwards, B. H., Tokach, S. K. (Polym. Mater. Sci. Eng. **49** [1983] 331/5).
[156] Szulenyi, F., Mohny, J. (Plasty Kauc. **20** [1983] 299/302).
[157] Wongkamolsesh, K., Kresta, J. E. (Polym. Mater. Sci. Eng. **49** [1983] 465/8).

[158] Ayrey, G., Hsu, S. Y., Poller, R. C. (Polym. Sci. Technol. [Plenum] **26** [1984] 171/87).
[159] Ball, P., Fuellmann, H., Schwalm, R., Heitz, W. (C. Mol. Chem. **1** [1984] 95/108; C.A. **103** [1985] No. 37803).
[160] Bats, J. P., Ducasse, Y., Lalande, R., Tauzia, M. (Eur. Polym. J. **20** [1984] 997/1001).

[161] Berlin, P. A., Bondarenio, S. P., Tiger, R. P., Entelis, S. G. (Khim. Fiz. **3** [1984] 722/30).
[162] Berlin, P. A., Tiger, R. P., Chirkov, Yu. N., Entelis, S. G. (Khim. Fiz. **3** [1974] 1448/54).
[163] Boitsov, E. N., Trub, E. P. (Kinetika Kataliz **25** [1984] 279/82).
[164] Camargo, R. E., Andrews, J. S., Macosko, C. W., Wellinhoff, S. T. (Polym. Prepr. Am. Chem. Soc. Div. Polym. Chem. **25** [1984] 294/5).
[165] Carlson, G. M., Neag, C. M., Kuo, C., Provder, T. (Advan. Urethane Sci. Technol. **9** [1984] 47/64).
[166] Carlson, G. M., Neag, C. M., Kuo, C., Provder, T. (Polym. Prepr. Am. Chem. Soc. Div. Polym. Chem. **25** [1984] 171/2).
[167] Castro, J. M., Macosko, C. W., Perry, S. J. (Polym. Commun. **25** [1984] 82/7; C.A. **100** [1984] No. 210510).
[168] Cook, J. R. (J. Text. Inst. **75** [1984] 191/5).
[169] Ebdon, J. R., Hourston, D. J., Klein, P. G. (Polymer **25** [1984] 1633/9).
[170] Guyot, A., Michel, A., Tran Van Hoang (Polym. Sci. Technol. [Plenum] **26** [1984] 155/69).

[171] Interox Chemicals Ltd. (Res. Discl. No. 244 [1984] 382/3).
[172] Matsusaka, K., Koyama, M. (Angew. Makromol. Chem. **125** [1984] 149/59).
[173] Nguyen, L. T., Suh, N. P. (Thermochim. Acta **76** [1984] 265/71).
[174] Pilati, F., Munari, A., Manaresi, P. (Polym. Commun. **25** [1984] 187/9 from C.A. **101** [1984] No. 91494).
[175] Robins, J., Edwards, B. H., Tokach, S. K. (Advan. Urethane Sci. Technol. **9** [1984] 65/76).
[176] Sebenik, A., Kavcic, M., Osredkar, U. (Hem. Ind. **38** [1984] 176/8).
[177] Solodovnik, P. I., Varlivans, V. (Latvijas PSR Zinatnu Akad. Vestis Kim. Ser. **1984** No. 6, pp. 687/96; C.A. **102** [1985] No. 85019).
[178] Trostyanskaya, E. B., Babaevskii, P. G., Kulik, S. G., Stepanova, M. I. (Vysokomol. Soedin. A **26** [1984] 1053/9).
[179] Tran Van Hoang, Michel, A., Guyot, A. (Polym. Degrad. Stab. **9** [1984] 73/87).
[180] Wang, L., Han, X., Zhang, Q., Zhang, Y. (Yingyong Huaxue **1** [1984] 68/71 from C.A. **102** [1985] No. 186428).

[181] Webb, D. D. (Proc. SPI 28th Ann. Tech. Mark. Conf., Westport, Conn., 1984, pp. 2/5).
[182] Wong, S., Frisch, K. C. (Polym. Mater. Sci. Eng. **50** [1984] 480/4).
[183] Yokoyama, T., Kinjo, N., Mukai, J. (J. Appl. Polym. Sci. **29** [1984] 1951/8).
[184] Yosomiya, R., Morimoto, K. (Polym. Bull. [Berlin] **12** [1984] 41/8).
[185] Byrne, C. A. (Polym. Prepr. Am. Chem. Soc. Div. Polym. Chem. **26** [1985] 10/1).
[186] Camargo, R. E., Macosko, C. W., Tirrell, M., Wellinghoff, S. T. (ACS Symp. Ser. No. 270 [1985] 27/51).
[187] Carlson, G. M., Provder, T., Neag, C. M. (12th Proc. Water-Borne Higher-Solids Coat. Symp., Hattiesburg, Miss., 1985, pp. 44/58).
[188] Cook, J. R., Fleischfresser, B. E., Wemyss, A. M., White, M. A. (J. Text. Inst. **76** [1985] 57/63).
[189] DeVido, J. P. (Mod. Paint Coat **75** [1985] 42/6).
[190] Fox, R. B., Bitner, J. L., Hinkley, J. A., Carter, W. (Polym. Eng. Sci. **25** [1985] 157/63).

[191] Hron, P., Aisman, M., Schatz, M. (Sb. Vys. Sk. Chem. Technol. Praze Oddil S **12** [1985] 55/63).

[192] Hronec, M., Cvengrosova, Z., Kizlink, J., Stolcova, M., Malik, L., Ilavsky, J., Sitek, J. (Oxid. Commun. **8** [1985/86] 51/64).
[193] Jirackova-Audouin, L., Verdu, J. (Eur. Polym. J. **21** [1985] 421/6).
[194] Kamatani, Y. (Polym. Mater. Sci. Eng. **52** [1985] 427/31).
[195] Keller, T. M. (J. Polym. Sci. Polym. Chem. Ed. **23** [1985] 2557/9).
[196] Kerle, E. J. (Res. Discl. No. 258 [1985] 547).
[197] Koyama, M., Matsusaka, K. (Nippon Kagaku Kaishi **1985** 249/54).
[198] Masiulanis, B., Siwek, P., Kwiatkowski, A. (Polimery [Warsaw] **30** [1985] 69/74).
[199] Matsuo, T., Kakei, M., Ikeda, S., Matsumoto, T., Yamashita, S. (Nippon Gomu Kyokaishi **58** [1985] 94/100).
[200] Matsuo, T., Kakei, M., Matsumoto, T., Yamashita, S. (Nippon Gomu Kyokaishi **58** [1985] 88/93).
[201] Mikheev, V. V., Svetlakov, N. V., Semenova, L. V., Gilmanov, R. R. (Lakokras. Mater. Ikh. Primen **1985** No. 5, pp. 10/1; C.A. **104** [1986] No. 150814).
[202] Ono, H. K., Jones, F. N., Pappas, S. P. (J. Polym. Sci. Polym. Letters Ed. **23** [1985] 509/15; C.A. **104** [1986] No. 34355).
[203] Provder, T., Carlson, G. M., Neag, C. M. (Polym. Prepr. Am. Chem. Soc. Div. Polym. Chem. **26** [1985] 19/20).
[204] Senger, J. S., Yilgor, I., McGrath, J. E., Patsiga, R. A. (Polym. Prepr. Am. Chem. Soc. Div. Polym. Chem. **26** [1985] 244/6).
[205] Smith, J. D. B. (Proc. 17th Electr. Electron. Insul. Conf., New York 1985, pp. 105/9).
[206] Sumkina, V. G., Palyutkin, G. M., Zharkov, V. V. (Kinetika Kataliz **26** [1985] 476/80).
[207] Webb, D. D. (J. Cell. Plast. **21** [1985] 208/12).
[208] Wongkamolsesh, K., Kresta, J. E. (ACS Symp. Ser. No. 270 [1985] 111/21).
[209] Yamada, A., Shiokaramatsu, Y., Yamashita, S. (Makromol. Chem. **186** [1985] 2275/82).
[210] Yamashita, S., Yamada, A., Ohata, M., Kohjiya, S. (Makromol. Chem. **186** [1985] 1373/8).
[211] Yilgor, I., MacGrath, J. E. (J. Appl. Polym. Sci. **30** [1985] 1733/9).

Patented Uses for Dibutyltin Dilaurate

Yngve, V., Carbide and Carbon Chemicals Corp., Vinyl Resin Compositions, U.S. 2307090 [1943]; C.A. **1943** 3532.

Rugeley, E. W., Quattlebaum, W. M., Carbide and Carbon Chemicals Corp., Stabilizers for Vinyl-Resin Filaments, U.S. 2344002 [1944]; C.A. **1944** 3392.

Johnson, E. W., Metal and Thermit Corp., Stabilized Chlorinated Esters of Fatty Acids, U.S. 2524 [1950]; C.A. **1951** 2498.

Churchill, J. W., Mathieson Chemical Corp., Alterungsschutzmittel für Polystyrol, Fr. 991331 [1949/51]; C. **1953** 3647.

Churchill, J. W., Mathieson Chemical Corp., Polymerized Dichlorostyrenes Containing Organotin Anticrazing Agents, U.S. 2643242 [1953]; C.A. **1953** 10904.

Sorenson, R. A., Nixon Nitration Works, Plastic Molding Powder, U.S. 2654716 [1953]; C.A. **1954** 2411.

Rossiter, W. T., Currie, C. C., Dow Corning Ltd., Organosiliconcompositionen, Brit. 713233 [1952/54]; C. **1955** 9922.

Higgins, N. A., E. I. du Pont de Nemours & Co., Polymers of Hydroxyacetic Acid and Its Ester, U.S. 2676945 [1954]; C.A. **1954** 11111.

Caldwell, J. R., Eastman Kodak Co., Organometallic Tin Compounds as Catalysts in the Preparation of Polyesters, U.S. 2720507 [1955]; C.A. **1956** 2205.

Barker, R. L., Pure Chemicals Ltd., Organotin Compounds and Their Use in Stabilization of Resinous Compositions, Brit. 761568 [1956]; C.A. **1957** 10561.

Rossiter, W. T., Currie, C. C., Dow Corning Corp., Stabilized, Corrosion-Inhibiting Polymeric Organosilicon Compositions, U.S. 2742368 [1956]; C.A. **1956** 11713.

Chemische Werke Albert, Hardening of Epoxide Resins, Brit. 783764 [1957]; C.A. **1958** 5883.

Di Giulio, E., Ciampa, G., Montecatini, Societa Generale Per l'Industria Mineraria e Chimica, Fibers from Highly Linear Propylene Polymers, Ital. 561304 [1957]; C.A. **1958** 13615.

Weinberg, E. L., Tomka, L. A., Metal & Thermit Corp., Antioxidant for Rubber, U.S. 2789107 [1957]; C.A. **1957** 10939.

Airs, R. S., Evans, H. C., Shell Research Ltd., High-Pressure Steam Turbine Oils, Brit. 833873 [1960]; C.A. **1960** 21745.

Squires, S., Goodier, K., Shell Research Ltd., Color-Stable Polystyrene, Brit. 881578 [1960]; C.A. **1962** 8940.

Hostettler, F., Union Carbide Corp., Foaming of Polyurethane, Ger. 1091324 [1960]; C.A. **1961** 25359.

Ito, T., Kyodo Yakuhin Co., Ltd., Heat-Stabilizer for Poly(Vinyl Chloride), Japan. 60-8337 [1960]; C.A. **57** [1962] 6146.

Union Chimique Belge S. A., Air-Drying Varnish Compositions, Brit. 827714 [1960]; C.A. **1961** 1027; Fr. 1177442 [1956/59].

Luz, W. D., Metallgesellschaft A.-G., Stabilization of Poly(Vinyl Chloride), Brit. 874574 [1960/61]; C.A. **57** [1962] 1080.

Horn, C. F., Vineyard, H., Union Carbide Corp., Polyamide Stabilization, Brit. 899896 [1959/62]; C.A. **57** [1962] 11391.

Tanaka, H., Tomioka, A., Odaira, A., Okamura, K., Satokawa, K., Yonetani, M., Daiwa Spinning Co., Ltd. and Osaka Kinzoku Kogyo Co., Ltd., Copolymer Spinning Solutions, Japan. 62-1475 [1962]; C.A. **60** [1964] 10869.

Nakana, T., Kubota, T., Umezawa, Y., Yokohama Rubber Co., Ltd., Improving the Heat Resistance of Rayon Tire Cords, U.S. 3108010 [1960/63]; C.A. **60** [1964] 751.

Strickland, A., Minnesota Mining & Manufacturing Co., Katalysator zur Synthese von Polyurethanen, Ger. Offen. 1159633 [1960/63]; C.A. **60** [1964] 9456.

Hostettler, F., Union Carbide Corp., Polyurethane Foams, U.S. 3194773 [1958/65]; C.A. **63** [1965] 8577.

Lazcano, C. S., Resinous Compositions and Stabilizers Therefore, Brit. 1008589 [1962/65]; C.A. **70** [1969] No. 97639.

Rhone-Poulenc S. A., Organic Compounds of Tin and Titanium, Fr. 1392648 [1964/65]; C.A. **63** [1965] 1816.

Wyandotte Chemicals Corp., Polyurethane Coating Compositions, Brit. 994348 [1961/65]; C.A. **63** [1965] 5904.

Oakes, V., Hutton, R. E., Tonge, B. L., Pure Chemicals Ltd., Vinyl Chloride Resin Stabilizers, Brit. 1047949 [1966]; C.A. **66** [1967] No. 19249.

Kusuhara, M., Oki, D., Tokyo Shibaura Electric Co., Ltd., Water-Repellent Treatment of Fibers and Paper, Japan. 69-32035 [1967/69]; C.A. **72** [1970] No. 112749.

Lewis, J. T., Malani, C., Midland Silicones Ltd., Mechanically Initiated Silane Crosslinking of Organic Polymers, Fr. 1577875 [1967/69]; C.A. **72** [1970] No. 91200.

Nelson, M. E., Dow Corning Corp., Treatment of Glass Fibers, Fr. 1577541 [1967/69]; C.A. **72** [1970] No. 91445.

Sterman, S., Marsden, J. G., Union Carbide Corp., Room-Temperature-Hardening Silicone Elastomers, Ger. Offen. 1913684 [1968/69]; C.A. **72** [1970] No. 4212.

Wessel, J. K., Dow Corning Corp., Room-Temperature Hardenable Organopolysiloxane Elastomer, Ger. Offen. 1907980 [1968/69]; C.A. **72** [1970] No. 13621.

Yamamoto, K., Nakatsuka, K., Mitsubishi Rayon Co., Ltd., Low-Molecular-Weight Polypropylene from High Molecular-Weight Polypropylene, Japan. 69-15185 [1965/69]; C.A. **72** [1970] No. 13232.

Beers, M. D., Smith, A. H., General Electric Co., Vulcanized Organopolysiloxanes, Ger. Offen. 2000396 [1969/70]; C.A. **73** [1970] No. 67499.

Cosan, Chemical Corp., Silicone Rubber Composition with Improved Vulcanization Characteristics, Fr. 1588111 [1968/70]; C.A. **73** [1970] No. 121397.

Critchfield, F. E., Lundberg, R. D., Union Carbide Corp., Catalytic Production of Solid, Linear Cyclic Ester Polymers, Fr. Demande 2026274 [1968/70]; C.A. **76** [1972] No. 127748.

Enoki, K., Yoshihara, Y., Kirai, Y., Japan Soda Co., Ltd., Stabilization of Ethylene Terephthalate Polyesters, Ger. Offen. 1954959 [1968/70]; C.A. **73** [1970] No. 26275.

Evans, E. M., Hughes, A., Shell Internationale Research Maatschappij N. V., Compressor Lubricant, Brit. 1176094 [1968/70]; C.A. **72** [1970] No. 69011.

Fukuda, T., Kubota, K., Japan Oils and Fats Co., Ltd., Easily-Burnable Resin, Japan. 70-30358 [1966/70]; C.A. **74** [1971] No. 64905.

Fuse, M., Todo, T., Fujimori Industry Co., Ltd., Removable Adhesive Tapes, Japan. 70-31686 [1967/70]; C.A. **75** [1971] No. 7014.

Goldman, G. K., Morris, L., Products Research and Chemical Corp., Nitrile-Substituted Polysiloxanes, U.S. 3531508 [1967/70]; C.A. **74** [1971] No. 4457.

Gruber, H., Degener, E., Schmelzer, H. G., Farbenfabriken Bayer A.-G., Elastic Polyurethanes, Ger. Offen. 1914365 [1969/70]; C.A. **74** [1971] No. 13990.

Gudgeon, H., Haggis, G. A., Yates, E. W., Imperial Chemical Industries Ltd., Polyols from 2,3-Tolylenediamines and Their Use for Preparing Polyurethane Foams, Ger. Offen. 2017038 [1969/70]; C.A. **74** [1971] No. 4246.

Hostettler, F., Union Carbide Corp., Polyurethane Foamed Resins, Brit. Amended 892136 [1957/70]; C.A. **76** [1972] No. 46929.

Imperial Chemical Industries, Ltd., Werkwijze ter Bereiding van een Hoog-Molekulaire Polyester, Neth. 6915283 [1969/70].

Ishida, S., Asahi Chemical Industry Co., Ltd., Poly(Oxymethylene), Ger. Offen. 1946077 [1968/70]; C.A. **72** [1970] No. 133404.

Ishida, S., Fujita, M., Kajihara, S., Tokushige, A., Aoyama, K., Asahi Chemical Industry Co., Ltd., Continuous Polymerization of Formaldehyde, Japan. 70-31470 [1966/70]; C.A. **74** [1971] No. 13781.

Ishida, S., Sato, K., Fujita, M., Fukuda, H., Mori, K., Asahi Chemical Industry Co., Ltd., Heat-Stabilized Thermoplastic Polyacetals, Japan. 70-20914 [1966/70]; C.A. **73** [1970] No. 88600.

Ishida, S., Sato, K., Fujita, M., Mori, K., Asahi Chemical Industry Co., Ltd., Molecular Weight Control in the Polymerization of Formaldehyde, Japan. 70-31469 [1966/70]; C.A. **74** [1971] No. 13557.

Ishida, S., Sato, K., Mori, K., Fujita, M., Asahi Chemical Industry Co., Ltd., Formaldehyde-N-Alkylacrylamide Copolymers, Japan. 70-14554 [1965/70]; C.A. **73** [1970] No. 56626.

Jayawant, M. D., E. I. du Pont de Nemours & Co., Polyurethanes Using Organotin and Time-Lapse Catalysts, Ger. Offen. 1956672 [1968/70]; C.A. **73** [1970] No. 67273.

Leebrick, J. R., Coasn Chemical Corp., Catalytic Hardening of Silicone Rubbers, Ger. Offen. 1803172 [1968/70]; C.A. **73** [1970] No. 99844.

Maier, J. E., Minnesota Mining and Manufacturing Co., Hardenable Epoxy-Amine Molding Compositions, Ger. Offen. 1950559 [1968/70]; C.A. **73** [1970] No. 4769.

Ono, H., Watanabe, K., Itami, A., Toyo Spinning Co., Ltd., Linear Polyurethanes, Japan. 70-03115 [1966/70]; C.A. **72** [1970] No. 112690.

Rudolph, K. H., Büchner, W., Töpsch, H., Noll, W., Goller, H., Farbenfabriken Bayer A.-G., Adhesion-Resistant Organopolysiloxane Coatings for Impregnation of Paper, Ger. Offen. 2015154 [1970]; C.A. **76** [1972] No. 47580.

Sakai, T., Fujioka, S., Okubo, J., Shinohara, Y., Toray Industries, Inc., Heat Stabilizers for Polysulfones, Japan. 70-25392 [1965/70]; C.A. **74** [1971] No. 77072.

Shima, T., Asami, A., Yamashiro, S., Teijin Ltd., Crystalline Polyamides Having at Least One Aliphatic Unsaturated Bond per 1000 Molecular Weight, Japan. 70-12150 [1967/70]; C.A. **73** [1970] No. 56625.

Smith, R. L., General Electric Co., Room-Temperature-Vulcanizing Silicone Rubber Foam, U.S. 3516951 [1967/70]; C.A. **73** [1970] No. 36312.

Sun Oil Company, Method of Annealing Aluminium, Brit. 1182800 [1967/70].

Tanaka, T., Ichigawa, W., Asahi Glass Co., Ltd., Fire-Retarding Polyurethane Resins, Japan. 70-13713 [1966/70]; C.A. **73** [1970] No. 88601.

Valle, P., Chomat, M., Societa Industrielle des Silicones, Silicone-Based Compositions for Foundry Core Molds, Fr. Addn. 94898 [1968/70]; C.A. **73** [1970] No. 99736.

Voigt, C. Söhne, Isocyanate-Containing Polymer Compositions, Ger. Offen. 1913911 [1969/70]; C.A. **73** [1970] No. 121204.

Yamamoto, T., Yukida, Y., Toray Industries, Inc., Polypropylene with Good Moldability, Japan. 70-12664 [1967/70]; C.A. **73** [1970] No. 88588.

Yamanouchi, S., Katsuki, H., Terazawa, T., Sumitomo Chemical Co., Ltd., Cross-Linked Poly(Vinyl Chloride), Ger. Offen. 1962848 [1969/70]; C.A. **73** [1970] No. 78164.

Barshtein, R. S., Li, P. Z., Gorbunova, V. G., Gazin, V. A., Martynov, S. P., Markina, E. E., Pertsov, L. D., Kalinkin, S. F., Shcherbakov, V. S., Polyester Plasticizers, U.S.S.R. 311930 [1967/71]; C.A. **76** [1972] No. 26012.

Castner, C. J., Russo, R. V., Berard, R. A., Celanese Corp., Modified Oxymethylene Polymers, Ger. Offen. 2129521 [1970/71]; C.A. **76** [1972] No. 114153.

Enoki, Y., Kiki, Y., Japan Soda Co., Ltd., Stabilization of Polyesters, Japan. 71-32067 [1968/71]; C.A. **77** [1972] No. 49557.

Fink, H. F., Koerner, G., Schmidt, G., Th. Goldschmidt, A.-G., Polishing Compositions Containing Wax and Organosilicon Compounds, Ger. Offen. 1929298 [1969/71]; C.A. **74** [1971] No. 77653.

Fulton, M., Midland Silicones Ltd., Vulcanizable Organopolysiloxane Compositions, U.S. 3607801 [1963/71]; C.A. **76** [1972] No. 15582.

Fulton, M., Johnes, J. D., Pearce, C. A., Midland Silicones Ltd., Cold Vulcanizing Polysiloxane Elastomer Compositions, Ger. Offen. 2018071 [1969/71]; C.A. **74** [1971] No. 127234.

Gibbons, J. P., Wondolowski, L., CPC International Inc., Polyurethane Resins Prepared from Alkoxylated Glucose Derivatives, U.S. 3586650 [1967/71]; C.A. **75** [1971] No. 64831.

Gibier-Rambaud, A., Naphthachimie, Polyether Polyol-Polyurethane-Bitumen Masses, Ger. Offen. 2036661 [1969/71]; C.A. **74** [1971] No. 127061.

Hartlage, J. V., Dow Corning Corp., Transparent Silicone Rubber Compositions Vulcanizable at Room Temperature, Ger. Offen. 2117027 [1970/71]; C.A. **76** [1972] No. 60733.

Japan Rayon Co., Ltd., Compositions for Imparting High Elastic Recovery to Extensible Knitted or Woven Fabrics, Brit. 1223574 [1968/71]; C.A. **74** [1971] No. 127489.

Katsibas, T., Reichold-Albert-Chemie A.-G., Aqueous Air- or Oven-Drying Lacquers Based on Dispersions or Solutions of Epoxy-Modified Polyurethane Resins, Ger. Offen. 2054468 [1969/71]; C.A. **75** [1971] No. 153026.

Leebrick, J. R., Cosan Chemical Corp., Organotin Curing Catalysts for Silicone Rubber Compositions, Brit. 1250498 [1968/71]; C.A. **76** [1972] No. 128461.

Leebrick, J. R., Cosan Chemical Corp., Tin Harada Complex Catalysts for the Preparation of Polyurethanes, Ger. Offen. 2122075 [1971]; C.A. **76** [1972] No. 100380.

Murai, K., Kanera, T., Saotome, K., Mitsui Toatsu Chemicals Co., Ltd., UV-Resistant Polyurethane, Japan. 71-38588 [1968/71]; C.A. **77** [1972] No. 6515.

Nitzsche, S., Hittmair, P., Kaiser, W., Wohlfarth, E., Wacker-Chemie GmbH, O-Aminosilyl Oximes Additives to Hardenable Polysiloxanes, Ger. Offen. 1941285 [1969/71]; C.A. **74** [1971] No. 127789.

Ogawa, K., Toray Industries, Inc., Polyester Compositions, Japan. 71-38707 [1968/71]; C.A. **77** [1972] No. 89421.

Ohtsuka, Y., Hiramatsu, T., Sekisui Chemical Co., Ltd., Vinyl Chloride Ethylene Copolymer Compositions, Japan. 71-42227 [1968/71]; C.A. **77** [1972] No. 115399.

Owen, W. J., Jomas, D. A., Midland Silicones Ltd., Silacyclobutane-Crosslinked Dimethylsiloxane Rubber, Ger. Offen. 2110871 [1970/71]; C.A. **76** [1972] No. 73517.

Polmanteer, K. E., Dow Corning Corp., Organosiloxane-Based Materials Hardenable to Elastomers at Room Temperature, Ger. Offen. 2063630 [1969/71]; C.A. **75** [1971] No. 152775.

Rice, D. M., Jefferson Chemical Co., Inc., Polyurethane Foam, Ger. Offen. 2049689 [1969/71]; C.A. **75** [1971] No. 64899.

Rozova, M. P., Compound for Coating Ceramic Capacitors, U.S.S.R 316712 [1964/71]; C.A. **76** [1972] No. 87333.

Ruch, D. J., Christie, H. W., Byerley, T. J., Midwest Research Institute, Flameproof, Heat-Resistant Polyurethane Foams, Ger. Offen. 1931763 [1969/71]; C.A. **74** [1971] No. 64968.

Sorkin, M. E., Dow Corning Corp., Rapid Thermosetting Organopolysiloxane-Containing Parting Emulsions Storage-Stable at Room Temperature, Ger. Offen. 2057121 [1969/71]; C.A. **75** [1971] No. 118900.

Suzuki, T., Ito, K., Nakataba, M., Shin-Etsu Chemical Industry Co., Ltd., Room-Temperature Vulcanizable Organopoly-Siloxane Compositions, Japan. 71-11982 [1966/71]; C.A. **76** [1972] No. 60741.

Umetsu, K., Fukatani, H., Tokyo Chemical Co., Ltd., Polymer Compositions Containing Chlorine for Measuring Radioactivity, Japan. 70-01258 [1966/71]; C.A. **76** [1972] No. 114204.

Wessel, J. K., Dow Corning Corporation, Method of Curing a Room Temperature Vulcanizable Silicone Composition to a Vulcanized Silicone Rubber, U.S. 3567493 [1968/71].

Colomb, H. O., Trecker, D. J., Brotherton, T. K., Union Carbide Corp., Radiation-Crosslinked Dinorborene Polymers, U.S. 3658669 [1968/72]; C.A. **77** [1972] No. 140945.

Duncan, J. S., Elmer, O. C., General Tire and Rubber Co., Composition for Lowering the Release Temperature of Phenol- and Lower Alkyl Substituted Phenol-Blocked Isocyanates, U.S. 3668186 [1970/72]; C.A. **77** [1972] No. 76117.

Fulton, M., Midland Silicones Ltd., Room Temperature-Vulcanizable Organopolysiloxane Compositions, Brit. 1269644 [1968/72]; C.A. **77** [1972] No. 21176.

Hirata, Y., Matsuda, F., Mitsui Toatsu Chemicals Co., Ltd., Poly(Vinyl Chloride) Plasticized with Butanetetracarboxylic Acid Esters, Japan. 72-18219 [1968/72]; C.A. **77** [1972] No. 153230.

Ishida, S., Ooshima, N., Mori, K., Matsumoto, K., Fujita, M., Asahi Chemical Industry Co., Ltd., Polymerizing Formaldehyde, Japan. 72-05109 [1968/72]; C.A. **77** [1972] No. 6136.

Kawabata, T., Ichii, R., Fukuda, K., Mitsui Toatsu Chemicals Co., Ltd., Catalysts to Give Polyisocyanates Containing Isocyanurate Rings, Japan. Kokai 72-34596 [1971/72]; C.A. **78** [1973] No. 125332.

Kent, E. W., General Latex and Chemicals (Canada) Ltd., Polyurethane Flocking Adhesives, Can. 902836 [1970/72]; C.A. **77** [1972] No. 115406.

Koga, S., Chisso Corp., Propylene-Ethylene Block Copolymers with Improved Workability, Japan. Kokai 72-34934 [1971/72]; C.A. **81** [1974] No. 26526.

Kojima, K., Tsujii, T., Fujita, T., Shinnippon Seitetsu Kagaku Kogyo Co., Ltd., Fire-Resistant Thermally Stable Styrene Resins, Japan. 72-04942 [1968/72]; C.A. **77** [1972] No. 127473.

Kondo, S., Kato, F., Isokawa, S., Mitsui Petrochemical Industries, Ltd., Katsuta Kako Co., Ltd., Polyethylene Compositions, Japan. 72-13305 [1967/72]; C.A. **77** [1972] No. 153207.

Leebrick, J. R., Cosan Chemical Corp., Curing Silicone Rubber Compositions Using Harada Complexes as Catalysts, U.S. 3661887 [1967/72]; C.A. **77** [1972] No. 49831.

Nitzsche, S., Hechtl, W., Hittmair, P., Wohlfarth, E., Wacker-Chemie GmbH, Diorganopolysiloxane Molding Materials Hardenable to Elastomers at Room Temperature, Ger. Offen. 2106651 [1971/72]; C.A. **77** [1972] No. 165819.

Nitzsche, S., Hittmair, P., Wohlfarth, E., Wacker-Chemie GmbH, Organopolysiloxane Elastomers, Ger. 1952756 [1969/72]; C.A. **78** [1973] No. 5229.

Richart, D. S., Polymer Corp., Reducing Sandiness in Vinyl Coatings by Use of an Organotin Compound, U.S. 3640747 [1966/72]; C.A. **76** [1972] No. 142512.

Schneider, M., Synthetic Resin Foam Material, Brit. 1283229 [1969/72]; C.A. **77** [1972] No. 115485.

Scott, H. G., Midland Silicones Ltd., Crosslinked Ethylene Graft Copolymer Films, Ger. Offen. 2151270 [1970/72]; C.A. **77** [1972] No. 35711.

Shima, T., Yamashiro, S., Ishizaki, T., Kawamura, T., Teijin Ltd., Adhesion of Polyester Fiber Rubber, Japan. 72-22119 [1968/72]; C.A. **78** [1973] No. 5234.

Snapp, T. C., Blood, A. E., Eastman Kodak Co., Preparing p-Dioxan-2-One Polymers Using an Organotin Catalyst, U.S. 3645941 [1970/72]; C.A. **76** [1972] No. 154453.

Tahara, M., Harata, Y., Japan Polystyrene Industry Co., Ltd., Stabilized Polystyrene Composition, Japan. 72-09744 [1968/72]; C.A. **77** [1972] No. 140979.

Wada, T., Imai, A., Inomata, H., Shin-Etsu Chemical Industry Co., Ltd., Room-Temperature-Hardening Organosilicon Compositions, Japan. 72-19616 [1968/72]; C.A. **77** [1972] No. 165828.

Wu, C., Union Carbide Corp., Reactions with Elemental Phosphorus to Produce Phosphorus Acid Ester Products, U.S. 3639531 [1966/72]; C.A. **76** [1972] No. 141699.

Wu, C., Union Carbide Corp., Flame-Retardant Organophosphorus Compositions, U.S. 3662029 [1966/72]; C.A. **77** [1972] No. 75988.

Aufdermarsh, C. A., du Pont de Nemours, E. I., and Co., Coating Material for the Production of the Primary Layer of a Two-Layer Adhesive for Binding Material for an Elastomer and a Polyester, Ger. Offen. 2316034 [1972/73]; C.A. **80** [1974] No. 48977.

Brown, R. A., Immont Corp., Polyurethane Production in the Presence of Boron Trifluoride, U.S. 3717604 [1970/73]; C.A. **78** [1973] No. 137648.

Buechner, W., Rudolph, K. H., Noll, W., Toepsch, H., Bayer A.-G., Adhesion-Reducing Agents for Separatory Paper, Ger. Offen. 2135673 [1971/73]; C.A. **78** [1973] No. 112979.

Dieter, J. A., Wolgemuth, L. G., Ryan, T. J., Atlantic Richfield Co., Catalytically Preparing Foam from a Nitrile Carbonate and a Nucleophilic Compound, U.S. 3746667 [1971/73]; C.A. **79** [1973] No. 147086.

Harayama, H., Morishima, Y., Sekisui Chemical Co., Ltd., Heat Foaming Resin Composition, Japan. 73-13340 [1969/73]; C.A. **80** [1974] No. 27953.

Hubbard, B. W., Kehr, C. L., W. R. Grace and Co., Coating Rigid Cores, U.S. 3767457 [1969/73]; C.A. **80** [1974] No. 109980.

Japanese Geon Co., Ltd., Alkylene Oxide Polymers, Brit. 1319350 [1969/73]; C.A. **79** [1973] No. 92859.

Jerabek, R. D., Marchetti, J. R., Zwack, R. R., PPG Industries, Inc., Cationically Depositable Polymeric Compositions, Ger. Offen. 2252536 [1971/73]; C.A. **80** [1974] No. 5023.

Johnson, C. E., Henderson, L. D., United States Dept. of the Navy, Nitrocellulose-Containing Adhesive Compositions, U.S. 3758325 [1972/73]; C.A. **80** [1974] No. 60566.

Kaburaki, K., Ishizawa, A., Tokyo Shibaura Electric Co., Ltd., Vinyl Carbazole Copolymer Compositions with Antistatic Properties, Japan. Kokai 73-16939 [1971/73]; C.A. **79** [1973] No. 67345.

Kacir, L., Narkis, M., Centre for Industrial Research (CIR), Ltd., Palram Plastic Works, Vinyl Chloride Polymer-Asbetos Molding Compositions, Ger. Offen. 2317984 [1972/73]; C.A. **80** [1974] No. 48701.

Marlin, L., Schwarz, E. G., Union Carbide Corp., Polyurethane Foam from a Heat Curable Froth, U.S. 3772224 [1969/73]; C.A. **80** [1974] No. 71685.

Munoz, E., Celanese Corp., Butyltin Laurate Boron Trifluoride Complex, Ger. Offen. 2228417 [1971/73]; C.A. **78** [1973] No. 111514.

Munoz, E., Celanese Corp., Trioxane Copolymers, Ger. Offen. 2228418 [1971/73]; C.A. **78** [1973] No. 137070.

Oakes, V., Hutton, R. E., Iles, B. R., AKZO GmbH, Tin Stabilizer Mixtures for PVC, Ger. Offen. 2307360 [1972/73]; C.A. **80** [1974] No. 27903.

Olstowski, F., Parrish, D. B., Dow Chemical Co., Articles from Nonelastomeric Polyurethane Compositions, Ger. Offen. 2243684 [1971/73]; C.A. **78** [1973] No. 148590.

Owen, W. J., Cooper, B. E., Dow Corning Ltd., Crosslinked Polyolefins, Ger. Offen. 2234717 [1971/73]; C.A. **78** [1973] No. 125509.

Scott, H. G., Dow Corning Corp., Crosslinked Silylvinyl Polymers, Ger. Offen. 2255116 [1971/73]; C.A. **79** [1973] No. 67135.

Scott, H. G., Thomas, B., Dow Corning Ltd., Polyethylene Foaming Compositions, Ger. Offen. 2310040 [1972/73]; C.A. **80** [1974] No. 71609.

Tsunawaki, K., Sasama, S., Watanabe, K., Nawata, K., Teijin Ltd., Diallyl Benzenedicarboxylates, Japan. Kokai 73-28444 [1971/73]; C.A. **79** [1973] No. 66032.

Yamamoto, T., Nippon Oils and Fats Co., Ltd., Heat Resistant Antistatic Agent, Japan. 73-10058 [1962/73]; C.A. **80** [1974] No. 109327.

Becker, R., Schimpfle, A., Heinatz, R., Wenk, R., Acceleration of Hardening Reactions of Polyurethane Systems Catalyzed with Organotin Compounds, Ger. [East] 109165 [1972/74]; C.A. **83** [1975] No. 98458.

Bosch, E., Roth, M., Gogolok, K., Wacker-Chemie GmbH, Strenghtening Binders for Building Materials, Ger. Offen. 2318494 [1973/74]; C.A. **82** [1975] No. 47241.

Ito, I., Inoue, Y., Sumitomo Chemical Co., Ltd., Poly(Vinyl Chloride) Plasticizers from Diglycerols, Japan. Kokai 74-99548 [1973/74]; C.A. **83** [1975] No. 148354.

Ito, K., Shin-Etsu Chemical Industry Co., Ltd., Catalyst Compositions for Condensation-Hardenable Silicone Rubbers, Japan. Kokai 74-134759 [1973/74]; C.A. **82** [1975] No. 157599.

Itsushiki, S., Fujikura Rubber Works, Ltd., Tsukahara, Y., Ogata, T., Fujikura Cable Works, Ltd., Room Temperature-Hardening Urethane Elastomer, Japan. Kokai 74-10245 [1972/74]; C.A. **81** [1974] No. 137017.

Jerabek, R. D., Marchetti, J. R., PPG Industries, Inc., Self-Crosslinkable Cationic Compositions for Electrodeposition, Fr. Demande 2211542 [1972/74]; C.A. **82** [1975] No. 113261.

Katsimbas, T., Reichold-Albert-Chemie A.-G., Water-Thinnable or Water-Dispersible Oil-Modified Polyurethanes, Ger. Offen. 1966836 [1968/74]; C.A. **81** [1974] No. 154791.

Marsushita, Y., Negi, Y., Masukawa, T., Ube-Nitto Chemical Industry Co., Ltd., Protective Resin Film for Goods, Japan. 74-11848 [1968/74]; C.A. **81** [1974] No. 154783.

Noma, T., Hitachi Shipbuilding and Engineering Co., Ltd., Thermal Insulation for Cryogenic Containers, U.S. 3802948 [1971/74]; C.A. **81** [1974] No. 121842.

Schimpfle, H. U., Becker, R., Lorenz, A., Heinatz, R., Wenk, R., Acceleration of Hardening Reactions of Polyurethane Systems Catalyzed with Organotin Compounds, Ger. [East] 109164 [1972/74]; C.A. **83** [1975] No. 80433.

Scott, H. G., Dow Corning Ltd., Crosslinked Ethylene-Vinyl Acetate Polymer, Ger. Offen. 2350876 [1972/74]; C.A. **82** [1975] No. 58834.

Takada, S., Niimi, S., Isogawa, M., Kanegafuchi Chemical Industry Co., Ltd., Fibrillated Poly(Vinyl Chloride)-Polyethylene Blends, Japan. Kokai 74-126903 [1973/74]; C.A. **83** [1975] No. 12033.

Tsunawaki, K., Santa, T., Watanabe, K., Aito, Y., Mitani, Y., Nawata, K., Teijin Ltd., Diallyl Naphthalenedicarboxylates, Japan. Kokai 74-45049 [1972/74]; C.A. **81** [1974] No. 120324.

Yanagisawa, I., Kojima, K., Tokieda, S., Oda, K., Mazaki, K., Nippon Steel Corp., Heat- and Fire-Resistant Styrene Resin Composition, Japan. Kokai 74-80159 [1972/74]; C.A. **82** [1975] No. 59067.

Yasuda, N., Yamashita, T., Ariyoshi, Y., Ajinomoto Co., Inc., N-Protected Amino Acid Esters, Japan. Kokai 74-133301 [1973/74]; C.A. **82** [1975] No. 171443.

Bargain, M., Rhone-Poulenc Industries, Silanes with Imide Group-Containing Compositions, Ger. Offen. 2504791 [1974/75]; C.A. **84** [1976] No. 18414.

Dickert, E. A., Cook Paint and Varnish Co., Phenolic Foam, U.S. 3872034 [1973/75]; C.A. **83** [1975] No. 44235.

Dumoulin, J., Rhone-Poulenc, S. A., Aqueous Dispersion of Organopolysiloxane and Allyl Alcohol-Acrylamide Copolymer as Paper Coating, U.S. 3865773 [1972/75]; C.A. **83** [1975] No. 45043.

Fang, J. C. S., Du Pont de Nemours, E. I., and Co., Polyester Polyurethane Coatings, Ger. Offen. 2428250 [1973/75]; C.A. **82** [1975] No. 172735.

Finelli, A. F., Goodyear Tire and Rubber Co., Polyurethane Prepared with 4,4-Diamino Diphenyl Disulfide, U.S. 3905944 [1973/75]; C.A. **83** [1975] No. 207370.

Hagiwara, T., Ohtsubo, T., Toyobo Co., Ltd., Polyamide Molding Compositions, Japan. Kokai 75-83452 [1973/75]; C.A. **84** [1976] No. 5907.

Hayasaka, K., Honny Chemicals Co., Ltd., Polymers for Artificial Leathers, Japan. Kokai 75-105796 [1974/75]; C.A. **84** [1976] No. 6269.

Hayasaka, K., Honny Chemicals Co., Ltd., Polymers for Artificial Leathers, Japan. Kokai 75-105797 [1974/75]; C.A. **84** [1976] No. 6268.

Hida, Y., Saeki, T., Shin-Etsu Chemical Industry Co., Ltd., Organopolysiloxane Compositions, Japan. Kokai 75-157447 [1974/75]; C.A. **84** [1976] No. 152426.

Horiuchi, Y., Katsube, T., Asahi Chemical Industry Co., Ltd., Thermoplastic Compositions with Improved Coating Receptance, Japan. Kokai 75-95345 [1973/75]; C.A. **83** [1975] No. 207107.

Hutton, R. E., Oakes, V., Iles, B. R., AKZO GmbH, Ger. Offen. 2455614 [1973/75]; C.A. **83** [1975] No. 180314.

Isayama, K., Hatano, I., Kanegafuchi Chemical Industry Co., Ltd., Oxylalkylene Polymers Containing Silicon Atoms at Chain Ends, Japan. Kokai 75-156599 [1974/75]; C.A. **85** [1976] No. 7037.

Iwazaki, K., Kokoku Chemical Industry Co., Ltd., Rigid Polyurethane Foams with Low Combustion Smokes, Japan. Kokai 75-78699 [1973/75]; C.A. **83** [1975] No. 132741.

Kan, P. T. Y., Cenker, M., BASF Wyandotte Corp., Carbodiimide Foams, Ger. Offen. 2458916 [1973/75]; C.A. **83** [1975] No. 165057.

Knox, W. B., Imperial Chemical Industries Ltd., Elastomer-Forming Compositions, Brit. 1401804 [1972/75]; C.A. **83** [1975] No. 207387.

Kobayashi, O., Ichihara, M., C. I. Kasei Ko., Ltd., Stabilizers for Vinyl Chloride Resins, Japan. Kokai 75-153056 [1974/75]; C.A. **84** [1976] No. 151501.

Kuehn, E., ICI America, Inc., Vinyl-Modified Polyurethane, Ger. Offen. 2425270 [1973/75]; C.A. **82** [1975] No. 112751.

Kumada, H., Murakami, Y., Nishiwaka, K., Dainippon Ink and Chemicals, Inc., Urethane Resin Varnishes, Japan. Kokai 75-144728 [1974/75]; C.A. **84** [1976] No. 61345.

Kume, S., Honma, M., Dainippon Ink and Chemicals, Inc., Dainippon Ink Research Institute, Isocyanate Composition, Japan. 75-15482 [1970/75]; C.A. **84** [1976] No. 75902.

Lander, H. L., Story Chemical Corp., Polymer Compositions, Ger. Offen. 2515775 [1974/75]; C.A. **84** [1976] No. 46273.

Laporte Industries Ltd., Polyurethane Foams, Belg. 818786 [1974/75]; C.A. **83** [1975] No. 116015.

Laporte Industries Ltd., Polyurethane Foams, Belg. 818787 [1974/75]; C.A. **83** [1975] No. 148383.

Maeda, S., Kondo, A., Nakanishi, H., Hodogaya Chemical Co., Ltd., Polymerizing Tetrahydrofuran, Japan. Kokai 75-126789 [1974/75]; C.A. **84** [1976] No. 74885.

Murayama, S., Nishizawa, H., Ebisawa, K., Tanuma, T., Tanaka, S., Hitachi Chemical Co., Ltd., Water-Soluble Alkyd Resin Coating Compositions, Japan. Kokai 75-139827 [1974/75]; C.A. **84** [1976] No. 61332.

Murayama, S., Nishizawa, H., Ebisawa, K., Tanuma, T., Tanaka, S., Hitachi Chemical Co., Ltd., Water-Soluble Acrylic Resin Coating Compositions, Japan. Kokai 75-139829 [1974/75]; C.A. **84** [1976] No. 61331.

Nakabayashi, S., Iimure, T., Umemoto, H., Nishida, T., Kansai Paint Co., Ltd., Photosensitive Compounds, Japan. Kokai 75-139721 [1974/75]; C.A. **85** [1976] No. 34755.

Nakatani, M., Imamura, T., Tanaka, M., Kubota Ltd., Stabilized Poly(Vinyl Chloride) Adhesive Compositions, Japan. Kokai 75-36528 [1973/75]; C.A. **83** [1975] No. 98463.

Ogino, A., Nakai, Y., Sakamoto, T., Takeda Chemical Industries, Ltd., Airdryable Unsaturated Polyesters, Japan. Kokai 75-90699 [1973/75]; C.A. **83** [1975] No. 194359.

Ono, K., Nishimura, T., Nomura, S., Idemitsu Kosan Co., Ltd., Impact-Resistant Rubber Compositions, Japan. Kokai 75-45847 [1973/75]; C.A. **83** [1975] No. 116579.

Prokai, B., Union Carbide Corp., Preparation of High Resilience Polyether Urethane Foam, U.S. 3905924 [1973/75]; C.A. **83** [1975] No. 207166.

Saam, J. C., Thomas, B., Dow Corning Corp., Crosslinkable Silane-Modified PVC, Ger. Offen. 2432006 [1973/75]; C.A. **83** [1975] No. 11362.

Schmalz, W., Michel, W., Schoen, M., Casella Farbwerke Mainkur A.-G., Thermosetting Coating Material as a Dispersion, Ger. Offen. 2414427 [1974/75]; C.A. **84** [1976] No. 61408.

Taller, R. A., Union Carbide Corp., Single Package 100 Percent Solids Urethane Coating Compositions, U.S. 3919174 [1974/75]; C.A. **84** [1976] No. 46292.

Thompson, D. R., E. I. du Pont de Nemours & Co., Addition Polymers with Polyethylenimine Terminal Groups, U.S. 3864379 [1968/75]; C.A. **82** [1975] No. 172769.

Tobolsky, A. V., Block Copolymers with Carbamate Linkages Between a Polyvinyl Chain and a Polymer Having Active Hydrogens, U.S. 3865898 [1973/75]; C.A. **83** [1975] No. 79965.

Vervloet, C., Shell Oil Co., Process for Preparing Polyurethane Products, U.S. 3905925 [1974/75]; C.A. **83** [1975] No. 207371.

Cenker, M., Robertson, E. J., Kan, P. T. Y., BASF Wyandotte Corp., Cellular Foams, Ger. Offen. 2545312 [1974/76]; C.A. **85** [1976] No. 47668.

Colomb, H. O., Trecker, D. J., Brotherton, T. K., Union Carbide Corp., Dinorbornenes, U.S. 3966797 [1968/76]; C.A. **85** [1976] No. 193574.

Dworkin, R. D., Eik, A., M and T Chemicals, Inc., Diorganotin Mercaptide Blowing Agent Activators for Cellular Vinyl Chloride Polymers, U.S. 3953385 [1973/76]; C.A. **85** [1976] No. 47644.

Goldin, G. S., Gershkokhen, S. L., Khludova, L. A., Shiryaev, V. I., Stepina, E. M., Hardening of Poly(imidosiloxanes), U.S.S.R. 509620 [1974/76]; C.A. **85** [1976] No. 34051.

Hayashi, N., Tanaka, S., Hitachi Chemical Co., Ltd., Esters and Alkyd Resins, Japan. Kokai 76-63803 [1974/76]; C.A. **85** [1976] No. 78995.

Ijichi, I., Morioko, A., Seki, K., Kai, S., Teranishi, H., Nitto Electric Industrial Co., Ltd., Pressure-Sensitive Adhesive Tapes, Japan. Kokai 76-81833 [1975/76]; C.A. **85** [1976] No. 144285.

Karasawa, Y., Narahara, T., Hitachi, Ltd., Thermosetting Resin Compositions, Japan. Kokai 76-89599 [1975/76]; C.A. **86** [1977] No. 6150.

Kume, S., Kuribayashi, A., Motohashi, S., Dainippon Ink and Chemicals, Inc., Polycarbamates, Japan. Kokai 76-04124 [1974/76]; C.A. **84** [1976] No. 181069.

Maslyuk, A. F., Biba, A. D., Romanenko, V. D., Lysobyk, S. E., Rositskii, A. A., Zholdakov, A. A., Volkova, V. V., Lacquer Composition, U.S.S.R. 502924 [1973/76]; C.A. **84** [1976] No. 152338.

Mine, K., Toray-Silicone Co., Ltd., Room-Temperature Curable Polyorganosiloxane Compositions, Japan. Kokai 76-30854 [1974/76]; C.A. **85** [1976] No. 34053.

Motohashi, A., Kogo, Y., Sankyo Organic Chemicals Co., Ltd., Heat-Resistant Poly(Vinyl Chloride) Resin Compositions, Japan. Kokai 76-81841 [1975/76]; C.A. **85** [1976] No. 144100.

Nakayama, T., Nakao, K., Tsukagoshi, I., Hitachi Chemical Co., Ltd., Adhesive Tapes, Japan. Kokai 76-16347 [1974/76]; C.A. **85** [1976] No. 6946.

Nakayama, T., Nakao, K., Tsukagoshi, I., Hitachi Chemical Co., Ltd., Adhesive Tapes, Japan. Kokai 76-42733 [1974/76]; C.A. **85** [1976] No. 79259.

Nakayama, T., Nakao, K., Tsukagoshi, I., Hitachi Chemical Co., Ltd., Adhesive Tapes Containing Crosslinkable Adhesives, Japan. Kokai 76-42734 [1974/76]; C.A. **85** [1976] No. 64320.

Nikolaev, A. F., Vorobev, O. L., Karkozov, V. G., Trizno, M. S., Kardashov, D. A., Yakovlev, A. D., Shibalovich, V. S., Kudryavtsev, B. B., Verkhoglyadova, T. Yu., et al., Lensovet Technological Institute, Leningrad, Epoxy Composition, Containing an Epoxy Resin, Dicyanamide, and a Metal Acetylacetonate, U.S.S.R. 535319 [1975/76]; C.A. **86** [1977] No. 73703.

NKF Kabel B. V., Cable Having at Least One Coating of a Plastic Containing a Crosslinked Polyolefin Modified with Silane, Neth. Appl. 7502714 [1975/76]; C.A. **86** [1977] No. 17728.

Ohtani, K., Oda, H., Furukawa Electric Co., Ltd., Crosslinked Polyolefins, Japan. Kokai 76-61550 [1974/76]; C.A. **85** [1976] No. 78986.

Olstowski, F., Parrish, D. B., Dow Chemical Co., Non-Elastomeric Polyurethane Compositions, U.S. 3996172 [1971/76]; C.A. **86** [1977] No. 56310.

Ono, K., Nishimura, T., Nomura, S., Nakai, M., Idemitsu Kosan Co., Ltd., Protein Reinforcement of Urethane Rubber, Japan. Kokai 76-39740 [1974/76]; C.A. **85** [1976] No. 34426.

Osterloh, R., Goethlich, L., Hartmann, H., BASF A.-G., Radiation-Hardenable Coatings, Ger. Offen. 2433908 [1974/76]; C.A. **84** [1976] No. 137385.

Piggott, K. E., Prolux Paint Manufactures Ltd., Coating Compositions, Ger. Offen. 2528377 [1974/76]; C.A. **84** [1976] No. 107273.

Pirck, D., Fuchs, G., Deutsche Texaco A.-G., Powder-Form Coating Composition with Hydroxyl Group-Containing Polymers and Masked Polyisocynates as Film-Forming Components, Ger. Offen. 2461416 [1974/76]; C.A. **85** [1976] No. 110193.

Saito, E., Nishimura, M., Fukazawa, H., Furukawa Electric Co., Ltd., Crosslinking of Poly(Vinyl Chloride), Japan. Kokai 76-132261 [1975/76]; C.A. **86** [1977] No. 73779.

Seki, M., Narahara, T., Hitachi, Ltd., Catalysts for Foam Production, Japan. Kokai 76-54698 [1974/76]; C.A. **85** [1976] No. 193596.

Smith, R. A., Suprenant, R. P., General Electric Co., Crosslinking Agents for Room Temperature Vulcanizable Silicone Rubber Compositions, U.S. 3957704 [1974/76]; C.A. **85** [1976] No. 64481.

Welte, R., Groegler, G., Bayer A.-G., Amidine-Metal Complexes, Ger. Offen. 2434185 [1974/76]; C.A. **84** [1976] No. 136685.

Akiba, I., Sekiguchi, T., Abe, M., Adeka Argus Chemical Co., Ltd., Poly(Vinyl Chloride) Electric Insulator Compositions, Japan. Kokai 77-77157 [1975/77]; C.A. **87** [1977] No. 152956.

Fukuoka, M., Kokan, S., Sumitomo Bakelite Co., Ltd., Vulcanization of Ethylene-Propene-Diene Rubber, Japan. Kokai 77-09054 [1975/77]; C.A. **86** [1977] No. 141425.

Hashimoto, H., Kato, M., Nishizawa, H., Showa Electric Wire and Cable Co., Ltd., Crosslinking of Silane-Modified Polyethylene, Japan. Kokai 77-69951 [1975/77]; C.A. **87** [1977] No. 118838.

Hashimoto, H., Kojima, T., Morita, M., Showa Electric Wire and Cable Co., Ltd., Crosslinking Composition, Japan. Kokai 77-56192 [1975/77]; C.A. **87** [1977] No. 85891.

Hashimoto, H., Kojima, T., Morita, M., Showa Electric Wire and Cable Co., Ltd., Crosslinked Polyolefin Moldings, Japan. Kokai 77-58760 [1975/77]; C.A. **87** [1977] No. 168844.

Hashimoto, H., Kojima, K., Morita, M., Suyama, S., Showa Electric Wire and Cable Co., Ltd., Joining of Cables, Japan. Kokai 77-80372 [1975/77]; C.A. **87** [1977] No. 185715.

Hashimoto, H., Nishikai, M., Kato, M., Nishizawa, H., Showa Electric Wire and Cable Co., Ltd., Room-Temperature-Curable Polyethylene Foam Molding, Japan. Kokai 77-71563 [1975/77]; C.A. **87** [1977] No. 152936.

Hashimoto, H., Suyama, S., Showa Electric Wire and Cable Co., Ltd., Crosslinked Polyethylene for Electric Cables, Japan. Kokai 77-80370 [1975/77]; C.A. **87** [1977] No. 185717.

Inomata, N., Nishijima, K., Kondo, Y., Tanaka, Y., Kyodo Chemical Co., Ltd., Heat Stabilizers for Chlorinated Resins, Japan. Kokai 77-80347 [1975/77]; C.A. **88** [1978] No. 74912.

Kogen, S., Fukuoka, M., Sumitomo Bakelite Co., Ltd., Japan. Kokai 77-124041 [1976/77]; C.A. **88** [1978] No. 90514.

Lampe, W. R., Marion Health and Safety, Inc., Silicone Rubber Composition for Making an Ear-Graft, Ger. Offen. 2166963 [1970/77]; C.A. **87** [1977] No. 119077.

M and T International B.V., Preparation of Cellular Vinyl Polymers, Brit. 1484087 [1973/77]; C.A. **88** [1978] No. 106232.

Mack, G. P., Organotin Mercapto Dicarboxylic Acid Esters and Compositions, U.S. 4058543 [1976/77]; C.A. **88** [1978] No. 51589.

Meyborg, H., Bayer A.-G., Catalysts for Isocyanate Polyaddition Reactions, Ger. Offen. 2601082 [1976/77]; C.A. **87** [1977] No. 85908.

Nakano, M., Ueshima, T., Maruyama, N., Mitsui Toatsu Chemicals, Inc., Poly(Vinyl Chloride) Films, Japan. Kokai 77-155655 [1976/77]; C.A. **88** [1978] No. 192094.

Ohta, S., Hitachi Chemical Co., Ltd., Crosslinking of Polyethylene, Japan. Kokai 77-09073 [1975/77]; C.A. **86** [1977] No. 156508.

Okado, R., Ichizuka, I., Tanaka, S., Nakayama, M., Asahi Denka Kogyo K. K., Coating Materials from Chlorinated Modified Polyisoprene, Japan. Kokai 77-29888 [1975/76]; C.A. **86** [1977] No. 191454.

Okamoto, M., Eidai Co., Ltd., Polyurethane-Coated Decorative Boards, Japan. Kokai 77-96648 [1976/77]; C.A. **88** [1978] No. 39172.

Okudaira, Y., Osanai, F., Yamada, T., Mitsubishi Plastics Industries, Ltd., Heat- and Weather-Resistant Poly(Vinyl Chloride) Compositions, Japan. Kokai 77-41661 [1975/77]; C.A. **87** [1977] No. 40263.

Schwindt, J., Bayer A.-G., Catalysts for Isocyanate Polyaddition Reactions, Ger. Offen. 2603834 [1976/77]; C.A. **87** [1977] No. 152725.

Suzuki, K., Saiki, N., Teijin, Ltd., High-Molecular-Weight Polyester Elastomers with Good Whitness, Japan. Kokai 77-06796 [1975/77]; C.A. **86** [1977] No. 156806.

Treadwell, K., Kushlefsky, B. G., Russo, R. V., M and T Chemicals, Inc., Polyurethane Foams, S. African 76-01743 [1976/77]; C.A. **87** [1977] No. 185479.

Vogt, H. C., Parekh, M., Patton, J. T., BASF Wyandotte Corp., Tin-Titanium Complexes as Esterification/Transesterification Catalysts, U.S. 4018708 [1975/77]; C.A. **86** [1977] No. 178187.

White, J. R., Krishnan, R. M., Wolfe, J. D., Goodyear Tire and Rubber Co., Mold Release Agent, U.S. 4038088 [1975/77]; C.A. **87** [1977] No. 103485.

Goller, H., Schulz, H. H., Leusner, B., Bayer A.-G., Pasty Masses Containing Crosslinker and Hardening Catalysts as Compounds for Room-Temperature-Vulcanizable Polysiloxane Elastomers, Ger. Offen. 2644193 [1976/78]; C.A. **88** [1978] No. 171549.

Hashimoto, H., Showa Electric Wire and Cable Co., Ltd., Silane-Crosslinked Polyolefin Moldings, Japan. Kokai Tokkyo Koho 78-125451 [1977/78]; C.A. **90** [1979] No. 104989.

Irie, S., Uesugi, K., Furukawa Electric Co., Ltd., Modified Polyolefin Compositions, Japan. Kokai Tokkyo Koho 78-92857 [1977/78]; C.A. **90** [1979] No. 7103.

Iwata, A., Fukishima, S., Une, S., Sekisui Chemical Co., Ltd., Light-Weight Polyurethane Foams, Japan. Kokai Tokkyo Koho 78-121086 [1977/78]; C.A. **90** [1979] No. 39735.

Kawabata, T., Asai, K., Saito, T., Nagahisa, S., Mitsui Nisso Urethane K. K., Water-Thinned Urethane Polymer Coating Materials for Shatterproofing of Bottles, Japan. Kokai Tokkyo Koho 78-139662 [1977/78]; C.A. **90** [1979] No. 139142.

Kawawada, S., Azuma, M., Sato, M., Hanawa, K., Kurimoto, H., Ogata, S., Kashiwazaki, S., Hitachi Cable, Ltd., Crosslinked Polyolefin Wire Sheats, Japan. Kokai 78-65342 [1976/78]; C.A. **89** [1978] No. 147645.

Kawawada, S., Azuma, M., Sato, M., Hanawa, K., Kurimoto, H., Ogata, S., Kashiwazaki, S., Hitachi Cable, Ltd., Crosslinked Polyolefin Wire Sheats, Japan. Kokai 78-65343 [1976/78]; C.A. **89** [1978] No. 147644.

Kenney, J. F., Treadwell, K., M and T Chemicals Inc., Rigid Polyurethane Foams, Fr. Demande 2390458 [1977/78]; C.A. **91** [1979] No. 158584.

Kita, A., Sakanaka, Y., Shimizu, A., Toyo Soda Mfg. Co., Ltd., Methyl Methacrylate-Grafted Neoprene Rubber Adhesives Having Good Storage Stability, Japan. Kokai Tokkyo Koho 78-132040 [1977/78]; C.A. **90** [1979] No. 105020.

Kita, A., Sakanaka, Y., Shimizu, A., Toyo Soda Mfg. Co., Ltd., Japan. Kokai Tokkyo Koho 78-132041 [1977/78]; C.A. **90** [1979] No. 122699.

Kawawada, S., Azumo, M., Sato, M., Hanawa, K., Kurimoto, H., Ogata, S., Kashiwazaki, S., Hitachi Cable, Ltd., Crosslinked Polyolefin Wire Sheats, Japan. Kokai 78-65344 [1976/78]; C.A. **89** [1978] No. 147643.

Kokan, S., Fukuoka, M., Sumitomo Bakelite Co., Ltd., Crosslinking of Thermoplastic Resins, Japan. Kokai 78-88050 [1977/78]; C.A. **89** [1978] No. 216345.

Kovacs, L., Szathmari, F., Budalakk Festek- es Mugyantagyar, Reactive Copolymer Solutions, Hung. Teljes 15876 [1976/78]; C.A. **91** [1979] No. 22557.

Nakanishi, K., Kurashiki Spinning Co., Ltd., Polyurethane Foam Insulators, Japan. Kokai 78-65395 [1976/78]; C.A. **89** [1978] No. 147815.

Nicks, P. F., Imperial Chemicals Industries Ltd., Coating Compositions, Brit. 1498408 [1974/78]; C.A. **89** [1978] No. 112563.

Nishizawa, H., Kato, M., Shimanuki, H., Nakamura, Y., Showa Electric Wire and Cable Co., Ltd., Extrusion and Crosslinking of Polyethylene, Japan. Kokai Tokkyo Koho 78-120796 [1977/78]; C.A. **90** [1979] No. 39734.

Ota, S., Yamada, M., Hitachi Chemical Co., Ltd., Crosslinking of Organic Polymers, Japan. Kokai Tokkyo Koho 78-138497 [1977/78]; C.A. **90** [1979] No. 122581.

Papa, A. J., Rollins, R. L., Critchfield, F. E., Union Carbide Corp., Phenolic Foam Modified with Phosphorus-Containing Isocyanate-Terminated Prepolymers, U.S. 4119584 [1977/78]; C.A. **90** [1979] No. 39690.

Saito, E., Hirukawa, H., Otani, K., Furukawa Electric Co., Ltd., Polyethylene Wire Sheats, Japan. Kokai Tokkyo Koho 78-145856 [1977/78]; C.A. **90** [1979] No. 138671.

Schaper, H., Phoenix Gummiwerke A.-G., Polyurethane Moldings, Ger. Offen. 2718173 [1977/78]; C.A. **90** [1979] No. 24531.

Sekmakas, K., Shah, R., De Soto, Inc., Electrodeposition of Aqueous Dispersions of Copolymers of Polyethylenically Unsaturated Epoxy Adducts Including Blocked Isocyanate Monomer, U.S. 4085161 [1976/78]; C.A. **89** [1978] No. 61133.

Stielau, M., One-Component Polyurethanes Expandable During Crosslinking by Moisture, Fr. Demande 2426703 [1978]; C.A. **92** [1980] No. 199313.

Yamada, M., Ota, S., Hitachi Chemical Co., Ltd., Improved Crosslinking Method for Silylated Polyolefins, Japan. Kokai Tokkyo Koho 78-132095 [1977/78]; C.A. **91** [1979] No. 21790.

Yamauchi, T., Kuroda, A., Hodogaya Chemical Co., Ltd., Hardening of Polyurethane Resin Mortar, Japan. Kokai 78-86734 [1977/78]; C.A. **89** [1978] No. 198622.

Chevron Research Co., Meta-Isocyanatobenzylisocyanate and Its Use in the Manufacture of Materials for a Polyurethane-Based Coating, Fr. Demande 2402646 [1977/79]; C.A. **91** [1979] No. 159153.

Dave, B., United States Steel Corp., Freely-Strippable Electrical Cable Insulation Composition, U.S. 4140818 [1977/79]; C.A. **90** [1979] No. 188207.

Eikawa, H., Otani, K., Nishiyama, H., Sasaki, Y., Furukawa Electric Co., Ltd., Molded Crosslinked Polyolefins, Japan. Kokai Tokkyo Koho 79-142255 [1978/79]; C.A. **92** [1980] No. 129978.

Emmons, W. D., Stevens, T. E., Rohm and Haas Co., Latex Compositions, S. African 77-07676 [1977/79]; C.A. **91** [1979] No. 109073.

Fujimoto, I., Isshiki, S., Kurita, Y., Sato, Y., Fujikura Cable Works, Ltd., Crosslinked Polyethylene, Japan. Kokai Tokkyo Koho 79-18857 [1977/79]; C.A. **90** [1979] No. 205404.

Hashimoto, H., Kato, M., Showa Electric Wire and Cable Co., Ltd., Fire-Resistant Crosslinked Molding, Japan. Kokai Tokkyo Koho 79-101862 [1978/79]; C.A. **92** [1980] No. 23583.

Horacek, H., Schoen, E., Reich, E., Volkert, O., BASF A.-G., Light-Resistant Polyurethane Integral Foams, Ger. Offen. 2825569 [1978/79]; C.A. **92** [1980] No. 94901.

Isagawa, M., Yamaguchi, N., Nikaido, T., Takiuchi, T., Matsushita Electric Works, Ltd., Vinyl Chloride Resin Compositions, Japan. Kokai Tokkyo Koho 79-103453 [1978/79]; C.A. **92** [1980] No. 23584.

Ivanow, S. A., Electrodepositable Resin, Span. 483678 [1979/80]; C.A. **93** [1980] No. 187862.

Kasahara, S., Kurihara, K., Murayama, H., Toyo Soda Mfg. Co., Ltd., Acrylic Resin-Coated PVC Sheets, Japan. Kokai Tokkyo Koho 79-26878 [1977/79]; C.A. **91** [1979] No. 22599.

Katayama, S., Nomura, H., Kurita, H., Shimabukuro, H., Tsutsumi, T., Nihon Hatsujo K. K., Hardenable Polymerizable Compositions and Polymers Derived from Them, Japan. Kokai Tokkyo Koho 79-03189 [1977/79]; C.A. **91** [1979] No. 21750.

Kleimann, H., Lienert, H. J., Meyborg, H., Groegler, G., Bayer A.-G., Foam Moldings Based on Polyurethane, Ger. Offen. 2737671 [1977/79]; C.A. **90** [1979] No. 169831.

Kojima, T., Kinoshita, S., Showa Electric Wire and Cable Co., Ltd., Coloring of Molded Polyolefins, Japan. Kokai Tokkyo Koho 79-71148 [1977/79]; C.A. **91** [1979] No. 176194.

Kokan, S., Fukuoka, M., Sumitomo Bakelite Co., Ltd., Crosslinking of Ethylenevinyl Alcohol Copolymers, Japan. Kokai Tokkyo Koho 79-20058 [1977/79]; C.A. **90** [1979] No. 188008.

Nikaido, T., Yamaguchi, N., Isagawa, M., Matsushita Electric Works, Ltd., Stable Poly(Vinyl Chloride) Composition, Japan. Kokai Tokkyo Koho 79-162745 [1978/79]; C.A. **92** [1980] No. 164776.

Roehr, C., Schlimper, R., Runge, J., Zwintzscher, R., Ehrhardt, H., VEB Leuna Werke "Walter Ulbricht", Stabilization of Polymers Against Thermo- and Photooxydative Decomposition, Ger. [East] 137312 [1975/79]; C.A. **92** [1980] No. 23627.

Salamon, M., Hagebaum, H. J., Bayer A.-G., Fiber Fill of Polyester Fibers, Brit. Appl. 2018320 [1978/79]; C.A. **92** [1980] No. 130557.

Satake, J., Arai, T., Shinogaki, K., Yamamoto, Y., Moriyasu, T., Nakazato, N., Nomura, T., Sumitomo Metal Industries, Ltd., Strippable Anticorrosive Coating Compositions, Japan. Kokai Tokkyo Koho 79-85233 [1977/79]; C.A. **92** [1980] No. 7949.

Sato, M., Hanawa, K., Ando, Y., Kurimoto, H., Ogata, S., Hitachi Cable, Ltd., Japan. Kokai Tokkyo Koho 79-76646 [1977/79]; C.A. **91** [1979] No. 176391.

Sato, M., Hanawa, K., Ando, Y., Kurimoto, H., Ogata, S., Hitachi Cable, Ltd., Crosslinking of Polyolefins and Electric Wires Insulated with Crosslinked Polyolefins, Japan. Kokai Tokkyo Koho 79-77657 [1977/79]; C.A. **91** [1979] No. 141869.

Sato, M., Hanawa, K., Shingyouchi, K., Kurimoto, H., Ogata, S., Hitachi Cable, Ltd., Crosslinking of Fire-Resistant Polyolefins and Fire-Resistant and Crosslinked Polyolefins for Insulating Electric Wires, Japan. Kokai Tokkyo Koho 79-76647 [1977/79]; C.A. **91** [1979] No. 158760.

Sudo, R., Isogai, T., Ijima, Y., Koyama, M., Hitachi, Ltd., Photocurable Epoxy Resin Adhesive Compositions for Joining Glass Materials, Japan. Kokai Tokkyo Koho 79-71133 [1977/79]; C.A. **91** [1979] No. 176195.

Sudo, R., Isogai, T., Iijima, Y., Koyama, M., Hitachi, Ltd., Photocurable Adhesive Compositions for Joining Glass, Japan. Kokai Tokkyo Koho 79-71134 [1977/79]; C.A. **91** [1979] No. 176169.

Watanabe, T., Mitsubishi Electric Corp., Molding of Liquid Rubber Compositions, Japan. Kokai Tokkyo Koho 79-90279 [1977/79]; C.A. **91** [1979] No. 194447.

Yamaguchi, N., Nikaido, T., Isagawa, M., Matsushita Electric Works, Ltd., Poly(Vinyl Chloride) Compositions, Japan. Kokai Tokkyo Koho 79-119548 [1978/79]; C.A. **92** [1980] No. 59746.

Zuppinger, P., Ciba-Geigy A.-G., Crosslinked High Polymers, Ger. Offen. 2825614 [1977/79]; C.A. **90** [1979] No. 153039.

AKZO N.V., Polyfunctional Isocyanates Free from Alkaline Groups and Urea, Belg. 878978 [1978/80]; C.A. **93** [1980] No. 27814.

Aragai, T., Takasuka, K., Matsunaga, T., Sekisui Chemical Co., Ltd., Foamed Concrete, Japan. Kokai Tokkyo Koho 80-42256 [1978/80]; C.A. **93** [1980] No. 119273.

Aragai, T., Takasuka, K., Matsunaga, T., Sekisui Chemical Co., Ltd., Foamed Concrete, Japan. Kokai Tokkyo Koho 80-42271 [1978/80]; C.A. **93** [1980] No. 119275.

Armstrong, G. H., Seefried, C. G., Van Cleve, R., Union Carbide Corp., High Resilience Flame-Retardant Polyurethane Foams Based on Polymer/Polyol Compositions, U.S. 4214055 [1978/80]; C.A. **93** [1980] No. 169432.

Asahi Chemical Industry Co., Ltd., Polyurethane Coating Compositions, Japan. Kokai Tokkyo Koho 80-54322 [1978/80]; C.A. **93** [1980] No. 116073.

Asahi Chemical Industry Co., Ltd., Vulcanization of Thermoplastic Rubber, Japan. Kokai Tokkyo Koho 80-73715 [1978/80]; C.A. **93** [1980] No. 151476.

Asahi Chemical Industry Co., Ltd., Resin Compositions, Japan. Kokai Tokkyo Koho 80-133419 [1979/80]; C.A. **94** [1981] No. 67381.

Asahi Glass Co., Ltd., Urethane Rubber Reaction Injection Molding Compositions for Automobile Bumpers, Japan. Kokai Tokkyo Koho 80-58215 [1978/80]; C.A. **93** [1980] No. 96569.

Ball, A., Minnesota Mining and Mfg. Co., Driographic Printing Plate, U.S. 4225663 [1974/80]; C.A. **95** [1981] No. 71055.

Blahak, J., Menk, H., Metzeler Kautschuk GmbH, Parting Agent for Use in Molding Polyurethanes, Ger. Offen. 2916700 [1979/80]; C.A. **94** [1981] No. 122691.

Brand, B. P., Ibbotson, A., Gazzard, E. G., Imperial Chemical Industries, Ltd., Iodoalkynyl Carbamates, Eur. Appl. 14032 [1979/80]; C.A. **94** [1981] No. 102865.

Bridgestone, Tire Co., Ltd., Compressed Polyurethane Foams, Japan. Kokai Tokkyo Koho 80-52329 [1978/80]; C.A. **93** [1980] No. 96205.

Britain, J. W., Ludwico, W. A., Mobay Chemical Corp., Mold Release Agent for Polyurethane Resins, Can. 1079903 [1975/80]; C.A. **93** [1980] No. 205736.

Brixius, D. W., Simms, J. A., du Pont de Nemours, E. I., and Co., Coating Compositions Comprising Isocyanate-Functional Polymers Containing a Terminal Thioalkyl Group, U.S. 4222909 [1977/80]; C.A. **94** [1981] No. 158473.

Carroll, W. G., Watts, A., Imperial Chemical Industries, Ltd., Liquid Polyisocyanate Compositions, Eur. Appl. 10850 [1978/80]; C.A. **93** [1980] No. 151096.

Cella, J. A., General Electric Co., 1,3-Silylcarbonyl Ethers, Ger. Offen. 2929226 [1978/80]; C.A. **93** [1980] No. 96541.

Cella, J. A., Mitchell, T. D., General Electric Co., Silyl Ethers of 1,3-Dicarbonyl Cyclic Compounds as Vulcanizing Agents for Silanol Terminated Polydiorganosiloxanes, Brit. Appl. 2026509 [1978/80]; C.A. **93** [1980] No. 27570.

Cenker, M., Kan, P. T., Schaaf, R. L., BASF Wyandotte Corp., Polyurethane Foams Prepared from Highly Stable Liquid Carbodiimide-Containing 4,4′-Diphenylmethane Diisocyanate, U.S. 4198489 [1978/80]; C.A. **93** [1980] No. 47805.

Cuscurida, M., Zimmermann, R. L., Ramey, B. J., Texaco Development Corp., Modified Diphenylmethane Diisocyanates Useful in Polyurethanes or Polyisocyanates, U.S. 4221877 [1976/80]; C.A. **93** [1980] No. 205545.

Daiichi Seiyaku Co., Ltd., One-Shot Polyurethane Compositions, Japan. Kokai Tokkyo Koho 80-99952 [1979/80]; C.A. **93** [1980] No. 221471.

Dainippon Ink and Chemicals, Inc., Poly(Vinyl Chloride) Compositions, Japan. Kokai Tokkyo Koho 80-102642 [1979/80]; C.A. **94** [1981] No. 16598.

Dainippon Ink and Chemicals, Inc., Poly(Vinyl Chloride) Resin Compositions, Japan. Kokai Tokkyo Koho 80-131036 [1979/80]; C.A. **94** [1981] No. 85139.

Denyer, R., Fortuin, M. S., Imperial Chemical Industries, Ltd., Dental Compositions Comprising a Selected Vinyl Urethane Prepolymer, Eur. Appl. 12535 [1978/80]; C.A. **93** [1980] No. 245508.

Deubzer, B., Brunner, E., Wilhelm, H., Sallersbeck, K., Wacker-Chemie GmbH, Wood Impregnated with Plastic, Ger. Offen. 2903376 [1979/80]; C.A. **93** [1980] No. 169947.

Deubzer, B., Brunner, E., Wilhelm, H., Sallersbeck, K., Wacker-Chemie GmbH, Polymer-Wood Composition, Ger. Offen. 2903452 [1979/80]; C.A. **93** [1980] No. 188032.

Disteldorf, J., Flakus, W., Schnurbusch, H., Chemische Werke Huels A.-G., Storage-Stable Urethane Acrylates, Ger. Offen. 2905205 [1979/80]; C.A. **93** [1980] No. 205564.

Dominguez, R. J. G., Rice, D. M., Texaco Development Corp., Polyurethane Elastomers, Ger. Offen. 3013069 [1979/80]; C.A. **94** [1981] No. 85457.

Dummer, G., Kratel, G., Niessner, P., Stohr, G., Wacker-Chemie GmbH, Increasing the Bulk Density of Silicon Dioxide, Ger. Offen. 2844459 [1978/80]; C.A. **93** [1980] No. 73443.

du Pont de Nemours, E. I., and Co., Polymers with Functional Isocyanate Groups and Thioalkyl Terminal Groups and Their Use in Preparing Coatings and Graft Polymers, Belg. 881844 [1980]; C.A. **94** [1981] No. 67421.

du Pont de Nemours, E. I., and Co., Moisture-Curable Coating Materials, Japan. Kokai Tokkyo Koho 80-84318 [1978/80]; C.A. **93** [1980] No. 206283.

Eikawa, H., Otani, K., Nishiyama, H., Sasaki, Y., Furukawa Electric Co., Ltd., Fire-Resistant Crosslinked Polyolefin Wire Sheath Compositions, Japan. Kokai Tokkyo Koho 80-45716 [1978/80]; C.A. **93** [1980] No. 72960.

Feltzin, J., Galvin, T. J., Kuehn, E., ICI Americas, Inc., In-Mold Coating Compositions Containing Functional Group Terminated Liquid Polymers, U.S. 4242415 [1978/80]; C.A. **94** [1981] No. 158004.

Ford Motor Co., Oligoester-Polyisocyanate Coatings, Japan. Kokai Tokkyo Koho 80-92772 [1979/80]; C.A. **93** [1980] No. 241314.

Fujikura Cable Works, Ltd., Crosslinked Vinyl Chloride Polymers, Japan. Kokai Tokkyo Koho 80-151049 [1979/80]; C.A. **94** [1981] No. 140698.

Funaki, M., Kuga, K., Atta, M., Asahi Glass Co., Ltd., Synthetic Resins Containing Fillers, Japan. Kokai Tokkyo Koho 80-45723 [1978/80]; C.A. **93** [1980] No. 48291.

Furukawa Electric Co., Ltd., Silane-Grafted Ethylene-Propene Rubber Compositions, Japan. Kokai Tokkyo Koho 80-71723 [1978/80]; C.A. **93** [1980] No. 151346.

Furukawa Electric Co., Ltd., Poly-α-Olefin Foams, Japan. Kokai Tokkyo Koho 80-75432 [1978/80]; C.A. **93** [1980] No. 187421.

Furukawa Electric Co., Ltd., Crosslinking of Poly-α-Olefins, Japan. Kokai Tokkyo Koho 80-129441 [1979/80]; C.A. **94** [1981] No. 85138.

GAF Corp., Copolymerizable UV Light Absorber (2-Cyano-3,3-diphenylacryloxy)-Alkylene Ethylenic Ethers, U.S. 4202834 [1979/80]; C.A. **94** [1981] No. 16591.

General Electric Co., Continuous Determination of Methyltriacetoxysilane in Siloxane, Japan. Kokai Tokkyo Koho 80-126848 [1979/80]; C.A. **94** [1981] No. 66728.

Gibard, A., Rhone-Poulenc Industries S. A., Organopolysiloxane Elastomer-Forming Compositions and Their Use in Mold Making, Eur. Appl. 10478 [1978/80]; C.A. **93** [1980] No. 96549.

Goodrich, J. E., Chevron Research Co., Rosin Acid Esters as Stabilizers for an Asphalt Premix, U.S. 4207231 [1979/80]; C.A. **93** [1980] No. 154983.

Goodrich, J. E., Chevron Research Co., Polymerizable Premix Composition for Preparation of Polyurethane Surfaces, U.S. 4237036 [1979/80]; C.A. **94** [1981] No. 32395.

Goodrich, J. E., Chevron Research Co., Polymerizable Premix Composition for Polyurethane Surfaces, U.S. 4238375 [1979/80]; C.A. **94** [1981] No. 66780.

Henderson's Industries, Ltd., Fire-Resistant Semirigid Polyurethane Foams with High Resilience, Japan. Kokai Tokkyo Koho 80-145720 [1979/80]; C.A. **94** [1981] No. 176117.

Hino, M., Oshima, T., Hayatsu, K., Yamamoto, M., Yasui, S., Sumitomo Chemical Co., Ltd., Plastic Resin Compositions for Electrocoating, Ger. Offen. 3000911 [1979/80]; C.A. **93** [1980] No. 169761.

Hitachi, Ltd., Rigid Polyurethane Foams with Skin Layers, Japan. Kokai Tokkyo Koho 80-127422 [1979/80]; C.A. **94** [1981] No. 48254.

Hitachi, Ltd., Coating of Magnesium and Its Alloy Products, Japan. Kokai Tokkyo Koho 80-148774 [1979/80]; C.A. **94** [1981] No. 107969.

Hitachi Cable, Ltd., Crosslinking of Polyolefins for Electric Insulators, Japan. Kokai Tokkyo Koho 80-82127 [1978/80]; C.A. **93** [1980] No. 169346.

Hitachi Chemical Co., Ltd., Glass Fiber-Reinforced Unsaturated Polyester-Polyurethane Blend Molding Compositions, Japan. Kokai Tokkyo Koho 80-90522 [1978/80]; C.A. **93** [1980] No. 221511.

Hitachi Chemical Co., Ltd., Unsaturated Polyester Compositions, Japan. Kokai Tokkyo Koho 80-155011 [1979/80]; C.A. **94** [1981] No. 176055.

Hitachi Chemical Co., Ltd., Polyurethane Bathtubs, Japan. Kokai Tokkyo Koho 80-163029 [1979/80]; C.A. **95** [1981] No. 8461.

Hochstrasser, U., Kertscher, E., Maillefer, S. A., Electrical Conductor Insulated by a Crosslinked Plastics Coating, Eur. Appl. 17624 [1979/80]; C.A. **94** [1981] No. 31799.

Honma, M., Shoji, A., Morita, T., Ishikawa, N., Dainippon Ink and Chemicals, Inc., Powder Coating Materials, Japan. Kokai Tokkyo Koho 80-70868 [1972/80]; C.A. **93** [1980] No. 96998.

Honny Chemicals Co., Ltd., Treatment of Poly(Vinyl Chloride) for Welding, Japan. Kokai Tokkyo Koho 80-90532 [1978/80]; C.A. **93** [1980] No. 222030.

Horacek, J., Cermak, J., Curing Method for Silicone Rubber, Czech. 184932 [1975/80]; C.A. **95** [1981] No. 116879.

Horacek, H., Nissen, D., Haardt, U., Reich, E., BASF A.-G., Compound Foam from Polyurethane Integral Foam and a Nonpolyurethane Foam, Ger. Offen. 2902255 [1979/80]; C.A. **94** [1981] No. 16746.

Hosaka, Y., Harita, Y., Kurokawa, M., Harada, T., Japan Synthetic Rubber Co., Ltd., Photosensitive Resin Printing Plate Supports, Japan. Kokai Tokkyo Koho 80-25046 [1978/80]; C.A. **93** [1980] No. 104841.

Hoy, K. L., Jones, A. P., Knopf, R. J., Union Carbide Corp., Shaped Polyurethane Hydrogel Articles, U.S. 4209605 [1978/80]; C.A. **93** [1980] No. 169361.

Ikeda, H., Tada, S., Kobayashi, T., Mitsui Polychemicals Co., Ltd., Hot-Melt Resin Compositions, Japan. Kokai Tokkyo Koho 80-40721 [1978/80]; C.A. **93** [1980] No. 73233.

Inaba, S., Hagiyama, H., Ichiki, M., Kamino, Y., Matsuki, N., Nagai, K., Suzuki, M., Hitachi Shipbuilding and Engineering Co., Ltd., Denitration Catalysts with a Porous Film, Japan. Kokai Tokkyo Koho 80-08875 [1978/80]; C.A. **92** [1980] No. 170023.

Isaka, T., Ishioka, M., Mitsubishi Petrochemical Co., Ltd., Crosslinked Polyolefin Films, Japan. Kokai Tokkyo Koho 80-09612 [1978/80]; C.A. **93** [1980] No. 27456.

Ishii, T., Kuratani, Y., Shimomura, K., Horioka, M., Yoshiki, S., Sekisui Chemical Co., Ltd., Metal Pipe Having a Synthetic Resin Lining, Japan. Kokai Tokkyo Koho 80-39360 [1978/80]; C.A. **93** [1980] No. 73239.

Ishimura, S., Kawamoto, H., Sumiya, Y., Asahi Chemical Industry Co., Ltd., Cationic Electrophoretic Coating Compositions, Japan. Kokai Tokkyo Koho 80-36233 [1978/80]; C.A. **93** [1980] No. 48638.

ITT Industries, Inc., Molded Polyurethane Components, Neth. Appl. 80-02637 [1979/80]; C.A. **94** [1981] No. 122776.

Jarre, W., Wurmb, R., BASF A.-G., Hydrophobic Polyurethane Foams, U.S. 4237237 [1977/80]; C.A. **94** [1981] No. 122760.

Kanegafuchi Chemical Industry Co., Ltd., Silyl Group-Containing Acrylic Polymer Coating Materials, Japan. Kokai Tokkyo Koho 80-129405 [1979/80]; C.A. **94** [1981] No. 67413.

Kanegafuchi Chemical Industry Co., Ltd., Waterproof Materials, Japan. Kokai Tokkyo Koho 80-160077 [1979/80]; C.A. **94** [1981] No. 193846.

Kao Soap Co., Ltd., Stable Quaternary Ammonium Chloride Compositions, Japan. Kokai Tokkyo Koho 80-07219 [1978/80]; C.A. **94** [1981] No. 71209.

Kaufmann, R., Mueller, J., Wegehaupt, K. H., Wacker-Chemie GmbH, Organopolysiloxanes, Ger. Offen. 2915751 [1979/80]; C.A. **94** [1981] No. 31538.

Kaufmann, R., Mueller, J., Wegehaupt, K. H., Wacker-Chemie GmbH, Organopolysiloxanes, Eur. Appl. 17985 [1979/80]; C.A. **94** [1981] No. 48270.

Keen, C. V., Dunlop, Ltd., Cellular Foaming Materials, Ger. Offen. 3018890 [1979/80]; C.A. **94** [1981] No. 85332.

Kempter, F. E., Schupp, E., BASF A.-G., Self-Crosslinking Cationic Binder, Ger. Offen. 2914331 [1979/80]; C.A. **94** [1981] No. 5016.

Kimura, H., Uesugi, K., Fukuda, A., Shibata, R., Furukawa Electric Co., Ltd, Silane-Crosslinked Polyolefin Resin Shaped Articles, Japan. Kokai Tokkyo Koho 80-40701 [1978/80]; C.A. **93** [1980] No. 73234.

Kinoshita, T., Izumi, T., Mitsui Petrochemical Industries, Ltd., Aqueous Emulsions of Amine-Terminated Polyurethanes, Japan. Kokai Tokkyo Koho 80-07829 [1978/80]; C.A. **93** [1980] No. 27332.

Kinyosha Co., Ltd., Coating Compositions for Swimming Pools, Japan. Kokai Tokkyo Koho 80-120665 [1979/80]; C.A. **94** [1981] No. 32290.

Kobayashi, Y., Tsuge, Y., Hirako, S., Kiniwa, H., Mitsubishi Chemical Industries Co., Ltd., Cationic Electrodeposition Process and Coating Composition, Brit. Appl. 2044280 [1980]; C.A. **95** [1981] No. 8946.

Kohkoku Chemical Industries Co., Ltd., Molded Polyurethane Foams, Japan. Kokai Tokkyo Koho 80-27241 [1978/80]; C.A. **93** [1980] No. 9240.

Kohkoku Chemical Industries Co., Ltd., Rigid Polyurethane Foams, Japan. Kokai Tokkyo Koho 80-92724 [1978/80]; C.A. **93** [1980] No. 240793.

Kohkoku Chemical Industries Co., Ltd., Insect Repellents for Polyurethane Foams, Japan. Kokai Tokkyo Koho 80-98217 [1979/80]; C.A. **94** [1981] No. 4717.

Kohkoku Chemical Industries Co., Ltd., Continuous Processes for Rigid Polyurethane Foam Laminates, Japan. Kokai Tokkyo Koho 80-105541 [1979/80]; C.A. **94** [1981] No. 4720.

Kohkoku Chemical Industries Co., Ltd., Rigid Polyurethane Foams Having Good Resistance to Water and Moisture, Japan. Kokai Tokkyo Koho 80-115422 [1979/80]; C.A. **94** [1981] No. 104527.

Kohkoku Chemical Industries Co., Ltd., Rigid Polyurethane Foams, Japan. Kokai Tokkyo Koho 80-147518 [1979/80]; C.A. **94** [1981] No. 140857.

Kojima, T., Konoshita, A., Takeuchi, K., Showa Electric Wire and Cable Co., Ltd., Water-Resistant Paper or Paper Laminates, Japan. Kokai Tokkyo Koho 80-40840 [1978/80]; C.A. **93** [1980] No. 28105.

Kopal, P., Polyurethane Foams, Czech. 182012 [1975/80]; C.A. **93** [1980] No. 169373.

Kubitza, W., Mennicken, G., Pedain, J., Bayer A.-G., Isocyanate Mixture and Its Use as Binders in Single-Component Lacqueurs, Ger. Offen. 2845514 [1978/80]; C.A. **93** [1980] No. 48620.

Lambert, R. J., Martens, J. A., Minnesota Mining and Mfg. Co., Plastic Film Label Material, Ger. Offen. 2944415 [1978/80]; C.A. **93** [1980] No. 133535.

Lee, Yu-Sum, B. F. Goodrich, Co., Alkylacrylate Polymer Composition and Laminate Structure Having this Composition as a Pressure-Sensitive Adhesive, Eur. Appl. 10753 [1978/80]; C.A. **93** [1980] No. 133578.

Lewarchik, R. J., Erikson, J. A., Birkmeyer, W. J., PPG Industries, Inc., Esters of Imidazolidinedione-Based Diepoxides and Coating Compositions Containing Them, U.S. 4228294 [1978/80]; C.A. **94** [1981] No. 17263.

Maekawa, I., Kagayama, A., Uchigasaki, I., Hitachi Chemical Co., Ltd., Thermosetting Resin Composition, Ger. Offen. 2948323 [1978/80]; C.A. **93** [1980] No. 115465.

Marx, M., Nissen, D., Jarre, W., BASF A.-G., Reinforced Foam Plastics, Ger. Offen. 2850610 [1978/80]; C.A. **93** [1980] No. 96240.

Massucco, A. A., Merrill, R. E., Arthur D. Little, Inc., Method of Preparing Screen Printing Stencils Using Novel Compounds and Compositions, U.S. 4209582 [1977/80]; C.A. **93** [1980] No. 248251.

Matsushita Electric Works, Ltd., PVC Molding Materials, Japan. Kokai Tokkyo Koho 80-147540 [1979/80]; C.A. **95** [1981] No. 8259.

Mayborg, H., Mormann, W., Illger, H. W., Bock, M., Bayer A.-G., Polyurethane Foams, Ger. Offen. 2914134 [1979/80]; C.A. **94** [1981] No. 31535.

McBrayer, R. L., BASF Wyandotte Corp., Microcellular Polyurethane Foams, U.S. 4220732 [1978/80]; C.A. **93** [1980] No. 240973.

Meyer, F., Kubens, R., Mehesch, H., Bergwerksverband GmbH, Bayer A.-G., Consolidating and Sealing Geological and Heaped Rock and Earth Formations, Eur. Appl. 16262 [1979/80]; C.A. **94** [1981] No. 89186.

Mitsubishi Chemical Industries Co., Ltd., Polyolefin Compositions, Japan. Kokai Tokkyo Koho 80-62945 [1978/80]; C.A. **93** [1980] No. 169297.

Mitsubishi Chemical Industries Co., Ltd., Modified Polybutadiene Sealing Compositions, Japan. Kokai Tokkyo Koho 80-78003 [1978/80]; C.A. **93** [1980] No. 151871.

Mitsubishi Petrochemical Co., Ltd., Grafted Ethylene Copolymers for Electric Insulators, Japan. Kokai Tokkyo Koho 80-74010 [1978/80]; C.A. **93** [1980] No. 187418.

Mitsubishi Petrochemical Co., Ltd., Extrusion Molding of Crosslinked Polyethylene, Japan. Kokai Tokkyo Koho 80-128441 [1979/80]; C.A. **94** [1981] No. 66902.

Mitsubishi Petrochemical Co., Ltd., Crosslinked Olefin Copolymer Foams, Japan. Kokai Tokkyo Koho 80-152724 [1979/80]; C.A. **94** [1981] No. 140695.

Mitsubishi Petrochemical Co., Ltd., Crosslinkable Ethylene Copolymer Compositions, Japan. Kokai Tokkyo Koho 80-155039 [1979/80]; C.A. **94** [1981] No. 157853.

Mitsubishi Petrochemical Co., Ltd., Crosslinkable Ethylene Copolymer Compositions, Japan. Kokai Tokkyo Koho 80-155040 [1979/80]; C.A. **94** [1981] No. 157854.

Mitsubishi Petrochemical Co., Ltd., Crosslinkable Ethylene Polymer Compositions, Japan. Kokai Tokkyo Koho 80-155045 [1979/80]; C.A. **94** [1981] No. 176059.

Mitsubishi Petrochemical Co., Ltd., Electrical Conductors Coated with Crosslinked Polyethylene Resin, Neth. Appl. 79-08254 [1978/80]; C.A. **93** [1980] No. 133330.

Mitsui Nisso Urethane K. K., Quickly Vulcanisable Urethane Rubber, Japan. Kokai Tokkyo Koho 80-98221 [1979/80]; C.A. **94** [1981] No. 4829.

Mitsui Petrochemical Industries, Ltd., Hydrocarbon Resin Grafted With Silane Compounds, Japan. Kokai Tokkyo Koho 80-73716 [1978/80]; C.A. **93** [1980] No. 169314.

Mitsui Toatsu Chemicals, Inc., Crosslinking of Polymers, Japan. Kokai Tokkyo Koho 80-75415 [1978/80]; C.A. **93** [1980] No. 169136.

Miyoshi, T., Takeuchi, K., Nukui, T., Oshida, H., Ajinomoto Co., Inc., Kawaguchi Chemical Industry Co., Ltd., Stabilized Halogen-Containing Resin Compositions, Japan. Kokai Tokkyo Koho 80-38828 [1978/80]; C.A. **93** [1980] No. 72956.

Nakada, H., Harada, M., Yamase, Y., Matsui, R., Nippon Soda Co., Ltd., Lacquer System for Cathodic Electroimmersion Lacquering, Ger. Offen. 2924343 [1979/80]; C.A. **94** [1981] No. 85809.

Neel, E., Perret, A., Cailho, R., Shell Internationale Research Maatschappij B. V., Catalytic Cracking of Hydrocarbon Oils, Ger. Offen. 2947710 [1978/80]; C.A. **94** [1981] No. 5755.

Nihon Hatsujo K. K., Polyurethane Foam Sealants, Japan. Kokai Tokkyo Koho 80-71777 [1978/80]; C.A. **93** [1980] No. 151868.

Nihon Tokushu Toryo Co., Ltd., Bulky Nonwoven Fabric, Japan. Kokai Tokkyo Koho 80-148266 [1979/80]; C.A. **94** [1981] No. 85642.

Nippon Oils and Fats Co., Ltd., Cationic Electrophoretic Coating Compositions, Japan. Kokai Tokkyo Koho 80-157667 [1979/80]; C.A. **94** [1981] No. 141326.

Nippon Paint Co., Ltd., Polyurethane Coatings, Japan. Kokai Tokkyo Koho 80-75419 [1978/80]; C.A. **93** [1980] No. 151848.

Nippon Soda Co., Ltd., Lacquer Composition for Cathodic Deposition, Neth. Appl. 79-04834 [1979/80]; C.A. **94** [1981] No. 176839.

Nippon Synthetic Chemical Industry Co., Ltd., Polyurethane Undercoating Compositions, Japan. Kokai Tokkyo Koho 80-155061 [1979/80]; C.A. **94** [1981] No. 141352.

Nissen, D., Stutz, H., Schuster, L., Marx, M., BASF A.-G., Transparent, Elastic Polyurethane-Urea Elastomers, Ger. Offen. 2837501 [1978/80]; C.A. **93** [1980] No. 27554.

Nitto Electric Industrial Co., Ltd., Primers for Concrete Substrates for Improved Adhesion to Polyurethane Sealants, Japan. Kokai Tokkyo Koho 80-99983 [1979/80]; C.A. **94** [1981] No. 5005.

Noethe, B., Siemens A.-G., Plastic Materials Based on Polyurethane, Especially for Sealing Electric Cables, Ger. Offen. 2847383 [1978/80]; C.A. **93** [1980] No. 74048.

Noethe, B., Siemens A.-G., Foamlike Filler Material for Longitudinal Sealing of Electrical or Optical Communication Cable with Plastic-Insulated Conducting Element, Ger. Offen. 2847386 [1978/80]; C.A. **93** [1980] No. 27502.

Noethe, B., Siemens A.-G., Filler Composition for Longitudinal Sealing of Electrical or Optical Communication Cables with Plastic-Insulated Conductors, Ger. Offen. 2487387 [1978/80]; C.A. **93** [1980] No. 48734.

Nomura, K., Kitagawa, T., Ono, H., Mitsubishi Electric Corp., Thermal Transfer Image Materials, Japan. Kokai Tokkyo Koho 80-39378 [1978/80]; C.A. **93** [1980] No. 85203.

Ohyabu, N., Hayakawa, K., Takeda, H., Asahi Chemical Industry Co., Ltd., Polyurethane Coating Composition, Japan. Kokai Tokkyo Koho 80-09644 [1978/80]; C.A. **93** [1980] No. 27883.

Otani, E., Tashiro, F., Hitachi Chemical Co., Ltd., Photocurable Coating Materials, Japan. Kokai Tokkyo Koho 80-36240 [1978/80]; C.A. **93** [1980] No. 73995.

Pospisil, V., Kakala, K., Kopal, P., Vulcanizing Agent, Czech. 182613 [1976/80]; C.A. **94** [1981] No. 67060.

Reilly, A., Sanok, J. L., Sheller-Globe Corp., Color Stable Integral Skin Foam, U.S. 4242463 [1979/80]; C.A. **94** [1981] No. 85112.

Ritz, J., Hotze, H., Mummenthey, H. D., Hoechst A.-G., Hardenable Copolymers, Ger. Offen. 2907997 [1979/80]; C.A. **93** [1980] No. 187841.

Rowlands, J. P., Polyols Modified by Polymers and Used in Polyurethane Production, Belg. 887514 [1980/81]; C.A. **95** [1981] No. 170470.

Rowton, R. L., Texaco Development Corp., High-Resilient Flexible Urethane Foams, U.S. 4239856 [1978/80]; C.A. **94** [1981] No. 85110.

Saint-Gobain Industries S. A., Preformed Polymeric Sheet for Use in Preparing a Glazing Laminate, Indian 147615 [1976/80]; C.A. **93** [1980] No. 221671.

Sanyo Chemical Industries, Ltd., Polyurethane Foam Moldings with a Skin, Japan. Kokai Tokkyo Koho 80-133417 [1980]; C.A. **94** [1981] No. 104543.

Schafer, H., Weber, C., Bayer A.-G., Reaction Injection Molding of Rubbers, U.S. 4218543 [1976/80]; C.A. **93** [1980] No. 205931.

Schaper, H., Phoenix A.-G., Polyurethane-Integral Foam Moldings for Motor Vehicle Exterior Parts, Ger. Offen. 2922769 [1979/80]; C.A. **94** [1981] No. 104589.

Schlatter, R., Grass, A., Swiss Aluminium, Ltd., Composite Sheet with Two Covering Layers and an Intermediate Core, Ger. Offen. 2842858 [1978/80]; C.A. **93** [1980] No. 9278.

Sekisui Chemical Co., Ltd., Aerated Concrete, Japan. Kokai Tokkyo Koho 80-62835 [1978/80]; C.A. **93** [1980] No. 191074.

Sekisui Chemical Co., Ltd., Composite Pipe, Japan. Kokai Tokkyo Koho 80-87528 [1978/80]; C.A. **93** [1980] No. 205805.

Sekisui Chemical Co., Ltd., Hot-Melt Adhesives, Japan. Kokai Tokkyo Koho 80-160074 [1979/80]; C.A. **94** [1981] No. 157993.

Sekisui Chemical Co., Ltd., PVC-Lined Pipes, Japan. Kokai Tokkyo Koho 80-161638 [1979/80]; C.A. **94** [1981] No. 176141.

Sekisui Chemical Co., Ltd., Hot-Melt Adhesive Compositions, Japan. Kokai Tokkyo Koho 80-165973 [1979/80]; C.A. **94** [1981] No. 209931.

Shin-Etsu Chemical Industry Co., Ltd., Moisture-Curable Sealing Compositions, Japan. Kokai Tokkyo Koho 80-115446 [1979/80]; C.A. **94** [1981] No. 48960.

Shin-Etsu Chemical Industry Co., Ltd., Room Temperature-Curable Rubber Compositions, Japan. Kokai Tokkyo Koho 80-129446 [1979/80]; C.A. **94** [1981] No. 122910.

Shintani, T., Kaneiwa, T., Sumitomo Bayer Urethane Co., Ltd., Thermally Stable Polyurethane Resin, Japan. Kokai Tokkyo Koho 80-31840 [1978/80]; C.A. **93** [1980] No. 72943.

Shiseido Co., Ltd., Skin Replica, Japan. Kokai Tokkyo Koho 80-106139 [1979/80]; C.A. **94** [1981] No. 4835.

Short, W. T., General Motors Corp., Moisture Curing Polyurethane Topcoat Paint Displaying Geometric Metamerism, U.S. 4199489 [1977/80]; C.A. **93** [1980] No. 48617.

Showa Electric Wire and Cable Co., Ltd., Manufacture of Silicone-Grafted Polyethylene, Japan. Kokai Tokkyo Koho 80-71708 [1978/80]; C.A. **93** [1980] No. 151342.

Showa Electric Wire and Cable Co., Ltd., Electrically Insulating Laminate Sheets, Japan. Kokai Tokkyo Koho 80-130016 [1979/80]; C.A. **94** [1981] No. 66861.

Showa Electric Wire and Cable Co., Ltd., Laminate Resin Sheets, Japan. Kokai Tokkyo Koho 80-132246 [1979/80]; C.A. **94** [1981] No. 66904.

Showa Electric Wire and Cable Co., Ltd., Polyolefin Nonwoven Fabrics, Japan. Kokai Tokkyo Koho 80-148268 [1979/80]; C.A. **94** [1981] No. 85641.

Stanley, H., Ray-Chaudhuri, D. K., National Starch and Chemical Corp., Blocked Isocyanate Diols, Preparation Thereof, and Polyurethanes Prepared Therefrom, Brit. Appl. 2038811 [1978/80]; C.A. **94** [1981] No. 209695.

Statton, L., Atlantic Richfield Co., Polyurethane Elastomers, U.S. 4202950 [1976/80]; C.A. **93** [1980] No. 96553.

Strolle, C. H., du Pont de Nemours, E. I. and Co., Urethane Enamel Coating Composition, U.S. 4215023 [1978/80]; C.A. **94** [1981] No. 85836.

Sumitomo Bakelite Co., Ltd., Copper Foil Laminates With Electric Insulators, Japan. Kokai Tokkyo Koho 80-127438 [1979/80]; C.A. **94** [1981] No. 66857.

Sumitomo Chemical Co., Ltd., Electrophoretic Coating Materials, Japan. Kokai Tokkyo Koho 80-69669 [1978/80]; C.A. **93** [1980] No. 133965.

Sumitomo Chemical Co., Ltd., Vinyl Chloride Elastomers, Japan. Kokai Tokkyo Koho 80-73743 [1978/80]; C.A. **93** [1980] No. 187553.

Takashi, S., Kiyotsugu, A., Toshihiko, K., Mitsui-Nisso Corp., A Thermosetting Polyurethane Resin Coating Agent, Eur. Appl. 19368 [1979/80]; C.A. **94** [1981] No. 85881.

Toa Gosei Chemical Industry Co., Ltd., Crosslinked Vinyl Chloride Resins, Japan. Kokai Tokkyo Koho 80-118927 [1979/80]; C.A. **94** [1981] No. 66706.

Tokyo Printing Ink Mfg. Co., Ltd., Separation Prevention in Urethane Polymer Components, Japan. Kokai Tokkyo Koho 80-133416 [1979/80]; C.A. **94** [1981] No. 66733.

Toppan Printing Co., Ltd., Resin Plates for Gravure Printing Plates, Japan. Kokai Tokkyo Koho 80-19754 [1976/80]; C.A. **93** [1980] No. 141018.

Toray Industries, Inc., Presensitized Lithographic Plate, Japan. Kokai Tokkyo Koho 80-124148 [1979/80]; C.A. **94** [1981] No. 112553.

Toray Silicone Co., Ltd., Oil- and Waterproofing Compositions for Synthetic Fibers, Japan. Kokai Tokkyo Koho 80-90682 [1978/80]; C.A. **93** [1980] No. 187715.

Toshiba Silicone Co., Ltd., Room-Temperature-Curable Compositions, Japan. Kokai Tokkyo Koho 80-60557 [1978/80]; C.A. **93** [1980] No. 115704.

Toshiba Silicone Co., Ltd., Room-Temperature-Curable Polysiloxane Compositions, Japan. Kokai Tokkyo Koho 80-94956 [1979/80]; C.A. **93** [1980] No. 240805.

Toyo Ink Mfg. Co., Ltd., Polyurethane Sheets, Japan. Kokai Tokkyo Koho 80-145753 [1979/80]; C.A. **94** [1981] No. 140676.

Toyo Linoleum Co., Ltd., Thermosetting Poly(Vinyl Chloride) Plastisol Compositions, Japan. Kokai Tokkyo Koho 80-137146 [1979/80]; C.A. **94** [1981] No. 104399.

Toyo Rubber Chemical Industry Co., Ltd., Polyurethane Cushions, Japan. Kokai Tokkyo Koho 80-140535 [1979/80]; C.A. **94** [1981] No. 85255.

Toyo Rubber Industry Co., Ltd., Porous Alloy Sinter, Japan. Kokai Tokkyo Koho 80-79804 [1978/80]; C.A. **93** [1980] No. 154510.

Toyo Rubber Industry Co., Ltd., Urethane Foam with Ionic Group as Deodorant, Japan. Kokai Tokkyo Koho 80-162341 [1979/80]; C.A. **95** [1981] No. 67203.

Toyobo Co., Ltd., Light-Polarizing Films, Japan. Kokai Tokkyo Koho 80-90546 [1978/80]; C.A. **93** [1980] No. 221662.

Toyobo Co., Ltd., Photosensitive Resin Compositions for Flexographic Plates, Japan. Kokai Tokkyo Koho 80-127551 [1979/80]; C.A. **94** [1981] No. 148400.

Toyobo Co., Ltd., Photosensitive Resin Compositions for Flexographic Plates, Japan. Kokai Tokkyo Koho 80-153936 [1979/80]; C.A. **94** [1981] No. 217626.

Toyoda Gosei Co., Ltd., PVC Automobile Weather Strips, Japan. Kokai Tokkyo Koho 80-73550 [1978/80]; C.A. **93** [1980] No. 169792.

Uesugi, K., Kimura, H., Furukawa Electric Co., Ltd., Silane Crosslinking of Polyolefin Electric Insulators, Japan. Kokai Tokkyo Koho 80-46951 [1978/80]; C.A. **93** [1980] No. 72958.

Ukima Gosei K. K., Leather Substitutes Having Good Softness, Japan. Kokai Tokkyo Koho 80-112381 [1979/80]; C.A. **94** [1981] No. 48470.

USM Corp., Powderd Polyurethane Adhesive, Fr. Demande 2448564 [1979/80]; C.A. **95** [1981] No. 63150.

Watanabe, T., Moriwaki, N., Nobuoka, M., Mitsubishi Electric Corp., Compositions for Radiowave Absorbers, Japan. Kokai Tokkyo Koho 80-36961 [1978/80]; C.A. **93** [1980] No. 105889.

Watanabe, T., Tajima, H., Moriwaki, N., Nobuoka, M., Matsushita Electric Industrial Co., Ltd., Electroconductive Composition, Japan. Kokai Tokkyo Koho 80-35476 [1978/80]; C.A. **93** [1980] No. 86908.

Wessling, R. A., Yats, L. D., Perry, W. O., Elms, W. J., Dow Chemical Co., A Water-Compatible Composition Based on an Epoxide Resin Modified by a Pyridinium Compound, Eur. Appl. 14851 [1979/80]; C.A. **93** [1980] No. 222006.

Williams, R. C., Rogemoser, D. R., Textron, Inc., Water-Dispersible Urethane Polymers and Aqueous Polymer Dispersions, Eur. Appl. 17199 [1979/80]; C.A. **94** [1981] No. 32293.

Wurstova, E., Silicon Agents for Antiadhesive and Hydrophobic Treatment of Paper, Czech. 181452 [1973/80]; C.A. **93** [1980] No. 48928.

Yamaguchi, N., Nikaido, T., Matsushita Electric Works, Ltd., Vinyl Chloride Resin Compositions, Japan. Kokai Tokkyo Koho 80-21411 [1978/80]; C.A. **92** [1980] No. 216277.

Yamaguchi, N., Nikaido, T., Matsushita Electric Works, Ltd., Vinyl Chloride Resin Molding Materials, Japan. Kokai Tokkyo Koho 80-43144 [1978/80]; C.A. **93** [1980] No. 47910.

Yamamoto, T., Waya, N., Nakao, S., Showa Kako K. K., Coated Stannic Acid as a Flame Retardant, Japan. Kokai Tokkyo Koho 80-07545 [1978/80]; C.A. **92** [1980] No. 147946.

Yatomi, T., Kajitani, N., Mitsui Nisso Urethane K. K., Scorch Prevention of Urethane Polymer Foams from Vinyl Copolymers-Containing Polyols, Japan. Kokai Tokkyo Koho 80-48215 [1978/80]; C.A. **93** [1980] No. 72896.

Yokohama Rubber Co., Ltd., Siloxane Primer Compositions, Japan. Kokai Tokkyo Koho 80-73772 [1978/80]; C.A. **93** [1980] No. 170000.

Yukuta, T., Yagura, K., Fuchigami, N., Bridgestone Tire Co., Ltd., Plastic Foam Fuel Tank and a Method of Producing the Same, Brit. Appl. 2038739 [1978/80]; C.A. **95** [1981] No. 8454.

AKZO N. V., Polyurethane Encasing Composition, Belg. 887935 [1979/81]; C.A. **95** [1981] No. 151913.

Allport, D. C., Haggis, G. A., Redman, R. P., Imperial Chemical Industries, Ltd., Polyurethane Elastomers, Brit. 1589562 [1976/81]; C.A. **95** [1981] No. 188443.

André, J. D., Killis, A., Le Nest, J. F., Cheradame, H. M., Etat Français, Solid Electrolyte Based on Macromolecular Material with Ionic Conductivity, Eur. Appl. 37776 [1980/81]; C.A. **96** [1982] No. 105624.

Arco, M. J., Minnesota Mining and Mfg. Co., One-Part Solvent-Free Thermosettable Blocked Prepolymer Composition Containing a Diene, Together with Chain Extender, Chain Terminator and a Dienophile, U.S. 4273909 [1980/81]; C.A. **96** [1982] No. 7751.

Asahi Chemical Industry Co., Ltd., Thermoplastic Crosslinked Elastomers, Japan. Kokai Tokkyo Koho 81-04605 [1979/81]; C.A. **95** [1981] No. 8590.

Asahi Chemical Industry Co., Ltd., Yellowing-Resistant, Waterproof Adhesives, Japan. Kokai Tokkyo Koho 81-22374 [1979/81]; C.A. **95** [1981] No. 26293.

Asahi Chemical Industry Co., Ltd., Urethane Polymer Coating Materials, Japan. Kokai Tokkyo Koho 81-26962 [1979/81]; C.A. **95** [1981] No. 63844.

Asahi Chemical Industry Co., Ltd., Photosensitive Resin Compositions Having Improved Water-Developability, Japan. Kokai Tokkyo Koho 81-60441 [1979/81]; C.A. **95** [1981] No. 212956.

Asahi Chemical Industry Co., Ltd., Electrophoretic Coating Materials, Japan. Kokai Tokkyo Koho 81-93728 [1979/81]; C.A. **95** [1981] No. 205522.

Asahi Chemical Industry Co., Ltd., Photocurable Coating Compositions, Japan. Kokai Tokkyo Koho 81-100816 [1980/81]; C.A. **96** [1982] No. 8243.

Asahi Chemical Industry Co., Ltd., Polymers with Terminal Functional Groups, Japan. Kokai Tokkyo Koho 81-104906 [1980/81]; C.A. **95** [1981] No. 188104.

Asahi Chemical Industry Co., Ltd., Manufacture of Polyacetals, Japan. Kokai Tokkyo Koho 81-167718 [1980/81]; C.A. **96** [1982] No. 123887.

Asahi Denka Kogyo K. K., Photocurable Coating Compositions, Japan. Kokai Tokkyo Koho 81-65022 [1979/81]; C.A. **95** [1981] No. 134515.

Asahi Denka Kogyo K. K., Photocurable Coating Compositions, Japan. Kokai Tokkyo Koho 81-65024 [1979/81]; C.A. **95** [1981] No. 154516.

Asahi Denka Kogyo K. K., Photocurable Coating Materials, Japan. Kokai Tokkyo Koho 81-67333 [1979/81]; C.A. **95** [1981] No. 171165.

Asah-Olin, Ltd., Polyurethane Moldings with Improved Bending Modulus, Japan. Kokai Tokkyo Koho 81-88418 [1979/81]; C.A. **95** [1981] No. 220990.

Blahak, J., Metzeler Kautschuk GmbH, Adhesive Polyurethane or Polyurea Elastomeric Films on Rubber or Vulcanized Rubber Molded Articles, Ger. Offen. 2932866 [1979/81]; C.A. **94** [1981] No. 176888.

Bluestein, B. A., General Electric Co., Room Temperature-Vulcanisable Silicone Rubber Compositions, U.S. 4304897 [1980/81]; C.A. **96** [1982] No. 36652.

Bovis, C., Trebuchen, P., Naphthachimie, S. A., Polymeric Coverings for Sports Surfaces, Brit. 1585029 [1977/81]; C.A. **95** [1981] No. 26296.

Bridgestone Tire Co., Ltd., Low-Smoke Emitting Poly(Isocyanate) Foam Resin, Japan. Kokai Tokkyo Koho 81-00817 [1979/81]; C.A. **95** [1981] No. 8213.

Burkhart, D. C., Harper, L. R., Sommerfeld, E. G., du Pont de Nemours, E. I., and Co., Epoxy-Urethanes Based on Copolyesters, U.S. 4267288 [1978/81]; C.A. **95** [1981] No. 63467.

Chang, D. Chi-Kung, du Pont de Nemours, E. I., and Co., High Solids Coating Composition of a Low Molecular Weight Acrylic Polymer and a Polyisocyanate Crosslinking Agent and a Substrate Coated Therewith, Eur. Appl. 29598 [1979/81]; C.A. **95** [1981] No. 117215.

Chattha, M. S., Van Oene, H., Ford Motor Co., Two Component Oligomeric Phosphate/Isocyanate Composition, U.S. 4259472 [1980/81]; C.A. **95** [1981] No. 8965.

Chisso Corp., Poly(Vinyl Chloride) Resins for Use in Fluidized Beds, Japan. Kokai Tokkyo Koho 81-106945 [1980/81]; C.A. **95** [1981] No. 221465.

Chisso Corp., Vinyl Chloride Copolymer Sheets for Vacuum Forming, Japan. Kokai Tokkyo Koho 81-142025 [1980/81]; C.A. **96** [1982] No. 86600.

Chisso Corp., PVC Coating Compositions, Japan. Kokai Tokkyo Koho 81-151744 [1980/81]; C.A. **96** [1982] No. 87177.

Chisso Corp., Vinyl Chloride Copolymer Calender Sheets, Japan. Kokai Tokkyo Koho 81-154038 [1980/81]; C.A. **96** [1982] No. 105437.

Chisso Corp., Silane-Modified Propylene Polymers, Japan. Kokai Tokkyo Koho 81-155213 [1979/81]; C.A. **96** [1982] No. 105184.

Christman, D. L., Kan, P. T., BASF Wyandotte Corp., Carbamylbiuret-Modified Polyisocyanates, U.S. 4271087 [1979/81]; C.A. **95** [1981] No. 188071.

Crose, J. M., Mobay Chemical Corp., Hydroxy-Functional Vinyl Copolymers, U.S. 4264755 [1975/81]; C.A. **95** [1981] No. 43927.

Cuscurida, M., Texaco Inc., Stabilizing Polyurea Polymer Polyols by Treating with a Secondary Amine, U.S. 4293470 [1980/81]; C.A. **96** [1982] No. 20864.

Cuscurida, M., Dominguez, R. J. G., Rice, D. M., Texaco Inc., RIM Elastomers with Improved Heat Distortion and Tear Properties, U.S. 4301110 [1980/81]; C.A. **96** [1982] No. 53574.

Cuscurida, M., Speranza, G. P., Texaco Inc., Polyurethane Foams Using a Polyurea Polymer Polyol, U.S. 4296213 [1979/81]; C.A. **96** [1982] No. 20821.

Daiichi Kogyo Seiyaku Co., Ltd., Baseball Practice Balls, Japan. Kokai Tokkyo Koho 81-161066 [1980/81]; C.A. **96** [1982] No. 124102.

Dainippon Ink and Chemicals, Inc., Resins for Electrophoretic Coating Compositions, Japan. Kokai Tokkyo Koho 81-86969 [1979/81]; C.A. **95** [1981] No. 152261.

Dainippon Ink and Chemicals, Inc.,Moisture-Curable, Stampable Sheets, Japan. Kokai Tokkyo Koho 81-149439 [1980/81]; C.A. **96** [1982] No. 86680.

Dainippon Ink and Chemicals, Inc., Kawamura Physical and Chemical Research Institute, Water-Curable Fire-Resistant Resin Compositions, Japan. Kokai Tokkyo Koho 81-76445 [1979/81]; C.A. **95** [1981] No. 188138.

Dainippon Toryo Co., Ltd., Antifouling Coating Materials, Japan. Kokai Tokkyo Koho 81-24073 [1979/81]; C.A. **95** [1981] No. 63807.

Dainippon Toryo Co., Ltd., Thermosetting Polymer Slurry Coating Materials, Japan. Kokai Tokkyo Koho 81-61403 [1979/81]; C.A. **95** [1981] No. 171117.

Dainippon Toryo Co., Ltd., Water-Thinned Thermosetting Coating Compositions, Japan. Kokai Tokkyo Koho 81-100818 [1980/81]; C.A. **96** [1982] No. 21377.

Dainippon Toryo Co., Ltd., Water-Thinned Thermosetting Coating Compositions, Japan. Kokai Tokkyo Koho 81-100819 [1980/81]; C.A. **95** [1981] No. 221442.

Dainippon Toryo Co., Ltd., Water-Thinned Coating Compositions, Japan. Kokai Tokkyo Koho 81-135564 [1980/81]; C.A. **96** [1982] No. 53991.

Dainippon Toryo Co., Ltd., Water-Thinned Coating Compositions, Japan. Kokai Tokkyo Koho 81-139562 [1980/81]; C.A. **96** [1982] No. 87149.

Dainippon Toryo Co., Ltd., Water-Thinned Coating Materials, Japan. Kokai Tokkyo Koho 81-149467 [1980/81]; C.A. **96** [1982] No. 105889.

Dainippon Toryo Co., Ltd., Water-Dispersible Thermosetting Coating Compositions, Japan. Kokai Tokkyo Koho 81-149469 [1980/81]; C.A. **96** [1982] No. 105904.

Dainippon Toryo Co., Ltd., Water Dispersed Thermosetting Coating Composition, Japan. Kokai Tokkyo Koho 81-151717 [1980/81]; C.A. **96** [1982] No. 124635.

Deiner, H., Mosch, F., Schilling, H., Chemische Fabrik Pfersee GmbH, Dimension Stabilizing of Textile Sheet Materials, Ger. Offen. 3014675 [1980/81]; C.A. **96** [1982] No. 36826.

Depetris, N., Bovis, C., Naphthachimie S. A., Material for Making Walls, S. African 80-06990 [1980/81]; C.A. **96** [1982] No. 200990.

Diefenbach, H., Dobbelstein, A., BASF A.-G., Aqueous Polymer Dispersion, Ger. Offen. 3002865 [1980/81]; C.A. **95** [1981] No. 134473.

Dominguez, R. G. J., Texaco Development Corp., Polyurethanes Molded by Injection with Reaction Having an Improved Catalytic Yield, Fr. Demande 2475456 [1980/81]; C.A. **95** [1981] No. 221104.

Dominguez, R. G. J., Rice, D. M., Texaco Development Corp., Elastomers Molded by Injection with Reaction Having Excellent Dimensional Stability at Increased Temperatures, Fr. Demande 2475455 [1980/81]; C.A. **95** [1981] No. 221105.

Dominguez, R. J. G., Rice, D. M., Texaco Development Corp., Catalyst System for Reaction Injection Molded Elastomers, U.S. 4273885 [1980/81]; C.A. **95** [1981] No. 99142.

du Pont de Nemours, E. I., and Co., Moisture-Curable Coating Materials, Japan. Kokai Tokkyo Koho 81-118409 [1980/81]; C.A. **96** [1982] No. 8285.

Eckardt, P., Hoechst A.-G., Porous Gypsum and Construction Elements Containing It, Ger. Offen. 2940785 [1979/81]; C.A. **95** [1981] No. 120036.

Efer, J., Kochmann, W., Pfeiffer, H. D., Rank, B., Schaefer, H., Thust, U., Trautner K., VEB Chemiekombinat Bitterfeld, Stabilization of Phosphoric Acid O,O-Dimethyl-O-2,2-dichlorovinyl Ester, Ger. [East] 151002 [1978/81]; C.A. **97** [1982] No. 127806.

Fischer, H., Plum, H., Hoechst A.-G., Copolymers Containing Hydroxyl Groups, Ger. Offen. 2942327 [1979/81]; C.A. **95** [1981] No. 8991.

Floyd, D. E., Henkel Corp., Polyamines Substituted Ammonium Compounds Prepared Therefrom and a Method for Producing a Cured Piece, Eur. Appl. 23158 [1979/81]; C.A. **94** [1981] No. 158460.

Fuji Kagakushi Kogyo Co., Ltd., Ink-Holding Porous Material, Japan. Kokai Tokkyo Koho 81-95934 [1979/81]; C.A. **96** [1982] No. 7771.

Fujikura Cable Works, Ltd., Crosslinkable Halogen-Containing Polymer Compositions and Crosslinking at Normal Temperature and Pressure, Japan. Kokai Tokkyo Koho 81-05854 [1979/81]; C.A. **94** [1981] No. 209757.

Fujikura Cable Works, Ltd., Crosslinked Polyolefin Electric Insulators, Japan. Kokai Tokkyo Koho 81-92946 [1979/81]; C.A. **95** [1981] No. 220824.

Fujikura Cable Works, Ltd., Polyethylene Cables, Japan. Kokai Tokkyo Koho 81-103818 [1980/81]; C.A. **95** [1981] No. 221004.

Fujikura Cable Works, Ltd., Polyethylene Cables, Japan. Kokai Tokkyo Koho 81-103819 [1980/81]; C.A. **96** [1982] No. 7781.

Fujikura Cable Works, Ltd., Neoprene Rubber Electric Insulators for Wire, Japan. Kokai Tokkyo Koho 81-163132 [1980/81]; C.A. **96** [1982] No. 124336.

Fujikura Cable Works, Ltd., Neoprene Rubber Electric Insulators for Wire, Japan. Kokai Tokkyo Koho 81-163133 [1980/81]; C.A. **96** [1982] No. 124338.

Fujikura Cable Works, Ltd., Crosslinking of Polyolefins and Olefin Copolymers, Japan. Kokai Tokkyo Koho 81-163142 [1980/81]; C.A. **97** [1982] No. 39828.

Fujikura Cable Works, Ltd., Crosslinking of Polyolefins and Olefin Copolymers, Japan. Kokai Tokkyo Koho 81-163143 [1980/81]; C.A. **97** [1982] No. 39829.

Fujikura Cable Works, Ltd., Molding of Crosslinked Polyolefins, Japan. Kokai Tokkyo Koho 81-167731 [1980/81]; C.A. **96** [1982] No. 144104.

Fujikura Cable Works, Ltd., Neoprene Rubber Compositions for Extrusion and Vulcanization, Japan. Kokai Tokkyo Koho 81-167739 [1980/81]; C.A. **96** [1982] No. 124339.

Fujikura Cable Works, Ltd., Toray Silicone Co., Ltd., Crosslinked Vinyl Chloride Polymers, Japan. Kokai Tokkyo Koho 81-74105 [1979/81]; C.A. **95** [1981] No. 170393.

Fujitsu Ltd., Radiation Resist Compositions, Japan. Kokai Tokkyo Koho 81-155939 [1980/81]; C.A. **97** [1982] No. 31275.

Furukawa Electric Co., Ltd., Polymer Compositions, Japan. Kokai Tokkyo Koho 81-143234 [1980/81]; C.A. **96** [1982] No. 53453.

Furukawa Electric Co., Ltd., Silane-Grafted Crosslinked Olefin Polymer Compositions, Japan. Kokai Tokkyo Koho 81-149453 [1980/81]; C.A. **96** [1982] No. 86491.

Gautier, A., Laisney, B., Letoffe, M., Rhone-Poulenc Industries S. A., Polysiloxane Compositions Curable into Elastomers at Ambient Temperature in the Presence of Water, Eur. Appl. 21859 [1979/81]; C.A. **94** [1981] No. 158142.

Gaughan, E. J., Stauffer Chemical Co., Sulfonylurea Herbicidal Antidotes, U.S. 4260824 [1979/81]; C.A. **95** [1981] No. 61815.

Gimpel, J., Feuerheld, K. H., Schenck, H. U., BASF A.-G., Masked Isocyanate Group-Containing Copolymers and Their Use in Electrodip Coating, Ger. Offen. 3017537 [1980/81]; C.A. **96** [1982] No. 21373.

Gimpel, J., Hartmann, H., Schenk, H. U., BASF A.-G., Varnish Binders and Their Use in Electrophoretic Varnishes, Eur. Appl. 21014 [1979/81]; C.A. **94** [1981] No. 176837.

Gimpel, J., Hartmann, H., Schenk, H. U., BASF A.-G., Varnish Binders and Their Use in Electrophoretic Varnishes, Eur. Appl. 21015 [1979/81]; C.A. **94** [1981] No. 141332.

Gras, R., Chemische Werke Huels A.-G., Coating Glass Surfaces with a Hard Plastic Protective Coating, Ger. Offen. 2938229 [1979/81]; C.A. **95** [1981] No. 8960.

Gras, R., Chemische Werke Huels A.-G., Coating Glass Surfaces with a Duroplastic Protective Coating, Ger. Offen. 2938309 [1979/81]; C.A. **94** [1981] No. 8963.

Haemer, L. F., Kimak, T., Congoleum Corp., Differential Gloss Products, U.S. 4298646 [1980/81]; C.A. **96** [1982] No. 87193.

Hahn, K., Horn, P., Marx, M., Weber, H., Weiss, W., Wurmb, R., BASF A.-G., Polyurethane Foams Modified with Melamine-Formaldehyde Pre-condensates, Ger. Offen. 3020091 [1980/81]; C.A. **96** [1982] No. 53466.

Hajimichael, A., Hopper, S., Smith, P. K., EMI Ltd., Polyurethane Encapsulant Material, Brit. Appl. 2068393 [1979/81]; C.A. **96** [1982] No. 124124.

Harmer, W. L., Minnesota Mining and Mfg. Co., Polyurethane Floor Varnish, U.S. 4273912 [1979/81]; C.A. **96** [1982] No. 8256.

Hertler, W. R., du Pont de Nemours, E. I., and Co., Coating Compositions Containing Polymers with Isocyano Groups and Nickel Oligomerization Catalysts, U.S. 4251421 [1979/81]; C.A. **94** [1981] No. 210482.

Hertler, W. R., du Pont de Nemours, E. I., and Co., Coating Compositions Containing Polymers with Isocyano Groups and Alkylboranes, U.S. 4251422 [1979/81]; C.A. **94** [1981] No. 176894.

Hitachi, Ltd., Urethane Foam Structure, Japan. Kokai Tokkyo Koho 81-32517 [1979/81]; C.A. **94** [1981] No. 209990.

Hitachi., Ltd., Polyurethane Foams with Skin, Japan. Kokai Tokkyo Koho 81-146734 [1980/81]; C.A. **96** [1982] No. 86602.

Hitachi Cable, Ltd., Moisture-Cured Polyolefin Insulator, Japan. Kokai Tokkyo Koho 81-126213 [1980/81]; C.A. **96** [1982] No. 53421.

Hitachi Chemical Co., Ltd., Molded Polyurethane, Japan. Kokai Tokkyo Koho 81-24156 [1979/81]; C.A. **95** [1981] No. 44343.

Hitachi Chemical Co., Ltd., Polyurethane Resin Compositions, Japan. Kokai Tokkyo Koho 81-34753 [1979/81]; C.A. **95** [1981] No. 25903.

Hitachi Chemical Co., Ltd., Photocurable Coating Materials, Japan. Kokai Tokkyo Koho 81-95364 [1979/81]; C.A. **95** [1981] No. 221458.

Hitachi Chemical Co., Ltd., Reinforced Polyurethane Elastomers, Japan. Kokai Tokkyo Koho 81-139552 [1980/81]; C.A. **96** [1982] No. 70250.

Hodakowski, L. E., Koleske, J. V., Union Carbide Corp., Novel Urethane-Acrylate and Radiation Curable Compositions, U.S. 4260703 [1979/81]; C.A. **95** [1981] No. 134502.

Horacek, H., Marx, M., Hobei, D., BASF A.-G., Polyurethane Moldings, Ger. Offen. 3014161 [1980/81]; C.A. **95** [1981] No. 205150.

Huebner, D. J., Weyenberg, D. R., Dow Corning Corp., Polysiloxane Emulsion Elastomers Reinforced with Emulsified Organosilicone Copolymers, U.S. 4288356 [1980/81]; C.A. **95** [1981] No. 188473.

Hughes, J., Keane, K. E., Imperial Chemical Industries, Ltd., Liquid Polyisocyanate Compositions, Brit. Appl. 2066813 [1979/81]; C.A. **96** [1982] No. 105225.

Hughes, J., Keane, K. E., Imperial Chemical Industries, Ltd., Liquid Polyisocyanate Compositions and Their Use, Eur. Appl. 31207 [1979/81]; C.A. **96** [1982] No. 7950.

Hughes, J., Keane, K. E., Imperial Chemical Industries, Ltd., Liquid Polyisocyanate Compositions and Their Use, Eur. Appl. 31650 [1979/81]; C.A. **95** [1981] No. 151626.

Hutt, J. W., Blanco, F. E., Products Research and Chemical Corp., Polyurethane Sealant System, U.S. 4284751 [1980/81]; C.A. **95** [1981] No. 171209.

Hynds, J., Imperial Chemical Industries Ltd., Manufacture of Polyurethane Foams, Brit. Appl. 2051838 [1979/81]; C.A. **95** [1981] No. 63212.

Idemitsu Kosan Co., Ltd., Laminated Sheets, Japan. Kokai Tokkyo Koho 81-51350 [1979/81]; C.A. **95** [1981] No. 82173.

Idemitsu Kosan Co., Ltd., Asphalt-Based Waterproofing Sheets, Japan. Kokai Tokkyo Koho 81-53286 [1979/81]; C.A. **95** [1981] No. 116991.

Idemitsu Kosan Co., Ltd., Curing of Liquid Polymers, Japan. Kokai Tokkyo Koho 81-84751 [1979/81]; C.A. **95** [1981] No. 170796.

Idemitsu Kosan Co., Ltd., Photocurable Liquid Rubber Coating Materials, Japan. Kokai Tokkyo Koho 81-86903 [1979/81]; C.A. **95** [1981] No. 152311.

Idemitsu Kosan Co., Ltd., Liquid Rubber Adhesives, Japan. Kokai Tokkyo Koho 81-90868 [1979/81]; C.A. **95** [1981] No. 188312.

Idstroem, B., Berol Kemi AB; Rigid Polyurethane Foams, Eur. Appl. 36403 [1980/81]; C.A. **96** [1982] No. 7542.

Iida, K., Kawamoto, T., Miyoshi, K., Asahi Chemical Industry Co., Ltd., Photosensitive Resinous Composition of Urethane, Fr. Demande 2477295 [1980/81]; C.A. **96** [1982] No. 77567.

Ilaria, J. E., Mobil Oil Corp., Coating Compositions Based on Epoxide Adducts and Polyurethane Prepolymers, Eur. Appl. 24811 [1979/81]; C.A. **95** [1981] No. 26729.

Ilaria, J. E., Mobil Oil Corp., Elastomeric Urethane Coating, U.S. 4282123 [1979/81]; C.A. **96** [1982] No. 124651.

Japan Ester Co., Ltd., Manufacture of Polyesters for Powder Coatings, Japan. Kokai Tokkyo Koho 81-82818 [1979/81]; C.A. **96** [1982] No. 87146.

Kallaur, M., Freeman Chemical Corp., Sheet Molding Compound, U.S. 4289684 [1980/81]; C.A. **95** [1981] No. 220865.

Kamatani, Y., Fujita, N., Takeda Chemical Industries, Ltd., Isocyanate Composition, Eur. Appl. 42701 [1980/81]; C.A. **96** [1982] No. 105168.

Kanebo NSC K. K., Bonding of Porous Materials, Japan. Kokai Tokkyo Koho 81-155269 [1980/81]; C.A. **96** [1982] No. 124861.

Kanebo Chemical Industry Co., Ltd., Idemitsu Kosan Co., Ltd., Waterproofing of Concrete, Japan. Kokai Tokkyo Koho 81-39253 [1979/81]; C.A. **95** [1981] No. 117212.

Kanegafuchi Chemical Industry Co., Ltd., Moisture-Curable Resin Compositions, Japan. Kokai Tokkyo Koho 81-67366 [1979/81]; C.A. **95** [1981] No. 170374.

Kinashi, T., Fujita, T., Kawabe, N., Toray Industries, Inc., Dry Planographic Printing Plate and Preparation Thereof, Brit. Appl. 2064803 [1979/81]; C.A. **96** [1982] No. 113495.

Kleeberg, W., Hellmann, K., Rubner, R., Wiedenmann, R., Siemens A.-G., Crosslinked Insulations and Sheaths of Cables and Cords, Eur. Appl. 41192 [1980/81]; C.A. **96** [1982] No. 105498.

Knopf, R. J., Union Carbide Corp., (N-Substituted Carbamoyloxy)alkanoyloxyhydrocarbyl Acrylate Esters, Eur. Appl. 37314 [1980/81]; C.A. **96** [1982] No. 53943.

Knopf, R. J., Hess, L. G., Union Carbide Corp., (N-Substituted Carbamoyloxy)alkyleneoxyhydrocarbyl Acrylate Esters, Eur. Appl. 36813 [1980/81]; C.A. **96** [1982] No. 182204.

Kohkoku Chemical Industries Co., Ltd., Rigid Plastic Foams, Japan. Kokai Tokkyo Koho 81-159250 [1980/81]; C.A. **96** [1982] No. 124167.

Kraemling, F., Mueller, A., Linden, L., Raedisch, H., Vereinigte Glaswerke GmbH, Plastic Interlayer for Multiple Glass Panes, Ger. Offen. 3032211 [1979/81]; C.A. **95** [1981] No. 44389.

Krbechek, L. O., Henkel Corp., Steroidal Carbamates, U.S. 4252731 [1980/81]; C.A. **94** [1981] No. 209079.

Kaogh, M. J., Union Carbide Corp., Water-Curable, Silane-Modified Alkylene Alkylacrylate Copolymer, U.S. 4291136 [1978/81]; C.A. **95** [1981] No. 205248.

Kuraray Co., Ltd., Greenhouse Films, Japan. Kokai Tokkyo Koho 81-161462 [1980/81]; C.A. **96** [1982] No. 123974.

Lampe, W. R., General Electric Co., Coating with a Silicone Rubber Vulcanizable at Ambient Temperature, Fr. Demande 2485404 [1980/81]; C.A. **96** [1982] No. 164338.

Lehner, A., Buethe, I., Heil, G., Lenz, W., Hartmann, H., BASF A.-G., Radiation-Hardenable Aqueous Binder Dispersions, Ger. Offen. 3005036 [1980/81]; C.A. **95** [1981] No. 171195.

Lehner, A., Gimpel, J., Buethe, I., Hartmann, H., Schenk, H. U., BASF A.-G., Coatings on Electrically Conductive Articles by Electrophoretic Plating of Anionic Polyurethanes from an Aqueous Dispersion, Ger. Offen. 3005034 [1980/81]; C.A. **95** [1981] No. 171127.

Lehner, A., Hartmann, H., Bachmann, R., Balz, W., Kohl, A., BASF A.-G., Magnetic Recording Tape, Ger. Offen. 3005009 [1980/81]; C.A. **96** [1982] No. 14507.

Lehner, A., Hartmann, H., Heil, G., Lenz, W., Buethe, I., Bachmann, R., BASF A.-G., Polymerizable Polyurethane Elastomers, Ger. Offen. 3005035 [1980/81]; C.A. **95** [1981] No. 170759.

Lin, K. C., Hammond, D. J., Owens-Corning Fiberglas Corp., Stable Aqueous Emulsion of Reactive Polysiloxane and Its Curing Agent, U.S. 4277382 [1979/81]; C.A. **95** [1981] No. 99404.

Loeschau, S., Niklas, M., Noack, J., Sachs, R., Trentsch, G., Wust, M., Stable Polyurethane Solutions, Ger. [East] 148460 [1976/81]; C.A. **95** [1981] No. 220565.

Loeschau, S., Niklas, M., Noack, J., Wust, M., Polyurethanes, Ger. [East] 148462 [1976/81]; C.A. **95** [1981] No. 188197.

Lorenz, D. H., Gruber, B. A., GAF Corp., Radiation Curable Coating Composition Comprising an Oligomer and a Copolymerizable Ultra-Violet Absorber, U.S. 4263366 [1979/81]; C.A. **95** [1981] No. 44912.

Maciejewski, J., Instytut Chemii Przemyslowej, Polysiloxane Compositions Absorbing Mechanical Energy, Ger. Offen. 3039692 [1979/81]; C.A. **95** [1981] No. 26436.

Maciejewski, J., Instytut Chemii Przemyslowej, Viscoelastic Polysiloxane Compositions, Pol. 109050 [1978/81]; C.A. **96** [1982] No. 124170.

McBrayer, R. L., BASF Wyandotte Corp., Microcellular Polyurethane Foam, Brit. Appl. 2075531 [1980/81]; C.A. **96** [1982] No. 105606.

McBrayer, R. L., Carver, T. G., BASF Wyandotte Corp., High Modulus Microcellular Polyurethane Foams, Brit. Appl. 2076001 [1980/81]; C.A. **96** [1982] No. 105478.

McDaniel, K. G., Rice, D. M., Cuscurida, M., Texaco Development Corp., Reaction Injection Molded Polyurethane, U.S. 4243760 [1979/81]; C.A. **95** [1981] No. 26415.

Meito Co., Ltd., Toshiba Silicone Co., Ltd., Waterproofing and Preservation of Wood, Japan. Kokai Tokkyo Koho 81-04408 [1979/81]; C.A. **94** [1981] No. 210562.

Mitsubishi Chemical Industries Co., Ltd., Urethane Polymers, Japan. Kokai Tokkyo Koho 81-30425 [1979/81]; C.A. **95** [1981] No. 26059.

Mitsubishi Chemical Industries Co., Ltd., Adhesives for Vulcanized Rubbers, Japan. Kokai Tokkyo Koho 81-55475 [1979/81]; C.A. **95** [1981] No. 99047.

Mitsubishi Chemical Industries Co., Ltd., Vulcanization of Hydroxy-Terminated Hydrocarbon Rubber, Japan. Kokai Tokkyo Koho 81-57819 [1979/81]; C.A. **95** [1981] No. 99157.

Mitsubishi Chemical Industries Co., Ltd., Solvent Recovery from Urethane Prepolymers, Japan. Kokai Tokkyo Koho 81-70019 [1979/81]; C.A. **95** [1981] No. 133804.

Mitsubishi Chemical Industries Co., Ltd., Adhesives for Vulcanized Rubber, Japan. Kokai Tokkyo Koho 81-95969 [1979/81]; C.A. **96** [1982] No. 886618.

Mitsubishi Chemical Industries Co., Ltd., Polyurethane Sealants, Japan. Kokai Tokkyo Koho 81-139577 [1980/81]; C.A. **96** [1982] No. 70577.

Mitsubishi Petrochemical Co., Ltd., Polyolefin Films for Surface Protection of Metal Plates, Japan. Kokai Tokkyo Koho 81-08256 [1979/81]; C.A. **94** [1981] No. 209958.

Mitsubishi Petrochemical Co., Ltd., Crosslinkable Polyolefin Compositions, Japan. Kokai Tokkyo Koho 81-08446 [1979/81]; C.A. **95** [1981] No. 8268.

Mitsubishi Petrochemical Co., Ltd., Crosslinkable Ethylene Copolymers, Japan. Kokai Tokkyo Koho 81-08447 [1979/81]; C.A. **94** [1981] No. 209796.

Mitsubishi Petrochemical Co., Ltd., Resin-Metal Composite, Japan. Kokai Tokkyo Koho 81-11246 [1979/81]; C.A. **94** [1981] No. 209846.

Mitsubishi Petrochemical Co., Ltd., Silane-Crosslinked Polyolefin Laminates, Japan. Kokai Tokkyo Koho 81-11248 [1979/81]; C.A. **94** [1981] No. 193208.

Mitsubishi Petrochemical Co., Ltd., Crosslinked-Polyethylene Blow-Molded Articles, Japan. Kokai Tokkyo Koho 81-48944 [1979/81]; C.A. **95** [1981] No. 116686.

Mitsubishi Petrochemical Co., Ltd., One-Shot Thermosetting Resin Compositions, Japan. Kokai Tokkyo Koho 81-65015 [1979/81]; C.A. **95** [1981] No. 134488.

Mitsubishi Petrochemical Co., Ltd., Pipes Coated with Crosslinked Polyolefin, Japan. Kokai Tokkyo Koho 81-65667 [1979/81]; C.A. **95** [1981] No. 117221.

Mitsubishi Petrochemical Co., Ltd., Crosslinked Polyethylene Resin Pipe, Japan. Kokai Tokkyo Koho 81-88446 [1979/81]; C.A. **95** [1981] No. 188314.

Mitsubishi Petrochemical Co., Ltd., Crosslinked Ethylene Copolymer Laminates, Japan. Kokai Tokkyo Koho 81-93542 [1979/81]; C.A. **95** [1981] No. 188328.

Mitsubishi Petrochemical Co., Ltd., Ethylene Copolymers with Unsaturated Silane, Japan. Kokai Tokkyo Koho 81-95912 [1979/81]; C.A. **95** [1981] No. 204894.

Mitsubishi Petrochemical Co., Ltd., Water-Crosslinkable Olefin Polymer Laminates, Japan. Kokai Tokkyo Koho 81-151561 [1980/81]; C.A. **96** [1982] No. 86695.

Mitsui Nisso Urethane K. K., Polyurethane Foams, Japan. Kokai Tokkyo Koho 81-88438 [1979/81]; C.A. **95** [1981] No. 204947.

Mitsui Nisso Urethane K. K., Thermoplastic Polyurethanes, Japan. Kokai Tokkyo Koho 81-127616 [1980/81]; C.A. **96** [1982] No. 20597.

Mitsui Nisso Urethane K. K., Polyurethane Foam Coatings, Japan. Kokai Tokkyo Koho 81-127669 [1980/81]; C.A. **96** [1982] No. 37039.

Miyaka, T., Takeda, K., Ikeda, A., Asahi Chemical Industry Co., Ltd., Coating Composition Containing a Nitrogen-Containing Acrylic Copolymer, an Amine-Modified Epoxy Compound and a Protected Polyisocyanate which can be Cathodically Deposited, Fr. Demande 2472590 [1979/81]; C.A. **95** [1981] No. 221407.

Moga-Gheorghe, S., Barbulescu, N., Costes, I. M., Rolea, G., Isoxazolyl-Phenyldiazene Derivatives, Rom. 77826 [1979/81]; C.A. **99** [1983] No. 139366.

Mohiuddin, G., International Telephone and Telegraph Corp., Molded Polyurethane Part, U.S. 4282285 [1979/81]; C.A. **95** [1981] No. 205064.

Monson, N. J., Minnesota Mining and Mfg. Co., Chip-Resistant Pigmented Polyurethane Protective Coating, U.S. 4254168 [1979/81]; C.A. **94** [1981] No. 141393.

Nagase, Y., Shibatani, K., Yamauchi, J., Omura, I., Kuraray Co., Ltd., Cement Compositions Used in Dentistry, Fr. Demande 2481112 [1980/81]; C.A. **96** [1982] No. 129820.

Narayan, T., Kan, P., BASF Wyandotte Corp., Liquid Carbodiimide- and Uretonimine-Isocyanurate-Containing Polyisocyanate Compositions and Microcellular Foams Made from Them, U.S. 4284730 [1980/81]; C.A. **95** [1981] No. 188349.

Nippon Carbide Industries Co., Inc., Submergible Vinyon Fibers, Japan. Kokai Tokkyo Koho 81-96911 [1979/81]; C.A. **95** [1981] No. 134293.

Nippon Carbide Industries Co., Inc., Compositions and Uses of Resin Fibers with Finely Divided Fiber Tips, Japan. Kokai Tokkyo Koho 81-120746 [1980/81]; C.A. **96** [1982] No. 8060.

Nippon Polyurethane Industry Co., Ltd., Coating Materials for Elastic Substrates, Japan. Kokai Tokkyo Koho 81-41264 [1979/81]; C.A. **95** [1981] No. 99422.

Nippon Soda Co., Ltd., Paint Composition for Electrolytic Coating by Cathodic Deposition, Fr. Demande 2459272 [1979/81]; C.A. **95** [1981] No. 99400.

Nippon Soflan Kako Co., Ltd., Polyurethane Binders, Japan. Kokai Tokkyo Koho 81-08420 [1979/81]; C.A. **95** [1981] No. 63135.

Nippon Synthetic Chemical Industry Co., Ltd., Aminated Polyester Powder Coating, Japan. Kokai Tokkyo Koho 81-59877 [1979/81]; C.A. **95** [1981] No. 152278.

Nippon Telegraph and Telephone Public Corp., Photoconductive Polymers for Electrophotographic Photosensitive Materials, Japan. Kokai Tokkyo Koho 81-47431 [1979/81]; C.A. **95** [1981] No. 195148.

Nissen, D., Schuster, L., Hutchison, J., Marx, M., Wurmb, R., BASF A.-G., Cellular Polyurethane-Polyurea Moldings, Ger. Offen. 2940738 [1979/81]; C.A. **95** [1981] No. 63426.

Nitto Chemical Industry Co., Ltd., Stabilized Halogen-Containing Resin Compositions, Japan. Kokai Tokkyo Koho 81-30453 [1979/81]; C.A. **95** [1981] No. 63222.

Nitto Electric Industrial Co., Ltd., Electrically Insulating Polyester Coating Materials, Japan. Kokai Tokkyo Koho 81-14528 [1979/81]; C.A. **94** [1981] No. 210442.

Nitto Electric Industrial Co., Ltd., Peelable Protective Coating Materials for Plastic Moldings, Japan. Kokai Tokkyo Koho 81-106927 [1980/81]; C.A. **95** [1981] No. 221464.

Noack, R., Schwetlick, K., Ermer, H., Heinrich, M., Herrmann, P., Storable Polyurethane-Single Component Lacquers, Ger. [East] 151466 [1979/81]; C.A. **96** [1982] No. 144674.

Park, K., Union Carbide Corp., Polycaprolactone Polyol Urethanes, Eur. Appl. 21825 [1979/81]; C.A. **94** [1981] No. 210465.

Park, R. S., Henderson's Industries, Ltd., Polyether-Based Polyurethane Foams Including a Flame-Retardant System Containing Antimony Trioxide, a Chlorinated Paraffin and Alumina Trihydrate, U.S. 4266042 [1977/81]; C.A. **95** [1981] No. 44256.

Parshina, A. M., Kiseleva, O. S., Flomina, E. E., Ogonyants, T. V., Stepina, E. M., Composition Based on Low-Molecular-Weight Polydimethylsiloxane Rubber, U.S.S.R. 887598 [1979/81]; C.A. **96** [1982] No. 144322.

Peters, E. N., Union Carbide Corp., Curable Resin Compositions Comprising Hydroxyl-Terminated Unsaturated Polyester Oligomer, a Polyisocyanate, an Ethylenically Unsaturated Monomer and a Catalyst, U.S. 4289682 [1979/81]; C.A. **95** [1981] No. 220864.

Phillips, B. A., Spitler, K. G., Keegan, R. E., Mobay Chemical Corp., Heat Sealing Polyurethane Foam, U.S. 4302272 [1979/81]; C.A. **96** [1982] No. 86718.

Plum, H., Hoechst A.-G., Hardenable Copolymers by Polymerization of Unsaturated Esters with Unsaturated Monomers, Ger. Offen. 3020524 [1979/81]; C.A. **96** [1982] No. 37062.

Rasshofer, W., Groegler, G., Findeisen, K., Bayer A.-G., Polyurethane Production Using Cyclic n-Hydroxyalkyl-Substituted Compounds Having Amidino Groups as Catalysts, Ger. Offen. 3015440 [1980/81]; C.A. **96** [1982] No. 20576.

Redman, R. P., Imperial Chemical Industries Ltd., Isocyanate Reactions, Eur. Appl. 39137 [1980/81]; C.A. **96** [1982] No. 54183.

Riley, R. L., Grabowsky, R. L., UOP Inc., Gas Separation Membranes, U.S. 4243701 [1977/81]; C.A. **94** [1981] No. 193408.

Rowlands, J. P., Polymer-Modified Polyols Useful in Preparing Polyurethanes, Neth. Appl. 81-00708 [1980/81]; C.A. **96** [1982] No. 20812.

Sanyo Chemical Industries, Ltd., Polyol Compositions for Polyurethane Foams, Japan. Kokai Tokkyo Koho 81-67331 [1979/81]; C.A. **95** [1981] No. 133872.

Schenk, W. N., Goodrich, B. F., Co., Liquid Hydroxyl Polymers for Adhesives, Belg. 887748 [1980/81]; C.A. **95** [1981] No. 204958.

Schimmel, K. F., Seiner, J. A., Dowbenko, R., Christenson, R. M., PPG Industries, Inc., Urethane Rheology Modifiers and Compositions Containing Them, U.S. 4298511 [1980/81]; C.A. **96** [1982] No. 36973.

Sekisui Chemical Co., Ltd., Adhesive Sheets, Japan. Kokai Tokkyo Koho 81-41282 [1979/81]; C.A. **95** [1981] No. 82004.

Sekisui Chemical Co., Ltd., Sealing Compositions, Japan. Kokai Tokkyo Koho 81-57868 [1979/81]; C.A. **95** [1981] No. 117240.

Sekisui Chemical Co., Ltd., Pressure-Sensitive Adhesives, Japan. Kokai Tokkyo Koho 81-67380 [1979/81]; C.A. **95** [1981] No. 170604.

Sekisui Chemical Co., Ltd., Adhesives, Japan. Kokai Tokkyo Koho 81-90869 [1979/81]; C.A. **96** [1982] No. 7762.

Sekisui Chemical Co., Ltd., Extrusion and Crosslinking of Polyolefins, Japan. Kokai Tokkyo Koho 81-166231 [1980/81]; C.A. **96** [1982] No. 123991.

Shanoski, H., General Tire and Rubber Co., Coating of Glass Fiber-Reinforced Polyester and Vinyl Ester Resin Moldings in the Mold, Fr. Demande 2485547 [1980/81]; C.A. **96** [1982] No. 201373.

Shin-Etsu Chemical Industry Co., Ltd., Organopolysiloxane Compositions Crosslinkable at Room Temperature, Ger. Offen. 3032625 [1979/81]; C.A. **94** [1981] No. 209796.

Showa Electric Wire and Cable Co., Ltd., Crosslinked Silyl-Modified Polyolefin Laminate, Japan. Kokai Tokkyo Koho 81-131631 [1980/81]; C.A. **96** [1982] No. 36305.

Singer, W., Nowak, M., Driscoll, A., Troy Chemical Corp., Tin-Containing Catalyst Compositions and Their Use in Catalyzing Urethane Preparation, Eur. Appl. 36289 [1980/81]; C.A. **96** [1982] No. 7543.

Smith, A. H., Beers, M. D., General Electric Co., Paintable One-Component Room Temperature Vulcanizable Systems, U.S. 4247445 [1979/81]; C.A. **94** [1981] No. 158512.

Smith, A. H., De Zuba, G. P., Mitchell, T. D., General Electric Co., Self-Bonding Room Temperature Vulcanizable Silicone Rubber Compositions, U.S. 4273698 [1979/81]; C.A. **95** [1981] No. 82257.

Strolle, C. H., Tronley, G. D., du Pont de Nemours, E. I., and Co., High-Solids Polyurethane Enamel Coating Compositions, Brit. Appl. 2064566 [1979/81]; C.A. **96** [1982] No. 105918.

Sugita, T., Sakaguchi, H., Tsuda, H., Suzuki, T., Mitsui-Nisso Corp., Rapid Curing Polyurethane Elastomer Prepared from a Diphenylmethanediisocyanate Based Liquid Prepolymer and a Curing Agent Containing a Poly(Tetramethylene Ether Glycol), a Diol and an Organometallic Catalyst, U.S. 4294951 [1980/81]; C.A. **96** [1982] No. 36622.

Sugita, T., Sakaguchi, H., Tsuda, H., Suzuki, T., Fukuda, K., Mitsui-Nisso Corp., Quick Hardening Polyurethane Elastomers and the Binding of the Elastomer into Metal, Ger. Offen. 3026366 [1979/81]; C.A. **94** [1981] No. 176454.

Sumika Color Co., Ltd., Stabilization of Polyurethanes, Japan. Kokai Tokkyo Koho 81-14551 [1979/81]; C.A. **95** [1981] No. 63182.

Sumika Color Co., Ltd., Stabilizers for Urethane Polymers, Japan. Kokai Tokkyo Koho 81-120724 [1980/81]; C.A. **96** [1982] No. 20848.

Sumitomo Bakelite Co., Ltd., Silane-Grafted Polyolefin Compositions, Japan. Kokai Tokkyo Koho 81-122817 [1980/81]; C.A. **96** [1982] No. 7280.

Svoboda, G. R., Freeman Chemical Corp., Composition for Coating Molded Articles, U.S. 4293659 [1980/81]; C.A. **96** [1982] No. 8302.

Szycher, M., Thermo Electron Corp., Aliphatic-Polyurethane Elastomers and its Use for Artificial Corneas, Fr. Demande 2477161 [1980/81]; C.A. **96** [1982] No. 21135.

Takeda Chemical Industries, Ltd., Semirigid Urethane Rubber Foams, Japan. Kokai Tokkyo Koho 81-28212 [1979/81]; C.A. **95** [1981] No. 116847.

Teijin Ltd., Aromatic Polyesters, Japan. Kokai Tokkyo Koho 81-32330 [1979/81]; C.A. **96** [1982] No. 35890.

Toa Gosei Chemical Industry Co., Ltd., Vinyl Chloride Resin Compositions, Japan. Kokai Tokkyo Koho 81-65037 [1979/81]; C.A. **95** [1981] No. 151643.

Toho Chemical Industry Co., Ltd., Hydrophilic Polyurethanes, Japan. Kokai Tokkyo Koho 81-32519 [1979/81]; C.A. **95** [1981] No. 8324.

Tokai Seiyu Kogyo Co., Ltd., Polyisocyanate Bisulfite Adducts, Japan. Kokai Tokkyo Koho 81-122340 [1980/81]; C.A. **96** [1982] No. 36219.

Toray Industries, Inc., Aqueous Polyurethane Emulsions, Japan. Kokai Tokkyo Koho 81-159213 [1980/81]; C.A. **96** [1982] No. 105599.

Toshiba Silicone Co., Ltd., Meito Co., Ltd., Silicone Rubber Paint, Japan. Kokai Tokkyo Koho 81-02358 [1979/81]; C.A. **95** [1981] No. 8943.

Toshiba Silicone Co., Ltd., Soiling-Resistant Sealants, Japan. Kokai Tokkyo Koho 81-76452 [1979/81]; C.A. **95** [1981] No. 171206.

Toyo Rubber Industry Co., Ltd., Yellowing-Resistant Polyurethane Foams, Japan. Kokai Tokkyo Koho 81-82812 [1979/81]; C.A. **95** [1981] No. 188147.

Toyota Motor Co., Ltd., Asahi Glass Co., Ltd., Polyurethane Containing Silicate Fillers, Japan. Kokai Tokkyo Koho 81-106915 [1980/81]; C.A. **95** [1981] No. 221102.

Voigt, H. U., Kabel- und Metallwerke Gutehoffnungshütte A.-G., Voltage Stabilization of Electrical Insulators for Electrical Cables, Ger. Offen. 2945216 [1979/81]; C.A. **95** [1981] No. 26331.

Voigt, H. U., Van Hove, C., Kabel- und Metallwerke Gutehoffnungshütte, Water-Resistant High Voltage Insulation for Electric Cables, Ger. Offen. 2935224 [1979/81]; C.A. **94** [1981] No. 176197.

Voronkov, M. G., Annenkova, V. Z., Khaliullin, A. K., Antonik, L. M., Irkutsk Institute of Organic Chemistry, U.S.S.R. 804682 [1979/81]; C.A. **95** [1981] No. 26115.

Vylet, J., Plicka, E., Karasek, O., Hlustik, K., Thermosetting Polyurethanes with Increased Storage Stability, Czech. 204880 [1979/81]; C.A. **99** [1983] No. 213575.

Ward, R. J., Union Carbide Corp., Adhesion Promoting Organic Silicon-Grafted Poly(diphenylene ethers), Can. 1113109 [1978/81]; C.A. **101** [1984] No. 92107.

Watson, S. L., Union Carbide Corp., Acrylated Urethanes, Eur. Appl. 21824 [1979/81]; C.A. **94** [1981] No. 176879.

Watson, S. L., Union Carbide Corp., Acrylated Urethane Polycarbonates, Eur. Appl. 25239 [1979/81]; C.A. **95** [1981] No. 44911.

Wellner, W., Botta, A., Gruber, H., Bayer A.-G., Polyureas by Reaction of Organic Polyisocyanates, Ger. Offen. 3018023 [1980/81]; C.A. **96** [1982] No. 164305.

Wick, G., AKZO N.V., Polyurethane Coatings for Membranes, Fr. Demande 2478110 [1980/81]; C.A. **96** [1982] No. 54025.

Williams, R. C., Rogemoser, D. R., Textron, Inc., Water-Dispersible Urethane Polymers, Aqueous Polymer Dispersion and Half-Esters Useful Therein, U.S. 4268426 [1979/81]; C.A. **95** [1981] No. 171116.

Willoughby, B. G., National Research Development Corp., Improvements in Polyurethane Catalysts, PCT Intern. Appl. 81-00411 [1979/81]; C.A. **95** [1981] No. 25897.

Wolfer, D., Schiller, A., Wacker-Chemie GmbH, Fire-Resistant Elastomers or Crosslinkable Compositions from Organopolysiloxanes after the Addition of a Crosslinking Agent, Ger. Offen. 3018549 [1980/81]; C.A. **96** [1982] No. 7980.

Wright, J. H., Lampe, W. R., Smith Jr., A. H., General Electric Co., Room-Temperature Vulcanizable Silicone Rubber Compositions with Sag Control, U.S. 4261758 [1979/81]; C.A. **95** [1981] No. 188455.

Yazawa, C., Koike, W., Oyaizu, Y., Ihara Chemical Industry Co., Ltd., Polyurethane Elastomer, S. African 80-03663 [1980/81]; C.A. **96** [1982] No. 124331.

Yoshitomi Pharmaceutical Industries, Ltd., Heat Stabilizer for ABS Resins, Japan. Kokai Tokkyo Koho 81-41241 [1979/81]; C.A. **95** [1981] No. 63263.

Adeka Argus Chemical Co., Stable Allyl Chloride Compositions, Japan. Kokai Tokkyo Koho 57 123126 [1981/82]; C.A. **98** [1983] No. 35121.

Adeka Argus Chemical Co., Ltd., Prevention of Discoloration of Dichloropropene Nematocides, Japan. Kokai Tokkyo Koho 57 126429 [1981/82]; C.A. **98** [1983] No. 12946.

Ahramjian, L., E. I. du Pont de Nemours & Co., Photocurable Polyurethane Film Coatings, Eur. Appl. 43073 [1980/82]; C.A. **96** [1982] No. 124707.

Ahramjian, L., E. I. du Pont de Nemours & Co., Photocurable Polyurethane Film Coatings, U.S. 4337130 [1980/82]; C.A. **97** [1982] No. 94117.

Alberino, L. M., Lockwood, R. J., Upjohn Co., Polyurethane from Polyisocyanate Blends, U.S. 4321333 [1981/82]; C.A. **97** [1982] No. 183386.

Asahi Chemical Industry Co., Ltd., Electrophoretic Coating Materials, Japan. Kokai Tokkyo Koho 57 123216 [1981/82]; C.A. **98** [1983] No. 18194.

Asahi Chemical Industry Co., Ltd., Electrophoretic Coating Materials, Japan. Kokai Tokkyo Koho 57 123217 [1981/82]; C.A. **98** [1983] No. 18195.

Asahi Chemical Industry Co., Ltd., Electrophoretic Coating Materials, Japan. Kokai Tokkyo Koho 57 123218 [1981/82]; C.A. **98** [1983] No. 18196.

Asahi Chemical Industry Co., Ltd., Photocurable Polyurethane Acrylate Coating Materials, Japan. Kokai Tokkyo Koho 57 165422 [1981/82]; C.A. **98** [1983] No. 127828.

Asahi Chemical Industry Co., Ltd., Yellowing-Free Polyurethane Composition, Japan. Kokai Tokkyo Koho 57 185315 [1981/82]; C.A. **98** [1983] No. 127115.

Asahi-Dow, Ltd., Stabilized Methylchloroform Compositions, Japan. Kokai Tokkyo Koho 82-116021 [1981/82]; C.A. **97** [1982] No. 215515.

Asahi Glass Co., Ltd., Energy-Absorbing Semirigid Polyurethane Foams, Japan. Kokai Tokkyo Koho 82-02340 [1980/82]; C.A. **96** [1982] No. 144171.

Asahi Glass Co., Ltd., Energy-Absorbing Polyurethane Foams, Japan. Kokai Tokkyo Koho 82-02341 [1980/82]; C.A. **96** [1982] No. 144172.

Asahi Glass Co., Ltd., Powdered Glass Fiber-Containing Urethane Polymers, Japan. Kokai Tokkyo Koho 82-59927 [1980/82]; C.A. **97** [1982] No. 110842.

Asahi Glass Co., Ltd., Silicic Acid Fillers and Polyurethane Containing the Fillers, Japan. Kokai Tokkyo Koho 82-59950 [1980/82]; C.A. **97** [1982] No. 110843.

Asahi Glass Co., Ltd., Polyol Compositions, Japan. Kokai Tokkyo Koho 57 135823 [1981/82]; C.A. **98** [1983] No. 65028.

Asahi Glass Co., Ltd., Polyurethane Sheets for Laminates with Glass, Japan. Kokai Tokkyo Koho 57 199649 [1981/82]; C.A. **99** [1983] No. 6620.

Asahi Glass Co., Ltd., Polyurethane Sheets for Lamination with Glass, Japan. Kokai Tokkyo Koho 57 199650 [1981/82]; C.A. **98** [1983] No. 180725.

Asahi Organic Chemicals Industry Co., Ltd., Fire-Resistant Polyurethane Foams, Japan. Kokai Tokkyo Koho 82-74323 [1980/82]; C.A. **97** [1982] No. 163990.

Asai Bussan Co., Ltd., Photocurable Resin Compositions, Japan. Kokai Tokkyo Koho 82-76016 [1980/82]; C.A. **97** [1982] No. 129185.

Balas, A., Palka, G., Potocki, A., Stelmasik, A., Buczkowski, P., Polytechnika Gdanska, Bonding of Polyurethane Elastomers with Other Materials, Pol. 112324 [1977/82]; C.A. **97** [1982] No. 128831.

Barsa, E. A., Stuber, F. A., Lin, Chung Yuan, Upjohn Co., Bis(cycloureas), a Composition Containing These Compounds and Their Use in Polyurethane Production, Belg. 891022 [1980/82]; C.A. **97** [1982] No. 94107.

Bayer A.-G., Weather-Resistant Polyurethane Coatings, Japan. Kokai Tokkyo Koho 57 175788 [1981/82]; C.A. **98** [1983] No. 145177.

Beers, M. D., General Electric Co., Devolatilized Room Temperature Vulcanizable Silicone Rubber Composition, U.S. 4356116 [1981/82]; C.A. **98** [1983] No. 35848.

Behula, F., Kopal, P., Polyurethane Moldings, Czech. 198741 [1978/82]; C.A. **97** [1982] No. 73738.

Bovis, C., Depetris, N., Naphthachimie, S. A., Structural Material with Good Mechanical and Thermal Resistance, Ger. Offen. 3043715 [1980/82]; C.A. **97** [1982] No. 39921.

Bridgestone Tire Co., Ltd., Urethane-Modified Polyisocyanurate Foams, Japan. Kokai Tokkyo Koho 57 119915 [1981/82]; C.A. **98** [1983] No. 161989.

Brixius, D. W., Simms, J. A., du Pont de Nemours, E. I., and Co., Isocyanate-Functional Polymers Containing a Terminal Monosulfide Group, U.S. 4351755 [1977/82]; C.A. **97** [1982] No. 218201.

Burgdoerfer, H. H., Schaepel, D., Schneider, G., von Bonin, W., von Gizycki, U., Bayer A.-G., Gel Padding and Its Use, Ger. Offen. 3103564 [1981/82]; C.A. **97** [1982] No. 199281.

Buysch, H. J., Krimm, H., Richter, W., Bayer A.-G., N,O-Disubstituted Urethanes, and Their Use as Raw Materials in the Manufacture of Isocyanates, Eur. Appl. 48927 [1980/82]; C.A. **97** [1982] No. 127294.

Canon, K. K., Electrostatographic Toners, Japan. Kokai Tokkyo Koho 57 124740 [1981/82]; C.A. **98** [1983] No. 188991.

Canon, K. K., Pressure-Fixing Type Electrostatographic Toners, Japan. Kokai Tokkyo Koho 57 185444 [1981/82]; C.A. **100** [1984] No. 94493.

Christman, D. L., Kan, P. T., BASF Wyandotte Corp., Carbamylbiuret-Modified Polyisocyanates, U.S. 4330636 [1979/82]; C.A. **97** [1982] No. 56105.

Christman, D. L., Kan, P. T., BASF Wyandotte Corp., Carbamylbiuret-Modified Polyisocyanates, U.S. 4331809 [1979/82]; C.A. **97** [1982] No. 73730.

Christman, D. L., Kan, P. T., BASF Wyandotte Corp., Carbamylbiuret-Modified Polyisocyanates, U.S. 4331810 [1979/82]; C.A. **97** [1982] No. 56966.

Christman, D. L., Kan, P. T., BASF Wyandotte Corp., Carbamylbiuret-Modified Polyisocyanates, U.S. 4332953 [1979/82]; C.A. **97** [1982] No. 93753.

Cuscurida, M., Dominguez, R. J. G., Texaco Inc., Catalyst System for RIM Elastomers, U.S. 4358547 [1981/82]; C.A. **98** [1983] No. 35883.

Cuscurida, M., Rice, D. M., McDaniel, K. G., Texaco Development Corp., Reaction Injection Molded Polyurethane, Brit. Appl. 2084163 [1980/82]; C.A. **97** [1982] No. 56964.

Cuscurida, M., Rice, D. M., McDaniel, K. G., Texaco Development Corp., Polyurethane Elastomers, Ger. Offen. 3034789 [1980/82]; C.A. **97** [1982] No. 24987.

Custom Coating, Inc., Polyurethane Backings for Carpets, Japan. Kokai Tokkyo Koho 82-77374 [1980/82]; C.A. **97** [1982] No. 164475.

Dainichi Nippon Cables, Ltd., Polyethylene Gas Pipes, Japan. Kokai Tokkyo Koho 57 127187 [1981/82]; C.A. **98** [1983] No. 35582.

Dainippon Ink and Chemicals, Inc., Curable Acrylic Coating Compositions, Japan. Kokai Tokkyo Koho 82-12058 [1980/82]; C.A. **96** [1982] No. 182916.

Dainippon Ink and Chemicals, Inc., Unsaturated Polyester Compositions, Japan. Kokai Tokkyo Koho 57 123211 [1981/82]; C.A. **98** [1983] No. 17544.

Dainippon Ink and Chemicals, Inc., Organic Isocyanate Derivatives, Japan. Kokai Tokkyo Koho 57 134486 [1981/82]; C.A. **98** [1983] No. 72139.

Dainippon Ink and Chemicals, Inc., Powder Coating Compositions, Japan. Kokai Tokkyo Koho 57 165461 [1981/82]; C.A. **98** [1983] No. 16545.

Dainippon Ink and Chemicals, Inc., Room Temperature-Curable Coating Compositions, Japan. Kokai Tokkyo Koho 57 167359 [1981/82]; C.A. **98** [1983] No. 145118.

Dainippon Toryo Co., Ltd., Colored Glass Flake-Containing Coating Materials, Japan. Kokai Tokkyo Koho 82-47366 [1980/82]; C.A. **97** [1982] No. 25215.

Diefenbach, H., Dobbelstein, A., Hille, H. D., BASF Farben und Fasern A.-G., Binders for Cathodically Depositable Coating Compositions, Ger. Offen. 3123536 [1981/82]; C.A. **98** [1983] No. 108969.

Disteldorf, J., Gras, R., Schnurbusch, H., Chemische Werke Huels A.-G., Powder Varnishes which Do Not Release Any Decomposition Products, Eur. Appl. 45998 [1980/82]; C.A. **96** [1982] No. 182929.

Dominguez, R. J. G., McCoy, D. R., Texaco Inc., Reaction-Injection Molding of Urethane Elastomers Using as the Catalyst System Methyldiethanolamine Dibutyltin Dilaurate and an Alkyltin Mercaptide, U.S. 4350779 [1981/82]; C.A. **97** [1982] No. 217755.

Dominguez, R. J. G., Rice, D. M., Texaco Development Corp., Reinforced Reaction Injection Molded Elastomers, Brit. Appl. 2083484 [1980/82]; C.A. **97** [1982] No. 40138.

Dominguez, R. J. G., Rice, D. M., Texaco Inc., Catalysts for Reaction-Injection Molding of Urethane Rubber, U.S. 4362824 [1981/82]; C.A. **98** [1983] No. 55359.

Dominguez, R. J. G., Rice, D. M., Zimmermann, R. L., Texaco Inc., Reaction-Injection Molding of Urethane Elastomers Using as the Catalyst System a Hydroxyalkylamine, Dibutyltin Dilaurate and an Alkyltin Mercaptide, U.S. 4350778 [1981/82]; C.A. **97** [1982] No. 217756.

Doshi, J. K., Wallenberg, S. A., De Soto, Inc., Two Component Polyurethane Coating Systems Having Extended Pot Life and Rapid Cure, U.S. 4341689 [1981/82]; C.A. **97** [1982] No. 146340.

Emanuel, P. R., Salman, S., Mohiuddin, G., ITT Industries, Inc., Molded Plastic Article, Fr. Demande 2486454 [1980/82]; C.A. **96** [1982] No. 182377.

Finkelmann, H., Rehage, G., Kreuzer, F. H., Consortium für Electrochemische Industrie GmbH, Crosslinked Organopolysiloxanes Exhibiting Liquid Crystal Properties, Ger. Offen. 3119459 [1981/82]; C.A. **98** [1983] No. 81903.

Fuji Fiber Glass K. K., Glass Fibers for Reaction Injection Molding, Japan. Kokai Tokkyo Koho 82-82040 [1980/82]; C.A. **97** [1982] No. 183439.

Fujikura Cable Works, Ltd., Vulcanization of Halogen-Containing Rubbers, Japan. Kokai Tokkyo Koho 82-05724 [1980/82]; C.A. **96** [1982] No. 182570.

Fujikura Cable Works, Ltd., Crosslinking of Chlorinated Polyethylene, Japan. Kokai Tokkyo Koho 82-08203 [1980/82]; C.A. **96** [1982] No. 182228.

Fujikura Cable Works, Ltd., Weathering-Resistant Wire-Covering Compositions, Japan. Kokai Tokkyo Koho 82-18805 [1977/82]; C.A. **97** [1982] No. 145963.

Fujikura Cable Works, Ltd., Crosslinking Chlorosulfonated Polyethylenes, Japan. Kokai Tokkyo Koho 82-28107 [1980/82]; C.A. **96** [1982] No. 182585.

Fujikura Cable Works, Ltd., Foamed-Polyethylene Insulated Cables, Japan. Kokai Tokkyo Koho 57 32444 [1974/82]; C.A. **98** [1983] No. 17575.

Fujikura Cable Works, Ltd., Polyolefin Rubber Wire Sheaths, Japan. Kokai Tokkyo Koho 82-49109 [1980/82]; C.A. **97** [1982] No. 25015.

Fujikura Cable Works, Inc., Polyethylene Cable Sheaths, Japan. Kokai Tokkyo Koho 82-82913 [1980/82]; C.A. **97** [1982] No. 199062.

Fujikura Cable Works, Ltd., Halogen-Containing Polymers Crosslinkable at Low Temperatures, Japan. Kokai Tokkyo Koho 82-96048 [1980/82]; C.A. **97** [1982] No. 183070.

Fujikura Cable Works, Ltd., Halogen-Containing Polymer-Crosslinking at Low Temperature, Japan. Kokai Tokkyo Koho 82-96049 [1980/82]; C.A. **97** [1982] No. 183071.

Fukayama, M., Ishida, T., Toray Silicone Co., Ltd., Room Temperature Curable Polyorganosiloxane Composition Containing Spindle-Shaped Calcium Carbonate, Eur. Appl. 64375 [1981/82]; C.A. **98** [1983] No. 144807.

Furukawa Electric Co., Ltd., Electric Cables, Japan. Kokai Tokkyo Koho 57 53612 [1975/82]; C.A. **99** [1983] No. 23713.

Furukawa Electric Co., Ltd., Electric Cables, Japan. Kokai Tokkyo Koho 57 53613 [1975/82]; C.A. **99** [1983] No. 23714.

Furukawa Electric Co., Ltd., Manufacture of Crosslinked Polyolefin Extrusion Moldings, Japan. Kokai Tokkyo Koho 82-61042 [1980/82]; C.A. **97** [1982] No. 110984.

Furukawa Electric Co., Ltd., Fire-Resistant Crosslinked Polyolefin Resin Compositions, Japan. Kokai Tokkyo Koho 57 145136 [1981/82]; C.A. **98** [1983] No. 108338.

Furukawa Electric Co., Ltd., PVC Cellular Sheets, Japan. Kokai Tokkyo Koho 57 180645 [1981/82]; C.A. **98** [1983] No. 144691.

General Electric Co., Room Temperature-Vulcanizable Silicone Rubber Compositions, Japan. Kokai Tokkyo Koho 82-36155 [1980/82]; C.A. **97** [1982] No. 128928.

Gruber, H., Kober, H., Bayer A.-G., Sealing Off Buildings, Eur. Appl. 52301 [1980/82]; C.A. **97** [1982] No. 74085.

Hammond, J. A., Colamco, Inc., Polyurethane Prepolymer Adhesives, Fr. Demande 2492834 [1980/82]; C.A. **97** [1982] No. 110824.

Hechtl, W., Garhammer, A., Wacker-Chemie GmbH, Reaction Products of Silicic Acid Esters with Organotin Compounds, and Their Application, Eur. Appl. 50358 [1980/82]; C.A. **97** [1982] No. 111138.

Henkel (Japan) Ltd., Printing Inks for Plastic Laminates, Japan. Kokai Tokkyo Koho 57 149323 [1981/82]; C.A. **98** [1983] No. 74031.

Hiraoka and Co., Ltd., Pile Carpets, Japan. Kokai Tokkyo Koho 82-101078 [1980/82]; C.A. **97** [1982] No. 217935.

Hirsch, B., Horn, G., Reuther, H., Frenzel, G., Silicon-Containing Dyes, Ger. [East] 154020 [1978/82]; C.A. **97** [1982] No. 129108.

Hitachi Cable, Ltd., Crosslinking of Vinyl Chloride-Olefin Copolymer, Japan. Kokai Tokkyo Koho 57 172909 [1981/82]; C.A. **98** [1983] No. 144469.

Hitachi Cable, Ltd., Vulcanization of Vinyl-Grafted Chlorosulfonated Polyethylene, Japan. Kokai Tokkyo Koho 57 172910 [1981/82]; C.A. **98** [1983] No. 127526.

Hitachi Cable, Ltd., Vulcanization of Vinyl Chloride-Grafted Chlorinated Polyethylene, Japan. Kokai Tokkyo Koho 57 172911 [1981/82]; C.A. **98** [1983] No. 144848.

Hitachi Cable, Ltd., Crosslinking of Vinyl Chloride-Grafted Ethylene-Vinyl Acetate Copolymer, Japan. Kokai Tokkyo Koho 57 172912 [1981/82]; C.A. **98** [1983] No. 144470.

Hitachi Cable, Ltd., Vulcanization of Ethylene-Propene-Vinyl Chloride Rubber, Japan. Kokai Tokkyo Koho 57 172913 [1981/82]; C.A. **98** [1983] No. 127525.

Hitachi Cable, Ltd., Crosslinkable Chlorinated Polyethylene Blends, Japan. Kokai Tokkyo Koho 57 209912 [1981/82]; C.A. **99** [1983] No. 24181.

Hitachi Chemical Co., Ltd., UV-Resistant Polyurethane Reaction Injection Moldings, Japan. Kokai Tokkyo Koho 82-63320 [1980/82]; C.A. **97** [1982] No. 128640.

Hitachi Chemical Co., Ltd., Stabilization of Polyurethane Resins Against Light, Japan. Kokai Tokkyo Koho 82-63321 [1980/82]; C.A. **97** [1982] No. 145732.

Hitachi Chemical Co., Ltd., Photocurable Coating of Oligourethane Acrylates, Japan. Kokai Tokkyo Koho 82-78414 [1980/82]; C.A. **97** [1982] No. 184082.

Hitachi Chemical Co., Ltd., Acrylic Urethane Oligomer Photosensitive Resin Compositions, Japan. Kokai Tokkyo Koho 82-78415 [1980/82]; C.A. **97** [1982] No. 183618.

Hitachi Chemical Co., Ltd., Modified Polyethylene for Lamination with Epoxy Resin Composites, Japan. Kokai Tokkyo Koho 82-87402 [1980/82]; C.A. **97** [1982] No. 199047.

Hitachi Chemical Co., Ltd., Polyurethane Foam Compacts, Japan. Kokai Tokkyo Koho 57 109835 [1980/82]; C.A. **98** [1983] No. 5086.

Hitachi Chemical Co., Ltd., Photohardenable Resin Composition, Japan. Kokai Tokkyo Koho 57 125209 [1981/82]; C.A. **98** [1983] No. 91146.

Hitachi Chemical Co., Ltd., Polyurethane Foam Moldings, Japan. Kokai Tokkyo Koho 57 126815 [1981/82]; C.A. **98** [1983] No. 35541.

Hitachi, Ltd., Polyurethane Foams, Japan. Kokai Tokkyo Koho 82-18719 [1980/82]; C.A. **97** [1982] No. 7233.

Hoffman, D. K., Dow Chemical Co., Addition Polymerizable Isocyanate-Polyol Anaerobic Adhesives, U.S. 4320221 [1980/82]; C.A. **96** [1982] No. 163867.

Hoffman, D. K., Frisch, K. C., Dow Chemical Co., Addition Polymerizable Isocyanate-Polyol Anaerobic Adhesives, PCT Intern. Appl. 82-02048 [1980/82]; C.A. **97** [1982] No. 128839.

Hoffmann, G., Bronstert, B., Hahn, H., Jun, Mong Jon, BASF A.-G., Photosensitive Multilayer Material and Adhesive Layers for It, Eur. Appl. 53258 [1980/82]; C.A. **98** [1983] No. 44250.

Hoffmann, G., Richter, P., BASF A.-G., Adhesives for Relief Printing Plates, Eur. Appl. 59385 [1981/82]; C.A. **98** [1983] No. 225306.

Holker, J. R., Lomax, G. R., Jeffries, R., Shirley Institute, Breathable, Non-Porous Polyurethane Film, Eur. Appl. 52915 [1980/82]; C.A. **97** [1982] No. 128595.

Hoshimitsu Chemical Co., Ltd., Photocurable Coating Materials, Japan. Kokai Tokkyo Koho 82-12021 [1980/82]; C.A. **96** [1982] No. 182910.

Hoshimitsu Chemical Co., Ltd., Nitrocellulose Emulsions, Japan. Kokai Tokkyo Koho 82-100144 [1980/82]; C.A. **97** [1982] No. 218140.

Hoshimitsu Chemical Co., Ltd., Water-Thinned Urethane Polymer Coating Materials, Japan. Kokai Tokkyo Koho 82-165420 [1981/82]; C.A. **98** [1983] No. 127791.

Howell, B. G., BICC PLC, Curable Polymer Compositions, Brit. Appl. 2099829 [1981/82]; C.A. **98** [1983] No. 90790.

Hughes, R. B., General Electric Co., Laminate Composition Comprising Polyetherimide Impregnated Fabric, U.S. 4313999 [1980/82]; C.A. **96** [1982] No. 163836.

Hughes, R. B., General Electric Co., High-Temperature Paint, U.S. 4322332 [1980/82]; C.A. **96** [1982] No. 219429.

Idemitsu Kosan Co., Ltd., Polyurethane Concrete, Japan. Kokai Tokkyo Koho 82-22163 [1980/82]; C.A. **96** [1982] No. 163726.

Idemitsu Kosan Co., Ltd., Resin Concrete, Japan. Kokai Tokkyo Koho 82-22164 [1980/82]; C.A. **96** [1982] No. 182441.

Idemitsu Kosan Co., Ltd., Resin Concrete, Japan. Kokai Tokkyo Koho 82-22166 [1980/82]; C.A. **97** [1982] No. 39831.

Idemitsu Kosan Co., Ltd., Polyurethane Compositions, Japan. Kokai Tokkyo Koho 82-49652 [1980/82]; C.A. **97** [1982] No. 128615.

Idemitsu Kosan Co., Ltd., Stabilizers for Urethane Polymers, Japan. Kokai Tokkyo Koho 82-49653 [1980/82]; C.A. **97** [1982] No. 73408.

Idemitsu Kosan Co., Ltd., Polyurethane Compositions, Japan. Kokai Tokkyo Koho 82-49654 [1980/82]; C.A. **97** [1982] No. 56700.

Idemitsu Kosan Co., Ltd., Polyurethane Compositions, Japan. Kokai Tokkyo Koho 82-49655 [1980/82]; C.A. **97** [1982] No. 39875.

Idemitsu Kosan Co., Ltd., Urethane Rubbers, Japan. Kokai Tokkyo Koho 82-80421 [1980/82]; C.A. **97** [1982] No. 128908.

Idemitsu Kosan Co., Ltd., Urethane Rubber for Solid Tires, Japan. Kokai Tokkyo Koho 82-80422 [1980/82]; C.A. **97** [1982] No. 164336.

Idemitsu Kosan Co., Ltd., Sealant Compositions, Japan. Kokai Tokkyo Koho 82-92078 [1980/82]; C.A. **97** [1982] No. 145970.

Idemitsu Kosan Co., Ltd., Crack-Resistant Urethane Rubber, Japan. Kokai Tokkyo Koho 82-121024 [1981/82]; C.A. **97** [1982] No. 217743.

Idemitsu Kosan Co., Ltd., Resin Concrete, Japan. Kokai Tokkyo Koho 82-153044 [1981/82]; C.A. **98** [1983] No. 108353.

Idemitsu Kosan Co., Ltd., Urethane Rubber Compositions with Good Fatigue Endurance, Japan. Kokai Tokkyo Koho 82-159812 [1981/82]; C.A. **98** [1983] No. 108657.

Idemitsu Kosan Co., Ltd., Liquid Diene Polymer Derivatives, Japan. Kokai Tokkyo Koho 57 170920 [1981/82]; C.A. **98** [1983] No. 162574.

Idemitsu Kosan Co., Ltd., Pressure-Sensitive Adhesives, Japan. Kokai Tokkyo Koho 57 209977 [1981/82]; C.A. **98** [1983] No. 216792.

Japan Ester Co., Ltd., Manufacture of Polyesters for Powder Coatings, Japan. Kokai Tokkyo Koho 57 126822 [1981/82]; C.A. **98** [1983] No. 36182.

Kanegafuchi Chemical Industry Co., Ltd., Silyl-Terminated Polymers, Japan. Kokai Tokkyo Koho 57 126823 [1981/82]; C.A. **98** [1983] No. 90849.

Kanegafuchi Chemical Industry Co., Ltd., Hardenable Silicone Compositions, Japan. Kokai Tokkyo Koho 57 145147 [1981/82]; C.A. **98** [1983] No. 73657.

Kanegafuchi Chemical Industry Co., Ltd., Vinyl Chloride Resin Foams, Japan. Kokai Tokkyo Koho 82-105430 [1980/82]; C.A. **97** [1982] No. 217381.

Kanegafuchi Chemical Industry Co., Ltd., Coating Materials Curable at Relatively Low Temperatures, Japan. Kokai Tokkyo Koho 57 172917 [1981/82]; C.A. **98** [1983] No. 145127.

Kanegafuchi Chemical Industry Co., Ltd., Moisture-Curable Silicone Rubber Compositions, Japan. Kokai Tokkyo Koho 57 182350 [1981/82]; C.A. **98** [1983] No. 162252.

Kanegafuchi Chemical Industry Co., Ltd., Room-Temperature-Curable Resin Compositions, Japan. Kokai Tokkyo Koho 57 190043 [1981/82]; C.A. **98** [1983] No. 162034.

Kato, Y., Furukawa, H., Kanegafuchi Chemical Industry Co., Ltd., Vinyl Resin Composition Containing Silyl Groups and a Paint Comprising Said Composition, Eur. Appl. 48461 [1980/82]; C.A. **97** [1982] No. 40460.

Keogh, M. J., Union Carbide Corp., Silicone-Modified Alkylene-Alkyl Acrylate Copolymers for Electric Insulators, Eur. Appl. 49155 [1980/81]; C.A. **97** [1982] No. 24667.

Keogh, M. J., Union Carbide Corp., Silicone-Modified Copolymers of Olefins and Alkyl Acrylates, U.S. 4328323 [1978/82]; C.A. **97** [1982] No. 39895.

Keogh, M. J., Union Carbide Corp., Flame-Retardant, Moisture-Curable Wire Insulation, U.S. 4353997 [1978/82]; C.A. **98** [1983] No. 5261.

Kitsuda, Y., Matsumura, M., Ohtsu, M., Matsushita Electric Works, Ltd., Epoxy Resin Composition, Ger. Offen. 3117960 [1980/82]; C.A. **96** [1982] No. 200825.

Kloubec, L., Stabilization of the Microcell Structure of Polyurethane Foams, Czech. 198727 [1978/82]; C.A. **97** [1982] No. 128925.

Koenig, K., Heitkämper, P., Bayer A.-G., Continuous Thermal Decomposition of Carbamic Acid Esters and Use of Mixtures of Carbamic Acid Esters and Isocyanates Obtained by this Process for Preparation of Isocyanates, Eur. Appl. 54817 [1980/82]; C.A. **97** [1982] No. 181949.

Kohno, K., Nishikawa, S., Hattori, Y., Kitao, K., Sunstar Giken K. K., Elastic Composition Hardenable at Room Temperature, Ger. Offen. 3215843 [1981/82]; C.A. **98** [1983] No. 127854.

Kuraray Co., Ltd., Pressure-Sensitives, Japan. Kokai Tokkyo Koho 82-47368 [1980/82]; C.A. **97** [1982] No. 39869.

Kurashiki Spinning Co., Ltd., Asahi Organic Chemicals Industry Co., Ltd., Polyurethane Foams, Japan. Kokai Tokkyo Koho 57 187312 [1981/82]; C.A. **98** [1983] No. 180460.

Kuroda, Y., Ito, S., Trashima, K., Miderikawa, A., Dainippon Ink Chemical Industry Co., Polyurethane Coatings for Magnetic Tapes, Eur. Appl. 59935 [1981/82]; C.A. **98** [1983] No. 5555.

Kwart, H., Varadhachary, S. N., Bonding Dissimilar Synthetic Polymeric Materials, U.S. 4333987 [1979/82]; C.A. **101** [1984] No. 5657.

Kyuya, Y., Sachio, I., Koichi, S., Takeda Chemical Industries, Ltd., Composition for Polyurethane Adhesives, Eur. Appl. 53359 [1980/82]; C.A. **97** [1982] No. 93662.

Lampe, W. R., Mitchell, T. D., Cella, J. A., General Electric Co., Silicone Products Business Division, Self-Bonding One-Component RTV Silicone Rubber Compositions, U.S. 4358575 [1981/82]; C.A. **98** [1983] No. 55374.

Lelu, A., Tauzia, J. M., Zilioli, F., Société Nationale des Poudres et Explosifs, Inhibitor Coating for Solid Propergols, Fr. Demande 2495133 [1980/82]; C.A. **98** [1983] No. 37094.

Lorant, I., Reti, J., Debreczy, Z., Szautin, B. V., Dinszburg, B. N., Bor-, Mubor-, es Cipoipari Kutato Intezet, Porous Leather Substitute, Hung. Teljes 23324 [1975/82]; C.A. **98** [1983] No. 180755.

Markovs, R. A., BASF Wyandotte Corp., Urethane Rubbers from Grafted Polyols for Reaction-Injection Molding, U.S. 4314038 [1981/82]; C.A. **96** [1982] No. 105600.

Matsui, H., Suzuki, T., Badische Petrochemical Co., Ltd., Polymerizable Composition Containing an Inorganic Filler for Preparation of Polymer Particles, Ger. Offen. 3210865 [1981/82]; C.A. **98** [1983] No. 55115.

Matsushita Electric Works, Ltd., Fire-Resistant Plastic Compositions, Japan. Kokai Tokkyo Koho 82-00168 [1980/82]; C.A. **96** [1982] No. 163696.

Matsushita Electric Works, Ltd., Epoxy Resin Potting Compositions, Japan. Kokai Tokkyo Koho 57 159813 [1981/82]; C.A. **98** [1983] No. 127136.

Matsuura, K., Yamaoka, N., Miyoshi, M., Nippon Oil Co., Ltd., Polyolefins Crosslinked by Silane Linkages, Ger. Offen. 3210192 [1981/82]; C.A. **98** [1983] No. 17611.

McEntire, E. E., Texaco Development Corp., Bis(dimethylaminopropyl)amine Derivatives as Polyurethane Catalysts, Can. 1132292 [1976/82]; C.A. **98** [1983] No. 55043.

McEntire, E. E., Dominguez, R. J. G., Texaco Development Corp., Catalyst for Producing Polyurethane Elastomers by Injection Molding, Ger. Offen. 3219349 [1981/82]; C.A. **98** [1983] No. 144818.

McEntire, E. E., Dominguez, R. J. G., Texaco, Inc., RIM Elastomers Using a Catalyst System which is a Polymer Containing Tertiary Amine Moieties, U.S. 4359540 [1981/82]; C.A. **98** [1983] No. 73661.

Mikami, T., Fuji Photo Film Co., Ltd., Pressure-Fixable Electrostatographic Tomer Material, Ger. Offen. 3215838 [1981/82]; C.A. **98** [1983] No. 188994.

Mitsubishi Petrochemical Co., Ltd., Crosslinkable Propylene Resin Compositions, Japan. Kokai Tokkyo Koho 82-12051 [1980/82]; C.A. **96** [1982] No. 200716.

Mitsubishi Petrochemical Co., Ltd., Silane-Vulcanization of Polyolefine Rubbers, Japan. Kokai Tokkyo Koho 82-23651 [1980/82]; C.A. **97** [1982] No. 7654.

Mitsubishi Petrochemical Co., Ltd., Antifoaming Agents, Japan. Kokai Tokkyo Koho 57 119807 [1981/82]; C.A. **98** [1983] No. 17541.

Mitsubishi Petrochemical Co., Ltd., Glass Bottles Coated with Resins for Prevention of Shattering, Japan. Kokai Tokkyo Koho 57 129844 [1981/82]; C.A. **98** [1983] No. 55723.

Mitsubishi Petrochemical Co., Ltd., Grafting of Vinylalkoxysilane on Propylene Polymers, Japan. Kokai Tokkyo Koho 57 147507 [1981/82]; C.A. **98** [1983] No. 73368.

Mitsubishi Petrochemical Co., Ltd., Crosslinked Polyethylene with Improved Creep Resistance, Japan. Kokai Tokkyo Koho 57 170913 [1981/82]; C.A. **98** [1983] No. 127165.

Mitsubishi Petrochemical Co., Ltd., Crosslinked Ethylene Polymer-Insulated Wire, Japan. Kokai Tokkyo Koho 57 208006 [1981/82]; C.A. **99** [1983] No. 6955.

Mitsui Nisso Urethane K. K., Adhesion of Urethane Rubber to Metals, Japan. Kokai Tokkyo Koho 57 126644 [1981/82]; C.A. **98** [1983] No. 5307.

Mitsui Nisso Urethane K. K., Thermoplastic Urethane Polymers, Japan. Kokai Tokkyo Koho 57 190012 [1981/82]; C.A. **98** [1983] No. 161717.

Mitsui Petrochemical Industries, Ltd., Rubber Electric Insulators, Japan. Kokai Tokkyo Koho 57 115439 [1981/82]; C.A. **98** [1983] No. 5319.

Mitsui Polychemicals Co., Ltd., Molding of Silane-Modified Ethylene Polymers, Japan. Kokai Tokkyo Koho 82-23650 [1980/82]; C.A. **97** [1982] No. 39832.

Mitsui Toatsu Chemicals, Inc., Prevention of Formation of Lumps in Vinyl Chloride Resin Articles, Japan. Kokai Tokkyo Koho 82-65744 [1981/82]; C.A. **97** [1982] No. 110876.

Mitsui Toatsu Chemicals, Inc., PVC Films for Greenhouses, Japan. Kokai Tokkyo Koho 57 123239 [1981/82]; C.A. **98** [1983] No. 35687.

Moravcik, A., Pach, L., Kacer, L., Buso, M., Agent Retarding the Release of Fertilizer into Soil, Czech. 210554 [1978/82]; C.A. **98** [1983] No. 71007.

Müller, K. F., Herber, S. J., Ciba-Geigy A.-G., Membrane-Modified Hydrogels and their Use as Active Agent Dispenser, Eur. Appl. 46136 [1980/82]; C.A. **97** [1982] No. 115323.

Myers, M. M., Tellefsen, W. J., Delta Oil Products Corp., Low Emission Foundry Binder System, U.S. 4311631 [1979/82]; C.A. **96** [1982] No. 124011.

Nagy, J., Demjen, Z., Palossy, L., Budapesti Muszaki Egyetem, Silicone Rubber-Based Paints for Artists, Hung. Teljes 23909 [1980/82]; C.A. **98** [1983] No. 217400.

Nakamura, T., Okamura, M., Moriguchi, Y., Hayase, T., Nippon Paint Co., Ltd., Coating Compositions Containing Polymers Having a Cellulose Constituent Crosslinked by Polyisocyanates, Fr. Demande 2497808 [1981/82]; C.A. **97** [1982] No. 199693.

Narayan, T., Patton, J. T., BASF Wyandotte Corp., Polyisocyanurate Dispersions Modified with Halogenated Alcohols, U.S. 4326043 [1981/82]; C.A. **97** [1982] No. 7636.

Nippon Chemiphar Co., Ltd., Photocurable Coating Materials for Poly(vinyl chloride) Sheets, Japan. Kokai Tokkyo Koho 57 172915 [1981/82]; C.A. **98** [1983] No. 127834.

Nippon Oils and Fats Co., Ltd., Refrigerator-Lubricating Oils, Japan. Kokai Tokkyo Koho 82-51795 [1980/82]; C.A. **97** [1982] No. 58358.

Nippon Oils and Fats Co., Ltd., Modified Wood Flour, Japan. Kokai Tokkyo Koho 57 205133 [1981/82]; C.A. **99** [1983] No. 71622.

Nippon Polyurethane Industry Co., Ltd., Urethane Polymers, Japan. Kokai Tokkyo Koho 57 123219 [1981/82]; C.A. **98** [1983] No. 17226.

Nippon Steel Corp., Daiichi Kogyo Co., Ltd., Coating Materials for Metals, Japan. Kokai Tokkyo Koho 82-21463 [1980/82]; C.A. **97** [1982] No. 7909.

Nippon Steel Corp., Nippon Paint Co., Ltd., Mitsubishi Chemical Industries Co., Ltd., Cationic Electrophoretic Coating Materials, Japan. Kokai Tokkyo Koho 57 134599 [1981/82]; C.A. **98** [1983] No. 73939.

Nippon Synthetic Chemical Industry Co., Ltd., Photocurable Acrylic Urethane Polymer Coating Materials, Japan. Kokai Tokkyo Koho 57 165416 [1981/82]; C.A. **98** [1983] No. 145164.

Nippon Synthetic Chemical Industry Co., Ltd., Photocurable Acrylic Urethane Polymer Coating Materials, Japan. Kokai Tokkyo Koho 57 165417 [1981/82]; C.A. **98** [1983] No. 145165.

Nippon Telegraph and Telephone Public Corp., Mitsubishi Electric Corp., Electromagnetic Wave Absorbers, Japan. Kokai Tokkyo Koho 57 195137 [1981/82]; C.A. **98** [1983] No. 180915.

Nippon Zeon Co., Ltd., Electrically Conductive Vinyl Chloride Copolymer Compositions, Japan. Kokai Tokkyo Koho 82-23646 [1980/82]; C.A. **97** [1982] No. 7306.

Nippon Zeon Co., Ltd., Soft Vinyl Chloride Resin Compositions, Japan. Kokai Tokkyo Koho 82-53545 [1980/82]; C.A. **97** [1982] No. 73416.

Nippon Zeon Co., Ltd., Electrically Conductive Vinyl Chloride Resin Compositions, Japan. Kokai Tokkyo Koho 82-63347 [1980/82]; C.A. **97** [1982] No. 128637.

Nippon Zeon Co., Ltd., Rigid Vinyl Chloride Polymer Compositions, Japan. Kokai Tokkyo Koho 57 200435 [1981/82]; C.A. **98** [1983] No. 199235.

NHK Spring Co., Ltd., Asphalt-Containing Polyurethane Foams, Japan. Kokai Tokkyo Koho 82-47336 [1980/82]; C.A. **97** [1982] No. 24735.

Noethe, B., Siemens A.-G., Sealing Materials for Electrical and Optical Communication Cables and Cable Connectors, U.S. 4348307 [1978/82]; C.A. **97** [1982] No. 184134.

O'Connor, J. M., Lickei, D. L., Rosin, M. L., Olin Corp., Modified Polyurethane Liquid Polymer Compositions, Eur. Appl. 52958 [1980/82]; C.A. **101** [1984] No. 73654.

Orton, M. L., Spurr, W. I., Imperial Chemical Industries PLC, Copolymerization of Unsaturated Urethane Monomers, Eur. Appl. 64809 [1981/82]; C.A. **98** [1983] No. 198886.

Otowa Kagaku Kogyo K. K., Fragrant Moldings, Japan. Kokai Tokkyo Koho 82-40558 [1980/82]; C.A. **97** [1982] No. 40162.

Peng, S. C., Thomson, D. M., Ford Motor Co., High Solids Paint Composition Comprising Hydroxy Functional Oligoesters and Hydroxy Functional Copolymers, U.S. 4322508 [1978/82]; C.A. **96** [1982] No. 219392.

Pernikis, R., Lazdina, B., Apsite, B. K., Karlivans, V., Barkane, R., Institute of Wood Pulp Chemistry, Academy of Sciences, Latvian S.S.S.R., Oligourethane Epoxides, U.S.S.R. 973551 [1981/82]; C.A. **98** [1983] No. 90131.

Pilot Ink Co., Ltd., Writing Tools, Japan. Kokai Tokkyo Koho 82-47697 [1980/82]; C.A. **97** [1982] No. 111435.

Plum, H., Hoechst A.-G., Copolymers and their Use, Eur. Appl. 56971 [1981/82]; C.A. **97** [1982] No. 184050.

Prajsnar, B., Treatment of Polyurethanes and Polyurethane Wastes, Ger. Offen. 3042804 [1980/82]; C.A. **94** [1982] No. 24827.

Rasshofer, W., Avar, G., Freitag, H. A., Groegler, G., Kopp, R., Bayer A.-G., Polymerization Catalyst for Polyurethanes, Ger. Offen. 3100977 [1981/82]; C.A. **97** [1982] No. 217309.

Rasshofer, W., Groegler, G., Kopp, R., Bayer A.-G., Tetrahydropyrimidines and their Use as Catalysts in Production of Polyurethane Plastics, Eur. Appl. 54876 [1980/82]; C.A. **97** [1982] No. 198718.

Reed, R., Chan, M. L., United States Dept. of the Navy, Additives to Prevent Gassing of Energetic Binder Containing Propellants, U.S. Appl. 353295 [1982/82]; C.A. **97** [1982] No. 200325.

Reed, W. N., Galvin, T. J., ICI Americas, Inc., Blister-Resistant Calcium Carbonate-Filled Polyisocyanurate Resin Molding Compositions, U.S. 4352906 [1981/82]; C.A. **97** [1982] No. 217451.

Ricoh Co., Ltd., Electrophotographic Plates, Japan. Kokai Tokkyo Koho 57 79947 [1980/82]; C.A. **98** [1983] No. 44171.

Ritz, J., Fischer, H., Plum, H., Hoechst A.-G., Curable Copolymers and their Use, Eur. Appl. 46561 [1980/81]; C.A. **97** [1982] No. 25231.

Rizk, S. D., Hsieh, H. W. S., Prendergast, J. J., Essex Chemical Corp., Silicon-Terminated Polyurethane Polymer, U.S. 4345053 [1981/82]; C.A. **97** [1982] No. 184132.

Roland, M., Heilfort, S., Adhesive for Joining Poly(vinyl chloride) Film to Carton or Paper, Ger. [East] 155522 [1980/82]; C.A. **97** [1982] No. 199287.

Sakamoto, M., Tamura, Y., Sakamoto, T., Synthetic Resin and its Use as Surface Material for Records, Ger. Offen. 3127667 [1981/82]; C.A. **97** [1982] No. 183681.

Sanns, F., Mobay Chemical Corp., Self-Granulating Reaction Product of 4,4′-Diphenylmethane Diisocyanate and Resorcinal, U.S. 4344892 [1981/82]; C.A. **98** [1982] No. 544655.

Sanyo Chemical Industries, Ltd., Primers for Urethane Polymer Thermal Insulators on Walls, Japan. Kokai Tokkyo Koho 82-61028 [1980/82]; C.A. **97** [1982] No. 129263.

Sanyo Chemical Industries, Ltd., Polyurethane Compositions, Japan. Kokai Tokkyo Koho 82-117524 [1981/82]; C.A. **97** [1982] No. 217315.

Sanyo Chemical Industries, Ltd., Soft Polyurethane Foams, Japan. Kokai Tokkyo Koho 57 162714 [1981/82]; C.A. **98** [1983] No. 90760.

Schaefer, W., Raedisch, H., Fuchs, R., Esser, G., Saint-Gobain Industries S. A., Process Utilizing Release Agent, U.S. 4331736 [1977/82]; C.A. **97** [1982] No. 56840.

Schaepel, D., Bayer A.-G., Active Agent-Containing Gel Material with a Depot Effect Based on a Polyurethane Matrix with High Molecular Weight Polyols, Ger. Offen. 3103499 [1981/82]; C.A. **98** [1983] No. 22084.

Schaepel, D., Bayer A.-G., Polyol Gels and their Use in Modeling Processes, Ger. Offen. 3103500 [1981/82]; C.A. **97** [1982] No. 145716.

Schimmel, K. F., Seiner, J. A., PPG Industries, Inc., Rheological Modifiers, Ger. Offen. 3150211 [1980/82]; C.A. **97** [1982] No. 128670.

Schimmel, K. F., Seiner, J. A., Christenson, R. M., Roger, M., Dowbenko, R., PPG Industries, Inc., Urethane Rheology-Modifying Agent and its Use, Ger. Offen. 3150157 [1980/82]; C.A. **97** [1982] No. 129190.

Schimmel, K. F., Seiner, J. A., Christenson, R. M., Roger, M., Dowbenko, R., PPG Industries, Inc., Urethane Rheology Modifiers and Coating Compositions Containing Them, U.S. 4327008 [1980/82]; C.A. **96** [1982] No. 219359.

Schwarz, D. S., B. F. Goodrich, Co., Adhesive Composition, U.S. 4336298 [1981/82]; C.A. **97** [1982] No. 128832.

Seiko Chemical Industry Co., Ltd., Acrylic Urethane Polymer Emulsion Coating Materials, Japan. Kokai Tokkyo Koho 82-61050 [1980/82]; C.A. **97** [1982] No. 129203.

Seiko Chemical Industry Co., Ltd., Water-Thinned Polyurethane Coating Compositions, Japan. Kokai Tokkyo Koho 57 133114 [1981/82]; C.A. **98** [1983] No. 55733.

Seiko Chemical Industry Co., Ltd., Water-Thinned Urethane Polymer Coating Materials, Japan. Kokai Tokkyo Koho 57 133115 [1981/82]; C.A. **98** [1983] No. 18200.

Sekisui Chemical Co., Ltd., Pressure-Sensitive Adhesives, Japan. Kokai Tokkyo Koho 57 109877 [1980/82]; C.A. **98** [1983] No. 5205.

Sekisui Chemical Co., Ltd., Pressure-Sensitive Adhesives, Japan. Kokai Tokkyo Koho 82-109878 [1980/82]; C.A. **97** [1983] No. 217583.

Sekisui Chemical Co., Ltd., Production of Porous Surgical Tapes, Japan. Kokai Tokkyo Koho 57 137375 [1981/82]; C.A. **98** [1983] No. 78175.

Sekisui Chemical Co., Ltd., Sustained-Release Transdermal Pharmaceuticals, Japan. Kokai Tokkyo Koho 82-139347 [1981/82]; C.A. **97** [1982] No. 203235.

Shanoski, H., General Tire and Rubber Co., One Component In-Mold Coating, U.S. 4331735 [1980/82]; C.A. **97** [1982] No. 40468.

Shone, E. B., Riches, K. M., Shell Internationale Research Maatschappij B. V., Marking Underwater Parts of Marine Structures, Brit. Appl. 2096019 [1981/82]; C.A. **98** [1983] No. 55754.

Showa Denko K. K., Electric Insulators Containing Voltage Stabilizers, Japan. Kokai Tokkyo Koho 57 39002 [1978/82]; C.A. **98** [1983] No. 108535.

Showa Electric Wire and Cable Co., Ltd., Crosslinked Polyolefin Fibers, Japan. Kokai Tokkyo Koho 82-05913 [1980/82]; C.A. **96** [1982] No. 124427.

Showa Electric Wire and Cable Co., Ltd., Polyolefin Nonwoven Fabrics, Japan. Kokai Tokkyo Koho 82-05956 [1980/82]; C.A. **96** [1982] No. 124493.

Showa Electric Wire and Cable Co., Ltd., Plastic Films for Electric Insulators, Japan. Kokai Tokkyo Koho 82-57409 [1980/82]; C.A. **97** [1982] No. 73622.

Showa Electric Wire and Cable Co., Ltd., Foamed Polymer, Japan. Kokai Tokkyo Koho 57 149338 [1981/82]; C.A. **98** [1983] No. 90650.

Showa Electric Wire and Cable Co., Ltd., Packings for Liquid Containers, Japan. Kokai Tokkyo Koho 57 149378 [1981/82]; C.A. **98** [1983] No. 73562.

Showa Electric Wire and Cable Co., Ltd., Insulated Electric Wires, Japan. Kokai Tokkyo Koho 57 151103 [1981/82]; C.A. **98** [1983] No. 90716.

Showa Electric Wire and Cable Co., Ltd., Paper Laminated Electric Insulators for Oil-Filled Cables, Japan. Kokai Tokkyo Koho 57 180813 [1981/82]; C.A. **98** [1983] No. 144696.

Showa Electric Wire and Cable Co., Ltd., Paper Laminated Electric Insulators for Oil-Filled Cables, Japan. Kokai Tokkyo Koho 57 180814 [1981/82]; C.A. **98** [1983] No. 162015.

Showa Electric Wire and Cable Co., Ltd., Wood Flour-Filled Crosslinkable Polyolefin Compositions, Japan. Kokai Tokkyo Koho 57 192446 [1981/82]; C.A. **98** [1983] No. 199226.

Showa Electric Wire and Cable Co., Ltd., Electric Insulating Paper Laminates, Japan. Kokai Tokkyo Koho 57 194413 [1981/82]; C.A. **98** [1983] No. 180705.

Showa Electric Wire and Cable Co., Ltd., Vulcanization of Ethylene-Olefin-Diene Rubber with Silanes, Japan. Kokai Tokkyo Koho 57 202337 [1981/82]; C.A. **98** [1983] No. 199615.

Smith, R. A., General Electric Co., Self-Binding, Solvent-Free, Cold-Hardening Silicone Rubber Composition and Articles Made from It, Ger. Offen. 3206474 [1981/82]; C.A. **98** [1983] No. 17909.

Sony Corp., Magnetic Recording Media, Neth. Appl. 81-05886 [1980/82]; C.A. **97** [1982] No. 129281.

Sony Corp., Nippon Polyurethane Industry Co., Ltd., Thermoplastic Polyurethane Resins, Neth. Appl. 81-05884 [1980/82]; C.A. **98** [1983] No. 18218.

Srapionyan, S. M., Atkarskii, A. A., Arutyunyan, B. G., Morozova, L. P., Karagezyan, G. N., Central Scientific-Research Institute for the Leather and Footwear Industry, "Masis" Industrial Shoe Enterprises, Erevan, Composition for Molding Footwear Soles, U.S.S.R. 905242 [1980/82]; C.A. **97** [1982] No. 24766.

Sumika Color Co., Ltd., Polyurethane Moldings, Japan. Kokai Tokkyo Koho 57 147545 [1981/82]; C.A. **98** [1983] No. 108341.

Sumika Color Co., Ltd., Toyota Motor Co., Ltd., Polyurethane Products, Japan. Kokai Tokkyo Koho 57 108154 [1980/82]; C.A. **98** [1983] No. 55059.

Sumimura, S., Amamiya, T., Toray Silicone Co., Ltd., Oil-Resistant Gasket Material and Packing Material, Eur. Appl. 50453 [1980/82]; C.A. **97** [1982] No. 25028.

Sumitomo Bakelite Co., Ltd., Photocurable Compositions Containing Polyisocyanate-Polythiol Adducts, Japan. Kokai Tokkyo Koho 82-80428 [1980/82]; C.A. **97** [1982] No. 163994.

Taylor, R. P., Philips, B. A., Mobay Chemical Corp., Elastomeric Molded Products, Eur. Appl. 44481 [1980/82]; C.A. **96** [1982] No. 182541.

Teijin Ltd., Dyeing of Fabrics, Japan. Kokai Tokkyo Koho 57 167469 [1981/82]; C.A. **98** [1983] No. 73776.

Teijin Ltd., Post-Photocuring of Coatings, Japan. Kokai Tokkyo Koho 57 192405 [1981/82]; C.A. **98** [1983] No. 181219.

Thiele, L., Wagner, K., Appenroth, S., Akademie der Wissenschaften der DDR, Storage-Stable, Tin Compound-Containing Polyol System for Preparing Polyurethanes, Ger. [East] 155620 [1980/82]; C.A. **97** [1982] No. 199107.

Timm, T., Hartwig, C., Gerlach, D., Phoenix A.-G., Coating of Fabric Strips with Polyurethane Compositions, Ger. Offen. 3042299 [1980/82]; C.A. **97** [1982] No. 57085.

Toa Gosei Chemical Industry Co., Ltd., Curable Polyurethanes, Japan. Kokai Tokkyo Koho 82-21417 [1980/82]; C.A. **96** [1982] No. 218722.

Toa Gosei Chemical Industry Co., Ltd., Crosslinking of Vinyl Chloride Polymer Compositions, Japan. Kokai Tokkyo Koho 82-25347 [1980/82]; C.A. **97** [1982] No. 24693.

Tokuyama Soda Co., Ltd., Polymer-Coated Inorganic Fillers, Japan. Kokai Tokkyo Koho 82-102989 [1980/82]; C.A. **97** [1982] No. 217380.

Toppan Printing Co., Ltd., Decorative Coatings, Japan. Kokai Tokkyo Koho 82-04791 [1980/82]; C.A. **97** [1982] No. 25245.

Toray Industries, Inc., Thermosetting Polyurethanes for Eyeglass Frames, Japan. Kokai Tokkyo Koho 82-22083 [1974/82]; C.A. **97** [1982] No. 164229.

Toray Industries, Inc., Composite for Leather Substitutes, Japan. Kokai Tokkyo Koho 82-39286 [1980/82]; C.A. **97** [1982] No. 24885.

Toray Industries, Inc., Negative Lithographic Plates, Japan. Kokai Tokkyo Koho 57 198465 [1981/82]; C.A. **100** [1984] No. 183210.

Toshiba Corp., Rigid Urethane Foams, Japan. Kokai Tokkyo Koho 57 125226 [1981/82]; C.A. **98** [1983] No. 55216.

Toshiba Silicone Co., Ltd., Room-Temperature-Curable Polyorganosiloxane Compositions, Japan. Kokai Tokkyo Koho 82-08247 [1980/82]; C.A. **96** [1982] No. 201080.

Toshiba Silicone Co., Ltd., Room-Temperature-Curable Polyorganosiloxane Compositions, Japan. Kokai Tokkyo Koho 82-42762 [1980/82]; C.A. **97** [1982] No. 40471.

Toshiba Silicone Co., Ltd., Sealing Compositions, Japan. Kokai Tokkyo Koho 82-76055 [1980/82]; C.A. **97** [1982] No. 184124.

Toshiba Silicone Co., Ltd., Moisture-Curable Siloxane Sealants, Japan. Kokai Tokkyo Koho 57 141447 [1981/82]; C.A. **98** [1983] No. 55769.

Toshiba Silicone Co., Ltd., Coating Compositions for Prevention of Dust, Japan. Kokai Tokkyo Koho 57 162763 [1981/82]; C.A. **98** [1983] No. 127817.

Toshiba Silicone Co., Ltd., Silicon-Containing Polyoxyalkylenes, Japan. Kokai Tokkyo Koho 57 164123 [1981/82]; C.A. **98** [1983] No. 108028.

Toyo Cloth Co., Ltd., Toyobo Co., Ltd., Leather Substitutes, Japan. Kokai Tokkyo Koho 57 101079 [1980/82]; C.A. **98** [1983] No. 17712.

Toyo Rubber Chemical Industry Co., Ltd., Polyurethane Foams, Japan. Kokai Tokkyo Koho 82-67623 [1980/82]; C.A. **97** [1982] No. 145747.

Toyo Rubber Industry Co., Ltd., High-Modulus Urethane Rubber, Japan. Kokai Tokkyo Koho 82-65720 [1980/82]; C.A. **97** [1982] No. 93749.

Toyo Rubber Industry Co., Ltd., Molded Plastic Foams, Japan. Kokai Tokkyo Koho 57 135127 [1981/82]; C.A. **98** [1983] No. 199388.

Toyo Rubber Industry Co., Ltd., Impact-Resistant Urethane Rubber, Japan. Kokai Tokkyo Koho 57 177015 [1981/82]; C.A. **98** [1983] No. 127499.

Toyo Soda Mfg. Co., Ltd., Vinyl Chloride-Grafted Ethylene-Vinyl Acetate Copolymer Transparent Resin Compositions, Japan. Kokai Tokkyo Koho 82-67646 [1980/82]; C.A. **97** [1982] No. 145692.

Toyo Soda Mfg. Co., Ltd., Transparent Resin Compositions, Japan. Kokai Tokkyo Koho 82-87451 [1980/82]; C.A. **97** [1982] No. 183406.

Toyo Soda Mfg. Co., Ltd., Ethylene-Vinyl Acetate-Vinyl Chloride Graft Copolymer Adhesives, Japan. Kokai Tokkyo Koho 57 167362 [1981/82]; C.A. **98** [1983] No. 127346.

Toyobo Co., Ltd., Magnetic Coating Compositions, Japan. Kokai Tokkyo Koho 57 165464 [1981/82]; C.A. **98** [1983] No. 145166.

Toyoda Gosei Co., Ltd., Transparent Soft Urethane Polymers, Japan. Kokai Tokkyo Koho 82-67622 [1980/82]; C.A. **97** [1982] No. 145445.

Toyoda Gosei Co., Ltd., Urethane Rubber Having Good Toughness, Japan. Kokai Tokkyo Koho 57 158221 [1981/82]; C.A. **98** [1983] No. 73672.

Tremblay, M., Canada, Minister of National Defence, Polyurethane Binders, Can. 1132140 [1979/82]; C.A. **98** [1983] No. 74895.

Vancleva, R., Union Carbide Corp., Polymer/Polyols and Polyurethanes Based Thereon, U.S. 4357430 [1981/82]; C.A. **97** [1982] No. 217346.

Varadhachary, S. N., Congoleum Corp., Bonding Dissimilar Synthetic Polymeric Materials and Their Products, U.S. 4337296 [1980/82]; C.A. **97** [1982] No. 93592.

Victor Co. of Japan, Ltd., Capacitive Disk Recording Materials, Japan. Kokai Tokkyo Koho 57 164139 [1981/82]; C.A. **98** [1983] No. 117186.

von Bittera, M., Dorn, H., Schapel, D., Stendel, W., Voegge, H., Bayer A.-G., Polyurethanes Containing Ectoparasiticides, Eur. Appl. 50784 [1980/82]; C.A. **97** [1982] No. 51225.

von Bittera, M., Federmann, M., von Gizycki, U., Schapel, D., Stenfdel, W., Voege, H., Bayer A.-G., Pet Collars Containing Ectoparasiticides, Eur. Appl. 50782 [1980/82]; C.A. **97** [1982] No. 51224.

Vylet, J., Plicka, E., Karasek, O., Solvent-Free Thermosetting Compositions with Low Viscosity and a Long Pot Life, Czech. 208040 [1979/82]; C.A. **98** [1983] No. 180485.

Wieczorrek, W., Mennicken, G., Bayer A.-G., Schatterproof Coating of Glass Surfaces, Ger. Offen. 3119151 [1981/82]; C.A. **98** [1983] No. 165843.

Wollensak, J. C., Ihrman, K. G., Ethyl Corp., Phenol Extended Polyurethanes Prepared by the RIM Process, U.S. 4314962 [1980/82]; C.A. **96** [1982] No. 218721.

Yamaba, M., Kojima, K., Kaya S., Asahi Glass Co., Ltd., Hardenable Fluorinated Copolymer, Fr. Demande 2488260 [1980/82]; C.A. **96** [1982] No. 201378.

Achilles Foam Board Co., Ltd., Rubberized Cloth, Japan. Kokai Tokkyo Koho 5838155 [1981/83]; C.A. **99** [1983] No. 89498.

Achilles Foam Board Co., Ltd., Urethane Polymer Foam Coating Materials, Japan. Kokai Tokkyo Koho 5876416 [1981/83]; C.A. **100** [1984] No. 23619.

Achilles Foam Board Co., Ltd., Rubber-Coated Fabrics, Japan. Kokai Tokkyo Koho 58116141 [1981/83]; C.A. **100** [1984] No. 8233.

Achilles Foam Board Co., Ltd., Laminates, Japan. Kokai Tokkyo Koho 58155941 [1982/83]; C.A. **100** [1984] No. 122391.

Adeka Argus Chemical Co., Ltd., Poly(Vinyl Chloride) Resin Compositions, Japan. Kokai Tokkyo Koho 5827734 [1981/83]; C.A. **99** [1983] No. 141026.

Annenkova, V. Z., Khaliullin, A. K., Bugun, L. G., Voronkov, M. G., Irkutsk Institute of Organic Chemistry, Adhesive, U.S.S.R. 990774 [1981/83]; C.A. **98** [1983] No. 144745.

Asahi Chemical Industry Co., Ltd., Water-Thinned Thermosetting Coating Materials, Japan. Kokai Tokkyo Koho 5808710 [1981/83]; C.A. **99** [1983] No. 39896.

Asahi Chemical Industry Co., Ltd., Water-Thinned Thermosetting Coating Compositions, Japan. Kokai Tokkyo Koho 5859251 [1981/83]; C.A. **99** [1983] No. 177565.

Asahi Chemical Industry Co., Ltd., Light-Stable Urethane Rubber, Japan. Kokai Tokkyo Koho 5883019 [1981/83]; C.A. **99** [1983] No. 213931.

Asahi Chemical Industry Co., Ltd., Composite Double-Base Propellants with Improved Strength, Japan. Kokai Tokkyo Koho 58125682 [1982/83]; C.A. **100** [1984] No. 54047.

Asahi Chemical Industry Co., Ltd., Composite Double-Base Propellants, Japan. Kokai Tokkyo Koho 58125683 [1982/83]; C.A. **100** [1984] No. 54046.

Asahi Chemical Industry Co., Ltd., Composite Double-Base Propellants with Improved Aging Resistance, Japan. Kokai Tokkyo Koho 58125684 [1982/83]; C.A. **100** [1984] No. 9522.

Asahi Chemical Industry Co., Ltd., Resin Compositions for Coating Materials, Japan. Kokai Tokkyo Koho 58160311 [1982/83]; C.A. **100** [1984] No. 122851.

Asahi Chemical Industry Co., Ltd., Polyoxyalkylene-Siloxanes, Japan. Kokai Tokkyo Koho 58174412 [1982/83]; C.A. **100** [1984] No. 104106.

Asahi Chemical Industry Co., Ltd., Resin Composition for Electrocoating Material, Japan. Kokai Tokkyo Koho 58176253 [1982/83]; C.A. **100** [1984] No. 176561.

Asahi Chemical Industry Co., Ltd., Photosensitive Coating Materials, Japan. Kokai Tokkyo Koho 58204060 [1982/83]; C.A. **101** [1984] No. 74412.

Asahi Glass Co., Ltd., Reaction Injection-Molding Compositions for Urethane Rubber, Japan. Kokai Tokkyo Koho 5832626 [1981/83]; C.A. **99** [1983] No. 71998.

Asahi Glass Co., Ltd., Polyurethane-Glass Laminate for Automobile Windows, Japan. Kokai Tokkyo Koho 5863444 [1981/83]; C.A. **99** [1983] No. 141261.

Asahi Glass Co., Ltd., Lamination with Polyurethane Films and Sheets, Japan. Kokai Tokkyo Koho 5863445 [1981/83]; C.A. **99** [1983] No. 213681.

Asahi Glass Co., Ltd., Urethane Polymer Laminates with Glass Plates, Japan. Kokai Tokkyo Koho 58119855 [1982/83]; C.A. **101** [1984] No. 52564.

Asahi Glass Co., Ltd., Anticorrosive Coating Materials with Good Weather Resistance, Japan. Kokai Tokkyo Koho 58133873 [1982/83]; C.A. **100** [1984] No. 105234.

Asahi Glass Co., Ltd., Antifouling Coating Materials, Japan. Kokai Tokkyo Koho 58133874 [1982/83]; C.A. **100** [1984] No. 105233.

Asahi Glass Co., Ltd., Dainippon Toryo Co., Ltd., Submersible Antifouling Paint, Japan. Kokai Tokkyo Koho 58136666 [1982/83]; C.A. **100** [1984] No. 105239.

Asahi Glass Co., Ltd., Dainippon Printing Co., Ltd., Coating Materials for Repairing Weathered Coatings, Japan. Kokai Tokkyo Koho 58137471 [1982/83]; C.A. **100** [1984] No. 105235.

Auto Chemical Industry Co., Ltd., Polyurethane-Type Insulating Coatings, Japan. Kokai Tokkyo Koho 58189269 [1982/83]; C.A. **101** [1984] No. 74442.

Barda, H. J., Saytech, Inc., Ester and Halogen Containing Polyols, PCT Intern. Appl. 83-04413 [1982/83]; C.A. **100** [1984] No. 140340.

Barnabeo, A. E., Union Carbide Corp., Polysiloxanes and Their Use in the Production of Silane-Modified Alkylene-Alkyl Acrylate Copolymers, U.S. 4408011 [1982/83]; C.A. **99** [1983] No. 214312.

Becker, W., Hoechst A.-G., Curing Agents for Elastic Epoxy Resins, Eur. Appl. 70536 [1981/83]; C.A. **98** [1983] No. 180513.

Beers, M. D., Smith, A. H., General Electric Co., Paintable One-Component Room Temperature Vulcanizable Systems, Can. 1147084 [1980/83]; C.A. **99** [1983] No. 106921.

Birtles, J. F., Brett, R. D., Roberts, W., British Ceramic Research Assoc., Ltd., Eur. Appl. 92988 [1982/83]; C.A. **100** [1984] No. 104922.

Bolgiano, N. C., Sigman, W. T., Armstrong World Industries, Inc., Surface Covering Material, Brit. Appl. 2107723 [1981/83]; C.A. **99** [1983] No. 39973.

Brinkmann, B., Fehr, E., Neffgen, B., Oechsner, W., Schering A.-G., Lechler Chemie GmbH, Hardenable Artificial Resin Mixtures and an Agent for Hardening Artificial Resin Mixtures from Prepolymeric Polyether-Urethane-Urea Amines and Epoxy Resins and/or Aromatic Carbamic Acid Aryl Esters, Ger. Offen. 3151591 [1981/83]; C.A. **99** [1983] No. 159466.

Bruylants, P. P., du Pont de Nemours, E. I., and Co., Acrylic Coating Composition with Multiple Uses, Belg. 896806 [1982/83]; C.A. **100** [1984] No. 87365.

Buethe, I., Heil, G., BASF A.-G., Elastic Radiation-Crosslinked Polyurethane Foams, Ger. Offen. 3127945 [1981/83]; C.A. **98** [1983] No. 180394.

Burton, B. L., Doorakian, G. A., Dow Chemical Co., Organophosphorus-Containing Compounds as Catalysts for Deplocking Blocked Isocyanates, U.S. 4370461 [1981/83]; C.A. **98** [1983] No. 145157.

Cargill, Inc., Polyester Powder Coating Materials, Japan. Kokai Tokkyo Koho 5825364 [1981/83]; C.A. **99** [1983] No. 89698.

Caroll, W. G., Farley, P., Imperial Chemical Industries PLC, Polymer-Modified Polyols, Eur. Appl. 79115 [1981/83]; C.A. **99** [1983] No. 71611.

Caroll, W. G., Farley, P., Marklow, R. J., Imperial Chemical Industries PLC, Polymer-Modified Polyols and their Use in the Manufacture of Polyurethane Products, Eur. Appl. 72096 [1981/83]; C.A. **98** [1983] No. 199591.

Carter, R. G., Miller, W. P., Watson, S. L., Union Carbide Corp., Acrylated Urethane Silicone Compositions, U.S. 4369300 [1979/83]; C.A. **98** [1983] No. 109038.

Chattha, M. S., Ford Motor Co., High Solids Urethane Coatings Prepared from a Polyisocyanate and a Polyhydroxy Oligomer, U.S. 4379906 [1981/83]; C.A. **98** [1983] No. 217340.

Chattha, M. S., Ford Motor Co., High Solids Urethane Coatings I., U.S. 4384103 [1981/83]; C.A. **99** [1983] No. 55172.

Chen, A. C. P., Guarino, J. P., Nagy, F. A., Mobil Oil Corp., Polyoxyalkylene Acrylate Carbamates for Radiation-Curable Coatings, Eur. Appl. 84227 [1981/83]; C.A. **99** [1983] No. 160023.

Chisso Corp., Lubricant Blooming-Resistant PVC Compositions, Japan. Kokai Tokkyo Koho 5853937 [1981/83]; C.A. **99** [1983] No. 159442.

Chisso Corp., Silane-Modified Propylene Polymer Compositions, Japan. Kokai Tokkyo Koho 58117244 [1982/83]; C.A. **100** [1984] No. 86637.

Chisso Corp., Crosslinked Moldings of Silane-Modified Propylene Polymers, Japan. Kokai Tokkyo Koho 58117245 [1982/83]; C.A. **100** [1984] No. 69488.

Christenson, R. M., Hockswender, T. R., PPG Industries, Inc., Polyurethane Crosslinkers and Their Use, U.S. 4403085 [1981/83]; C.A. **99** [1983] No. 196756.

Daicel Chemical Industries, Ltd., Magnetic Coating Materials for Recording Media, Japan. Kokai Tokkyo Koho 58161132 [1982/83]; C.A. **100** [1984] No. 104761.

Daiichi Kogyo Seiyaku Co., Ltd., Polyurethane Resins, Japan. Kokai Tokkyo Koho 58189221 [1982/83]; C.A. **100** [1984] No. 210958.

Dainichi Nippon Cables, Ltd., Crosslinking of Poly(Vinyl Chloride) at Relatively Low Temperatures, Japan. Kokai Tokkyo Koho 5811517 [1981/83]; C.A. **99** [1983] No. 23567.

Dainichi Nippon Cables, Ltd., Crosslinking Polyolefins Containing a Vinyl Silicon Compound with a Silanol Condensation Catalyst, Japan. Kokai Tokkyo Koho 5813612 [1981/83]; C.A. **99** [1983] No. 38973.

Dainichi Nippon Cables, Ltd., Crosslinking Polyolefins Containing a Vinyl Silicon Compound with a Silanol Condensation Catalyst after being Irradiated with UV Beams in an Inert Atmosphere, Japan. Kokai Tokkyo Koho 5813614 [1981/83]; C.A. **98** [1983] No. 216239.

Dainichi Nippon Cables, Ltd., Ethylene Polymer Pipes, Japan. Kokai Tokkyo Koho 5837391 [1981/83]; C.A. **99** [1983] No. 177140.

Dainichi Nippon Cables, Ltd., Polyethylene Pipes for Drinking Water, Japan. Kokai Tokkyo Koho 5838727 [1981/83]; C.A. **99** [1983] No. 123781.

Dainichi Nippon Cables, Ltd., Water-Crosslinkable Olefin Polymer Pipes, Japan. Kokai Tokkyo Koho 5857582 [1981/83]; C.A. **99** [1983] No. 177190.

Dainichi Nippon Cables, Ltd., Olefin Polymer Pipe Joints, Japan. Kokai Tokkyo Koho 5857590 [1981/83]; C.A. **99** [1983] No. 177189.

Dainichi Nippon Cables, Ltd., Olefin Polymer Pipes for Hot Water, Japan. Kokai Tokkyo Koho 5865387 [1981/83]; C.A. **100** [1984] No. 35461.

Dainichi Nippon Cables, Ltd., Olefin Polymer Pipes for Hot Water, Japan. Kokai Tokkyo Koho 5865388 [1981/83]; C.A. **100** [1984] No. 35460.

Dainichi Nippon Cables, Ltd., Two-Part Adhesives, Japan. Kokai Tokkyo Koho 58189276 [1982/83]; C.A. **100** [1984] No. 8332.

Dainichi Nippon Cables, Ltd., Pipes, Japan. Kokai Tokkyo Koho 58211092 [1982/83]; C.A. **100** [1984] No. 157914.

Dainichiseika Color and Chemicals Mfg. Co., Ltd., Coloring Composition for Polyurethane, Japan. Kokai Tokkyo Koho 58104945 [1981/83]; C.A. **99** [1983] No. 213549.

Dainichiseika Color and Chemicals Mfg. Co., Ltd., Curable Coating Materials, Japan. Kokai Tokkyo Koho 58198571 [1982/83]; C.A. **100** [1984] No. 122925.

Dainichiseika Color and Chemicals Mfg. Co., Ltd., Polyurethane Coating Compositions, Japan. Kokai Tokkyo Koho 58201856 [1982/83]; C.A. **100** [1984] No. 176612.

Dainippon Ink and Chemicals, Inc., Thermosetting Coating Materials, Japan. Kokai Tokkyo Koho 5889611 [1981/83]; C.A. **99** [1983] No. 214283.

Dainippon Ink and Chemicals, Inc., Composite Coatings, Japan. Kokai Tokkyo Koho 58189071 [1982/83]; C.A. **100** [1984] No. 105260.

Dainippon Ink and Chemicals, Inc., Thermosetting Powder Coating Compositions, Japan. Kokai Tokkyo Koho 58213062 [1982/83]; C.A. **100** [1984] No. 176583.

Dainippon Printing Co., Ltd., Binders for Thermal Recording Materials, Japan. Kokai Tokkyo Koho 58199192 [1982/83]; C.A. **101** [1984] No. 81725.

Dainippon Toryo Co., Ltd., Anticorrosive Coating Materials for Chromate-Treated Galvanized Steel Plates, Japan. Kokai Tokkyo Koho 5840372 [1981/83]; C.A. **99** [1983] No. 124184.

Dainippon Toryo Co., Ltd., Aqueous Coating Compositions, Japan. Kokai Tokkyo Koho 5852370 [1981/83]; C.A. **99** [1983] No. 141626.

Dainippon Toryo Co., Ltd., Aqueous Dispersion Thermosetting Coating Compositions, Japan. Kokai Tokkyo Koho 5867762 [1981/83]; C.A. **99** [1983] No. 214238.

Dainippon Toryo Co., Ltd., Aqueous Dispersion Thermosetting Coating Compositions, Japan. Kokai Tokkyo Koho 5867763 [1981/83]; C.A. **99** [1983] No. 196714.

Das, S. K., Kania, C. M., PPG Industries, Inc., Thermosetting Cationic Acrylic Latex Containing Blocked Isocyanates, Fr. Demande 2513647 [1981/83]; C.A. **99** [1983] No. 72278.

Dawdy, T. H., Lord Corp., Mounting Adhesive System, Ger. Offen. 3304816 [1982/83]; C.A. **99** [1983] No. 177214.

Dawdy, T. H., Lord Corp., Compositions and Methods for Improving Adhesion to Plastic Substrates, U.S. 4397707 [1982/83]; C.A. **99** [1983] No. 159675.

Denki Kagaku Kogyo K. K., Chloroprene Polymer-Urethane Adhesives, Japan. Kokai Tokkyo Koho 5821470 [1981/83]; C.A. **99** [1983] No. 71860.

Disalvo, A. L., Tsai, C. C., Stauffer Chemical Co., Compositions for Forming Poly(Oxazolidone-Urethane) Thermosets, Eur. Appl. 77174 [1981/83]; C.A. **99** [1983] No. 39365.

Disalvo, A. L., Yu Jun Shen, A., Stauffer Chemical Co., Compositions for Forming Poly-(Oxazolidone-Urethane) Thermosets, Eur. Appl. 77175 [1981/83]; C.A. **99** [1983] No. 23614.

Dominguez, R. J. G., Texaco Development Corp., Polyurea-Containing Reaction Injection Molded Elastomers, Eur. Appl. 92672 [1982/83]; C.A. **100** [1984] No. 52918.

Ford Motor Co., Solvent-Based Thermosetting Coating Compositions, Japan. Kokai Tokkyo Koho 58113217 [1981/83]; C.A. **100** [1984] No. 35951.

Ford Motor Co., Solvent-Based Thermosetting Coating Compositions, Japan. Kokai Tokkyo Koho 58113218 [1981/83]; C.A. **100** [1984] No. 53354.

Ford Motor Co., Thermosetting Coating Materials for Baking at Lower Temperatures, Japan. Kokai Tokkyo Koho 58113265 [1981/83]; C.A. **100** [1984] No. 70034.

Fukumura, M., Shibata, H., Doi, S., Isaka, T., Inoue, T., Sekisui Kaseihin Kogyo K. K., Crosslinked Polyethylene Foams, Belg. 894652 [1981/83]; C.A. **99** [1983] No. 6527.

Furukawa Electric Co., Ltd., Moisture-Curable Electric Insulators, Japan. Kokai Tokkyo Koho 5866211 [1981/83]; C.A. **99** [1983] No. 177607.

Furukawa Electric Co., Ltd., Silane-Crosslinked Polyolefin Compositions, Japan. Kokai Tokkyo Koho 5867744 [1981/83]; C.A. **99** [1983] No. 176938.

Furukawa Electric Co., Ltd., Preparation of Crosslinked Chlorinated Polyethylene, Japan. Kokai Tokkyo Koho 58129016 [1982/83]; C.A. **100** [1984] No. 52577.

Furukawa Electric Co., Ltd., Silane-Crosslinked Polyolefins in Cable Sheathing, Japan. Kokai Tokkyo Koho 58210927 [1982/83]; C.A. **100** [1984] No. 140524.

Gaughan, E. J., Stauffer Chemical Co., Thiophosphoryl Carbamate Herbicide Antidotes, U.S. 4420323 [1982/83]; C.A. **100** [1984] No. 98330.

Genz, J., Heitz, W., Bayer A.-G., Dialkyl and Diaryl Carbonates, Ger. Offen. 3203190 [1982/83]; C.A. **99** [1983] No. 139320.

Genz, J., Heitz, W., Bayer A.-G., Dialkyl Carbonates, Eur. Appl. 85347 [1982/83]; C.A. **99** [1983] No. 194429.

Graziano, F. D., Kuziemka, E. J., Materila Sciences Corp., Organopolysiloxane Coating Compositions, U.S. 4369268 [1981/83]; C.A. **98** [1983] No. 109036.

Grunzinger, R. E., Minnesota Mining and Mfg. Co., Storable One-Part Reactive Liquid Thermal Curing Polymeric Coating Compositions, Eur. Appl. 82618 [1981/83]; C.A. **99** [1983] No. 106906.

Hajimichael, A., Hopper, S., Smith, P. K., EMI Ltd., Encapsulant Material Suitable For Use under Conditions of Prolonged Vacuum, U.S. 4383099 [1981/83]; C.A. **99** [1983] No. 54767.

Hanisch, F., Winter, R., Kabelmetall Electro GmbH, Mixtures Crosslinkable by the Action of Moisture, Ger. Offen. 3207713 [1982/83]; C.A. **99** [1983] No. 177006.

Harima Chemicals, Inc., Photocurable Resin, Japan. Kokai Tokkyo Koho 58160313 [1982/83]; C.A. **100** [1984] No. 53385.

Hedlund, J. L., Puligandla, J., International Business Machines Corp., Corrosion Inhibiting Compositions for Metals, Eur. Appl. 96180 [1982/83]; C.A. **100** [1984] No. 89751.

Hicks, D. D., Celanese Corp., Co-Reactive Urethane Surfactants and Stable Aqueous Epoxy Dispersions, U.S. 4423201 [1982/83]; C.A. **100** [1984] No. 105214.

Hitachi Chemical Co., Ltd., Cellular Polyolefin Manufacture, Japan. Kokai Tokkyo Koho 5829635 [1981/83]; C.A. **99** [1983] No. 89339.

Hitachi Chemical Co., Ltd., Cellular Polyolefin Leather-Substitute Manufacture, Japan. Kokai Tokkyo Koho 5829636 [1981/83]; C.A. **99** [1983] No. 206426.

Hitachi Chemical Co., Ltd., Adhesives for Fiber Reinforced Plastics, Japan. Kokai Tokkyo Koho 58134171 [1982/83]; C.A. **100** [1984] No. 35518.

Hoffman, D. K., Dow Chemical Co., Addition Polymerizable Adduct of a Polymeric Monoahl and an Unsaturated Isocyanate, U.S. 4394491 [1980/83]; C.A. **99** [1983] No. 141012.

Holubka, J. W., Ford Motor Co., Ford-Werke A.-G., Ford (France) S.A., Alcohol-Diblocked Diisocyanate Diurea Oligomers and Coating Compositions Based Thereon, Eur. Appl. 84261 [1981/83]; C.A. **99** [1983] No. 177571.

Holubka, J. W., Ford Motor Co., Diblocked Diisocyanate Diurea Oligomers and Coating Compositions Comprising Them, U.S. 4396753 [1981/83]; C.A. **99** [1983] No. 141671.

Holubka, J. W., Ford Motor Co., Diblocked Diisocyanate Urea Urethane Oligomers and Coating Compositions Comprising Them, U.S. 4409381 [1981/83]; C.A. **100** [1984] No. 8711.

Holubka, J. W., Dickie, R. A., Ford Motor Co., Coating Composition Comprising Chain-Extendable Crosslinkable Polyol and Diblocked Diisocyanate, U.S. 4403086 [1981/83]; C.A. **99** [1983] No. 214309.

Hosoi, N., Fuji Photo Film Co., Ltd., Electrostatographic Toner Material, Ger. Offen. 3245802 [1981/83]; C.A. **100** [1984] No. 43040.

Howell, B. G., BICC PLC, Polymer Compositions, Brit. Appl. 2101138 [1981/83]; C.A. **98** [1983] No. 144061.

Hoy, K. L., Hoy, R. C., Union Carbide Corp., Thixotropic Thickeners for Coatings, Eur. Appl. 96882 [1982/83]; C.A. **100** [1984] No. 87364.

Huber, H., Haensel, E., Geier, G., Dynamit Nobel A.-G., Crosslinkable Resin Mixtures, Ger. Offen. 3220866 [1982/83]; C.A. **100** [1984] No. 104529.

Hyzak, D. L., Pallos, F. M., Stauffer Chemical Co., Halogenated Allylthioisopropyl N-Methylcarbamates as Herbicide Extenders, U.S. 4422869 [1981/83]; C.A. **100** [1984] No. 116480.

Ichinomiys, T., Seiki, R., Imai, T., Seikoh Chemical Co., Ltd., Casting a Polyurethane Sheet, Ger. Offen. 3300424 [1982/83]; C.A. **99** [1983] No. 213791.

Idemitsu Kosan Co., Ltd., Pressure-Sensitive Adhesive Materials, Japan. Kokai Tokkyo Koho 5807471 [1981/83]; C.A. **99** [1983] No. 23695.

Idemitsu Kosan Co., Ltd., Tough and Antivibration Three-Layer Laminates, Japan. Kokai Tokkyo Koho 5825954 [1981/83]; C.A. **99** [1983] No. 39522.

Idemitsu Kosan Co., Ltd., Resin Concrete, Japan. Kokai Tokkyo Koho 5826068 [1981/83]; C.A. **99** [1983] No. 123755.

Idemitsu Kosan Co., Ltd., Resin Concretes, Japan. Kokai Tokkyo Koho 5874560 [1981/83]; C.A. **99** [1983] No. 141272.

Idemitsu Kosan Co., Ltd., Puncture-Preventing Sealants in Tires, Japan. Kokai Tokkyo Koho 58108284 [1981/83]; C.A. **100** [1984] No. 8265.

Idemitsu Kosan Co., Ltd., Polymer Concrete, Japan. Kokai Tokkyo Koho 58120562 [1982/83]; C.A. **99** [1983] No. 177208.

Idemitsu Kosan Co., Ltd., Polymer Concrete, Japan. Kokai Tokkyo Koho 58120563 [1982/83]; C.A. **99** [1983] No. 196297.

Idemitsu Kosan Co., Ltd., Polyurethane Composition, Japan. Kokai Tokkyo Koho 58168618 [1982/83]; C.A. **100** [1984] No. 35273.

Idemitsu Kosan Co., Ltd., Polyurethane Compositions, Japan. Kokai Tokkyo Koho 58189222 [1982/83]; C.A. **100** [1984] No. 192988.

Idemitsu Kosan Co., Ltd., Rubber Blende with Asphalt, Japan. Kokai Tokkyo Koho 58189243 [1982/83]; C.A. **100** [1984] No. 104895.

Idemitsu Kosan Co., Ltd., Pressure-Sensitive Adhesives, Japan. Kokai Tokkyo Koho 58196227 [1982/83]; C.A. **100** [1984] No. 122241.

Idemitsu Kosan Co., Ltd., Liquid Resin Composition for Adhesive Film, Japan. Kokai Tokkyo Koho 58198512 [1982/83]; C.A. **100** [1984] No. 8198.

Iimure, T., Nippon Paint Co., Ltd., Photocurable Coating Compositions, Brit. Appl. 2107334 [1981/83]; C.A. **99** [1983] No. 39907.

Inoue, MTP K. K., Polyurethane Foam-Reinforcement-Skin Composite Moldings, Japan. Kokai Tokkyo Koho 5822154 [1981/83]; C.A. **99** [1983] No. 106615.

Japan Synthetic Rubber Co., Ltd., Lignin-Containing Rubber Compositions with Good Processability, Japan. Kokai Tokkyo Koho 5879036 [1981/83]; C.A. **99** [1983] No. 196453.

Japan Synthetic Rubber Co., Ltd., Lamination of Halogenated Polymers with Diene Rubbers, Japan. Kokai Tokkyo Koho 5896550 [1981/83]; C.A. **100** [1984] No. 35469.

Johnson, C., Dendor, P., Hodgkins, H. L., United States Dept. of the Navy, U.S. 4388126 [1980/83]; C.A. **99** [1983] No. 90500.

Kabel- und Metallwerke Gutehoffnungshütte A.-G., Composite Pipe, Belg. 8950243 [1981/83]; C.A. **99** [1983] No. 89390.

Kanegafuchi Chemical Industry Co., Ltd., Room Temperature-Curing Compositions, Japan. Kokai Tokkyo Koho 5802326 [1981/83]; C.A. **99** [1983] No. 39343.

Kanegafuchi Chemical Industry Co., Ltd., Room-Temperature Curable Compositions, Japan. Kokai Tokkyo Koho 5802342 [1981/83]; C.A. **98** [1983] No. 21516.

Kanegafuchi Chemical Industry Co., Ltd., Moisture-Curable Sealant Compositions, Japan. Kokai Tokkyo Koho 5817154 [1981/83]; C.A. **99** [1983] No. 89742.

Kanegafuchi Chemical Industry Co., Ltd., Siloxane-Modified Polyether Rubber, Japan. Kokai Tokkyo Koho 5842619 [1981/83]; C.A. **99** [1983] No. 141381.

Kanegafuchi Chemical Industry Co., Ltd., Dispersion Coating Compositions, Japan. Kokai Tokkyo Koho 5889661 [1981/83]; C.A. **100** [1984] No. 8665.

Kanegafuchi Chemical Industry Co., Ltd., One-Component Curable Composition, Japan. Kokai Tokkyo Koho 58111855 [1981/83]; C.A. **99** [1983] No. 196751.

Kanegafuchi Chemical Industry Co., Ltd., Room-Temperature Vulcanizable Polyoxyalkyne Having Terminal Silyl Groups, Japan. Kokai Tokkyo Koho 58132022 [1982/83]; C.A. **99** [1984] No. 21934.

Kanegafuchi Chemical Industry Co., Ltd., Stabilized Single-Liquid Hardenable Compositions, Japan. Kokai Tokkyo Koho 58136606 [1982/83]; C.A. **99** [1983] No. 213569.

Kanegafuchi Chemical Industry Co., Ltd., Thermosetting Coating Materials, Japan. Kokai Tokkyo Koho 58157810 [1982/83]; C.A. **100** [1984] No. 105242.

Keogh, M. J., Union Carbide Corp., Masterbatch Composition Comprising a Matrix Having a Polysiloxane Dispersed in It, U.S. 4369289 [1978/83]; C.A. **98** [1983] No. 144733.

Keogh, M. J., Union Carbide Corp., Polysiloxane Having Combined Organotitanates, U.S. 4404349 [1978/83]; C.A. **100** [1984] No. 7436.

Koda, Y., Ona, I., Takeda, A., Toray Silicone Co., Ltd., Fluorosilicone-Containing Compositions for Treatment of Fibers, U.S. 4417024 [1981/83]; C.A. **100** [1984] No. 87213.

Kollmeier, H. J., Klietsch, B. J., Lammerting, H., Langenhagen, R. D., Th. Goldschmidt A.-G., Highly Elastic Cold-Hardened Polyurethane Foams, Ger. 3215317 [1982/83]; C.A. **99** [1983] No. 159467.

Konishiroku Photo Industry Co., Ltd., Magnetic Recording Tapes, Japan. Kokai Tokkyo Koho 58222436 [1982/83]; C.A. **100** [1984] No. 211896.

Kordemenos, P. I., Ford Motor Co., Ford-Werke A.-G., Ford (France) S.A., Aqueous Compositions Comprising a Blocked Isocyanate Crosslinking Agent, Eur. Appl. 83232 [1981/83]; C.A. **99** [1983] No. 106908.

Kubitza, W., Mennicken, G., Bayer A.-G., Use of Clear Varnishes from Organic Polyisocyanates for Coating Sheet Structures from Poly(Vinyl Chloride), Ger. Offen. 3202166 [1982/83]; C.A. **99** [1983] No. 160090.

Kuraray Co., Ltd., Shin-Etsu Chemical Industry Co., Ltd., Membranes for Separating Liquid Mixtures, Japan. Kokai Tokkyo Koho 5858104 [1981/83]; C.A. **99** [1983] No. 39530.

Kurashiki Spinning Co., Ltd., Asahi Organic Chemicals Industry Co., Ltd., Fire-Resistant Laminates, Japan. Kokai Tokkyo Koho 58136432 [1982/83]; C.A. **100** [1984] No. 52811.

Legue, N. R., Shapiro, M., Synthetic Surfaces, Inc., Adhesive Consisting Essentially of a Ricinoleate Urethane Polyol and a Chlorinated Poly-(Vinyl Chloride), Brit. Appl. 2111512 [1981/83]; C.A. **99** [1983] No. 177215.

Leifert, E., BASF Farben und Fasern A.-G., Binder for Cathode Electrodip Lacquers, Ger. Offen. 3124168 [1981/83]; C.A. **98** [1983] No. 127798.

Letoffe, M., Rhone-Poulenc Specialities Chimiques, Moisture-Curable Organopolysiloxane Compositions Prepared by Condensing Organosiloxanes Having Silicon-Bound Hydroxyl Groups, Alkoxysilylated Crosslinking Agents and a Substituted Hydroxylamine, Eur. Appl. 70786 [1981/83]; C.A. **99** [1983] No. 23843.

Maass, G., Sattlegger, H., Lücking, H. J., Bayer A.-G., Room Temperature Vulcanizable Organopolysiloxane Molding Compositions Storable in the Absence of Water, Ger. Offen. 3135185 [1981/83]; C.A. **98** [1983] No. 180905.

Makhlouf, J. M., McCollum, G. J., Kerr, P. R., PPG Industries, Inc., Unsaturated Urea-Urethane Polymers and Their Further Treatment, Ger. Offen. 3248132 [1981/83]; C.A. **99** [1983] No. 123554.

Markusch, P. H., Potter, T. A., Mobay Chemical Corp., Isocyanate-Terminated Prepolymers with Low Free Monomer Contents, U.S. 4413111 [1982/83]; C.A. **100** [1984] No. 52512.

Matsui, M., Momose, C., Terasaki, K., Dainichi Nippon Cables, Ltd., Adhesive for Jointing Polyolefin Articles, Eur. Appl. 83780 [1981/83]; C.A. **99** [1983] No. 123836.

Matsushita Electric Industrial Co., Ltd., Ozone Removal, Japan. Kokai Tokkyo Koho 58183928 [1982/83]; C.A. **100** [1984] No. 90642.

Matsushita Electric Works, Ltd., Radically Curable Urethane Polymer Acrylates, Japan. Kokai Tokkyo Koho 58219214 [1982/83]; C.A. **100** [1984] No. 193692.

Matsushita Electric Works, Ltd., Radically Curable Acrylic Urethane Polymer Coating Materials, Japan. Kokai Tokkyo Koho 58219215 [1982/83]; C.A. **100** [1984] No. 193693.

Matsushita Electric Works, Ltd., Radical-Polymerizable Urethane Compounds, Japan. Kokai Tokkyo Koho 58219216 [1982/83]; C.A. **101** [1984] No. 74443.

McBrayer, R. L., BASF Wyandotte Corp., Reinforced Reaction Injection Molding of Polyurethanes, Brit. Appl. 2104535 [1981/83]; C.A. **98** [1983] No. 216623.

McBrayer, R. L., BASF Wyandotte Corp., A Microcellular Polyurethane Foam with Improved Green Strength, U.S. 4388420 [1980/83]; C.A. **99** [1983] No. 71926.

McGary, C. W., Pascarella, V. J., Rhodes, D. R., Taller, R. A., Warner-Lambert Co., A Polyurethane Elastomer and an Improved Hypoallergenic Polyurethane Flexible Glove Prepared from It, Eur. Appl. 89780 [1982/83]; C.A. **100** [1984] No. 12698.

McGary, C. W., Pascarella, V. J., Taller, R. A., Rhodes, D. R., Anglin, P. E., Daugherty, C. W., Warner-Lambert Co., Powder Coating Process for Forming Soft Flexible Polyurethane Film, Eur. Appl. 89181 [1982/83]; C.A. **100** [1984] No. 39643.

McGary, C. W., Rhodes, D. R., Pascarella, V. J., Warner-Lambert Co., Crystalline Grindable Polyurethane Prepolymers, U.S. 4403084 [1982/83]; C.A. **99** [1983] No. 213935.

Mikami, T., Fuji Photo Film Co., Ltd., Electrostatographic Toner Materials, Ger. Offen. 3226176 [1981/83]; C.A. **98** [1983] No. 170370.

Mikami, T., Hosoi, N., Fuji Photo Film Co., Ltd., Electrostatographic Toner Material, Ger. Offen. 3236689 [1981/83]; C.A. **99** [1983] No. 113699.

Mitsubishi Electric Corp., Cork Sealants, Japan. Kokai Tokkyo Koho 5896672 [1981/83]; C.A. **100** [1984] No. 8720.

Mitsubishi Petrochemical Co., Ltd., Water-Crosslinkable Flexible Sheets, Japan. Kokai Tokkyo Koho 5829663 [1981/83]; C.A. **99** [1983] No. 71887.

Mitsubishi Petrochemical Co., Ltd., Thermoplastic Rubber, Japan. Kokai Tokkyo Koho 58132032 [1982/83]; C.A. **100** [1984] No. 35631.

Mitsubishi Petrochemical Co., Ltd., Crosslinkable Exandable Propylene Polymer Compositions, Japan. Kokai Tokkyo Koho 58134131 [1982/83]; C.A. **100** [1984] No. 23254.

Mitsubishi Petrochemical Co., Ltd., Crosslinkable Expandable Propylene Resin Compositions, Japan. Kokai Tokkyo Koho 58134132 [1982/83]; C.A. **100** [1984] No. 52582.

Mitsubishi Petrochemical Co., Ltd., Low-Shrinkage Unsaturated Polyester Resin Compositions, Japan. Kokai Tokkyo Koho 58173114 [1982/83]; C.A. **100** [1984] No. 211018.

Mitsubishi Rayon Co., Ltd., Thermosetting Resin Compositions, Japan. Kokai Tokkyo Koho 58145723 [1982/83]; C.A. **100** [1984] No. 52589.

Mitsuboshi Belting Ltd., Adhesives for Waterproof Sheets, Japan. Kokai Tokkyo Koho 5887033 [1981/83]; C.A. **100** [1984] No. 8273.

Mitsui-Nisso Corp., Urethane Polymer Undercoating Materials for Automobiles, Japan. Kokai Tokkyo Koho 58119381 [1982/83]; C.A. **100** [1984] No. 53353.

Mitsui-Nisso Corp., Polyurethane-Siloxanes, Japan. Kokai Tokkyo Koho 58217515 [1982/83]; C.A. **100** [1984] No. 157186.

Mitsui-Nisso Urethane K. K., Thermoplastic Urethane Polymers, Japan. Kokai Tokkyo Koho 5805325 [1981/83]; C.A. **99** [1983] No. 23144.

Mitsui Polychemicals Co., Ltd., Poly(Vinyl Chloride) Compositions, Japan. Kokai Tokkyo Koho 58168646 [1982/83]; C.A. **101** [1984] No. 8185.

Mitsui Toatsu Chemicals, Inc., Granular Fire Extinguishers, Japan. Kokai Tokkyo Koho 5878676 [1981/83]; C.A. **99** [1983] No. 107605.

Mitsui Toatsu Chemicals, Inc., Granular Fire Extinguishers, Japan. Kokai Tokkyo Koho 58175575 [1982/83]; C.A. **100** [1984] No. 70788.

Miyai, S., Nakajima, K., Somezawa, M., Sony Corp., Electron Beam-Curable Resin, PCT Intern. Appl. 83-00696 [1981/83]; C.A. **99** [1983] No. 24165.

Miyake, J., Yamazaki, K., Kamatani, Y., Takeda Chemical Industries, Ltd., One-Can Type Pressure-Sensitive Adhesive Composition, Eur. Appl. 81103 [1981/83]; C.A. **99** [1983] No. 89385.

Miyake, J., Yamazaki, K., Kamatani, Y., Takeda Chemical Industries, Ltd., Two-Package Type Pressure-Sensitive Adhesive Composition, Eur. Appl. 81693 [1981/83]; C.A. **99** [1983] No. 89392.

Murakami, R., Uyeda, Y., Nishijima, K., Yanagihara, T., Nippon Paint Co., Ltd., Cationic Electrophoretic Rust Protection Coating, Ger. Offen. 3315285 [1982/83]; C.A. **100** [1984] No. 53315.

Naoi, Takashi, Kakimi, F., Mikami, T., Fuji Photo Film Co., Ltd., Electrostatographic Toner Material, Ger. Offen. 3245801 [1981/83]; C.A. **99** [1983] No. 203577.

Naples, G., Textron, Inc., Storage-Stable One-Component Urethanes and Their Use, U.S. 4381388 [1981/83]; C.A. **99** [1983] No. 39954.

Nippon Carbide Industries Co., Inc., Sealants, Japan. Kokai Tokkyo Koho 58176289 [1982/83]; C.A. **100** [1984] No. 122947.

Nippon Oils and Fats Co., Ltd., Double-Base Propellants with Improved Mechanical Properties and Combustibility, Japan. Kokai Tokkyo Koho 58194791 [1982/83]; C.A. **100** [1984] No. 105977.

Nippon Paint Co., Ltd., Electrophoretic Coating Compositions, Japan. Kokai Tokkyo Koho 58136667 [1982/83]; C.A. **100** [1984] No. 70038.

Nippon Paint Co., Ltd., Electrophoretic Paint Composition, Japan. Kokai Tokkyo Koho 58136668 [1982/83]; C.A. **100** [1984] No. 105205.

Nippon Paint Co., Ltd., Cationic Electrodeposition Composition, Japan. Kokai Tokkyo Koho 58185662 [1982/83]; C.A. **100** [1984] No. 211776.

Nippon Polyurethane Industry Co., Ltd., Thermosetting Coating Compositions, Japan. Kokai Tokkyo Koho 5801753 [1981/83]; C.A. **99** [1983] No. 24185.

Nippon Polyurethane Industry Co., Ltd., Thermosetting Coating Compositions, Japan. Kokai Tokkyo Koho 58113260 [1981/83]; C.A. **100** [1984] No. 87392.

Nippon Polyurethane Industry Co., Ltd., Polyurethane Molded Products, Japan. Kokai Tokkyo Koho 58162623 [1982/83]; C.A. **100** [1984] No. 193004.

Nissen, D., Hickmann, E., BASF A.-G., Cellular Polyurethane or Polyurethane-Urea Moldings, Ger. Offen. 3126436 [1981/83]; C.A. **98** [1983] No. 127404.

Nissen, D., Marx, M., Schmidt, H. U., BASF A.-G., Molding Compositions Based on Polyurethane or Polyurethane-Polyurea Elastomers with Improved Light- and Weather-Resistance, Ger. Offen. 3215908 [1982/83]; C.A. **100** [1984] No. 8216.

Nissen, D., Neumann, P., Marx, M., Eilingsfeld, H., BASF A.-G., Cellular Polyurethane-Polyurea Moldings and Alkyl-Substituted Phenylenediamines, Ger. Offen. 3126435 [1981/83]; C.A. **98** [1983] No. 144716.

NHK Spring Co., Ltd., Waterproof Sheets, Japan. Kokai Tokkyo Koho 5849241 [1981/83]; C.A. **99** [1983] No. 57966.

Pawlowski, C. E., Dow Chemical Co., Carbamic Acid Derivatives, U.S. 4387058 [1981/83]; C.A. **99** [1983] No. 122064.

Pedain, J., Wellner, W., Koenig, K., Gruber, H., Bayer A.-G., Use of Single- or Multicomponent Systems as or for Producing Laminates for Glass Constructions, Ger. Offen. 3200430 [1982/83]; C.A. **99** [1983] No. 196765.

Potter, W. D., Smith and Nephew Associated Co. PLC, Dental Prosthesis Coating with Polyurethane Lining Material, Eur. Appl. 91285 [1982/83]; C.A. **99** [1983] No. 218649.

Potter, W. D., Kiamil, S. B., White, N. D., Smith and Nephew Associated Co. PLC, Surgical Bandages Containing Urethane Polymers, Eur. Appl. 94222 [1982/83]; C.A. **100** [1984] No. 74014.

Rasshofer, W., Kopp, R., Paul, R., Bayer A.-G., Polyurethane Foams and Storage-Stable Preliminary Products for Their Production, Ger. Offen. 3141117 [1981/83]; C.A. **99** [1983] No. 54577.

Reed, R., Chan, M. L., United States Dept. of the Navy, Propellant Binders Cure Catalyst, U.S. 4379903 [1982/83]; C.A. **98** [1983] No. 218231.

Rizk, S. D., Shah, N. B., Essex Chemical Corp., "One-Pot" Moisture-Curable Two-Component Coating Composition, U.S. 4367313 [1981/83]; C.A. **98** [1983] No. 109033.

Rowlands, J. P., Interchem International S. A., Polymer-Modified Polyols Useful in Polyurethane Manufacture, U.S. 4374209 [1980/83]; C.A. **98** [1983] No. 161742.

Sannopuko, K. K., Thickening Agents for Aqueous Systems, Japan. Kokai Tokkyo Koho 58213074 [1982/83]; C.A. **100** [1984] No. 140901.

Sanyo Chemical Industries, Ltd., Urethane Polymer Foams with Low Levels of Water Absorption, Japan. Kokai Tokkyo Koho 5811518 [1981/83]; C.A. **99** [1983] No. 23501.

Sanyo Chemical Industries, Ltd., Polyurethanes, Japan. Kokai Tokkyo Koho 58162626 [1982/83]; C.A. **100** [1984] No. 52617.

Sanyo Chemical Industries, Ltd., Toyota Motor Co., Ltd., Urethane Polymer Foams, Japan. Kokai Tokkyo Koho 5829816 [1981/83]; C.A. **99** [1983] No. 89065.

Saruyama, T., Toray Silicone Co., Ltd., Room Temperature-Vulcanizable Polyorganosiloxane Compositions Containing Alcoholic and Carboxylic Organofunctionality, Eur. Appl. 81119 [1981/83]; C.A. **99** [1983] No. 72023.

Schulze, K., Dipox Kurt Schulze K.-G., Free-Flowing Additive for Crosslinking Organic Polymers, Ger. Offen. 3150808 [1981/83]; C.A. **99** [1983] No. 106333.

Sekisui Chemical Co., Ltd., Crosslinking of Polyolefins with Silanes, Japan. Kokai Tokkyo Koho 5893704 [1981/83]; C.A. **100** [1984] No. 7808.

Sekisui Chemical Co., Ltd., Polyurethane Adhesives, Japan. Kokai Tokkyo Koho 58213072 [1982/83]; C.A. **100** [1984] No. 176098.

Sekisui Chemical Co., Ltd., Polyurethane Adhesives, Japan. Kokai Tokkyo Koho 58215474 [1982/83]; C.A. **100** [1984] No. 157879.

Sekisui Chemical Co., Ltd., Urethane Adhesives, Japan. Kokai Tokkyo Koho 58217576 [1982/83]; C.A. **100** [1984] No. 193213.

Shin-Etsu Chemical Industry Co., Ltd., Release Agents for Polyurethane Molding, Japan. Kokai Tokkyo Koho 5845297 [1981/83]; C.A. **99** [1983] No. 106434.

Shin-Etsu Chemical Industry Co., Ltd., Aqueous Silicone Emulsions, Japan. Kokai Tokkyo Koho 5869250 [1981/83]; C.A. **100** [1984] No. 8219.

Shin-Etsu Chemical Industry Co., Ltd., Silicone Rubber Coating Materials, Japan. Kokai Tokkyo Koho 58101153 [1981/83]; C.A. **100** [1984] No. 23603.

Shin-Etsu Chemical Industry Co., Ltd., Finishing Agents for Textiles, Japan. Kokai Tokkyo Koho 58126377 [1982/83]; C.A. **100** [1984] No. 35798.

Shinto Paint Co., Ltd., Anticorrosive Coatings, Japan. Kokai Tokkyo Koho 5873800 [1981/83]; C.A. **99** [1983] No. 177566.

Showa Denko, K. K., Silane-Crosslinked Ethylene-Diene Copolymer, Japan. Kokai Tokkyo Koho 58173106 [1982/83]; C.A. **101** [1984] No. 92043.

Showa Electric Wire and Cable Co., Ltd., Powdered Silicone-Grafted Polyolefin, Japan. Kokai Tokkyo Koho 5801724 [1981/83]; C.A. **99** [1983] No. 23497.

Showa Electric Wire and Cable Co., Ltd., Flame Retardant Compositions for Silicone-Grafted Polyolefins, Japan. Kokai Tokkyo Koho 5876443 [1981/83]; C.A. **100** [1984] No. 7778.

Showa Electric Wire and Cable Co., Ltd., Silane-Crosslinked Polyolefin Hoses, Japan. Kokai Tokkyo Koho 5877989 [1981/83]; C.A. **100** [1984] No. 35440.

Showa Electric Wire and Cable Co., Ltd., Swelling-Resistant Polyolefin Films for Oil-Filled Electric Cables, Japan. Kokai Tokkyo Koho 5889324 [1981/83]; C.A. **99** [1983] No. 196278.

Showa Electric Wire and Cable Co., Ltd., Laminates of Metal and Silanol-Crosslinked Polyolefin, Japan. Kokai Tokkyo Koho 5896548 [1981/83]; C.A. **100** [1984] No. 8044.

Showa Electric Wire and Cable Co., Ltd., Oil-Impregnable Insulating Paper Laminates, Japan. Kokai Tokkyo Koho 58103714 [1981/83]; C.A. **100** [1984] No. 8094.

Showa Electric Wire and Cable Co., Ltd., Flame-Resistant Crosslinked Polyolefin Composition, Japan. Kokai Tokkyo Koho 58132013 [1982/83]; C.A. **99** [1983] No. 213566.

Showa Electric Wire and Cable Co., Ltd., Silyl-Modified High-Density Polyethylene Laminates, Japan. Kokai Tokkyo Koho 58166047 [1982/83]; C.A. **100** [1984] No. 104633.

Simroth, D. W., Critchfield, F. E., Shook, E. G., Union Carbide Corp., Polymer/Polyol Compositions Having Improved Combustion Resistance and a Process for Preparing Polyurethanes, Eur. Appl. 95653 [1982/83]; C.A. **100** [1984] No. 121802.

Singh, B., Novak, R. W., American Cyanamid Co., Blocked Isocyanate, U.S. 4374771 [1982/83]; C.A. **98** [1983] No. 181259.

Skoulchi, M. M., National Starch and Chemical Corp., Monomeric Carbamic Ester Photoinitiators, Brit. Appl. 2100722 [1981/83]; C.A. **98** [1983] No. 143995.

Smith, N., Sommerford Plastics Ltd., Marking a Trafficway Surface, Brit. Appl. 2113234 [1982/83]; C.A. **100** [1984] No. 8569.

Steinbach, H. H., Schnurrbusch, K., Rieder, M., Bayer A.-G., Impregnating Agent for Wooden Surfaces, Ger. Offen. 3222963 [1982/83]; C.A. **100** [1984] No. 87520.

Stresinka, J., Malcovsky, E., Mokry, J., Valent, V., Polyesterpolyols, Czech. 201783 [1978/83]; C.A. **99** [1983] No. 54500.

Sumitomo Chemical Co., Ltd., Fire-Resistant Polyurethane Sealing Compositions, Japan. Kokai Tokkyo Koho 58157882 [1982/83]; C.A. **100** [1984] No. 176631.

Sumitomo Chemical Co., Ltd., Polyurethane Sealants, Japan. Kokai Tokkyo Koho 58160385 [1982/83]; C.A. **100** [1984] No. 122943.

Takeda Chemical Industries, Ltd., Polyurethane Coating Compositions, Japan. Kokai Tokkyo Koho 58108262 [1981/83]; C.A. **100** [1984] No. 35937.

Takeda Chemical Industries, Ltd., Quick-Drying Urethane Polymer Coating Materials, Japan. Kokai Tokkyo Koho 58109528 [1981/83]; C.A. **100** [1984] No. 53347.

Tanaka, M., Kamatani, Y., Nasu, K., Takeda Chemical Industries, Ltd., One-Pack Type Thermosetting Polyurethane Coating Composition, Eur. Appl. 81712 [1981/83]; C.A. **99** [1983] No. 89737.

Teijin Ltd., Bleeding-Resistant Leather Substitutes, Japan. Kokai Tokkyo Koho 5841975 [1981/83]; C.A. **99** [1983] No. 141232.

Teijin Ltd., Photosensitive Resin Compositions for Flexographic Plates, Japan. Kokai Tokkyo Koho 5876828 [1981/83]; C.A. **99** [1983] No. 113720.

Teijin Ltd., A Method for Making a Printing Plate, Japan. Kokai Tokkyo Koho 5880640 [1981/83]; C.A. **101** [1984] No. 63683.

Theodore, A. N., Chattha, M. S., Ford Motor Co., High Solids Urethane Coatings with Enhanced Flexibility and Impact Strength, U.S. 4376187 [1981/83]; C.A. **99** [1983] No. 6964.

Toa Gosei Chemical Industry Co., Ltd., Hardening Composition, Japan. Kokai Tokkyo Koho 58134113 [1982/83]; C.A. **100** [1984] No. 86690.

Toray Industries, Inc., Flame-Retardant Styrene Polymer Compositions, Japan. Kokai Tokkyo Koho 5865741 [1981/83]; C.A. **99** [1983] No. 196013.

Toray Industries, Inc., Light-Sensitive Compositions for Lithographic Use, Japan. Kokai Tokkyo Koho 58152236 [1982/83]; C.A. **99** [1983] No. 203609.

Toray Silicone Co., Ltd., Room Temperature Setting Organopolysiloxane Compositions, Japan. Kokai Tokkyo Koho 58149947 [1982/83]; C.A. **100** [1984] No. 122195.

Toshiba Corp., Quick-Curing Urethane Polymer Foams, Japan. Kokai Tokkyo Koho 5829815 [1981/83]; C.A. **99** [1983] No. 89066.

Toshiba Corp., Rigid Urethane Foam, Japan. Kokai Tokkyo Koho 58145736 [1982/83]; C.A. **100** [1984] No. 193163.

Toshiba Corp., Information Recording Material, Japan. Kokai Tokkyo Koho 58154749 [1982/83]; C.A. **100** [1984] No. 86725.

Toshiba Corp., Capacitance Variation Type Video Desks, Japan. Kokai Tokkyo Koho 58200954 [1981/83]; C.A. **100** [1984] No. 201054.

Toyo Rubber Chemical Industry Co., Ltd., Urethane Rubber Foam Manufacture, Japan. Kokai Tokkyo Koho 5859231 [1981/83]; C.A. **99** [1983] No. 141387.

Toyo Rubber Industry Co., Ltd., Safety Tires, Japan. Kokai Tokkyo Koho 5853035 [1975/83]; C.A. **100** [1984] No. 158063.

Toyo Rubber Industry Co., Ltd., Yellowing-Resistant Polyurethane Products, Japan. Kokai Tokkyo Koho 5887115 [1981/83]; C.A. **99** [1983] No. 177283.

Toyo Rubber Industry Co., Ltd., Vibration-Damping Coating Materials, Japan. Kokai Tokkyo Koho 58210965 [1982/83]; C.A. **100** [1984] No. 140936.

Toyobo Co., Ltd., Low-Friction Block Copolyester Compositions, Japan. Kokai Tokkyo Koho 5811547 [1982/83]; C.A. **99** [1983] No. 39320.

Toyobo Co., Ltd., Polyurethane Compositions, Japan. Kokai Tokkyo Koho 5825355 [1981/83]; C.A. **99** [1983] No. 106679.

Toyobo Co., Ltd., Adhesives, Japan. Kokai Tokkyo Koho 58217572 [1982/83]; C.A. **100** [1984] No. 176131.

Union Carbide Corp., Moisture-Curable Silane-Modified Alkylene-Alkyl Acrylate Copolymer Compositions, Japan. Kokai Tokkyo Koho 58187445 [1982/83]; C.A. **100** [1984] No. 192990.

Unitika Ltd., Heat-Resistant Adhesives, Japan. Kokai Tokkyo Koho 5827767 [1981/83]; C.A. **99** [1983] No. 123756.

Van Elven, A., Overmars, H. G. J., Nederlandse Centrale Organisatie voor Toegepast-Natuurwetenschappelijk Oderzoek, Poly(Etherurethane) Particles Having Absorptive Activity, as well as a Method for the Extraction or Separation of Metals from Liquids by Employing Said Particles, Eur. Appl. 92295 [1982/83]; C.A. **100** [1984] No. 52508.

Voigt, H., Schmidtchen, H. M., Kabel- und Metallwerke Gutehoffnungshütte A.-G., Moisture Crosslinkable Polymeric or Elastomeric Composition, Indian 150981 [1978/83]; C.A. **99** [1983] No. 123630.

von Bittera, M., Bayer A.-G., Membrane for Measuring Active Ingredient Release and Evaluating Skin Absorption In Vitro, Ger. Offen. 3212735 [1982/83]; C.A. **100** [1984] No. 39590.

Watson, P. J. H., Deborah Coatings, Ltd., Coating Composition, Brit. Appl. 2105356 [1981/83]; C.A. **98** [1983] No. 199974.

Watson, P. J. H., Deborah Coatings, Ltd., Coating Composition, Brit. Appl. 2106526 [1981/83]; C.A. **98** [1983] No. 217349.

Weber, C., Wirtz, H., Seel, K., Bayer A.-G., Elastic Moldings, Ger. Offen. 3147736 [1981/83]; C.A. **99** [1983] No. 72020.

Wick, G., AKZO GmbH, Polyurethanes for Hemodialysis Membranes, Ger. Offen. 3147025 [1981/83]; C.A. **99** [1983] No. 146147.

Wolf, E., Schmitt, F., Chemische Werke Huels A.-G., Storable Polyurethane-Single Component Stoving Varnish, Ger. Offen. 3221558 [1982/83]; C.A. **100** [1984] No. 122910.

Yamabe, M., Higaki, H., Shinohara, T., Tanabe, H., Nakayama, S., Dainippon Toryo Co., Ltd., Asahi Glass Co., Ltd., Corrosion-Preventing Coating, Ger. Offen. 3303828 [1982/83]; C.A. **100** [1984] No. 23630.

Yamano, N., Arimatsu, Y., Katsuo K., Tani, K., Mitamura, H., Saitoh, M., Kamatani, H., Toyobo Co., Ltd., Urethane Polymers Containing a Tertiary Nitrogen Atom, Brit. Appl. 2105734 [1981/83]; C.A. **99** [1983] No. 23153.

Zalucha, D. J., Sexsmith, F. H., Howard, D. D., Nulph, M. L., Lord Corp., Thermally-Responsive Polymeric Materials, U.S. 4409383 [1981/83]; C.A. **100** [1984] No. 23231.

Zecher, W., Merten, R., Bayer A.-G., Ethene-1,2-dicarbamic Acid Esters, Ger. Offen. 3204129 [1982/83]; C.A. **100** [1984] No. 6048.

Zoellner, R., Godthardt, L. P., Andres, K. H., Schmidt, W., Bayer A.-G., Eur. Appl. 77964 [1981/83]; C.A. **99** [1983] No. 89746.

Achilles Corp., Adhesive Sheets with Good Creep Resistance, Japan. Kokai Tokkyo Koho 59157165 [1983/85]; C.A. **101** [1984] No. 231700.

Asahi Chemical Industry Co., Ltd., Abrasion-Resistant Coatings, Japan. Kokai Tokkyo Koho 5951908 [1982/84]; C.A. **101** [1984] No. 132586.

Asahi Chemical Industry Co., Ltd., Polyoxymethylene Molding Compositions, Japan. Kokai Tokkyo Koho 50129247 [1983/84]; C.A. **101** [1984] No. 193339.

Asahi Glass Co., Ltd., Soil Repellents, Japan. Kokai Tokkyo Koho 5933315 [1982/84]; C.A. **101** [1984] No. 74280.

Asahi Glass Co., Ltd., Polyurethane Elastomers, Japan. Kokai Tokkyo Koho 5947222 [1982/84]; C.A. **101** [1984] No. 56314.

Asahi Glass Co., Ltd., Polyurethane Elastomers, Japan. Kokai Tokkyo Koho 5980426 [1982/84]; C.A. **101** [1984] No. 112239.

Asahi Glass Co., Ltd., Automobile Exterior Materials, Japan. Kokai Tokkyo Koho 59117518 [1983/84]; C.A. **101** [1984] No. 172832.

Asahi Glass Co., Ltd., Bilayer Safety Glass Laminates, Japan. Kokai Tokkyo Koho 59133048 [1983/84]; C.A. **101** [1984] No. 231649.

Asahi Glass Co., Ltd., Thermoplastic Polyurethane Sheet or Film and Laminated Glass, Japan. Kokai Tokkyo Koho 59135216 [1983/84]; C.A. **102** [1985] No. 63004.

Asahi Glass Co., Ltd., Polyurethane Reaction Injection Molding Compositions, Japan. Kokai Tokkyo Koho 59145129 [1983/84]; C.A. **102** [1985] No. 25927.

Asahi Glass Co., Ltd., Fluorine-Containing Releases, Japan. Kokai Tokkyo Koho 59157190 [1983/84]; C.A. **102** [1985] No. 114702.

Asahi Glass Co., Ltd., Urethane Rubbers, Japan. Kokai Tokkyo Koho 59159816 [1983/84]; C.A. **102** [1985] No. 8050.

Asahi Glass Co., Ltd., Reaction Injection Molding, Japan. Kokai Tokkyo Koho 59191712 [1983/84]; C.A. **102** [1985] No. 63420.

Asahi Glass Co., Ltd., Polyurethane Elastomers, Japan. Kokai Tokkyo Koho 59227921 [1983/84]; C.A. **102** [1985] No. 150748.

Bard, R. C., Inc., Preparation of Polymers for Gaskets Around Fistulas, Japan. Kokai Tokkyo Koho 5927910 [1982/84]; C.A. **101** [1984] No. 28326.

Barsa, E. A., Sherwood, P. W., Upjohn Co., N-Isocyanatoalkanoyl-Substituted Cyclic Ureas, U.S. 4448816 [1983/84]; C.A. **101** [1984] No. 131840.

Battey, P. K., Talbot, P. D., Belzona Molecular Metalife, Ltd., Polyurethane Composition, Brit. 2121813 [1982/84]; C.A. **100** [1984] No. 122937.

Battice, D. R., Dow Corning Corp., High Resiliency Polyurethane Foams Containing Grafted Polyols, U.S. 4477601 [1983/84]; C.A. **102** [1985] No. 25589.

Beis, D., Prevotat, J., SOGECAN, Thermoplastic Compositions Containing a Polymer Grafted by a Silane, Fr. Demande 2546172 [1983/84]; C.A. **102** [1985] No. 221740.

Blanco, F. E., Products Research and Chemical Corp., Unsaturated Ureide Polymers, and Cured Elastomers Produced from Them, Eur. Appl. 104035 [1982/84]; C.A. **101** [1984] No. 24840.

Blanco, F. E., Products Research and Chemical Corp., Unsaturated Ureide Polymers, and Cured Elastomers Produced from Them, U.S. 4426506 [1982/84]; C.A. **100** [1984] No. 87039.

Buchan, I. A., Smith and Nephew Associated Co. PLC, Medical Device Suitable for the Prophylaxis of Pressure Sores, Eur. Appl. 122035 [1983/84]; C.A. **102** [1985] No. 226004.

Bueltjer, U., Eimer, D., Zeitler, G., Reich, E., Siekmeyer, A., Weckert, H. J., BASF A.-G., Flat Composite Articles of Polyurethane and Fibers, Based on Polyurethane Rigid Foams, and Their Use, Fr. Demande 2545039 [1983/84]; C.A. **102** [1985] No. 205060.

Cassata J. C., Chattha, M. S., Ford Motor Co., Ford-Werke A.-G., Ford (France) S.A., Crosslinked Flow-Control Additives for High-Solids Paints, Eur. Appl. 117658 [1983/84]; C.A. **102** [1985] No. 8307.

Celanese, Corp., Dispersing Agents for Aqueous Epoxy Resin, Japan. Kokai Tokkyo Koho 5943050 [1982/84]; C.A. **101** [1984] No. 39956.

Chang, W. H., Ambrose, R. R., McKeough, D. T., Porter, S., Johnston, B. K., PPG Industries, Inc., Thermosetting High-Solids Solvent-Based Polyester-Urethane Two-Component Coating Compositions, U.S. 4485228 [1983/84]; C.A. **102** [1985] No. 115271.

Chisso Corp., Silane-Crosslinked Cellular Polypropylene, Japan. Kokai Tokkyo Koho 5951925 [1982/84]; C.A. **101** [1984] No. 56072.

Colpitts, R., Polythetics, Inc., Composite Denture Combining Soft Polyurethane and Hard Polymer Components, Brit. Appl. 2130886 [1982/84]; C.A. **101** [1984] No. 137070.

Cuscurida, M., Texaco Inc., Polyester Polyols Made with Polyester Polycarbonates and Polyurethanes Therefrom, U.S. 4435527 [1983/84]; C.A. **100** [1984] No. 175994.

Cuscurida, M., Speranza, G. Ph., Texaco Development Corp., Polyurethane Polymer Polyols Made with Aromatic Nitrogen-Containing Polyols and Polyurethanes From Them, Eur. Appl. 116758 [1983/84]; C.A. **102** [1985] No. 7626.

Daicel Chemical Industries, Ltd., Mitsui-Nisso Corp., Polyurethane Resins, Japan. Kokai Tokkyo Koho 59179514 [1983/84]; C.A. **102** [1985] No. 96446.

Daimer, W., Gmoser, J., Schipfer, R., Vianova Kunstharz A.-G., Self-Crosslinking, Cathodically Depositable Binders Based on Modified Alkylphenol-Formaldehyde Condensation Product, Austrian 375953 [1983/84]; C.A. **102** [1985] No. 26497.

Daiichi Kogyo Seiyaku Co., Ltd., Finishing of Fabrics for Improved Transparency, Japan. Kokai Tokkyo Koho 59157381 [1983/84]; C.A. **101** [1984] No. 212635.

Daiichi Kogyo Seiyaku Co., Ltd., Nippon Steel Corp., Coating Compositions for Metals, Japan. Kokai Tokkyo Koho 59197466 [1983/84]; C.A. **102** [1985] No. 151032.

Daiichi Kogyo Seiyaku Co., Ltd., Nippon Steel Corp., Polyurethane Coating Composition for Metals, Japan. Kokai Tokkyo Koho 59197469 [1983/84]; C.A. **102** [1985] No. 151031.

Dainichi Nippon Cables, Ltd., Water-Curable Polyolefin Pipes, Japan. Kokai Tokkyo Koho 59220326 [1983/84]; C.A. **102** [1985] No. 133262.

Dainippon Ink and Chemicals, Inc., Resin Composition for Powder Paint, Japan. Kokai Tokkyo Koho 5930869 [1982/85]; C.A. **101** [1984] No. 92938.

Dainippon Ink and Chemicals, Inc., Liquid Developer for Electrophotography, Japan. Kokai Tokkyo Koho 5934540 [1982/84]; C.A. **101** [1984] No. 31135.

Dainippon Ink and Chemicals, Inc., Curable Polyurethane Compositions, Japan. Kokai Tokkyo Koho 5962621 [1982/84]; C.A. **101** [1984] No. 74500.

Dainippon Ink and Chemicals, Inc., Weather-Resistant Resin Moldings, Japan. Kokai Tokkyo Koho 5993732 [1982/84]; C.A. **101** [1984] No. 193821.

Dainippon Ink and Chemicals, Inc., Thermosetting Polyurethanes, Japan. Kokai Tokkyo Koho 59108023 [1982/84]; C.A. **102** [1985] No. 8330.

Dainippon Ink and Chemicals, Inc., Glass Bottle Reinforcing Compositions, Japan. Kokai Tokkyo Koho 59195566 [1983/84]; C.A. **102** [1985] No. 189700.

Dainippon Printing Co., Ltd., Thermal Transfer Sheet, Japan. Kokai Tokkyo Koho 59198195 [1983/84]; C.A. **102** [1985] No. 176601.

Dainippon Printing Co., Ltd., Adhesive Compositions, Japan. Kokai Tokkyo Koho 59215372 [1983/84]; C.A. **102** [1985] No. 150619.

Dainippon Toryo Co., Ltd., High-Solids, Nonaqueous Resin Dispersions, Japan. Kokai Tokkyo Koho 59179550 [1983/84]; C.A. **102** [1985] No. 63705.

Dainippon Toryo Co., Ltd., Coating Formation, Japan. Kokai Tokkyo Koho 59222267 [1983/84]; C.A. **102** [1985] No. 222260.

Daito Kasei Kogyo Co., Ltd., Seiko Seisakusho Co. Ltd., Sotojima Masaharu, Ink for Writing Materials, Japan. Kokai Tokkyo Koho 59170166 [1983/84]; C.A. **102** [1985] No. 186812.

Dawdy, T. H., Lord Corp., Epoxy-Modified Assembly Adhesive with Improved Heat Resistance, Ger. Offen. 3333006 [1982/84]; C.A. **101** [1984] No. 92400.

Deguchi, J., Inoue, T., Mitsubishi Petrochemical Co., Ltd., Crosslinked Products from Propylene Resin, Ger. Offen. 3346267 [1982/84]; C.A. **101** [1984] No. 111992.

Dell, J. D., Minnesota Mining and Mfg. Co., Film-Forming Composition Containing an Antimicrobial Agent, Eur. Appl. 100591 [1982/84]; C.A. **100** [1984] No. 161797.

Dinsch, St., Klein, B., Boettger, T., VEB Synthesewerk Schwarzheide, Polyurethanes in Solution, Ger. [East] 160743 [1981/84]; C.A. **101** [1984] No. 152874.

Dirlikov, St. K., Schneider, C. J., Dow Chemical Co., Polyurethanes Derived from 1,4-Lactones of 3,6-Anhydro-2,3,4,5,6-Pentahydroxyhexanoic Acid, U.S. 4438226 [1983/84]; C.A. **100** [1984] No. 192955.

Dirlikov, St. K., Schneider, C. J., Dow Chemical Co., Polyurethanes from 1:4—3:6 Dianhydrohexitols, U.S. 4443563 [1983/84]; C.A. **101** [1984] No. 24146.

DiSalvo, A. L., Tsai, Ch. Ch., Stauffer Chemical Co., Modified Poly(Oxazolidone/Urethane) Compositions, Eur. Appl. 127976 [1983/84]; C.A. **102** [1985] No. 114601.

Ellerstein, St. M., Lee, S. A., Thiokol Corp., Radiation-Curable Coating for Photographic Laminate, Eur. Appl. 104057 [1982/84]; C.A. **100** [1984] No. 218977.

Esselborn, E., Fock, J., Th. Goldschmidt A.-G., Polymers with Pendant Long-Chain Alkyl and/or Aryl Realdues, Ger. Offen. 3230771 [1982/84]; C.A. **100** [1984] No. 175560.

Ford Motor Co., Crosslinkable Primers for Metal Coating, Japan. Kokai Tokkyo Koho 59136318 [1983/84]; C.A. **101** [1984] No. 232036.

Ford Motor Co., Crosslinked Flow Regulators for High-Solids Paints, Japan. Kokai Tokkyo Koho 59161431 [1983/84]; C.A. **102** [1985] No. 26542.

Frentzel, R. L., Rua, L., Pacheco, A., Olin Corp., Carboxylic Acid-Containing Monoether or Polyether Polyol Addition Products, and Polyurethane Prepolymers Derived Therefrom, Eur. Appl. 119349 [1983/84]; C.A. **102** [1985] No. 46724.

Frentzel, R. L., Rua, L., Pacheco, A. L., Olin Corp., Carboxylic Acid-Containing Polyurethane Prepolymers and Their Aqueous Dispersions, U.S. 4460738 [1983/84]; C.A. **101** [1984] No. 131354.

Frisch, K. C., Baumann, H., Schaum-Chemie Wilhelm Bauer GmbH und Co. KG, Polyurea Foams Prepared from Isocyanate, Water, and a Lower Alkanol, U.S. 4454251 [1982/84]; C.A. **101** [1984] No. 73781.

Fritsch, H., Koerner, G., Quilitsch, H., Schamberg, E., Th. Goldschmidt A.-G., Inorganic Waterproofing Materials, Ger. Offen. 3312911 [1983/84]; C.A. **102** [1985] No. 66503.

Fuji Fiber Glass, K. K., Glass Fiber-Reinforced Polyurethane Compositions in Reaction Injection Molding, Japan. Kokai Tokkyo Koho 5986636 [1982/84]; C.A. **101** [1984] No. 111984.

Fujikura Cable Works, Ltd., Corrosion-Inhibiting Coatings Containing Isocyanate Polymers, Japan. Kokai Tokkyo Koho 5956583 [1982/84]; C.A. **101** [1984] No. 41984.

Fujikura Cable Works, Ltd., Manufacture of Crosslinked Polyethylene-Insulated Electric Wires, Japan. Kokai Tokkyo Koho [1982/84]; C.A. **101** [1984] No. 112565.

Fujikura Cable Works, Ltd., Electric Cables, Japan. Kokai Tokkyo Koho 5999618 [1982/84]; C.A. **101** [1984] No. 173132.

Fukuda, T., Sakamoto, S., Toray Industries, Inc., Urethanized Acrylic Resin Material for Plastic Lens, U.S. 4487904 [1983/84]; C.A. **102** [1985] No. 114830.

Furukawa Electric Co., Ltd., Crosslinked Polyethylene-Insulated Wire, Japan. Kokai Tokkyo Koho 5949657 [1976/84]; C.A. **102** [1985] No. 168031.

Furukawa Electric Co., Ltd., Extrusion-Coating of Silane-Grafted Ethylene Polymers, Japan. Kokai Tokkyo Koho 59129130 [1983/84]; C.A. **101** [1984] No. 193259.

Furukawa Electric Co., Ltd., Extrusion of Silane-Crosslinkable Polyolefins, Japan. Kokai Tokkyo Koho 59169832 [1983/84]; C.A. **102** [1985] No. 25868.

Furukawa Electric Co., Ltd., Molding of Silane-Crosslinked Polyolefins, Japan. Kokai Tokkyo Koho 59232125 [1983/84]; C.A. **102** [1985] No. 151052.

General Electric Co., One-Shot Silicone Rubber Compositions Curable at Room Temperature, Japan. Kokai Tokkyo Koho 59142250 [1983/84]; C.A. **102** [1985] No. 47183.

Geuskens, G., Bastin, P., Debie, B., Gromen, M., Polart, J., Delaunois, G., Cableries et Corderies du Hainaut, S. A., Société Anon. Fabricable – Kabelfabrik Van Huizingen N.V., Polyolefins Crosslinkable by Grafting a Hydrolyzable Composition of Silane, Belg. 899506 [1984/84]; C.A. **102** [1985] No. 132690.

Gilch, H. G., von Voithenberg, H., Albert, K. H., Bostik Ltd., Adhesive Compositions, Brit. Appl. 2137638 [1983/84]; C.A. **102** [1985] No. 7968.

Gould, F. E., Johnston, Ch. W., Tyndale Plains-Hunter Ltd., Polyurethane Polyene Compositions, U.S. 44543309 [1980/84]; C.A. **101** [1984] No. 131762.

Haemer, L. F., Kimak, Th., Congoleum Corp., Differential Gloss Products, PCT Intern. Appl. 8400719 [1982/84]; C.A. **101** [1984] No. 8863.

Ham, N. M., Miyazaki, T., Ciba-Geigy Corp., Castable Polyurethane Systems, U.S. 4476292 [1984/84]; C.A. **102** [1985] No. 132934.

Harashima, A., Nakamura, T., Kishimoto, K., Toray Silicone Co., Ltd., Reactive Injection-Molded Products, Eur. Appl. 116966 [1983/84]; C.A. **101** [1984] No. 231481.

Hayase, Sh., Onishi, Y., Toshiba Corp., Photocurable Silicon Compound Composition, Eur. Appl. 104590 [1982/84]; C.A. **101** [1984] No. 8523.

Hicks, D. D., Celanese Corp., Epoxide Resin Aqueous Dispersant Comprising the Reaction Product of Diisocyanate, Diol and Polyether Glycol Monoether, U.S. 4446256 [1982/84]; C.A. **101** [1984] No. 56084.

Hirose, T., Isayama, K., Kawakubo, F., Takanoo, M., Yukimoto, S., Kanegafuchi Chemical Industry Co., Ltd., Eur. Appl. 108946 [1982/84]; C.A. **101** [1984] No. 132061.

Hitachi, Ltd., Metal Oxide Film Formation, Japan. Kokai Tokkyo Koho 59199506 [1983/84]; C.A. **102** [1985] No. 134365.

Hitachi Cable, Ltd., Resin Coating on Optical Fibers, Japan. Kokai Tokkyo Koho 5905203 [1982/84]; C.A. **100** [1984] No. 160934.

Hitachi Cable, Ltd., Crosslinked Polyolefin Wire Coatings, Japan. Kokai Tokkyo Koho 5980439 [1982/84]; C.A. **101** [1984] No. 112603.

Hitachi Cable, Ltd., Crosslinked Electric Insulators, Japan. Kokai Tokkyo Koho 59102931 [1982/84]; C.A. **101** [1984] No. 172648.

Hitachi Cable, Ltd., Crosslinking of Polyolefins, Japan. Kokai Tokkyo Koho 59133299 [1983/84]; C.A. **101** [1984] No. 193152.

Hitachi Cable, Ltd., Crosslinking of Polyolefins, Japan. Kokai Tokkyo Koho 59138233 [1983/84]; C.A. **101** [1984] No. 193268.

Hitachi Cable, Ltd., Recording Discs, Japan. Kokai Tokkyo Koho 59138248 [1983/84]; C.A. **102** [1985] No. 47005.

Hitachi Cable, Ltd., Fire-Resistant Silane-Grafted Polyolefins, Japan. Kokai Tokkyo Koho 59217750 [1983/84]; C.A. **102** [1985] No. 132968.

Hitachi Chemical Co., Ltd., Photoresist Composition, Japan. Kokai Tokkyo Koho 5946642 [1982/84]; C.A. **102** [1985] No. 36759.

Hitachi Chemical Co., Ltd., Cellular, Crosslinked Polyolefin Sheets, Japan. Kokai Tokkyo Koho 5974133 [1982/84]; C.A. **101** [1984] No. 112058.

Hitachi Chemical Co., Ltd., Photosensitive Resin Composition, Japan. Kokai Tokkyo Koho 59116652 [1982/84]; C.A. **102** [1985] No. 70229.

Hitachi Chemical Co., Ltd., Photosensitive Resin Composition, Japan. Kokai Tokkyo Koho 59120613 [1982/84]; C.A. **101** [1984] No. 172470.

Hitachi Chemical Co., Ltd., Low-Temperature-Curable Traffic Paints, Japan. Kokai Tokkyo Koho 59142261 [1983/84]; C.A. **102** [1985] No. 80471.

Hitachi Chemical Co., Ltd., Fire-Resistant Photosensitive Resin Compositions, Japan. Kokai Tokkyo Koho 59149917 [1983/84]; C.A. **102** [1985] No. 7701.

Hitachi Chemical Co., Ltd., Crosslinked Cellular Propylene Polymers, Japan. Kokai Tokkyo Koho 59230038 [1983/84]; C.A. **102** [1985] No. 168026.

Hoffman, D. K., Harris, R. F., Tefertiller, N. B., Rains, R. C., Dow Chemical Co., Stable Dispersions of Polymers in Polyfunctional Compounds Having a Plurality of Active Hydrogens and Polyurethanes Produced from Them, U.S. 4460715 [1979/84]; C.A. **101** [1984] No. 193125.

Holladay, H. Ph., Malon, R. F., Zampini, A., Monsanto Co., Multicomponent Membrane for Gas Separation, Eur. Appl. 107636 [1982/84]; C.A. **101** [1984] No. 92432.

Holubka, J. W., Ford Motor Co., Ltd., Ford-Werke A.-G., Ford (France) S.A., Crosslinkable Coatings, Eur. Appl. 114090 [1983/84]; C.A. **102** [1985] No. 8321.

Holubka, J. W., Ford Motor Co., Ltd., Ford-Werke A.-G., Ford (France) S.A., Coating Composition Comprising Bis-Diene Oligomers and Bis-Dieneophile Oligomers, Eur. Appl. 114091 [1983/84]; C.A. **102** [1985] No. 8334.

Hube, H., Metz, O., Anton, E., Plaschil, E., VEB Chemische Werke Buna, Flameproofing and Reinforcement of Low-Molecular-Weight Diene Polymers, Ger. [East] 160722 [1980/84]; C.A. **101** [1984] No. 74124.

Hughes, D. W., Dow Chemical Co., Thiabicyclononanediisocyanates and Polymers Made from Them, U.S. 4458061 [1982/84]; C.A. **101** [1984] No. 152498.

Idemitsu Kosan Co., Ltd., Coating Compositions, Japan. Kokai Tokkyo Koho 5936161 [1982/84]; C.A. **101** [1984] No. 56592.

Idemitsu Kosan Co., Ltd., Liquid Diene Polymer Compositions, Japan. Kokai Tokkyo Koho 5938224 [1982/84]; C.A. **101** [1984] No. 8548.

Idemitsu Kosan Co., Ltd., Polyurethane Compositions, Japan. Kokai Tokkyo Koho 5938225 [1982/84]; C.A. **101** [1984] No. 8547.

Idemitsu Kosan Co., Liquid Diene Copolymers, Japan. Kokai Tokkyo Koho 5975912 [1982/84]; C.A. **101** [1984] No. 131744.

Idemitsu Kosan Co., Ltd., Liquid Diene Type Polymers Containing Amino Groups, Japan. Kokai Tokkyo Koho 59124903 [1982/84]; C.A. **101** [1984] No. 153311.

Idemitsu Kosan Co., Ltd., Liquid Diene Copolymers, Japan. Kokai Tokkyo Koho 59159808 [1983/84]; C.A. **102** [1985] No. 47479.

Idemitsu Kosan Co., Ltd., Liquid Diene Polymer Compositions, Japan. Kokai Tokkyo Koho 59166523 [1983/84]; C.A. **102** [1985] No. 63430.

Idemitsu Kosan Co., Ltd., Liquid Diene Polymer Derivatives, Japan. Kokai Tokkyo Koho 59227905 [1983/84]; C.A. **102** [1985] No. 168411.

Ihrman, K. G., Ethyl Corp., 1-Methyl-2,6-Diamino-3-Isopropylbenzene, U.S. 4440952 [1982/84]; C.A. **101** [1984] No. 7782.

Institute for Production and Development Science, Silicone Compositions, Japan. Kokai Tokkyo Koho 5918759 [1982/84]; C.A. **101** [1984] No. 39670.

Jacobs, W., Parekh, G. G., Blank, W. J., American Cyanamid Co., Hydroxyalkyl Carbamate-Containing Resins for Cathodic Electrodeposition, U.S. 4484994 [1984/84]; C.A. **102** [1985] No. 80444.

Kalyanji, P. U., Minnesota Mining and Mfg. Co., Fluorochemical Polyesters and Fibrous Substrates Treated Therewith, Eur. Appl. 113217 [1982/84]; C.A. **102** [1985] No. 47288.

Kanegafuchi Chemical Industry Co., Ltd., Reactive Emulsions, Japan. Kokai Tokkyo Koho 5906219 [1982/84]; C.A. **101** [1984] No. 92933.

Kanegafuchi Chemical Industry Co., Ltd., Silicon-Containing Rubbers, Japan. Kokai Tokkyo Koho 5912932 [1982/84]; C.A. **100** [1984] No. 211418.

Kanegafuchi Chemical Industry Co., Ltd., Crosslinked Elastomers, Japan. Kokai Tokkyo Koho 5925837 [1982/84]; C.A. **101** [1984] No. 8520.

Kanegafuchi Chemical Industry Co., Ltd., Curable Elastic Compositions, Japan. Kokai Tokkyo Koho 59168014 [1983/84]; C.A. **102** [1985] No. 26131.

Kanegafuchi Chemical Industry Co., Ltd., High-Molecular-Weight Alkylene Oxide Polymers, Japan. Kokai Tokkyo Koho 59230024 [1984/84]; C.A. **102** [1985] No. 186467.

Kao Corp., Crosslinkable Polyester Hot-Melt Adhesives, Japan. Kokai Tokkyo Koho 59172573 [1983/84]; C.A. **102** [1985] No. 96632.

Kao Corp., Hot-Melt Adhesives, Japan. Kokai Tokkyo Koho 59172574 [1983/84]; C.A. **102** [1985] No. 47043.

Kao Corp., Hot-Melt Adhesives, Japan. Kokai Tokkyo Koho 59172575 [1983/84]; C.A. **102** [1985] No. 47042.

Kao Corp., Adhesives, Japan. Kokai Tokkyo Koho 59172577 [1983/84]; C.A. **102** [1985] No. 47041.

Kao Corp., Crosslinkable Polyester Hot-Melt Adhesives, Japan. Kokai Tokkyo Koho 59174673 [1983/84]; C.A. **102** [1985] No. 63284.

Katsuta Kako Co., Ltd., Stabilizers for Halogen-Containing Resins, Japan. Kokai Tokkyo Koho 5938250 [1982/84]; C.A. **101** [1984] No. 92144.

Keogh, M. J., Union Carbide Corp., Compositions Based on Water-Curable Silane-Modified Copolymers of Alkylene-Alkyl Acrylates, Can. 1177195 [1982/84]; C.A. **102** [1985] No. 133654.

Keogh, M. J., Union Carbide Corp., Water-Curable, Silane-Modified Alkyl Acrylate Copolymers, U.S. 4434272 [1978/84]; C.A. **100** [1984] No. 175875.

Keogh, M. J., Union Carbide Corp., Compositions Based on a Polysiloxane and an Organotitanate and Their Use in the Preparation of Water-Curable, Silane-Modified Alkene-Alkyl Acrylate Copolymers, U.S. 4446279 [1978/84]; C.A. **101** [1984] No. 74425.

Keogh, M. J., Wallace, S. L., Brown, G. D., Union Carbide Corp., Compositions Based on Alkylene-Alkyl Acrylate Copolymers and Silanol Condensation Catalysts, and Their Use in the Production of Covered Wires and Cable, Eur. Appl. 129121 [1983/84]; C.A. **102** [1985] No. 150375.

Klietsch, B. J., Kollmeier, H. J., Lammerting, H., Langenhagen, R. D., Th. Goldschmidt A.-G., Highly Elastic Cold-Hardenable Polyurethane Foams, Ger. Offen. 3234462 [1982/84]; C.A. **100** [1984] No. 193009.

Kobunshi Oyo Gijutsu Kenkyu Kumiai, Plastic Film Laminates with Low Gas Permeability, Japan. Kokai Tokkyo Koho 5959438 [1982/84]; C.A. **101** [1984] No. 92952.

Kobunshi Oyo Gijutsu Kenkyu Kumiai, Coated Vinyl Films with Low Gas Permeability, Japan. Kokai Tokkyo Koho 5959440 [1982/84]; C.A. **101** [1984] No. 73964.

Kobunshi Oyo Gijutsu Kenkyu Kumiai, Coated Poly(Ethylene Terephthalate) Films with Low Gas Permeability, Japan. Kokai Tokkyo Koho 5959441 [1982/84]; C.A. **101** [1984] No. 132048.

Kritchevsky, G. R., Celanese Corp., Isocyanate-Coupled Reinforced Oxymethylene Polymers Using an Improved Catalyst System, U.S. 4469842 [1983/84]; C.A. **101** [1984] No. 172485.

Kuraray Co., Ltd., Synthetic Tufted Carpets with Improved Resilience, Japan. Kokai Tokkyo Koho 59163485 [1983/84]; C.A. **101** [1984] No. 193630.

Lee, F. T. H., Green, J., FMC Corp., Polyurethane Coating Composition with a Curative Containing Polyhydroxyalkylphosphine Oxide, U.S. 4456742 [1983/84]; C.A. **101** [1984] No. 132675.

Lee, F. T. H., Green, J., FMC Corp., Polyurethane Adhesive Composition and a Curative System Containing a Poly(Hydroxyalkyl)Phosphine Oxide, U.S. 4456743 [1983/84]; C.A. **101** [1984] No. 153179.

Lee, K. W., Pfeffer, J. R., Piccirilli, R., Chang, W. H., PPG Industries, Inc., Film-Forming Resins Containing Alkoxy Silane Groups, U.S. 4429082 [1982/84]; C.A. **100** [1984] No. 211778.

Lee, S. A., Ellerstein, St. M., Morton Thiokol, Inc., Radiation-Curable Coating for Photographic Laminate, S. African 8401613 [1984/84]; C.A. **102** [1985] No. 195080.

Liang, T. M., Spitler, K. G., Mobay Chemical Corp., System for Production of Polyurethanes, U.S. 4448903 [1983/84]; C.A. **101** [1984] No. 39392.

Liang, T. M., Spitler, K. G., Mobay Chemical Corp., System for the Production of Polyurethanes, U.S. 4477602 [1983/84]; C.A. **102** [1984] No. 7724.

Liechti, W., Togo Management A.-G., Moisture-Reactive Single-Component Polyurethane Adhesive Sealing Compositions, Ger. Offen. 3416773 [1984/84]; C.A. **102** [1985] No. 80521.

Lin, I. S., Gromelski, St. J., Werner, J., Brown, M. J., Chakrabarti, P. M., GAF Corp., Diamine-Diol Chain Extender Blends for Reaction-Injection Molding, Eur. Appl. 102203 [1982/84]; C.A. **100** [1984] No. 211433.

Loch, W., Schupp, E., Osterloh, R., Ahlers, K., BASF Farben und Fasern A.-G., Varnish Vehicle Self-Crosslinking in Heat and Its Use, Ger. Offen. 3311517 [1983/84]; C.A. **102** [1985] No. 47424.

Loch, W., Schupp, E., Osterloh, R., Ahlers, K., BASF Farben und Fasern A.-G., Thermosetting Coating and Its Use, Ger. Offen. 3311518 [1983/84]; C.A. **102** [1985] No. 47426.

Lock, M. R., Frisch, K. C., Dow Chemical Co., Urethane Coatings from Aliphatic Aromatic Diisocyanates, U.S. 4465713 [1982/84]; C.A. **101** [1984] No. 193790.

Mabuchi, A., Toyoda Gosei Co., Ltd., Nonyellowing, Semirigid Polyurethane Foam, Ger. Offen. 3409402 [1983/84]; C.A. **102** [1985] No. 79935.

Malysa, M., Stanko, K., Rak, R., Grohs, R., Centralne Laboratorium Przemyslu Obuwniczego, Molds and Patterns of Gypsum-Based Molding Sand Mix for Pressureless Casting of Low-Melting Alloys Capable of Reproducing Mold Cavity Surface Configurations with a High Accuracy, Pol. 126805 [1980/84]; C.A. **102** [1985] No. 136306.

Matsumoto, T., Diethylene Glycol Bis(Alkyl Phthalate) Plasticizers, Japan. Kokai Tokkyo Koho 5915436 [1982/84]; C.A. **100** [1984] No. 193040.

Matsushita, Electric Works, Ltd., Radical-Polymerizable Prepolymers, Japan. Kokai Tokkyo Koho 5904615 [1982/84]; C.A. **101** [1984] No. 8825.

Matsushita Electric Works, Ltd., Photocurable Urethane Acrylate Prepolymers, Japan. Kokai Tokkyo Koho 5974112 [1982/84]; C.A. **101** [1984] No. 74474.

Matsushita Electric Works, Ltd., Photocurable Urethane Acrylate Prepolymers, Japan. Kokai Tokkyo Koho 5974113 [1982/84]; C.A. **101** [1984] No. 92961.

Matsushita Electric Works, Ltd., Radically Curable Polyurethane Prepolymers, Japan. Kokai Tokkyo Koho 5974116 [1982/84]; C.A. **101** [1984] No. 132647.

Matsushita Electric Works, Ltd., Radically Polymerizable Prepolymers, Japan. Kokai Tokkyo Koho 59179619 [1983/84]; C.A. **102** [1985] No. 47499.

Matsushita Electric Works, Ltd., Radically Polymerizable Prepolymers, Japan. Kokai Tokkyo Koho 59184218 [1983/84]; C.A. **102** [1985] No. 63315.

Matsushita Electric Works, Ltd., Light-Curable Urethane Resins, Japan. Kokai Tokkyo Koho 59184219 [1983/84]; C.A. **102** [1985] No. 63765.

Matsushita Electric Works, Ltd., Photocurable Polyurethanes, Japan. Kokai Tokkyo Koho 59221322 [1983/84]; C.A. **102** [1985] No. 133712.

Matsushita Electric Works, Ltd., Kyoto Densen K. K., Electrically Insulated Wire, Japan. Kokai Tokkyo Koho 59134509 [1983/84]; C.A. **102** [1985] No. 7892.

Metikes, M. M. M. A. W. M., Molding Composition, Brit. Appl. 2127838 [1982/84]; C.A. **101** [1984] No. 74012.

Meyer, F., Bergwerksverband GmbH, Sealing and Strengthening Water-Bearing Geological Formations by Polyurethane-Resin-Forming Compositions, U.S. 4454252 [1975/84]; C.A. **101** [1984] No. 96698.

Miller, L. I., du Pont de Nemours, E. I., and Co., Additive for Alkyd Resin Coating Compositions, U.S. 4442256 [1982/84]; C.A. **101** [1984] No. 8794.

Mitsubishi Chemical Industries Co., Ltd., Water-Borne Coating Compositions, Japan. Kokai Tokkyo Koho 5920359 [1982/84]; C.A. **101** [1984] No. 25099.

Mitsubishi Chemical Industries Co., Ltd., Polyurethane Coating Materials, Japan. Kokai Tokkyo Koho 5927967 [1982/84]; C.A. **101** [1984] No. 25097.

Mitsubishi Monsanto Chemical Co., Powdered PVC Compositions, Japan. Kokai Tokkyo Koho 5974146 [1982/84]; C.A. **101** [1984] No. 131828.

Mitsubishi Petrochemical Co., Ltd., Crosslinked Adhesives, Japan. Kokai Tokkyo Koho 5924767 [1982/84]; C.A. **101** [1984] No. 73931.

Mitsubishi Petrochemical Co., Ltd., Crosslinked Adhesive Films for Wood, Japan. Kokai Tokkyo Koho 5924769 [1982/84]; C.A. **101** [1984] No. 73932.

Mitsubishi Petrochemical Co., Ltd., Modified Propylene Polymers, Japan. Kokai Tokkyo Koho 5936115 [1982/84]; C.A. **101** [1984] No. 7829.

Mitsubishi Petrochemical Co., Ltd., Phosphorus-Containing Acrylate Resin Modified with Urethane, Japan. Kokai Tokkyo Koho 59187020 [1983/84]; C.A. **103** [1985] No. 55481.

Mitsubishi Petrochemical Co., Ltd., Moldable Propylene Polymer Foam Beads, Japan. Kokai Tokkyo Koho 59223733 [1983/84]; C.A. **102** [1985] No. 150394.

Mitsubishi Rayon Co., Ltd., Epoxy Resin Compositions, Japan. Kokai Tokkyo Koho 59219320 [1983/84]; C.A. **102** [1985] No. 150411.

Mitsubishi Rayon Co., Ltd., Fiber-Reinforced Thermosetting Resin Molding Compositions, Japan. Kokai Tokkyo Koho 59219347 [1983/84]; C.A. **102** [1985] No. 133063.

Mitsui-Nisso Corp., Thermoplastic Polyurethanes, Japan. Kokai Tokkyo Koho 5904614 [1982/84]; C.A. **100** [1984] No. 192937.

Mitsui-Nisso Corp., Polyurethanes, Japan. Kokai Tokkyo Koho 59166522 [1983/84]; C.A. **102** [1985] No. 63749.

Mitsui Toatsu Chemicals, Inc., Vinyl Chloride Resin Compositions, Japan. Kokai Tokkyo Koho 5920342 [1982/84]; C.A. **101** [1984] No. 24543.

Mitsui Toatsu Chemicals, Inc., Copolymers for High-Refractive-Index Lenses, Japan. Kokai Tokkyo Koho 5996109 [1982/84]; C.A. **101** [1984] No. 172330.

Mitsui Toatsu Chemicals, Inc., PVC Agricultural Cover Films with Low Toxicity, Japan. Kokai Tokkyo Koho 59145234 [1983/84]; C.A. **101** [1984] No. 231675.

Monsanto Co., Hollow-Fiber Membranes for Gas Separation, Japan. Kokai Tokkyo Koho 5973006 [1982/84]; C.A. **101** [1984] No. 39577.

Moriarity, Th. C., PPG Industries, Inc., Curable Compositions Containing Urethane Curing Agents, U.S. 4452930 [1983/84]; C.A. **101** [1984] No. 132593.

Moriarity, Th. C., PPG Industries, Inc., Low Temperature Urethane Curing Agents, U.S. 4452963 [1983/84]; C.A. **101** [1984] No. 132594.

Mueller, K. F., Heiber, S. J., Plankl, W. L., Ciba-Geigy A.-G., Silicon-Containing Polymers Having an Increased Oxygen Permeability, Eur. Appl. 109355 [1982/84]; C.A. **101** [1984] No. 131839.

Murphy, J. R., Babiec, J. S., Meyers, J. A., Atlantic Richfield Co., Polyurethane Foam Game Ball and Composition, U.S. 4454253 [1983/84]; C.A. **101** [1984] No. 74156.

Narisawa, Sh., Funaki, M., Kojima, H., Kuga, K., Asahi Glass Co., Ltd., Polyurethane Elastomer by Reaction Injection Molding, Eur. Appl. 112557 [1982/84]; C.A. **101** [1984] No. 112240.

Natarajan, K., Zimmerman, D., Celanese Corp., Isocyanate-Coupled Reinforced Polyoxymethylenes, U.S. 4480071 [1983/84]; C.A. **102** [1985] No. 25744.

National House Industrial Co., Ltd., Meisei Churchill Co., Ltd., One-Component Liquid Sealants, Japan. Kokai Tokkyo Koho 59196383 [1983/84]; C.A. **102** [1985] No. 133733.

Neos Co., Ltd., Internal Release Agents for Polyurethane Reaction Injection Molding, Japan. Kokai Tokkyo Koho 59213716 [1983/84]; C.A. **102** [1985] No. 205050.

Neto, A. P., Dow Quimica S. A., Crosslinkable Composition, from Polyol Polyethers with Increased Molecular Weight and Crosslinking Agents, Producing Flexible Polyurethane Foam and Quilts, Braz. Pedido 8306762 [1983/84]; C.A. **102** [1985] No. 46746.

Nippon Carbide Industries Co., Inc., Ultraviolet Radiation-Shielding Agricultural Vinyl Films, Japan. Kokai Tokkyo Koho 5941346 [1982/84]; C.A. **101** [1984] No. 73939.

Nippon Carbide Industries Co., Inc., Resins for Sealing, Japan. Kokai Tokkyo Koho 5964614 [1982/84]; C.A. **101** [1984] No. 132686.

Nippon Kayaku Co., Ltd., Urethane Poly(meth)acrylates, Japan. Kokai Tokkyo Koho 59217718 [1983/84]; C.A. **102** [1985] No. 222255.

Nippon Kayaku Co., Ltd., Ester Diol Urethane Poly(meth)acrylates, Japan. Kokai Tokkyo Koho 59232114 [1983/84]; C.A. **102** [1985] No. 151051.

Nippon Oils and Fats Co., Ltd., Thermosetting Resin Coating Compositions, Japan. Kokai Tokkyo Koho 5996177 [1982/84]; C.A. **102** [1985] No. 47464.

Nippon Oils and Fats Co., Ltd., Cationic Electrodeposition Paint Compositions, Japan. Kokai Tokkyo Koho 59135269 [1983/84]; C.A. **102** [1985] No. 80429.

Nippon Oils and Fats Co., Ltd., Thermosetting Resins for Dispersion of Pigments, Japan. Kokai Tokkyo Koho 59174620 [1983/84]; C.A. **102** [1985] No. 133682.

Nippon Oils and Fats Co., Ltd., Cationic Electrodeposited Paint Compositions, Japan. Kokai Tokkyo Koho 59176365 [1983/84]; C.A. **102** [1985] No. 47462.

Nippon Paint Co., Ltd., Primer Compositions for Cationic Electrodeposition, Japan. Kokai Tokkyo Koho 5911378 [1982/84]; C.A. **100** [1984] No. 176572.

Nippon Paint Co., Ltd., Coating Materials, Japan. Kokai Tokkyo Koho 5945359 [1982/84]; C.A. **101** [1984] No. 112512.

Nippon Paint Co., Ltd., Cationic Electrophoretic Coating Compositions, Japan. Kokai Tokkyo Koho 59199779 [1983/84]; C.A. **102** [1985] No. 151014.

Nippon Soda Co., Ltd., Polyurethane Elastomers, Japan. Kokai Tokkyo Koho 59117522 [1982/84]; C.A. **102** [1985] No. 8043.

Nippon Soda Co., Ltd., Polyurethane Elastomers, Japan. Kokai Tokkyo Koho 59210928 [1983/84]; C.A. **102** [1985] No. 168171.

Nippon Synthetic Chemical Industry Co., Ltd., Room-Temp.-Curable Siloxanes, Japan. Kokai Tokkyo Koho 59184260 [1983/84]; C.A. **102** [1985] No. 132998.

Nippon Synthetic Chemical Industry Co., Ltd., Adhesive Compositions, Japan. Kokai Tokkyo Koho 59223774 [1983/84]; C.A. **102** [1985] No. 186325.

Nippon Telegraph and Telephone Public Corp., Idemitsu Kosan Co., Ltd., Curable Liquid Resin Compositions for Optical Fiber Coating Materials, Japan. Kokai Tokkyo Koho 59227914 [1983/84]; C.A. **102** [1985] No. 168412.

Nippon Telegraph and Telephone Public Corp., Idemitsu Kosan Co., Ltd., Curable Liquid Resin Compositions for Optical Fiber Coating Materials, Japan. Kokai Tokkyo Koho 59227915 [1983/84]; C.A. **102** [1985] No. 168413.

Nishinippon Electric Wire and Cable Co., Ltd., Heat-Shrinkable Polyolefin Tubes, Japan. Kokai Tokkyo Koho 59215812 [1983/84]; C.A. **102** [1985] No. 114848.

Nissan Motor Co., Ltd., Binders for Polydiene-Type Composite Propellants, Japan. Kokai Tokkyo Koho 5978996 [1982/84]; C.A. **101** [1984] No. 113386.

Nissan Motor Co., Ltd., Thickeners for Polydiene Composite Propellants, Japan. Kokai Tokkyo Koho 59128291 [1983/84]; C.A. **102** [1985] No. 48268.

Nitto Electric Industrial Co., Ltd., Protective Covering Films, Japan. Kokai Tokkyo Koho 5952616 [1982/84]; C.A. **101** [1984] No. 92909.

Nitto Electric Industrial Co., Ltd., Waterproofing Sheets Using Elastic Sealants, Japan. Kokai Tokkyo Koho 5958076 [1982/84]; C.A. **101** [1984] No. 153606.

Nitto Electric Industrial Co., Ltd., Radiochemically-Curable Release Agents, Japan. Kokai Tokkyo Koho 5959776 [1982/84]; C.A. **101** [1984] No. 132045.

Nitto Electric Industrial Co., Ltd., Elastic Sealant, Japan. Kokai Tokkyo Koho 59176378 [1983/84]; C.A. **102** [1985] No. 133729.

Nitto Electric Industrial Co., Ltd., Elastic Sealant, Japan. Kokai Tokkyo Koho 59176379 [1983/84]; C.A. **102** [1985] No. 133730.

Nitto Electric Industrial Co., Ltd., Elastic Sealant, Japan. Kokai Tokkyo Koho 59176380 [1983/84]; C.A. **102** [1985] No. 133731.

Nitto Electric Industrial Co., Ltd., Elastic Sealant, Japan. Kokai Tokkyo Koho 59176381 [1983/84]; C.A. **102** [1985] No. 133732.

Nitto Electric Industrial Co., Ltd., Photosetting Hydrosol Compositions, Japan. Kokai Tokkyo Koho 59179612 [1983/84]; C.A. **102** [1985] No. 115239.

Otsuka Chemical Co., Ltd., Siloxane Compositions, Japan. Kokai Tokkyo Koho 5912964 [1982/84]; C.A. **100** [1984] No. 211444.

Otsuka Chemical Co., Ltd., Fire-Resistant Thermal Insulating Sheets, Japan. Kokai Tokkyo Koho 5926987 [1982/84]; C.A. **101** [1984] No. 42641.

Otsuka Chemical Co., Ltd., Fireproof Thermal Insulation Sheets, Japan. Kokai Tokkyo Koho 5935938 [1982/84]; C.A. **101** [1984] No. 8399.

Paxton, L. D., Madison, R. A., Dunbar, J. E., Dow Chemical Co., 8-Quinolinyl Carbamates and Their Use as Urinary Tract Antimicrobials, U.S. 4472404 [1982/84]; C.A. **102** [1985] No. 7099.

Peerman, D., Henkel Corp., Polyester Polyols for Polyurethanes from Dimethylol Tricyclo Compounds and Caprolactone, U.S. 4438225 [1983/84]; C.A. **100** [1984] No. 193350.

Piccirilli, R., Chang, W. H., PPG Industries, Inc., Polymeric Organo Functional Allanes as Reactive Modifying Materials, U.S. 4468492 [1983/84]; C.A. **101** [1984] No. 193854.

Plum, H., Hoechst A.-G., Polymers Containing Primary Amino Groups and Their Use, Ger. Offen. 3229047 [1982/84]; C.A. **100** [1984] No. 140904.

Pogozelski, V. F., Schmidle, C. J., Congoleum Corp., Gloss Retentive Surface Covering, Can. 1175717 [1981/84]; C.A. **102** [1985] No. 115281.

Pollak, R. B., Lee Pharmaceuticals, Endodontic Filling and Sealing Composition, U.S. 4449938 [1982/84]; C.A. **101** [1984] No. 116770.

Polythetics, Inc., Dentures from Soft Polyurethane and Hard Acrylic Polymers, Japan. Kokai Tokkyo Koho 5995209 [1982/84]; C.A. **101** [1984] No. 137067.

Rasshofer, W., Schaefer, H., Paul, R., Beuth, J., Bayer A.-G., Foam Moldings with a Closed Outer Skin, Ger. Offen. 3231399 [1982/84]; C.A. **100** [1984] No. 211090.

Reichel, C. J., Levis, W. W., Hartmann, R. J., BASF Wyandotte Corp., Flexible Polyurethane Foams Having High Indentation Load Deflection Prepared from Polyol Blends, Eur. Appl. 116309 [1983/84]; C.A. **101** [1984] No. 231478.

Reichel, C. J., Patton, J. T., Narayan, T., BASF Wyandotte Corp., Haloalkoxymethylmelamine Polymer Fire Retardants for Polyurethanes, Brit. Appl. 2133412 [1982/84]; C.A. **101** [1984] No. 212216.

Reichel, C. J., Patton, J. T., Narayan, T., BASF Wyandotte Corp., Aminoplast Resin Dispersions for Flame-Resistant Polyurethanes, U.S. 4454254 [1982/84]; C.A. **101** [1984] No. 92251.

Saitama Gomu Kogyo K. K., Water-Swellable Sealing Compositions, Japan. Kokai Tokkyo Koho 59164332 [1983/84]; C.A. **102** [1985] No. 63775.

Salisbury, W. C., Ex-Cell-O Corp., Polyurethane RIM System, Eur. Appl. 102709 [1982/84]; C.A. **101** [1984] No. 73895.

Samigullin, F. K., Kafengauz, I. M., Korzyuk, E. L., Frolov, Yu. M., Gladkovskii, G. A., Nepyshnevskii, V. M., Simonovskii, F. I., All-Union Scientific-Research Institute of Synthetic Resins, Linear Polyurethanes, U.S.S.R. 1085988 [1982/84]; C.A. **101** [1984] No. 56002.

Sanyo Chemical Industries, Ltd., Skin-Forming Agents for Polyurethane Foams, Japan. Kokai Tokkyo Koho 5918723 [1982/84]; C.A. **101** [1984] No. 8354.

Sanyo Chemical Industries, Ltd., Polyurethanes, Japan. Kokai Tokkyo Koho 59117519 [1982/84]; C.A. **101** [1984] No. 172341.

Sanyo Chemical Industries, Ltd., Polyurethanes, Japan. Kokai Tokkyo Koho 59117520 [1982/84]; C.A. **101** [1984] No. 152883.

Sanyo Chemical Industries, Ltd., Urethane-Vinyl Resins, Japan. Kokai Tokkyo Koho 59232110 [1983/84]; C.A. **103** [1985] No. 7803.

Sanyo Chemical Industries, Ltd., Inoue MTP K. K., In-Mold Coatings for Polyurethane Foams, Japan. Kokai Tokkyo Koho 5918733 [1982/84]; C.A. **100** [1984] No. 193723.

Schaefer, W., Meiners, H. J., Seel, K., Reichmann, W., Wagner, K., Findeisen, K., Bayer A.-G., Microcellular Molding Compositions, Ger. Offen. 3242925 [1982/84]; C.A. **101** [1984] No. 92086.

Schaerer, A. J., Zeley, A., Amrotex, A. G., Flexible, Closed-Cell Foamed Insulation from a Polyolefin, Ger. Offen. 3310295 [1983/84]; C.A. **102** [1985] No. 25928.

Schatz, M., Dolezal, V., Vondracek, P., Protective Insulating Material for Pipelines, Czech. 224161 [1980/84]; C.A. **101** [1984] 172716.

Sekisui Chemical Co., Ltd., Crosslinkable Alkoxysilane Compositions, Japan. Kokai Tokkyo Koho 5971364 [1982/84]; C.A. **101** [1984] No. 172384.

Sekisui Chemical Co., Ltd., Adhesive Tapes, Japan. Kokai Tokkyo Koho 5996181 [1982/84]; C.A. **101** [1984] No. 231618.

Sekisui Chemical Co., Ltd., Adhesive Tapes, Japan. Kokai Tokkyo Koho 5998184 [1982/84]; C.A. **101** [1984] No. 231620.

Showa Electric Wire and Cable Co., Ltd., Fire-Retardant Silane-Crosslinked Polyolefin Compositions, Japan. Kokai Tokkyo Koho 5964653 [1982/84]; C.A. **101** [1984] No. 131816.

Showa Union Gosei Co., Ltd., Flame-Resistant and Modified Polyurethane Foams, Japan. Kokai Tokkyo Koho 59120616 [1982/84]; C.A. **101** [1984] No. 172469.

Simroth, D. W., Critchfield, F. E., Shook, E. G., Union Carbide Corp., Polymer/Polyol Compositions Having Improved Combustion Resistance, U.S. 4463102 [1982/84]; C.A. **101** [1984] No. 193127.

Singer, W., Versfelt, Ch. C., Troy Chemical Corp., Haloalkynes and Their Use as Fungicides, Brit. Appl. 2138292 [1983/84]; C.A. **102** [1985] No. 57826.

Stankiewicz, A., Schallschluck GmbH und Co. K.G., Polyurethane Foam with Noise-Reducing Properties, Ger. Offen. 3316652 [1983/84]; C.A. **102** [1985] No. 186255.

Starnes, W. H., Bell Telephone Laboratories, Inc., Stabilization of Vinyl Chloride Polymers, U.S. 4443586 [1977/84]; C.A. **101** [1984] No. 39383.

Stutz, H., Eckert, G., BASF A.-G., Crosslinked Polyurethane Ionomer Dispersions, Ger. Offen. 3233605 [1982/84]; C.A. **101** [1984] No. 56566.

Sumika Color Co., Ltd., Polyurethane Molded Materials, Japan. Kokai Tokkyo Koho 5971359 [1982/84]; C.A. **101** [1984] No. 172382.

Sumika Color Co., Ltd., Polyurethane Molded Materials, Japan. Kokai Tokkyo Koho 5971360 [1982/84]; C.A. **101** [1984] No. 172383.

Sumitomo Bakelite Co., Ltd., Extrusion of Silane-Grafted Polyolefins, Japan. Kokai Tokkyo Koho 59102930 [1982/84]; C.A. **101** [1984] No. 132022.

Sumitomo Bakelite Co., Ltd., Plastic Pipe for Hot-Water Heating Systems, Japan. Kokai Tokkyo Koho 59155682 [1983/84]; C.A. **102** [1985] No. 25995.

Sumitomo Bakelite Co., Ltd., Composite Laminates for Printed Circuit Boards, Japan. Kokai Tokkyo Koho 59209859 [1983/84]; C.A. **102** [1985] No. 96698.

Sumitomo Electric Industries, Ltd., Electric Wire Coated with Crosslinked Polyolefins, Japan. Kokai Tokkyo Koho 5929922 [1976/84]; C.A. **101** [1984] No. 231667.

Sekisui Chemical Co., Ltd., Polyurethane Adhesive Compositions, Japan. Kokai Tokkyo Koho 59105069 [1982/84]; C.A. **102** [1985] No. 7858.

Sekisui Chemical Co., Ltd., Catalysts for Crosslinking Alkoxysilane-Grafted Polymers, Japan. Kokai Tokkyo Koho 59122538 [1982/84]; C.A. **102** [1985] No. 25632.

Sekisui Chemical Co., Ltd., Urethane-Type Adhesives, Japan. Kokai Tokkyo Koho 59129279 [1983/84]; C.A. **101** [1984] No. 231642.

Sekisui Chemical Co., Ltd., Urethane-Type Adhesives, Japan. Kokai Tokkyo Koho 59131683 [1983/84]; C.A. **101** [1984] No. 231639.

Sekisui Chemical Co., Ltd., Polyurethane Adhesives, Japan. Kokai Tokkyo Koho 59230076 [1983/84]; C.A. **102** [1985] No. 168007.

Sekisui Chemical Co., Ltd., Crosslinking Polyethylene, Japan. Kokai Tokkyo Koho 59232108 [1983/84]; C.A. **102** [1985] No. 185996.

Sekisui Kaseihin Kogyo, K. K., Silanol-Crosslinked Cellular Polyethylene, Japan. Kokai Tokkyo Koho 5958038 [1982/84]; C.A. **101** [1984] No. 39474.

Sekisui Kaseihin Kogyo, K. K., Crosslinked Resin Foams, Japan. Kokai Tokkyo Koho 5962642 [1982/84]; C.A. **101** [1984] No. 39381.

Shimizu, Ch., Toshiba Silicone Co., Ltd., Dustproof Film Forming Material, U.S. 4476278 [1984/84]; C.A. **102** [1985] No. 8320.

Shin-Etsu Chemical Industry Co., Ltd., Room-Temperature-Vulcanizable Silicone Rubber Compositions, Japan. Kokai Tokkyo Koho 5901566 [1982/84]; C.A. **100** [1984] No. 158057.

Shin-Etsu Chemical Industry Co., Ltd., Room-Temperature-Surface-Hardenable Organopolysiloxane Compositions, Japan. Kokai Tokkyo Koho 5989356 [1982/84]; C.A. **101** [1984] No. 193867.

Shin-Etsu Chemical Industry Co., Ltd., Thickening Agents for Liquid Siloxane Compositions, Japan. Kokai Tokkyo Koho 59166560 [1983/84]; C.A. **102** [1985] No. 79784.

Showa Electric Wire and Cable Co., Ltd., Adhesive Primer for Urethane Rubber, Japan. Kokai Tokkyo Koho 5911382 [1982/84]; C.A. **101** [1984] No. 92396.

Sumitomo Electric Industries, Ltd., Electric Wire Coated with Crosslinked Polyolefins, Japan. Kokai Tokkyo Koho 5929923 [1976/84]; C.A. **101** [1984] No. 231666.

Sunstar Giken K. K., Room Temperature-Curable Resin Composition for Elastic Sealing Agent, Japan. Kokai Tokkyo Koho 5978211 [1982/84]; C.A. **101** [1984] No. 172337.

Sunstar Giken K. K., Moisture-Curable Adhesive Compositions, Japan. Kokai Tokkyo Koho 59113076 [1982/84]; C.A. **102** [1985] No. 79915.

Sunstar Giken K. K., Flexible PVC Laminates, Japan. Kokai Tokkyo Koho 59232854 [1983/84]; C.A. **102** [1985] No. 221821.

Takeda Chemical Industries, Ltd., Isocyanate Terminated Urethane Prepolymer Compositions, Japan. Kokai Tokkyo Koho 5925812 [1983/84]; C.A. **100** [1984] No. 211817.

Takeda Chemical Industries, Ltd., Primer for Polyolefin Resin, Japan. Kokai Tokkyo Koho 59124937 [1983/84]; C.A. **102** [1985] No. 63039.

Tateosian, L. H., Royer, J. R., Eden, G. T., Dentsply Research and Development Corp., U.S. 4425094 [1981/84]; C.A. **100** [1984] No. 161826.

Taylor, R. P., Dewhurst, J. E., Abouzahr, S. M., Mobay Chemical Corp., Process and Composition for Urethane Elastomer Moldings, U.S. 4442235 [1983/84]; C.A. **101** [1984] No. 56326.

Tetenbaum, M. T., Crowley, B. C., NL Industries, Inc., Modified Polyurethane Dispersible in Water, Belg. 899570 [1983/84]; C.A. **102** [1985] No. 25587.

Thayer, L. E., Dow Corning Corp., Siloxane-Poly(Vinyl Alcohol) Coatings, Eur. Appl. 117607 [1983/84]; C.A. **101** [1984] No. 212837.

Toa Gosei Chemical Industry Co., Ltd., Ionizing Radiation-Curable Magnetic Coating Materials, Japan. Kokai Tokkyo Koho 59129267 [1983/84]; C.A. **101** [1984] No. 173217.

Toho Chemical Industry Co., Ltd., Aqueous Resin Coating Materials, Japan. Kokai Tokkyo Koho 59157101 [1983/84]; C.A. **102** [1985] No. 47430.

Tokai Rubber Industries, Ltd., Capsules for Fixing, Japan. Kokai Tokkyo Koho 59180000 [1983/84]; C.A. **102** [1985] No. 80526.

Toray Industries, Inc., Processing of Photoimaging Laminates, Japan. Kokai Tokkyo Koho 5917552 [1982/84]; C.A. **102** [1985] No. 36780.

Toray Industries, Inc., Resin for Plastic Lenses with High Refractive Indexes, Japan. Kokai Tokkyo Koho 59133211 [1983/84]; C.A. **101** [1984] No. 193037.

Toray Silicone Co., Ltd., Internal Mold Release Agents for Polyurethane Reaction Injection Molding, Japan. Kokai Tokkyo Koho 5938044 [1982/84]; C.A. **101** [1984] No. 8325.

Toshiba Chemical Products Co., Ltd., Conductive Paste, Japan. Kokai Tokkyo Koho 59206459 [1983/84]; C.A. **102** [1985] No. 222313.

Toshiba Corp., Tin Oxide or Hydroxide Cation-Exchanger for Radioactive Cobalt in Nuclear Reactor Cooling Water, Japan. Kokai Tokkyo Koho 5987050 [1982/84]; C.A. **101** [1984] No. 179683.

Toshiba Silicone Co., Ltd., Dust-Preventing Coating Compositions, Japan. Kokai Tokkyo Koho 59105057 [1982/84]; C.A. **101** [1984] No. 173206.

Toshiba Silicone Co., Ltd., Polysiloxane Compositions Curable at Room Temperature, Japan. Kokai Tokkyo Koho 59124953 [1983/84]; C.A. **102** [1985] No. 8075.

Toshiba Silicone Co., Ltd., Antidusting Coating Materials, Japan. Kokai Tokkyo Koho 59126470 [1983/84]; C.A. **101** [1984] No. 212806.

Toshiba Silicone Co., Ltd., Takamatsu Yushi K. K., Method for Collecting Glass Fiber, Japan. Kokai Tokkyo Koho 5957931 [1982/84]; C.A. **101** [1984] No. 135769.

Toyo Cloth Co., Ltd., Leather Substitutes by Transfer Process, Japan. Kokai Tokkyo Koho 5942111 [1976/84]; C.A. **102** [1985] No. 133143.

Toyoda Gosei Co., Ltd., Adhesives for PVC-Metal Composites, Japan. Kokai Tokkyo Koho 5941374 [1982/84]; C.A. **101** [1984] No. 8404.

Toyo Ink Mfg. Co., Ltd., Overprint Varnishes, Japan. Kokai Tokkyo Koho 59168074 [1983/84]; C.A. **102** [1985] No. 63790.

Toyo Kohan Co., Ltd., Tokyo, Quality Control of Red Oxide in Manufacture of Weather-Resistant PVC, Japan. Kokai Tokkyo Koho 59202245 [1983/84]; C.A. **102** [1985] No. 115235.

Toyo Rubber Chemical Industry Co., Ltd., Polyurethane Foams, Japan. Kokai Tokkyo Koho 5945133 [1982/84]; C.A. **101** [1984] No. 92164.

Toyo Rubber Industry Co., Ltd., Puncture-Sealing Compositions for Tires, Japan. Kokai Tokkyo Koho 5915949 [1974/84]; C.A. **101** [1984] No. 112269.

Toyo Rubber Industry Co., Ltd., Polyurethanes, Japan. Kokai Tokkyo Koho 59102917 [1982/84]; C.A. **101** [1984] No. 173214.

Toyo Rubber Industry Co., Ltd., Photosensitive Composition, Japan. Kokai Tokkyo Koho 59111638 [1982/84]; C.A. **102** [1985] No. 36783.

Toyo Rubber Industry Co., Ltd., Photosensitive Coating Compositions, Japan. Kokai Tokkyo Koho 59113069 [1982/84]; C.A. **102** [1985] No. 63729.

Toyo Rubber Industry Co., Ltd., Epoxy-Modified Polyurethane, Japan. Kokai Tokkyo Koho 59179517 [1983/85]; C.A. **102** [1985] No. 132997.

Toyo Rubber Industry Co., Ltd., Toyobo Co., Ltd., Preparation of Modified Polyols, Japan. Kokai Tokkyo Koho 59152290 [1983/84]; C.A. **102** [1985] No. 47169.

Toyo Rubber Industry Co., Ltd., Toyobo Co., Ltd., Two-Component Coating Compositions, Japan. Kokai Tokkyo Koho 59152962 [1983/84]; C.A. **102** [1985] No. 26509.

Toyo Soda Mfg. Co., Ltd., Tear Strength Improvement of Vinyl Chloride Graft Copolymers, Japan. Kokai Tokkyo Koho 59179552 [1983/84]; C.A. **102** [1985] No. 150346.

Toyo Soda Mfg. Co., Ltd., Vinyl Chloride Graft Copolymer Compositions for Films and Sheets, Japan. Kokai Tokkyo Koho 59179553 [1983/84]; C.A. **102** [1985] No. 96502.

Toyo Soda Mfg. Co., Ltd., Improving the Tear Strength of Vinyl Chloride Graft Copolymers, Japan. Kokai Tokkyo Koho 59184248 [1983/84]; C.A. **102** [1985] No. 96520.

Toyobo Co., Ltd., Magnetic Recording Medium, Japan. Kokai Tokkyo Koho 59154633 [1983/84]; C.A. **102** [1985] No. 54962.

Toyobo Co., Ltd., Magnetic Recording, Japan. Kokai Tokkyo Koho 59154634 [1983/84]; C.A. **102** [1985] No. 54963.

Toyota Motor Co., Ltd., Polyurethane Foams, Japan. Kokai Tokkyo Koho 5980420 [1982/84]; C.A. **101** [1984] No. 193476.

Toyota Motor Co., Ltd., Release Agent for Urethane Moldings, Japan. Kokai Tokkyo Koho 59209819 [1983/84]; C.A. **102** [1985] No. 114722.

Tsutsui, K., Ricoh Co., Ltd., Lithographic Printing Plate, Ger. Offen. 3422378 [1983/84]; C.A. **103** [1985] No. 30338.

Ube Industries, Ltd., Vinyl Chloride Resin Molding Materials, Japan. Kokai Tokkyo Koho 5996153 [1982/84]; C.A. **101** [1984] No. 152950.

Ueno, R., Tsuchiya, H., Itoh, S., Yamamoto, I., Kabushiki Kaisha Ueno Seiyaku Oyo Kenkyujo, Benzyl Esters of Aromatic Hydroxy Carboxylic Acids, Eur. Appl. 117502 [1983/84]; C.A. **102** [1985] No. 45629.

United States Dept. of the Navy, Fluorinated Polyurethanes, U.S. Appl. 602256 [1984/84]; C.A. **101** [1984] No. 231432.

Unitika Ltd., Thermosetting Polyester Adhesives, Japan. Kokai Tokkyo Koho 5945376 [1982/84]; C.A. **101** [1984] No. 73947.

Valko, J. T., PPG Industries, Inc., β-Hydroxyurethane Low-Temperature Curing Agents, U.S. 4435559 [1982/84]; C.A. **100** [1984] No. 140950.

Vylet, J., Plicka, E., Karasek, O., Alkanolamine Carbamate Chain Extenders for Thermoreactive Systems, Czech. 208984 [1979/84]; C.A. **100** [1984] No. 193072.

Wolf, D., Polyurethane Adhesive, Czech. 212991 [1980/84]; C.A. **100** [1984] No. 193281.

Yokohama Rubber Co., Ltd., Single Liquid-Type Moisture-Curable Urethane Prepolymers, Japan. Kokai Tokkyo Koho 5901522 [1982/84]; C.A. **100** [1984] No. 211838.

Yokohama Rubber Co., Ltd., Urethane Prepolymer Compositions, Japan. Kokai Tokkyo Koho 5978227 [1982/84]; C.A. **101** [1984] No. 112576.

Zampini, A., Monsanto Co., Permeation-Modified Membrane, U.S. 4484935 [1983/84]; C.A. **102** [1985] No. 63347.

Abramov, S. A., Petrov, E. A., Gommen, R. A., Gladkovskii, G. A., Spirin, I. S., Vedeneeva, G. F., Iudina, L. P., Naber, B., All-Union Scientific-Research Institute of Synthetic Resins; VEB Synthesewerk Schwarzheide, Flexible Polyurethane Foams, Ger. [East] 230534 [1982/85]; C.A. **106** [1987] No. 19459.

Agger, R. T., Crabtree, A., Hardy, A., Emhart Australia Pty, Ltd.; Bostik GmbH; Bostik S.A.; Bostik Ltd.; USM Corp., Heat Curable Compositions, PCT Intern. Appl. 8502860 [1983/85]; C.A. **104** [1986] No. 6827.

Ambrose, R. R., Chang, W. H., McKeough, D. T., Peffer, J. R., PPG Industries, Inc., High-Solide-Polyurethane Polyol Coating Compositions, U.S. 4543405 [1984/85]; C.A. **104** [1986] No. 52184.

Ansel, R. E., De Soto, Inc., Radiation-Curable Coatings Containing Reactive Pigment Dispersants, U.S. 4496686 [1984/85]; C.A. **102** [1985] No. 150979.

Asahi Chemical Industry Co., Ltd., Photocurable Resin Compositions Resistant to Sandblasting, Japan. Kokai Tokkyo Koho 6010242 [1983/85]; C.A. **103** [1985] No. 38783.

Asahi Chemical Industry Co., Ltd., Photopolymerizable Composition, Japan. Kokai Tokkyo Koho 6024542 [1983/85]; C.A. **104** [1986] No. 120048.

Asahi-Eckart, Ltd., Composite Metal Powder, Japan. Kokai Tokkyo Koho 121201 [1983/85]; C.A. **103** [1985] No. 199793.

Asahi Glass Co., Ltd., Reaction Injection Molding, Japan. Kokai Tokkyo Koho 6004519 [1983/85]; C.A. **102** [1985] No. 186484.

Asahi Glass Co., Ltd., Polyurethane Elastomers, Japan. Kokai Tokkyo Koho 6032815 [1983/85]; C.A. **102** [1985] No. 186475.

Asahi Glass Co., Ltd., Polyurethane Elastomers, Japan. Kokai Tokkyo Koho 6053521 [1983/85]; C.A. **103** [1985] No. 124933.

Asahi Glass Co., Ltd., Coating Materials, Japan. Kokai Tokkyo Koho 6088078 [1983/85]; C.A. **103** [1985] No. 161944.

Bagaglio, G., van Assche, J., Watts, A., Imperial Chemical Industries PLC, Stable Polyol Composition and Polyurethane Manufacture, Braz. Pedido 8501123 [1984/85]; C.A. **105** [1986] No. 79502.

Bagaglio, G., van Assche, J., Watts, A., Imperial Chemical Industries PLC, Polyol Compositions Containing Mold Release Agent, Eur. Appl. 155116 [1984/85]; C.A. **104** [1986] No. 34956.

Balle, G., Paul, R., Rasshofer, W., Bayer A.-G., Cellular Moldings, Ger. Offen. 3347573 [1983/85]; C.A. **103** [1985] No. 216526.

Barnabeo, A. E., Union Carbide Corp., Water-Curable, Azidosulfonyl Silane-Modified Ethylene Polymers, Eur. Appl. 150773 [1984/85]; C.A. **104** [1986] No. 19980.

Barnabeo, A. E., Union Carbide Corp., Water-Curable, Azide Sulfonyl Silane-Modified, Alkylene-Alkyl Acrylate Copolymers, U.S. 4514545 [1984/85]; C.A. **103** [1985] No. 6899.

Bauman, T. M., Lee, Ch. L., Dow Corning Corp., Nonslumping Foamable Polyorganosiloxane Compositions Containing Silica and Fibers, U.S. 4529741 [1984/85]; C.A. **103** [1985] No. 179484.

Bauman, T. M., Lee, Ch. L., Dow Corning Corp., Nonslumping, Foamable Compositions Containing Siloxane Graft Polymers, U.S. 4548958 [1984/85]; C.A. **104** [1986] No. 7290.

Bauman, T. M., Lee, Ch. L., Rabe, J. A., Dow Corning Corp., Foamable Water-Based Silicone Aerosols, U.S. 4559369 [1984/85]; C.A. **104** [1986] No. 131238.

Bergstrom, C., Neste Oy, Polyolefin Foam Prepared with Water and Crosslinked with a Silane, Belg. 900940 [1983/85]; C.A. **102** [1985] No. 204969.

Bergstrom, C., Brenner, J., Neste Oy, Crosslinked Polyolefin Articles, Belg. 901442 [1984/85]; C.A. **103** [1985] No. 196849.

Bezwada, R. S., American Cyanamid Co., Storage Stable, One Package, Heat Curable Polyurea/Urethane Coating Compositions, U.S. 4523003 [1984/85]; C.A. **103** [1985] No. 89185.

Blum, R., Buensch, H., Druschke, W., Mueller, H., BASF A.-G., Storage-Stable, Heat-Activatable, Liquid Mixture of Polyols and/or Polyamines and Polyisocyanates for Polyurethane Adhesives, Ger. Offen. 3325735 [1983/85]; C.A. **102** [1985] No. 150431.

Blum, R., Frank, W., Horn, P., Osterloh, R., Welz, M., BASF A.-G., Room Temperature Storage-Stable Thermosetting Mixtures based on Compounds with Reactive Hydrogen Atoms and Polyisocyanates, Ger. Offen. 3343124 [1983/85]; C.A. **103** [1985] No. 179105.

Bravet, J. L., Colmon, D., Daude, G., Moncheaux, M. J., Saint-Gobain Vitrage, Safety Glass Laminates, Eur. Appl. 132198 [1983/85]; C.A. **103** [1985] No. 55040.

Bravet, J. L., Colomon, D., Daude, G., Moncheaux, M. J., Saint-Gobain Vitrage, Film of Transparent Plastic Material with High Optical Quality, Fr. Demande 2548956 [1983/85]; C.A. **103** [1985] No. 55041.

Booth, R. G., Bartlett, D. L., International Paint PLC, Solvent-Resistant Coating Composition, Eur. Appl. 147984 [1983/85]; C.A. **103** [1985] No. 197490.

Byk-Chemie GmbH, Adducts for Dispersing Agents for Coating of Solids, Japan. Kokai Tokkyo Koho 60166318 [1984/85]; C.A. **104** [1986] No. 90591.

Cadek, V., Vulcanizing Agent for Room-Temperature Vulcanization of Silicone Rubber, Czech. 227634 [1982/85]; C.A. **104** [1986] No. 111180.

Cavitt, M. B., Wassberg, N. L., Dow Chemical Co., Advanced Epoxy Resins Having Improved Impact Resistance when Cured, U.S. 4552814 [1984/85]; C.A. **104** [1986] No. 208912.

Cenegy, L. F., Protecting a Substrate with a Multi-Density Composite Polyurethane, U.S. 4507336 [1982/85]; C.A. **102** [1985] No. 205175.

Chang, E. Y. C., American Cyanamid Co., Catalyst System for Polyurethane Compositions, Eur. Appl. 154180 [1984/85]; C.A. **104** [1986] No. 69625.

Change, W. H., Ambrose, R. R., McKeough, D. T., Porter, S., Eslinger, D. R., PPG Industries, Inc., Thermosetting High-Solids Solvent-Based Polyester-Urethane One-Component Coating Compositions, Eur. Appl. 140186 [1983/85]; C.A. **103** [1985] No. 106440.

Chiyode Kagaku Kenkyusho, Lubricant Corrosion Inhibitor, Japan. Kokai Tokkyo Koho 6018590 [1983/85]; C.A. **102** [1985] No. 206414.

Cuscurida, M., Texaco, Inc., Polymer Polyols from Alkylene Oxide Adducts of Alkanolamines, U.S. 4518778 [1983/85]; C.A. **103** [1985] No. 38135.

Daicel Chemical Industries, Ltd., Resins for Magnetic Coatings, Japan. Kokai Tokkyo Koho 6006719 [1983/85]; C.A. **102** [1985] No. 222267.

Daiichi Kogyo Seiyaku Co., Ltd., Nippon Steel Corp., Coating Compositions for Metals, Japan. Kokai Tokkyo Koho 6002355 [1983/85]; C.A. **102** [1985] No. 186785.

Daikyo Chemical Co., Ltd., Stabilizing Electrically Conductive Halogen-Containing Resins, Japan. Kokai Tokkyo Koho 6023436 [1983/85]; C.A. **103** [1985] No. 7277.

Dainippon Ink and Chemicals, Inc., Caulking and Sealing Compounds, Japan. Kokai Tokkyo Koho 6023471 [1983/85]; C.A. **103** [1985] No. 55570.

Dainippon Ink and Chemicals, Inc., Preparation of Photosensitive Urethane-Polyester Resins, Japan. Kokai Tokkyo Koho 6032818 [1983/85]; C.A. **103** [1985] No. 30296.

Dainippon Printing Co., Ltd., Adhesive Compositions, Japan. Kokai Tokkyo Koho 6044570 [1983/85]; C.A. **103** [1985] No. 105923.

Dainippon Printing Co., Ltd., Ionizing Radiation-Curable Urethane Compositions, Japan. Kokai Tokkyo Koho 6090211 [1983/85]; C.A. **103** [1985] No. 161290.

Dainippon Printing Co., Ltd., Electrolytically Dissociating Radiochemically Hardenable Urethane Compositions, Japan. Kokai Tokkyo Koho 60101107 [1983/85]; C.A. **104** [1986] No. 70423.

Dainippon Printing Co., Ltd., Shin-Etsu Chemical Industry Co., Ltd., Printable Siloxane Release Coatings for Adhesive Tapes, Japan. Kokai Tokkyo Koho 6094485 [1983/85]; C.A. **103** [1985] No. 106007.

Den Hartog, H. C., Walus, A. N., du Pont de Nemours, E. I., and Co., Coating Composition of an Acrylic Polymer Having Primary Amine Groups, U.S. 4529765 [1984/85]; C.A. **103** [1985] No. 106410.

De Soto, Inc., Electron Beam-Curable Coating Compositions, Japan. Kokai Tokkyo Koho 60210674 [1984/85]; C.A. **104** [1986] No. 170258.

Disteldorf, J., Fiakus, W., Chemische Werke Huels A.-G., Moisture-Hardening Lacquer from Isocyanate Resin, Ger. Offen. 3322723 [1983/85]; C.A. **102** [1985] No. 133711.

Doebler, K. P., Graef, K., Jaeger, K. E., Johannsen, F., Ritz, J., Saatweber, D., Herberts GmbH, Ger. Offen. 3333073 [1983/85]; C.A. **103** [1985] No. 38748.

Dow Corning Corp., Perennatorwerk Alfred Hagen GmbH, Organopolysiloxane Composition Curable Into an Elastomer and Its Use, Belg. 901479 [1984/85]; C.A. **103** [1985] No. 38556.

Druetzler, T. W., Sherwin-Williams Co., Isocyanate-Functional Prepolymers and Coating Materials Based on Them, U.S. 4560494 [1984/85]; C.A. **104** [1986] No. 226461.

Eisenbach, C. D., Guenter, C., Bayer A.-G., Hexaurethanediols, Ger. Offen. 3330946 [1983/85]; C.A. **102** [1985] No. 220599.

Eisenbach, C. D., Guenter, C., Bayer A.-G., Tetraurethanediols, Ger. Offen. 3330947 [1983/85]; C.A. **103** [1985] No. 104733.

Falardeau, E. R., Frisch, K. C., Lock, M. R., Dow Chemical Co., Silica-Bonded Tin Urethane Catalysts, U.S. 4507410 [1984/85]; C.A. **103** [1985] No. 6854.

Falk, R. A., Ciba-Geigy Corp., Di-Perfluoroalkyl Carbamyl Group-Containing Acrylates and Methacrylates, U.S. 4540805 [1984/85]; C.A. **104** [1986] No. 69290.

Fesman, G., Lin, R. Y., Rehder, R. A., Stauffer Chemical Co., Flame-Retardant Mixture for Polyurethane Materials, Eur. Appl. 138204 [1983/85]; C.A. **103** [1985] No. 72143.

Field, J. R., Kocher, H. S., Martellock, A. C., Xerox Corp., Fusing System with Unblended Silicone Oil, U.S. 4515884 [1982/85]; C.A. **103** [1985] No. 45795.

Fuji Photo Film Co., Ltd., Method for Fabrication of Microencapsulated Toners, Japan. Kokai Tokkyo Koho 60120367 [1983/85]; C.A. **104** [1986] No. 13022.

Fuji Xerox Co., Ltd., Electrophotographic Photoreceptor, Japan. Kokai Tokkyo Koho 6004945 [1983/85]; C.A. **102** [1985] No. 212636.

Fujii, H., Koide, T., Isobe, K., Aoki, H., Dainippon Printing Co., Ltd., Shin-Etsu Chemical Industry Co., Ltd., Printable Release Paper, Japan. Kokai Tokkyo Koho 60155451 [1984/85]; C.A. **104** [1986] No. 7408.

Fujikura Cable Works, Ltd., Coating of Crosslinked Polyethylenes on Electric Wires, Japan. Kokai Tokkyo Koho 6006045 [1978/85]; C.A. **102** [1985] No. 222210.

Fujioka, S., Ishitobi, T., Toray Industries, Inc., Radically-Curing Compositions, Japan. Kokai Tokkyo Koho 60163914 [1984/85]; C.A. **104** [1986] No. 188300.

Fukuda, T., Saito, M., Sakamoto, S., Toray Industries, Inc., Plastic Lenses with High Refractive Index, Japan. Kokai Tokkyo Koho 60249101 [1984/85]; C.A. **104** [1986] No. 187790.

Fukuda, T., Sakamoto, S., Toray Industries, Inc., Resin Material for Plastic Lens, Eur. Appl. 134861 [1983/85]; C.A. **103** [1985] No. 7241.

Fukuda, T., Sakamoto, S., Wada, S., Toray Industries, Inc., Plastic Lenses with High Refractive Index, Japan. Kokai Tokkyo Koho 60217301 [1984/85]; C.A. **104** [1986] No. 208494.

Funaki, M., Atsuta, M., Kuga, K., Asahi Glass Co., Ltd., Treated Siliceous Fillers for a Polyurethane, Japan. Kokai Tokkyo Koho 60260653 [1985/85]; C.A. **105** [1986] No. 7421.

Furukawa Electric Co., Ltd., Extruded Silane-Crosslinked Polyolethylenes, Japan. Kokai Tokkyo Koho 6024916 [1983/85]; C.A. **102** [1985] No. 205605.

Furukawa Electric Co., Ltd., Polyolefin Rubber-Insulated Wires, Japan. Kokai Tokkyo Koho 6093710 [1983/85]; C.A. **103** [1985] No. 217041.

Furukawa, H., Kato, Y., Kanegafuchi Chemical Industry Co., Ltd., Aminosilane-Modified Resins, Japan. Kokai Tokkyo Koho 60197733 [1984/85]; C.A. **104** [1986] No. 131644.

Ganster, O., Knipp, U., Luckas, B., Bayer A.-G., Polyurethane Molding Compositions Containing Urea Groups, Ger. Offen. 3407931 [1984/85]; C.A. **104** [1986] No. 89740.

Grape, W., Saykowski, F., Schlak, O., Wuerminghausen, T., Bayer A.-G., Stable Silicone Emulsions, Ger. Offen. 3323909 [1983/85]; C.A. **102** [1985] No. 115287.

Gras, R., Riemer, H., Wolf, E., Chemische Werke Huels A.-G., Mat Powder Coating Containing Uretdione Group-Containing Adducts of Isophorone Diisocyanate, Ger. Offen. 3328133 [1983/85]; C.A. **103** [1985] No. 23826.

Guan, Yunlin., Xu, Meixuan., Chen, Tonghui., Tianjin University, Polyurethane-Based Binder, Faming Zhuanli Shenqing Gongkai Shuomingshu 85100023 [1985]; C.A. **105** [1986] No. 192442.

Hachitsuka, T., Sugitoge, T., Emoto, K., Mizumura, Y., Toyobo Co., Ltd., Thermoplastic Polyurethane-Polyureas, Japan. Kokai Tokkyo Koho 60147427 [1984/85]; C.A. **104** [1986] No. 70414.

Hasegawa, M., Kobayashi, H., Sunazuka, H., Yoshino, A., Matsuda, T., Shingo, Y., Fujikura, Ltd., Flame-Retardant Crosslinked Composition and Flame-Retardant Cable Using It, Brit. Appl. 2156825 [1984/85]; C.A. **104** [1986] No. 151023.

Hashimoto, H., Hanai, S., Showa Electric Wire and Cable Co., Ltd., Silane-Crosslinked Ethylene Copolymer Moldings, Japan. Kokai Tokkyo Koho 60250033 [1984/85]; C.A. **105** [1986] No. 25406.

Herold, J., Oezelli, R. N., Scheer, H., Henkel KG, Elastic Binding Between Glass and Lacquered Sheets, Suitable for Transfer of Force, Ger. Offen. 3322442 [1983/85]; C.A. **102** [1985] No. 186242.

Hirata, F., Fujinami, K., Nasu, K., Takeda Chemical Industries, Ltd., Polymers with Isocyanate Pendant Groups, Japan. Kokai Tokkyo Koho 60233110 [1984/85]; C.A. **105** [1986] No. 7054.

Hirukawa, H., Otani, K., Furukawa Electric Co., Ltd., Crosslinking of Polyolefins, Japan. Kokai Tokkyo Koho 60186515 [1984/85]; C.A. **104** [1986] No. 89871.

Hirukawa, H., Otani, K., Furukawa Electric Co., Ltd., Crosslinking of Silane-Grafted Polyolefins, Japan. Kokai Tokkyo Koho 60186558 [1984/85]; C.A. **104** [1986] No. 69757.

Hirukawa, H., Otani, K., Furukawa Electric Co., Ltd., Heat-Shrinkable Polyolefin Tubes, Japan. Kokai Tokkyo Koho 60262829 [1984/85]; C.A. **104** [1986] No. 226069.

Hitachi Chemical Co., Ltd., Adhesive Films, Japan. Kokai Tokkyo Koho 6013872 [1983/85]; C.A. **102** [1985] No. 221851.

Hitachi Chemical Co., Ltd., Thermal-Transfer Recording Materials, Japan. Kokai Tokkyo Koho 6036189 [1983/85]; C.A. **103** [1985] No. 45887.

Hitachi Chemical Co., Ltd., Isocyanuric Ring-Containing Esters, Japan. Kokai Tokkyo Koho 60123478 [1983/85]; C.A. **104** [1986] No. 69621.

Hitachi Cable, Ltd., Chlorinated Polyethylene Compositions, Japan. Kokai Tokkyo Koho 60118709 [1983/85]; C.A. **104** [1986] No. 69688.

Hitachi Cable, Ltd., Crosslinking Halogen-Containing Polymers, Japan. Kokai Tokkyo Koho 60124613 [1983/85]; C.A. **103** [1985] No. 179212.

Hoffmann, D. K., Dow Chemical Co., Polyoxyalkylene Methacrylate Copolymer Dispersions, U.S. 4513124 [1980/85]; C.A. **103** [1985] No. 38130.

Holker, J. R., Lomax, G. R., Shirley Institute, Breathable, Non-Poromeric Polyurethane Films, Brit. UK Pat. Appl. 2157703 [1984/85]; C.A. **104** [1986] No. 208273.

Horiguchi, S., Yamakita, H., Mimura, M., Bayer Gohsei-Silicone Co., Ltd., Polysiloxane Compositions for Food Models, Japan. Kokai Tokkyo Koho 60221458 [1984/85]; C.A. **104** [1986] No. 208498.

Houze, E. C., Vasta, J. A., du Pont de Nemours, E. I., and Co., Acrylic Polyurethane Coating Compositions, U.S. 4503175 [1983/85]; C.A. **102** [1985] No. 205560.

Hughes, A., Topham, A., Baxenden Chemical Co., Ltd., Blocked Isocyanates, Eur. Appl. 159117 [1984/85]; C.A. **104** [1986] No. 90586.

Human Industry Corp., Flame-Retardant Polyurethane Foams, Japan. Kokai Tokkyo Koho 60123521 [1983/85]; C.A. **104** [1986] No. 20594.

Human Industry Corp., Flame-Retardant Polyurethane Foams, Japan. Kokai Tokkyo Koho 60123539 [1983/85]; C.A. **104** [1986] No. 35286.

Ikeda, A., Kaneko, S., Ai, H., Asahi Chemical Industry Co., Ltd., Photopolymerizing Composition, Japan. Kokai Tokkyo Koho 60138540 [1983/85]; C.A. **103** [1985] No. 224434.

Ikemoto, Y., Murai, T., Isobe, T., Daicel Chemical Industries, Ltd., Epoxy Urethane Resins, Japan. Kokai Tokkyo Koho 60219218 [1984/85]; C.A. **104** [1986] No. 187455.

Imperial Chemical Industries PLC, Photopolymerizable Compositions, Japan. Kokai Tokkyo Koho 6032801 [1983/85]; C.A. **103** [1985] No. 6855.

Inoue, S., Nishibori, S., Kinoshita, H., Komori, H., Yoshimura, M., Daiichi Kogyo Seiyaku Co., Ltd., Flame-Retarding Plasticizers, Japan. Kokai Tokkyo Koho 60212435 [1984/85]; C.A. **104** [1986] No. 187519.

Irie, S., Kimura, H., Furukawa Electric Co., Ltd., Water-Crosslinkable Polymer Compositions, Japan. Kokai Tokkyo Koho 60258246 [1984/85]; C.A. **105** [1986] No. 7410.

Isayama, K., Hirose, T., Iwahara, T., Kawakubo, F., Kanegafuchi Chemical Industry Co., Ltd., Curable Phenolic Resin Blende, Eur. Appl. 159605 [1984/85]; C.A. **104** [1986] No. 89859.

Isayama, K., Hirose, T., Kawakubo, F., Kanegafuchi Chemical Industry Co., Ltd., Moisture-Curable Silane-Terminated Acrylic Polymers, Fr. Demande 2564846 [1984/85]; C.A. **104** [1986] No. 208516.

Ishikawa, I., Nagata, K., Tomosada, T., Sanyo Chemical Industries, Ltd., Polyurethane Manufacture, Japan. Kokai Tokkyo Koho 60197718 [1984/85]; C.A. **104** [1986] No. 225765.

Isobe, T., Murai, T., Ikemoto, Y., Daicel Chemical Industries, Ltd., Urethane Epoxy Resins, Japan. Kokai Tokkyo Koho 60219217 [1984/85]; C.A. **104** [1986] No. 208244.

Isoda, T., Watanabe, Y., Nakayama, N., Aoki, M., Ricoh Co., Ltd., Carrier Particles for Use in a Two-Component Dry-Type Developer, Ger. Offen. 3436410 [1983/85]; C.A. **103** [1985] No. 96324.

Ito, K., Suzuki, T., Mitsui Toatsu Chemicals, Inc., Fast-Drying Polyurethane Compositions for Coatings, Japan. Kokai Tokkyo Koho 60215073 [1984/85]; C.A. **104** [1986] No. 188257.

Ito, Y., Daito, T., Honny Chemicals Co., Ltd., Adhesives for Polyester Substrates, Japan. Kokai Tokkyo Koho 60135472 [1983/85]; C.A. **104** [1986] No. 187710.

Jansa, J., Hynek, V., Mohr, P., Budin, J., Novy, J., Polackova, V., Polyamide of Polyester Sewing Monofilament, Czech. 220661 [1981/85]; C.A. **105** [1986] No. 7913.

Kajita, Y., Seki, I., Yagyu, H., Hitachi Cable, Ltd., Crosslinking of Chlorinated Polyethylene Compositions, Japan. Kokai Tokkyo Koho 60144315 [1984/85]; C.A. **104** [1986] No. 51545.

Kamatani, Y., Sakamoto, T., Takeda Chemical Industries, Ltd., Curing Polyurethane Compounds, Eur. Appl. 144841 [1983/85]; C.A. **103** [1985] No. 143504.

Kamatani, Y., Sakamoto, T., Fujita, N., Takeda Chemical Industries, Ltd., Moisture-Curable Polyurethane Compositions, Japan. Kokai Tokkyo Koho 60206820 [1984/85]; C.A. **104** [1986] No. 187446.

Kanegafuchi Chemical Industry Co., Ltd., Room-Temperature-Curable Sealants, Japan. Kokai Tokkyo Koho 6006747 [1984/85]; C.A. **102** [1985] No. 205582.

Kanegafuchi Chemical Industry Co., Ltd., Weather-Resistant Delustered Agricultural Sheets, Japan. Kokai Tokkyo Koho 6032809 [1983/85]; C.A. **102** [1985] No. 205167.

Kapps, M., Wiedermann, R., Adam, N., Weigand, E., Bayer A.-G., Hard, Closed-Cell, Flame-Resistant Polyurethane Foams, Ger. Offen. 3402310 [1984/85]; C.A. **103** [1985] No. 216302.

Kawakubo, F., Takanoo, M., Yukimoto, S., Isayama, K., Kanegafuchi Chemical Industry Co., Ltd., Moisture-Cured Polymers, Eur. Appl. 159715 [1984/85]; C.A. **104** [1986] No. 130793.

Kidd, P. D., Sterrett, T. L., Sybron Corp., Urethane Modified Orthodontic Adhesive, U.S. 4554336 [1983/85]; C.A. **104** [1986] No. 174691.

Kinumura, A., Yamaguchi, K., Shiraishi, M., Fuji, K., Sekisui Chemical Co., Ltd., Pressure-Sensitive Adhesive Tapes, Japan. Kokai Tokkyo Koho 60252682 [1984/85]; C.A. **104** [1986] No. 226037.

Klaar, K., Kuehnel, W., Spielau, P., Weiss, R., Dynamit Nobel A.-G., Crosslinked Moldings Based on Polyethylene, Ger. Offen. 3306909 [1983/85]; C.A. **102** [1985] No. 114628.

Kodera, T., Dainippon Printing Co., Ltd., Ultraviolet or Electron Beam-Curable Polymer Film, Japan. Kokai Tokkyo Koho 60263140 [1984/85]; C.A. **104** [1986] No. 216579.

Kojima, H., Narisawa, S., Funski, M., Asahi Glass Co., Ltd., Urethane Rubbers, Japan. Kokai Tokkyo Koho 60245621 [1984/85]; C.A. **105** [1986] No. 7753.

Kojima, H., Shibata, S., Asahi Glass Co., Ltd., Reaction-Injection Molding of Plastics, Japan. Kokai Tokkyo Koho 60244511 [1984/85]; C.A. **104** [1986] No. 187694.

Koleske, J. V., Osborn, C. L., Union Carbide Corp., Blends of Cyclic Vinyl Ether-Containing Compounds and Urethane Acrylates, Eur. Appl. 155704 [1984/85]; C.A. **104** [1986] No. 52157.

Kopp, R., Freitag, H. A., Bayer A.-G., Polyurethanes in Foams if Desired, Ger. Offen. 3329452 [1983/85]; C.A. **103** [1985] No. 7206.

Kotleba, J., Svitkova, O., Gendiar, J., Novak, L., Cmolik, J., Mixed Agent with Regulating and Lubricating Properties for Poly(Vinyl Chloride) Processing, Czech. 219651 [1980/85]; C.A. **105** [1986] No. 209920.

Kumasaka, S., Tada, S., Fujii, O., Kuga, S., Human Industry Corp., Flame Retardant Polyurethane Foams, Japan. Kokai Tokkyo Koho 60133018 [1983/85]; C.A. **104** [1986] No. 89771.

Kumasaka, S., Tada, S., Wakabayashi, H., Human Industry Corp., Solutions for Regenerating Polyurethane Foams, Japan. Kokai Tokkyo Koho 60190437 [1984/85]; C.A. **104** [1986] No. 69892.

Kumasaka, S., Tada, S., Wakabayashi, H., Human Industry Corp., Regenerated Asphalt Foam Solution, Japan. Kokai Tokkyo Koho 60190438 [1984/85]; C.A. **104** [1986] No. 187720.

Kumada, K., Fukuda, M., Sanyo Chemical Industries, Ltd., Polyurethanes Based on Epoxy-Containing Polymeric Polyols, Ger. Offen. 3432827 [1983/85]; C.A. **103** [1985] No. 23284.

Kumada, K., Fukuda, M., Sanyo Chemical Industries, Ltd., Polyurethane Moldings with High Modulus, Japan. Kokai Tokkyo Koho 60235816 [1984/85]; C.A. **105** [1986] No. 7335.

Kumada, K., Fukuda, M., Sanyo Chemical Industries, Ltd., Polyols Containing Dispersed Vinyl Polymers for Polyurethanes, Japan. Kokai Tokkyo Koho 60252612 [1984/85]; C.A. **105** [1986] No. 79868.

Kuo, A. L., Goddard, E. D., Ritscher, J. S., Union Carbide Corp., Internal Mold Release for Reaction Injection Molded Polyurethanes, Eur. Appl. 138130 [1983/85]; C.A. **103** [1985] No. 124468.

Kurashige, Y., Sakuma, K., Tanaka, T., Idemitsu Petrochemical Co., Ltd., Concrete Protecting Materials, Japan. Kokai Tokkyo Koho 60243162 [1984/85]; C.A. **104** [1986] No. 208696.

Kurashige, Y., Tanaka, T., Ito, K., Idemitsu Petrochemical Co., Ltd., Laminated Panels for Vibration Dampers, Japan. Kokai Tokkyo Koho 60190350 [1984/85]; C.A. **104** [1986] No. 20501.

Kurita, A., Fujimoto, T., Toshiba Silicone Co., Ltd., Vulcanization Accelerator Composition for Silicone Rubber, Japan. Kokai Tokkyo Koho 60202153 [1984/85]; C.A. **104** [1986] No. 70137.

Kuriyama, S., Nakajima, S., Sanyo Chemical Industries, Ltd., Silylated Polyurethane, Japan. Kokai Tokkyo Koho 60133019 [1983/85]; C.A. **104** [1986] No. 34847.

Kuriyama, S., Nakajima, S., Sanyo Chemical Industries, Ltd., Polymeric Polyols, Japan. Kokai Tokkyo Koho 60147422 [1984/85]; C.A. **104** [1986] No. 69287.

Kuriyama, S., Nakajima, S., Sanyo Chemical Industries, Ltd., Silyl-Containing Urethane Prepolymers, Japan. Kokai Tokkyo Koho 60213722 [1984/85]; C.A. **104** [1986] No. 208936.

Larson, W. K., Minnesota Mining and Mfg. Co., Sulfopolyester-Containing Photopolymer Systems, Eur. Appl. 146326 [1983/85]; C.A. **103** [1985] No. 125192.

Lehner, A., Balz, W., Velic, M., Kopke, H., Grau, W., Baur, R., BASF A.-G., Magnetic Recording Support Binding Material, Ger. Offen. 3341698 [1983/85]; C.A. **103** [1985] No. 80781.

Lehner, A., Kopke, H., Roller, H., Balz, W., Grau, W., Koester, E., Sommermann, F., BASF A.-G., Magnetic Recording Support from Nonmagnetic Support Material, Ger. Offen. 3341699 [1983/85]; C.A. **103** [1985] No. 80782.

Leiffer, J. L., Stopps, W. E., St. Clair, T. I., Watkins, V. E., Kelly, T. P., United States National Aeronautics and Space Administration, Structural Pressure-Sensitive Silicone Adhesives, U.S. 569536 [1984/85]; C.A. **102** [1985] No. 133396.

Lewarchik, R. J., Noren, G. K., Metcalfe, R., Bonin, D. J., Poklacki, E. S., De Soto, Inc., Low Temperature-Curing Polyester Urethane Systems, U.S. 4530977 [1984/85]; C.A. **103** [1985] No. 125204.

Lewarchik, R. J., Thompson, J. M., De Soto, Inc., Thermally Cured Topcoats on Vacuum Deposited Metal, Eur. Appl. 136667 [1983/85]; C.A. **103** [1985] No. 55486.

Lister, J. W., Interox Chemicals, Ltd., Crosslinkable Polymeric Coating Compositions, Brit. Appl. 2160534 [1984/85]; C.A. **105** [1986] No. 62299.

Loehr, G., Mondt, J., Graeff, H., Hoechst A.-G., Crosslinkable Fluoropolymer Coatings, Ger. Offen. 3347655 [1983/85]; C.A. **103** [1985] No. 197491.

Lucast, D. H., Shubkin, R. L., Filbey, A. H., Wollensak, J. C., Ethyl Corp., Polyurethanes from Dialkyl Diaminobenzenes, U.S. 4526905 [1982/85]; C.A. **103** [1985] No. 161665.

Maekawa, I., Omori, E., Uchigasaki, I., Hida, H., Yokoyama, N., Tanno, S., Nakano, F., Ito, T., Hitachi Chemical Co., Ltd., Hitachi, Ltd., Moisture-Resistant Coatings for Circuit Boards, Japan. Kokai Tokkyo Koho 60186569 [1984/85]; C.A. **104** [1986] No. 150942.

Maeno, K., Murachi, T., Toyoda Gosei Co., Ltd., Waterproofing Aerosol Sprays, Ger. Offen. 3521250 [1984/85]; C.A. **104** [1986] No. 150990.

Maki, H., Kawamura, T., Uchikata, H., Daiichi Kogyo Seiyaku Co., Ltd., Anticorrosive Polyurethane Coating Composition, Ger. Offen. 3504228 [1984/85]; C.A. **104** [1986] No. 170232.

Malcovsky, E., Stresinka, J., Mokry, J., Modified Polyurethane Foams, Czech. 219168 [1981/85]; C.A. **103** [1985] No. 216419.

Martens, J., Williams, B. H., Minnesota Mining and Mfg. Co., Improved Processable Radiation Curable Poly(Vinyl Chloride) Resin Compositions, Eur. Appl. 132032 [1983/85]; C.A. **102** [1985] No. 221661.

Martineu, P., Guillet, A., Conservatome, Crosslinking of Grafted Polyolefins, Fr. Demande 2558472 [1984/85]; C.A. **104** [1986] No. 110782.

Masar, B., Cefelin, P., Sequenced Block Copolymers of Synthetic Peptides with Linear Structure, Czech. 217739 [1981/85]; C.A. **106** [1987] No. 19213.

Matsukura, N., Suzuki, Y., Fujiwara, H., Ashida, T., Jujo Paper Co., Ltd., Release Compositions for Openings of Paper Containers, Japan. Kokai Tokkyo Koho 60147458 [1984/85]; C.A. **104** [1986] No. 35149.

Matsuyama, A., Ozawa, H., Hirose, S., Mitsui Toatsu Chemicals, Inc., Crosslinkable Polymers, Japan. Kokai Tokkyo Koho 60233114 [1984/85]; C.A. **104** [1986] No. 187066.

Meckel, W., Bayer A.-G., Polyurethanes with Aromatic Amino End Groups and Their Use, Ger. Offen. 3401753 [1984/85]; C.A. **103** [1985] No. 216307.

Meier, H. M., Dhein, R., Winkel, J., Klein, G., Kloeker, W., Bayer A.-G., Arylamine Derivatives Polymerizable in Unsaturated Resins and Their Use as Hardening Accelerators, Ger. Offen. 3345104 [1983/85]; C.A. **103** [1985] No. 196855.

Meshii, M., Shigeta, S., Ueda, F., Yanagida, Y., Dainippon Ink and Chemicals, Inc., Coating Materials for Optical Fibers, Japan. Kokai Tokkyo Koho 60181170 [1984/85]; C.A. **104** [1986] No. 131632.

Miller, L. I., du Pont de Nemours, E. I., and Co., Additive for Coating Compositions from Alkyd Resins, Belg. 901300 [1983/85]; C.A. **104** [1986] No. 7257.

Mitsubishi Chemical Industries Co., Ltd., Urethane Resin Compositions, Japan. Kokai Tokkyo Koho 6076525 [1983/85]; C.A. **103** [1985] No. 105728.

Mitsubishi Monsanto Chemical Co., Adhesion-Preventing Flexible PVC Films, Japan. Kokai Tokkyo Koho 6006734 [1983/85]; C.A. **102** [1985] No. 221831.

Mitsubishi Petrochemical Co., Ltd., Electric Wire Coated with Crosslinked Propylene Resins, Japan. Kokai Tokkyo Koho 60105109 [1983/85]; C.A. **103** [1985] No. 197064.

Mitsui-Nisso Corp., Stable Polyurethane Moldings, Japan. Kokai Tokkyo Koho 6015418 [1983/85]; C.A. **103** [1985] No. 7204.

Mitsui-Nisso Corp., Improvement of the Heat Resistance of Ultraflexible Polyurethane Elastomers, Japan. Kokai Tokkyo Koho 6067524 [1983/85]; C.A. **103** [1985] No. 72440.

Mitsui-Nisso Corp., Lining of Polyurethane Elastomers, Japan. Kokai Tokkyo Koho 6082171 [1983/85]; C.A. **103** [1985] No. 106448.

Mitsui-Nisso Corp., Polyurethane Foams, Japan. Kokai Tokkyo Koho 6096629 [1983/85]; C.A. **104** [1986] No. 20568.

Miyama, M., Onishi, M., Toray Silicone Co., Ltd., Room Temperature-Vulcanizable Siloxane Coatable After Curing, Japan. Kokai Tokkyo Koho 60158254 [1984/85]; C.A. **104** [1986] No. 35656.

Miyama, M., Onishi, M., Toray Silicone Co., Ltd., Organopolysiloxane Compositions Coatable After Room-Temperature-Setting, Japan. Kokai Tokkyo Koho 60166344 [1984/85]; C.A. **104** [1986] No. 52181.

Mochida Syoko, K. K., Fluorinated Rubber Laminates, Japan. Kokai Tokkyo Koho 6097846 [1983/85]; C.A. **103** [1985] No. 106134.

Mollett, Ch. Ch., Sneddon, G. R., Albright and Wilson Ltd., Composition and Method of Deinking of Recycled Cellulosic Material, Brit. Appl. 2158836 [1984/85]; C.A. **104** [1986] No. 151160.

Mollett, Ch. Ch., Sneddon, G. R., Albright and Wilson Ltd., Composition and Method for Deinking Recycled Cellulosic Material, Eur. Appl. 163444 [1984/85]; C.A. **104** [1986] No. 131797.

Morimoto, T., Nishiuchi, K., Wada, K., Otsuka Chemical Co., Ltd., Metamorphic Alkali Metal Titanates, Ger. Offen. 3442270 [1983/85]; C.A. **104** [1986] No. 80423.

Moroishi, Y., Sakai, I., Tawara, S., So, I., Nitto Electric Industrial Co., Ltd., One-Package Elastic Sealants, Japan. Kokai Tokkyo Koho 60202180 [1984/85]; C.A. **104** [1986] No. 188353.

Murai, T., Isobe, T., Ikemoto, Y., Daicel Chemical Industries, Ltd., Polyurethane Epoxides, Japan. Kokai Tokkyo Koho 60219215 [1984/85]; C.A. **104** [1986] No. 208246.

Murai, T., Isobe, T., Ikemoto, Y., Daicel Chemical Industries, Ltd., Polyurethane Epoxides Japan. Kokai Tokkyo Koho 60219216 [1984/85]; C.A. **104** [1986] No. 208245.

Murai, T., Isobe, T., Ikemoto, Y., Daicel Chemical Industries, Ltd., Epoxy Urethane Resins, Japan. Kokai Tokkyo Koho 60260614 [1984/85]; C.A. **104** [1986] No. 187474.

Nagai, T., Baba, S., Mizuno, M., Toyo Rubber Industry Co., Ltd., Manufacture of Porous Ceramic Materials, Japan. Kokai Tokkyo Koho 60239376 [1984/85]; C.A. **104** [1986] No. 211929.

Nakamoto, H., Fukushima, H., Niimoto, M., Mitsubishi Rayon Co., Ltd., Ultraviolet-Setting Resin Compositions for Preparation of Replicas of Optical Disks, Japan. Kokai Tokkyo Koho 60206815 [1984/85]; C.A. **105** [1986] No. 200576.

Nakamura, M., Nagai, H., Takeuchi, S., Deguchi, A., Sumitomo Bayer Urethane Co., Ltd., Polyurethane Moldings, Japan. Kokai Tokkyo Koho 60168715 [1984/85]; C.A. **104** [1986] No. 6927.

Nichias Corp., Cellular Polyurethanes, Japan. Kokai Tokkyo Koho 60110717 [1983/85]; C.A. **103** [1985] No. 197068.

Nippon Paints Co., Ltd., Photosensitive Compounds for Coatings, Japan. Kokai Tokkyo Koho 6023411 [1984/85]; C.A. **103** [1985] No. 7812.

Nippon Polyurethane Industry Co., Ltd., One-Component Liquid Polyurethanes for Thick Coatings, Japan. Kokai Tokkyo Koho 60108476 [1983/85]; C.A. **103** [1985] No. 197487.

Nippon Soda Co., Ltd., Polyurethanes with Inorganic Fillers, Japan. Kokai Tokkyo Koho 6071625 [1983/85]; C.A. **103** [1985] No. 142893.

Nippon Steel Corp., Sanyo Chemical Industries, Ltd., Sealants for Steel Segment Joints, Japan. Kokai Tokkyo Koho 6076588 [1983/85]; C.A. **103** [1985] No. 89193.

Nippon Zeon Co., Ltd., Electrically Conductive Vinyl Chloride Resin Compositions, Japan. Kokai Tokkyo Koho 6020948 [1983/85]; C.A. **103** [1985] No. 23332.

Nishino, K., Miyashita, H., Fukui, T., Takeda Chemical Industries, Ltd., One-Package Polyurethane In-Mold Coating Compositions, Japan. Kokai Tokkyo Koho 60221437 [1984/85]; C.A. **104** [1986] No. 170264.

Nitto Electric Industrial Co., Ltd., Elastic Sealants, Japan. Kokai Tokkyo Koho 6028482 [1983/85]; C.A. **103** [1985] No. 23871.

Nitto Electric Industrial Co., Ltd., Elastic Sealants, Japan. Kokai Tokkyo Koho 6053583 [1983/85]; C.A. **103** [1985] No. 55574.

Nitto Electric Industrial Co., Ltd., Elastic Sealants, Japan. Kokai Tokkyo Koho 6053584 [1983/85]; C.A. **103** [1985] No. 55573.

Nitto Electric Industrial Co., Ltd., One-Package Elastic Sealants, Japan. Kokai Tokkyo Koho 6076589 [1983/85]; C.A. **103** [1985] No. 89192.

NTN-Rulon Industries Co., Ltd., Mitsubishi Petrochemical Co., Ltd., Transparent Friction Liners for Retention of Magnetic Tapes, Japan. Kokai Tokkyo Koho 60108433 [1983/85]; C.A. **103** [1985] No. 124710.

O'Donnell, T. W., Olson, D. R., General Electric Co., Polymerizable 3-Aroyloxyphenyl Carbamates and Their Use, U.S. 4520074 [1984/85]; C.A. **103** [1985] No. 55561.

Oji-Yuka Synthetic Paper Co., Ltd., Heat-Shrinkable Vinylsilane-Grafted Propylene Polymer Tubes, Japan. Kokai Tokkyo Koho 6021233 [1983/85]; C.A. **102** [1985] No. 221866.

Okada, Y., Igarashi, Y., Kureha Chemical Industry Co., Ltd., Composite Laminate Film, Eur. Appl. 152102 [1984/85]; C.A. **104** [1986] No. 20409.

Osborn, C. L., Koleske, J. V., Drake, K., Union Carbide Corp., Compositions Useful in Curable Coatings, Eur. Appl. 133908 [1983/85]; C.A. **103** [1985] No. 38785.

Ott, G., Dobbelstein, A., Geist, M., Schoen, G., Ahlers, K., BASF Farben und Fasern A.-G., Water-Dispersible Binders from Modified Epoxide-Amine Adducts, Ger. Offen. 3409189 [1984/85]; C.A. **104** [1986] No. 70380.

Ottaviani, R. A., Short, W. T., Hart, D. J., General Motors Corp., Sag Control of High-Solids Polyurethane Clearcoats by Urea Thixotrope/Silica Systems, U.S. 4528319 [1984/85]; C.A. **103** [1985] No. 143516.

Ozaki, Y., Iwata, H., Sekisui Chemical Co., Ltd., Ethylene Polymer Foams, Japan. Kokai Tokkyo Koho 60219235 [1984/85]; C.A. **106** [1987] No. 5956.

Ozaki, Y., Iwata, H., Sekisui Chemical Co., Ltd., Ethylene Polymer Foams, Japan. Kokai Tokkyo Koho 60219236 [1984/85]; C.A. **104** [1986] No. 131118.

Ozaki, Y., Matsumura, A., Nakamura, H., Toyo Rubber Industry Co., Ltd., Aqueous Polyurethane Dispersions, Japan. Kokai Tokkyo Koho 60197720 [1984/85]; C.A. **104** [1986] No. 69651.

Ozono, M., Miyabayashi, T., Three Bond Co., Ltd., Photocurable Rubber Elastic Compositions, Japan. Kokai Tokkyo Koho 60163911 [1984/85]; C.A. **104** [1986] No. 90265.

Parekh, G. G., Jacobs, W., Blank, W. J., American Cyanamid Co., Carbamate Compounds and Compositions, Eur. Appl. 152820 [1984/85]; C.A. **104** [1986] No. 90624.

Perrey, H., Matner, M., Bayer A.-G., Heat Sensitizers for Polymer Latexes, Ger. Offen. 3330197 [1983/85]; C.A. **102** [1985] No. 204936.

Piccirilli, R. M., McKeough, D. T., Chang, W. H., PPG Industries, Inc., Polymer Microparticles, Brit. Appl. 2157702 [1984/85]; C.A. **104** [1986] No. 208883.

Piotrowski, B., Buening, R., Hanisch, H., Janser, B., Dynamit Nobel A.-G., Silane-Modified Compositions from Vinylidene Fluoride Polymers for Crosslinked Molding, Ger. Offen. 3327596 [1983/85]; C.A. **102** [1985] No. 167808.

Rathousky, J., Kruchna, O., Soil-Repelling Composition for Building Materials and Structures which Facilitates Washing, Czech. 220522 [1981/85]; C.A. **104** [1986] No. 208941.

Rathousky, J., Kruchna, O., Colored, Soil-Repellant, and Easily Washable Coating for Concrete, Czech. 225692 [1982/85]; C.A. **104** [1986] No. 38837.

Richter, R., Mueller, P., Hombach, R., Dollhausen, M., Avar, G., Freitag, H. A., Bayer A.-G., Tin Carboxylate-Sulfonyl Isocyanate Adducts as Reversibly Blocked Catalysts for Polyurethane Manufacture, Ger. Offen. 3326566 [1983/85]; C.A. **102** [1985] No. 185988.

Ricoh Co., Ltd., Lithographic Printing Original Plates, Japan. Kokai Tokkyo Koho 6085995 [1983/85]; C.A. **103** [1985] No. 62624.

Ricoh Co., Ltd., Photolithographic Plates, Japan. Kokai Tokkyo Koho 6090338 [1983/85]; C.A. **103** [1985] No. 113389.

Rifi, M. R., Union Carbide Corp., Water-Curable, Silane-Modified Chlorosulfonated Olefinic Polymers, U.S. 4493924 [1983/85]; C.A. **102** [1985] No. 114850.

Rizk, S. D., Hsieh, H. W. S., Essex Specialty Products, Inc., Thermosetting One-Package Polyurethane Compositions, Eur. Appl. 153135 [1984/85]; C.A. **104** [1986] No. 35674.

Rukavina, T. G., PPG Industries, Inc., Cyanoethyl Acrylate-Acrylic Acid Copolymer, U.S. 4554318 [1984/85]; C.A. **104** [1986] No. 130922.

Sakaguchi, H., Ichii, R., Sugita, T., Mitsui-Nisso Corp., Urethane Rubber for Floor Covering, Japan. Kokai Tokkyo Koho 60134077 [1983/85]; C.A. **104** [1986] No. 7051.

Sakaguchi, H., Ichii, R., Sugita, T., Mitsui-Nisso Corp., Surface Protection of Cellular Plastics, Japan. Kokai Tokkyo Koho 60147329 [1984/85]; C.A. **104** [1986] No. 52178.

Sakai, K., Shimokawa, Y., Toray Industries, Inc., Waterless Lithographic Printing Plate Material, Japan. Kokai Tokkyo Koho 60139482 [1983/85]; C.A. **104** [1986] No. 139343.

Sakuma, K., Kurashige, Y., Idemitsu Petrochemical Co., Ltd., Glossy Liquid Diene Polymer Compositions, Japan. Kokai Tokkyo Koho 60149625 [1984/85]; C.A. **104** [1986] No. 20820.

Sakuma, K., Kurashige, Y., Idemitsu Petrochemical Co., Ltd., Liquid Diene Polymer Compositions, Japan. Kokai Tokkyo Koho 60235826 [1984/85]; C.A. **104** [1986] No. 226182.

Sakurai, Y., Otsuka, Y., Toyo Rubber Industry Co., Ltd., Polyurethane Coating Materials, Japan. Kokai Tokkyo Koho 60149667 [1984/85]; C.A. **104** [1986] No. 111487.

San Nopon Ltd., Sanyo Chemical Industries, Ltd., Thickeners with Excellent Coloring and Leveling Properties, Japan. Kokai Tokkyo Koho 6049022 [1983/85]; C.A. **103** [1985] No. 72679.

Sano, T., Kawabata, K., Miya, Y., Daichi Kogyo Seiyaku Co., Ltd., Adhesion Promoters for Polyolefins, Japan. Kokai Tokkyo Koho 60250043 [1984/85]; C.A. **105** [1986] No. 44875.

Sanyo Chemical Industries, Ltd., Accelerators for Moisture-Curable Polymers, Japan. Kokai Tokkyo Koho 6013850 [1983/85]; C.A. **102** [1985] No. 222221.

Sanyo Chemical Industries, Ltd., Curable Resin Compositions, Japan. Kokai Tokkyo Koho 6018538 [1983/85]; C.A. **103** [1985] No. 55477.

Sanyo Chemical Industries, Ltd., Polyurethanes, Japan. Kokai Tokkyo Koho 6026022 [1983/85]; C.A. **102** [1985] No. 222276.

Saykowski, F., Wuerminghausen, Th., Sattlegger, H., Achtenberg, Th., Bayer A.-G., Room Temperature Vulcanizable Silicone Pastes, Ger. Offen. 3336135 [1983/85]; C.A. **103** [1985] No. 106072.

Schatz, M., Vondracek, P., Aisman, M., Hradec, M., Polycondensation Crosslinking of Linear Polymers, Czech. 218338 [1980/85]; C.A. **103** [1985] No. 142961.

Schenectady Chemicals, Inc., Blocked Isocyanates, Japan. Kokai Tokkyo Koho 60166319 [1984/85]; C.A. **104** [1986] No. 111490.

Schmidt, H. U., Werner, F., BASF A.-G., Polyurethane-Polyurea Moldings with Improved Mold Release Properties, Ger. Offen. 3405680 [1984/85]; C.A. **104** [1986] No. 51508.

Schoenbaechler, M., Saur, W., Gurit-Essex A.-G., Chemically Curable Two-Component Polyurethane Compositions, Ger. Offen. 3407031 [1984/85]; C.A. **104** [1986] No. 89778.

Schupp, E., Lock, W., Osterloh, R., Ahlers, K., BASF Farben und Fasern A.-G., Synthetic Resin Containing Basic Nitrogen-Containing Groups and Its Use, Ger. Offen. 3325061 [1983/85]; C.A. **102** [1985] No. 150998.

Sekisui Chemical Co., Ltd., Silane-Crosslinked Two-Layered Polyethylene Pipes, Japan. Kokai Tokkyo Koho 6002351 [1983/85]; C.A. **102** [1985] No. 186205.

Sekisui Chemical Co., Ltd., Crosslinkable Polypropylene, Japan. Kokai Tokkyo Koho 6018510 [1983/85]; C.A. **102** [1985] No. 221354.

Sekisui Chemical Co., Ltd., Flame-Retardant Foams, Japan. Kokai Tokkyo Koho 6055036 [1983/85]; C.A. **103** [1985] No. 23392.

Sekisui Chemical Co., Ltd., Antistatic Plastic Sheets, Japan. Kokai Tokkyo Koho 6061258 [1983/85]; C.A. **103** [1985] No. 38408.

Sekisui Chemical Co., Ltd., Crosslinked Polyolefin Pipes, Japan. Kokai Tokkyo Koho 6084332 [1983/85]; C.A. **103** [1985] No. 105976.

Sekisui Chemical Co., Ltd., Crosslinkable Polyolefin Compositions, Japan. Kokai Tokkyo Koho 6084346 [1983/85]; C.A. **103** [1985] No. 124487.

Sekisui Chemical Co., Ltd., Glass Laminates, Japan. Kokai Tokkyo Koho 6096551 [1983/85]; C.A. **103** [1985] No. 182613.

Sekisui Chemical Co., Ltd., Photocurable, Electrically Conductive Coating Compositions, Japan. Kokai Tokkyo Koho 6099173 [1983/85]; C.A. **103** [1985] No. 143499.

Sekisui Chemical Co., Ltd., Photocurable Antistatic Coatings, Japan. Kokai Tokkyo Koho 6099176 [1983/85]; C.A. **103** [1985] No. 161929.

Shibuno, T., Terayama, A., Nitto Electric Industrial Co., Ltd., PVC Plastisols with Stable Viscosity, Japan. Kokai Tokkyo Koho 60203660 [1984/85]; C.A. **104** [1986] No. 187501.

Shin-Etsu Chemical Industry Co., Ltd., Discoloration Inhibitor for Building Materials and Pavement Indicator, Japan. Kokai Tokkyo Koho 6051103 [1983/85]; C.A. **103** [1985] No. 41645.

Shiromizu, T., Iwata, M., Chikasawa, T., Fujikura Ltd., Crosslinked Polyethylene Cables, Japan. Kokai Tokkyo Koho 60144333 [1984/85]; C.A. **104** [1986] No. 51725.

Shiromizu, T., Maeda, K., Iwata, M., Fujikura Ltd., Crosslinked Polyethylene Insulated Electric Cables, Japan. Kokai Tokkyo Koho 60139713 [1983/85]; C.A. **104** [1986] No. 70410.

Short, W. T., Ottaviani, R. A., General Motors Corp., Urea Flow Control Agents for Urethane Paint Prepared by Reaction of an Isocyanate-Terminated Prepolymer and an Ethanolamine, U.S. 4522986 [1984/85]; C.A. **103** [1985] No. 72673.

Showa Electric Wire and Cable Co., Ltd., Polyolefin Tubes Shrinkable at Room Temperature, Japan. Kokai Tokkyo Koho 60110730 [1983/85]; C.A. **103** [1985] No. 143052.

Showa Electric Wire and Cable Co., Ltd., Toyo Soda Mfg. Co., Ltd., Extrusion of Vinylsilane-Grafted Crosslinkable Polyolefins, Japan. Kokai Tokkyo Koho 6099153 [1983/85]; C.A. **103** [1985] No. 124604.

Smith, R. A., General Electric Co., Completely Solventless Two-Component RTV Silicone Composition, Can. 1186833 [1982/85]; C.A. **103** [1985] No. 72385.

Smith, St. B., Thermocell Development, Ltd., High Tear-Strength Flexible Urethane Foam Composition, U.S. 4525490 [1984/85]; C.A. **103** [1985] No. 88867.

Smith, St. B., Thermocell Development, Ltd., Sprayable Urethane Resin Composition, U.S. 4543366 [1984/85]; C.A. **104** [1986] No. 111503.

Sogabe, Y., Mizoguchi, I., Uchida, K., Harada, K., Achilles Corp., Double-Stick Adhesive Sheets, Japan. Kokai Tokkyo Koho 60181183 [1984/85]; C.A. **104** [1986] No. 6940.

Sogabe, Y., Uchida, K., Harada, K., Achilles Corp., Cellular Polymer Sheets, Japan. Kokai Tokkyo Koho 60149642 [1984/85]; C.A. **104** [1986] No. 35148.

Sony Corp., Electron-Beam Curing Resin, Japan. Kokai Tokkyo Koho 6065017 [1983/85]; C.A. **103** [1985] No. 63775.

Sony Corp., Polyisocyanate Compounds, Japan. Kokai Tokkyo Koho 6065018 [1983/85]; C.A. **103** [1985] No. 55547.

Sorio, A. A., Gale, G. M., Union Carbide Corp., One-Extrusion Method of Making a Shaped Crosslinkable Extruded Polymeric Product, PCT Intern. Appl. 8504618 [1984/85]; C.A. **104** [1986] No. 187683.

Speranza, G. P., Cuscurida, M., Dominguez, R. J. G., Texaco Inc., RIM Elastomers Made from Terephthalate Polyester Polyol Derived Polymer Polyols, U.S. 4540768 [1984/85]; C.A. **103** [1985] No. 216713.

Steinbach, H. H., Rieder, M., Bayer A.-G., Release Agent for Polyurethane Molding Compositions, Ger. Offen. 3401484 [1984/85]; C.A. **103** [1985] No. 197037.

Such, Ch. H., Dulux Australis Ltd., Metallic Coating Compositions, Brit. Appl. 2156828 [1984/85]; C.A. **104** [1986] No. 208884.

Suda, K., Yagi, M., Takagi, I., Kawakami Paint Mfg. Co., Ltd., Powder Coating Process for Precoated Steel, Japan. Kokai Tokkyo Koho 60257878 [1984/85]; C.A. **105** [1986] No. 44856.

Sugano, Sh., Hagiwara, K., Miura, I., Honto, A., Nagami Chemical Industrial Co., Ltd., Toshiba Silicone Co., Ltd., Primer Composition, Eur. Appl. 136074 [1983/85]; C.A. **102** [1985] No. 72287.

Sutton, T. R., Hughes, D. W., Dow Chemical Co., Reaction Injection Molded Polyurethanes, Polyureas and Polyureas-Urethanes Employing Thiapolycyclic Polyisocyanates, U.S. 4495309 [1984/85]; C.A. **102** [1985] No. 186466.

Szabat, J. F., Mobay Chemical Corp., Fire-Resistant Flexible Polyurethane Foam, U.S. 4546117 [1983/85]; C.A. **104** [1986] No. 20246.

Takahashi, M., Wakabayashi, K., Dainippon Ink and Chemicals, Inc., Additives for Polyurethane Foams, Japan. Kokai Tokkyo Koho 60202114 [1984/85]; C.A. **104** [1986] No. 169500.

Takiyama, E., Morita, K., Showa Highpolymer Co., Ltd., Thermosetting Polymers, Japan. Kokai Tokkyo Koho 60197719 [1984/85]; C.A. **104** [1986] No. 225766.

Tanaka, M., Ohashi, H., Shin-Etsu Chemical Industry Co., Ltd., Fiber Finishing Agents, Japan. Kokai Tokkyo Koho 60259680 [1984/85]; C.A. **105** [1986] No. 44716.

Tatsuta, Electric Wire and Cable Co., Ltd., Crosslinked Vinyl Chloride Resin Moldings, Japan. Kokai Tokkyo Koho 6042429 [1983/85]; C.A. **103** [1985] No. 72285.

Tawara, S., Sakai, I., Moroishi, Y., So, I., Nitto Electric Industrial Co., Ltd., One-Package Elastic Sealants, Japan. Kokai Tokkyo Koho 60202181 [1984/85]; C.A. **104** [1986] No. 188352.

Taylor, R. P., Cekoric, M. E., Dewhurst, J. E., Abouzahr, S. M., Mobay Chemical Corp., Internal Mold Release Agent for Use in Reaction Injection Molding, U.S. 4519965 [1984/85]; C.A. **103** [1985] No. 38542.

Theodore, A. N., Chattha, M. S., Ford Motor Co., Nonaqueous Dispersions Based on Capped Stabilizers and Reactants Comprising Polyfunctional Monomers II, U.S. 4528317 [1983/85]; C.A. **103** [1985] No. 161933.

Theodore, A. N., Chattha, M. S., Ford Motor Co., Nonaqueous Dispersions Based on Capped Stabilizers and Polyfunctional Monomers, U.S. 4533695 [1983/85]; C.A. **103** [1985] No. 161920.

Toa, Gosei Chemical Industry Co., Ltd., Curable Condensation Compositions, Japan. Kokai Tokkyo Koho 60127317 [1983/85]; C.A. **104** [1986] No. 89769.

Toa Nenryo Kogyo K. K., Crosslinked Resin Composition from Modified Polyolefin and Silane Catalyst, Japan. Kokai Tokkyo Koho 6013804 [1983/85]; C.A. **103** [1985] No. 72157.

Tochu Plastic Kogya K. K., Metal-Polyethylene Composite Sheets, Japan. Kokai Tokkyo Koho 6046239 [1983/85]; C.A. **103** [1985] No. 7487.

Tokuyama Sekisui Industry Co., Ltd., Electroconductive Graft PVC Compositions, Japan. Kokai Tokkyo Koho 6076552 [1984/85]; C.A. **103** [1985] No. 72176.

Toray Industries, Inc., Waterless Lithographic Plates, Japan. Kokai Tokkyo Koho 6009799 [1983/85]; C.A. **102** [1985] No. 212733.

Toray Industries, Inc., Resin of High Refractive Index for Lens Use, Japan. Kokai Tokkyo Koho 6011513 [1983/85]; C.A. **103** [1985] No. 23288.

Toray Industries, Inc., Photosensitive Compositions, Japan. Kokai Tokkyo Koho 6015428 [1983/85]; C.A. **103** [1985] No. 142885.

Toray Industries, Inc., Preparation of Dampening-Free Lithographic Plate, Japan. Kokai Tokkyo Koho 6028654 [1983/85]; C.A. **103** [1985] No. 45857.

Toray Industries, Inc., Preparation of Dampening-Free Lithographic Plate, Japan. Kokai Tokkyo Koho 6028655 [1983/85]; C.A. **103** [1985] No. 45856.

Toray Industries, Inc., Plastic Lens with High Refractive Index, Japan. Kokai Tokkyo Koho 6051706 [1983/85]; C.A. **103** [1985] No. 88787.

Toray Industries, Inc., Radically Curable Oligomers, Japan. Kokai Tokkyo Koho 60118712 [1983/85]; C.A. **105** [1986] No. 7026.

Toshiba Silicone Co., Ltd., Room Temperature-Curable Silicone Rubber Compositions, Japan. Kokai Tokkyo Koho 6018544 [1983/85]; C.A. **103** [1985] No. 7606.

Toshiba Silicone Co., Ltd., Silicone Rubber Compositions, Japan. Kokai Tokkyo Koho 6018545 [1983/85]; C.A. **102** [1985] No. 222010.

Toshiba Silicone Co., Ltd., Silicone Compositions Curable at Room Temperature, Japan. Kokai Tokkyo Koho 6049062 [1983/85]; C.A. **103** [1985] No. 72735.

Toshiba Silicone Co., Ltd., Coating Compositions, Japan. Kokai Tokkyo Koho 6049066 [1983/85]; C.A. **103** [1985] No. 55527.

Toyo Rubber Chemical Industry Co., Ltd., Tachikawa Spring Co., Ltd., Polyurethane Foams, Japan. Kokai Tokkyo Koho 60130630 [1983/85]; C.A. **104** [1986] No. 6649.

Toyo Rubber Industry Co., Ltd., Electron Beam-Curable Thermosetting Resins, Japan. Kokai Tokkyo Koho 6042450 [1983/85]; C.A. **103** [1985] No. 23362.

Toyo Rubber Industry Co., Ltd., Magnetic Coating Materials, Japan. Kokai Tokkyo Koho 6090272 [1983/85]; C.A. **103** [1985] No. 161946.

Toyo Rubber Industry Co., Ltd., Magnetic Coating, Japan. Kokai Tokkyo Koho 60120765 [1983/85]; C.A. **104** [1986] No. 80761.

Toyobo Co., Ltd., Thermoplastic Polyurethanes, Japan. Kokai Tokkyo Koho 60115616 [1983/85]; C.A. **104** [1986] No. 6342.

Toyoda Gosei Co., Ltd., Abrasion-Resistant Frames for Sliding Automobile Windows, Japan. Kokai Tokkyo Koho 6092363 [1983/85]; C.A. **103** [1985] No. 89179.

Ueno, H., Nippon Dry Chemical Co., Ltd., Powdery Fire Extinguisher for Metal Fire, Japan. Kokai Tokkyo Koho 60188180 [1984/85]; C.A. **104** [1986] No. 91614.

Ui, M., Tokai Rubber Industries, Ltd., Adhesives for Anchor Bolts, Japan. Kokai Tokkyo Koho 60156780 [1984/85]; C.A. **104** [1986] No. 51749.

Umpleby, J. D., BP Chemicals Ltd., Silane-Crosslinkable Polyolefin Compositions, Eur. Appl. 149903 [1984/85]; C.A. **103** [1985] No. 161392.

Umpleby, J. D., BP Chemicals Ltd., Crosslinkable Composition Comprising a Thermoplastic Silyl-Modified Polymer and a Supported Silanol Condensation Catalyst, Eur. Appl. 150595 [1984/85]; C.A. **103** [1985] No. 196917.

von Bittera, M., Schaepel, D., von Gizycki, U., Rupp, R., Bayer A.-G., Materials with Self-Adhering Surfaces and Their Use, Ger. Offen. 3341555 [1983/85]; C.A. **103** [1985] No. 88809.

Wakabayashi, T., Yoshida, S., Asahi Glass Co., Ltd., Optical Elements, Japan. Kokai Tokkyo Koho 60194401 [1984/85]; C.A. **104** [1986] No. 90075.

Wanat, St., American Hoechst Corp., Catalyzed Urethanation of Polymers, Brit. Appl. 2155487 [1984/85]; C.A. **104** [1986] No. 89253.

Weaver, W. R., Libbey-Owens-Ford Co., Molding of Seals on Windshilds, Fr. Demande 2553083 [1983/85]; C.A. **103** [1985] No. 124813.

Werner, F., Marx, M., Horn, P., Schmidt, H. U., BASF A.-G., Cellular Polyurethane-Polyurea Moldings, Ger. Offen. 3405679 [1984/85]; C.A. **104** [1986] No. 51510.

Werner, F., Marx, M., Schmidt, H. U., BASF A.-G., Polyurethane-Polyurea Moldings with Improved Demolding Properties and an Internal Mold Release Agent, Ger. Offen. 3405875 [1984/85]; C.A. **104** [1986] No. 51509.

Wood, L. L., Wazolek, W. R., Fulmer, G. E., W. R. Grace and Co., Crosslinked Polyurethane Emulsion, Ger. Offen. 3433029 [1983/85]; C.A. **103** [1985] No. 55488.

Wuerminghausen, Th., Saykowski, F., Sattlegger, H., Achtenberg, Th., Bayer A.-G., Single-Component Silicone Pastes, Ger. Offen. 3323912 [1983/85]; C.A. **102** [1985] No. 150763.

Yagyu, H., Watanabe, K., Shibayama, M., Hitachi Cable, Ltd., Moisture-Curable Electric Insulation, Japan. Kokai Tokkyo Koho 60198006 [1984/85]; C.A. **104** [1986] No. 150350.

Yamada, K., Ohayashi, Sh., Mitsui, A., Nippon Steel Chemical Co., Ltd., Room-Temperature-Curable Silicone Elastomer Compositions, Japan. Kokai Tokkyo Koho 60210663 [1984/85]; C.A. **104** [1986] No. 187944.

Yamamoto, A., Kansai Paint Co., Ltd., Colored Coating Compositions for Heat-Resistant Glass, Japan. Kokai Tokkyo Koho 60188476 [1984/85]; C.A. **104** [1986] No. 111507.

Yamazaki, Y., Mabuchi, A., Yano, K., Toyoda Gosei Co., Ltd., Glass Guide Rails for Glass Windows of Motor Vehicles, Ger. Offen. 3503479 [1984/85]; C.A. **104** [1986] No. 52176.

Yasui, S., Nagata, K., Sanyo Chemical Industries, Ltd., Polyurethanes, Japan. Kokai Tokkyo Koho 60161415 [1984/85]; C.A. **104** [1986] No. 69940.

Yokota, M., Goshima, T., Fujioka, Sh., Toray Industries, Inc., Hard Contact Lenses, Japan. Kokai Tokkyo Koho 60146219 [1984/85]; C.A. **104** [1986] No. 39781.

Yoshihara, M., Morikawa, K., Nitto Electric Industrial Co., Ltd., Coating Materials for Optical Glass Fibers, Japan. Kokai Tokkyo Koho 60171249 [1984/85]; C.A. **104** [1986] No. 154364.

Yoshihara, M., Morikawa, K., Nitto Electric Industrial Co., Ltd., Coating Materials for Optical Glass Fibers, Japan. Kokai Tokkyo Koho 60251157 [1984/85]; C.A. **104** [1986] No. 191667.

Yoshihara, M., Nakajima, T., Morikawa, K., Komada, M., Yamamoto, K., Nitto Electric Industrial Co., Ltd., Coating Materials for Optical Glass Fibers, Japan. Kokai Tokkyo Koho 60251152 [1984/85]; C.A. **104** [1986] No. 211881.

1.4.1.2.1.5.4.1.7 Dibutyltin Biscarboxylates $(C_4H_9)_2Sn(OOCR)_2$ with R = $C_{11}H_{23}$, $C_{13}H_{27}$, $C_{15}H_{31}$, $C_{17}H_{35}$, $C_{20}H_{41}$, and $C_{21}H_{43}$

$(C_4H_9)_2Sn(OOC(CH_2)_7C(CH_3)_3)_2$

The synthesis of this compound is not described in the literature. It is used as a catalyst for the synthesis of polyurethane elastomers [43].

Dibutyltin Dimyristinate, $(C_4H_9)_2Sn(OOCC_{13}H_{27})_2$

A synthesis of this compound is not described in the literature. The compound melts between 40 and 42°C and decomposes above 250°C with formation of $C_{13}H_{27}COC_{13}H_{27}$ [24]. Dibutyltin dimyristate was tested as a stabilizer for polyvinyl chloride against heat and γ radiation from ^{60}Co but it was less effective than other dibutyltin dicarboxylates [10].

Dibutyltin Dipalmitate, $(C_4H_9)_2Sn(OOCC_{15}H_{31})_2$

The synthesis of this compound is not described in the literature. It can be separated from other butyltin compounds by thin layer chromatography [31].

The compound is used as an anthelminthicum for the removal of Raillietina cesticillus and Ascaridia galli from chickens [2], but with only low activity [14], and as a thermooxidative stabilizer for di-2-ethylhexyldodecanedioate and esters of pentaerythritol with monocarboxylic acids at 180 and 220°C [44].

Dibutyltin Distearate, $(C_4H_9)_2Sn(OOCC_{17}H_{35})_2$

$(C_4H_9)_2Sn(OOCC_{17}H_{35})_2$ is prepared from $(-(C_4H_9)_2SnO-)_n$ and $C_{17}H_{35}COOH$, with heating to drive off the water in 91% [20] or 95 to 98% yield [9], or from $(-(C_4H_9)_2SnO-)_n$ and $(C_{17}H_{35}CO)_2O$ [32]. A melting point of 48 to 50°C [9] or 49 to 50°C [20] is given.

^{119}Sn Mössbauer spectrum (mm/s at 74 K): $\delta = 1.45$ relative to SnO_2, $\Delta = 3.30$ [11], and $\delta = -1.34$ relative β-Sn, $\Delta = 3.56$ [12].

The IR spectrum of dibutyltin distearate was measured and assigned in CCl_4 solution and in Nujol mull (Table 22) [30].

Paper chromatography [8, 37], ESR spectroscopy [34], and Mössbauer spectroscopy [15] are used for the separation and determination of dibutyltin distearate from other organotin compounds, from residues or from polyvinyl chloride. A special method, using potentiometric titration with Sb electrodes was developed for the determination of RCOO groups in stannyl esters including $(C_4H_9)_2Sn(OOCC_{17}H_{35})_2$ [16].

The toxicity of $(C_4H_9)_2Sn(OOCC_{17}H_{35})_2$ is not high. The LD_{50} for rats is more than 4000 mg/kg bodyweight [36]. The acute oral toxicity to deer mice was found to be $LD_{50} = 460$ to 1600 mg/kg [45]. The compound was used as an anthelminticum [1, 2, 14]. The toxicity of a PVC material used in blood transfusion, which contains $(C_4H_9)_2Sn(OOCC_{17}H_{35})_2$ as a stabilizer was tested by an optical method [39].

Table 22
Infrared Spectrum of $(C_4H_9)_2Sn(OOCC_{17}H_{35})_2$.
Wave numbers in cm^{-1}.

Nujol	CCl_4	assignment
1587 (s)	1595 (s)	$\nu_{as}(OCO)$
1464 (s)	1592 (s)	$\delta_{as}(CH_2)$
1406 (s)	1406 (s)	$\delta_s(CH_2 \leftarrow CH_2COO)$
1385 (s)	1385 (s)	$\nu_s(OCO)$, $\delta_s(CH_3 \leftarrow R)$
1342		
1316		$\nu_s(OCO)$
1299	1290	$\delta_w(CH_2)$
1282	1282	$\delta_s(CH_2)$
1263		
725 (s)	725 (s)	$\delta_r(CH_2)$
694 (s)		$\delta_r(CH_2 \leftarrow CH_2Sn)$
	671 (s)	
606	621	$\nu_{as}(SnC)$
476 (w)	476 (w)	
460 (w)	460 (w)	

Dibutyltin distearate is used as a stabilizer for polyvinyl chloride [4, 6, 17, 18, 19, 32, 35, 41, 46] and halogenated resins [7, 42] against heat and light. Mechanistic studies of the stabilization of PVC in the presence of phosphites showed that $(C_4H_9)_2Sn(OOCC_{17}H_{35})_2$ forms synergistic mixtures with the phosphites in the thermal decomposition of PVC [22]. $(C_4H_9)_2Sn(OOCC_{17}H_{35})_2$ is useful as a stabilizer for polyesters [26], for methylchloroform compositions [40], and for lubricants [44]. Dibutyltin distearate catalyzes the conversion of haloalkane nitriles to triazines [13], the reaction between tosylhydrazine and phenyl isocyanate [27, 28], and the condensation of polydimethylsiloxane rubber with tetraethoxysiloxane [3, 5]. Polypropylene fibers are improved by incorporating $(C_4H_9)_2Sn(OOCC_{17}H_{35})_2$ [23] or a product of the reaction between $(C_4H_9)_2Sn(OOCC_{17}H_{35})_2$ and a nickel complex [21, 25]. Dibutyltin distearate is also used as an additive for polyamide molding compositions [29].

$(C_4H_9)_2Sn(OOCC_{17}H_{35}\text{-}i)_2$

The compound is mentioned in a patent as an ingredient in a liquid stabilizer mixture for chlorine-containing thermoplasts [33].

$(C_4H_9)_2Sn(OOCCH(CH_3)C_{18}H_{37})_2$ and $(C_4H_9)_2Sn(OOCC_{21}H_{43})_2$

Both compounds are mentioned in a patent as stabilizers for PVC [38].

References:

[1] Cita, C., Dr. Salsbury's Laboratories (Brit. 711563 [1951/54]; C. **1955** 5375).
[2] Kerr, K. B., Walde, A. W. (Exptl. Parasitol. **5** [1956] 560/70).

[3] Baranovskaya, N. B., Berlin, A. A., Zakharova, M. Z., Mizikin, A. I. (Khim. Prakt. Primen. Kremneorg. Soedin. Tr. 2nd Konf., Leningrad 1958 [1961], pp. 88/95; C.A. **1960** 7206).
[4] Berlin, A. A., Popova, Z. V., Yanovskii, D. M. (Zh. Prikl. Khim. **33** [1960] 871/7; J. Appl. Chem. [USSR] **33** [1960] 870/5).
[5] Novikov, A. S., Nudelman, Z. N. (Kauch. Rezina **19** No. 12 [1960] 3/7; Soviet Rubber Technol. **19** No. 12 [1960] 5/7).
[6] Popovam, Z. V., Yanovskii, D. M. (Mezhdunar. Simp. Makromol. Khim. Dokl. Avtoreferaty, Moscow 1960, Vol. 3, pp. 372/9 from C.A. **1961** 9940).
[7] Ito, T., Kyodo Pharmaceutical Co. (Japan. 61-284 [1961]; C.A. **1961** 20518).
[8] Gasparic, J., Cee, A. (J. Chromatog. **8** [1962] 393/8).
[9] Sheverdina, N. I., Abramova, L. V., Paleeva, I. E., Kocheshkov, K. A. (Khim. Prom. [Kiev] **1962** 707/8; C.A. **59** [1963] 8776).
[10] Shteding, M. M., Karpov, V. L. (Vysokomol. Soedin. **4** [1962] 1806/11; C.A. **59** [1963] 2989).

[11] Aleksandrov, A. Yu., Delyagin, N. N., Mitrofanov, K. P., Polak, L. S., Shpinel, V. S. (Dokl. Akad. Nauk SSSR **148** [1963] 126/8; Dokl. Phys. Chem. Proc. Acad. Sci. USSR **148/153** [1963] 1/3).
[12] Gol'danskii, V. I., Makarov, E. F., Stukan, R. A., Trukhtanov, V. A., Khrapov, V. V. (Dokl. Akad. Nauk SSSR **151** [1963] 357/60; Dokl. Phys. Chem. Proc. Acad. Sci. USSR **148/153** [1963] 598/601).
[13] Emerson, W. E., Dorfman, E., Hooker Chemical Corp. (Fr. 1574807 [1967/69]; C.A. **72** [1970] No. 67772).
[14] Tareeva, A. I., Borodina, G. M. (Farmakol. Toksikol. [Moscow] **30** [1967] 207/9; C.A. **67** [1967] No. 20351).
[15] Aleksandrov, A. Yu., Baldokhin, Yu. V., Braginskii, R. P., Gol'danskii, V. I., Korytko, L. A., Leshchenko, S. S., Finkel, G. E. (Khim. Vysokikh Energ. **2** [1968] 331/7; High Energy Chem. [USSR] **2** [1968] 285/90).
[16] Groagova, A., Pribyl, M. (Z. Anal. Chem. **234** [1968] 423/8).
[17] Stapfer, C. H., Carlisle Chemical Works, Inc. (Fr. 1537462 [1966/68]; C.A. **71** [1969] No. 13830).
[18] Troitskaya, L. C., Troitskii, B. B. (Plasticheskie Massy **1968** No. 9, pp. 12/5; Soviet Plast. **1968** No. 9, pp. 13/6).
[19] Hoch, S., Tenneco Chemicals Inc. (Ger. 1801274 [1968/69]; C.A. **71** [1969] No. 13811).
[20] Mikhailov, G. D., Chegolya, A. S. (Sin. Volokna **1969** 18/21; C.A. **74** [1971] No. 3711).

[21] Senda, K., Ichikawa, A., Nakajima, E., Sasaki, M., Hirose, M., Mitsubishi Rayon Co., Ltd. (Japan. 69-26179 [1966/69]; C.A. **72** [1970] No. 101804).
[22] Troitskaya, L. S., Troitskii (Izv. Akad. Nauk SSSR Ser. Khim. **1969** 2141/8; Bull. Acad. Sci. [USSR] Div. Chem. Sci. **1969** 1997/2003).
[23] Yamamoto, K., Nakatsuka, K., Mitsubishi Rayon Co., Ltd. (Japan. 69-15185 [1965/69]; C.A. **72** [1970] No. 13232).
[24] Matsuda, S., Matsuda, H., Yamai, Y., Ninimiya, K. (Kogyo Kagaku Zasshi **73** [1970] 1007/9).
[25] Yamamoto, T., Yukida, Y., Toray Industries, Inc. (Japan. 70-12665 [1967/70]; C.A. **73** [1970] No. 88659).
[26] Ogawa, K., Toray Industries, Inc. (Japan. 71-38707 [1968/71]; C.A. **77** [1972] No. 89421).
[27] Grekov, A. P., Otroshko, G. V. (Zh. Org. Khim. **10** [1974] 1905/8; J. Org. Chem. [USSR] **10** [1974] 1916/8).
[28] Grekov, A. P., Otroshko, G. V. (Kratk. Tezisy Vses. Soveshch. Probl. Mekh. Geteroliticheskikh Reakts. **1974** 19/20 from C.A. **85** [1976] No. 77131).

[29] Hagiwara, T., Ohtsubo, T., Toyobo Co., Ltd. (Japan. 75-83452 [1973/75]; C.A. **84** [1976] No. 5907).
[30] Morikawa, T. (Kagaku To Kogyo [Osaka] **49** [1975] 98/113; C.A. **84** [1976] No. 5795).
[31] Moshkovskaya, M. V., Nefedov, V. D., Molchanova, N. G., Zhuravlev, V. E. (Tr. Estestvennonauchn. Inst. Permsk. Gos. Univ. **13** [1975] 210/3; C.A. **86** [1977] No. 121468).
[32] Dworkin, R. D., Ejk, A. J., M and T Chemicals, Inc. (Ger. 2626554 [1975/76]; C.A. **86** [1977] No. 140253).
[33] Mueller, H., Ciba-Geigy A.-G. (Ger. 2546900 [1974/76]; C.A. **85** [1976] No. 79038).
[34] Stegmann, H. B., Uber, W., Scheffler, K. (Z. Anal. Chem. **286** [1977] 59/64).
[35] Sato, K., Kogo, Y., Kiryu, M., Sankyo Organic Chemicals Co., Ltd. (Japan. 78-134052 [1977/78]; C.A. **90** [1979] No. 138626).
[36] Smith, P. J., Luijten, J. G. A., Klimmer, O. R. (Intern. Tin Res. Inst. Publ. No. 538 [1978]).
[37] Vasundhara, T. S., Parihar, D. B. (Z. Anal. Chem. **294** [1979] 408).
[38] Watanabe, N., Tawada, H., Sato, T., Onoda, K., Miyoshi Oil and Fat Co., Ltd. (Japan. 79-03861 [1977/79]; C.A. **90** [1979] No. 187962).
[39] Eskov, A. P., Kayumov, R. I., Luzhetskii, A. S., Gurilev, O. M., Ryazanov, Yu. N., Smelik, G. I., Arefev, I. M. (Gig. Sanit. **1985** No. 1, pp. 62/4; C.A. **102** [1985] No. 107541).
[40] Asahi-Dow Ltd. (Japan. Kokai Tokkyo Koho 82-116021 [1981/82]; C.A. **97** [1982] No. 215515).
[41] Nitto Chemical Industry Co., Ltd. (Japan. Kokai Tokkyo Koho 82-00150 [1980/82]; C.A. **96** [1982] No. 218765).
[42] Sankyo Organic Chemicals Co., Ltd. (Japan. Kokai Tokkyo Koho 57147537 [1981/82]; C.A. **98** [1983] No. 90486).
[43] Bats, J. P., Ducasse, Y., Lalande, R., Tauzia, M. (Eur. Polym. J. **20** [1984] 997/1001).
[44] Hronec, M., Cvengrosova, Z., Kizlink, J., Stolcova, M., Malik, L., Ilavsky, J., Sitek, J. (Oxid. Commun. **8** [1985/86] 51/64).
[45] Schafer, E. W., Bowles, W. A. (Arch. Environ. Contam. Toxicol. **14** [1985] 111/29).
[46] Troitskii, B. B., Troitskaya, L. S., Yakhnov, A. S., Novikova, M. A., Razuvaev, G. A. (Vysokomol. Soedin. B **27** [1985] 49/53).

1.4.1.2.1.5.4.1.8 Dibutyltin Biscarboxylates, $(C_4H_9)_2Sn(OOCR)_2$ with R = Cycloalkyl

$(C_4H_9)_2Sn(OOCC_5H_9\text{-c})_2$

The compound is prepared from $(C_4H_9)_2SnCl_2(NH_3)_2$ and cyclo-C_5H_9COOH in C_6H_6 in 98% yield, free of any tributyltin impurities. The refractivity index n_D^{20} is 1.4960 [1].

$(C_4H_9)_2Sn(OOCC_6H_{11}\text{-c})_2$

The compound is used as a catalyst for the synthesis of polyurethanes [2].

References:

[1] Viveen, W. J. C., Schröder, A., Koninklijke Industrieele Maatschappij vorheen Noury & van der Lande N. V. (Ger. Offen. 1167836 [1962/64]).
[2] Bats, J. P., Ducasse, Y., Lalande, R., Tauzia, M. (Eur. Polym. J. **20** [1984] 997/1001).

1.4.1.2.1.5.4.2 Dibutyltin Biscarboxylates, $(C_4H_9)_2Sn(OOCR)_2$ with R = Substituted Alkyl or Cycloalkyl

The compounds belonging to this section are listed in Table 23. They were prepared by the following methods.

Method I: $(C_4H_9)_2SnCl_2$ and RCOOM (M = Na or K, 1:2 mole ratio).
No. 5 is obtained in high yields by the reaction between $(C_4H_9)_2SnCl_2$ and $CHCl_2COONa$ [17] in CH_2Cl_2-H_2O at 20°C using the interfacial method [21]. This method has also been used for the synthesis of No. 62 from $(C_4H_9)_2SnCl_2$ and $C_6H_5CH(OH)COONa$ in petroleum ether-H_2O at 0°C [21]. The reaction between $(C_4H_9)_2SnCl_2$ and $(C_6H_5)_2C(OH)COOK$ in refluxing C_2H_5OH for 1 h yields No. 63 [66].

Method II: $(-(C_4H_9)_2SnO-)_n$ and RCOOH (1:2 mole ratio).
The condensation reaction between $(-(C_4H_9)_2SnO-)_n$ and the free acid is the method most frequently used; it is usually carried out in an organic solvent under reflux and is completed by removal of the water formed by azeotropical distillation. Thus, C_6H_6 is the solvent for the synthesis of Nos. 3 and 25 [32], Nos. 10 to 13 [19], Nos. 24 and 26 to 31 [72], No. 32 [23], No. 33 [22], Nos. 37, 38, and 51 [53], and of No. 50 (in the presence of catalytic amounts of $CH_3C_6H_4SO_3H$-4) [2]. A mixture of C_6H_6 and C_2H_5OH (3:1 v/v) was used in the preparation of Nos. 53 and 54 [76] and Nos. 55 to 57 [77]. Nos. 42 [15, 18], 52 [27], 64, 65 [24], and 71 [65] were synthesized in refluxing $C_6H_5CH_3$. No. 2 was prepared in THF solution [68] and No. 73 in $CH_3COC_4H_9$-i [11]. Heating of stoichiometric mixtures of $(-(C_4H_9)_2SnO-)_n$ and the appropriate RCOOH derivative under vacuum without a solvent leads to No. 17 (70 to 100°C) [44], Nos. 19 and 20 (120 to 130°C) [7], and No. 58 (160 to 190°C) [67]. The conditions for the preparation of No. 63 could not be ascertained [33].

Method III: Transesterification of $(C_4H_9)_2Sn(OOCCH_3)_2$ with a twofold molar amount of $CH_2ClCOOH$ in C_6H_6 yields No. 4 [20].

Method IV: $(C_4H_9)_2SnH_2$ and RCOOH.
Addition of $(C_4H_9)_2SnH_2$ (88% purity) to the appropriate RCOOH derivative dissolved in dioxane causes H_2 evolution within 5 to 10 min (complete after 6 to 16 h) leading to formation of Nos. 1 and 4 to 6. Depending on the acid to hydride ratio and the acid itself, $RCOO(C_4H_9)_2SnOSn(C_4H_9)_2OOCR$ compounds are formed as byproducts [12].

Method V: $RCOO(C_4H_9)_2SnOSn(C_4H_9)_2OH$ or $RCOO(C_4H_9)_2SnOSn(C_4H_9)_2OOCR$ and RCOOH.
$CF_3COO(C_4H_9)_2SnOSn(C_4H_9)_2OH$ reacts with CF_3COOH (1:3 mole ratio) in boiling C_6H_6 within 1 h to give No. 1. Under the same conditions, the tetrabutyl-1,3-diacyloxydistannoxanes (R = CH_2Cl, $CHCl_2$, CCl_3) react with the twofold molar amount of RCOOH yielding Nos. 4 to 6 [57]. No. 5 is also obtained by the reaction of 4-$NO_2C_6H_4COO(C_4H_9)_2SnOSn(C_4H_9)_2OOCC_6H_4NO_2$-4 with $CHCl_2COOH$ (1:2 mole ratio) in boiling $C_6H_5CH_3$ [26].

Method VI: Addition of $(-(C_4H_9)_2SnO-)_n$ to a C_6H_6 or $C_6H_5CH_3$ solution containing the product of the reaction of $HSCH_2CH_2COOH$ and $HSCH_2CH_2COOC_8H_{17}$-i with 2-HOC_6H_4CHO [51]; $HSCH_2CH_2COOH$ and $C_{17}H_{35}COOCH_2CH(OH)CH_2COOH$ with $(C_2H_5)_2CHCHO$ [56]; $HSCH_2CH_2COOH$ and $HSCH_2COOC_{12}H_{25}$ with $C_2H_5OOCCH_2COCH_3$; $HSCH_2CH_2COOH$ and $HSCH_2CH_2COOC_8H_{17}$-i with $C_{11}H_{23}CHO$; or $HSCH_2CH_2COOH$ and $HSCH_2COOC_{12}H_{25}$ with $C_{11}H_{23}CHO$ [51] leads to the formation of Nos. 45 to 49, respectively.

Table 23

$(C_4H_9)_2Sn(OOCR)_2$ Compounds with R = Substituted Alkyl or Cycloalkyl.
Further information on compounds preceded by an asterisk is given at the end of the table.
Explanations, abbreviations, and units on p. X.

No.	OOCR group method of preparation (yield in %)	properties and remarks	Ref.
1	$OOCCF_3$ IV [12] V (47 [57])	m.p. 57°	[57]
		with $H_2O \rightarrow$ $CF_3COO(C_4H_9)_2SnOSn(C_4H_9)_2OH$	
		used for the preparation of conducting SnO_2 films on glass	[70]
		studies on toxicity [55, 64] (LD_{50} = 53.6 mg/kg rat) and mutagenic properties [60]	[55,60, 64]
2	$OOCC_3F_7$ II [68]	catalyst for polyurethane formation	[68,71]
3	$OOCC_7F_{15}$ II	m.p. 63 to 65° b.p. 137°/0.1	[32]
4	$OOCCH_2Cl$ III (76 [20]) IV (81 [12]) V (41 [57])	m.p. 87 to 88° [20, 57], 87 to 89° [12], 90.3° [49]	[12,20, 49,57]
		b.p. 165 to 166°/5 (dec.)	[20]
		^{119}Sn-γ (78 to 80 K): δ = 1.46 [39], 1.60 [13], or −0.64 (α-Sn) [27], Δ = 3.56 [27, 39], 3.65 [13, 31]	[13,27, 31,39]
		thermolysis at 240° → $(C_4H_9)_2SnCl_2$ and products containing no chlorine	[49]
		with $(C_4H_9)_2Sn(OCH_3)_2 \rightarrow CH_2ClCOOCH_3$, $CH_2ClCOO(C_4H_9)_2SnOSn(C_4H_9)_2OCH_3$	[20,25]
		with $H_2O \rightarrow CH_2ClCOOH$, $CH_2ClCOO(C_4H_9)_2SnOSn(C_4H_9)_2OH$	[57]
		tested as a fungicide and bactericide	[1,6]
		used as a gelation catalyst for polyurethane foams	[48]
5	$OOCCHCl_2$ Ib (70 [17], 88 [21]) IV (33 [12]) V (64 [57], 90 [26])	m.p. 112 to 113° [57], 112 to 114° [12, 21, 26]	[12,21, 26,57]
		^{119}Sn-γ (78 to 80 K): δ = 1.54 [39], −0.56 (α-Sn) [27], Δ = 3.73 [27, 39]	[27,39]
		with $H_2O \rightarrow CHCl_2COOH$, $CHCl_2COO(C_4H_9)_2SnOSn(C_4H_9)_2OOCCHCl_2$	[57]
		with $KOOC(CH_2)_4COOK \rightarrow$ $(-(C_4H_9)_2SnOOC(CH_2)_4COO-)_n$	[17]

Table 23 (continued)

No.	OOCR group method of preparation (yield in %)	properties and remarks	Ref.
6	$OOCCCl_3$ IV (9 [12]) V (3 [57])	m.p. 113° ^{119}Sn-γ (78 to 80 K): δ = 1.5 [16, 39], (in C_2H_5OH) [16], 1.65 [13], Δ = 3.8 [13, 31], 3.9 (in C_2H_5OH) [16], 4.0 [16, 39] with $H_2O \rightarrow CCl_3COOH$, $CCl_3COO(C_4H_9)_2SnOSn(C_4H_9)_2OH$	[57] [13, 16, 31, 39] [57]
7	$OOCCHClCH_3$	no preparation reported used as a gelation catalyst for polyurethane foams	 [48]
8	$OOCCH_2CH_2Cl$	no preparation reported thermolysis at 240° → $(C_4H_9)_2SnCl_2$, $CH_2{=}CHCOOH$ use like No. 7	 [49] [48]
9	$OOCCH_2CH_2CH_2Cl$	no preparation reported use like No. 7	 [48]
10	$OOC(CH_2)_3CH_2Cl$ II	^{119}Sn-γ (77 K): δ = −1.20 (β-Sn), Δ = 3.41 structure discussion	[19]
11	$OOC(CH_2)_5CH_2Cl$ II	^{119}Sn-γ (77 K): δ = −1.27 (β-Sn), Δ = 3.35 structure discussion	[19]
12	$OOC(CH_2)_7CH_2Cl$ II	^{119}Sn-γ (77 K): δ = −1.15 (β-Sn), Δ = 3.39 structure discussion	[19]
13	$OOC(CH_2)_{11}CH_2Cl$ II	^{119}Sn-γ (77 K): δ = −1.40 (β-Sn), Δ = 2.89 structure discussion	[19]
14	$OOCCH_2OH$	no preparation reported tested as a fungicide and bactericide	 [6]
15	$OOCCH(OH)CH_3$	no preparation reported used as an additive for the preparation of siloxane elastomers curing at room temperature	 [5]
16	$OOC(CH_2)_{10}CH(OH)C_6H_{13}$	no preparation reported stabilizer and lubricant for PVC	 [62, 63]
17	$OOC(CH_2)_7CH(OH)CH(OH)$-$(CH_2)_5CH_2OH$ II	stabilizer for PVC	[44]

Table 23 (continued)

No.	OOCR group method of preparation (yield in %)	properties and remarks	Ref.
18	$OOC(CH(OH))_4CH_2OH$ d-gluconic acid	no preparation reported anthelmintic activity against Raillietina cesticillus and Ascaridia galli	 [3]
19	$OOC(CH_2)_7CH\text{-}CH(CH_2)_7CH_3$ (CH-CH bridged by O: \ / O) II [7]	stabilizer for PVC catalyst for the preparation of polyacetal resins	[7,8,9] [29]
20	$OOC(CH_2)_7CH\text{-}CHCH_2\text{-}$ (CH-CH bridged by O: \ / O) $CH(OH)(CH_2)_5CH_3$ II	stabilizer for PVC	[7]
21	$OOCCH_2CH_2OCH_3$ not ascertained [41]	thermolysis (%) $\rightarrow$ C_4H_{10} (6.2), CH_3OH (1.6), $CH_2{=}CHCOOCH_3$ (1.6), $CH_3OCH_2CH_2COOC_4H_9$ (12.7), $Sn(C_4H_9)_4$ (13.5) stabilizer for PVC	[40] [41]
22	$OOCCH_2CH_2OC_4H_9$ not ascertained [41]	thermolysis (%) $\rightarrow$ C_4H_{10} (5.9), C_4H_9OH (7.8), $CH_2{=}CHCOOC_4H_9$ (7.8), $C_4H_9OCH_2CH_2COOC_4H_9$ (12.0), $(C_4H_9)_3SnOOCCH_2CH_2OC_4H_9$ (5.7) stabilizer for PVC	[40] [41]
23	$OOCCH_2CH_2OC_{10}H_{21}$ not ascertained	stabilizer for PVC	[41]
24	$OOCCH_2OC_6H_5$ II [72]	m.p. 128° ^{17}O NMR (saturated solution in $CHCl_3$, 330 K, neat D_2O as external reference): 249 ± 3 (OCO) IR (Nujol): $\nu_{as}(OCO)$ 1610, $\nu_s(OCO)$ 1330 UV: 215 (sh), 262.5 (sh), 270, 276.5 in C_6H_{12}; 217.5, 262.5 (sh), 270, 276 in CH_3OH monomeric in the solid state with intramolecular O$\rightarrow$Sn coordination	[72] [75] [72] [72,75]
25	$OOCCH_2OC_6F_5$ II	m.p. 138 to 139°	[32]
26	$OOCCH_2OC_6H_4Cl\text{-}2$ II	m.p. 156° IR (Nujol): $\nu_{as}(OCO)$ 1615, $\nu_s(OCO)$ 1325 structure like No. 24	[72]

Table 23 (continued)

No.	OOCR group method of preparation (yield in %)	properties and remarks	Ref.
27	$OOCCH_2OC_6H_4Cl$-4 II	m.p. 110° IR (Nujol): ν_{as}(OCO) 1640, ν_s(OCO) 1340 UV (C_6H_{12}): 231.9, 273.75 (sh), 281.25, 288.75 structure like No. 24	[72]
28	$OOCCH_2OC_6H_3Cl_2$-2,4 II	m.p. 166° IR (Nujol): ν_{as}(OCO) 1610, ν_s(OCO) 1320 structure like No. 24	[72]
29	$OOCCH_2OC_6H_2Cl_3$-2,4,5 II	m.p. 158° IR (Nujol): ν_{as}(OCO) 1610, ν_s(OCO) 1360 structure like No. 24	[72]
30	$OOCCH_2OC_6H_4CH_3$-2 II	m.p. 117 to 120° IR (Nujol): ν_{as}(OCO) 1600, ν_s(OCO) 1340 structure like No. 24	[72]
31	$OOCCH_2OC_6H_3(CH_3$-2)Cl-4 II	m.p. 122° IR (Nujol): ν_{as}(OCO) 1608, ν_s(OCO) 1338 structure like No. 24 herbicide	[72] [30]
32	$OOCCH_2ON{=}C(CH_3)C_6H_5$ II	m.p. 95 to 103° tested as a bactericide, fungicide, and herbicide	[23]
33	$OOCCH_2ON{=}CHC_6H_4OCH_3$-4 II	m.p. 84 to 87° tested as a bactericide, fungicide, and anthelmintic stabilizer for polymers	[22]
34	$OOCCH_2SCH_3$	no preparation reported improves the moldability of polypropylene	 [42, 43]
35	$OOCCH_2SC_8H_{17}$-i	no preparation reported heat stabilizer for PVC	 [69]
36	$OOCCH_2SCH_2COOCH_2CH{=}CH_2$	no preparation reported heat and color stabilizer for PVC	 [10]
37	$OOCCH_2SCH_2C_6H_5$ II [53]	m.p. 59 to 60° with HCl → $(C_4H_9)_2SnCl_2$, $C_6H_5CH_2SCH_2COOH$ poor stabilizer for PVC	[53] [46, 53]

Table 23 (continued)

No.	OOCR group method of preparation (yield in %)	properties and remarks	Ref.
38	$OOCCH_2SC_6H_5$ II [53]	m.p. 89.5 to 90.5°	[53]
		^{17}O NMR (saturated solution in $CHCl_3$, 330 K, neat D_2O as external reference): 272 (OCO)	[75]
		with HCl → $(C_4H_9)_2SnCl_2$, $C_6H_5SCH_2COOH$ poor stabilizer for PVC	[46, 53]
39	$OOCCH_2S\text{-}Sb(SCH_2COOC_8H_{17})_2$	no preparation reported heat stabilizer for PVC and vinyl chloride – vinyl acetate copolymer	[50]
40	$OOCCH_2CH_2SH$	no preparation reported investigations on its mode of action as stabilizer for PVC	[14]
41	$OOCCH_2CH_2SCH_2CH_2COOC_2H_5$ not ascertained [41]	tested as heat stabilizer for PVC	[41, 52]
42	$OOCCH_2CH_2SCH_2CH_2COOC_4H_9$ II	stabilizer for polyolefins	[15, 18]
43	$OOCCH_2CH_2SCH_2CH_2COOC_8H_{17}$ not ascertained	stabilizer for PVC	[41]
44	$OOCCH_2CH_2SCH_2CH_2COOC_{12}H_{25}$ not ascertained	stabilizer for PVC	[41]
45	$OOCCH_2CH_2SCH(C_6H_4OH\text{-}2)S\text{-}CH_2CH_2COOC_8H_{17}\text{-}i$ VI	clear, yellow liquid stabilizer for PVC	[51]
46	$OOCCH_2CH_2SCH(CH(C_2H_5)_2)\text{-}SCH_2CH_2COOCH_2CH(OH)CH_2\text{-}OOCC_{17}H_{35}$ VI	stabilizer for PVC	[56]
47	$OOCCH_2CH_2SC(CH_3)(CH_2COO\text{-}C_2H_5)SCH_2COOC_{12}H_{25}$ VI	light yellow liquid stabilizer for PVC	[51]
48	$OOCCH_2CH_2SCH(C_{11}H_{23})SCH_2\text{-}COOC_8H_{17}\text{-}i$ VI	yellow liquid stabilizer for PVC	[51]
49	$OOCCH_2CH_2SCH(C_{11}H_{23})S\text{-}CH_2COOC_{12}H_{25}$ VI	light yellow liquid stabilizer for PVC	[51]
50	$OOCCH(CH_3)CH_2SC_{12}H_{25}$ II	stabilizer for Cl-containing resins	[2]

Table 23 (continued)

No.	OOCR group method of preparation (yield in %)	properties and remarks	Ref.
51	$OOCCH_2CH(C_6H_5)SCOCH_3$ II [53]	m.p. 132 to 133° with HCl → $(C_4H_9)_2SnCl_2$, $CH_3OCS\text{-}CH(C_6H_5)CH_2COOH$ poor stabilizer for PVC	[53] [46, 53]
52	$OOCCH(CH_3)NHCOCH_3$ II	m.p. 76°	[27]
53	$OOCCH(CH_2C_3H_7\text{-}i)NHCOCH_3$ II (75)	m.p. 185° ^1H NMR ($CDCl_3$): 0.5 to 1.88 (m, C_4H_9), 2.01 (s, CH_3CO), 2.34 (m, br, CH of C_3H_7-i), 4.5 (m, br, CHN), 6.65 (d, NH) ^{119}Sn-γ (80 K): $\delta = 1.42$, $\Delta = 3.46$ IR (KBr): ν(NH) 3360 (s), ν(CO)/N 1665 (s), ν_{as}(OCO) 1590 (s), ν_s(OCO) 1390 (s), ν(SnC) 570 (s), ν(SnO) 465 (m) distorted *trans* octahedral structure with bidentate OCO groups in solution and in the solid state	[76]
54	$OOCCH(CH_2C_6H_5)NHCOCH_3$ II (83)	m.p. 160° ^1H NMR ($CDCl_3$): 0.56 to 1.6 (m, C_4H_9), 2.00 (s, CH_3CO), 4.93 (m, CH), 5.9 (m, NH, CH_2), 7.3 (m, C_6H_5) ^{119}Sn-γ (80 K): $\delta = 1.33$, $\Delta = 3.38$ IR (KBr/$CHCl_3$): ν(NH) 3360 (s), ν(C=O)/N 1670 (s), ν_{as}(OCO) 1585 (s), ν_s(OCO) 1400 (m), ν(SnC) 590 (m), ν(SnO) 495 (m) structure like No. 53	[76]
*55	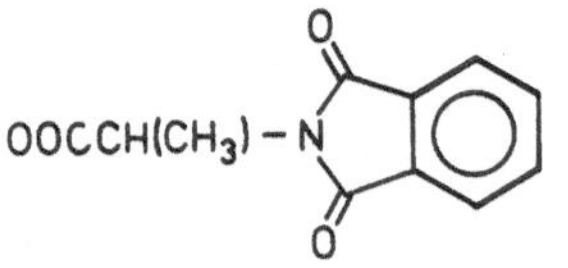 II (60)	3-H_2O solvate m.p. 170 to 173° ^1H NMR ($CDCl_3$): 0.80 to 1.7 (m, C_4H_9, CH_3), 4.65 (q, br, CH), 7.65 (m, C_6H_4) ^{119}Sn-γ (80 K): $\delta = 1.37$, $\Delta = 3.52$ IR (KBr): ν_s(C=O)/N 1776, ν_{as}(C=O)/N + ν_{as}(OCO) 1720 (sh), 1705 (s, br), and 1605 (s, br), ν_s(OCO) 1380 (s, br), ν(SnC) 570 (w, br), ν(SnO) 490 (m, br), ν(SnN) 435 (m, br)	[77]

Table 23 (continued)

No.	OOCR group method of preparation (yield in %)	properties and remarks	Ref.
*56	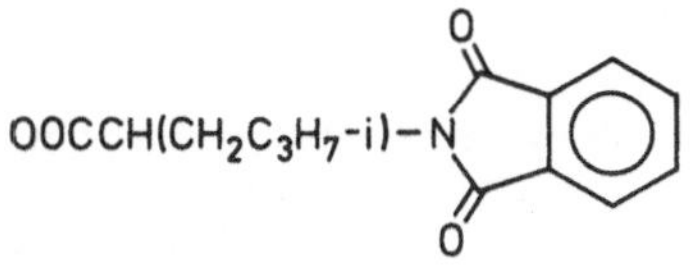 II (70)	m.p. 200 to 206° ^{1}H NMR ($CDCl_3$): 0.98 to 1.5 (m, C_4H_9, $CH_2C_3H_7$-i), 4.45 to 4.90 (CH), 7.75 (m, C_6H_4) ^{119}Sn-γ (80 K): δ = 1.16, Δ = 3.40 IR (KBr): ν_s(C=O)/N 1780 (m), ν_s(C=O)/N + ν_{as}(OCO) 1730 (s), 1710 (sh), and 1600 (s), ν_s(OCO) 1392 (s), ν(SnC) 585 (m), ν(SnO) 490 (s), ν(SnN) 460 (m)	[77]
*57	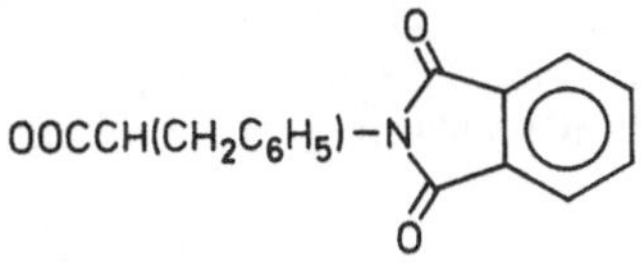 II (75)	4-H_2O solvate m.p. 208 to 210° ^{1}H NMR ($CDCl_3$): 0.85 (t, H-δ), 1.10 to 1.66 (m, H-α,β,γ), 3.37 (d, CH_2), 4.67 (t, CH), 7.10 (s, C_6H_5), 7.75 (m, C_6H_4) ^{119}Sn-γ (80 K): δ = 1.33, Δ = 3.45 IR (KBr): ν_s(C=O)/N 1780 (m), ν_s(C=O)/N + ν_{as}(OCO) 1730 (sh), 1720 (s), and 1600 (s), ν_s(OCO) 1390 (s), ν(SnC) 570 (m), ν(SnO) 495 (s), ν(SnN) 460 (s, sh)	[77]
58	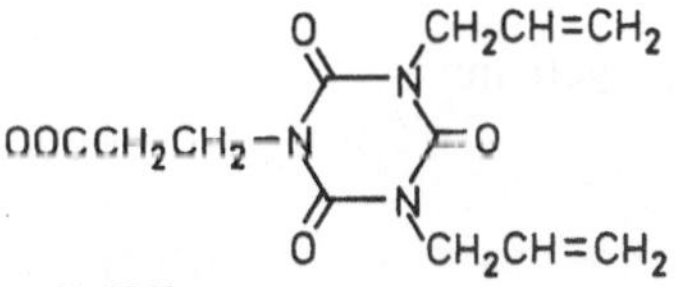 II (99)	m.p. 59 to 61° n_D^{20} = 1.5281 catalyst for polyurethane synthesis	[67]
59	$OOCCH_2C_6H_5$	no preparation reported analysis by titration in nonaqueous medium	[28]
60	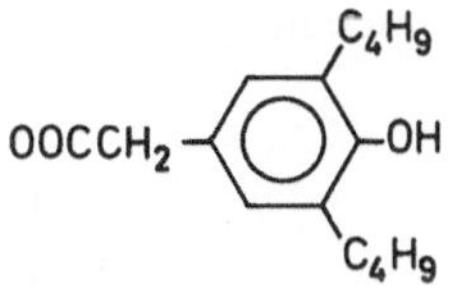	no preparation reported improves the moldability of propylene block copolymers	[58]

Table 23 (continued)

No.	OOCR group method of preparation (yield in %)	properties and remarks	Ref.
61	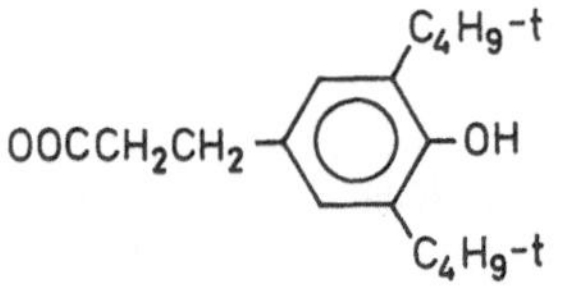	no preparation reported stabilizer for PVC	[38, 45, 47, 61]
62	$OOCCH(C_6H_5)OH$ I (100 [21])	m.p. 165 to 166° tested as anthelmintic against Raillietina cesticillus and Ascaridia galli from chicken	[21] [3]
*63	$OOCC(C_6H_5)_2OH$ I [66] II (86 [33])	m.p. 139 to 146° [65], 182° [66] IR (Nujol): ν(OH) 3404 (s), $ν_{as}$(OCO) 1720 (s) and 1630 (vs, br), $ν_s$(OCO) 1344 (vs, br), $ν_{as}$(SnC) 540 (m) and 532 (m), $ν_s$(SnC) 507 (m), ν(SnO) 290 (s)	[33, 66]
64	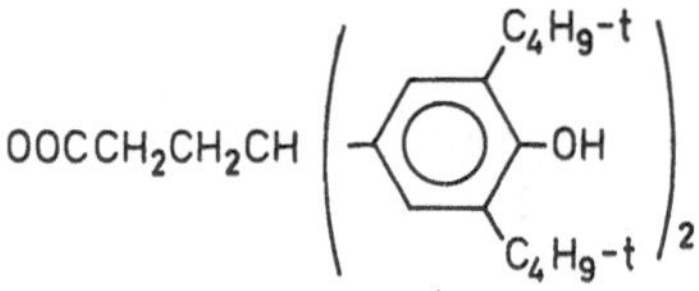 II	m.p. 60 to 70° antioxidant for polyolefins	[24]
65	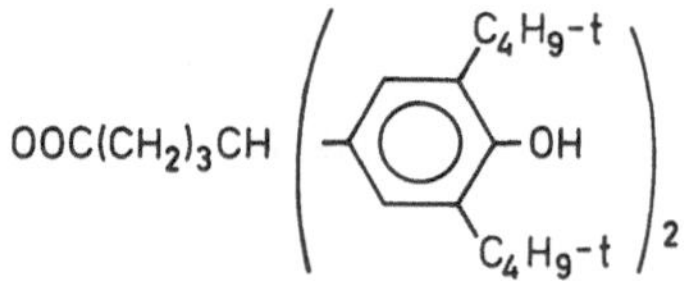 II	m.p. 56 to 62° antioxidant for polyolefins	[24]
66	$OOCCH_2COOCH_3$	no preparation reported investigations on its mode of action as stabilizer for PVC	[14]
*67	$OOCCH_2COOC_2H_5$ special	stabilizer for halogen-containing resins component of room temperature vulcanizable siloxane compositions	[4] [78]
68	$OOCCH_2CH_2COOC_2H_5$ not ascertained	stabilizer for PVC	[41]

Table 23 (continued)

No.	OOCR group method of preparation (yield in %)	properties and remarks	Ref.
69	$OOCCH_2CH_2COOC_8H_{17}$ not ascertained [41]	stabilizer for PVC characterization by gel permeation chromatography study of the kinetic of diffusion and equilibrium concentration in nonplasticized PVC at 100°	[41] [71] [72]
70	$OOCCH_2CH_2COOC_{12}H_{25}$	no preparation reported additive to polypropylene fibers to improve dyeability	 [34,35]
71	$OOCCH_2CH_2COOCH_2CH_2N(CH_3)_2$ II (61)	catalyst for promoting reactions of isocyanates with organic compounds possessing active hydrogen	[65]
72	$OOCCH_2CH_2CONHC_{12}H_{25}$	no preparation reported additive to polypropylene fibers to improve dyeability	 [34,35]
73	$OOCCH_2CH_2CON(CH_2CH_2CN)_2$ II	retards discoloration in vinyl halide-acrylonitrile copolymers subjected to high temperatures or UV light	[11]
74	$OOC(CH_2)_4COOH$	no preparation reported improves curing of siloxane elastomers at room temperature	 [5]
75	$OOC(CH_2)_4COOC_4H_9$	no preparation reported improves the moldability and spinability of isotactic polypropylene study of its effect on the thermal oxydation stability, color, and transparency of polypropylene	 [42]
76	$OOC(CH_2)_8COOC_4H_9$	no preparation reported improves the moldability and spinability of isotactic polypropylene	 [42]
*77	$OOCCH(CH_3)CH_2Si(C_2H_5)_3$ special	b.p. 200°/1.5 D = 1.0730 at 20°, n_D^{20} = 1.4798	[54,59]
78	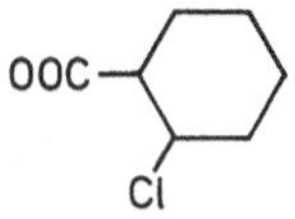	no preparation reported gelation catalyst for polyurethane foams	 [48]

* Further information:

$(C_4H_9)_2Sn(OOCCHR'NC_8H_4O_2)_2$ (R' = CH_3, $CH_2C_3H_7$-i, $CH_2C_6H_5$, Table **23**, Nos. **55** to **57**). The thermogravimetric analysis of No. 55 or 57 shows the loss of three or four water molecules, respectively, present in the crystal lattice between 20 and 120°C. The other decomposition products could not be identified. The spectroscopic data support a distorted octahedral structure (Formula I, R = C_4H_9) for the monomers. Cryoscopic molecular weight determinations in C_6H_6 show all three compounds to be dimeric at low temperature (Formula II, R = C_4H_9) [77].

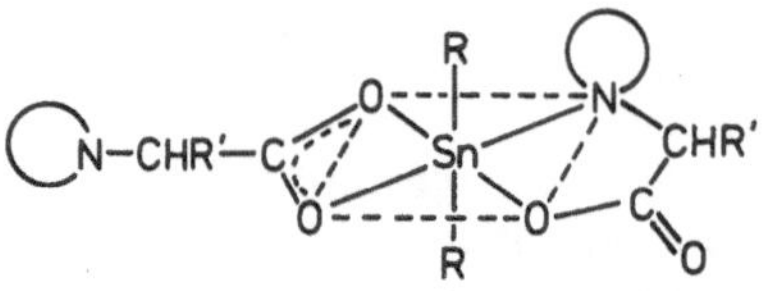

I

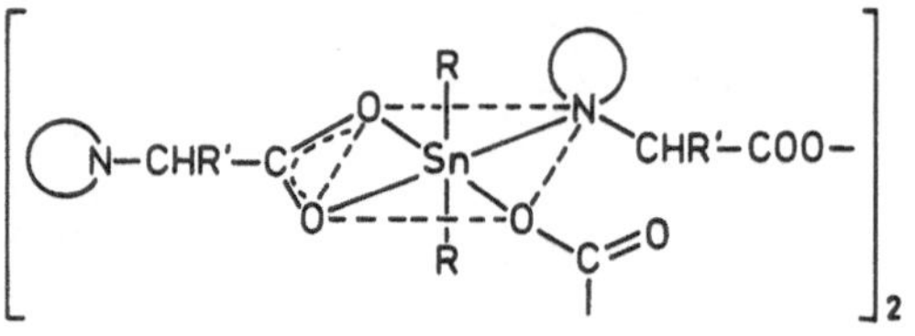

II

$(C_4H_9)_2Sn(OOCC(C_6H_5)_2OH)_2$ (Table **23**, No. **63**). The IR spectrum is interpreted in terms of a distorted octahedral *cis* or *trans* arrangement of the C_4H_9 groups with coordinating OH oxygen atoms and nonchelating carbonyl groups (Formulas III and IV, R = C_4H_9) [66].

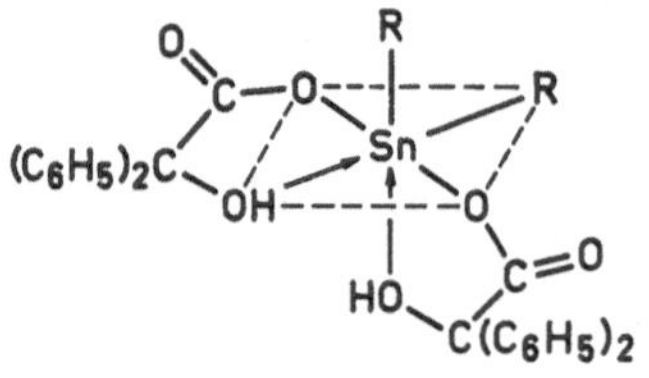

III

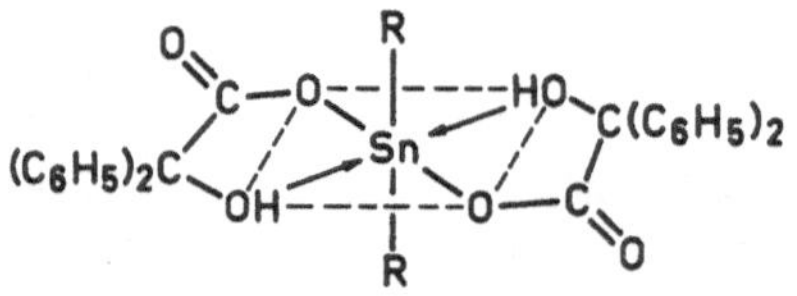

IV

$(C_4H_9)_2Sn(OOCCH_2COOC_2H_5)_2$ (Table **23**, No. **67**) has been obtained by treating $C_2H_5OOCCH_2COOC_2H_5$ with $(C_4H_9)_2SnCl_2$ in $C_6H_5CH_3$ [4].

$(C_4H_9)_2Sn(OOCCH(CH_3)CH_2Si(C_2H_5)_3)_2$ (Table **23**, No. **77**) is formed in a 64% yield by the reaction of $(C_2H_5)_3SiCH_2CH(CH_3)COOSi(CH_3)_3$ with $(-(C_4H_9)_2SnO-)_n$ (2:1 mole ratio) at 120°C for 2 h along with $(CH_3)_3SiOSi(CH_3)_3$ [54, 59].

References:

[1] Farbwerke Hoechst A.-G. (Brit. 797073 [1953]; C.A. **1959** 22714).
[2] Leistner, W. E., Hecker, A. C., Argus Chemical Laboratory, Inc. (U.S. 2680107 [1954]; C.A. **1955** 4713).
[3] Kerr, K. B., Walde, A. W. (Experim. Parasitol. **5** [1956] 560/70).
[4] Mack, G. P., Parker, E., Carlisle Chemical Works, Inc. (U.S. 2745819 [1956]; C.A. **51** [1957] 9214).
[5] Polmanteer, K. E., Dow Corning Corp. (Ger. 1019462 [1957]; C.A. **1960** 10378).
[6] Brückner, H., Härtel, K., Farbwerke Hoechst A.-G. (Ger. 1025198 [1958]; C.A. **1960** 12468).
[7] Verity-Smith, H., Pure Chemicals Ltd. (Brit. 791119 [1958]; C.A. **1958** 16201).
[8] Rybakova, N. A., Popova, Z. V., Yanovskii, D. M., Zilberman, E. N., Sharetskii, A. M. (U.S.S.R. 117676 [1959]; C.A. **1959** 19881).

[9] Berlin, A. A., Popova, Z. V., Yanovskii, D. M. (Zh. Prikl. Khim. **33** [1960] 871/7; J. Appl. Chem. [USSR] **33** [1960] 870/5).
[10] Thinius, K., Schroeder, E. (Ger. [East] 21727 [1959/61]; C.A. **57** [1962] 1080).

[11] Lynn, J. W., Walter, A. T., Union Carbide Corp. (U.S. 3053870 [1958/62]; C.A. **58** [1963] 3457).
[12] Sawyer, A. K., Kuivila, H. G. (J. Org. Chem. **27** [1962] 610/4).
[13] Aleksandrov, A. Yu., Delyagin, N. N., Mitrofanov, K. P., Polak, L. S., Shpinel, V. S. (Dokl. Akad. Nauk SSSR **148** [1963] 126/8; Dokl. Phys. Chem. Proc. Acad. Sci. USSR **148/153** [1963] 1/3).
[14] Frye, A. H., Horst, R. W., Paliobagis, M. A. (Polym. Prepr. Am. Chem. Soc. Div. Polym. Chem. **4** [1963] 260/82).
[15] Hechenbleikner, I., Homberg, O. A., Deutsche Advance Produktion GmbH (Belg. 622648 [1961/63]; C.A. **58** [1963] 14227).
[16] Aleksandrov, A. Yu., Dorfman, Ya. G., Lependina, O. L., Mitrofanov, K. P., Plotnikova, M. V., Polak, L. S., Temkin, A. Ya., Shpinel, V. S. (Russ. J. Phys. Chem. **38** [1964] 1185/8).
[17] Frankel, M., Gertner, D., Wagner, D., Zilkha, A. (J. Appl. Polym. Sci. **9** [1965] 3383/8).
[18] Hechenbleikner, I., Homberg, O. A., Carlisle Chemical Works, Inc. (U.S. 3209017 [1961/65]; C.A. **63** [1965] 16382).
[19] Krizhanskii, L. M., Okhlobystin, O. Yu., Popov, A. V., Rogozev, B. I. (Dokl. Akad. Nauk SSSR **160** [1965] 1121/3; Dokl. Phys. Chem. Proc. Acad. Sci. USSR **160/165** [1965] 142/4).
[20] Zemlyanskii, N. N., Panov, E. M., Shamagina, O. P., Kocheshkov, K. A. (Zh. Obshch. Khim. **35** [1965] 1029/31; J. Gen. Chem. [USSR] **35** [1965] 1034/5).

[21] Frankel, M., Gertner, D., Wagner, D., Zilkhan, A. (J. Organometal. Chem. **9** [1967] 83/8).
[22] Nakanishi, M., Inamasu, S., Yoshitomi Pharmaceutical Industries, Ltd. (Japan. 67-16296 [1964/67]; C.A. **68** [1968] No. 105366).
[23] Nakanishi, M., Inamasu, S., Yoshitomi Pharmaceutical Industries, Ltd. (Japan. 67-21334 [1965/67]; C.A. **69** [1968] No. 27527).
[24] Nakanishi, M., Kuriyama, T., Oe, T., Yoshitomi Pharmaceutical Industries, Ltd. (Japan. 67-13302 [1964/67]; C.A. **68** [1968] No. 49767).
[25] Zemlyanskii, N. N., Gol'dshtein, I. P., Gur'yanova, E. N., Syutkina, O. P., Panov, E. M., Slovokhtova, N. A., Kocheshkov, K. A. (Izv. Akad. Nauk SSSR Ser. Khim. **1967** 728/35; Bull. Acad. Sci. [USSR] Div. Chem. Sci. **1967** 707/12).
[26] Frankel, M., Wagner, D., Gertner, D., Zilkha, A. (Israel J. Chem. **6** [1968] 817/21).
[27] Gol'danskii, V. I., Khrapov, V. V., Okhlobystin, O. Yu., Rochev, V. Ya. (in: Gol'danskii, V. I., Herber, R. H., Chemical Application of Mössbauer Spectroscopy, New York 1968, pp. 336/76).
[28] Groagova, A., Pribyl, M. (Z. Anal. Chem. **234** [1968] 423/8).
[29] Asahi Kasei Kogyo Kabushi Kaisha (Brit. 1151927 [1965/69]: C.A. **71** [1969] No. 13755).
[30] Babin, V. V., Denisenkova, R. N., Ugryumov, E. P., Shcheglov, Yu. V., Bliznyuk, N. K., Streltsov, R. V., Kolomiets, A. F., North-Caucasus Scientific-Research Institute of Phytopathology and All-Union Scientific-Research Institute of Phytopathology (U.S.S.R. 250603 [1968/69]; C.A. **72** [1970] No. 77819).

[31] Kothekar, V., Shpinel, V. S. (Zh. Strukt. Khim. **10** [1969] 37/42; J. Struct. Chem. [USSR] **10** [1969] 33/7).
[32] La Liberte, B. R., Reiff, H. F., Davidsohn, W. E. (Org. Prep. Proced. **1** [1969] 173/6).
[33] Mikhailov, G. D., Chegolya, A. S. (Sin. Volokna **1969**, 18/21; C.A. **74** [1971] No. 3711).
[34] Senda, K., Ichikawa, A., Ozeki, T., Nakajima, E., Sakai, M., Nishikawa, T., Mitsubishi Rayon Co., Ltd. (Japan. 69-13589 [1966/69]; C.A. **72** [1970] No. 13783).

[35] Senda, K., Ichikawa, A., Oseki, T., Nakajima, E., Sakai, M., Yasui, A., Hirose, M., Mitsubishi Rayon Co., Ltd. (Japan. 69-28872 [1966/69]; C.A. **72** [1970] No. 122813, **73** [1970] No. 131931, **73** [1970] No. 57071).
[36] Frankel, M., Migdal, S., Gertner, D., Zilkha, A. (Israel J. Chem. **8** [1970] 647/50).
[37] Kresta, J., Jadrnik, B. (Chem. Prumysl **20** [1970] 222/3; C.A. **73** [1970] No. 99299).
[38] Seki, T., Hiyama, Y., Sato, Y., Nitto Chemical Industrial Co., Ltd. (U.S. 3511803 [1966/70]; C.A. **73** [1970] No. 26283).
[39] Smith, P. J. (Organometal. Chem. Rev. A **5** [1970] 373/402).
[40] Yamaji, Y., Moriguchi, J., Matsuda, H., Matsuda, S. (Kogyo Kagaku Zasshi **73** [1970] 1013/7; C.A. **73** [1970] No. 87985).

[41] Yamaji, Y., Moriguchi, J., Matsuda, H., Matsuda, S. (Kogyo Kagaku Zasshi **73** [1970] 2218/21).
[42] Yamamoto, T., Yukida, Y., Toray Industries, Inc. (Japan. 70-12664 [1967/70]; C.A. **73** [1970] No. 88588).
[43] Yamamoto, T., Yukida, Y., Toray Industries, Inc. (Japan. 70-12665 [1967/70]; C.A. **73** [1970] No. 88659).
[44] Ghatge, N. D., Vernekar, S. P. (Ind. Eng. Chem. Prod. Res. Develop. **10** [1971] 214/6).
[45] Ibbotson, A., Grindley, P. R., Imperial Chemical Industries Ltd. (Brit. 1239923 [1967/71]; C.A. **75** [1971] No. 118934).
[46] Rockett, B. W., Hadlington, M., Poyner, W. R. (J. Polym. Sci. Polym. Letters **9** [1971] 371/4).
[47] Seki, T., Hiyama, Y., Sato, Y., Nitto Chemical Industrial Co., Ltd. (U.S. 3591551 [1966/71]; C.A. **75** [1971] No. 118958).
[48] Takeya, M., Nitto Kasei Co., Ltd. (Japan. 71-38988 [1969/71]; C.A. **77** [1972] No. 6512).
[49] Yamai, Y., Ninamiya, K., Matsuda, S. (Kogyo Kagaku Zasshi **74** [1971] 1181/4; C.A. **77** [1972] No. 63909).
[50] Yukitomi, M., Hiramatsu, T., Shinkawa, Y., Ito, T., Kyoto Pharmaceutical Industries, Ltd. (Japan. 72-06106 [1968/72]; C.A. **77** [1972] No. 140965).

[51] Coates, H., Collins, J. D., Siddiqui, I. H., Albright and Wilson Ltd. (Ger. Offen. 2256613 [1971/73]; C.A. **79** [1973] No. 105823).
[52] Matsuda, S., Yamaji, Y., Moriguchi, J., Nippon Su-Homa Co., Ltd. (Japan. 73-39013 [1970/73]; C.A. **81** [1974] No. 50648).
[53] Rockett, B. W., Hadlington, M., Poyner, W. R. (J. Appl. Polym. Sci. **17** [1973] 3457/64).
[54] Voronkov, M. G., Mirskov, R. G., Ishchenko, O. S., Korotaeva, I. M. (Zh. Obshch. Khim. **43** [1973] 1198/9; J. Gen. Chem. [USSR] **43** [1973] 1191).
[55] Belyaeva, N. N., Bystrova, T. A., Arkhangelskii, V. V., Panteleimonov, V. A. (Gig. Sanit. **1974** No. 9, pp. 96/8; C.A. **82** [1975] No. 68959).
[56] Coates, H., Collins, J. D., Siddiqui, I. H., Albright and Wilson Ltd. (Ger. Offen. 2359346 [1972/74]; C.A. **82** [1975] No. 17899).
[57] Peruzzo, V., Tagliavini, G. (J. Organometal. Chem. **66** [1974] 437/45).
[58] Imai, K., Matsuda, K., Koga, S., Chisso Corp. (Japan. 75-17217 [1971/75]; C.A. **84** [1976] No. 165698).
[59] Voronkov, M. G., Mirskov, R. G., Ishchenko, O. S., Sitnikova, S. P. (Zh. Obshch. Khim. **45** [1975] 2634/8; J. Gen. Chem. [USSR] **45** [1975] 2595/9).
[60] Belyaeva, N. N., Bystrova, T. A., Revazova, Yu. A., Arkhangel'skii, V. I. (Gig. Sanit. **1976** No. 5, pp. 10/4, C.A. **85** [1976] No. 104765).

[61] Glushkova, L. V., Skripko, L. A., Belova, S. Yu., Efimov, A. A., Medvedv, A. I., Romanchenko, T. S., Ushakova, R. S. (U.S.S.R. 516696 [1975/76]; C.A. **85** [1976] No. 95266).

[62] King, L. F., Exxon Research and Engineering Co. (Can. 1013099 [1973/77]; C.A. **87** [1977] No. 185577).
[63] Nitto Kasei Co., Ltd. (Japan. Kokai Tokkyo Koho 80-142041 [1979/80]; C.A. **94** [1981] No. 157837).
[64] Arkhangel'skii, V. I. (Gig. Sanit. **1981** No. 7, pp. 18/9; C.A. **95** [1981] No. 144856).
[65] Bechara, I. S., Mascioli, R. L., Air Products and Chemicals, Inc. (Eur. Appl. 60974 [1981/82]; C.A. **98** [1983] No. 107559).
[66] Mesubi, M. A. (Spectrochim. Acta A **38** [1982] 989/91).
[67] Gordetsov, A. S., Noskov, N. M., Tasalova, M. E., Pavlova, L. A., Karlik, V. M., Kuzina, V. I., Dergunov, Yu. I. (Zh. Prikl. Khim. **56** [1983] 2635/8; J. Appl. Chem. [USSR] **56** [1983] 2453/7; C.A. **100** [1984] No. 210021).
[68] Robins, J., Edwards, B. H., Tokach, S. K. (Polym. Mater. Sci. Eng. **49** [1983] 331/5; C.A. **101** [1984] No. 6275).
[69] Adeka Argus Chemical Co., Ltd. (Japan. Kokai Tokkyo Koho 59120646 [1982/84]; C.A. **102** [1985] No. 7674).
[70] Nippon Sheet Glass Co., Ltd. (Japan. Kokai Tokkyo Koho 5957914 [1982/84]; C.A. **101** [1984] No. 156364).
[71] Robins, J., Edwards, B. H., Tokach, S. K. (Advan. Urethane Sci. Technol. **9** [1984] 65/76; C.A. **102** [1985] No. 7130).
[72] Chattopadhyay, T. K., Kumar, A. K., Majee, B. (Indian J. Chem. A **24** [1985] 713/5).
[73] Jirackova-Audouin, L., Ranceze, D., Verdu, J. (Analusis **13** [1985] 59/64).
[74] Jirackova-Audouin, L., Verdu, J. (Eur. Polym. J. **21** [1985] 421/6).
[75] Lycka, A., Holecek, J. (J. Organometal. Chem. **294** [1985] 179/82).
[76] Sandhu, G. K., Gupta, R., Sandhu, S. S., Parish, R. V. (Polyhedron **4** [1985] 81/7).
[77] Sandhu, G. K., Gupta, R., Sandhu, S. S., Parish, R. V., Brown, K. (J. Organometal. Chem. **279** [1985] 373/84).
[78] Wengrovius, J. H., Niedrach, L. W., General Electric Co. (U.S. 4554310 [1984/85]; C.A. **104** [1986] No. 110834).

1.4.1.2.1.5.4.3 Dibutyltin Biscarboxylates, $(C_4H_9)_2Sn(OOCR)_2$ with R = Alkenyl or Cycloalkenyl

1.4.1.2.1.5.4.3.1 Dibutyltin Biscarboxylates, $(C_4H_9)_2Sn(OOCR)_2$ with HOOCR = Unsaturated Monocarboxylic Acid

The compounds belonging to this section are listed in Table 24. The following methods of preparation have been used.

Method I: $(\text{-}(C_4H_9)_2SnO\text{-})_n$ and RCOOH (1:2 mole ratio) or RCO-O-OCR (1:1 mole ratio).
The reaction between $(\text{-}(C_4H_9)_2SnO\text{-})_n$ and CH_2=CHCOOH at 20°C for 3 h affords an oil which does not crystallize at low temperature. Purification by molecular distillation is accompanied by extensive thermal polymerization yielding only 13% of pure No. 1 [36]. No. 2 is the product of the reaction of $(\text{-}(C_4H_9)_2SnO\text{-})_n$ with CH_2=$C(CH_3)COOH$ [32], conducted in H_2O or in an organic solvent [3, 5, 14, 26], or in refluxing $C_6H_5CH_3$ for 3 h [12]. No conditions are reported for the synthesis of No. 3 from $(\text{-}(C_4H_9)_2SnO\text{-})_n$ and CH_3CH=CHCOOH [38]. Treatment of $(\text{-}(C_4H_9)_2SnO\text{-})_n$ with a 5% excess of the appropriate acid in refluxing C_6H_6 with azeotropic removal of the water formed, affords Nos. 5 to 10 [35]. No. 5 has also been obtained by

reacting $(-(C_4H_9)_2SnO-)_n$ and *trans*-cinnamic acid for 3 h. The product crystallizes on cooling [36]. No. 11 has been prepared by dissolving $(-(C_4H_9)_2SnO-)_n$ in a hot solution (60 to 80°C) of oleic acid anhydride, $C_{17}H_{33}CO\text{-}O\text{-}OCC_{17}H_{33}$, in $C_6H_5CH_3$ [8]. Nos. 13 to 19 are obtained by heating $(-(C_4H_9)_2SnO-)_n$ with the appropriate acid in $C_6H_5CH_3$ under reflux [28]. The reaction between $(-(C_4H_9)_2SnO-)_n$ and 3,5-(t-$C_4H_9)_2$-4-HO-C_6H_2CH=C(CN)COOH in refluxing C_6H_5OH gives crystalline No. 22 [18].

Method II: The "in situ" reaction of $(C_4H_9)_2Sn(OOCCH_2CN)_2$, obtained from $(-(C_4H_9)_2SnO-)_n$ and $NCCH_2COOH$ in C_6H_6, with $CH_3COC_4H_9$-i [17], 4-HOC_6H_4CHO [22], or 3,5-(t-$C_4H_9)_2$-4-HO-C_6H_2CHO [17, 22] in the presence of small amounts of $C_4H_9NH_2$, C_5H_5N, or $C_5H_{11}N$ leads to Nos. 20, 21, or 22, respectively.

Table 24
$(C_4H_9)_2Sn(OOCR)_2$ Compounds with HOOCR = Unsaturated Monocarboxylic Acid.
Explanations, abbreviations, and units on p. X.

No.	OOCR group method of preparation (yield in %)	properties and remarks	Ref.
1	OOCCH=CH$_2$ I (13 [36])	colorless, viscous liquid b.p. ≈120°/0.1 (with extensive polymerization) with $(C_4H_9)_2SnCl_2$ → $(C_4H_9)_2Sn(Cl)OOCCH{=}CH_2$ (90%)	[36]
		used as a crosslinking agent for the preparation of butadiene-styrene copolymer latex [33], or as a catalyst for the preparation of photosensitive compositions for printing plates [37]	[33, 37]
2	OOCC(CH$_3$)=CH$_2$ I (85 [3, 5, 14, 26]), (72 [12]), (84 [32])	m.p. 34° [12], 50° [3, 5, 14, 26, 29], 51 to 52.5° [32]	[3, 5, 12, 14, 26, 29, 32]
		^{119}Sn-γ (78 K): δ = −0.80 (α-Sn) [16], 1.40 [9], 1.45 [11]; Δ = 3.50 [9, 16], 3.90 [11]	[9, 11, 16]
		IR: ν(C=C) 1650, δ(CH)/Sn 775, ν(SnC) 435 and 525 [32]; spectrum depicted in the range 4000 to 500 [10]	[10, 32]
		tested as a biocide	[2, 14, 24, 26]
		used for the preparation of flame resistant, transparent polyester moldings [19], of benzene durable rubber [29], of vulcanized elastomers [4], and as a catalyst in polyacetal resin preparation [21]	[4, 19, 21, 29]

Table 24 (continued)

No.	OOCR group method of preparation (yield in %)	properties and remarks	Ref.
3	OOCCH=CHCH$_3$ I [38]	tested as anthelmintic against Raillietina cesticillus and Ascaridia galli	[2]
		tested as a thermo-oxydative stabilizer for ester-type synthetic lubricants	[38]
		light and heat stabilizer for vinyl chloride copolymers	[23]
4	OOCCH=CHCl	no preparation reported	
		gelation catalyst for polyurethane foams	[25]
5	OOCCH=CHC$_6$H$_5$ I (88 [35], 95 [36])	m.p. 77 to 78° [35], 80 to 81° [36]	[35, 36]
		with (C$_4$H$_9$)$_2$SnCl$_2$ $\rightarrow$ (C$_4$H$_9$)$_2$Sn(Cl)OOCCH=CHC$_6$H$_5$ (86%)	[36]
		catalyzes the reaction between CH$_3$C$_6$H$_3$(NCO)$_2$-2,6 and HO(CH$_2$CH$_2$O)$_3$H	[35]
		stabilizer for PVC	[20]
6	OOCCH=CHC$_6$H$_4$Cl-4 I (92)	m.p. 149 to 150° catalytic effect like No. 5	[35]
7	OOCCH=CHC$_6$H$_4$OCH$_3$-4 I (85)	m.p. 89 to 90° ^{119}Sn-γ (? K): δ = 1.35, Δ = 3.38 catalytic effect like No. 5	[35]
8	OOCCH=CHC$_6$H$_4$N(CH$_3$)$_2$-4 I (85)	m.p. 164 to 165° ^{119}Sn-γ (? K): δ = 1.34, Δ = 3.18 catalytic effect like No. 5	[35]
9	OOCCH=CHC$_6$H$_4$NO$_2$-4 I (94)	m.p. 137 to 138° ^{119}Sn-γ (? K): δ = 1.50, Δ = 3.80 catalytic effect like No. 5	[35]
10	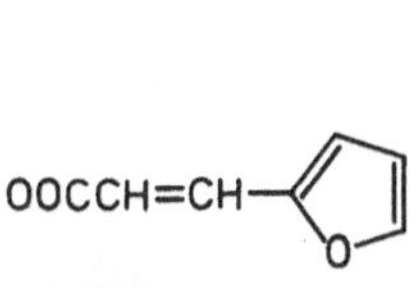 I (91)	m.p. 60 to 61° ^{119}Sn-γ (? K): δ = 1.38, Δ = 3.32 catalytic effect like No. 5	[35]
11	OOC(CH$_2$)$_7$CH=CH(CH$_2$)$_7$CH$_3$ I (>95 [8])	yellow liquid n$_D$ = 1.4812 to 1.4816	[8]
		detection in polymers by ESR [31], TLC [6, 34], or PC [7]	[6, 7, 31, 34]

Table 24 (continued)

No.	OOCR group method of preparation (yield in %)	properties and remarks	Ref.
11 (continued)		tested as an anthelmintic like No. 3 [2], or in cats suffering from dipylidiasis [15]	[2, 15]
		stabilizer for PVC [13], or ester-type synthetic lubricants [38]	[13, 38]
		catalyst for the preparation of silicone rubber vulcanizable at ambient temperature [27], and of polyurethane foam with high elasticity [30]	[27, 30]
		component of coatings for metals, glass, and ceramics	[1]
12	$OOC(CH_2)_8CH{=}CH_2$	no preparation reported tested as an anthelmintic like No. 3	[2]
13	$OOCCH{=}CHCOC_6H_5$ *trans* I (100)	m.p. 111° heat stabilizer and photodegradation promotor for PVC	[28]
14	$OOCCH{=}CHCOC_6H_4Cl$-4 I	m.p. 192° effect on PVC like No. 13	[28]
15	$OOCCH{=}CHCOC_6H_4OCH_3$-4 I	m.p. 130° effect on PVC like No. 13	[28]
16	$OOCCH{=}CHCOC_6H_4CH_3$-4 I	m.p. 162° effect on PVC like No. 13	[28]
17	$OOCCH{=}CHCOC_6H_2(CH_3)_3$-2,4,6 *trans* I	viscous liquid effect on PVC like No. 13	[28]
18	$OOCCH{=}CHCOC_6H_3(CH_3\text{-}3)Cl$-6 I	m.p. 148° effect on PVC like No. 13	[28]
19	$OOCCH{=}C(COC_6H_5)_2$ I	m.p. 150° effect on PVC like No. 13	[28]
20	$OOCC(CN){=}C(CH_3)C_4H_9$-i II	m.p. 123° stabilizer for PVC	[17]
21	$OOCC(CN){=}CHC_6H_4OH$-4 II	stabilizer for PVC	[22]
22	$OOCC(CN){=}CH\text{-}C_6H_2(OH\text{-}4)\text{-}(C_4H_9\text{-}t)_2$-3,5 I [18] II [17, 22]	m.p. 181° (dec.) stabilizer for PVC	[17, 18] [17, 22]

References:

[1] Dörfelt, C., Wolff, W., Farbwerke Hoechst A.-G. (Ger. 1084101 [1957]; C.A. **56** [1962] 4503).
[2] Kerr, K. B., Walde, A. W. (Exptl. Parasitol. **5** [1956] 560/70).
[3] Kochkin, D. A., Kotrelev, V. N., Kalinina, S. P., Kuznetsova, G. I., Laine, L. V., Chervova, L. V., Borisova, A. I., Borisenko, V. V. (Vysokomol. Soedin. **1** [1959] 1507/12; Polym. Sci. [USSR] [1959] 30/8).
[4] Montermoso, J. C., Andrews, T. M., Marinelli, L. P., Laliberte, B. R. (Proc. Intern. Rubber Conf. Preprint Papers, Washington, D.C., 1959, pp. 526/8).
[5] Shostakovskii, M. F., Kalinina, S. P., Kotrelev, V. N., Kochkin, D. A., Kuznetsova, G. I., Laine, L. V., Borisova, A. I., Borisenko, V. V. (Mezhdunar. Simp. Makromol. Khim. Dokl. Avtoreferaty, Moscow 1960, Vol. 1, pp. 160/6).
[6] Türler, M., Högl, O. (Mitt. Gebiete Lebensmittelunters. Hyg. **52** [1961] 123/30).
[7] Gasparic, J., Cee, A. (J. Chromatog. **8** [1962] 393/8).
[8] Sheverdina, N. I., Abramova, L. V., Palleva, I. E., Kocheshkov, K. A. (Khim. Prom. **1962** 707/8; C.A. **59** [1963] 8776).
[9] Aleksandrov, A. Yu., Delyagin, N. N., Mitrofanov, K. P., Polak, L. S., Shpinel, V. S. (Dokl. Akad. Nauk SSSR **148** [1963] 126/8; Dokl. Phys. Chem. Proc. Acad. Sci. USSR **148/153** [1963] 1/3).
[10] Cummins, R. A., Dunn, P. (Rept. Defence Stand. Lab. Australia No. 266 1963).

[11] Aleksandrov, A. Y., Okhlobystin, O. Yu., Polak, L. S., Shpinel, V. S. (Dokl. Akad. Nauk SSSR **157** [1964] 934/7; Dokl. Phys. Chem. Proc. Acad. Sci. USSR **154/159** [1964] 768/71).
[12] Dunn, P., Norris, T. (Rept. Defence Stand. Lab. Australia No. 269 [1964]; C.A. **61** [1961] 3134).
[13] Gelfman, Y. A., Lauris, I. V., Kuskova, V. P. (Sb. Tr. Vses. Nauchno Issled. Inst. Novykh Stroit. Mater. **1966** No. 14, pp. 54/8; C.A. **68** [1968] No. 79010).
[14] Kochkin, D. A., Azerbaev, I. N. (Vestn. Akad. Nauk. Kaz. SSR **22** [1966] 53/61).
[15] Tareeva, A. I., Borodina, G. M. (Farmakol. Toksikol. [Moscow] **30** [1967] 207/9).
[16] Gol'danskii, V. I., Khrapov, V. V., Okhlobystin, O. Yu., Rochev, V. Ya. (in: Gol'danskii, V. I., Herber, R. H., Chemical Application of Mössbauer-Spectroscopy, New York 1968, pp. 336/76).
[17] Matsuda, H., Kohara, M., Takikawa, I., Fukushima T., Sekisui Chemical Co., Ltd. (Japan. 68-21290 [1965/68]; C.A. **70** [1969] No. 58628).
[18] Matsuda, N., Ohara, M., Takigawa, I., Fukushima, T., Sekisui Chemical Co., Ltd. (Japan. 68-22577 [1966/68]; C.A. **70** [1969] No. 96955).
[19] Ralohlo, K., Alfos, F., Schnell, H., Prater, K., Farbenfabriken Bayer A.-G. (Ger. 1266497 [1965/68]; C.A. **69** [1968] No. 3448).
[20] Stapfer, C. H., Carlisle Chemical Works, Inc. (Fr. 1537462 [1966/68]; C.A. **71** [1969] No. 13830).

[21] Asahi Kasei Kogyo Kabushiki Kaisha (Brit. 1151927 [1965/69]; C.A. **71** [1969] No. 13755).
[22] Kanai, M., Matsuda, N., Ohara, M., Takigawa, I., Fukishima, T., Sekisui Chemical Co., Ltd. (Japan. 69-28086 [1965/69]; C.A. **72** [1970] No. 32731).
[23] Mack, G. P., M and T. Chemicals Inc. (Ger. 1300700 [1959/69]; C.A. **71** [1969] No. 82157).
[24] Kochkin, D. A., Novoderzhkina, I. S., Voronkov, N. A., Zubov, P. I., Azerbaev, I. N. (Fizol. Opt. Aktiv. Polim. Veshchestva Tr. 2nd Vses. Simp., Riga 1969 [1971], pp. 89/102).
[25] Takeya, M., Nitto Kasei Co., Ltd. (Japan. 71-38988 [1969/71]; C.A. **77** [1972] No. 6512).
[26] Kochkin, D. A., Azerbaev, I. N. (Dokl. 4th Vses. Konf. Khim. Atsetilena, Alma Ata 1972, Vol. 3, pp. 209/16; C.A. 79 [1973] No. 126823).
[27] Warren, W. R., General Electric Co. (Ger. Offen. 2146863 [1972]; C.A. **77** [1972] No. 35983).

[28] Grover, P. N., Wirth, H. O., Stroh, V. K., Ciba-Geigy A.-G. (Ger. Offen. 2516168 [1974/75]; C.A. **84** [1976] No. 60560).
[29] Kochkin, D. A. (Compt. Rend. 4th Congr. Intern. Corrosion Marine Salissures, Antibes/ Juan-les-Pins 1976 [1977], pp. 281/4; C.A. **88** [1978] No. 1049).
[30] Russo, R. V., M and T Chemicals, Inc. (Ger. Offen. 2658271 [1975/77]; C.A. **87** [1977] No. 85783).

[31] Stegmann, H. B., Uber, W., Scheffler, K. (Z. Anal. Chem. **286** [1977] 59/64).
[32] Rzaev, Z. M., Yusifov, G. A., Kochkin, D. A. (Issled. Obl. Sint. Polim. Monomernykh Prod. **1979** 79/81; C.A. **92** [1980] No. 147591).
[33] Sahajpal, V. K., Sarver, L. D., Deegan, C. C., Borg-Warner Corp. (U.S. 4133788 [1978/79] 122506; C.A. **90** [1979] No. 122506).
[34] Vasundhara, T. S., Parihar, D. B. (Z. Anal. Chem. **294** [1979] 408).
[35] Aliev, I. M., Noskov, N. M., Tasilova, M. E., Klyuchinskii, S. A., Dergunov, Yu. I., Zavgorodnii, V. S., Rogozev, B. I., Petrov, A. A. (Zh. Obshch. Khim. **52** [1982] 1866/71; J. Gen. Chem [USSR] **52** [1982] 1654/9).
[36] Ayrey, G., Humphrey, M. J., Poller, R. C. (Eur. Polym. J. **18** [1982] 693/7).
[37] Toyo Rubber Industry Co., Ltd. (Japan. Kokai Tokkyo Koho 59113430 [1982/84]; C.A. **102** [1985] No. 36784).
[38] Hronec, M., Cvengrosova, Z., Kizlink, J., Stolcova, M., Malik, L., Ilavsky, J., Sitek, J. (Oxid. Commun. **8** [1985/86] 51/64).

1.4.1.2.1.5.4.3.2 Dibutyltin Biscarboxylates of the Type $(C_4H_9)_2Sn(OOC\text{-}CX{=}CX\text{-}COYR)_2$ with X = H, Cl, or Br and Y = O or S

The compounds belonging to this section are listed in Table 25. They all are derivatives of maleic acid, *cis*-HOOCCH=CHCOOH (Z-form), with only one exception being No. 13 which is expressed as a derivative of fumaric acid, *trans*-HOOCCH=CHCOOH (E-form), in the patent [35]. The compounds are prepared by the following two methods.

Method I: $(C_4H_9)_2SnCl_2$ and HOOC-CX=CX-COYR or KOOC-CX=CX-COYR (1:2 mole ratio).
The reaction of $(C_4H_9)_2SnCl_2$ with the appropriate maleic acid monoester, in the presence of a base to neutralize the HCl formed, yields Nos. 6, 11, 13 [103], and 20 [103] (in aqueous NaOH) [112].
The reactions between the labeled compounds $(CH_3CH_2CH_2{}^{14}CH_2)_2SnCl_2$ or $(C_4H_9)_2{}^{113}SnCl_2$ and $KOOCCH{=}CHCOOCH_3$, or between $(C_4H_9)_2SnCl_2$ and KOOCCH=CHCOO$^{14}CH_3$, in refluxing CH_3COCH_3 for 1 h lead to the appropriately labeled No. 2 [13].

Method II: $(-(C_4H_9)_2SnO-)_n$ and HOOCCX=CXCOYR (1:2 mole ratio).
Treatment of $(-(C_4H_9)_2SnO-)_n$ with the appropriate monoester of maleic acid, halogenated maleic acid, or monothiomaleic acid followed by separation of the H_2O either by heating the reaction mixtures to ≧90°C, by distillation of the solvent-water azeotrope, or by means of a Dean-Stark water separator leads to No. 2 (C_6H_6, reflux/1 h) [22, 24], or to labeled No. 2, $(C_4H_9)_2Sn(OO^{14}CCH{=}CH^{14}COOCH_3)_2$, [126, 127], No. 3 [38] (110°C) [2, 3], No. 4 [38], No. 5 [38] (110°C) [2, 3], No. 6 [38], (90 to 100°C) [123], (110°C) [2, 3, 4], (117°C) [8], (C_6H_6, reflux/3 h) [96], No. 8 (110°C) [2, 3], No. 9 (C_6H_6, reflux/3 h) [96], No. 10 [38] (90 to 100°C) [123], No. 11 (90 to 100°C) [123], No. 14 [38], No. 15 ($C_6H_5CH_3$, reflux) [29], (90 to 100°C) [123], Nos. 17, 27, and 44 (C_6H_6, reflux) [41], No. 18 (heating) [104], Nos. 19, 22, 23, and 29 [34], No. 20 (90 to 100°C) [123], (C_6H_6, reflux) [18], No. 21 (C_6H_6, reflux) [18], No. 25 [34, 38], Nos. 26

and 28 (C_6H_6, reflux) [40], No. 31 ($C_6H_5CH_3$, 90 to 95°C, reduced pressure) [5, 7], No. 32 (100 to 110°C/1 h) [125], No. 33 [34], (90 to 100°C) [123], (100 to 110°C/1 h) [125], Nos. 34 to 38 (100 to 110°C) [125], Nos. 41 and 42 (120°C) [23], No. 43 (90 to 100°C) [123], No. 45 ($C_6H_5CH_3$) [69, 70], No. 46 ($C_6H_5CH_3$) [69], Nos. 48, 49, and 51 (heating/reduced pressure) [124], Nos. 50 and 52 (100 to 110°C/1 h, than 120°C/10 Torr) [113], and Nos. 53 and 54 ($C_6H_5CH_3$) [10, 12].

Table 25
Dibutyltin Biscarboxylates of the Type $(C_4H_9)_2Sn(OOCCX{=}CXCOYR)_2$ with X = H, Cl, or Br and Y = O or S.
Further information on compounds preceded by an asterisk is given at the end of the table.
Explanations, abbreviations, and units on p. X.

No.	group R method of preparation (yield in %)	properties and remarks	Ref.
$(C_4H_9)_2Sn(OOCCH{=}CHCOOR)_2$ compounds			
*1	H	no preparation reported	
		stabilizer for radiation-curable inks and vapor barrier coatings	[119]
*2	CH_3 I (98 [13]) II [22, 24, 126, 127]	oil, decomposes on distillation	[22, 24]
		n_D^{23} = 1.4930	
		IR (liquid film): ν_{as}(OCO)/Sn 1587 (s), ν_s(OCO)/Sn 1350 (m); indication for bidentate OCO groups	[24]
*3	C_2H_5 II [2, 3], (61 [38])	D = 1.34 at 20° [2, 3], n_D^{20} = 1.4949 [76], 1.5005 [2, 3, 38]	[2, 3, 38, 76]
		stabilizer for PVC	[2, 3, 74, 76, 87, 128]
		stabilizer for polyacetal resins	[33]
		stabilizer for acrylic fibers	[118]
4	C_3H_7 II (73 [38])	n_D^{20} = 1.4960	[38]
		stabilizer for PVC	[34, 37]
5	C_3H_7-i II [2, 3], (73 [38])	m.p. 37°	[2]
		D = 1.32 at 20° [3], n_D^{20} = 1.4930 [38], 1.4982 [2, 3]	[2, 3, 38]
		stabilizer for PVC	[2, 3, 117, 135]
*6	C_4H_9 I (≧80 [103]) II [2, 3, 4, 8, 96], (76 [38], 89 [123])	light yellow oil	[8, 96]
		D = 1.26 at 20° [3, 8], n_D^{20} = 1.4865 [76, 94], 1.4919 [38], 1.4927 [2, 3, 8], 1.4943 [123], n_D^{25} = 1.4939 [96]	[2, 3, 8, 38, 76, 94, 96, 123]
		identification by paper- [6], thin layer- [17], and gel permeation chromatography [137]	[6, 17, 137]

Table 25 (continued)

No.	group R method of preparation (yield in %)	properties and remarks	Ref.
		IR spectrum depicted (1725 to 487)	[77]
		LD_{50} (acute oral) = 120 mg/kg rat	[36, 56, 92]
		stabilizing effect under further information	
7	C_4H_9-i	no preparation reported stabilizer for PVC	[66, 106]
8	C_5H_{11} II	D = 1.23 [3], n_D^{20} = 1.4906 [2,3] stabilizer for PVC	[2, 3]
9	C_5H_{11}-i II	not distillable oil n_D^{25} = 1.4933 stabilizer for PVC	[96]
*10	C_8H_{17} II (89 [38], 99 [123])	n_D^{20} = 1.4825 [76, 94], 1.4839 [123], 1.4842 [38]	[38, 76, 94, 123]
		IR spectrum depicted (1750 to 487)	[77]
		stabilizing effect under further information	
*11	$CH_2CH(C_2H_5)C_4H_9$ I (≧80 [103]) II (98 [123])	m.p. 22 to 24°, n_D^{25} = 1.4821	[123]
		identification by PC	[6]
		stabilizing and catalytic effects under further information	
*12	$(CH_2)_5C_3H_7$-i	no preparation reported	
*13	C_8H_{17}-i I (≧80 [103], 94 [112])	^{119}Sn-γ (80 K): δ = 1.44 [102, 133], Δ = 3.62 [102], 3.66 [133] stabilizing and catalytic effects under further information	[102, 133]
14	C_9H_{19} II (89 [38])	n_D^{20} = 1.4915	[38]
		^{119}Sn-γ (78 K): δ = 1.38, Δ = 3.66	[48]
		LD_{50} (acute oral) = 170 mg/kg rat [36, 92], 3200 mg/kg mouse [92]	[36, 92]
15	$C_{12}H_{25}$ II [29], (96 [123])	m.p. 26 to 27°	[123]
		identification by PC	[6]
		stabilizer for PVC [39, 123] and polypropylene [31]	[31, 39, 123]
		PVC stabilization mechanisms	[29]
16	$C_{16}H_{33}$ II [40]	identification by PC	[6]
		stabilizer for PVC [34] and halogen-containing resin compositions [89]	[34, 89]
17	$CH_2CH(C_6H_{13})C_8H_{17}$ II	stabilizer for PVC	[41]

Table 25 (continued)

No.	group R method of preparation (yield in %)	properties and remarks	Ref.
18	$C_{18}H_{37}$ II [104]	identification by PC stabilizer for PVC	[6] [104]
19	$C_{20}H_{41}$ II	stabilizer for PVC	[34]
20	C_6H_{11}-c I (≧80 [103]) II (70 [18], 88 [123])	m.p. 70 to 71° [123], 71 to 73° [18], 73 to 74° [76] quantitative analysis like No. 10 stabilizer for PVC effect on oxidative thermal degradation of PVC [63] studied by UV [62]	[18,76,123] [43] [18, 20, 34, 37, 50, 76, 87, 103, 123, 124] [62, 63]
21	C_6H_{10}-c-CH_3-2 II	stabilizer for PVC	[18]
22	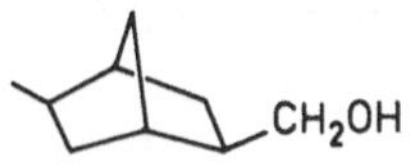 II	stabilizer for PVC	[34]
23	$CH_2C_6H_5$ II [34]	stabilizer for PVC studies of the effect on photooxydation of stabilized PVC by means of IR and UV [130], or ESR [98, 140]	[19, 21, 34, 37, 39, 80, 122] [98, 130, 140]
24	$CH(C_3H_7\text{-}i)C_6H_5$	no preparation reported stabilizer for PVC studies of the effect on photodegradation of stabilized PVC by means of TLC	 [82]
25	$CH_2CH{=}CH_2$ II [34], (73 [38])	n_D^{20} = 1.5063 stabilizer for PVC	[38] [34, 124]
26	$CH_2CH{=}CHCH_3$ II	stabilizer for PVC	[40]
27	$(CH_2)_8CH{=}CHC_8H_{17}$ II [41]	stabilizer for PVC studies on the photodegradation of stabilized PVC by means of weatherometer tests	[41] [57]

Table 25 (continued)

No.	group R method of preparation (yield in %)	properties and remarks	Ref.
28	$CH_2CH{=}CHC_6H_5$ II	stabilizer for PVC	[40]
29	C_6H_9 (hexynyl) II	stabilizer for PVC	[34]
30	CH_2CH_2OH	no preparation reported catalyst for the preparation of polyurethane lacquers	 [114]
31	$CH_2CH(OH)CH_3$ II (100 [5, 7])	D = 1.374 at 20°, n_D^{20} = 1.5060 stabilizer for PVC	[5, 7] [5, 9]
32	$CH_2CH_2OCH_3$ II (87)	n_D^{20} = 1.5050 stabilizer for PVC	[125]
33	$CH_2CH_2OC_2H_5$ II [34], (85 [125], 97 [123])	n_D^{20} = 1.4971 [76], 1.4985 [123, 125] stabilizer for PVC studies on the thermal degradation of stabilized PVC	[76, 123, 125] [34, 35, 76, 87, 123] [87]
34	$CH_2CH_2OC_3H_7$ II (87)	n_D^{20} = 1.4980 stabilizer for PVC	[125]
35	$CH_2CH_2OC_4H_9$ II (91)	n_D^{20} = 1.4935 stabilizer for PVC	[125]
36	$CH_2CH_2OCH_2CH(C_2H_5)C_4H_9$ II (97)	n_D^{20} = 1.4800 stabilizer for PVC	[125]
37	$CH_2CH_2OC_6H_{11}$-c II (87)	n_D^{20} = 1.5025 stabilizer for PVC	[125]
38	$CH_2CH_2OC_6H_5$ II (97 [125])	n_D^{20} = 1.5400 stabilizer for PVC	[125] [35, 125]
39	$CH_2CH_2OCH_2CH_2OH$	no preparation reported stabilizer for PVC	 [35]
40	$CH(CH_3)CH_2OCH_2CH(OH)CH_3$	no preparation reported stabilizer for PVC	 [35]
41	$CH_2CH_2OOCC(CH_3){=}CH_2$ II	antimicrobial agent stabilizer for PVC formation of homopolymers and copolymers with $CH_2{=}C(CH_3)COOCH_3$	[23]
42	$CH(CH_3)CH_2OOCCH{=}CH_2$ II	properties like No. 41	[23]

Table 25 (continued)

No.	group R method of preparation (yield in %)	properties and remarks	Ref.
43	$CH_2CH_2SC_2H_5$ II (96)	n_D^{20} = 1.5183 stabilizer for PVC	[123]
44	$(CH_2)_8CH{=}CHCH_2CH(OH)C_6H_{13}$ II	stabilizer for PVC	[41]
45	$CH_2C(CH_2OOCC_{11}H_{23})_2CH_3$ II	stabilizer for PVC	[69, 70]
46	$CH_2C(CH_2OOCCH{=}CHCOO$- C_8H_{17}-i$)_2CH_3$ II	stabilizer for PVC	[69]
47	$(CH_2CH(CH_3)O)_{40}OCC_{11}H_{23}$	no preparation reported antistatic agent for thermoplastics	 [26]
$(C_4H_9)_2Sn(OOCCCl{=}CClCOOR)_2$ compounds			
48	C_4H_9 II	stabilizer for PVC	[124]
49	C_6H_{11}-c II	stabilizer for PVC	[124]
50	$CH_2CH(C_2H_5)C_4H_9$ II (95 [113])	yellow, viscous liquid stabilizer for PVC	[113] [113, 124]
$(C_4H_9)_2Sn(OOCCBr{=}CBrCOOR)_2$ compounds			
51	C_4H_9 II	stabilizer for PVC	[124]
52	$CH_2CH(C_2H_5)C_4H_9$ II (92 [113])	dark yellow, viscous liquid stabilizer for PVC	[113] [113, 124]
$(C_4H_9)_2Sn(OOCCH{=}CHCOSR)_2$ compounds			
53	$CH(COOC_4H_9)CH_2COOC_4H_9$ II	yellow oil stabilizer for PVC	[10, 12]
54	$CH_2COOC_8H_{17}$ II	yellow oil stabilizer for PVC	[10,12]

* Further information:

$(C_4H_9)_2Sn(OOCCH{=}CHCOOH)_2$ (Table **25**, No. **1**). The treatment of $(-(C_4H_9)_2SnO-)_n$ with maleic acid or its anhydride does not lead to No. 1 but to the well-defined products $(C_4H_9)_2SnOOCCH{=}CHCOO \cdot H_2O$ and $(-(C_4H_9)_2SnOOCCH{=}CHCOO-)_n$ (n = 3 or 4) [24]. Mössbauer spectroscopic investigations on the effect of dibutyltin dimaleate as an irradiation stabilizer for polyethylene show that the optimum stabilizing effect occurs in the 100 to 300 Mrad range and that irradiation causes Sn-C bond splitting yielding butyl radicals, which hinder the radiolysis of the polymer [11].

$(C_4H_9)_2Sn(OOCCH{=}CHCOOCH_3)_2$ (Table **25**, No. **2**) reacts with equimolar amounts of $(C_4H_9)_2SnCl_2$ to give an 80% yield of $(C_4H_9)_2Sn(Cl)OOCCH{=}CHCOOCH_3$ [115]. The Diels-Alder reaction between No. 2 and $CH_2{=}C(CH_3)C(CH_3){=}CH_2$ leads to a gummy product from which no pure compound could be isolated, whereas the reaction with cyclopentadiene in C_6H_6 at room temperature for 12 h readily gives the expected adduct (Formula I) [22, 25].

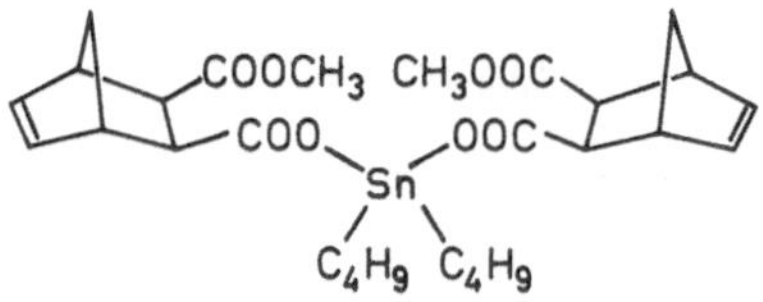

I

The acute oral toxicity of No. 2 for rats has been estimated to be LD_{50} = 62 to 1330 mg/kg [56, 92]. No. 2 is about one-tenth as effective as $(C_6H_5)_3SnOOCCH_3$ (fentin acetate) in controlling potato haulm blight (Phytophthora infestans) in laboratory tests on detached leaflets, but is less phytotoxic [47].

The compound is used as a stabilizer for PVC [57, 83, 104, 121]. Extensive studies on the mechanisms of stabilization by using ^{14}C- or ^{113}Sn-labeled No. 2 were conducted [14, 15, 16, 111, 126, 127]. It is also used as a stabilizer for acrylic fibers [65], polystyrene compositions [52], and for polyisocyanate containing coatings [110].

$(C_4H_9)_2Sn(OOCCH{=}CHCOOC_2H_5)_2$ (Table **25**, No. **3**). The thermostabilizing effect of No. 3 on irradiated polyethylene has been investigated by Mössbauer spectroscopy. The butyl radicals formed on irradiation add to free valences arising in the polymer chain, and the stable Sn^{II} compounds inhibit processes of thermal oxidation of the polymer [27]. For mechanistic studies on the stabilizing effect of No. 3 on the thermal degradation of PVC, see [76, 87].

$(C_4H_9)_2Sn(OOCCH{=}CHCOOC_4H_9)_2$ (Table **25**, No. **6**) is used as a stabilizer for PVC (2, 3, 4, 8, 29, 49, 51, 68, 72, 76, 81, 87, 88, 93, 94, 96, 97, 100, 107, 108, 109, 123, 124, 129, 131, 138], chlorinated esters of fatty acids [1], chlorinated paraffin oils [53], halogen-containing polyesters [78], and acrylonitrile polymers [60]. Studies on its stabilizing effect for polymers, especially for PVC, mostly in comparison to other organotin stabilizers, are conducted in [29, 53, 76, 87, 94, 96, 107, 123, 124, 129, 131, 138]. The compound is also used for the stabilization of rigid polyurethane foam [120] and as an additive in the preparation of fire-resistant thermoplastic polystyrene [61] or thermoplastic polyacryl fibers [55, 71, 86].

$(C_4H_9)_2Sn(OOCCH{=}CHCOOC_8H_{17})_2$ (Table **25**, No. **10**) can be determined quantitatively by potentiometric tritration with CH_3ONa in DMF, THF, or C_5H_5N using Sb and calomel electrodes [43]. The compound is an effective stabilizer for PVC [42, 57, 68, 76, 81, 94, 99, 101, 123, 129, 132, 134] and chlorinated paraffin oil [53]. The thermal degradation [87, 94] and photodegradation [57, 132] of thus stabilized PVC have been studied.

$(C_4H_9)_2Sn(OOCCH{=}CHCOOCH_2CH(C_2H_5)C_4H_9)_2$ (Table **25**, No. **11**) is used as a stabilizer for PVC [28, 30, 44, 45, 46, 58, 75, 84, 91, 103, 113, 123, 124], chloroparaffin oil [53], other chlorinated resins [28, 64], and for ester type synthetic lubricants [136]. It also serves as a catalyst in the preparation of polyurethane foams [79].

$(C_4H_9)_2Sn(OOCCH{=}CHCOO(CH_2)_5C_3H_7\text{-i})_2$ (Table **25**, No. **12**). The studies on the electrochemical behavior of the compound in 50% (v/v) and 80% (v/v) H_2O-C_2H_5OH media (pH 7.3) show two very close one-electron steps at ca. −0.8 V and a two-electron step at ca. −1.4 V *vs.* the saturated calomel electrode. The differential pulse anodic stripping voltammetry has been used for determining the compound in water down to ppb levels [139].

$(C_4H_9)_2Sn(OOCCH{=}CHCOOC_8H_{17}\text{-i})_2$ (Table **25**, No. **13**) stabilizes PVC [32, 34, (*trans* form) 35, 37, 54, 95, 103], acrylic ester copolymers [90], and quarternary ammonium chlorides in detergents, cosmetics, or pharmaceuticals [105]. The compound is used as a catalyst in the preparation of polyurethane foams with high elasticity [67, 73, 85]. The thermal degradation [102], photochemical degradation (artificial sunlight) [116], and the degradation by γ irradiation of PVC, stabilized by No. 13 [133], has been investigated by means of ^{119}Sn Mössbauer spectroscopy.

References:

[1] Johnson, E. W., Metal and Thermit Corp. (U.S. 2524528 [1950]; C.A. **1951** 2498/9).
[2] Koninklijke Industrieele Maatschappij voorheen Noury & van der Lande N.V. (Neth. 87074 [1957]; C.A. **1959** 5132).
[3] Koninklijke Industrieele Maatschappij voorheen Noury & van der Lande N.V. (Brit. 787930 [1957]; C.A. **1958** 7774).
[4] Koninklijke Industrieele Maatschappij voorheen Noury & van der Lande N.V. (Fr. 1133139 [1955/57]; C. **1958** 14445).
[5] Mack, G. P., Parker, E., Carlisle Chem. Works, Inc. (U.S. 2938013 [1960]; C.A. **1960** 20334).
[6] Gasparic, J., Cee, A. (J. Chromatog. **8** [1962] 393/8).
[7] Mack, G. P., Parker, E., Carlisle Chem. Works, Inc. (U.S. 3019247 [1957/62]; C.A. **57** [1962] 7309).
[8] Miskowiec, J. (Polimery [Warsaw] **7** [1962] 255/6; C.A. **58** [1963] 11528).
[9] Kauder, O. S., S. A. Argus Chemical N.V. (Belg. 620371 [1963]; C.A. **59** [1963] 6581; Brit. 973019).
[10] Metalorgana Ets. (Brit. 938961 [1963]; C.A. **60** [1964] 3007).

[11] Aleksandrov, A. Yu., Berlyant, S. M., Karpov, V. L., Leshchenko, S. S., Okhlobystin, O. Yu., Finkel, E. E., Shpinel, V. S. (Vysokomol. Soedin. **6** [1964] 2105/7; Polym. Sci. [USSR] **6** [1964] 2334/6).
[12] Cramer, C. R., de Bruijn, W. F. L. (U.S. 3126400 [1958/64]; C.A. **61** [1958] 3148).
[13] Frye, A. H., Horst, R. W. (Intern. J. Appl. Radiat. Isotop. **15** [1964] 169/74).
[14] Frye, A. H., Horst, R. W., Paliobagis, M. A. (J. Polym. Sci. A **2** [1964] 1801/14).
[15] Frye, A. H., Horst, R. W., Paliobagis, M. A. (J. Polym. Sci. A **2** [1964] 1765/84).
[16] Frye, A. H., Horst, R. W., Paliobagis, M. A. (J. Polym. Sci. A **2** [1964] 1785/99).
[17] van der Heide, R. F. (Z. Lebensm. Untersuch. Forsch. **124** [1964] 348/50).
[18] Gloskey, C. R., M & T Chemicals Inc. (Brit. 1009368 [1962/65]; C.A. **64** [1966] 8240).
[19] M & T Chemicals Inc. (Belg. 661478 [1965]; C.A. **64** [1966] 19904).
[20] M & T Chemicals Inc. (Belg. 661480 [1965]; C.A. **64** [1966] 19904).

[21] M & T Chemicals Inc. (Neth. Appl. 297476 [1965]; C.A. **64** [1966] 14365).
[22] Mufti, A. S., Poller, R. C. (Polymer **7** [1966] 641/2).
[23] Chas. Pfizer & Co., Inc. (Brit. 1089428 [1965/67]; C.A. **68** [1968] No. 13889).

[24] Mufti, A. S., Poller, R. C. (J. Chem. Soc. C **1967** 1362/4).
[25] Mufti, A. S., Poller, R. C. (J. Chem. Soc. C **1967** 1767/8).
[26] Takeda, T., Iino, K., Ando, M., Kawakami, Y., Seki, T., Japan Telegram and Telephon Corp. and Nitto Chemical Industrial Co., Ltd. (Japan. 67-24044 [1964/67]; C.A. **68** [1968] No. 87885).
[27] Aleksandrov, A. Yu., Baldokhin, Yu. V., Braginskii, R. P., Gol'danskii, V. I., Korytko, L. A., Leshchenko, S. S., Finkel, E. E. (Khim. Vysokikh Energ. **2** [1968] 331/7; High Energy Chem. [USSR] **2** [1968] 285/90; C.A. **69** [1968] No. 59783).
[28] Freeze, J. T., Fikes, A. L., Pfizer, Chas., and Co., Inc. (Fr. 1540230 [1966/69]; C.A. **71** [1969] No. 4175).
[29] Klemchuk, P. P. (Advan. Chem. Ser. No. 85 [1968] 1/17).
[30] Koninklijke Industrieele Maatschappij voorheen Noury & van der Lande N.V. (Neth. Appl. 6615781 [1966/68]; C.A. **69** [1968] No. 44398; Fr. 1542351; Brit. 1169770).

[31] Shimoi, M., Koda, Y., Hattori, H., Toyo Rayon Co., Ltd. (Japan. 68-03012 [1965/68]; C.A. **69** [1968] No. 44396).
[32] Weisfeld, L. B., Thacker, G. A., Giamundo, L. (Advan. Chem. Ser. No. 85 [1968] 38/44).
[33] Asahi Kasei Kogyo Kabushiki Kaisha (Brit. 1151927 [1965/69]; C.A. **71** [1969] No. 13755).
[34] Hoch, S., Tenneco Chemicals Inc. (Ger. Offen. 1801274 [1968/69]; C.A. **71** [1969] No. 13811).
[35] Kauder, O. S., Argus Chemical Corp. (U.S. 3483159 [1965/69]; C.A. **72** [1970] No. 44559).
[36] Klimmer, O. R. (Arzneimittel-Forsch. **19** [1969] 934/9).
[37] Mayo, W. E., Billiton M & T Chemische Industrie N.V. (Ger. Offen. 1917847 [1969]; C.A. **72** [1970] No. 13400).
[38] Mikhailov, G. D., Chegolya, A. S. (Sin. Volokna **1969** 18/21; C.A. **74** [1971] No. 3711).
[39] Gibbons, A. J., Ringwood, R. C., M and T Chemicals Inc. (Brit. 1060067 [1962/70]; C.A. **76** [1972] No. 26030).
[40] Hoch, S., Tenneco Chemicals, Inc. (Ger. Offen. 2005290 [1969/70]; C.A. **73** [1970] No. 110531).

[41] Hoch, S., Tenneco Chemicals, Inc. (Ger. Offen. 2005291 [1969/70]; C.A. **73** [1970] No. 110525).
[42] Kapisinska, V. (Plast. Hmotya Kaucuk **7** [1970] 236/7; C.A. **73** [1970] No. 99517).
[43] Kapisinska, V., Caplovic, J. (Chem. Prumysl **20** [1970] 487/8).
[44] Klimsch, P., Kühnert, P. (Ger. [East] 71359 [1968/70]; C.A. **73** [1970] No. 78136).
[45] Klimsch, P., Kühnert, P. (Ger. [East] 71360 [1968/70]; C.A. **73** [1970] No. 67304).
[46] Klimsch, P., Kühnert, P. (Ger. [East] 71361 [1968/70]; C.A. **73** [1970] No. 67298).
[47] McIntosh, A. H. (Ann. Appl. Biol. **66** [1970] 115/8).
[48] Smith, P. J. (Organometal. Chem. Rev. A **5** [1970] 373/402).
[49] Kidooka, S., Sotani, T., Sankyo Organic Chemicals Co., Ltd. (Japan. 71-32066 [1968/71]; C.A. **77** [1972] No. 49555).
[50] Oakes, V., Pure Chemicals, Ltd. (Ger. Offen. 2102534 [1970/71]; C.A. **75** [1971] No. 152553).

[51] Seki, T., Yano, K., Tahara, O., Mototani, H., Nitto Kasai Co., Ltd. (Japan. 71-29385 [1968/71]; C.A. **77** [1972] No. 35671).
[52] Tahara, M., Harata, Y., Japan Polystyrene Industry Co., Ltd. (Japan. 72-09744 [1968/72]; C.A. **77** [1972] No. 140979).
[53] Viska, J., Sulc, J., Galle, A. (Chem. Prumysl **22** [1972] 341/4; C.A. **78** [1973] No. 6235).
[54] Yukitomi, M., Hiramatsu, T., Shinkawa, Y., Ito, T., Kyoto Pharmaceutical Industries, Ltd. (Japan. 72-06106 [1968/72]; C.A. **77** [1972] No. 140965).

[55] Kurioka, S., Kozuka, A., Yasumoto, T., Kubota, A., Ohtoshi, N., Kanegafuchi Chemical Industry Co., Ltd. (Japan. 73-29501 [1970/73]; C.A. **81** [1974] No. 27051).
[56] Luijten, J. G. A., Klimmer, O. R. (Tin Res. Inst. Publ. No. 501 [1973] 1/18).
[57] Oki, Y., Mori, F. (Kobunshi Kagaku **30** [1973] 737/41; C.A. **81** [1974] No. 64496).
[58] Kulas, F. R., Thorshaug, N. P. (Polym. Eng. Sci. **14** [1974] 366/70).
[59] Moore, K. G., Golightly, D. S., Haughton, M. H., Albright and Wilson, Ltd. (Brit. 1378851 [1970/74]; C.A. **82** [1975] No. 157283).
[60] Szita, J., Koroscil, A., American Cyanamid Co. (Ger. Offen. 2402957 [1973/74]; C.A. **82** [1975] No. 5330).

[61] Tamai, T., Kojima, S., Nakamoto, F., Taijin Chemicals, Ltd. (Japan. Kokai 74-125448 [1973/74]; C.A. **82** [1975] No. 112754).
[62] Troitskii, B. B., Troitskaya, L. S., Denisova, V. N., Novikova, M. A., Dubova, Z. B., Kuznetsov, V. A. (Tr. Khim. Khim. Tekhnol. **1974** 56/7; C.A. **83** [1975] No. 79910).
[63] Troitskii, B. B., Troitskaya, L. S., Denisova, V. N., Dubova, Z. B., Novikova, M. A. (Tr. Khim. Khim. Tekhnol. **1974** 120/2; C.A. **83** [1975] No. 80181).
[64] Itsukaichi, Y., Kondo, T., Sankyo Organic Chemicals Co., Ltd. (Japan. Kokai 75-33243 [1973/75]; C.A. **83** [1975] No. 98347).
[65] Kozuka, K., Kurioka, S., Yasumoto, T., Kobayashi, S., Otoshi, N., Kubota, A., Kanegafuchi Chemical Industry Co., Ltd. (Fr. Demande 2236970; C.A. **83** [1975] No. 165650).
[66] Malyshev, L. N., Zavarova, T. B. (Plast. Massy **1975** 42/3; C.A. **83** [1975] No. 180196).
[67] Russo, R. V. (Cell. Plast. Transp. 18th Ann. Tech. Conf. Cell. Plast. Div. Soc. Plast. Ind., Detroit 1975, pp. 30/4; C.A. **85** [1976] No. 79168).
[68] Tomidokoro, T., Kurita, N., Adeka Argus Chemical Co., Ltd. (Japan. Kokai 75-139839 [1974/75]; C.A. **84** [1976] No. 75134).
[69] Coates, H., Collins, J. D., Siddiqui, I. H., Albright and Wilson, Ltd. (U.S. 3933743 [1972/76]; C.A. **84** [1976] No. 136516).
[70] Coates, H., Collins, J. D., Siddiqui, I. H., Albright and Wilson, Ltd. (U.S. 3978102 [1972/76]; C.A. **86** [1977] No. 17627).

[71] Minagawa, M., Ito, M., Sekiguchi, T., Adeka Argus Chemical Co., Ltd. (Japan. Kokai 76-61546 [1974/76]; C.A. **85** [1976] No. 79097).
[72] Minagawa, M., Nakahara, Y., Adeka Argus Chemical Co., Ltd. (Japan. Kokai 76-73046 [1972/76]; C.A. **85** [1976] No. 109501).
[73] Russo, R. V., (J. Cell. Plast. **12** [1976] 203/7).
[74] Sakamoto, K., Uchida, J., Nitto Kasei Co., Ltd. (Japan. Kokai 76-53546 [1974/76]; C.A. **85** [1976] No. 95206).
[75] Sato, K., Murata, K., Yazawa, S., Sankyo Organic Chemicals Co., Ltd. (Japan. Kokai 76-93949 [1975/76]; C.A. **86** [1977] No. 30528).
[76] Troitskii, B. B., Troitskaya, L. S., Denisova, V. N., Novikova, M. A. (Fiz. Khim. Osn. Sint. Pereab. Polim. [Gorkiy] **1976** No. 1, pp. 58/61; C.A. **87** [1977] No. 53972).
[77] Choe, S. G., Rynag, Y. B., Chu, S. I. (Hwahak Kwa Hwahak Kongop **20** [1977] 283/9; C.A. **88** [1978] No. 135814).
[78] Hayashi, S., Ohkubo, K., Toshiba Chemical K. K. (Japan. Kokai 77-59655 [1975/77]; C.A. **87** [1977] No. 152925).
[79] Kitzler, J., Hajek, K. (Czech. 170373 [1974/77]; C.A. **88** [1978] No. 172057).
[80] Motohashi, A., Kogo, Y., Tsukahara, Y., Sankyo Organic Chemicals Co., Ltd. (Japan. Kokai 77-80349 [1975/77]; C.A. **87** [1977] No. 185540).

[81] Oda, S., Inoue, T., Yoshitomi Pharmaceutical Industries, Ltd. (Japan. Kokai 77-128942 [1976/77]; C.A. **88** [1978] No. 171085).

[82] Oki, Y., Mori, F., Koyama, M. (Kobunshi Ronbunshu **34** [1977] 43/7; C.A. **86** [1977] No. 122184).
[83] Okudaira, H., Osanai, F., Yamada, T., Mitsubishi Plastics Industries, Ltd. (Japan. Kokai 77-154854 [1976/77]; C.A. **89** [1978] No. 7088).
[84] Rabek, J. F., Canback, G., Ranby, B. (J. Appl. Polym. Sci. **21** [1977] 2211/23).
[85] Russo, R. V., M and T Chemicals, Inc. (Ger. Offen. 2658271 [1975/77]; C.A. **87** [1977] No. 85783).
[86] Sekiguchi, T., Ogasawara, T., Adeka Argus Chemical Co., Ltd. (Japan. Kokai 77-77155 [1975/77]; C.A. **87** [1977] No. 185532).
[87] Troitskii, B. B., Troitskaya, L. S., Denisova, V. N., Novikova, M. A., Luzinova, Z. B. (Eur. Polym. J. **13** [1977] 1033/41).
[88] Ito, M., Nishimura, A., Adeka Argus Chemical Co., Ltd. (Japan. Kokai Tokkyo Koho 78-102357 [1977/78]; C.A. **90** [1979] No. 55746).
[89] Kitano, Y., Nitto Kasei Co., Ltd. (Japan. Kokai Tokkyo Koho 78-94359 [1977/78]; C.A. **90** [1979] No. 7101).
[90] Kovacs, L., Szathmari, F., Budalakk Festek es Mugyantagyar (Hung. Teljes 15876 [1976/78]; C.A. **91** [1979] No. 22557).

[91] Minagawa, M., Sekiguchi, T., Inoue, T., Kurita, N., Adeka Argus Chemical Co., Ltd. (Japan. Kokai 78-59744 [1976/78]; C.A. **89** [1978] No. 111465).
[92] Smith, P. J., Luijten, J. G. A., Klimmer, O. R. (Intern. Tin Res. Inst. Publ. No. 538 [1978] 1/20).
[93] Tanaka, Y., Kano, T., Fukuda, A., Kyodo Chemical Co., Ltd. (Japan. Kokai Tokkyo Koho 78-111346 [1977/78]; C.A. **90** [1979] No. 55763).
[94] Troitskii, B. B., Troitskaya, L. S., Denisova, V. N., Luzinova, Z. B. (Polym. J. **10** [1978] 377/85).
[95] Belyusova, M., Durmis, J., Karvas, M., Hanus, M. (Czech. 178233 [1975/79]; C.A. **92** [1980] No. 1229990).
[96] Ghatge, N. D., Vaidya, S. V. (Kautschuk Gummi Kunstst. **32** [1979] 254/7).
[97] Ichise, M., Kitano, Y., Nitto Chemical Industry Co., Ltd. (Japan. Kokai Tokkyo Koho 79-107947 [1978/79]; C.A. **92** [1980] No. 111702).
[98] Mori, F., Koyama, M., Oki, Y. (Angew. Makromol. Chem. **75** [1979] 123/35).
[99] Saito, S., Konuki, M., Sekiguchi, T., Kawaken Fine Chemicals Co., Ltd.; Adeka Argus Chemical Co., Ltd. (Japan. Kokai Tokkyo Koho 79-143457 [1978/79]; C.A. **92** [1980] No. 129977).
[100] Sekiguchi, T., Tsuruga, K., Adeka Argus Chemical Co., Ltd. (Japan. Kokai Tokkyo Koho 79-68851 [1977/79]; C.A. **91** [1979] No. 141676).

[101] Stanczyk, W., Kowalski, J., Chojnowski, J. (Polimery [Warsaw] **24** [1979] 111/4).
[102] Allen, W. D., Brooks, J. S., Clarkson, R. W. (J. Organometal. Chem. **199** [1980] 299/310).
[103] Ceprini, M. Q., Collins, J. D., Hoch, S., Wodd, D. A., Albright and Wilson, Ltd. (U.S. 4231949 [1978/80]; C.A. **94** [1981] No. 84306).
[104] Dainippon Ink and Chemicals, Inc. (Japan. Kokai Tokkyo Koho 80-71741 [1978/80]; C.A. **93** [1980] No. 133379).
[105] Kao Soap Co., Ltd. (Japan. Kokai Tokkyo Koho 80-07219 [1978/80]; C.A. **94** [1981] No. 71209).
[106] Minsker, K. S., Kolesov, S. V., Kotsenko, L. M. (Vysokomol. Soedin. A **22** [1980] 2253/8; Polym. Sci. [USSR] **22** [1980] 2471/7; C.A. **94** [1981] No. 66593).
[107] Park, G. S., Hoang, T. V. (Eur. Polym. J. **16** [1980] 779/83).
[108] Sankyo Organic Chemicals Co., Ltd. (Japan. Kokai Tokkyo Koho 80-165936 [1979/80]; C.A. **94** [1981] No. 157882).

[109] Sekiguchi, T., Tsuriga, K., Kimura, M., Adeka Argus Chemical Co., Ltd. (Japan. Kokai Tokkyo Koho 80-31860 [1978/80]; C.A. **93** [1980] No. 47862).
[110] Toyobo Co., Ltd. (Japan. Kokai Tokkyo Koho 80-80473 [1978/80]; C.A. **94** [1981] No. 5026).

[111] Ayrey, G., Hsu, S. Y., Poller, R. C. (Org. Coat. Appl. Polym Sci. Proc. **46** [1981] 630/4).
[112] Collins, J. D., Wood, D. A., Tenneco Chemicals, Inc. (U.S. 4292252 [1978/81]; C.A. **95** [1981] No. 204160).
[113] Kitzlink, J. G., Poller, R. C. (Eur. Polym. J. **17** [1981] 639/40).
[114] Noack, R., Schwetlick, K., Ermer, H., Heinrich, M., Herrmann, P. (Ger. [East] 151466 [1979/81]; C.A. **96** [1982] No. 144674).
[115] Ayrey, G., Humphrey, M. J., Poller, R. C. (Eur. Polym. J. **18** [1982] 693/7).
[116] Brooks, J. S., Clarkson, R. W., Allen, D. W., Mellor, M. T. J., Williamson, A. G. (Polym. Degrad. Stab. **4** [1982] 359/63).
[117] Kamas, F., Svoboda, P., Kralicek, J., Kral, I. (Czech. 200061 [1978/82]; C.A. **98** [1983] No. 108434).
[118] Kanegafuchi Chemical Industry Co., Ltd. (Japan. Kokai Tokkyo Koho 82-82516 [1980/82]; C.A. **97** [1982] No. 217901).
[119] Nowak, M. T. (Radiat. Curing **9** [1982] 29/30).
[120] Tsybukko, N. N., Martinovich, F. S., Satsura, V. M., Mandrikova, A. I. (U.S.S.R. 958432 [1979/82]; C.A. **98** [1983] No. 73396).

[121] Beck, H., Frassek, K. H., Verhulst, E. M., Giezen, E. A., AKZO GmbH (Ger. Offen. 3136931 [1981/83]; C.A. **99** [1983] No. 39339).
[122] Gibbons, A. J., Ringwood, R. C., M and T Chemicals, Inc. (U.S. 4418169 [1962/83]; C.A. **100** [1984] No. 35384).
[123] Kizlink, J., Caplovic, J., Obertova, L. (Plasty Kaucuk **20** [1983] 170/3).
[124] Kizlink, J., Jakubcek, E., Dimitrova, J. (Ropa Uhlie **25** [1983] 565/7).
[125] Kizlink, J., Spirk, E., Novak, L., Kotleba, J., Kopal, F. (Plasty Kaucuk **20** [1983] 269/73).
[126] Ayrey, G., Hsu, S. Y., Poller, R. C. (Polym. Sci. Technol. [Plenum] **26** [1984] 171/87).
[127] Ayrey, G., Hsu, S. Y., Poller, R. C. (J. Polym. Sci. Polym. Chem. Ed. **22** [1984] 2871/86).
[128] Dainippon Ink and Chemicals, Inc. (Japan. Kokai Tokkyo Koho 59 96150 [1982/84]; C.A. **101** [1984] No. 212174).
[129] Jamil, F. A. (J. Petrol. Res. **3** [1984] 45/56).
[130] Matsusaka, K., Koyama, M. (Angew. Makromol. Chem. **125** [1984] 149/59).

[131] Tran Van Hoang, Michel, A., Guyot, A. (Polym. Degrad. Stab. **9** [1984] 89/102).
[132] Allen, D. W., Brooks, J. S., Unwin, J., McGuinness, J. D. (Chem. Ind. [London] **1985** 524/5).
[133] Brooks, J. S., Allen, D. W., Unwin, J. (Polym. Degrad. Stab. **10** [1985] 79/94).
[134] Dainippon Ink and Chemicals, Inc. (Japan. Kokai Tokkyo Koho 60 130629 [1983/85]; C.A. **103** [1985] No. 216390).
[135] Goegh, T., Durmis, J., Salko, J., Karvas, M., Masek, J., Voeroesova, M. (Czech. 216431 [1980/85]; C.A. **103** [1985] No. 105826).
[136] Hronec, M., Cvengrosova, Z., Kizlink, J., Stolcova, M., Malik, L., Ilavsky, J., Sitek, J. (Oxid. Commun. **8** [1985/86] 51/64).
[137] Jirackova-Audouin, L., Ranceze, D., Verdu, J. (Analusis **13** [1985] 59/64).
[138] Jirackova-Audouin, L., Verdu, J. (Eur. Polym. J. **21** [1985] 421/6).
[139] Kitamura, H., Sugimae, A., Nakamoto, M. (Bull. Chem. Soc. Japan **58** [1985] 2641/7).
[140] Koyama, M., Matusaka, K. (Nippon Kagaku Kaishi **1985** 249/54; C.A. **102** [1985] No. 132916).

1.4.1.2.1.5.4.3.3 Dibutyltin Biscarboxylates, $(C_4H_9)_2Sn(OOCR)_2$ with R = Cycloalkenyl

The compounds belonging to this section are summarized in Table 26. They are prepared by the following general methods.

Method I: The reaction between $(-(C_4H_9)_2SnO-)_n$ and the appropriate 1,2,3,6-tetrahydrophthalic acid derivative RCOOH (1:2 molar ratio) yields No. 1 (petroleum ether, reflux/1 h with removal of water) [3], or No. 2 ($CH_3COC_4H_9$-i, reflux/2 h) [1].

Method II: The Diels-Alder reaction between $(C_4H_9)_2Sn(OOCCH{=}CHCOOR')_2$ and cyclopentadiene (1:4 mole ratio) in C_6H_6 at room temperature for 12 h affords No. 4 (R' = CH_3) [2, 3], No. 6 (R' = C_4H_9), No. 7 (R' = C_8H_{17}), No. 8 (R' = $CH_2CH(C_2H_5)C_4H_9$), No. 9 (R' = $C_{12}H_{23}$), No. 10 (R' = C_6H_{11}-c), and No. 11 (R' = $CH_2CH_2SC_2H_5$) [6].

Method III: No. 5 has been obtained by the reaction of $(-(C_4H_9)_2SnO-)_n$ with the dimethyl ester of chlorendic acid (1:2 molar ratio) [5].

Table 26
Dibutyltin Biscarboxylates, $(C_4H_9)_2Sn(OOCR)_2$ with R = Cycloalkenyl.
Explanations, abbreviations, and units on p. X.

No.	OOCR group method of preparation (yield in %)	properties and remarks	Ref.
1	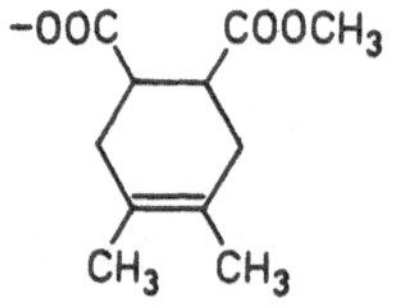 I	clear, viscous liquid decomposes on distillation	[3]
2	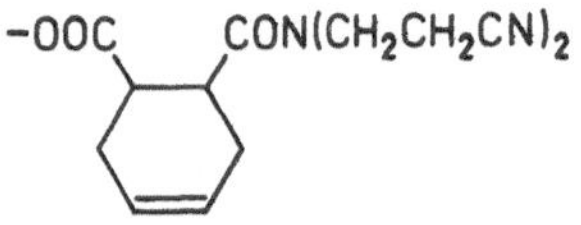 I	stabilizer for retarding discoloration in vinyl halide-acrylonitrile copolymers subjected to high temperatures or UV light	[1]
3	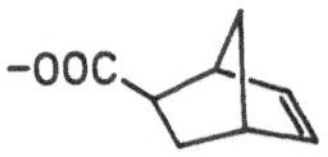	no preparation reported heat stabilizer for PVC	[4]
4	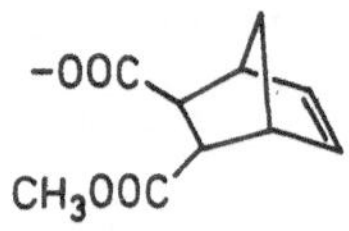 II (64)	m.p. 56 to 58°	[2, 3]
5	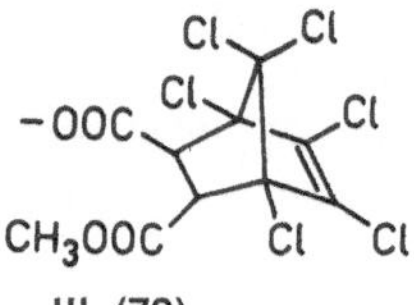 III (79)	m.p. 58 to 59.5°	[5]

Table 26 (continued)

No.	OOCR group method of preparation (yield in %)	properties and remarks	Ref.
6	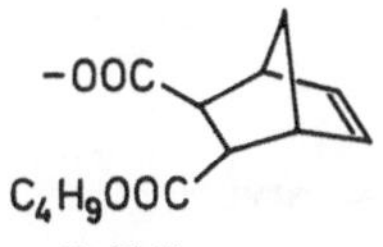 II (91)	n_D^{20} = 1.4962, 1.4988 stabilizer for PVC	[6]
7	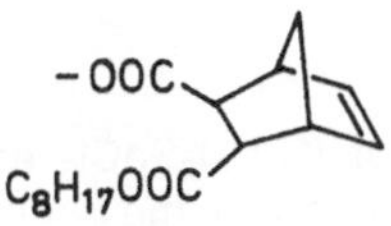 II (92)	n_D^{20} = 1.4855, 1.4895 stabilizer for PVC	[6]
8	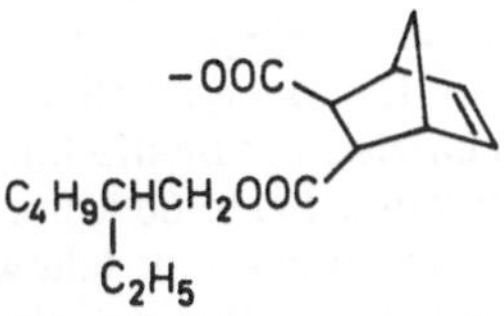 II (93)	n_D^{20} = 1.4956, 1.4982 stabilizer for PVC	[6]
9	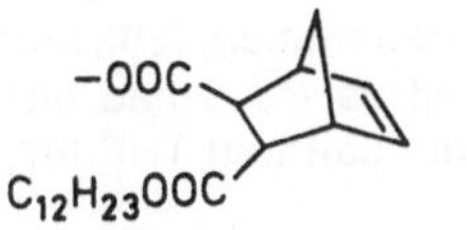 II (88)	n_D^{20} = 1.4888, 1.4930 stabilizer for PVC	[6]
10	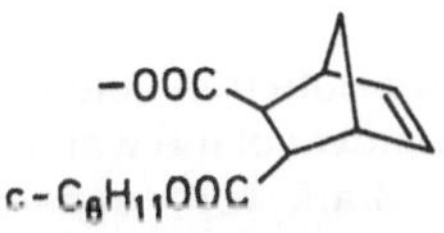 II (92)	m.p. 85 to 87°, 86 to 88° stabilizer for PVC	[6]
11	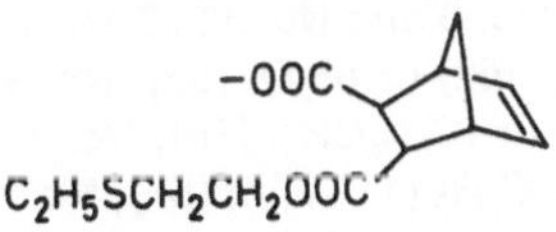 II (86)	n_D^{20} = 1.5180, 1.5222 stabilizer for PVC	[6]

References:

[1] Lynn, J. W., Walter, A. T., Union Carbide Corp. (U.S. 3053870 [1958/62]; C.A. **58** [1963] 3457).
[2] Mufti, A. S., Poller, R. C. (Polymer **7** [1966] 641/2).
[3] Mufti, A. S., Poller, R. C. (J. Chem. Soc. C **1967** 1767/8).

[4] Sato, K., Mori, E., Ihara, K., Sankyo Organic Chemicals Co., Ltd. (Japan. 78-41350 [1976/78]; C.A. **89** [1978] No. 111408).
[5] Dunyamaliev, A. D., Novoderezhkina, I. S., Rzaev, Z. M., Shakhtakhtinskii, T. N. (Issled. Obl. Sint. Polim. Monomernykh Prod. **1979** 41/4; C.A. **92** [1980] No. 164572).
[6] Kizlink, J., Caplovic, J., Obertova, L. (Plasty Kaucuk **20** [1983] 170/3; C.A. **100** [1984] No. 7666).

1.4.1.2.1.5.4.4 Dibutyltin Biscarboxylates, $(C_4H_9)_2Sn(OOCR)_2$ with R = Aryl

The compounds belonging to this section are listed in Table 27. They are prepared by the following general methods.

Method I: a. $(C_4H_9)_2SnCl_2$ and RCOOH (1:2 mole ratio).
The method of splitting off HCl from the mixtures of $(C_4H_9)_2SnCl_2$ and the appropriate free acid, dissolved in C_6H_6, and its withdrawal from the equilibrium by a suitable HCl acceptor has been used for the synthesis of Nos. 3 and 5 ($N(C_2H_5)_3$) [51], No. 37 (Na_2CO_3) [44], or No. 43 (NH_3) [13, 15, 17].

b. $(C_4H_9)_2SnCl_2$ and RCOOM (1:2 mole ratio; M = K, Na, or NH_4).
Most of the metathetic reactions between $(C_4H_9)_2SnCl_2$ and the alkali or ammonium salts of the appropriate RCOOH have been carried out by the interfacial technique. A solution of $(C_4H_9)_2SnCl_2$ in petroleum ether (60 to 80°C) is added to an aqueous solution of the appropriate RCOOM compound, followed by vigorous stirring for 0.5 h. The product is then isolated from the organic phase by evaporating the organic solvent, or it already precipitates during the reaction (M, temperature, isolation): No. 1 (K, 20°C, evaporation) [18], No. 3 (Na, 0°C, evaporation) [23], Nos. 16, 26, 27, and 30 (Na, 0°C, precipitation) [23], and No. 18 (Na, ether as organic solvent) [60]. No. 18 as well as No. 9 has also been prepared in a one-phase reaction using the sodium salt and THF [67, 71] or CH_3OH as the solvent, respectively [51].
For the preparation of No. 35 from $(C_4H_9)_2SnCl_2$ and 4-CH_2=$CHC_6H_4COONH_4$, no conditions are reported [16].

Method II: $(-(C_4H_9)_2SnO-)_n$ and RCOOH (1:2 mole ratio).
The dehydration between $(-(C_4H_9)_2SnO-)_n$ and the appropriate free acid in boiling water or in an organic solvent under reflux, and with removal of the water formed by means of distillation, azeotropic distillation, a Dean-Stark water separator, or a chemical H_2O acceptor, has been used for the preparation of No. 1 [34, 55], (H_2O) [56], (hot C_6H_6, in the presence of $Mg(ClO_4)_2$) [22]; Nos. 2, 29, and 30 ($C_6H_5CH_3$) [76]; No. 5 (C_6H_6) [52], ($C_6H_5CH_3$) [76]; No. 6 and No. 38 ($C_6H_5CH_3$) [52]; No. 7 (H_2O) [65]; Nos. 8 and 26 (C_6H_6 or $C_6H_5CH_3$) [62] or (H_2O) [65]; No. 9 (C_6H_6) [33] or (H_2O) [59]; No. 11 (C_6H_6 or $C_6H_5CH_3$) [62], or ($C_6H_5CH_3$) [76]; No. 12 (C_6H_6, 4 h) [61]; Nos. 17, 19, 20, 24, 31, and 32 (1:1, C_6H_6-C_2H_5 (1:1 v/v)) [57]; No. 18 [34], (C_6H_6) [30]; No. 21 ($C_6H_5CH_3$) [38, 43]; No. 22 ($C_6H_5CH_3$) [38]; No. 23 (C_6H_6, 8 h) [80]; No. 27 (C_6H_6) [30, 52, 62], ($C_6H_5CH_3$) [28, 62], or ($C_2H_5OCH_2CH_2OC_2H_5$) [33]; Nos. 28 and 34 (C_6H_6) [30]; No. 33 [34], or ($C_6H_5CH_3$) [76]; No. 35 (C_6H_6) [16]; No. 45 ($CH_3COC_4H_9$-i) [10].

Method III: Alkoxy-carboxy ligand exchange (1:2 mole ratio).
Heating a mixture of $(C_4H_9)_2Sn(OC_4H_9\text{-t})_2$ and C_6H_5COOH, dissolved in C_6H_6, in a bath with boiling water for 1.5 h, leads to the formation of No. 1 [22]. The ligand exchange between $(C_4H_9)_2Sn(OCH_3)_2$ and 2-HSC_6H_4COOH, 2-$C_8H_{17}SC_6H_4COOH$, or 2-$C_6H_5COC_6H_4COOH$ at 120°C affords Nos. 14, 15, or 39 within 1 h [24].

Method IV: The transesterification of $(C_4H_9)_2Sn(OOCCH_3)_2$ with the twofold molar amount of $4\text{-}NH_2C_6H_4COOH$ leads to No. 18 [63]; with the twofold amount of the 1,2-anhydride of trimellitic acid in o-xylene at 130 to 140°C, No. 44 is obtained [60].

Method V: $(C_4H_9)_2SnH_2$ and RCOOH or RCO-OO-OCR.
$(C_4H_9)_2SnH_2$ reacts with C_6H_5COOH (1:2.5 mole ratio), $2\text{-}ClC_6H_4COOH$ (1:1.25 mole ratio), and $4\text{-}CH_3C_6H_4COOH$ (1:2.5 mole ratio) in dioxane with evolution of H_2 (start after 5 to 10 min, completion within 6 to 16 h) and formation of No. 1 (80%), No. 3 (47.3%), or No. 33 (39%), along with $RC_6H_4COO(C_4H_9)_2SnSn(C_4H_9)_2OOCR$ (R = H (13%), 2-Cl (49%), or 4-NH_2 (64%)), respectively. The relative amounts were calculated from the results of the bromine analysis which is indicative for the amount of the byproduct and the total H_2 gas evolved [11]. No. 1 has been obtained in a 52% yield when starting from an 1:2 molar ratio for hydride/acid, using ether as a solvent [14]. The reaction between equimolar amounts of $(C_4H_9)_2SnH_2$ and $C_6H_5CO\text{-}OO\text{-}OCC_6H_5$ in C_6H_6 at 50 to 60°C/2 h affords 85% of No. 1 [19].

Method VI: Heating a 1:4 molar mixture of $Cl(C_4H_9)_2SnOSn(C_4H_9)_2Cl$ and C_6H_5COOH to 180°C for 0.5 h yields No. 1 [9].

Table 27
Dibutyltin Biscarboxylates, $(C_4H_9)_2Sn(OOCR)_2$ with R = Aryl.
Further information on compounds preceded by an asterisk is given at the end of the table.
Explanations, abbreviations, and units on p. X.

No.	OOCR group method of preparation (yield in %)	properties and remarks	Ref.
*1	$OOCC_6H_5$ Ib (95 [18]) II [55, 56], (77 [34], 85 [22]) III (85 [22]) V (52 [14], 80 [11], 85 [19]) VI (84 [9])	m.p. 63 to 65° [34], 65 to 67° [9, 18], 66 to 68° [19, 51], 67° [73], 68 to 71° [14], 69°, 69 to 69.5° [22], 69 to 70° [65]	[9, 14, 18, 19, 22, 34, 51, 65, 73]
		b.p. ca. 200° (bath)/0.005	[9]
		μ = 1.76 D; mixture of *cis* and *trans* O→Sn dative bonds	[51]
		^{17}O NMR (neat D_2O as external standard, ±3 ppm, 330 K): 275 (70% (v/v) in $CHCl_3$), 280 (50% (v/v) in $CHCl_3$); bidentate carboxyl groups with four equalized Sn⋯O bonds	[77]
		^{119}Sn-γ: δ = −0.48 (α-Sn), Δ = 3.44 or −0.53 (α-Sn) and 3.56 [26], δ = 1.57, Δ = 3.56 [40] or 1.62 and 3.44 [20, 40] at 77 to 78 K; δ = 1.46, Δ = 3.67 at 80 K [51]	[20, 26, 40, 51]
		IR on p. 293	[51]
2	$OOCC_6H_4F\text{-}4$ II	m.p. 69°	[76]
		with $\overline{OCH_2CH}CH_2OC_6H_5$ → $(C_4H_9)_2Sn(OCH(CH_2OC_6H_5)CH_2OOCC_6H_4F\text{-}4)_2$; kinetics studied	

Table 27 (continued)

No.	OOCR group method of preparation (yield in %)	properties and remarks	Ref.
*3	$OOCC_6H_4Cl$-2 Ia [51] Ib (83 [23]) V (47 [11]) special [25]	m.p. 78 to 80.5° [51], 84 to 85° [23, 25] ^{119}Sn-γ (80 K): δ = 1.44, Δ = 3.56; octahedral structure with bidentate ligands and *trans* C-Sn-C bonds IR (KBr): ν_{as}(OCO) 1590 and 1555, ν_s(OCO) 1385 and 1350	[23, 25, 51] [51]
*4	$OOCC_6H_4Cl$-3 special	m.p. 35 to 40° ^{1}H NMR (CCl_4): 7.30, 7.40, 8.07, and 8.10 (C_6H_4)	[47]
5	$OOCC_6H_4Cl$-4 Ia [51] II [52, 76]	m.p. 125 to 127° [51], 127° [76] ^{119}Sn-γ (80 K): δ = 1.34, Δ = 3.26; octahedral structure with *trans* C-Sn-C arrangement with $OCH_2CHCH_2OC_6H_5$ → $(C_4H_9)_2$Sn-$(OCH(CH_2OC_6H_5)CH_2OOCC_6H_4Cl\text{-}4)_2$; kinetics studied improves the processibility of crystalline polypropylene catalyzes formation and polymerization of urethanes	[51, 76] [51] [76] [52] [33]
6	$OOCC_6H_3Cl_2$-2,4 II	improves the processibility of crystalline polypropylene	[52]
7	$OOCC_6H_4Br$-3 II	m.p. 77 to 78° catalyst for peptide synthesis	[65]
8	$OOCC_6H_4Br$-4 II [62, 65]	m.p. 141° [62], 143 to 143.5° [65] with $OCH_2CHCH_2OC_6H_5$ → $(C_4H_9)_2$Sn-$(OCH(CH_2OC_6H_5)CH_2OOCC_6H_4Br\text{-}4)_2$; kinetics studied catalyst for peptide synthesis	[62, 65] [62] [65, 69]
*9	$OOCC_6H_4OH$-2 Ib [51] II [33], (87 [59])	m.p. 69 to 72° [33], 74° [59], 76 to 78° [51] ^{119}Sn-γ (80 K): δ = 1.47, Δ = 3.60 IR: ν(OH) 3114, ν_{as}(OCO) 1620 in KBr [51]; ν(C=O) 1635 in CCl_4 [59]	[33, 51, 59] [51] [51, 59]
10	$OOCC_6H_4OH$-4	no preparation reported catalyst for the preparation and polymerization of urethanes stabilizer for halogen-containing resins, e.g. PVC, against heat and discoloration	 [33] [48]

Table 27 (continued)

No.	OOCR group method of preparation (yield in %)	properties and remarks	Ref.
11	$OOCC_6H_4OCH_3$-4 II [62, 76]	m.p. 71° [62], 71 to 71.5° [65] with $OCH_2CHCH_2OC_6H_5 \rightarrow (C_4H_9)_2Sn$-$(OCH(CH_2OC_6H_5)CH_2OOCC_6H_4OCH_3$-$4)_2$; kinetics studied catalyst for peptide formation	[62, 65] [62, 76] [65, 69]
*12	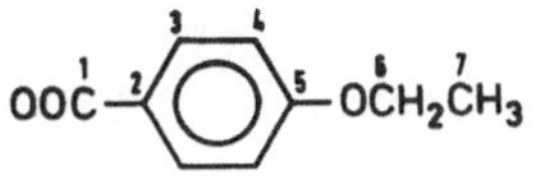 II [61]	^{13}C NMR (50% in $CDCl_3$): 13.5 (C-δ), 25.5 (C-α [58, 61], J(Sn,C) = 586 [58], 598 [61]), 26.4 (C-γ, J(Sn,C) = 98), 26.8 (C-β, J(Sn,C) = 20) [58, 61]; 14.7 (C-7), 63.7 (C-6), 114.0 (C-4), 122.8 (C-2), 132.5 (C-3), 162.9 (C-5) [58], 175.6 (C-1) [58, 61]	[58, 61]
13	$OOCC_6H_4OCH_2CH(OH)$-CH_2Cl-2	prepared from No. 9 and OCH_2CHCH_2Cl (see further information of No. 9)	[59]
14	$OOCC_6H_4SH$-2 III	m.p. 135°, heat resistant UV absorber (200 to 400 nm range) used as impregnation agent	[24]
15	$OOCC_6H_4SC_8H_{17}$-2 III	n_D^{20} = 1.5992 properties like No. 14	[24]
16	$OOCC_6H_4SO_2NH_2$-4 Ib (100)	m.p. 225°	[23]
17	$OOCC_6H_4NH_2$-2 II [57]	m.p. 102 to 103° ^1H NMR and IR spectroscopic results indicate bidentate carboxylate groups and exclude N → Sn coordination with $OCH_2CH_2 \rightarrow$ inexpensive oxyethylation products containing only 11% Sn	[57] [7]
18	$OOCC_6H_4NH_2$-4 Ib [67], (50 [60], 64 [71]) II [30], (62 [34]) IV [63]	m.p. 118 to 119° [30], 123 to 124° [60], 124 to 126° [34], 127 to 128° [63] ^1H NMR: 0.9, 1.5, 2.6, 3.80, 6.78, and 7.80; spectrum depicted IR: ν(NH) 3300 [63], 3360 to 3330, ν(OCO) 1710 and 1265 [71]; spectrum depicted (4000 to 800) [63] with No. 44 (after heat/vacuum treatment) → polymeric Sn-containing acid imides	[30, 34, 60, 63] [63] [63, 71] [60]

References on p. 295

Table 27 (continued)

No.	OOCR group method of preparation (yield in %)	properties and remarks	Ref.
18 (continued)		with $C_6H_4(NH_2)_2$ and 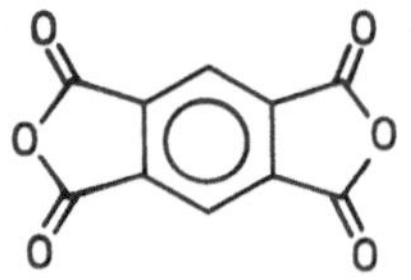 (after heat/vacuum treatment) → copolymeric acid imides	[67, 71]
		used for modification of polyurethane elastomers	[63]
19	$OOCC_6H_4NHCH_3$-2 II	colorless crystals m.p. 69 to 70°	[57]
20	$OOCC_6H_4NHC_6H_5$-2 II	light yellow substance m.p. 45°	[57]
21	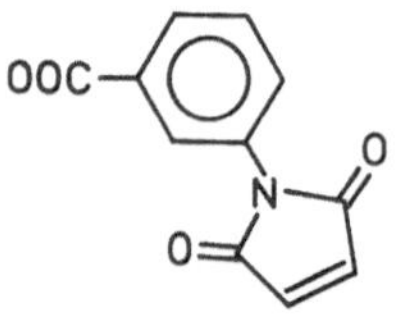 II (80 [38, 43])	m.p. 163.5 to 164° IR data emission spectral analysis	[38]
22	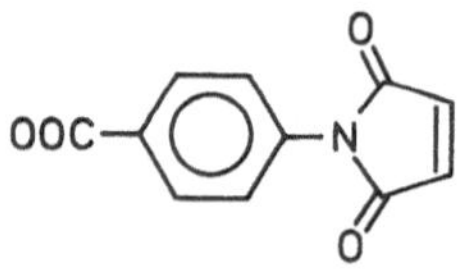 II (65)	m.p. 209 to 210° IR data emission spectral analysis	[38]
23	$OOCC_6H_4N{=}CHC_6H_5$-2 II (98)	orange crystals m.p. 178° IR (Nujol): ν_{as} (C=O) 1695, ν(C=N) 1600, ν(C=C) 1570, ν(N → Sn) 450 six-coordinated structure with N → Sn dative bonds suggested	[80]
24	$OOCC_6H_3(NH_2$-2)Cl-5 II [57]	light yellow crystals m.p. 123°	[57]
		curing agent for epoxy resins	[78]
25	$OOCC_6H_4NO_2$-2	no preparation reported	
		catalyst for preparation and polymerization of urethanes	[33]

References on p. 295

Table 27 (continued)

No.	OOCR group method of preparation (yield in %)	properties and remarks	Ref.
26	$OOCC_6H_4NO_2$-3	m.p. 115° [62], 117 to 118° [65],	[23, 62, 65]
	Ib (84 [23])	120 to 122° [23]	
	II [62, 65]	with $\overline{OCH_2CH}CH_2OC_6H_5 \rightarrow (C_4H_9)_2Sn$- $(OCH(CH_2OC_6H_5)CH_2OOCC_6H_4NO_2-3)_2$; kinetics studied	[62, 76]
		catalyst for peptide formation	[65]
*27	$OOCC_6H_4NO_2$-4	m.p. 205 to 206° [30], 214° [62], 218° [28, 29],	[23, 25, 28,
	Ib (87 [23])	218 to 219° [23, 25]	29, 30, 62]
	II [30, 33, 52, 62], (83 [28])	with $\overline{OCH_2CH}CH_2OC_6H_5 \rightarrow (C_4H_9)_2Sn$- $(OCH(CH_2OC_6H_5)CH_2OOCC_6H_4NO_2-4)_2$; kinetics studied	[62, 76]
	special [25, 29]	catalyst for the preparation and polymerization of urethanes	[33]
		improves the processibility of polypropylene	[52]
28	$OOCC_6H_3(Cl$-2$)NO_2$-4	m.p. 101 to 102°	[30]
	II		
29	$OOCC_6H_3(NO_2$-3$)OCH_3$-4	m.p. 148°	[76]
	II	with $\overline{OCH_2CH}CH_2OC_6H_5 \rightarrow (C_4H_9)_2Sn$- $(OCH(CH_2OC_6H_5)CH_2OOCC_6H_3(NO_2-3)$- OCH_3-4$)_2$; kinetics studied	
30	$OOCC_6H_3(NO_2)_2$-3,5	m.p. 187 to 188° [23], 196° [76]	[23, 76]
	Ib (89 [23])	with $\overline{OCH_2CH}CH_2OC_6H_5 \rightarrow (C_4H_9)_2Sn$- $(OCH(CH_2OC_6H_5)CH_2OOCC_6H_3$- $(NO_2)_2$-3,5$)_2$; kinetics studied	[76]
	II [76]		
31	$OOCC_6H_4As(C_6H_5)_2$-2	colorless crystals	[57]
	II	m.p. 79°	
32	$OOCC_6H_4As(C_6H_4CH_3$-4$)_2$-2	colorless crystals	[57]
	II	m.p. 65°	
33	$OOCC_6H_4CH_3$-4	m.p. 64 to 65° [34], 68° [76]	[34, 76]
	II [76], (75 [34])	with $\overline{OCH_2CH}CH_2OC_6H_5 \rightarrow (C_4H_9)_2Sn$- $(OCH(CH_2OC_6H_5)CH_2OOCC_6H_4CH_3)_2$; kinetics studied	[76]
	V (39 [11])		
34	$OOCC_6H_4C_4H_9$-t-4	m.p. 95 to 96°	[30]
	II		
35	$OOCC_6H_4CH{=}CH_2$-4	copolymerizes with styrene	[16]
	Ib	fungicidal agent	
	II		
36	$OOCC_6H_2(OH$-4$)(C_4H_9$-t$)_2$-3,5	no preparation reported	
		stabilizer for PVC	[45]

References on p. 295

Table 27 (continued)

No.	OOCR group method of preparation (yield in %)	properties and remarks	Ref.
37	$OOCC_6H_2(OH\text{-}2)(CH_3\text{-}4)SCH_3\text{-}5$ Ia	bactericide and fungicide	[44]
38	$OOCC_6H_4COCH_3\text{-}4$ II	improves processibility of polypropylene	[52]
39	$OOCC_6H_4COC_6H_5\text{-}2$ III	m.p. 65°, heat resistant UV absorber (200 to 400 nm range) used as impregnation agent	[24]
40	$OOCC_6H_4COOC_2H_5\text{-}2$	no preparation reported additive to crystalline polypropylene, improving dyeability and weather resistance	[36, 37]
41	$OOCC_6H_4COOC_4H_9\text{-}2$	no preparation reported additive to acrylic fibers, preventing devitrification during dyeing and steaming	[49]
42	$OOCC_6H_4COOC_8H_{17}\text{-}2$	no preparation reported additive to acrylic fibers, preventing delustering during treatment with boiling H_2O	[54]
43	$OOCC_6H_4COOCH_2CH(C_2H_5)\text{-}C_4H_9\text{-}4$ Ia	$n_D^{20} = 1.5180$ stabilizer for PVC	[13, 15, 17]
*44	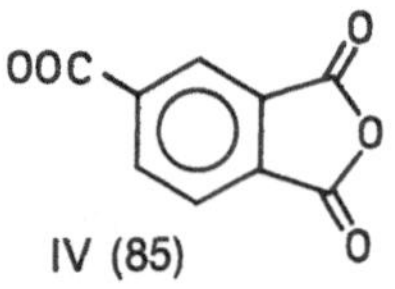 IV (85)	m.p. 138 to 140°	[60]
45	$OOCC_6H_4CON(CH_2CH_2CN)_2\text{-}2$ II	stabilizer for vinyl halide-acrylonitrile copolymers, preventing discoloration by heat or UV light	[10]
46	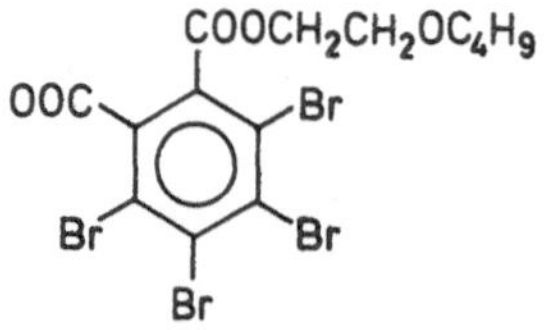	no preparation reported additive for thermochromic compositions useful in display panels for timers and computers	[50]

References on p. 295

* Further information:

$(C_4H_9)_2Sn(OOCC_6H_5)_2$ (Table **27**, No. **1**). The C_6H_5COO groups can be determined by non-aqueous titration with CH_3ONa in C_5H_5N using thymolphthalein as visual indicator, or an Sb electrode for potentiometric indication (error $< 0.5\%$) [27]. A rapid determination of nonagram amounts of the compound using molecular emission cavity analysis is described in [53].

The IR spectra of No. 1 show that the OCO absorptions are the same in Nujol mull as in $CHCl_3$ solution, indicating chelation rather than bridging in the solid [51]:

ν in cm^{-1}	KBr disk	solid film	Nujol mull	$CHCl_3$ solution
$\nu_{as}(OCO)$	1600	1598	1603	1603
	1558	1560	1567	1567
$\nu_s(OCO)$	1362	1365	1363	1360

Equimolar quantities of No. 1 and $(C_4H_9)_2SnH_2$ react at room temperature within 5 to 18 h with evolution of H_2 and formation of $C_6H_5COO(C_4H_9)_2SnSn(C_4H_9)_2OOCC_6H_5$ (48%) [8]. The compound reacts with an excess of $\overline{OCH_2C}HCH_2OC_6H_5$ at 165 to 185°C within 5 to 8 h in an inert atmosphere to give the insertion product $(C_4H_9)_2Sn(OCH(CH_2OC_6H_5)CH_2OOCC_6H_5)_2$ [56]. This reaction was studied kinetically and mechanistically and the results were compared with those obtained for $(C_4H_9)_2Sn(OOCC_6H_4X)_2$ compounds (X = 4-NO_2, 3-NO_2, 4-Br, 4-OCH_3), also in aprotic, strongly polar solvents [56, 73, 76]. The reaction between the title compound and $V(C_5H_5)_2$ at 20°C for 1 d in a sealed tube using $C_6H_5CH_3$ as a solvent, affords C_5H_6 (84%), $C_5H_5V(OOCC_6H_5)_2$, $(C_4H_9)_3SnOOCC_6H_5$ and $Sn(C_4H_9)_2$, $Sn(OOCC_6H_5)_2$ (10 to 20%), and Sn and No. 1 (10 to 20%) as decomposition products of the intermediate $[C_4H_9SnOOCC_6H_5]$ [64]. No. 1 forms in C_6H_6 an 1:1 complex with $C_6H_5CH_2NH_2$ with a formation constant of $K_1 = 2.4 \pm 0.1$ L/mol at 27°C (determined by ^1H NMR), which after addition of $CH_3COOC_2H_5$ changes into a triple associate [70]. An 1:1 complex is also formed with o-phenanthroline in C_6H_6 with replacing of $C{=}O \rightarrow Sn$ coordination by $N \rightarrow Sn$ dative bonds [51]. A water-plasticizable polymer is obtained from the reaction between No. 1 and resorcinol diglycidyl ether [74].

No. 1 has been tested as a fungicide [2, 5], e.g., against potato haulm blight, Phytophthora infestans [12], as an anthelmintic against the cestode Raillietina cesticillus and the nematode Ascaridia galli from chickens [3], as well as a bactericide [2, 5]. White wheat seeds (Triticum aestivum) treated with 2% of No. 1 and offered as feed to house mice, get repel 90% of the animals [79].

The title compound has been used as a stabilizer and property improver of PVC [42, 55], polypropylene [35, 36, 37, 46], polyolefins [41], siloxane elastomers [4], and silicone rubbers [81]. It works as a catalyst in the preparation of polyacetal resins [32], in peptide synthesis [65, 66, 69], in the reaction between 4-methyl-m-phenylene diisocyanate and triethylene glycol (comparison with other dialkyltin biscarboxylates) [68], in the polymerization of phenylglycidic ester [72], or in the aminolysis of $CH_3COOC_6H_4NO_2$-4 or $CH_3COOC_6H_3(NO_2)_2$-2,4 by $NH_2CH_2C_6H_5$, $NH(CH_2C_6H_5)_2$, or $NH(C_3H_7\text{-}i)_2$ [75].

$(C_4H_9)_2Sn(OOCC_6H_4Cl\text{-}2)_2$ (Table **27**, No. **3**) is formed in a 92% yield in the 1:2 molar reaction between 4-$NO_2C_6H_4COO(C_4H_9)_2SnOSn(C_4H_9)_2OOCC_6H_4NO_2$-4 and 2-$ClC_6H_4COOH$ in refluxing $C_6H_5CH_3$, along with $(C_4H_9)_2Sn(OOCC_6H_4NO_2\text{-}4)_2$ (96% yield). Equimolar amounts of $(\text{-}(C_4H_9)_2SnO\text{-})_n$, 2-$ClC_6H_4COOH$, and 4-$NO_2C_6H_4COOH$ react in refluxing $C_6H_5CH_3$ to give 92% No. 3 and 98% of $(C_4H_9)_2Sn(OOCC_6H_4NO_2\text{-}4)_2$. $(C_4H_9)_2Sn(Cl)OOCC_6H_4Cl$-2, obtained from $Cl(C_4H_9)_2SnOSn(C_4H_9)_2Cl$ and 2-ClC_6H_4COOH, reacts with 4-$NO_2C_6H_4COONa$ in refluxing

References on p. 295

C_6H_6 to give 74% No. 3 and 85% $(C_4H_9)_2Sn(OOCC_6H_4NO_2\text{-}4)_2$. 91% of No. 3 and 98% of $(C_4H_9)_2Sn(OOCC_6H_4NO_2\text{-}4)_2$ were obtained by the interfacial reaction between a solution of 4-$NO_2C_6H_4COONa$ and 2-ClC_6H_4COONa in water and a solution of $(C_4H_9)_2SnCl_2$ in CH_2Cl_2 at room temperature for 0.5 h [25]. No. 3 forms a 1:1 complex with o-phenanthroline in which the O→Sn coordination in the six-coordinate compound is replaced by N→Sn coordination [51]. The compound is used as a catalyst for the synthesis and polymerization of urethanes [33].

$(C_4H_9)_2Sn(OOCC_6H_4Cl\text{-}3)_2$ (Table **27**, No. **4**) is formed by oxidation of $(C_4H_9)_2Sn(CH{=}CH_2)_2$ with 3-$ClC_6H_4CO\text{-}OOH$ in C_6H_6 along with $(C_4H_9)_2Sn(CH{=}CH_2)\overline{CHCH_2O}$ [47].

$(C_4H_9)_2Sn(OOCC_6H_4OH\text{-}2)_2$ (Table **27**, No. **9**). The IR spectroscopic data indicate a six-coordinate structure with monodentate carboxylate groups and coordinating hydroxyl oxygens [51, 59]. Whereas in [51], a *trans* C-Sn-C arrangement is deduced from the quadrupole splitting value (Δ = 3.60 mm/s), the title compound is formulated as a *cis* complex in [59].

The reaction of No. 9 with $\overline{OCH_2CHCH_2Cl}$ (1:20 mole ratio) at 80 to 90 °C has an autocatalytic character and leads to No. 13, $(C_4H_9)_2Sn(OOCC_6H_4OCH_2CH(OH)CH_2Cl\text{-}2)_2$, and $(C_4H_9)_2Sn\text{-}(OC_6H_4COOCH_2CH(OH)CH_2Cl\text{-}2)_2$, indicating an equilibrium between two isomeric forms of No. 9, $(C_4H_9)_2Sn(OOCC_6H_4OH\text{-}4)_2$ and $(C_4H_9)_2Sn(OC_6H_4COOH\text{-}2)_2$, in solvents capable of inducing proton transfer as excess of $\overline{OCH_2CHCH_2Cl}$ or also CH_3SOCH_3 [59]. No. 9 forms a 1:1 complex with o-phenanthroline in which the HO→Sn coordination is replaced by N→Sn coordination [51].

The title compound is an effective fungicide against Phytophthora infestans causing potato haulm blight [7, 39] and controls peach leaf curl and powdery mildew of cucumber [7].

No. 9 is used in stabilizer compositions [31], as a heat stabilizer for PVC [6, 12], or as light and oxidation stabilizer for dichlorobutadiene resins [1]. It catalyzes the formation of polyacetal resins [32], and the formation and polymerization of urethanes [33].

$(C_4H_9)_2Sn(OOCC_6H_4OC_2H_5\text{-}4)_2$ (Table **27**, No. **12**) is monomeric in dilute $CHCl_3$ solution, and shows a $^1J(Sn,C)$ value of 586 [58] or 598 Hz [61] indicating five-coordinate Sn and equivalent carboxyl carbon shifts of δ = 175.6 ppm, thus suggesting a structure of Formula I with rapid bidentate-unidentate interchange of the coordinating groups [58, 61].

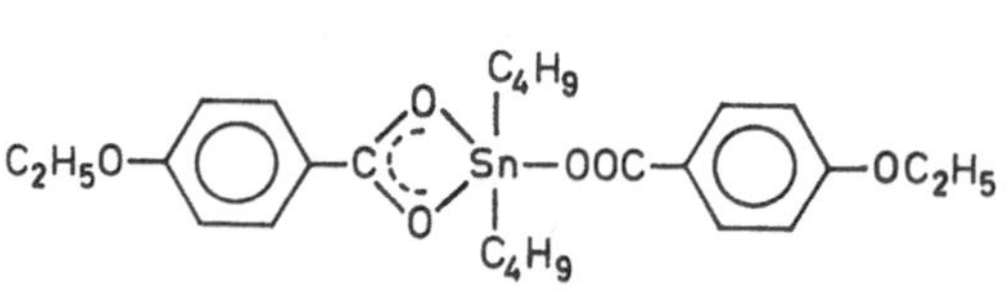

I

$(C_4H_9)_2Sn(OOCC_6H_4NO_2\text{-}4)_2$ (Table **27**, No. **27**). In addition to the special methods for its formation already described for No. 3, the title compound also results from the reactions between 4-$NO_2C_6H_4COO(C_4H_9)_2SnOSn(C_4H_9)_2OOCC_6H_4NO_2$-4 and $CHCl_2COOH$ in refluxing $C_6H_5CH_3$ (96% yield), or between $Cl(C_4H_9)_2SnOOC(CH_2)_4COOSn(C_4H_9)_2Cl$ and 4-$NO_2C_6H_4$-$COONa$ in $CH_3COOC_2H_5$ at room temperature (80% yield) [25]. A quantitative yield is obtained by treatment of $(4\text{-}NO_2C_6H_4COO(C_4H_9)_2SnOSn(C_4H_9)_2\text{-})_2O$ with the sixfold molar amount of 4-$NO_2C_6H_4COOH$ in refluxing $C_6H_5CH_3$ for 9 h [29].

$(C_4H_9)_2Sn(OOCC_6H_3C_2O_3\text{-}3,4)_2$ (Table **27**, No. **44**) reacts with $(C_4H_9)_2Sn(OOCC_6H_4NH_2\text{-}4)_2$ to give a polymeric acid amide of Formula II which after heat/vacuum treatment changes into a polymeric acid imide (Formula III) [60].

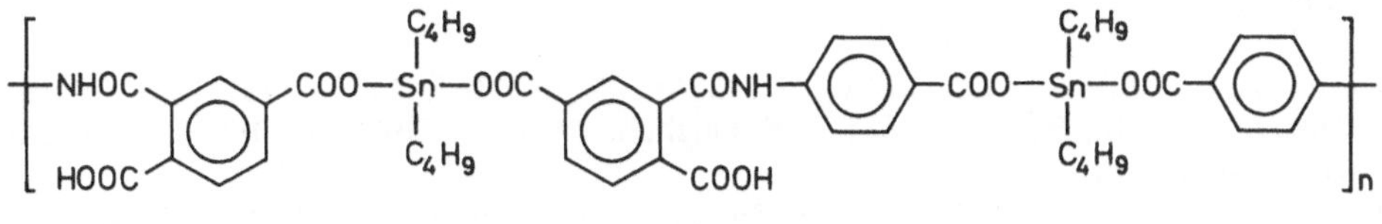

II

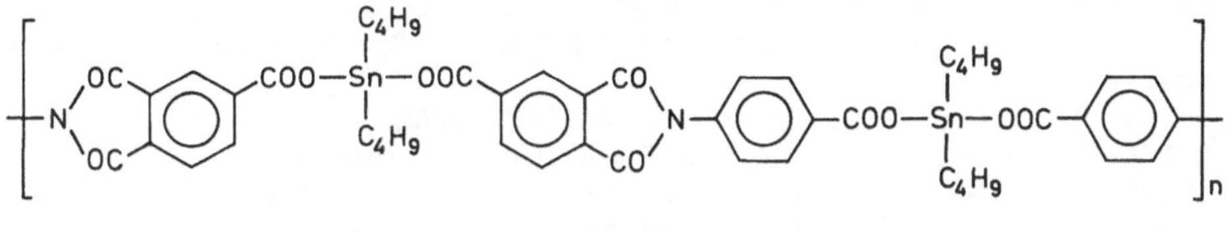

III

References:

[1] Rowland, G. P., Reid, R. J., Firestone Tire and Rubber Co. (U.S. 2445739 [1948]; C.A. **1949** 440).
[2] Farbwerke Hoechst A.-G. (Brit. 797073 [1953]; C.A. **1959** 22714).
[3] Kerr, K. B., Walde, A. W. (Experim. Parasitol. **5** [1956] 560/70).
[4] Polmanteer, K. E., Dow Corning Corp. (Ger. 1019462 [1957]; C.A. **1960** 10378).
[5] Brückner, H., Härtel, K., Farbwerke Hoechst A.-G. (Ger. 1025198 [1958]; C.A. **1960** 12468).
[6] Berlin, A. A., Popova, Z. V., Yanovskii, D. M. (Zh. Prikl. Khim. **33** [1960] 871/7; J. Appl. Chem. [USSR] **33** [1960] 870/5).
[7] Farbwerke Hoechst A.-G. (Brit. 849220 [1960]; C.A. **1961** 17503).
[8] Sawyer, A. K., Kuivila, H. G. (J. Am. Chem. Soc. **82** [1960] 5958/9).
[9] Alleston, D. L., Davies A.-G. (J. Chem. Soc. **1962** 2050/4).
[10] Lynn, J. W., Walter, A. T., Union Carbide Corp. (U.S. 3053870 [1958/62]; C.A. **58** [1963] 3457).

[11] Sawyer, A. K., Kuivila, H. G. (J. Org. Chem. **27** [1962] 610/4).
[12] Shteding, M. N., Karpov, V. L. (Vysokomol. Soedin. **4** [1962] 1806/11; Polym. Sci. [USSR] **4** [1962] 561/6).
[13] Koninklijke Industrieele Maatschappij voorheen Noury and van der Lande N.V. (Fr. 1320473 [1961/63]; C.A. **59** [1963] 8788/9).
[14] Sawyer, A. K., Kuivila, H. G., Metal and Thermit Corp. (U.S. 3083217 [1960/63]; C.A. **59** [1963] 7559).
[15] Koninklijke Industrieele Maatschappij voorheen Noury and van der Lande N.V. (Neth. 109491 [1962/64]; C.A. **62** [1964] 9173).
[16] Leebrink, J. R., M and T Chemicals, Inc. (Brit. 952490 [1960/64]; C.A. **61** [1964] 8340).
[17] Viveen, W. J. C., Schröder, A., Koninklijke Industrieele Maatschappij voorheen Noury and van der Lande N.V. (Ger. Offen. 1167836 [1962/64]).

[18] Frankel, M., Gertner, D., Wagner, D., Zilkha, A. (J. Appl. Polym. Sci. **9** [1965] 3383/8).
[19] Vyazankin, N. S., Bychkov, V. T. (Zh. Obshch. Khim. **35** [1965] 684/7; J. Gen. Chem. [USSR] **35** [1965] 685/7).
[20] Herber, R. H., Stöckler, H. A. (Tech. Rept. Ser. Intern. At. Energy Agency **50** [1966] 110/20).

[21] Mussell, D. R., Dow Chemical Co. (U.S. 3284290 [1963/66]; C.A. **66** [1967] No. 18296).
[22] Vyazankin, N. S., Bychkov, V. T. (Zh. Obshch. Khim. **36** [1966] 1684/7; J. Gen. Chem. [USSR] **36** [1966] 1681/3).
[23] Frankel, M., Gertner, D., Wagner, D., Zilkha, A. (J. Organometal. Chem. **9** [1967] 83/8).
[24] Dynamit Nobel A.-G. (Fr. 1539228 [1967/68]; C.A. **71** [1969] No. 80957).
[25] Frankel, M., Wagner, D., Gertner, D., Zilkha, A. (Israel J. Chem. **6** [1968] 817/21).
[26] Gol'danskii, V. I., Khrapov, V. V., Oklobystin, O. Yu., Rochev, V. Ya. (in: Gol'danskii, V. I., Herber, R. H., Chemical Application of Mössbauer Spectroscopy, New York 1968, pp. 336/76).
[27] Groagova, A., Pribyl, M. (Z. Anal. Chem. 234 [1968] 423/8).
[28] Migdal, S., Gertner, D., Zilkha, A. (Eur. Polym. J. **4** [1968] 465/72).
[29] Migdal, S., Gertner, D., Zilkha, A. (Can. J. Chem. **46** [1968] 2409/13).
[30] Saruto, K., Gono, T., Suenobu, Y., Yoshitomi Pharmaceutical Industries, Ltd. (Japan. 68-07941 [1964/68]; C.A. **69** [1968] No. 87186).

[31] Stapfer, C. H., Carlisle Chemical Works, Inc. (Fr. 1537462 [1966/68]; C.A. **71** [1969] No. 13830).
[32] Asahi Kasei Kogyo Kabushiki Kaisha (Brit. 1151927 [1965/69]; C.A. **71** [1969] No. 13755).
[33] Fuchsman, C. H., Brown, J. C., Ferro Corp. (Brit. 1163857 [1967/69]; C.A. **71** [1969] No. 103290).
[34] Mikhailov, G. D., Chegolya, A. S. (Sin. Volokna **1969** 18/21; C.A. **74** [1971] No. 3711).
[35] Senda, K., Ichikawa, A., Nakajima, E., Sasaki, M., Hirose, M., Mitsubishi Rayon Co., Ltd. (Japan. 69-26179 [1966/69]; C.A. **72** [1970] No. 101804).
[36] Senda, K., Ichikawa, A., Oseki, T., Nakajima, E., Sakai, M., Nishikawa, T., Mitsubishi Rayon Co., Ltd. (Japan. 69-13589 [1966/69]; C.A. **72** [1970] No. 13783).
[37] Senda, K., Ichikawa, A., Oseki, T., Nakajima, E., Sakai, M., Yasui, A., Hirose, M., Mitsubishi Rayon Co., Ltd. (Japan. 69-28872 [1966/69]; C.A. **72** [1970] No. 122813, **73** [1970] No. 131931, **73** [1970] No. 57071).
[38] Zykova, S. K., Ivanov, V. S., Naumova, I. V. (Zh. Obshch. Khim. **39** [1969] 718; J. Gen. Chem. [USSR] **39** [1969] 686).
[39] McIntosh, A. H. (Ann. Appl. Biol. **66** [1970] 115/8).
[40] Smith, P. J. (Organometal. Chem. Rev. A **5** [1970] 373/402).

[41] Yukidsa, Y., Kuboda, Y., Yamamoto, T., Toray Industries, Inc. (Japan. 70-34452 [1966/70]; C.A. **75** [1971] No. 37284).
[42] Yukitomi, M. (Plast. Age **1970** 111/8; C.A. **72** [1970] No. 112097).
[43] Zykova, S. K., Naumova, I. V., Ivanov, V. S. (Metody Polouch. Khim. Reaktivov Prep. **1970** No. 22, pp. 32/5; C.A. **77** [1972] No. 5580).
[44] Fuchsman, C. H., Meek, W. H., Ferro Corp. (U.S. 3574693 [1969/71]; C.A. **75** [1971] No. 20626).
[45] Seki, T., Hiyama, Y., Sato, Y., Nitto Chemical Industrial Co., Ltd. (U.S. 3591551 [1966/71]; C.A. **75** [1971] No. 118958).
[46] Suzuki, R., Tahara, T., Mototani, H., Nitto Kaisei Co., Ltd. (U.S. 3674769 [1969/72]; C.A. **77** [1972] No. 102655).
[47] Ayrey, G., Parsonage, J. R., Poller, R. C. (J. Organometal. Chem. **56** [1973] 193/8).

[48] Ueno, R., Kashibara, W., Ueno Pharmaceutical Applied Research Laboratory (Japan. 73-23332 [1969/73]; C.A. **80** [1974] No. 121856).
[49] Kozuka, N., Kurioka, S., Yasumoto, T., Kobayashi, S., Ohtoshi, N., Kanegafuchi Chemical Industry Co., Ltd. (Japan. 73-29500 [1970/73]; C.A. **81** [1974] No. 27052).
[50] Murakami, T., Muneoka, M., Suzuki, S., Nitto Electric Industrial Co., Ltd. (Japan. Kokai 73-40679 [1971/73]; C.A. **79** [1973] No. 99219).

[51] Naik, D. V., May, J. C., Curran, C. (J. Coord. Chem. **2** [1973] 309/15).
[52] Suzuki, R., Tahara, T., Takubo, T., Nitto Kaisei Co., Ltd. (Japan. 73-22181 [1969/73]; C.A. **80** [1974] No. 71674).
[53] Akpofure, C. O., Belcher, R., Bogdanski, S. L., Townshend, A. (Anal. Letters **8** [1975] 921/9).
[54] Kozuka, K., Kurioka, S., Yasumoto, T., Kobayashi, S., Otoshi, N., Kubota, A., Kanegafuchi Chemical Industry Co., Ltd. (Fr. Demande 2236970 [1973/75]; C.A. **83** [1975] No. 165650).
[55] Dworkin, R. D., Ejk, A. J., M and T Chemicals, Inc. (Ger. Offen. 2626554 [1975/76]; C.A. **86** [1977] No. 140253).
[56] Klebanov, M. S., Shologon, I. M., Novikova, T. V. (Zh. Obshch. Khim. **47** [1977] 1078/81; J. Gen. Chem. [USSR] **47** [1977] 987/9).
[57] Sandhu, S. S., Sindhu, J. K., Sandhu, G. K. (Indian J. Chem. A **15** [1977] 654/5).
[58] Domazetis, G., Magee, R. J., James, B. D. (J. Organometal. Chem. **148** [1978] 339/54).
[59] Klebanov, M. S., Shoshina, L. V., Shologon, I. M., Nikonova, L. P. (Zh. Obshch. Khim. **48** [1978] 138/41; J. Gen. Chem. [USSR] **48** [1978] 117/20).
[60] Koton, M. M., Kiseleva, T. M., Dergacheva, E. N., Nikolaeva, S. N. (Zh. Obshch. Khim. **48** [1978] 1561/3; J. Gen. Chem. [USSR] **48** [1978] 1430/2).

[61] Domazetis, G., Magee, R. J., James, B. D. (J. Inorg. Nucl. Chem. **41** [1979] 1547/53).
[62] Klebanov, M. S., Shologon, I. M. (Zh. Obshch. Khim. **49** [1979] 812/6; J. Gen. Chem. [USSR] **49** [1979] 704/8).
[63] Inagaki, S., Yamada, E., Okamoto, H., Furukawa, J. (Nippon Gomu Kyokaishi **53** [1981] 745/50; C.A. **94** [1980] No. 104635).
[64] Latyaeva, V. N., Lineva, A. N., Zimina, S. V., Gordetsov, A. S., Lergunov, Yu. I. (Zh. Obshch. Khim. **51** [1981] 1101/6; J. Gen. Chem. [USSR] **51** [1981] 921/6).
[65] Litvinenko, L. M., Garkusha-Bozhko, I. P., Oleinik, N. M., Klebanov, M. S., Nesterenko, Yu. A. (Zh. Org. Khim. **17** [1981] 307/11; J. Org. Chem. [USSR] **17** [1981] 255/60).
[66] Oleinik, N. M., Garkusha-Bozhko, I. P., Litvinenko, L. M. (Ukr. Khim. Zh. **47** [1981] 723/8; Soviet Progr. Chem. **47** No. 7 [1981] 50/4).
[67] Pathak, C. P., Samant, S., Murty, M. V. R., Babu, G. N. (Org. Coat. Appl. Polym. Sci. Proc. **46** [1982] 154/7).
[68] Aliev, I. M., Noskov, N. M., Tasilova, M. E., Klyuchinskii, S. A., Dergunov, Yu. I., Zavgorodnii, V. S., Rogozev, B. I., Petrov, A. A. (Zh. Obshch. Khim. **52** [1982] 1866/71; J. Gen. Chem. [USSR] **52** [1982] 1654/9).
[69] Garkusha-Bozhko, I. P., Oleinik, N. M., Litvinenko, L. M. (Zh. Org. Khim. **18** [1982] 2340/50; J. Org. Chem. [USSR] **18** [1982] 2068/77).
[70] Litvinenko, L. M., Kapkan, L. M., Oleinik, N. M., Garkusha-Bozhko, I. P., Pekhtereva, T. M. (Zh. Org. Khim. **18** [1982] 114/9; J. Org. Chem. [USSR] **18** [1982] 97/101).

[71] Babu, G. N., Pathak, C. P., Samant, S. (Polym. Sci. Technol. [Plenum] **21** [1983] 373/81).
[72] Klebanov, M. S., Shoshina, L. V., Shologon, I. M. (Dokl. Akad. Nauk SSSR **268** [1983] 1410/2; Dokl. Chem. Proc. Acad. Sci. USSR **268/273** [1983] 64/6).
[73] Klebanov, M. S., Shoshina, L. V., Shologon, I. M. (Zh. Obshch. Khim. **53** [1983] 1131/7; J. Gen. Chem. [USSR] **53** [1983] 1004/9).

[74] Lagunov, V. A., Polozenko, V. I., Sinani, A. B., Stepanov, V. A. (Fiz. Tverd. Tela [Leningrad] **25** [1983] 1816/21).

[75] Litvinenko, L. M., Oleinik, N. M., Garkusha-Bozhko, I. P. (Zh. Org. Khim. **19** [1983] 2353/8; J. Org. Chem. [USSR] **19** [1983] 2056/61).

[76] Klebanov, M. S., Shoshina, L. V., Shologon, I. M. (Zh. Obshch. Khim. **54** [1984] 2052/7; J. Gen. Chem. [USSR] **54** [1984] 1833/7).

[77] Lycka, A., Holecek, J. (J. Organometal. Chem. **294** [1985] 179/82).

[78] Popov, L. K., Plyashechnik, N. I., Sitkin, A. I., Ochneva, V. A., Batizat, V. P., Kovaleva, N. M., Zaitseva, N. P., Krasilnikov, F. S., Zorina, R. K., Kuznetsova, L. I. (Plast. Massy **1985** 41/2).

[79] Schafer, E. W., Bowles, W. A. (Arch. Environ. Contam. Toxicol. **14** [1985] 111/29).

[80] Shanker, R., Sakuntala, E. N. (Syn. React. Inorg. Metal-Org. Chem. **15** [1985] 779/88).

[81] Shin-Etsu Chemical Industry Co., Ltd. (Japan. Kokai Tokkyo Koho 60 04555 [1983/85]; C.A. **102** [1985] No. 221999).

1.4.1.2.1.5.4.5 Dibutyltin Biscarboxylates, $(C_4H_9)_2Sn(OOCR)_2$ with R = Heterocycle

The compounds belonging to this section are listed in Table 28 and are prepared by the following methods.

Method I: No. 4 precipitates from a reaction mixture obtained by mixing CH_3OH solutions containing $(C_4H_9)_2SnCl_2$ and 2-$HOOCC_5H_4N$ in a 1:2 mole ratio [3].

Method II: $(-(C_4H_9)_2SnO-)_n$ and RCOOH (1:2 mole ratio).
By this method have been prepared No. 1 [2], (C_6H_6, reflux, 3 h) [1], or (C_2H_5OH-C_6H_6, heating in a water-bath, 2 h) [7]; No. 2 (C_6H_6, reflux, 3 h) [1]; No. 3 (C_2H_5OH-C_6H_6, heating in a water-bath, 2 h) [7]; No. 4 (CH_3OH-C_6H_6 (1:5, v/v), reflux, 2 h) [6]; No. 5 (C_6H_6, reflux) [4]; Nos. 6 and 7 ($C_6H_5CH_3$, reflux, 4 h) [5]. The water formed in these condensation reactions is removed by azeotropic distillation.

Table 28
Dibutyltin Biscarboxylates, $(C_4H_9)_2Sn(OOCR)_2$ with R = Heterocycle.
Further information on compounds preceded by an asterisk is given at the end of the table.
Explanations, abbreviations, and units on p. X.

No.	OOCR group method of preparation (yield in %)	properties and remarks	Ref.
1	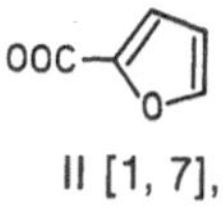 II [1, 7], (61 [2])	dirty white substance	[7]
		m.p. 47 to 48° [7], 96 to 98° [2], 118° [1]	[1, 2, 7]
		^{1}H NMR ($CDCl_3$): 0.80 to 1.73 (C_4H_9), 6.50 (H-4), 7.15 and 7.20 (d, H-3), 7.58 (H-5)	[7]
		IR: ν_{as}(OCO) 1540 (m), ν_s(OCO) 1415 (m), ν(SnC) 540 (m), ν(SnO) 468 (m)	
		UV (solid): λ = 215	
		six-coordinate *trans* structure suggested	
		useful as a fungicide	[1]

Table 28 (continued)

No.	OOCR group method of preparation (yield in %)	properties and remarks	Ref.
2	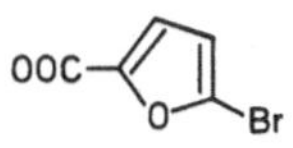 II	m.p. 120° useful as a fungicide	[1]
3	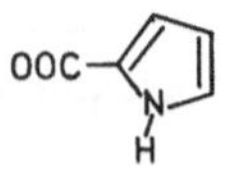 II	brown solid m.p. 108 to 109° ^{1}H NMR ($CDCl_3$): 0.78 to 1.74 (C_4H_9), 6.25 (m, H-4), 6.82 (m, H-3), 7.19 (H-5) IR: ν_{as}(OCO) 1530 (s), ν_s(OCO) 1420 (s) six-coordinate *trans* structure suggested	[7]
*4	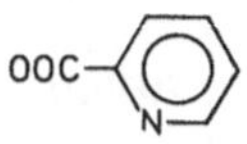 I [3] II [6]	m.p. 198 to 200° (dec.) [3], 198.5 to 200° [6, 8] ^{1}H NMR ($CDCl_3$): 0.74, 0.84, and 1.35 (C_4H_9), 7.96 to 8.87 and 9.28 (ligand) ^{119}Sn NMR ($CDCl_3$): -312 ± 2, ^{1}J(Sn, C) = 711, ^{2}J(Sn, C) = 33.5, ^{3}J(Sn, C) = 122.1, ^{2}J(Sn, H) = 74 ± 2 ^{119}Sn-γ (80 K): δ = 1.45, Δ = 4.35 IR and R spectra in Table 29, p. 300 UV: λ_{max} ($\varepsilon \cdot 10^{-3}$) = 264 (10.1) and 238 (3.8, sh) in $CHCl_3$; 263 (9.0) and 221 (11.9) in CH_3CN	[3, 6, 8] [6] [8] [3] [6]
5	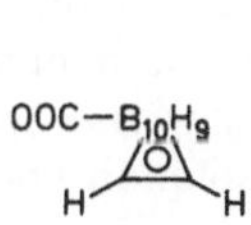 II (67)	orange needles m.p. 187 to 189° ^{1}H NMR: 0.8 to 1.75 (C_4H_9), 4.22 (s, C_5H_5), 4.40 and 5.85 (t's, C_5H_4) IR (KBr): ν_{as}(OCO) 1582, ν_s(OCO) 1320; six-coordinate structure with bidentate, chelating OCO groups deduced only poor heat stabilizer for PVC	[4]
6	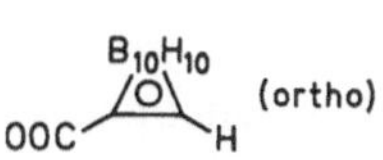 II (70)	m.p. 142 to 143°	[5]
7	 II (68)	m.p. 220 to 222°	[5]

References on p. 300

* Further information:

$(C_4H_9)_2Sn(OOCC_5H_4N\text{-}2)_2$ (Table **28**, No. **4**). The IR and Raman frequencies of the compound are listed in Table 29 [6]. In KBr the assignments ν_{as}(OCO) 1666 and 1624, and ν_s(OCO) 1383 and 1348 are reported [3].

Table 29
IR and Raman Spectra of solid $(C_4H_9)_2Sn(OOCC_5H_4N\text{-}2)_2$ [6].
Wave numbers in cm^{-1}.

IR CsBr	Raman powder	assignment	IR CsBr	Raman powder	assignment
3102 (vw,br)		$\nu_{as}(CH)_{ring}$	1381 (ms)	1382(28)	ν_s(OCO)
3085 (vw,br)	3088(36)		1344 (s)	1345(27)	
	3074(15)	$\nu_s(CH)_{ring}$	1254 (m)	1260(7)	$\delta_s(CH)_{ring}$
	3055(9)		1236 (m)		
2988 (ms)		$\nu_{as}(CH_3)$	1167 (m)	1173(16)	
2953 (ms)	2954(25)	$\nu_{as}(CH_2)$	1158 (m)	1165(17)	ν(ring)
	2921(43)	$\nu_s(CH_3)$	1148 (sh)	1154(52)	
2891 (m)	2893(29)	$\nu_s(CH_2)$	1087 (mw)	1086(16)	ν_{as}(ring)
	2868(19)		1048 (m)	1047(27)	$\delta_s(CH)_{ring}$
	2858(21)	$\nu_s(CH_2)$/Sn		1037(20)	ν(ring)
	2851(22)		1009 (ms)	1009(100)	ν_s(ring)
1669 (s)		ν_{as}(OCO)	842 (m)	847(19)	$\pi(CH)_{ring}$
1653 (sh)	1655(16)			824(8)	$\nu_s(OCO)_{ring}$
			754 (ms)		$\varrho\omega(CH)_{ring}$
1626 (s)	1623(7)		699 (ms)	699(11)	δ(OCO)
1598 (ms)	1596(27)	ν_{as}(C=C, C=N)	690 (ms)		
1581 (s)	1576(33)		631 (ms)	631(22)	δ_s(OCO)
1563 (s)	1566(25)		596 (w)	597(71)	ν_s(SnC)
1464 (m)	1471(13)	ν_s(C=C, C=N)	424 (m)	425(20)	ν_{as}(SnO)
1440 (m)	1440(29)			391(4)	ν(SnN)

From the summary of spectroscopic data, it is concluded that the solid compound has six-coordinate Sn with *trans* butyl groups and two bidentate ligands each bonded through O and N [3, 6, 8]. There is also intermolecular C=O → Sn bridging indicated, giving some oligomeric character to the solid. The IR and Raman signals assigned to intramolecular N → Sn and intermolecular C=O → Sn coordination are missing from $CHCl_3$ or CH_3CN solution spectra, indicating an essentially tetrahedral, nonbridging structure in solution [6].

References:

[1] Nakanishi, M., Tsuda, A., Yoshitomi Pharmaceutical Industries, Ltd. (Japan. 64-15690 [1961/64]; C.A. **62** [1965] 6513).

[2] Mikhailov, G. D., Chegolya, A. S. (Sin. Volokna **1969** 18/21; C.A. **74** [1971] No. 3711).

[3] Naik, D. V., Curran, C. (Inorg. Chem **10** [1971] 1017/20).
[4] Morris, D. R., Rocket, B. W. (J. Organometal. Chem. **35** [1972] 179/84).
[5] Mironov, V. F., Pechurina, S. Ya., Grigos, V. I. (Dokl. Akad. Nauk SSSR **230** [1976] 865/8; Dokl. Chem. Proc. Acad. Sci. USSR **226/231** [1976] 619/22).
[6] Howard, W. F., Nelson, W. H. (J. Mol. Struct. **53** [1979] 165/77).
[7] Sandhu, G. K., Boparai, N. S., Sandhu, S. S. (Syn. React. Inorg. Metal-Org. Chem. **10** [1980] 535/51).
[8] Howard, W. F., Crecely, R. W., Nelson, W. H. (Inorg. Chem. **24** [1985] 2204/8).

1.4.1.2.1.5.5 Dibutyltin Oxalate, $(C_4H_9)_2\overline{SnOOC\text{-}COO}$, and Dibutyltin Dicarboxylates of the $(C_4H_9)_2\overline{SnOOC\text{-}R\text{-}COO}$ Type

1.4.1.2.1.5.5.1 Dibutyltin Oxalate, $(C_4H_9)_2\overline{SnOOC\text{-}COO}$

Dibutyltin oxalate is formed as a monohydrate in the reaction of $(C_4H_9)_2SnCl_2$ with excess of HOOC-COOH · $2H_2O$ in H_2O-C_2H_5OH (1/1) at room temperature [2], or of $(C_4H_9)_2Sn(OOCCH_3)_2$ with a solution of HOOC-COOH in a 1/1 mixture of xylene and dioxane under reflux (95% yield). It is an insoluble and infusable compound, decomposing at 195°C [1]. The solubility product is 1.70×10^{-9}, 1.24×10^{-9}, or 1.22×10^{-9} in water, in 25% C_2H_5OH, or in 25% C_2H_5OH containing 2.5 M CH_3COOH, respectively. The precipitation of the salt, which may be followed amperometrically by measuring the reduction wave-height at −0.9 V, can be used as analytical method for dibutyltin compounds [2]. A rapid determination of ng amounts of the title compound itself is possible by means of molecular emission cavity analysis [4]. As well as other insoluble organotin compounds, the oxalate stabilizes fiber-forming acrylonitrile polymers against thermal and light degradation [3].

References:

[1] Andrews, T. M., Bower, F. A., LaLiberte, B. R., Montermoso, J. C. (J. Am. Chem. Soc. **80** [1958] 4102/4).
[2] Haasova, L., Pribyl, M. (Z. Anal. Chem. **249** [1970] 35/8).
[3] Palethorpe, G. (U.S. 3642628 [1968/72]; C.A. **76** [1972] No. 155461).
[4] Akpofure, C. O., Belcher, R., Bogdanski, S. L., Townshend, A. (Anal. Letters **8** [1975] 921/9).

1.4.1.2.1.5.5.2 Dibutyltin Maleate, $(C_4H_9)_2\overline{SnOOCCH{=}CHCOO}$

This compound is described in the literature as monomeric, oligomeric, or polymeric. References in which dibutyltin maleate is described as definitely trimeric, tetrameric, oligomeric, or polymeric are not quoted in this section.

1.4.1.2.1.5.5.2.1 Synthesis, Properties, and Reactions

Dibutyltin maleate is prepared from $(\text{-}(C_4H_9)_2SnO\text{-})_n$ and maleic acid by dehydration in quantitative yields [16] or from $(\text{-}(C_4H_9)_2SnO\text{-})_n$ and maleic anhydride [25]; thus in water after 1 h reflux [18], in THF at 50°C [23], in cyclohexane after 1 h at 75°C in 90 to 98% yield [15], in iso-octane at 80°C [8, 11, 19], in toluene between 60 and 100°C in yields up to 98% [1, 4, 22], or after passing a mixture of dibutyltin oxide and maleic anhydride through a pin mill and heating the mixture for 12 h at 40°C [9].

Melting points for the yellow solid are 101°C [21], 103 to 105°C [4], 107 to 110°C [18, 23], 112 to 114°C [16], 125 to 135°C [8], 128 to 129°C [29], 134°C [9], and 136 to 138°C [15].

Analytical methods have been developed for the determination of Sn and other elements in organotin stabilizers including dibutyltin maleate. A review on the existing procedures for the determination of Sn, including a new method which can be used for rapid microanalysis, is given in [27]. Nanogram amounts of dibutyltin maleate can be determined rapidly using molecular emission cavity analysis [24]; polarographic methods [2, 26], and ESR spectroscopy are also used for this purpose [30]. The separation and determination of dibutyltin maleate from other organotin compounds, from residues, or from polyvinyl chloride and other plastic material can be accomplished by thin layer chromatography [3, 7, 13, 28, 31].

^{119}Sn Mössbauer spectrum (in mm/s): $\delta = -0.65$ (α-Sn), $\Delta = 3.49$ (78 K) [12]; $\delta = 1.18$, $\Delta = 3.93$ (80 K) [14]; $\delta = 1.38$, $\Delta = 3.74$ (? K) [20]; $\delta = 1.45$, $\Delta = 3.49$ (80 K) [17]; $\delta = 1.50$, $\Delta = 3.50$ (78 K) [5].

The IR spectrum is depicted between 4000 and 500 cm^{-1} [6].

Dibutyltin maleate, containing monomeric and oligomeric forms, reacts with cyclopentadiene at 0°C with formation of the Diels-Alder-adduct 1,2,3,6-tetrahydro-3,6-methanophthalate in a polymeric form [10].

References:

[1] Gloskey, C. R., Metal and Thermit Corp. (U.S. 2838554 [1958]; C.A. **1958** 16212).

[2] Baker Allen, R. (Diss. Univ. New Hampshire 1959, pp. 1/70; Diss. Abstr. **20** [1959] 897).

[3] Türler, M., Högl, O. (Mitt. Gebiete Lebensmittelunters. Hyg. **52** [1961] 123/30).

[4] Sheverdina, N. I., Abramova, L. V., Paleeva, I. E., Kocheshkov, K. A. (Khim. Prom. **1962** 707/8; C.A. **59** [1963] 8776).

[5] Aleksandrov, A. Yu., Delyagin, N. N., Mitrofanov, K. P., Polak, L. S., Shpinel, V. S. (Dokl. Akad. Nauk SSSR **148** [1963] 126/8; Dokl. Phys. Chem. Proc. Acad. Sci. USSR **148/153** [1963] 1/3).

[6] Cummins, R. A., Dunn, P. (Dept. Defence Stand. Lab. Australia No. 266 [1963] 1/106).

[7] Neubert, G. (Z. Anal. Chem. **203** [1964] 265/72).

[8] Carlisle Chemical Works, Inc. (Neth. Appl. 66-04233 [1965/66]; C.A. **66** [1967] No. 76156).

[9] Oakes, V., Pure Chemicals Ltd. (Brit. 1043609 [1964/66]; C.A. **66** [1967] No. 28899).

[10] Mufti, A. S., Poller, R. C. (J. Chem. Soc. C **1967** 1767/8).

[11] Oakes, V., Pure Chemicals Ltd. (Ger. 1246735 [1965/67]; C.A. **68** [1968] No. 49766).

[12] Gol'danskii, V. I., Khrapov, V. V., Okhlobystin, O. Yu., Rochev, V. Ya. (in: Gol'danskii, V. I., Herber, R. H., Chemical Application of Mössbauer-Spectroscopy, New York 1968, pp. 336/76).

[13] Huber H., Wimmer, J. (Kunststoffe **58** [1968] 786/8).

[14] Fitzsimmons, B. W., Seeley, N. J., Smith, A. W. (J. Chem. Soc. A **1969** 143/6).

[15] Hirshman, J. L., Breza, E. J., M en T. Billiton, Chemische Industrie N.V. (Ger. Offen. 1908883 [1968/69]; C.A. **72** [1970] No. 12884).

[16] Mikhailov, G. D., Chegolya, A. S. (Sin. Volokna **1969** 18/21; C.A. **74** [1971] No. 3711).

[17] Smith, P. J. (Organometal. Chem Rev. A **5** [1970] 373/402).

[18] Shibuya, A., Tokyo Fine Chemical Co., Ltd. (Japan. 70-29648 [1967/70]; C.A. **74** [1971] No. 23009).

[19] Carlisle Chemical Works, Inc. (Brit. 1099106 [1965/71]; C.A. **75** [1971] No. 140984).
[20] Maddock, A. G., Platt, R. H. (J. Chem. Soc. A **1971** 1191/5).
[21] Minsker, K. S., Fedoseyeva, G. T., Zavarova, T. B., Krats, E. O. (Vysokomol. Soedin. A **13** [1971] 2265/78; Polym. Sci. [USSR] A **13** [1971] 254460).
[22] Miyake, M., Itsukaichi, Y., Kasuya, T., Sankyo Organic Chemicals Co., Ltd. (Japan. 71-04366 [1968/71]; C.A. **75** [1971] No. 6101).
[23] Kiyoto, K., Narita, A., Miki, A., Tokyo Fine Chemical Co., Ltd. (Japan. 73-30622 [1970/73]; C.A. **80** [1974] No. 27384).
[24] Akpofure, C. O., Belcher, R., Bogdanski, S. L., Townshend, A. (Anal. Letters **8** [1975] 921/9).
[25] Dworkin, R. D., Ejk, A. J., M and T Chemicals, Inc. (Ger. Offen. 2626554 [1975/76]; C.A. **86** [1977] No. 140253).
[26] Fleet, B., Fouzder, N. B. (J. Electroanal. Chem. Interfacial Electrochem. **63** [1975] 69/78).
[27] Marr, I. L. (Talanta **22** [1975] 387/94).
[28] Kasarinova, N., Kozitskaya, L. (Ukr. Khim. Zh. **42** [1976] 526/8; Soviet Progr. Chem. **42** No. 7 [1976] 77/8; C.A. **85** [1976] No. 95066).
[29] Troitskii, B. B., Troitskaya, L. S., Denisova, V. N., Novikova, M. A. (Fiz. Khim. Osn. Sint. Pererab. Polim. **1** [1976] 58/61; C.A. **87** [1977] No. 53972).
[30] Stegmann, H. B., Uber, W., Scheffler, K. (Z. Anal. Chem. **268** [1977] 59/64).
[31] Novitskaya, L. P., Dregval, G. F., Brodskaya, N. M. (Gig. Sanit. **1979** No. 6, pp. 48/51).

1.4.1.2.1.5.5.2.2 Toxicity and Biocidal Uses

Because of the use of dibutyltin maleate as a stabilizer for polyvinyl chloride, the toxicity of this compound and some other organotin stabilizers was tested during their entry through skin covers [9]. Tests on rats and rabbits showed tin in the brain [4]. The approximate lethal dose of dibutyltin maleate for deer mice was found to be 470 mg/kg bodyweight [11]. Other tests on rats in a trace element-controlled environment showed that dibutyltin maleate added to purified amino acid diets enhanced the growth of the rats, demonstrating that tin is a hitherto unrecognized essential trace element [7].

Dibutyltin maleate shows a high anthelmintic activity against tapeworms of chickens [2, 3, 6]. Intensive investigations showed a remarkable activity on the adult and immature forms of the avian cestodes Raillietina tetragona, Raillietina echinobothrida, Raillietina cesticillus, Choanotaenia infundibulum and Hymenolepis carioca, but no action on nematodes like Ascaridia styphlocerca, Subulura brumpti and Acuaria spiralis [5].

Patents claim the use of dibutyltin maleate as a veterinary composition effective in the control of protozoal and helminthic infections and in enhancing the meat production capacity and maturation of fowl and domesticated animals [1], as an additive to plasticized PVC to inhibit the growth of microorganisms on the surfaces of the plastic during immersion in water [10], and as an ingrediant of an asphalt based antifouling marine paint [8].

References:

[1] Kerr, K. B., Walde A. W., Dr. Salsburys Laboratories (U.S. 2702776 [1953/55]; C.A. **1955** 7816).
[2] Kerr, K. B., Walde, A. W. (Experim. Parasitol. **5** [1956] 560/70).
[3] Edgar, S. A., Teer, P. A. (Poultry Sci. **36** [1957] 329/34).

[4] Turchini, J., van Kien, L. K., Castel, P., Tuong, T. (Therapie **13** [1958] 873/6).
[5] Graber, M., Gras, G. (Rev. Elev. Med. Vet. Pays Trop. **16** [1963] 427/38).
[6] Graber, M., Gras, G. (Rev. Elev. Med. Vet. Pays Trop. **19** [1966] 7/14).
[7] Schwarz, K., Milne, D. B., Vinyard, E. (Biochem. Biophys. Res. Commun. **40** [1970] 22/9).
[8] Okamoto, T., Murayama, K., Sato, S., Tokyo Electric Power Co., Inc.; Sankyo Yuka Kogyo Co., Ltd. (Japan. 74-38690 [1969/74]; C.A. **83** [1975] No. 29975).
[9] Statsek, N. K., Kucherenko, T. V. (Gig. Primeneniya Polim. Mater. **1976** 61/2; C.A. **86** [1976] No. 184164).
[10] Spielau, P., Vohwinkel, H., Dynamit Nobel A.-G. (Ger. Offen. 3014291 [1980/81]; C.A. **96** [1981] No. 7650).
[11] Schafer, E. W., Bowles, W. A. (Arch. Environ. Contam. Toxicol. **14** [1985] 111/29).

1.4.1.2.1.5.5.2.3 Uses

Dibutyltin maleate is prepared in industrial scale for the use as a stabilizer for polyvinyl chloride [1, 2, 7, 13, 19]. It is manufactured under different trade names like Advastab T-340, Irgastab DBTM, Irgastab T 290, Mark 645, or Thermolite 13.

The compound stabilizes PVC against degradation by heat [3, 4, 5, 8, 11, 12, 20, 22 to 30, 32 to 35, 38, 41, 44], by light, or UV-irradiation [11, 15, 17, 31, 35, 37, 40, 41], by γ irradiation [6, 21], or under environmental and weathering conditions [16, 18, 29, 36].

The mechanism of the stabilization reaction was studied using several methods [8, 12, 26, 27, 30, 31, 32, 35, 41], including kinetic measurements on the dehydrochlorination of PVC [9, 10, 39, 42]. Dibutyltin maleate is complexing the labile chlorine atoms [9] and neutralizing the hydrogen chloride formed during the decomposition of PVC [27]. A second effect is substitution of chlorine atoms by bridging maleate ligands [9]. A combination of dibutyltin maleate and dibutyltin bis(iso-octyl-thioglycollate) shows a high efficiency in stabilizing PVC against heat and light [22, 26, 31, 32, 35].

Dibutyltin maleate is also used as a stabilizer for polyepichlorohydrin [14] and as a catalyst for the preparation of glass fibers with isocyanates on the surface [43].

Patents in which the use of dibutyltin maleate is claimed are listed at the end of the reference list according to the year of the issue of the patent, see p. 306.

References:

[1] Gay, P. J. (Tin Its Uses No. **29** [1953] 6).
[2] Hedges, E. S. (Metall **9** [1955] 23/5).
[3] Luijten, J. G. A., Pezarro, S. (Brit. Plast. **30** [1957] 183/6).
[4] Berlin, A. A., Popova, Z. V., Yanovskii, D. M. (Zh. Prikl. Khim. **33** [1960] 871; J. Appl. Chem. [USSR] **33** [1960] 870/5).
[5] Popova, Z. V., Yanovskii, D. M. (Mezhdunar. Simp. Makromol. Khim. Avtoreferaty, Moscow **1960**, Vol. 3, pp. 372/9; C.A. **1961** 9940).
[6] Shteding, M. N., Karpov, V. L. (Vysokomol. Soedin. **4** [1962] 1806/11; Polym. Sci. [USSR] **4** [1962] 561/6).
[7] Kinoshita, Y., Muraoka, K. (Shokuhin Eiseigaku Zasshi **4** [1963] 78/85; C.A. **59** [1963] 9232).
[8] Gelfman, Ya. A., Lauris, I. V., Kuskova, V. P. (Sb. Tr. Vses. Nauchn. Issled. Inst. Novykh Stroit. Mater. **1966** No. 14, pp. 54/8; C.A. **68** [1968] No. 79010).

[9] Klemchuk, P. P. (Advan. Chem. Ser. **85** [1968] 1/17).
[10] Terman, L. M., Sedelnikova, V. N., Dyachkovskaya, O. S., Malysheva, I. P. (Sin. Issled. Eff. Khim. Polim. Mater. **1969** No. 3, pp. 170/6; C.A. **75** [1971] No. 64783).

[11] Krats, E. O., Zavarova, T. B., Fedoseeva, G. T., Minsker, K. S. (Vysokomol. Soedin. A **13** [1971] 899/905; Polym. Sci. [USSR] A **13** [1971] 1013/21).
[12] Minsker, K. S., Fedoseeva, G. T., Zavarova, T. B., Krats, E. O. (Vysokomol. Soedin. A **13** [1971] 2265/78; Polym. Sci. [USSR] A **13** [1971] 2544/60).
[13] Vaiman, E. Ya., Pakshver, A. B., Fikhman, V. D. (Khim. Volokna **1971** 25/7; C.A. **75** [1971] No. 78024).
[14] Nakamura, Y., Oka, S., Mori, K., Tamura, K. (Nippon Gomu Kyokaishi **46** [1973] 507/13; C.A. **79** [1973] No. 105992).
[15] Oki, Y., Mori, F. (Kobunshi Kagaku **30** [1973] 737/41; C.A. **81** [1974] No. 64496).
[16] Levy, G. L., Heffner, M. H., Gross, R. C. (Tech. Papers Soc. Plast. Eng. **21** [1975] 46/50).
[17] Scott, G., Tahan, M. (Eur. Polym. J. **11** [1975] 535/9).
[18] Szabo, E., Lally, R. E. (Polym. Eng. Sci. **15** [1975] 277/80).
[19] Nass, L. I. (Encycl. PVC **1976** 295/384).
[20] Troitskii, B. B., Troitskaya, L. S., Denisova, V. N., Novikova, M. A. (Fiz. Khim. Osn. Sint. Pererab. Polim. [Gorkyi] No. 1 [1976] 58/61; C.A. **87** [1977] No. 53972).

[21] Kim, Bong Heup, Kang, Dou Yol, Lee, Jae In (Chongi Hakhoechi **26** [1977] 185/90; C.A. **87** [1977] No. 24039).
[22] Troitskii, B. B., Troitskaya, L. S., Denisova, V. N., Novikova, M. A., Luzinova, Z. B. (Eur. Polym. J. **13** [1977] 1033/41).
[23] Scott, G., Tahan, M., Vyvoda, J. (Eur. Polym. J. **14** [1978] 913/23).
[24] Miyazawa, T., Massaya, T. (Jigyo Gaiyo-Nagano-ken Seimitsu Kogyo Shikenjo **1979/80** 103/6; C.A. **95** [1981] No. 63042).
[25] Stanczyk, W., Kowalski, J., Chojnowski, J. (Polimery [Warsaw] **24** [1979] 111/4; C.A. **91** [1979] No. 124260).
[26] Cooray, B. B., Scott, G. (Polym. Degrad. Stab. **2** [1980] 35/51).
[27] Cooray, B. B., Scott, G. (Eur. Polym. J. **16** [1980] 169/77).
[28] Cooray, B. B., Scott, G. (Eur. Polym. J. **16** [1980] 1145/51).
[29] Kolawole, E. G., Mbamali, G. E., Olayemi, J. Y. (J. Appl. Polym. Sci. **25** [1980] 2133/8).
[30] Minsker, K. S., Kolesov, S. V., Kotsenko, L. M. (Vysokomol. Soedin. A **22** [1980] 2253/8; Polym. Sci. [USSR] A **22** [1980] 2471/7).

[31] Cooray, B. B., Scott, G. (Eur. Polym. J. **17** [1981] 229/32).
[32] Cooray, B. B., Scott, G. (Eur. Polym. J. **17** [1981] 379/84).
[33] Cooray, B. B., Scott, G. (Eur. Polym. J. **17** [1981] 385/7).
[34] Protic, P. M., Lukic, P. L. (Polimeri [Zagreb] **2** [1981] 115/8; C.A. **96** [1982] No. 86396).
[35] Bellenger, V., Verdu, J., Carette, L. B., Vymalalova, Z., Vymazal, Z. (Polym. Degrad. Stab. **4** [1982] 303/12).
[36] Erben, F., Veseley, R. (Angew. Makromol. Chem. **114** [1983] 95/104).
[37] Ghatge, N. D., Mahajan, S. S., Vaidya, S. V. (Indian J. Technol. **21** [1983] 161/4).
[38] Kizlink, J., Jakubcek, E., Dimitrova, J. (Ropa Uhlie **25** [1983] 565/7; C.A. **100** [1984] No. 104347).
[39] Guyot, A., Michel, A., Van Hong, Tran (Polym. Sci. Technol. [Plenum] **26** [1984] 155/69).
[40] Ho, Ben Y. K. (J. Vinyl Technol. **6** [1984] 162/6).

[41] Matsusaka, K., Koyama, M. (Angew. Makromol. Chem. **125** [1984] 149/59).
[42] Van Hong, Tran, Michel, A., Guyot, A. (Polym. Degrad. Stab. **9** [1984] 89/102).
[43] Yosomiya, R., Morimoto, K. (Polym. Bull. [Berlin] **12** [1984] 41/8).
[44] Yassin, A. A., Sabaa, M. W., Mohamed, N. A. (Polym. Degrad. Stab. **13** [1985] 225/47).

Patented Uses for Dibutyltin Maleate

Johnson, E. W., Metal and Thermit Corp., Stabilized Chlorinated Esters of Fatty Acids, U.S. 2524528 [1950]; C.A. **1951** 2489.

Churchill, J. W., Mathieson Chemical Corp., Alterungsschutzmittel für Polystyrole, Fr. 991331 [1949/51]; C. **1953** 3647.

Harding, J., Union Carbide and Carbon Corp., Heat and Light-Stabilized Vinyl Resins, U.S. 2597987 [1952]; C.A. **1952** 8893.

Walker, A. T., Fremon, G. H., Union Carbide and Carbon Corp., Solutions of Acrylonitrile Copolymers and Related Synthetic Products, U.S. 2603620 [1952]; C.A. **1953** 880.

Churchill, J. W., Mathieson Chemical Corp., Polymerized Dichlorostyrenes Containing Anticrazing Agents, U.S. 2643242 [1953]; C.A. **1953** 10904.

Sorenson, R.A., Nixon Nitration Works, Plastic Molding Powder, U.S. 2654716 [1953]; C.A. **1954** 2411.

Barker, R.L., Pure Chemicals Ltd., Organotin Compounds and their Use in Stabilization of Resinous Compositions, Brit. 761568 [1956]; C.A. **1957** 10561.

Chemische Werke Albert, Hardening of Epoxide Resins, Brit. 783764 [1957]; C.A. **1958** 5883.

Weinberg, E. L., Tomka, L. A., Metal and Thermit Corp., Organotin-Compound Nonstaining, Nondiscoloring Antioxidants for Rubber, U.S. 2789107 [1957]; C.A. **1957** 10939.

Squires, S., Goodier, K., Shell Research Ltd., Color-Stable Polystyrene, Brit. 881578 [1960]; C.A. **56** [1962] 8940.

Lazcano, C.S., Resinous Compositions and Stabilizers therefore, Brit. 1008589 [1962/65]; C.A. **70** [1969] No. 97639.

Penneck, R. J., Pinner, S. H., B. X. Plastics Ltd., Stabilizing Systems for Chlorine-Containing Polymers, Brit. 1001344 [1965]; C.A. **63** [1965] 13507.

Horrocks, J. A., Bakelite Xylonite Ltd., Stabilizers for Polymeric Compositions, Brit. 1061747 [1967]; C.A. **67** [1967] No. 12144.

Horrocks, J. A., Bakelite Xylonite Ltd., Organotin Stabilizers for Polymers, Brit. 1069165 [1967]; C.A. **67** [1967] No. 32778.

Kubo, M., Fukumoto, O., Toyo Rayon Co., Ltd., Heat-Stable and Dyable Polyolefins, Japan. 67-4275 [1967]; C.A. **67** [1967] No. 44419.

Freeze, J. T., Fikes, A. L., Pfizer, Chas., and Co., Inc., Organotin Stabilizers for Halogen-Containing Polymers, Fr. 1540230 [1966/68]; C.A. **71** [1969] No. 4175.

Aijima, I., Fukuma, N., Sakurai, H., Chiaya, M., Okamoto, T., Tsuchii, Y., Itsumi, H., Asahi Chemical Industry Co., Ltd., Polypropylene Composite Fibers, Japan. 69-17567 [1966/69]; C.A. **72** [1970] No. 13750.

Asahi Kasei Kogyo Kabushiki Kaisha, Catalyst for the Preparation of Polyacetal Resins, Brit. 1151927 [1965/69]; C.A. **71** [1969] No. 13755.

Hoshino, A., Nakajima, A., Nakamura, A., Asahi Dow Ltd., Stabilizing Vinylidene Chloride-Vinyl Chloride Copolymer Compositions, Japan. 69-16218 [1965/69]; C.A. **72** [1970] No. 22320.

Japan Carbide Industries Co., Inc., Chlorinated Poly(Vinyl Chloride) Compositions for Shaping, Brit. 1171205 [1966/69]; C.A. **72** [1970] No. 67736.

Mago, W. E., Billiton M. en T. Chemische Industrie N. V., Stabilisierung von halogenhaltigen Polymeren, Ger. 1917847 [1969]; C.A. **72** [1970] No. 13400.

Senda, K., Ichikawa, A., Oseki, T., Nakajima, E., Sakai, M., Yasui, A., Hirose, M., Mitsubishi Rayon Co., Ltd., Polypropylene Fibers with Good Dyeability, Japan 69-28872 [1966/69]; C.A. **72** [1970] No. 122813.

Brook, J. W., Hoch, S., Tenneco Chemicals, Inc., Heat and Light Stabilizers for Poly(Vinyl Chloride), Ger. Offen. 2006711 [1969/70]; C.A. **73** [1970] No. 121255.

Broyde, B., Organometallics which Increase the Sensitivity of Negative-Working Photoresists, Ger. Offen. 1949502 [1968/70]; C.A. **73** [1970] No. 30651.

Chapman, A. C., Hoye, P. A. T., Frank, G., Albright and Wilson (Mfg.) Ltd., Light-Stabilized Poly(Vinyl Chloride) Compositions, Brit. 1203442 [1967/70]; C.A. **73** [1970] No. 88697.

Enoki, K., Yoshihara, Y., Kirai, Y., Japan Soda Co., Ltd., Stabilization of Ethylene Terephthalate Polyesters, Ger. Offen. 1954959 [1968/70]; C.A. **73** [1970] No. 26275.

Imperial Chemical Industries Ltd., Katalysator zur Darstellung von Polyestern, Neth. Appl. 69-15283 [1969/70].

Ishida, S., Sato, K., Mori, K., Fujita, M., Asahi Chemical Industry Co., Ltd., Formaldehyde-N-Alkylacrylamide Copolymers, Japan. 70-14554 [1965/70]; C.A. **73** [1970] No. 56626.

Ishida, S., Asahi Chemical Industry Co., Ltd., Poly(oxymethylene), Ger. Offen. 1946077 [1968/70]; C.A. **72** [1970] No. 133404.

Monsanto Co., Stabilizers for Polymers, Brit. 1180511 [1967/70]; C.A. **72** [1970] No. 122443.

Moore, E., Wear, G. H., General Tire and Rubber Co., Imparting Self-Extinguishing Properties to Shock-Resistant Poly(Vinyl Chloride) Resins, Ger. Offen. 2007706 [1969/70]; C.A. **73** [1970] No. 121258.

Nakata, T., Kawamata, K., Osaka Soda Co., Ltd., Organotin-Phosphorus Ester Catalysts for Alkene Oxide Polymerization, Ger. Offen. 1941690 [1968/70]; C.A. **72** [1970] No. 133372.

Sakai, T., Fujioka, S., Okubo, J., Shinohara, Y., Toray Industries, Inc., Heat Stabilizers for Polysulfones, Japan. 70-25392 [1965/70]; C.A. **74** [1971] No. 77072.

Speitel, R., ARFA Röhrenwerke A. G., Poly(Vinyl Chloride) Plastisol or Organosol, Swiss 493573 [1967/70]; C.A. **73** [1970] No. 100153.

Yamanouchi, S., Katsuki, H., Terazawa, T., Sumitomo Chemical Co., Ltd., Cross-Linked Poly(Vinyl Chloride), Ger. Offen. 1962848 [1969/70]; C.A. **73** [1970] No. 78164.

Jones, P. W., Fisons Ltd., Embossing by Selective Expansion of Thermoplastic Materials, Ger. Offen. 2128964 [1970/71]; C.A. **76** [1972] No. 100719.

Kawaguchi, H., Kusunogi, M., Ogawa, M., Hayashi, S., Owari, S., Showa Denko K. K., Poly(Vinyl Chloride)-Chlorinated Polyethylene Compositions, Japan. 71-37904 [1968/71]; C.A. **77** [1972] No. 89429.

Ogawa, K., Toray Industries, Inc., Polyester Compositions, Japan. 71-38707 [1968/71]; C.A. **77** [1972] No. 89421.

Ohtsuka, Y., Hiramatsu, T., Sekisui Chemical Co., Ltd., Vinyl Chloride-Ethylene Copolymer Compositions, Japan. 71-42227 [1968/71]; C.A. **77** [1972] No. 115399.

Bruyere, G., Solvay et Cie., Stabilization of Plasticized Poly(Vinyl Chloride), Ger. Offen. 2137373 [1970/72]; C.A. **76** [1972] No. 154895.

Crescenzo, F. G., Dow, R. L., United States Dept. of the Navy, Gas Generator Compositions, U.S. 3639183 [1965/72]; C.A. **76** [1972] No. 88051.

Hirata, Y., Matsuda, F., Mitsui Toatsu Chemicals Co., Ltd., Poly(Vinyl Chloride) Plasticized with Butanetetracarboxylic Acid Esters, Japan. 72-18219 [1968/72]; C.A. **77** [1972] No. 153230.

Kakiuchi, H., Nakatsuka, R., Taniyama, M., Ishibashi, T., Sumitomo Bakelite Co., Ltd., Vinyl Chloride Resin Compositions, Japan. 72-33489 [1969/72]; C.A. **78** [1973] No. 125344.

Kaneko, S., Hisanaga, N., Watanabe, Y., Kawaguchi, T., Takeshima, A., Toa Gosei Chemical Industry Co., Ltd., Stabilized Butadiene-Derivative Compositions, Japan. 72-22090 [1969/72]; C.A. **78** [1973] No. 17084.

Kondo, S., Kato, F., Isokawa, S., Mitsui Petrochemical Industries, Ltd., Katsuta Kao Co., Ltd., Polyethylene Compositions, Japan. 72-13305 [1967/72]; C.A. **77** [1972] No. 153207.

Lozanou, M., Pfizer, Chas., and Co., Inc., Lubricating Stabilizers for Chlorine-Containing Polymers, U.S. 3644246 [1969/72]; C.A. **77** [1972] No. 35730.

Plum, H., Schering A.-G., Diisobutyltin Maleates as Stabilizers for Poly(Vinyl Chloride), Ger. 1494342 [1962/72]; C.A. **76** [1972] No. 127999.

Harayama, H., Morishima, Y., Sekisui Chemical Co., Ltd., Heat Foaming Resin Composition, Japan. 72-13340 [1969/72]; C.A. **80** [1974] No. 27953.

Kaburaki, K., Ishizawa, A., Tokyo Shibaura Electric Co., Vinyl Carbazole Copolymer Compositions with Antistatic Properties, Japan. Kokai 73-16939 [1971/73]; C.A. **79** [1973] No. 67345.

Morton, J. F., Schlaffer, J. C., Eitzel, R. J., Fowler, J. A., Congoleum Industries, Inc., Resinous Surface Covering, U.S. 3745040 [1971/73]; C.A. **79** [1973] No. 105966.

Nagasawa, K., Itsukaichi, Y., Noda, Y., Murata, K., Sankyo Organic Chemicals Co., Ltd., Halogen-Containing Resin Compositions, Japan 73-38766 [1970/73]; C.A. **81** [1974] No. 26681.

Yamamoto, T., Nippon Oils and Fats Co., Ltd., Heat Resistant Antistatic Agent, Japan. 73-10058 [1962/73]; C.A. **80** [1974] No. 109327.

Abe, K., Morita, Y., Hirai, T., Inoue, S., Toray Industries, Inc., Vinyl Chloride Resin Composition, Japan. 74-13496 [1970/74]; C.A. **83** [1975] No. 60564.

Kabuki K., Ishizawa, A., Sato, K., Tokyo Shibaura Electric Co., Ltd., Tokyo Optical Co., Ltd., Antistatic Polymers, Japan. Kokai 74-05439 [1972/74]; C.A. **82** [1975] No. 87097.

Kanai, Z., Honma, M., Shouji, A., Ishikawa, N., Dainippon Ink and Chemicals, Inc., Resin Powder Coating Compositions, Japan. Kokai 74-97024 [1972/74]; C.A. **82** [1975] No. 113241.

Matsuo, T., Nakata, T., Nippon Zeon Co., Ltd.; Osaka Soda Co., Ltd., Alkylene Sulfide Polymer, Japan. 74-28915 [1970/74]; C.A. **82** [1975] No. 73688.

Ohta, T., Resin Coating of Aluminum, Japan. Kokai 74-100110 [1972/74]; C.A. **82** [1975] No. 157964.

Szita, J., Koroscil, A., American Cyanamid Co., Stabilizing Halogen-Containing Acrylonitrile Polymers, Ger. Offen. 2402957 [1973/74]; C.A. **82** [1975] No. 5330.

Tsujimura, K., Kitayama, H., Iino, A., Asahi Dow, Ltd., Stabilizing Vinylidene Chloride Copolymers, Japan. 74-16110 [1970/74]; C.A. **82** [1975] No. 31942.

Hutton, R. E., Oakes, V., Iles, B. R., AKZO GmbH, Stabilization of Halogen-Containing Polymers, Ger. Offen. 2455614 [1973/75]; C.A. **83** [1975] No. 180314.

Ide, F., Kishida, K., Hasegawa, A., Kaneda, M., Mitsubishi Rayon Co., Ltd., Vinyl Chloride Resin Compositions, Japan. Kokai 75-123761 [1974/75]; C.A. **84** [1976] No. 45362.

Ide, F., Kishida, K., Hasegawa, A., Kaneda, M., Mitsubishi Rayon Co., Ltd., Vinyl Chloride Resin Compositions with Good Workability, Japan. Kokai 75-123763 [1974/75]; C.A. **84** [1976] No. 45359.

Ide, F., Kishida, K., Hasegawa, A., Kaneda, M., Mitsubishi Rayon Co., Ltd., Vinyl Chloride Resin Compositions with Good Workability, Japan. Kokai 75-123764 [1974/75]; C.A. **84** [1976] No. 45361.

Kobayashi, O., Ichihara, M., C.I. Kasei Co., Ltd., Stabilizers for Vinyl Chloride Resins, Japan. Kokai 75-153056 [1974/75]; C.A. **84** [1976] No. 151501.

Masaharu, Y., Shigenobu, I., Tsuyoshi, T., Hiroki, I., Showa Denko K. K., Resin Composition of High Adhesiveness, Ger. Offen. 2456594 [1973/75]; C.A. **83** [1975] No. 132903.

McInerney, E. J., General Electric Co., Polyolefin Composition Resistant to Hot Alkaline Solutions, U.S. 3878266 [1973/75]; C.A. **83** [1975] No. 115894.

Miyoshi, H., Sakagami, K., Takada, K., Horioka, M., Sekisui Chemical Co., Ltd., Heat-Resistant Vinyl Chloride Resin Compositions, Japan. Kokai 75-72948 [1973/75]; C.A. **83** [1975] No. 180393.

Miyoshi, H., Takada, K., Imada, M., Sekisui Chemical Co., Ltd., Japan. Kokai 75-88154 [1973/75]; C.A. **84** [1976] No. 5913.

Morita, Y., Hirai, T., Inoue, S., Toray Industries, Inc., Improved ABS Resin Composition, Japan. 75-17107 [1970/75]; C.A. **83** [1975] No. 180504.

Nakamura, J., Konuma, H., Kokurya, S., Uejima, T., Tsuge, C., Showa Denko K. K., Impact-Resistant Thermoplastic Resin Compositions, Japan. Kokai 75-148462 [1974/75]; C.A. **84** [1976] No. 106577.

Palm, R. A., BASF A.-G., Blowing Agents for Chlorine-Containing Polymers, Ger. 1669762 [1968/75]; C.A. **83** [1975] No. 165217.

Dworkin, R. D., Ejk, A. J., M and T Chemicals, Inc., Diorganotin Carboxylates, Ger. Offen. 2626554 [1975/76]; C.A. **86** [1977] No. 140253.

Ikuma, S., Saito, Y., Mitsubishi Monsanto Chemical Co., Fire-Resistant Resin Compositions, Japan. Kokai 76-28851 [1974/76]; C.A. **85** [1976] No. 6739.

Ikuma, S., Saito, Y., Mitsubishi Monsanto Chemical Co., Fire-Resistant Resin Composition, Japan. Kokai 76-28852 [1974/76]; C.A. **85** [1976] No. 6738.

Ikuma, S., Saito, Y., Iwanaga, A., Mitsubishi Monsanto Chemical Co., Weather- and Fire-Resistant Styrene Resin Compositions, Japan. Kokai 76-62847 [1974/76]; C.A. **85** [1976] No. 79092.

Itokuma, S., Saito, Y., Mitsubishi Monsanto Chemical Co., Weathering-Resistant Thermoplastic Molding Compositions having Good Impact Strength, Japan. Kokai 76-133349 [1975/76]; C.A. **86** [1977] No. 90993.

Kokuryo, S., Arai, F., Kurosawa, S., Showa Denko K. K., Heat-Resistant Norbornene Derivative Ring-Opening Polymer Compositions, Japan. Kokai 76-112867 [1975/76]; C.A. **86** [1977] No. 90947.

Matsuzaki, T., Takubo, T., Mototani, H., Saisho, T., Nitto Kasei Co., Ltd., Stabilized Poly(Vinyl Chloride) Compositions having Good Processability, Japan. Kokai 76-07051 [1974/76]; C.A. **84** [1976] No. 151624.

Matsuzaki, T., Takubo, T., Mototani, H., Saisho, T., Nitto Kasei Co., Ltd., Poly(Vinyl Chloride) Compositions Having Good Processability and Heat and Light Resistance, Japan. Kokai 76-20251 [1974/76]; C.A. **84** [1976] No. 165727.

Shibahara, T., Sasaki, M., Eguchi, H., Habara, T., Iwamura, T., Toray Industries, Inc., Preliminary Purification of Dimethyl Terephthalate, Japan. Kokai 76-39640 [1974/76]; C.A. **85** [1976] No. 123617.

Shibahara, T., Sasaki, M., Eguchi, H., Iwamura, T., Toray Industries, Inc., Esters, Japan. Kokai 76-39619 [1974/76]; C.A. **85** [1976] No. 108426.

Ito, S., Asaoka, T., Sankyo Organic Chemicals Co., Ltd., Heat- and Weather-Resistant Vinyl Chloride Resin Compositions with Improved Processability, Japan. Kokai 77-26552 [1975/77]; C.A. **87** [1977] No. 6903.

Kitamura, M., Hida M., Aoki T., Mitsubishi Plastics Industries, Ltd., Weather-Resistant Laminated Sheets, Japan. Kokai 77-57275 [1975/77]; C.A. **87** [1977] No. 69416.

Minagawa, G., Sekiguchi, T., Ogasawara, T., Adeka Argus Chemical Co., Ltd., Lubricants for Thermoplastic Resins, Japan. Kokai 77-81358 [1975/77]; C.A. **87** [1977] No. 202600.

Okumura, K., Kabushiki Kaisha Settsu Chemical Shoji, Protective Coating for a Car Body, Japan. Kokai 77-96990 [1976/77]; C.A. **87** [1977] No. 186227.

Dworkin, R. D., Ejk, A. J., M and T Chemicals, Inc., Stabilizing Vinyl Chloride Polymers, U.S. 4085077 [1976/78]; C.A. **89** [1978] No. 111125.

Green, J., Versnel, J., Wolford, L. T., Cities Service Co., Fire-Resistant Polystyrene Compositions Modified by Rubber, Belg. 858536 [1976/78]; C.A. **88** [1978] No. 153677.

Ito, Y., Sato, K., Kiryu, M., Sankyo Organic Chemicals Co., Ltd., Heat Stabilizers for ABS Resin-PVC Blends, Japan. Kokai 78-66954 [1976/78]; C.A. **89** [1978] No. 164425.

Iwata, A., Ozaki, Y., Egi, S., Sekisui Chemical Co., Ltd., Tokuyama Soda Co., Ltd., Chlorinated Poly(Vinyl Chloride) Foams, Japan. Kokai 78-64276 [1976/78]; C.A. **89** [1978] No. 147646.

Kametani, M., Lio, Y., Nitto Chemical Industry Co., Ltd., Dialkylaminoethyl Acrylates and Methacrylates, Japan. Kokai Tokkyo Koho 78-144522 [1977/78]; C.A. **90** [1979] No. 169290.

Kawasumi, H., Isobe, Y., Morita, K., Hayashi, H., Ito, M., Toa Gosei Chemical Industry Co., Ltd., Manufacture of Vinyl Chloride Polymers, Japan. Kokai 78-14788 [1976/78]; C.A. **89** [1978] No. 7045.

Levek, R. P., Williams, D. O., Great Lakes Chemical Corp., Stabilized Fire-Resistant Styrene Polymer Mixtures, Ger. Offen. 2733695 [1976/78]; C.A. **88** [1978] No. 137462.

Maeda, K., Takabayashi, Y., Nippon Carbide Industries Co., Ltd., Weather-Resistant Vinyl Chloride Resin Laminated Sheets, Japan. Kokai 78-65377 [1976/78]; C.A. **89** [1978] No. 147817.

Motohashi, A., Murayama, R., Kizaki,Y., Kamiyama, S., Sankyo Organic Chemicals Co., Ltd., Heat Stabilizers for ABS Resins Containing Fire-Proofing Agents, Japan. Kokai Tokkyo Koho 78-134051 [1977/78]; C.A. **90** [1979] No. 138625.

Nikaido, T., Data, H., Hasegawa, S., Matsushita Electric Works, Ltd., Rigid PVC Molding Compositions, Japan. Kokai 78-67757 [1976/78]; C.A. **89** [1978] No. 147641.

Nikaido, T., Date, H., Hasegawa, S., Matsushita Electric Works, Ltd., Impact Modifiers for Rigid Poly(Vinyl Chloride) Compositions, Japan. Kokai 78-82852 [1976/78]; C.A. **89** [1978] No. 180866.

Nikaido, T., Date, H., Hasegawa, S., Matsushita Electric Works, Ltd., Vinyl Chloride Resin Compositions, Japan. Kokai Tokkyo Koho 78-128653 [1977/78]; C.A. **90** [1979] No. 88344.

Pillar, W. O., ARCO Polymers, Inc., Expandable Styrene Polymer Particles, Ger. Offen. 2732377 [1976/78]; C.A. **89** [1978] No. 44701.

Pillar, W. O., ARCO Polymers, Inc., Molding Expandable Styrene Polymer Particles, U.S. 4113672 [1976/78]; C.A. **90** [1979] No. 24273.

Sato, K., Tsukahara, Y., Ihara, K., Sankyo Organic Chemicals Co., Ltd., Stable Resin Blends, Japan. Kokai 78-66957 [1976/78]; C.A. **89** [1978] No. 147637.

Imada, K., Eguchi, Y., Shin-Etsu Chemical Industry Co., Ltd., Fire-Resistant Poly(Vinyl Chloride) Compositions, Japan. Kokai Tokkyo Koho 79-114558 [1978/79]; C.A. **92** [1980] No. 59730.

Kano, T., Fukuda, A., Tanaka, Y., Kyodo Chemical Co., Ltd., Heat Stabilizers for Halogen-Containing Polymer Blends, Japan. Kokai Tokkyo Koho 79-138047 [1978/79]; C.A. **92** [1980] No. 129960.

Kogo, Y., Kiryu, M., Ilnuma, K., Sankyo Organic Chemicals Co., Ltd., Stabilizing Compositions for Halogen-Containing Polymers, Japan. Kokai Tokkyo Koho 79-145751 [1978/79]; C.A. **92** [1980] No. 129986.

Kuegler, F., Schneider, H., Lentia GmbH Chem. und Pharm. Erzeugnisse—Industriebedarf, Flame-Resistant Polypropylene Fibers, Eur. Appl. 5496 [1978/79]; C.A. **92** [1980] No. 165145.

Takashina, N., Shiokawa, S., Fujita, R., Usuda, Y., Mitsubishi Gas Chemical Co., Inc., Poly(Vinyl Chloride) Molding Compounds, Japan. Kokai Tokkyo Koho 79-154445 [1978/79]; C.A. **92** [1980] No. 147908.

Yamaguchi, N., Nikaido, T., Isagawa, M., Matsushita Electric Works, Ltd., Rigid Poly(Vinyl Chloride) Having Good Impact Strength, Japan. Kokai Tokkyo Koho 79-138049 [1978/79]; C.A. **92** [1980] No. 129965.

Cerefice, S. A., Paschke, E. E., Standard Oil Co. (Indiana), Alkyl Alkoxymethylbenzoates, U.S. 4220753 [1978/80]; C.A. **94** [1981] No. 4810.

Dainippon Ink and Chemicals, Inc., Stabilizers for Halogen-Containing Resins, Japan. Kokai Tokkyo Koho 80-71741 [1978/80]; C.A. **93** [1980] No. 13379.

Dainippon Ink and Chemicals, Inc., Poly(Vinyl Chloride) Compositions, Japan. Kokai Tokkyo Koho 80-102642 [1979/80]; C.A. **94** [1981] No. 16598.

Erben, F., Svoboda, P., Vesely, R., Stabilizing Polymers, Czech. 185378 [1975/80]; C.A. **95** [1981] No. 116508.

Fields, E. K., Cerefice, S. A., Paschke, E. E., Standard Oil Co. (Indiana), Monomers for Poly(methylenebenzoate) Polymers from Toluic Acid Compounds, U.S. 4182847 [1977/80]; C.A. **93** [1980] No. 47450.

Hariki, Y., Tateno, I., Harasawa, I., Nippon Carbide Industries Co., Inc., Cover Films for Ginger Cultivation, Japan. Kokai Tokkyo Koho 80-23919 [1978/80]; C.A. **93** [1980] No. 48078.

Hariki, Y., Tateno, I., Harasawa, I., Nippon Carbide Industries Co., Inc., Film for Cultivation of Beans, Japan. Kokai Tokkyo Koho 80-23920 [1978/80]; C.A. **93** [1980] No. 27450.

Inoe, M., Ueno, M., Nippon Zeon Co., Ltd., Vinyl Chloride Resin Compositions, Japan. Kokai Tokkyo Koho 80-23144 [1978/80]; C.A. **93** [1980] No. 9008.

Kao Soap Co., Ltd., Stable Quarternary Ammonium Chloride Compositions, Japan. Kokai Tokkyo Koho 80-07219 [1978/80]; C.A. **94** [1981] No. 71209.

Kobayashi, S., C. I. Kasei Co., Ltd., Antistatic Agents for Poly(Vinyl Chloride), Japan. Kokai Tokkyo Koho 80-03457 [1978/80]; C.A. **92** [1980] No. 199325.

Kyowa Chemical Industry Co., Ltd., Thermoplastic Resin Compositions and Prevention of Degradation, Japan. Kokai Tokkyo Koho 80-80445 [1978/80]; C.A. **93** [1980] No. 169183.

Martin, K. J., Voelker, M. J., Ryan, R. J., RCA Corp., Conductive Molding Composition, Ger. Offen. 3000447 [1979/80]; C.A. **93** [1980] No. 221515.

Matsushita Electric Works, Ltd., Vinyl Chloride Resin Molding Materials, Japan. Kokai Tokkyo Koho 80-147541 [1979/80]; C.A. **94** [1981] No. 122598.

Mitsubishi Chemical Industries Co., Ltd., Polyolefin Compositions, Japan. Kokai Tokkyo Koho 80-62945 [1978/80]; C.A. **93** [1980] No. 169297.

Mitsubishi Plastics Industries, Ltd., Heat-Shrinkable Plastic Films, Japan. Kokai Tokkyo Koho 80-73548 [1978/80]; C.A. **93** [1980] No. 240770.

Mitsubishi Plastics Industries, Ltd., Improving the Weather Resistance of Vinyl Chloride Resin Moldings, Japan. Kokai Tokkyo Koho 80-84331 [1978/80]; C.A. **93** [1980] No. 241306.

Mitsubishi Plastics Industries, Ltd., Coated PVC Sheets with Improved Weatherability, Japan. Kokai Tokkyo Koho 80-115431 [1978/80]; C.A. **94** [1981] No. 32330.

Mitsubishi Plastics Industries, Ltd., Heat-Stabilized Poly(Vinyl Chloride) Compositions Having Low Levels of Plate Out, Japan. Kokai Tokkyo Koho 80-125143 [1979/80]; C.A. **94** [1981] No. 66713.

Mitsubishi Plastics Industries, Ltd., Photocurable Epoxy Coating Compositions, Japan. Kokai Tokkyo Koho 80-127468 [1978/80]; C.A. **94** [1981] No. 85816.

Mitsubishi Plastics Industries, Ltd., Epoxy Compositions, Japan. Kokai Tokkyo Koho 80-147525 [1979/80]; C.A. **94** [1981] No. 123244.

Nippon Carbide Industries Co., Inc., Cultivation of Liliaceae Green Vegetables, Japan. Kokai Tokkyo Koho 80-10203 [1977/80]; C.A. **93** [1980] No. 73282.

Nitto Electric Industrial Co., Ltd., Adhesive Tapes for Covering Metals, Japan. Kokai Tokkyo Koho 80-127479 [1979/80]; C.A. **94** [1981] No. 48511.

Nowak, M. T., Celanese Corp., Radiation-Curable Coating Composition Having Utility as a Moisture Barrier Film, U.S. 4201642 [1978/80]; C.A. **93** [1980] No. 97015.

Nowak, M. T., Rybny, C. B., Celanese Corp., Ultraviolet Curable Self-Pigmented Coating Composition, U.S. 4194955 [1978/80]; C.A. **93** [1980] No. 9688.

Osaka Soda Co., Ltd., Transparent Crosslinked Chlorinated Polyethylene Compositions, Japan. Kokai Tokkyo Koho 80-125139 [1979/80]; C.A. **94** [1981] No. 48319.

Sanyo Chemical Industries, Ltd., Polyurethane Foam Spray Coating Materials, Japan. Kokai Tokkyo Koho 80-98218 [1979/80]; C.A. **95** [1981] No. 26732.

Sumitomo Chemical Co., Ltd., Vinyl Chloride Elastomers, Japan. Kokai Tokkyo Koho 80-73743 [1978/80]; C.A. **93** [1980] No. 187553.

Toyoda Gosei Co., Ltd., Automobile Glass Runs, Japan. Kokai Tokkyo Koho 80-157661 [1979/80]; C.A. **94** [1981] No. 158481.

Toyoda Gosei Co., Ltd., Coatings for Automobile Glass Runs, Japan. Kokai Tokkyo Koho 80-157662 [1979/80]; C.A. **94** [1981] No. 158482.

Yamaguchi, N., Nikaido, T., Matsushita Electric Works, Ltd., Rigid Poly(Vinyl Chloride) Molding Compositions, Japan. Kokai Tokkyo Koho 80-34239 [1978/80]; C.A. **93** [1980] No. 47855.

Yamaguchi, N., Nikaido, T., Matsushita Electric Works, Ltd., Rigid Poly(Vinyl Chloride) Molding Compositions with Good Impact Strength, Japan. Kokai Tokkyo Koho 80-34240 [1978/80]; C.A. **93** [1980] No. 115420.

Yamamoto, T., Waya, N., Nakao, S., Showa Kako K. K., Coated Stannic Acid as a Flame Retardant, Japan. Kokai Tokkyo Koho 80-07545 [1978/80]; C.A. **92** [1980] No. 147946.

Yoshiga, N., Nakamura, H., Ohmura, M., Mitsubishi Plastics Industries, Ltd., Heat-Shrinkable Poly(Vinyl Chloride) Film, Ger. Offen. 2920986 [1978/80]; C.A. **93** [1980] No. 27492.

Boboli, E., Malasnicki, W. L., Pietraszkiewicz, O., Rzeszowski, J., Sosinska, H., Lato, C., Instytut Przemyslu Orgaicznego, Polymeric Di-n-alkyltin Maleates in the Form of Nondusty Powders as Thermal Stabilizers of Poly(Vinyl Chloride) and Other Chlorinated Polymers, Pol. 108827 [1977/81]; C.A. **96** [1982] No. 69881.

Boussely, J., Pigerol, C., De Cointet De Fillian, P., Omnium Financier Aquitaine pour l'Hygiène et la Santé, Compositions of Organostannic Derivates and Dihydropyridines Useful for the Stabilization of Vinylic Resins, Eur. Appl. 27439 [1979/81]; C.A. **96** [1982] No. 7555.

Dainippon Ink and Chemicals, Inc., Polyurethane Coating Compositions, Japan. Kokai Tokkyo Koho 81-98266 [1980/81]; C.A. **95** [1981] No. 205564.

Dainippon Ink and Chemicals, Inc., Coating Process, Japan. Kokai Tokkyo Koho 81-115664 [1980/81]; C.A. **96** [1982] No. 21409.

Efer, J., Kochmann, W., Pfeiffer, H. D., Rank B., Schaefer, H., Thust, U., Trautner, K., VEB Chemiekombinat Bitterfeld, Stabilization of Phosphoric Acid O,O-Dimethyl O-2,2 Dichlorovinyl Ester, Ger. [East] 151002 [1978/81]; C.A. **97** [1982] No. 127806.

Hahn, K., Hinselmann, K., Halbritter, K., Rebafka, W., Weber, H., BASF A.-G., Fire-Resistant Styrene-Based Foams, Ger. Offen. 2932303 [1979/81]; C.A. **94** [1981] No. 157919.

Hahn, K., Hinselmann, K., Halbritter, K., Rebafka, W., Weber, H., BASF A.-G., Flame-Retardant Polystyrene Foams, U.S. 4298702 [1979/81]; C.A. **96** [1982] No. 20907.

Horak, Z., Rosik, L., Seiner, F., Vecerka, F., Vitr, J., Impact-Resistant, Fire-Resistant Polymer Compositions, Czech. 190219 [1977/81]; C.A. **97** [1982] No. 24711.

Kamada, K., Osaka, N., Kaneda, M., Mitsubishi Rayon Co., Ltd., Vinyl Chloride Polymer Composition, Eur. Appl. 40543 [1980/81]; C.A. **96** [1982] No. 69995.

Kanegafushi Chemical Industry Co., Ltd., Vinyl Chloride Copolymers Having Good Heat Stability, Japan. Kokai Tokkyo Koho 81-86911 [1979/81]; C.A. **95** [1981] No. 170056.

Kishida, K., Hasegawa, A., Sugimori, M., Mitsubishi Rayon Co., Ltd., Delustered Thermoplastic Resin Compositions, Brit. Appl. 2057466 [1979/81]; C.A. **95** [1981] No. 188133.

Kyodo Chemical Co., Ltd., Halogen-Containing Resins with Antistatic Properties, Japan. Kokai Tokkyo Koho 81-02335 [1979/81]; C.A. **95** [1981] No. 26089.

Kyodo Chemical Co., Ltd., Antistatic Compositions of Halogen-Containing Resins, Japan. Kokai Tokkyo Koho 81-17045 [1979/81]; C.A. **95** [1981] No. 98825.

Matsushita Electric Industrial Co., Ltd., Lubricating Oils, Japan. Kokai Tokkyo Koho 81-131696 [1980/81]; C.A. **96** [1982] No. 71649.

Matsushita Electric Works, Ltd., Crosslinking Agents for Epoxy Resins, Japan. Kokai Tokkyo Koho 81-159218 [1980/81]; C.A. **96** [1982] No. 123941.

Mitsubishi Plastics Industries, Ltd., Plastic Sheets with Metallic Luster, Japan. Kokai Tokkyo Koho 81-133164 [1980/81]; C.A. **96** [1982] No. 36314.

Mitsubishi Plastics Industries, Ltd., PVC Films with Whitened Pattern, Japan. Kokai Tokkyo Koho 81-136829 [1980/81]; C.A. **96** [1980] No. 86604.

Mitsubishi Rayon Co., Ltd., Impact Modifiers for Thermoplastic Compositions, Japan. Kokai Tokkyo Koho 81-166217 [1980/81]; C.A. **96** [1982] No. 123985.

Miyata, S., Kuroda, M., Kyowa Chemical Industry Co., Ltd., Inhibiting the Thermal or UV Degradation of Thermoplastic Resins, Ger. Offen. 3019632 [1980/81]; C.A. **96** [1982] No. 36327.

Miyata, S., Kuroda, M., Kyowa Chemical Industry Co., Ltd., Inhibiting the Thermal or Ultraviolet Degradation of Thermoplastic Resin Composition Having Stability to Thermal or Ultraviolet Degradation, U.S. 4299759 [1980/81]; C.A. **96** [1982] No. 36333.

Nippon Carbide Industries Co., Inc., Films for Growing Lettuce and Chinese Cabbage, Japan. Kokai Tokkyo Koho 81-46375 [1977/81]; C.A. **96** [1982] No. 105462.

Nippon Carbide Industries Co., Inc., Films for Agricultural Uses, Japan. Kokai Tokkyo Koho 81-68327 [1979/81]; C.A. **95** [1981] No. 133882.

Nippon Telegraph and Telephone Public Corp., Sumitomo Bakelite Co., Ltd., Hitachi Chemical Co., Ltd., Recycle of Plastics, Japan. Kokai Tokkyo Koho 81-23774 [1978/81]; C.A. **95** [1981] No. 151875.

Sanyo Chemical Industries, Ltd., Polyurethane Foams, Japan. Kokai Tokkyo Koho 81-67329 [1979/81]; C.A. **95** [1981] No. 151834.

Sanyo Chemical Industries, Ltd., Polyol Compositions for Polyurethane Foams, Japan. Kokai Tokkyo Koho 81-67331 [1979/81]; C.A. **95** [1981] No. 133872.

Adeka Argus Chemical Co., Ltd., Discoloration- and Flame-Resistant Thermoplastics, Japan. Kokai Tokkyo Koho 57 73049 [1980/82]; C.A. **98** [1983] No. 73287.

Adeka Argus Chemical Co., Ltd., Prevention of Discoloration of Dichloropropene Nematocides, Japan. Kokai Tokkyo Koho 57 126429 [1981/82]; C.A. **98** [1983] No. 12946.

Asahi-Dow Ltd., Stabilized Methylchloroform Compositions, Japan. Kokai Tokkyo Koho 82-116021 [1981/82]; C.A. **97** [1982] No. 215515.

Horak, Z., Voeroes, F., Rosik, L., Virt, J., Vecerka, F., Foamed Polymer with Low Flammability, Czech. 201748 [1978/82]; C.A. **99** [1983] No. 71694.

Japan Synthetic Rubber Co., Ltd., Self Extinguishing Thermoplastic Compositions, Japan. Kokai Tokkyo Koho 82-98540 [1980/82]; C.A. **97** [1982] No. 217373.

Kanegafushi Chemical Industry Co., Ltd., Fire-Resistant Resin Compositions, Japan. Kokai Tokkyo Koho 82-14638 [1980/82]; C.A. **96** [1982] No. 218720.

Kanegafushi Chemical Industry Co., Ltd., Fire- and Impact Resistant Thermoplastic Molding Compositions, Japan. Kokai Tokkyo Koho 82-80445 [1980/82]; C.A. **97** [1982] No. 145769.

Kanegafushi Chemical Industry Co., Ltd., Modified Poly(Vinyl Chloride) Compositions, Japan. Kokai Tokkyo Koho 82-98537 [1980/82]; C.A. **97** [1982] No. 199076.

Kanegafushi Chemical Industry Co., Ltd., Vinyl Chloride Resin Foams, Japan. Kokai Tokkyo Koho 82-105430 [1980/82]; C.A. **97** [1982] No. 217381.

Kanegafushi Chemical Industry Co., Ltd., Poly(Vinyl Chloride) Composites with Improved Flow, Japan. Kokai Tokkyo Koho 57 155212 [1981/82]; C.A. **98** [1983] No. 90496.

Kanegafushi Chemical Industry Co., Ltd., Poly(Vinyl Chloride) Beads Impregnated with Copolymers, Japan. Kokai Tokkyo Koho 57 158215 [1981/82]; C.A. **98** [1983] No. 108356.

Kanegafushi Chemical Industry Co., Ltd., Video Disk Resin Molding Compositions, Japan. Kokai Tokkyo Koho 57 165441 [1981/82]; C.A. **98** [1983] No. 144655.

Kanegafushi Chemical Industry Co., Ltd., Modified Vinyl Chloride Polymers with Good Melt Flow, Japan. Kokai Tokkyo Koho 57 179214 [1981/82]; C.A. **98** [1983] No. 144431.

Kanegafushi Chemical Industry Co., Ltd., Resin Compositions for Molding of Disks, Japan. Kokai Tokkyo Koho 57 180650 [1981/82]; C.A. **98** [1983] No. 161714.

Kitsuada, Y., Matsamura, M., Ohtsu, M., Matsushita Electric Works, Ltd., Epoxy Resin Composition, Ger. Offen. 3117960 [1980/82]; C.A. **96** [1982] No. 200825.

Krejci, B., Zemba, J., Kriz, M., Plastisols from Vinyl Chloride Polymers with Increased Thermal Stability, Czech. 210406 [1980/82]; C.A. **98** [1983] No. 90516.

Macho, V., Porubsky, J., Chochlac, S., Szulenyi, F., Pleva, S., Konc, J., Hard Vinyl Chloride Polymers and Copolymers with Decreased Brittleness, Czech. 189418 [1977/82]; C.A. **97** [1982] No. 73457.

Matsushita Electric Works, Ltd., Epoxy Resin Electric Insulators, Japan. Kokai Tokkyo Koho 82-65722 [1980/82]; C.A. **97** [1982] No. 110853.

Matsushita Electric Works, Ltd., Epoxy Resin Potting Compositions, Japan. Kokai Tokkyo Koho 57 126842 [1981/82]; C.A. **98** [1983] No. 35540.

Matsushita Electric Works, Ltd., Epoxy Potting Compounds, Japan. Kokai Tokkyo Koho 57 139117 [1981/82]; C.A. **98** [1983] No. 73349.

Mitsubishi Plastics Industries, Ltd., Films Having a Whitened Pattern, Japan. Kokai Tokkyo Koho 82-02339 [1980/82]; C.A. **96** [1982] No. 144002.

Mitsubishi Rayon Co., Ltd., Processing Aids for Vinyl Chloride Polymers, Japan. Kokai Tokkyo Koho 82-23645 [1980/82]; C.A. **97** [1982] No. 7303.

Mitsubishi Rayon Co., Ltd., Processability Improvement of PVC Resin Compositions, Japan. Kokai Tokkyo Koho 82-74347 [1980/82]; C.A. **97** [1982] No. 164033.

Mitsubishi Rayon Co., Ltd., Impact-Resistant Poly(Vinyl Chloride) Compositions, Japan. Kokai Tokkyo Koho 82-102940 [1980/82]; C.A. **97** [1982] No. 217306.

Nippon Carbide Industries Co., Ltd., Fruit Cultivation UV-Shielding Films, Japan. Kokai Tokkyo Koho 57-122725 [1981/82]; C.A. **98** [1983] No. 17736.

Shin-Etsu Chemical Industry Co., Ltd., Vinyl Chloride Resin Compositions for Video Disks, Japan. Kokai Tokkyo Koho 57 127939 [1981/82]; C.A. **98** [1983] No. 55242.

Tijin Ltd., Heat-Shielding Sheets, Japan. Kokai Tokkyo Koho 82-70649 [1980/82]; C.A. **97** [1982] No. 145918.

Tokuyama Soda Co., Ltd., Heat Stabilizers for Chloride-Containing Polymers, Japan. Kokai Tokkyo Koho 82-40537 [1980/82]; C.A. **97** [1982] No. 24725.

Tokuyama Soda Co., Ltd., Heat Stabilizers for Chlorine-Containing Resins, Japan. Kokai Tokkyo Koho 82-83544 [1980/82]; C.A. **97** [1982] No. 199055.

Tokuyama Soda Co., Ltd., Poly(Vinyl Chloride)-Poly(Vinyl Pyrrolidone) Blend Compositions, Japan. Kokai Tokkyo Koho 57 187340 [1981/82]; C.A. **98** [1983] No. 161777.

Toyo Soda Mfg. Co., Ltd., Vinyl-Chloride-Grafted Ethylene-Vinyl Acetate Copolymer Transparent Resin Compositions, Japan. Kokai Tokkyo Koho 82-67646 [1980/82]; C.A. **97** [1982] No. 145692.

Toyo Soda Mfg. Co., Ltd., Transparent Resin Compositions, Japan. Kokai Tokkyo Koho 82-87451 [1980/82]; C.A. **97** [1982] No. 183406.

Toyo Soda Mfg. Co., Ltd., Ethylene-Vinyl Acetate-Vinyl Chloride Graft Copolymer Adhesives, Japan. Kokai Tokkyo Koho 57 167362 [1981/82]; C.A. **98** [1983] No. 127346.

Toyoda Gosei Co., Ltd., Adhesives for Siloxane-Contaminated Poly(Vinyl Chloride) Moldings, Japan. Kokai Tokkyo Koho 82-16084 [1980/82]; C.A. **96** [1982] No. 218915.

Yonezawa, K., Furukawa, H., Azuma, M., Kanegafuchi Chemical Industry Co., Ltd., Curable Vinyl Polymer Having a Silyl Group, U.S. 4334036 [1978/82]; C.A. **97** [1982] No. 94119.

Agency of Industrial Sciences and Technology, Modified Vinyl Chloride Polymer Products, Japan. Kokai Tokkyo Koho 58 141229 [1982/83]; C.A. **100** [1984] No. 211200.

Agency of Industrial Sciences and Technology, Crosslinking of Chlorine-Containing Plastics, Japan. Kokai Tokkyo Koho 58 179252 [1982/83]; C.A. **100** [1984] No. 140252.

Agency of Industrial Sciences and Technology, Chlorine-Containing Plastic Compositions, Japan. Kokai Tokkyo Koho 58 179253 [1982/83]; C.A. **100** [1984] No. 140251.

Datta, P., Poliniak, E. S., RCA Corp., High Density Information Disk Lubricants, U.S. 4410748 [1982/83]; C.A. **100** [1984] No. 9816.

Horak, Z., Rosik, L., Seiner, F, Vecerka, F., Virt, J., Impact-Resistant Polystyrene with Reduced Flammability and Antistatic Properties, Czech. 201903 [1978/83]; C.A. **99** [1983] No. 213567.

Matsushita Electric Industrial Co., Ltd., Electroconductive Recording Disk Molding Compositions, Japan. Kokai Tokkyo Koho 58 70440 [1981/83]; C.A. **100** [1984] No. 69267.

Matsushita Electric Industrial Co., Ltd., Electroconductive Recording Disk Molding Compositions, Japan. Kokai Tokkyo Koho 58 70442 [1981/83]; C.A. **100** [1984] No. 7838.

Matsushita Electric Industrial Co., Ltd., Electroconductive Recording Disk Molding Compositions, Japan. Kokai Tokkyo Koho 58 70443 [1981/83]; C.A. **100** [1984] No. 7837.

Matsushita Electric Works, Ltd., Plasticizing Flame Retardants for Laminated Circuit Boards, Japan. Kokai Tokkyo Koho 58 142847 [1982/83]; C.A. **100** [1984] No. 69303.

Millick, W. H., Hercofina, Partially Hydrolized, DMT Process Residue, and a Useful Propylene Oxide Derivative of this Residue, U.S. 4394286 [1981/83]; C.A. **99** [1983] No. 106246.

Mitsubishi Rayon Co., Ltd., Delustering Agent for Thermoplastics, Japan. Kokai Tokkyo Koho 58 29856 [1981/83]; C.A. **99** [1983] No. 123567.

Reed, J. O., Mathis, R. D., Phillips Petroleum Co., Polymer Stabilization, U.S. 4411853 [1982/83]; C.A. **100** [1984] No. 23280.

Rosik, L., Horak, Z., Seiner, F., Vecerka, F., Virt, J., Polystyrene with Lower Flammability and Improved Light Aging Resistance, Czech. 201731 [1978/83]; C.A. **100** [1984] No. 157707.

Showa Denko K. K., Olefin Polymer Mixture with Adhesion to Polyurethanes, Japan. Kokai Tokkyo Koho 58 191706 [1982/83]; C.A. **100** [1984] No. 193007.

Sumika Color Co., Ltd., Chalking-Resistant PVC Resin Compositions, Japan. Kokai Tokkyo Koho 58 21438 [1981/83]; C.A. **99** [1983] No. 54567.

Sumitomo Electric Industries, Ltd., Polyolefin Composition for Electric Field Relaxing Tapes, Japan. Kokai Tokkyo Koho 58 19808 [1981/83]; C.A. **99** [1983] No. 39337.

Taoka Chemical Co., Ltd., Acrylic Adhesive Compositions, Japan. Kokai Tokkyo Koho 58 21469 [1981/83]; C.A. **99** [1983] No. 89311.

TDK Corp., Optical Recording Medium, Japan. Kokai Tokkyo Koho 58 171733 [1982/83]; C.A. **100** [1984] No. 43150.

Voeroesova, M., Krejci, B., Hanzalik, S., Kriz, M., Stabilizers for the Thermal Stabilization of Polymers Containing Halogen, Czech. 211881 [1980/83]; C.A. **100** [1984] No. 35378.

Adeka Argus Chemical Co., Ltd., Halogen-Containing Resin Compositions, Japan. Kokai Tokkyo Koho 59 196351 [1983/84]; C.A. **102** [1985] No. 133016.

Bill, R., Wolf, R., Sandoz A.-G., Flame-Proofing Compositions, Brit. Appl. 2139633 [1983/84]; C.A. **102** [1985] No. 63120.

Bill, R., Wolf, R., Sandoz-Patent-GmbH, Flame-Inhibiting Compositions, Ger. Offen. 3416447 [1983/84]; C.A. **102** [1985] No. 79817.

Daicel Chemical Industries, Ltd., Flame-Resistant Resin Compositions, Japan. Kokai Tokkyo Koho 59 28343 [1976/84]; C.A. **101** [1984] No. 321475.

Denki Kagaku Kogyo K. K., Resin Compositions for Video Disks, Japan. Kokai Tokkyo Koho 59 187037 [1983/84]; C.A. **102** [1985] No. 96522.

Japan Synthetic Rubber Co., Ltd., Resin Compositions Adhering to Polyurethane Coatings, Japan. Kokai Tokkyo Koho 59 47250 [1982/84]; C.A. **101** [1984] No. 112079.

Hitachi Cable, Ltd., Recording Discs, Japan. Kokai Tokkyo Koho 59 138248 [1983/84]; C.A. **102** [1985] No. 47005.

Japan Exlan Co., Ltd. Polyester, Japan. Kokai Tokkyo Koho 59105023 [1982/84]; C.A. **101** [1984] No. 171938.

Kanegafuchi Chemical Industry Co., Ltd., Curable Resin Compositions, Japan. Kokai Tokkyo Koho 5974149 [1982/84]; C.A. **101** [1984] No. 153139.

Kishida, K., Hasegawa, A., Sugimori, M., Mitsubishi Rayon Co., Ltd., Delustered Thermoplastic Resin Composition, U.S. 4464513 [1979/84]; C.A. **101** [1984] No. 172468.

Matsushita Electric Industrial Co., Ltd., Electrically Conductive Video Disks, Japan. Kokai Tokkyo Koho 59113537 [1982/84]; C.A. **101** [1984] No. 231647.

Mitsubishi Rayon Co., Ltd., Poly(Vinyl Chloride) Compositions, Japan. Kokai Tokkyo Koho 5933342 [1982/84]; C.A. **101** [1984] No. 39358.

Mitsubishi Rayon Co., Ltd., Impact-Resistant Thermoplastic Resin Compositions, Japan. Kokai Tokkyo Koho 59149915 [1983/84]; C.A. **102** [1985] No. 46736.

Mitsui Toatsu Chemicals, Inc., Vinyl Chloride Resin Compositions for Synthetic Wood, Japan. Kokai Tokkyo Koho 5953546 [1982/84]; C.A. **101** [1984] No. 92172.

Nippon Zeon Co., Ltd., Rigid Resin Compositions, Japan. Kokai Tokkyo Koho 5968361 [1982/84]; C.A. **101** [1984] No. 131746.

Nippon Zeon Co., Ltd., Rigid Resin Compositions, Japan. Kokai Tokkyo Koho 5998154 [1982/84]; C.A. **101** [1984] No. 172419.

Paschke, E. E., Cerefice, S. A., Standard Oil Co. (Indiana), Poly(p-methylene-benzoate) from p-Hydroxymethylbenzoic Acid, U.S. 4431798 [1982/84]; C.A. **100** [1984] No. 192580.

Suzuki, N., Hayakawa, F., Dainippon Printing Co., Ltd., Plastic Card Having a Metallic Luster, U.S. 4479995 [1981/84]; C.A. **102** [1985] No. 103745.

Teijin Ltd., Treatment of Chlorine-Containing Synthetic Fibers, Japan. Kokai Tokkyo Koho 59100771 [1982/84]; C.A. **101** [1984] No. 231902.

Tochu Plastic Kogyo K. K., Plastic Sheets for Electromagnetic Shielding, Japan. Kokai Tokkyo Koho 59184239 [1983/84]; C.A. **102** [1985] No. 96672.

Toyo Soda Mfg., Co., Ltd., Tear Strength Improvement of Vinyl Chloride Graft Copolymers, Japan. Kokai Tokkyo Koho 59179552 [1983/84]; C.A. **102** [1985] No. 150346.

Toyo Soda Mfg., Co., Ltd., Vinyl Chloride Graft Copolymer Compositions for Films and Sheets, Japan. Kokai Tokkyo Koho 59179553 [1983/84]; C.A. **102** [1985] No. 96502.

Toyo Soda Mfg. Co., Ltd., Improving the Tear Strength of Vinyl Chloride Graft Copolymers, Japan. Kokai Tokkyo Koho 59184248 [1983/84]; C.A. **102** [1985] No. 96520.

Asahi, T., Suzuki, Y., Shimizu, A., Uno, S., Toyo Soda Mfg. Co., Ltd., Vinyl Chloride Resin Foams, Japan. Kokai Tokkyo Koho 60149636 [1984/85]; C.A. **104** [1986] No. 51735.

DiMarco, L. A., RCA Corp., Solvent-Cast Capacitive Density Information Disks, U.S. 4515830 [1983/85]; C.A. **103** [1985] No. 79576.

Harasawa, I., Hariki, Y., Nippon Carbide Industries Co., Inc., Plastic Films for Growth of Plants, Japan. Kokai Tokkyo Koho 6056462 [1976/85]; C.A. **104** [1986] No. 208353.

Higashijima, M., Sasaki, H., Takana, H., Osaka Soda Co., Ltd., Resin-Coated Fabrics, Japan. Kokai Tokkyo Koho 60176 [1984/85]; C.A. **103** [1985] No. 216843.

Imai, T., Nochimori, S., Kimura, S., Japan Synthetic Rubber Co., Ltd., Fire-Resistant Graft Polymers, Japan. Kokai Tokkyo Koho 60 255809 [1984/85]; C.A. **105** [1986] No. 25409.

Ishikawa, I., Nagata, K., Tomosada, T., Sanyo Chemical Industries, Ltd., Polyurethanes for Reaction-Injection Molding, Japan. Kokai Tokkyo Koho 60 206819 [1984/85]; C.A. **104** [1986] No. 208486.

Kanegafuchi Chemical Industry Co., Ltd., Crosslinked PVC Blends, Japan. Kokai Tokkyo Koho 60 49052 [1983/85]; C.A. **103** [1985] No. 23519.

Kitano, Y., Yano, K., Matsumaru, T., Nitto Chemical Industry Co., Ltd., Stabilized Halogen-Containing Polymer Compositions, Japan. Kokai Tokkyo Koho 60 177053 [1984/85]; C.A. **104** [1986] No. 208280.

Kobayashi, T., Arakawa, H., Nippon Zeon Co., Ltd., Vinyl Chloride Resin Compositions, Japan. Kokai Tokkyo Koho 60 173036 [1984/85]; C.A. **104** [1986] No. 130836.

Kobayashi, T., Arakawa, H., Nakayama, T., Watanabe, J., Nippon Zeon Co., Ltd., Flexible Vinyl Chloride Polymer Compositions, Japan. Kokai Tokkyo Koho 60 170648 [1984/85]; C.A. **104** [1986] No. 150085.

Kobayashi, T., Arakawa, H., Ohira, T., Nippon Zeon Co., Ltd., Vinyl Chloride Polymer Blends, Ger. Offen. 3505361 [1984/85]; C.A. **104** [1986] No. 34930.

Kogo, Y., Marushima, N., Sankyo Organic Chemicals Co., Ltd., Stabilization of Chlorinated Polyethylene, Japan. Kokai Tokkyo Koho 60 184535 [1984/85]; C.A. **104** [1986] No. 130856.

Labib, M. E., Poll, R. F., Wang, C. C., RCA Corp., Capacitance Electronic Disc Molding Compositions, U.S. 4522747 [1984/85]; C.A. **103** [1985] No. 105810.

Mark, V., General Electric Co., Aromatic Carbonates, Ger. Offen. 3445553 [1983/85]; C.A. **104** [1986] No. 50646.

Mark, V., General Electric Co., Aromatic Carbonates, Ger. Offen. 3455555 [1983/85]; C.A. **104** [1986] No. 88288.

Mitsubishi Rayon Co., Ltd., Impact Modifiers for Thermoplastic Resins, Japan. Kokai Tokkyo Koho 60 32849 [1983/85]; C.A. **103** [1985] No. 7287.

Mitsubishi Rayon Co., Ltd., Impact-Resistant Thermoplastic Molding Compositions, Japan. Kokai Tokkyo Koho 60 112811 [1983/85]; C.A. **103** [1985] No. 179362.

Miyazaki, M., Sekisui Chemical Co., Ltd., Adhesive Sheets for Copying, Japan. Kokai Tokkyo Koho 60 217284 [1984/85]; C.A. **104** [1986] No. 70038.

Nakamura, T., Takahashi, S., Nippon Zeon Co., Ltd., Vinyl Chloride Polymer Compositions, Japan. Kokai Tokkyo Koho 60 163954 [1984/85]; C.A. **104** [1986] No. 169476.

Nara, S., Nishiyama, S., Yukitomi, M., Matsunaga, K., Toppan Printing Co., Ltd., Chemical Embossing of Plastic Surfaces, Japan. Kokai Tokkyo Koho 60 34447 [1977/85]; C.A. **104** [1986] No. 90682.

Narita, T., Mitsubishi Rayon Co., Ltd., Vinyl Chloride Polymer Compositions, Japan. Kokai Tokkyo Koho 60 161449 [1984/85]; C.A. **104** [1986] No. 90028.

Nippon Carbide Industries Co., Inc., UV-Absorbing Mulching Films, Japan. Kokai Tokkyo Koho 60 75215 [1983/85]; C.A. **103** [1985] No. 88833.

Nippon Zeon Co., Ltd., Expandable Vinyl Chloride Resin Compositions, Japan. Kokai Tokkyo Koho 60 08330 [1983/85]; C.A. **102** [1985] No. 204931.

Nippon Zeon Co., Ltd., Heat-Resistant Resin Compositions, Japan. Kokai Tokkyo Koho 60 49051 [1983/85]; C.A. **103** [1985] No. 72136.

Okano, K., Narita, T., Mitsubishi Rayon Co., Ltd., Methyl Methacrylate Polymers for Polymer Alloys, Japan. Kokai Tokkyo Koho 60 240714 [1984/85]; C.A. **104** [1986] No. 187528.

Okano, K., Narita, T., Mitsubishi Rayon Co., Ltd., Vinyl Chloride Polymer Alloys, Japan. Kokai Tokkyo Koho 60 240 [1984/85]; C.A. **104** [1986] No. 169549.

Ono, K., Mitsubishi Rayon Co., Ltd., Fire-Resistant Styrene Resin Compositions, Japan. Kokai Tokkyo Koho 60 208343 [1984/85]; C.A. **105** [1986] No. 25178.

Schepers, H. A. J., Debets, W. A. M., Stamicarbon B. V., Thermoplastic Molding Compound, Eur. Appl. 140423 [1983/85]; C.A. **103** [1985] No. 72162.

Stamicarbon, B. V., Poly(Vinyl Chloride)-Based Composition, Neth. Appl. 83-03139 [1983/85]; C.A. **103** [1985] No. 72155.

Suzuki, N., Hayakawa, F., Dainippon Printing Co., Ltd., Taihei Chemicals, Ltd., Plastic Card Having Metallic Luster, Can. 1188969 [1981/85]; C.A. **103** [1985] No. 216619.

Tachibana, H., Nagata, M., Ogoshi, Z., Kanegafuchi Chemical Industry Co., Ltd., Thermoplastic Blend Molding Compositions, Japan. Kokai Tokkyo Koho 60 208346 [1984/85]; C.A. **104** [1986] No. 90113.

Tachibana, H., Nagata, M., Ogoshi, Z., Kanegafuchi Chemical Industry Co., Ltd., Transparent Thermoplastic Molding Compositions with Heat, Fire, and Impact Resistance, Japan. Kokai Tokkyo Koho 60 210653 [1984/85]; C.A. **104** [1986] No. 130807.

Tachibana, H., Nagata, M., Ogoshi, Z., Kanegafuchi Chemical Industry Co., Ltd., Heat- and Impact-Resistant Heat-Stable Thermoplastic Molding Compositions, Japan. Kokai Tokkyo Koho 60 212446 [1984/85]; C.A. **104** [1986] No. 131101.

Tachibana, H., Nagata, M., Ogoshi, Z., Kanegafuchi Chemical Industry Co., Ltd., Heat-, Fire-, and Impact-Resistant Polymer Compositions, Japan. Kokai Tokkyo Koho 60 212462 [1984/85]; C.A. **104** [1986] No. 187518.

Tachibana, H., Ogoshi, Z., Hagata, M., Kanegafuchi Chemical Industry Co., Ltd., Heat- and Fire-Resistant Thermoplastic Polymer Blends, Japan. Kokai Tokkyo Koho 60 248757 [1984/85]; C.A. **104** [1986] No. 208525.

Tachibana, H., Ogoshi, Z., Nagata, M., Kanegafuchi Chemical Industry Co., Ltd., Heat-, Fire-, and Impact-Resistant Thermoplastic Polymer Blends, Japan. Kokai Tokkyo Koho 60 248758 [1984/85]; C.A. **104** [1986] No. 208524.

Tachibana, H., Ogoshi, Z., Nagata, M., Kanegafuchi Chemical Industry Co., Ltd., Heat- and Fire-Resistant Thermoplastic Blends, Japan. Kokai Tokkyo Koho 60 248759 [1984/85]; C.A. **104** [1986] No. 208523.

Tachibana, H., Ogoshi, Z., Nagata, M., Kanegafuchi Chemical Industry Co., Ltd., Heat-, Fire- and Impact-Resistant Thermoplastic Compositions, Japan. Kokai Tokkyo Koho 60 248761 [1984/85]; C.A. **104** [1986] No. 208339.

Takahashi, S., Nakamura, T., Nippon Zeon Co., Ltd., Hard Resin Compositions, Japan. Kokai Tokkyo Koho 60 168741 [1984/85]; C.A. **104** [1986] No. 110775.

Yamamoto, K., Takenaka, T., Sumitomo Chemical Co., Ltd., Adhesive Composition with Enhanced Storage Stability, U.S. 4515917 [1983/85]; C.A. **103** [1985] No. 7349.

Yamazaki, T., Shimomura, T., Kamiide, K., Koei Chemical Co., Ltd., Japan. Kokai Tokkyo Koho 60 142990 [1983/85]; C.A. **104** [1986] No. 111463.

1.4.1.2.1.5.5.3 Other Dibutyltin Dicarboxylates of the $(C_4H_9)_2\overline{SnOOC\text{-}R\text{-}COO}$ Type

The compounds belonging to this section are presented in Table 30 and are prepared by the methods described below.

General remark: All the compounds are formulated as monomers, although they are oligomeric or even polymeric in part.

Method I: Metathetic reactions between $(C_4H_9)_2SnX_2$ with X = Cl, Br, or I and MOOC-R-COOM with M = K, Na, or $[N(C_2H_5)_4]$ (1:1 mole ratio).
Preferably, the interfacial technique has been used, neutralizing aqueous solutions of the appropriate dicarboxylic acid with the base MOH and adding them to rapidly stirred solutions of $(C_4H_9)_2SnX_2$ in suitable organic solvents (X, M, solvents (PE = petroleum ether), temperature/time): No. 2 (Cl, Br, or I, Na, CCl_4-H_2O, 25°C/15 min [27]; Cl, Na, PE-H_2O, 20°C/20 min [11]); No. 5 (Cl, Na, PE-H_2O, 0°C/1 h) [11]; No. 7 (Cl, Br, or I, Na, CCl_4-H_2O, 25°C/15 min [27]; Cl, Na, PE-H_2O, 0°C/20 min [11, 26]); No. 30 (Cl, K, PE-H_2O, 0°C/1 h) [11]; No. 32 (Cl, Na, PE-H_2O, 0°C/1 h) [11]; Nos. 42 [38, 42] and 43 (Cl, Br, or I, Na, $CHCl_3$-H_2O, 25°C/1 to 20 min) [39, 43, 44].
No. 7 has also been obtained from $(C_4H_9)_2SnCl_2$ and $NaOOC(CH_2)_4COONa$ only in H_2O at 80°C/2 h [19, 25]. No. 41 precipitates on mixing methanolic solutions of $(C_4H_9)_2SnCl_2$ and of the $[N(C_2H_5)_4]^+$ salt of pyridine-2,6-dicarboxylic acid in a 1:1 mole ratio at room temperature [34].

Method II: $(\text{-}(C_4H_9)_2SnO\text{-})_n$ and HOOC-R-COOH or $\overline{OOC\text{-}R\text{-}CO}$ (1:1 mole ratio).
The reaction between $(\text{-}(C_4H_9)_2SnO\text{-})_n$ and the appropriate free dicarboxylic acid or dicarboxylic acid anhydride has been used for the synthesis of Nos. 4 and 6 (acid) [20], No. 9 (acid) [21], No. 29 (anhydride) [48] (for these four reactions are no further conditions available); of No. 10 [4], No. 11 [17], No. 28 [41], and No. 40 [59], all with the free acid in refluxing C_6H_6; of No. 12 [5], Nos. 13, 20, 21 [5, 7, 8, 10, 12], Nos. 14 and 17 [9, 13], Nos. 16 and 19 [5, 7, 8, 12], Nos. 18 and 39 [7, 8, 12], and Nos. 23 and 24 [53], all with the free acid in refluxing $C_6H_5CH_3$; as well as of No. 2 [4], No. 26 [23], No. 27 [23, 30], No. 29 [23], No. 31 [45], Nos. 33, 34, 35 [22, 36], No. 36 [22, 33, 36], and No. 37 [22, 32, 36], all with the dicarboxylic acid anyhdride in refluxing $C_6H_5CH_3$.

Method III: Transesterification of $(C_4H_9)_2Sn(OOCCH_3)_2$ (1:1 mole ratio).
The reactions of $(C_4H_9)_2Sn(OOCCH_3)_2$ with itaconic acid, $HOOCCH_2CH(=CH_2)COOH$ (C_6H_6, reflux/1.5 h), with adipic acid, $HOOC(CH_2)_4COOH$ or 3-methyladipic acid, $HOOCCH_2CH(CH_3)CH_2CH_2COOH$ (xylene, slow temperature rise to 136°C), or with terephthalic acid, $HOOCC_6H_4COOH$-4 ($C_6H_5CH_3$, reflux) with destillative removal of CH_3COOH lead to Nos. 3, 7, 8, and 32, respectively [4].
No. 7 also has been prepared by the reactions of $(C_4H_9)_2Sn(OOCR')_2$ (R' = H, CH_3, or $CHCl_2$) with $KOOC(CH_2)_4COOK$ in petroleum ether-water with yields of 80%, 89%, or 65%, respectively [11].

Method IV: Treatment of $(C_4H_9)_2Sn(OCH_3)_2$ with equal molar amounts of $NCH_3(CH_2COOH)_2$ in dioxane-C_6H_6 under reflux affords No. 22 in excellent yields [52].

Method V: Diels-Alder addition has been used for the synthesis of Nos. 28 and 29, starting from crude dibutyltin maleate which was reacted either with cyclopentadiene (excess C_5H_6, 0°C/12 h [16]; C_6H_6, 20°C/4.5 h [23, 30]), or with hexachlorocyclopentadiene (xylene, reflux/18 h) [23].

Table 30
Dibutyltin Dicarboxylates of the $(C_4H_9)_2\overline{SnOOC\text{-}R\text{-}COO}$ Type.
Further information on compounds preceded by an asterisk is given at the end of the table.
Explanations, abbreviations, and units on p. X.

No.	OOC-R-COO group method of preparation (yield in %)	properties and remarks	Ref.
1	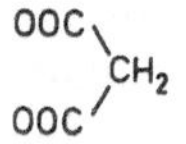	no preparation reported stabilizer for polystryrene; prevents crazing	 [1]
2	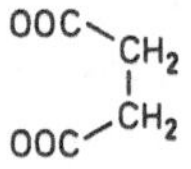 I (69 [27], 77 [11]) II (95 [4])	m.p. 187 to 187.5° (tetrameric) [4], 235° (polymeric) [11]; softening range 187 to 252° (n ≧ 7) [27] catalyst for transesterification reactions	[4, 11, 27] [40]
3	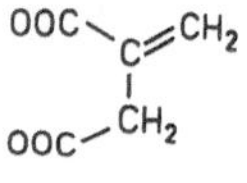 III	no cyclic monomer, but waxy or plastic, nonelastic and fully saturated products are obtained	[4]
4	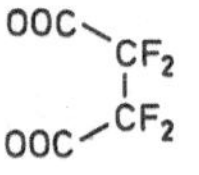 II	m.p. 204 to 206° (dimeric)	[20]
5	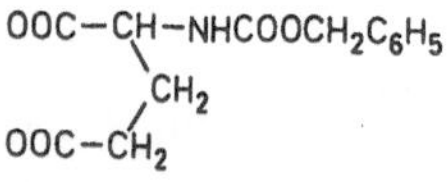 I (82)	m.p. 163 to 165° (polymeric)	[11]
6	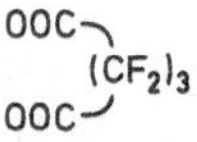 II	m.p. 188 to 190° (dimeric)	[20]

References on p. 329

Table 30 (continued)

No.	OOC-R-COO group method of preparation (yield in %)	properties and remarks	Ref.
7	$OOC\text{–}$ $(CH_2)_4$ $OOC\text{–}$ I (34 to 78 [26, 27], 85 [11], 93 [19, 25]) III (65 to 90 [11], 85 [4])	m.p. 132 to 133° (prepared in xylene-water, Method I) [11], 136 to 137° (trimeric) [4]; softening between 153 and 224° (dependent on the kind of dibutyltin dihalide [27], or on the organic solvent and the isolation procedure used in Method I [26]); 220° (prepared in petroleum ether-water, Method I) [11]; 215 or 230° (Method III) [11]	[4, 11, 26, 27]
		degradation at $400°/10^{-6}$, examined by MS	[14]
		tested as a fungicide against Phytophthora infestans, causing potato haulm blight	[28]
8	$OOC-CH_2CHCH_3$ $OOC-CH_2CH_2$ III (78)	m.p. 143 to 144.5° (dimeric)	[4]
9	$OOC\text{–}$ $(CH_2)_7$ $OOC\text{–}$ II (100 [21])	m.p. 60 to 62°	[21]
		internal lubricant [24] and heat stabilizer for PVC [35]	[24, 35]
10	$OOC\text{–}$ $(CH_2)_8$ $OOC\text{–}$ II (96 [4])	m.p. 78 to 82° (heptameric)	[4]
		catalyst for polyoxymethylene preparation	[29]
		additive to stabilizer compositions	[18]
11	$OOC-CH_2S$ CH_2 $OOC-CH_2S$ II	m.p. 146 to 148° stabilizer for PVC	[17]
12	$OOC-CH_2S$ $C(CH_3)_2$ $OOC-CH_2S$ II	m.p. 58 to 60° stabilizer for halogen-containing resins	[5]
13	$OOC-CH_2S$ CHC_6H_5 $OOC-CH_2S$ II	m.p. 157 to 162° stabilizer for halogen-containing resins	[5, 7, 8, 10, 12]

References on p. 329

Table 30 (continued)

No.	OOC-R-COO group method of preparation (yield in %)	properties and remarks	Ref.
14	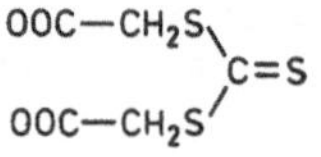 II (93)	m.p. 192 to 196° stabilizer for polyolefins	[9, 13]
15	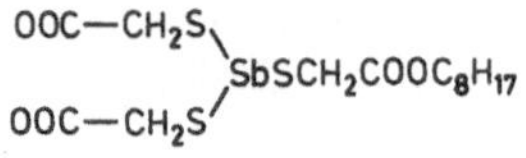	no preparation reported stabilizer for PVC and vinyl chloride-vinyl acetate copolymers	[37]
16	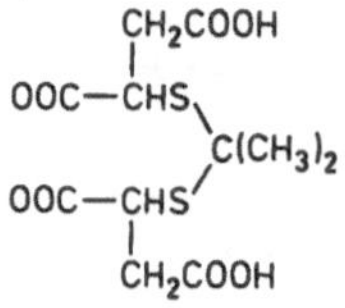 II	stabilizer for halogen-containing resins	[5, 7, 8, 12]
17	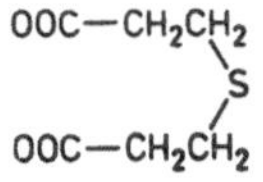 II [9, 13]	m.p. 84 to 92° stabilizer for polyolefins	[9, 13] [6, 9, 13]
18	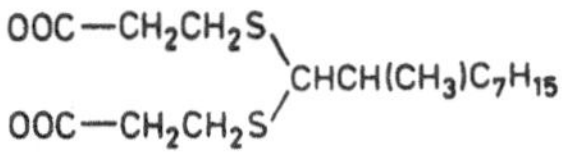 II	stabilizer for halogen-containing resins	[7, 8, 12]
19	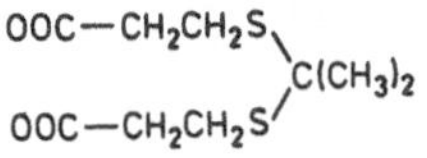 II	m.p. 58 to 60° (wrong C.A. citation; belongs to No. 12) [12], 76 to 78° [5, 7, 8] stabilizer for halogen-containing resins	[5, 7, 8, 12]
20	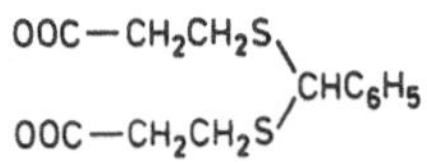 II	m.p. 79 to 89° stabilizer for halogen-containing resins	[5, 7, 8, 10, 12]
21	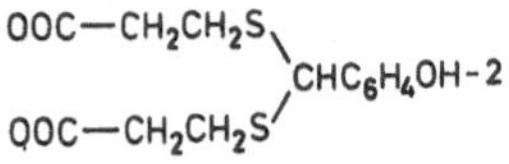 II	m.p. 131 to 134° (erroneous -C_6H_4OH-3 in [8]) stabilizer for halogen-containing resins	[5, 7, 8, 10, 12]

References on p. 329

Table 30 (continued)

No.	OOC-R-COO group method of preparation (yield in %)	properties and remarks	Ref.
22	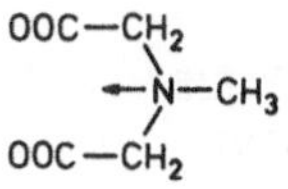 IV (93 [52])	m.p. 166°, stable against boiling H_2O	[52]
		^{1}H NMR ($CDCl_3$, 30°): 2.66 (CH_3N, $^{3}J(Sn,H)$ = 14), 3.50 (CH_2N)	
		^{119}Sn-γ (77 K): δ = 1.54, Δ = 3.88	[54]
		IR: ν(CO) 1670 and 1608 (in KBr) or 1690 (3% in $CHCl_3$)	
		from spectroscopic data, six-coordinate Sn with intramolecular N→Sn and intermolecular C=O→Sn coordination suggested	[52, 54]
23	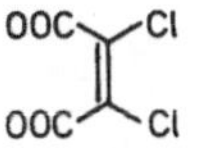 II (83 [53])	m.p. 155°	[53]
		IR (KBr): ν_{as}(OCO) 1650 and 1600, ν_s(OCO) 1400, ν(CCl) 700	
		less suitable as a heat stabilizer for PVC than non-halogenated maleic acid derivatives, due to lower reactivity in Diels-Alder reactions	[53, 56]
24	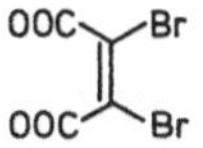 II (80 [53])	m.p. 147°	[53]
		IR (KBr): ν_{as}(OCO) 1650 and 1600, ν_s(OCO) 1350, ν(CBr) 665	
		stabilizing effect for PVC like No. 23	[53, 56]
25	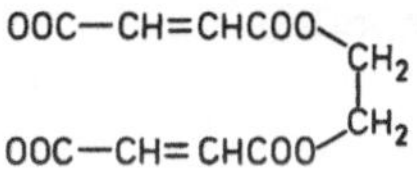	no preparation reported	
		n_D^{20} = 1.5185	[46]
		studies on the stabilizing effect in thermal degradation of PVC in comparison to other organotin maleates and thioglycolates	[46, 47]
26	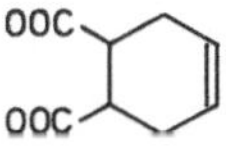 II	m.p. 72 to 74.5°	[23]
		fungicide, especially against Alternaria solani on tomatoes, and Erysiphe polygoni on beans	
27	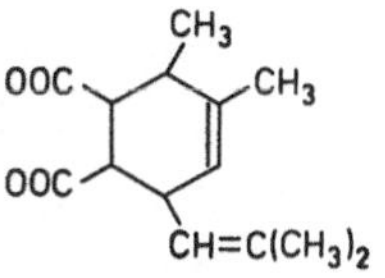 II	fungicide like No. 26	[23, 30]

References on p. 329

Table 30 (continued)

No.	OOC-R-COO group method of preparation (yield in %)	properties and remarks	Ref.
28	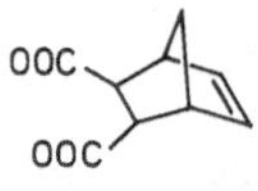 II (94 [41]) V [16, 23, 30]	m.p. 202 to 203° [41], 222 to 227° [23, 30], 239 to 242° (polymeric) [16] copolymerizes with styrene fungicide against Alternaria solani on tomatoes	[16, 23, 30, 41] [41] [23, 30]
29	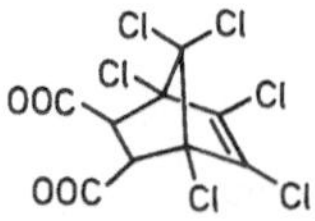 II [23], (82 [48]) V [23]	m.p. 147 to 148.5° [48], 160 to 170° [23] fungicide like No. 26	[23, 48] [23]
30	 I (81)	m.p. > 300° (polymeric)	[11]
31	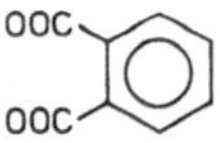 II [45]	no physical properties reported stabilizer for PVC [45], or polypivalolactone [31] additive for room temperature curable silicone resin compositions [49, 51, 58, 60], polyether adhesives [50], polyurethanes [57], or ABS compositions containing 45% nitrile [61] anticrazing agent for polydichlorostyrene	 [31, 45] [49, 50, 51, 57, 58, 60, 61] [1, 2, 3]
*32	 I (70 [11]) III (100 [4]) special [15]	m.p. >245° [4], or >300° (polymeric chains) [11, 15]	[4, 11, 15]

References on p. 329

Table 30 (continued)

No.	OOC-R-COO group method of preparation (yield in %)	properties and remarks	Ref.
33	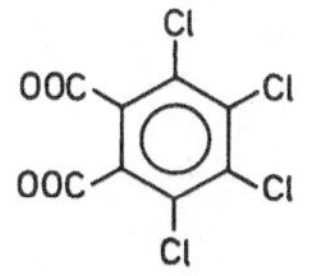 II (~100)	m.p. 200 to 202° fungicide like No. 28	[22, 36]
34	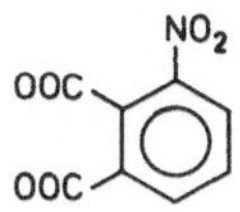 II (~100)	m.p. 145 to 147° fungicide against Erysiphe polygony on beans	[22, 36]
35	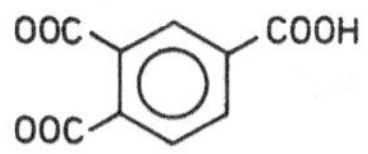 II (~100)	m.p. 237 to 240° fungicide like No. 28	[22, 36]
36	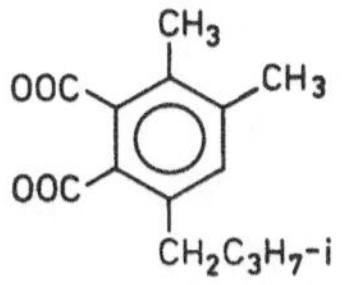 II (~ 100)	m.p. 89 to 90° fungicide against Alternaria solani on tomatoes, Cercospora beticola on sugar beets, or Colletotrichum lagenarium on cucumbers	[22, 33, 36]
37	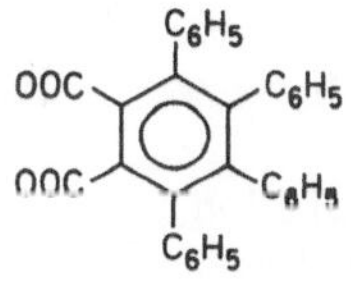 II (~100)	m.p. 223 to 225° fungicide	[22, 32, 36]
*38	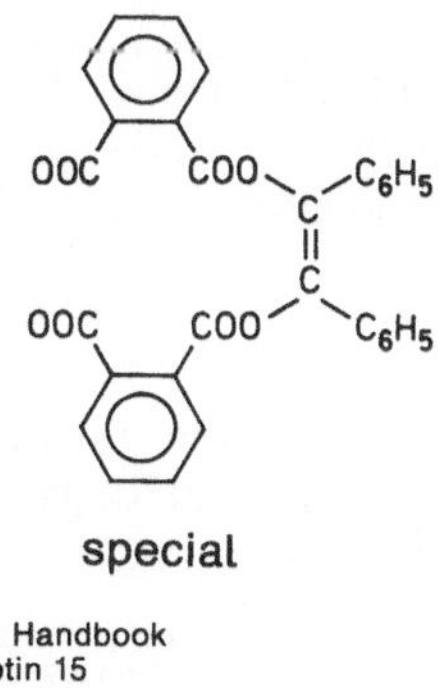 special	m.p. 125 to 127° ^{1}H NMR ($CDCl_3$): 0.7 to 2.0 (C_4H_9), 7.0 to 8.2 (C_6H_4, C_6H_5) IR: ν(C=C) 1710	[55]

References on p. 329

Table 30 (continued)

No.	OOC-R-COO group method of preparation (yield in %)	properties and remarks	Ref.
39	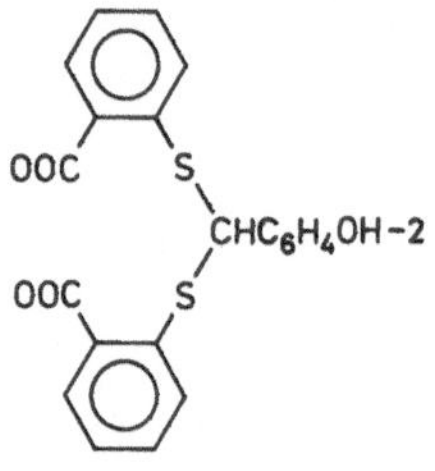 II	stabilizer for halogen-containing resins	[7, 8, 12]
40	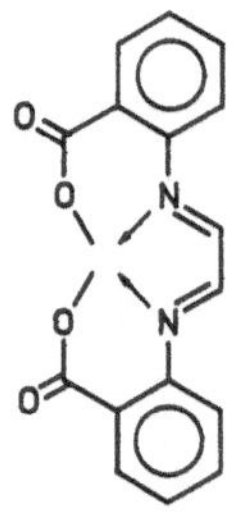 II (97)	m.p. 167° ^{1}H NMR: 6.9 to 7.8 (m, C_6H_4), 8.6 (s, CH=N) IR: ν_{as}(OCO) 1680, ν(C=N) 1610, ν(C=C) 1580	[59]
41	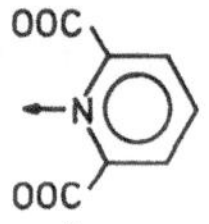 I	m.p. 253 to 255° ^{119}Sn-γ (80 K): δ = 1.46, Δ = 4.07 IR (KBr): ν_{as}(OCO) 1685, ν_s(OCO) 1335; planar tridentate ligand	[34]
42	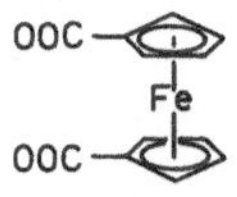 I (20 to 40)	assumed as an alternative choice to linear oligomers study of the synthesis as a function of several reaction variables	[38, 42]
43	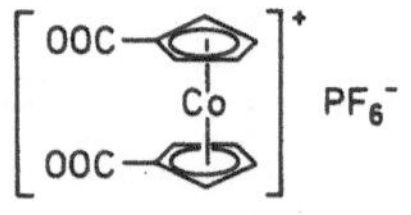 I (30 to 49)	assumed as an alternative choice to linear oligomers study of the synthesis as a function of several reaction variables thermograms between 0 and 600°	[39, 43, 44]

* Further information:

$(C_4H_9)_2\overline{SnOOC-C_6H_4-COO}$-4 (Table **30**, No. **32**) is also formed in the interfacial reaction of $Cl(C_4H_9)_2SnSSn(C_4H_9)_2Cl$ with $NaOOCC_6H_4COONa$-4 in petroleum ether-water at 0°C, with 12 h stirring in a 16% yield, along with $(-(C_4H_9)_2SnS-)_3$ [15].

$(C_4H_9)_2\overline{SnOOC\text{-}C_{28}H_{18}O_4\text{-}COO}$ (Table **30**, No. **38**) is formed in the reaction between $(C_4H_9)_2\overline{SnOC(C_6H_5){=}C(C_6H_5)O}$ (Table 14, No. 43) and phthalic acid anhydride in C_6H_6 at 55°C for 1 h, along with $(C_4H_9)_2\overline{SnOOCC_6H_4COO}$-2 (this table, No. 31) and $C_6H_5COCOC_6H_5$ [55].

References:

[1] Churchill, J. W., Mathieson Chemical Corp. (Fr. 991331 [1949/51]; C. **1953** 3647).
[2] Churchill, J. W., Mathieson Chemical Corp. (U.S. 2643242 [1953]; C.A. **1953** 10904).
[3] Churchill, J. W., Mathieson Chemical Corp. (Can. 506310 [1949/54]).
[4] Andrews, T. M. Bower, F. A., LaLiberte, B. R., Montermoso, J. C. (J. Am. Chem. Soc. **80** [1958] 4102/4).
[5] Hechenbleikner, I., Bresser, R. E., Homberg, O. A., Deutsche Advance Produktion GmbH (Belg. 616326 [1961/62]).
[6] Thinius, K., Walther, H., Bergheer, H., Institut für Chemie und Technologie der Plaste (Ger. 1126604 [1958/62]; C.A. **57** [1962] 6147).
[7] Hechenbleikner, I., Bresser, R. E., Homberg, O. A., Carlisle Chemical Works, Inc. (Belg. 630452 [1962/63]; C.A. **61** [1964] 8339).
[8] Hechenbleikner, I., Bresser, R. E., Homberg, O. A., Carlisle Chemical Works, Inc. (U.S. 3078290 [1961/63]; C.A. **58** [1963] 13993).
[9] Hechenbleikner, I., Homberg, O. A., Deutsche Advance Produktion GmbH (Belg. 622648 [1961/63]; C.A. **58** [1963] 14227).
[10] Deutsche Advance Produktion GmbH (Brit. 1004663 [1961/65]; C.A. **63** [1965] 1655).

[11] Frankel, M., Gertner, D., Wagner, D., Zilkha, A. (J. Appl. Polym. Sci. **9** [1965] 3383/8).
[12] Hechenbleikner, I., Bresser, R. E., Homberg, O. A., Carlisle Chemical Works, Inc. (U.S. 3217004 [1963/65]; C.A. **67** [1967] No. 90939).
[13] Hechenbleikner, I., Homberg, O. A., Carlisle Chemical Works, Inc. (U.S. 3209017 [1961/65]; C.A. **63** [1965] 16382).
[14] Bruck, S. D. (J. Polym. Sci. Polym. Chem. Ed. **5** [1967] 2158/9).
[15] Migdal, S., Gertner, D., Zilkha, A. (Can. J. Chem. **45** [1967] 2987/92).
[16] Mufti, A. S., Poller, R. C. (J. Chem. Soc. C **1967** 1767/8).
[17] Thinius, K., Schlimper, R. (Ger. [East] 54106 [1963/67]; C.A. **67** [1967] No. 74190).
[18] Stapfer, C. H., Carlisle Chemical Works, Inc. (Fr. 1537462 [1966/68]; C.A. **71** [1969] No. 13830).
[19] Boboli, E., Rajewski, M., Kowalski, M., Pazgan, A., Lao, W., Institut Przemyslu Organicznego (Fr. 1580291 [1968/69]).
[20] LaLiberte, B. R., Reiff, H. F., Davidsohn, W. E. (Org. Prep. Proced. **1** [1969] 173/6).

[21] Mikhailov, G. D., Chegolya, A. S. (Sin. Volokna **1969** 18/21; C.A. **74** [1971] No. 3711).
[22] Mineri, P. P., Tenneco Chemicals, Inc. (U.S. 3454611 [1967/69]; C.A. **71** [1969] No. 81537).
[23] Mineri, P. P., Tenneco Chemicals, Inc. (U.S. 3479380 [1967/69]).
[24] Stapfer, C. H., Carlisle Chemical Works, Inc. (Fr. 1578260 [1967/69]; C.A. **72** [1970] No. 112263).
[25] Boboli, E., Rajewski, M., Kowalski, M., Pazgan, A., Lato, C., Institut Przemyslu Organicznego (Pol. 59754 [1967/70]; C.A. **74** [1971] No. 31850).
[26] Carraher, C. E., Dammeier, R. L. (Makromol. Chem. **135** [1970] 107/12).
[27] Carraher, C. E., Dammeier, R. L. (Polym. Prep. Am. Chem. Soc. Div. Polym. Chem. **11** [1970] 606/12).
[28] McIntosh, A. H. (Ann. Appl. Biol. **66** [1970] 115/8).

[29] Ishida, S., et al., Asahi Chemical Industry Co., Ltd. (Ger. Offen. 1946077 [1968/70]; C.A. **72** [1970] No. 133404).
[30] Mineri, P. P. (U.S. 3517088 [1969/70]; C.A. **73** [1970] No. 56241).

[31] Nagato, S., Suzuki, J., Saitama, H. M., Daicell Co., Ltd. (U.S. 3510449 [1966/70]; C.A. **73** [1970] No. 26272).
[32] Mineri, P. P., Tenneco Chemicals, Inc. (Ger. Offen. 1933173 [1969/71]; C.A. **74** [1971] No. 88141).
[33] Mineri, P. P., Tenneco Chemicals, Inc. (U.S. 3595963 [1969/71]; C.A. **75** [1971] No. 129942).
[34] Naik, D. V., Curran, C. (Inorg. Chem. **10** [1971] 1017/20).
[35] Stapfer, C. H., Shah, A. C., Cincinnati Milacron Chemicals, Inc. (Ger. Offen. 2057234 [1969/71]; C.A. **75** [1971] No. 64893).
[36] Mineri, P. P., Tenneco Chemicals, Inc. (Brit. 1273029 [1969/72]; C.A. **77** [1972] No. 621133).
[37] Yukitomi, M., Hiramatsu, T., Shinkawa, Y., Ito, T., Kyoto Pharmaceutical Industries, Ltd. (Japan. 72-06106 [1968/72]; C.A. **77** [1972] No. 140965).
[38] Carraher, C. E., Lessek, P. J. (Am. Chem. Soc. Div. Org. Coatings Plastics Chem. Papers **33** [1973] 420/6).
[39] Carraher, C. E., Peterson, G. F., Sheats, J. E. (Am. Chem. Soc. Div. Org. Coatings Plastics Chem. Papers **33** [1973] 427/33).
[40] Tsunawaki, K., Sasama, W., Watanabe, K., Nawata, K., Teijin Ltd. (Japan. Kokai 73-28444 [1971/73]; C.A. **79** [1973] No. 66032).

[41] Aslanov, I. A., Kochkin, D. A., Koton, M. M. (Dokl. Akad. Nauk SSSR **216** [1974] 319/20; Dokl. Chem. Proc. Acad. Sci. USSR **214/219** [1974] 313/4).
[42] Carraher, C. E., Lessek, P. J. (Angew. Makromol. Chem. **38** [1974] 55/66).
[43] Carraher, C. E., Peterson, G. F., Sheats, J. E., Kirsch, T. (J. Makromol. Sci. Chem. A **8** [1974] 1009/22).
[44] Carraher, C. E., Peterson, G. F., Sheats, J. E. Kirsch, T. (Makromol. Chem. **175** [1974] 3089/96).
[45] Dworkin, R. D., Ejk, A. J., M and T Chemicals, Inc. (Ger. Offen. 2626554 [1975/76]; C.A. **86** [1977] No. 140253).
[46] Troitskii, B. B., Troitskaya, L. S., Denisova, V. N., Novikova, M. A. (Fiz. Khim. Osn. Sint. Pererab. Polim. [Gorkyi] No. 1 [1976] 58/61).
[47] Troitskii, B. B., Troitskaya, L. S., Denisova, V. N., Novikova, M. A., Luzinova, Z. B. (Eur. Polym. J. **13** [1977] 1033/41).
[48] Dunyamaliev, A. D., Novoderezhkina, I. S., Rzaev, Z. N., Shakhtakhtinskii, T. N. (Issled. Obl. Sint. Polim. Monomernykh Prod. **1979** 41/4).
[49] Kanegafuchi Chemical Industry Co., Ltd. (Japan. Kokai Tokkyo Koho 80-56154 [1978/80]; C.A. **93** [1980] No. 115435).
[50] Kanegafuchi Chemical Industry Co., Ltd. (Japan. Kokai Tokkyo Koho 80-56153 [1978/80]; C.A. **93** [1980] No. 169291).

[51] Tani, N., Mita, T., Isayama, K., Kanegafuchi Chemical Industry Co., Ltd. (Japan. Kokai Tokkyo Koho 80-21453 [1978/80]; C.A. **93** [1980] No. 27906).
[52] Tzschach, A., Jurkschat, K., Zschunke, A., Mügge, C. (J. Organometal. Chem. **193** [1980] 299/305).
[53] Kizlink, J. G., Poller, R. C. (Eur. Polym. J. **17** [1981] 639/40).
[54] Korecz, L., Saghier, A. A., Burger, K., Tzschach, A., Jurkschat, K. (Inorg. Chim. Acta **58** [1982] 243/9).
[55] Davies, A. G., Hawari, J. A. A. (J. Chem. Soc. Perkin Trans. I **1983** 875/82).
[56] Kizlink, J., Jakubcek, E. (Ropa Uhlie **25** [1983] 565/7; C.A. **100** [1984] No. 104347).

[57] Matsushita Electric Works, Ltd. (Japan. Kokai Tokkyo Koho 5974115 [1982/84]; C.A. **101** [1984] No. 132646).
[58] Takase, J., Hirose, T., Isayama, K., Kanegafuchi Chemical Industry Co., Ltd. (U.S. 4444974 [1982/84]; C.A. **101** [1984] No. 92973).
[59] Sakuntala, E. N., Vasanta, E. N. (Z. Naturforsch. **40b** [1985] 1173/6).
[60] Sekisui Chemical Co., Ltd. (Japan. Kokai Tokkyo Koho 6001253 [1983/85]; C.A. **103** [1985] No. 7246).
[61] Shiobara, T., Abe, H., Sekisui Chemical Co., Ltd. (Japan. Kokai Tokkyo Koho 60195152 [1984/85]; C.A. **104** [1986] No. 110826).

1.4.1.2.1.5.6 Dibultyltin Oxycarboxylates, $(C_4H_9)_2SnOOCRO$

$(C_4H_9)_2SnC_3H_4O_3$

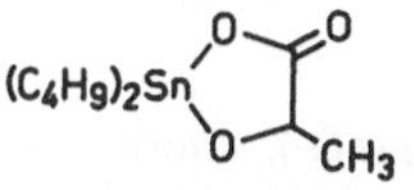

The cyclic lactic acid derivative is formed in the reaction between $(C_4H_9)_2SnCl_2$ and $CH_3CH(OH)COOH$ in refluxing ether and in the presence of C_5H_5N as HCl acceptor and has been shown to be a stabilizer for resin compositions, especially those containing $-CH_2CHCl-$ units [1].

$(C_4H_9)_2SnC_8H_6O_3$

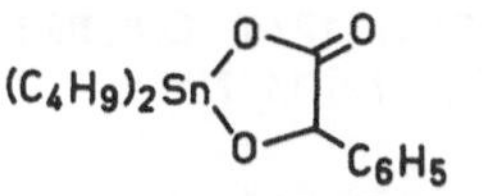

The reaction between $(-(C_4H_9)_2SnO-)_n$ and mandelic acid, $C_6H_5CH(OH)COOH$, in refluxing C_6H_6 leads to the formation of the title compound in a yield of 93%. It melts at 194 to 195°C.

^{119}Sn Mössbauer spectrum (77 K): $\delta = 1.36$, $\Delta = 3.71$ mm/s. IR spectrum (Nujol): $\nu_{as}(OCO)$ 1613 (vs) and $\nu_s(OCO)$ 1362 (s), leaving all other bands in the 1650 to 450 cm^{-1} range unassigned [3].

$(C_4H_9)_2SnC_7H_4O_3$

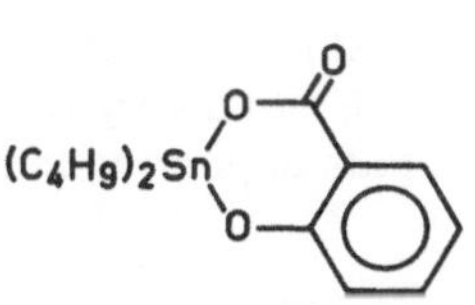

The addition of salicylic acid, $2-HOC_6H_4COOH$, to a suspension of $(-(C_4H_9)_2SnO-)_n$ in C_6H_6 and subsequent refluxing of the reaction mixture with azeotropic distillation of the water produced in the condensation reaction causes the formation of the title compound in a yield of 94%. The product melts at 226 to 229°C.

^{119}Sn Mössbauer spectrum (77 K): $\delta = 1.26$; $\Delta = 3.18$ mm/s. IR spectrum (Nujol): $\nu_{as}(OCO)$ 1512 (vs), $\nu_s(OCO)$ 1403 (s), $\delta(OCO)$ 650 (s) from the otherwise unassigned bands between 1650 and 200 cm^{-1}.

The low value of $\nu_{as}(OCO)$ suggests that association occurs via C=O→Sn coordination, whereas the appearance of an additional band at 1630 (m) cm^{-1} in the $CHCl_3$ spectrum points to a free carbonyl group of monomeric species in solution. Evidence for the association in the solid is also given by the value of the ratio $\Delta/\delta = 2.52$ of the Mössbauer spectrum.

The title compound reacts with $SnCl_2$ in CH_3COCH_3 under reflux for 3 h to give $Sn^{II}OOCC_6H_4O$-2, or with catechol, HOC_6H_4OH-2, in refluxing C_6H_6 within 1 h to yield $(C_4H_9)_2SnOC_6H_4O$-2 [3].

$(C_4H_9)_2SnC_7H_5NO_3$

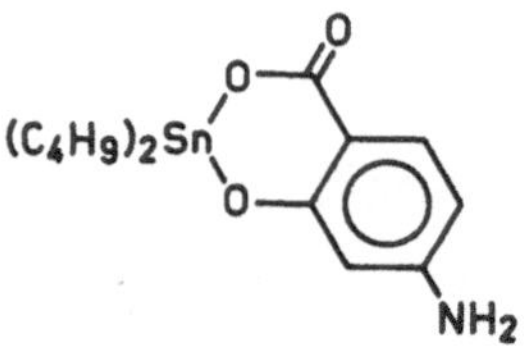

The cyclization of 4-aminosalicylic acid with $(-(C_4H_9)_2SnO-)_n$ affords the title compound in a yield of 93%. It is useful as a heat stabilizer for poly(vinyl chloride) [2].

References:

[1] Weinberg, E. L., Metal & Thermit Corp. (Brit. 753998 [1956]; C.A. **1957** 6683).
[2] Pinkowski, N. J., M and T Chemicals Inc. (Can. 992551 [1976]; C.A. **86** [1977] No. 43819).
[3] Honnick, W. D., Zuckerman, J. J. (Inorg. Chem. **18** [1979] 1437/43).

1.4.1.2.1.5.7 Dibutyltin Carbonic Acid Esters, $(C_4H_9)_2Sn(OOCOR)_2$, and Dibutyltin Acylimines, $(C_4H_9)_2Sn(OCR{=}NR')_2$

$(C_4H_9)_2Sn(OOCOCH_3)_2$

Dibutyltin methyl carbonate, the only representative of this class, seems to be formed by passing dry CO_2 through $(C_4H_9)_2Sn(OCH_3)_2$. Heat is evolved, and a very viscous liquid results which has the composition $(C_4H_9)_2Sn(OCH_3)_2 \cdot 1.84\ CO_2$, thus appearing to be largely the title compound.

^{1}H NMR spectrum: $\delta = 3.61$ (OCH_3) ppm. IR spectrum: ν(C=O) 1325 cm^{-1} [1].

$(C_4H_9)_2Sn(OCR{=}N\text{-}N{=}CHC_5H_4FeC_5H_5)_2$ with R = C_6H_5, 4-ClC_6H_4, 2-$NO_2C_6H_4$, HOC_6H_4, C_5H_4N

The red-violet ferrocenyl derivatives have been prepared by heating 1:2 molar mixtures of $(-(C_4H_9)_2SnO-)_n$ and the appropriate $C_5H_5FeC_5H_4CH{=}N\text{-}NHCOR$ derivative in C_6H_6 under reflux for 8 h.

The melting points, IR and ^{1}H NMR spectroscopic data of the compounds are given below:

R	C_6H_5	4-ClC_6H_4	2-$NO_2C_6H_4$	HOC_6H_4	C_5H_4N
m.p. (in °C)	125 (dec.)	193	208	176	185 (dec.)
IR (in cm^{-1})					
ν(C=N-N=C)	1590 (vs)	1600 (s)	1610 (s)	1610 (s)	1610 (s)
ν(N=CO)	1500 (vs)	1500 (sh)	1530 (s)	1515 (s)	1520 (vs)
ν(CO)	1230 (s)	1235 (s)	1240 (m)	1250 (m)	1240 (s)
ν(SnC)	595 (m)	560 (s)	605 (m)	605 (w)	605 (s)
^{1}H NMR (in ppm)					
solvent	$CDCl_3$	$CDCl_3$	$CDCl_3$	CCl_4	$CDCl_3$
δ (m, C_4H_9)	1.26	1.16	1.5	1.2	1.23
δ (s, C_5H_5)	3.9	3.93	4.04	3.93	3.93
δ (br, C_5H_4)	4.26, 4.83	4.09, 4.46	4.26, 4.60	4.33, 4.56	4.37, 4.86
δ (s, CH=N)	7.01	7.01	7.23	7.16	7.23
δ(OH)				7.86	
δ(C_6H_4, C_6H_5)/H-2 .	7.33	7.23		6.76	7.93
H-3 .	8.16	8.03		7.26	8.66
H-4 .	7.33		8.33		
H-5 .	8.16	8.03		7.26	8.66
H-6 .	7.33	7.23		6.76	7.93

The position of the OH substituent at the phenyl ring is reported to be 2-HOC_6H_4 [2], but should be 4-HOC_6H_4 according to the NMR data. These data also indicate that the C_5H_4N group is bonded with the C-4 atom to the acylimine carbon atom.

The complexes are suggested to have a six-coordinate Sn atom, with *trans*-C_4H_9 groups and with the monobasic bidentate ligands coordinating the Sn atom by the azomethine nitrogen (Formula I) [2].

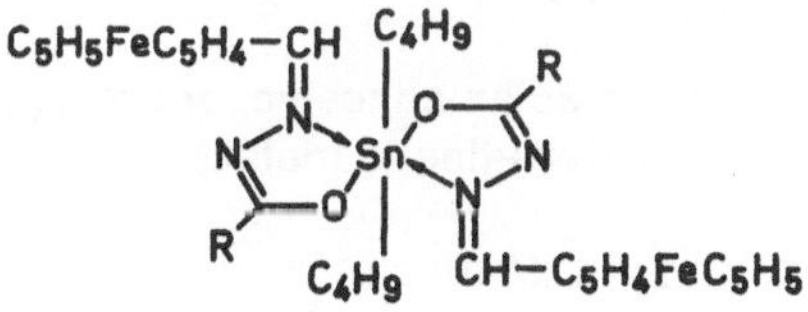

I

References:

[1] Davies, A. G., Harrison, P. G. (J. Chem. Soc. C **1967** 1313/7).
[2] Patil, S. R., Kantak, U. N., Sen, D. N. (Inorg. Chim. Acta **68** [1983] 1/6).

1.4.1.2.1.5.8 Dibutyltin Peroxides, $(C_4H_9)_2Sn(OOR)_2$ or $(C_4H_9)_2\overline{SnOORO}$, and Dibutyltin Peroxycarboxylates, $(C_4H_9)_2Sn(OOCOR)_2$

The compounds belonging to this section are listed in Table 31 and are prepared by the following methods.

Method I: No. 1 has been obtained by the addition of a solution of $(C_4H_9)_2SnCl_2$ in petroleum ether to an aqueous solution of t-C_4H_9OOK (prepared from t-C_4H_9OOH and KOH at < 10°C) at room temperature for 1 h [2].

Method II: Almost quantitative yields of No. 1 are obtained by the reaction of $(\text{-}(C_4H_9)_2SnO\text{-})_n$ with t-C_4H_9OOH in refluxing C_6H_6 for 3 h, with azeotropic removal of H_2O [3].

Method III: Reaction of $(C_4H_9)_2Sn(OCH_3)_2$ with peroxy compounds. $(C_4H_9)_2Sn(OCH_3)_2$ (prepared in situ from $(C_4H_9)_2SnCl_2$ and CH_3ONa in CH_3OH) reacts with the twofold molar amount of t-C_4H_9OOH yielding No. 1 [1]. Nos. 3 to 11 are obtained by the addition of H_2O_2 (98%) to ethereal solutions containing $(C_4H_9)_2Sn(OCH_3)_2$ and the appropriate aldehyde in a 1:1 molar ratio, keeping the mixtures for 12 h at −10°C and then removing the solvent and the CH_3OH formed under reduced pressure. In the case of No. 12 and No. 13, the ketone $CH_3COC_2H_5$ or c-$C_5H_{10}CO$ has been used instead of an aldehyde [7]. The peroxyacid derivative No. 14 is formed immediately and quantitatively after mixing benzene solutions containing $(C_4H_9)_2Sn(OCH_3)_2$ and $C_5H_{11}COOOH$ in a 1:2 mole ratio. Rapid decomposition prevents isolation [8].

Table 31
Dibutyltin Peroxides, $(C_4H_9)_2Sn(OOR)_2$ or $(C_4H_9)_2\overline{SnOORO}$,
and Dibutyltin Peroxycarboxylates, $(C_4H_9)_2Sn(OOCOR)_2$.
Further information on compounds preceded by an asterisk is given at the end of the table.
Explanations, abbreviations, and units on p. X.

No.	peroxy group method of preparation (yield in %)	properties and remarks	Ref.
1	OOC_4H_9-t I [2] II (94 [3]) III [1]	viscous liquid b.p. 100° (bath)/< 0.001, $n_D^{25} = 1.4652$ with H_2O in air → white, infusible, non-peroxidic solid	[2] [1]
2	$OOC(CH_3)_2C_6H_5$ II (90 [3])	used as an adhesive for pressureless bonding of polyolefins to metals	[4, 5, 6]
*3	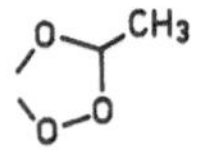 III (129 ?)	m.p. 125° (dec.)	[7]
*4	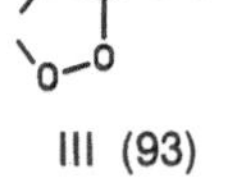 III (93)	m.p. 126° (dec.)	[7]

Table 31 (continued)

No.	peroxy group method of preparation (yield in %)	properties and remarks	Ref.
*5	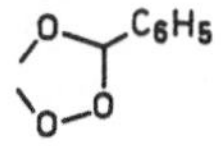 III (79)	m.p. 114° (dec.) thermolysis at 130° for 10 min → O_2, $(-(C_4H_9)_2SnO-)_n$, C_6H_5CHO, and C_6H_5COOH tested as a catalyst for styrene polymerization	[7]
*6	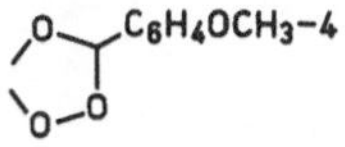 III (69)	m.p. 103° (dec.)	[7]
*7	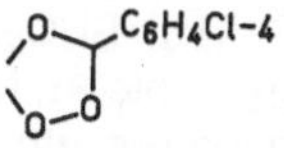 III (79)	m.p. 110 to 113° (dec.)	[7]
*8	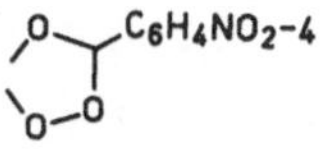 III (83)	m.p. 121 to 123° (dec.)	[7]
*9	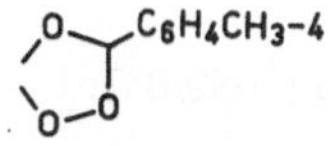 III (86)	m.p. 117° (dec.)	[7]
*10	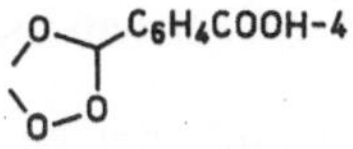 III (37)	m.p. 123 to 125° (dec.)	[7]
*11	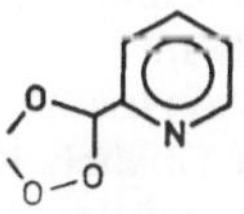 III (81)	m.p. 91° (dec.)	[7]
*12	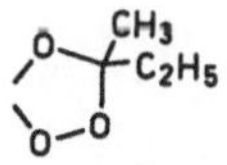 III (93)	m.p. 120° (dec.)	[7]
*13	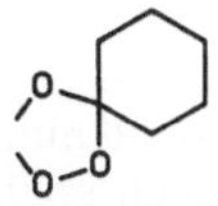 III (89)	m.p. 133° (dec.)	[7]

References on p. 336

Table 31 (continued)

No.	peroxy group method of preparation (yield in %)	properties and remarks	Ref.
14	$OOCOC_5H_{11}$ III	not isolated dec. (in C_6H_6, 20°, 20 min; second-order reaction) → O_2 and $(C_4H_9)_2Sn(OOCC_5H_{11})_2$ (1:1 mole ratio) IR (C_6H_6): ν(OOCO) 1680; spectrum depicted (4000 to 900 range)	[8]

* Further information:

$(C_4H_9)_2\overline{SnOOCR'R''O}$ (R' = H and R" = CH_3, C_2H_5, C_6H_5, 4-$CH_3OC_6H_4$, 4-ClC_6H_4, 4-$NO_2C_6H_4$, 4-$CH_3C_6H_4$, 4-$HOOCC_6H_4$, or 2-C_5H_4N; R' = CH_3, R" = C_2H_5; R = R" = $(CH_2)_5$; Table **31**, Nos. **3** to **13**, respectively). In solution, these peroxides are predominantly cyclic monomers, although appreciable quantities of oligomeric macrocycles or linear polymers may be present. They decompose thermally by first-order kinetics (k = 10^{-2} to 10^{-3} s^{-1} at 50°C), with an activation energy of 22.9 kcal/mol for No. 5 [7].

References:

[1] Alleston, D. L., Davies, A. G. (J. Chem. Soc. **1962** 2465/71).
[2] Mageli, O. L., Harrison, J. B., Wallace and Tiernan Inc. (U.S. 3152156 [1959/64]; C.A. **61** [1964] 16093).
[3] Dahlmann, J., Deutsche Akademie der Wissenschaften zu Berlin (Ger. Offen. 1931 232 [1968/70]; C.A. **72** [1970] No. 111616).
[4] Reicherdt, W., Wunsch, K., Dahlmann, J., Deutsche Akademie der Wissenschaften zu Berlin (Fr. 2036424 [1969/70]; C.A. **75** [1971] No. 99083).
[5] Reicherdt, W., Wunsch, K., Dahlmann, J., Stedtler, L., Deutsche Akademie der Wissenschaften zu Berlin (Brit. 1189629 [1969/70]; C.A. **73** [1970] No. 15710).
[6] Reicherdt, W., Wunsch, K., Stedtler, L., Dahlmann, J., Deutsche Akademie der Wissenschaften zu Berlin (Ger. Offen. 1906419 [1968/70]; C.A. **74** [1971] No. 127790).
[7] Dannley, R. L., Aue, W. A., Shubber, A. K. (J. Organometal. Chem. **38** [1972] 281/6).
[8] Malkov, V. D., Maslennikov, V. P., Vyshinskii, N. N., Aleksandrov, Yu. A. (Zh. Obshch. Khim. **44** [1974] 2708/12; J. Gen. Chem. [USSR] **44** [1974] 2661/5).

1.4.1.2.1.5.9 Dibutyltin Dihalogenates

$(C_4H_9)_2Sn(OClO_3)_2$

Aqueous solutions of $(C_4H_9)_2Sn(OClO_3)_2$ have been prepared by titrating weighed amounts of $(C_4H_9)_2SnCl_2$ with standard $AgClO_4$ solutions. To prevent decomposition, such solutions have to be stored in dark bottles at low temperatures. They have been used for studies on complex formation between $[(C_4H_9)_2Sn]^{2+}$ and F^- ions in a constant ionic medium (1 M $NaClO_4$) at 25°C by potentiometric and solubility methods, leading from the mononuclear complex $[(C_4H_9)_2SnF]^+$, via $(C_4H_9)_2SnF_2$, to the mononuclear complex $[(C_4H_9)_2SnF_3]^-$.

Reference:

Magno, F., Bontempelli, G., Mazzocchin, G. A., Pilloni, G. (J. Organometal. Chem. **67** [1974] 33/42).

1.4.1.2.1.5.10 $(C_4H_9)_2Sn(ONO_2)_2$, $(C_4H_9)_2SnON(C_6H_4CH_3\text{-}4)COCON(C_6H_4CH_3\text{-}4)O$, and Other Compounds Containing the $(C_4H_9)_2Sn(ON)_2$ Unit

$(C_4H_9)_2Sn(ONO_2)_2$

Dibutyltin dinitrate has been prepared by extraction of $AgNO_3$ from a Soxhlet thimble with a solution of $(C_4H_9)_2SnCl_2$ in dry CH_3OH [19] in a yield of 66% [4]. The reaction between equivalent amounts of $(\text{-}(C_4H_9)_2SnO\text{-})_n$ and HNO_3 in C_6H_6 affords $(C_4H_9)_2Sn(OH)NO_3$ which on heating under reduced pressure also yields the title compound [5], which melts at 103.5 to 104.5°C [4].

$(C_4H_9)_2Sn(ONO_2)_2$ forms a 1:1 complex with 1,10-phenanthroline (Formula I) in dry C_2H_5OH [4] as well as with 2-aminobenzthiazole (Formula II) or benzimidazol (Formula III) in C_6H_6 or $CHCl_3$. If the last two reactions are conducted in CH_3COCH_3, 1:1 complexes of HNO_3 and $O_2NO(C_4H_9)_2SnOSn(C_4H_9)_2ONO_2$ are formed, which under otherwise severe anhydrous conditions, is explained by the detection of mesityl oxide, $(CH_3)_2C{=}CHCOCH_3$, the H_2O-elimination product of acetone [19].

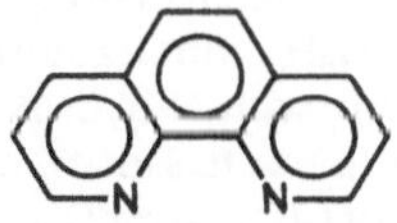

I

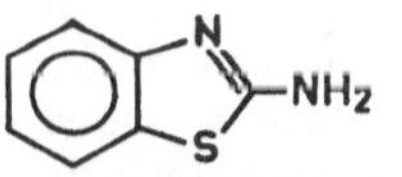

II

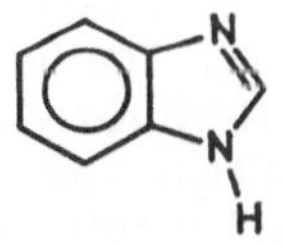

III

The compound has been tested as an anthelmintic against Raillietina cesticillus and Ascaridia galli from chickens [2].

$(C_4H_9)_2SnON(C_6H_4CH_3\text{-}4)COCON(C_6H_4CH_3\text{-}4)O$

The reaction between $HONH(C_6H_4CH_3\text{-}4)COCONH(C_6H_4CH_3\text{-}4)OH$ and $(\text{-}(C_4H_9)_2SnO\text{-})_n$ (1:1 mole ratio) in C_6H_6 affords, after azeotropic removal of water, the title compound in yields of

50 to 60%. It melts at 205°C. Molecular weight estimations indicate the compound to be a tetrameric macrocycle. Comparative studies of the spectroscopic results suggest a twofold bidentate ligand, the C=O → Sn coordination causing the tin to become six-coordinate.

^{1}H NMR spectrum (in $CDCl_3$): δ(ppm) = 0.86 (CH_3/ring), 0.93 to 2.25 (C_4H_9), 6.85 to 7.12 (C_6H_4). IR spectrum: ν(C=O) 1575 cm^{-1} [29].

Other Compounds Containing the $(C_4H_9)_2Sn(ON)_2$ Unit

The compounds are listed in Table 32 and are prepared by the following general methods.

Method I: Reaction of $(C_4H_9)_2SnCl_2$ with hydroxylamines or oximes or their alkali salts, and with $AgON=C(CN)_2$ or $AgON=N(O)C_6H_5$ (1:2 mole ratio).
Nos. 7, 8, 10, and 12 to 16 have been prepared by slow addition of a solution of $N(C_2H_5)_3$ in C_6H_6 to a 1:2 molar mixture of $(C_4H_9)_2SnCl_2$ and the appropriate N-acyl-N-arylhydroxylamine in the same solvent and heating to reflux for 4 h [21, 25]. No. 8 has also been obtained by addition of an aqueous NaOH solution to a solution of $(C_4H_9)_2SnCl_2$ and $C_6H_5CON(C_6H_5)OH$ in dioxane-water (75% v/v) until pH 8.69 is reached [23].
No. 19 precipitates immediately after mixing CH_3OH solutions of $(C_4H_9)_2SnCl_2$ and of the potassium salt of 3,5-diphenyl-4-methyl-1-hydroxy-2-pyrazole N-oxide [16]. No. 22 or No. 23 is formed by heating $(C_4H_9)_2SnCl_2$ with $C_6H_5CH=NONa$ or $(CH_3)_2C=NONa$ (1:3 mole ratio), respectively [6]. The reaction of $(C_4H_9)_2SnCl_2$ with $AgON=C(CN)_2$ in CH_3CN [11], or $AgON=N(O)C_6H_5$ in ether (exothermic) [15] yields No. 30 or 31, respectively.

Method II: Reaction between $(-(C_4H_9)_2SnO-)_n$ and hydroxylamines or oximes (1:2 mole ratio).
Nos. 2 to 5 [18], No. 6 [14], No. 8 [14, 17], Nos. 9 and 13 [14], No. 16 [24], and Nos. 23, 25, 28, and 29 [20] have been prepared by the reaction of $(-(C_4H_9)_2SnO-)_n$ with the appropriate N-acylhydroxylamine or oxime in C_6H_6 with azeotropic removal of water. The heating of mixtures of $(-(C_4H_9)_2SnO-)_n$ with N-hydroxysuccinimide or 1-N-hydroxy-1,2,3-benzotriazole under vacuum leads to No. 17 or 18 [22, 26]. $C_6H_5CH_3$ has been used as a solvent in the reaction of $(-(C_4H_9)_2SnO-)_n$ with $C_6H_5CH=NOH$ [8, 9, 10] or $(CH_3)_2C=NOH$ yielding No. 22 or 23 [8]. No conditions are reported for the synthesis of No. 11 from $(-(C_4H_9)_2SnO-)_n$ and $4-NO_2C_6H_4CON(C_6H_5)OH$ [28].

Method III: Reaction of $(C_4H_9)_2Sn(OR')_2$ (R′ = CH_3, C_2H_5, i-C_3H_7) with $(C_2H_5)_2NOH$ or oximes (1:2 mole ratio).
Azeotropic removal of C_2H_5OH from refluxing solutions of $(C_4H_9)_2Sn(OC_2H_5)_2$ and $(C_2H_5)_2NOH$ or $CH_3CH=NOH$ in C_6H_6 affords No. 1 [12, 13], or No. 20, respectively [20]. Nos. 22, 23, 24, and 27 are obtained by treatment of $(C_4H_9)_2Sn(OC_3H_7\text{-}i)_2$ with the appropriate oxime in C_6H_6 under reflux and removal of the i-C_3H_7OH-C_6H_6 azeotrope [20]. $(C_4H_9)_2Sn(OCH_3)_2$ has been used in the reacton with $C_3H_7CH=NOH$ (no conditions), or with $(CH_3)_2NCH_2CH_2CH_2C(CH_3)=NOH$ (at 110°C), yielding No. 21 [1] or No. 26, respectively [3].

References on p. 345

Table 32
Compounds containing the $(C_4H_9)_2Sn(ON)_2$ Unit.
Further information on compounds preceded by an asterisk is given at the end of the table.
Explanations, abbreviations, and units on p. X.

No.	ON group method of preparation (yield in %)	properties and remarks	Ref.
1	$ON(C_2H_5)_2$ III [12], (40 [13])	yellow liquid, b.p. 103 to 106°/1 molecular weight corresponds to a degree of association of n ≈ 3	[13]
2	$ONHCOCH_3$ II	monomeric in C_6H_6 IR discussion; *cis*-octahedral geometry with C=O→Sn coordination suggested potentiometric estimation of stepwise formation constants	[18]
3	$ONHCOC_3H_7$-i II (77)	m.p. 110°, monomeric in C_6H_6 IR and structure discussion like No. 2 potentiometric estimation of stepwise formation constants	[18]
4	$ONHCOC_6H_5$ II (98)	yellow solid, m.p. 68°, monomeric in C_6H_6 IR and structure discussion like No. 2 potentiometric estimation of stepwise formation constants	[18]
5	$ONHCOC_6H_4Cl$-4 II (92)	colorless solid, m.p. 73°, monomeric in C_6H_6 IR and structure discussion like No. 2	[18]
6	$ON(C_2H_5)COC_6H_4Cl$-4 II	colorless crystals, m.p. 100° intramolecular C=O→Sn coordination suggested	[14]
7	$ON(C_6H_5)COCH_3$ I	m.p. 110 to 112° ^{1}H NMR (sat. sol. in $CDCl_3$): 0.96 (t, H-δ), 1.43 and 1.70 (m's, H-α, -β, -γ), 1.93 (s, CH_3CO), 2.03 (s, ?), 7.4 (s, C_6H_5) ^{119}Sn-γ (77 K): δ = 1.28, Δ = 3.09 IR (KBr): ν(C=O) 1568 (vs), ν(N-O) 966 (m) and 908 (m), ν(SNC) 600 (vs), 590 (vs), and 510 (s), ν(SnO) 483 (s) and 400 (m, br), along with unassigned bands in the 1700 to 300 range UV (CH_3OH): $\lambda_{max}(\log \varepsilon)$ = 253(3.64) spectroscopic results suggest a highly distorted *trans*-octahedral structure with bidentate, C=O→Sn coordinating ligands (angle C-Sn-C = 130° calculated)	[25]

References on p. 345

Table 32 (continued)

No.	ON group method of preparation (yield in %)	properties and remarks	Ref.
*8	$ON(C_6H_5)COC_6H_5$ I [21, 25], ($\sim$80 [23]) II [14], (81 [17])	m.p. 106 to 107° [14], 107 to 108° [25], 108° [21], 120° [23]	[14, 21, 23, 25]
		^{1}H NMR ($CDCl_3$): 0.92 (t, H-δ), 1.55 (m, H-α, -β, -γ) [21, 25], 7.11 [17] or 7.16 (m, C_6H_5) [21, 25]	[17, 21, 25]
		^{119}Sn-γ (77 K): δ = 1.31, Δ = 3.29 [25], or δ = 1.34, Δ = 3.30 [17]	[17, 25]
		IR and R in Table 33 on p. 344 [17]; complete IR spectrum listed [25]	[17, 23, 25]
		UV (CH_3OH): λ_{max}(log ε) = 266(4.09)	[25]
		with $(C_4H_9)_2Sn(SCN)_2$ (C_6H_6, reflux/1 h) → $(C_4H_9)_2Sn(SCN)ON(C_6H_5)COC_6H_5$	[14]
9	$ON(C_6H_5)COC_6H_4Cl$-4 II	m.p. 98 to 100°	[14]
		intramolecular C=O → Sn coordination suggested	
10	$ON(C_6H_5)COC_6H_4I$-2 I [21, 25]	m.p. 116°	[21, 25]
		^{1} NMR ($CDCl_3$): 0.96 (t, H-δ), 1.58 and 1.80 (m's, H-α, -β, -γ), 7.10 (m, C_6H_5, H-3, -4, -5/C_6H_4I-2), 7.65 and 7.75 (s's or d, H-6/C_6H_4I-2)	
		^{119}Sn-γ (77 K): δ = 1.37, Δ = 3.33	[25]
		IR (KBr): ν(C=O) 1540 (vs), ν(N-O) 915 (s) and 905 (s, sh), ν(SnC) 590 (m) and 530 (m), and ν(SnO) 480 (s) and 405 (s), along with unassigned bands in the 1700 to 300 range	
		UV (CH_3OH): λ_{max}(log ε) = 261(4.24)	
		trans-octahedral structure like No. 7 suggested (angle C-Sn-C = 137.2° calculated)	
		cis-octahedral structure deduced	[21]
11	$ON(C_6H_5)COC_6H_4NO_2$-4 II	no physical properties reported	[28]
		with $(C_4H_9)_2SnX_2$ (X = Cl, Br, I, SCN) → $(C_4H_9)_2Sn(X)ON(C_6H_5)COC_6H_4NO_2$-4	
12	$ON(C_6H_5)COC_6H_3(NO_2)_2$-3,5 I	m.p. 60 to 62° (somewhat impure)	[25]
		^{1}H NMR ($CDCl_3$): 0.96 (unresolved d or t, H-δ), 7.40 (s), 8.40 (d), 8.90 (d), and 9.10 (s) (H aryl)	
		^{119}Sn-γ (77 K): δ = 1.37 ± 0.05, Δ = 3.53 ± 0.10	
		IR (KBr): ν(C=O) 1522 (vs), ν(N-O) 961 (m) and 890 (s), ν(SnC) 595 (m) and 490 (w, br)?, or ν(SnO) 490 (w, br)?	
		trans-octahedral structure like No. 7 suggested (angle C-Sn-C = 143.5° calculated)	

References on p. 345

Table 32 (continued)

No.	ON group method of preparation (yield in %)	properties and remarks	Ref.
13	$ON(C_6H_4Cl\text{-}4)COC_6H_5$ I [21, 25] II [14]	m.p. 100° [21, 25], 102 to 103° [14]	[14, 21, 25]
		^1H NMR ($CDCl_3$): 0.92 (t, H-δ), 1.64 (m, H-α, -β, -γ), 7.10 (m, C_6H_4, C_6H_5)	[21, 25]
		^{119}Sn-γ (77 K): δ = 1.34, Δ = 3.23	[25]
		IR (KBr): ν(C=O) 1552 (vs), ν(N-O) 930 (s) and 925 (s), ν(SnC) 595 (m) and 500 (m), ν(SnO) 455 (m) and 405 (w), along with unassigned bands in the 1700 to 300 range UV (CH_3OH): λ_{max}(log ε) = 273(4.29) *trans*-octahedral structure like No. 7 suggested (angle C-Sn-C = 134.2° calculated)	
		cis-octahedral structure deduced	[21]
14	$ON(C_6H_4CH_3\text{-}2)COC_6H_5$ I [21, 25]	m.p. 83° ^1H NMR ($CDCl_3$): 0.92 (t, H-δ), 1.48 and 1.78 (m's, H-α, -β, -γ), 2.26 (s, CH_3/tolyl), 7.10 (m, C_6H_4, C_6H_5)	[21, 25]
		^{119}Sn-γ (77 K): δ = 1.32, Δ = 3.34	[25]
		IR (KBr): ν(C=O) 1537 (vs), ν(N-O) 917 (s) and 907 (vs), ν(SnC) 594 (m) and 489 (m), ν(SnO) 445 (w) and 421 (w), along with unassigned bands in the 1700 to 300 range UV (CH_3OH): λ_{max}(log ε) = 256(4.26) *trans*-octahedral structure like No. 7 suggested (angle C-Sn-C = 137.5° calculated)	
		cis-octahedral structure deduced	[21]
15	$ON(C_6H_4CH_3\text{-}3)COC_6H_5$ I [21, 25]	m.p. 65° ^1H NMR ($CDCl_3$): 0.93 (unresolved d or t, H-δ), 1.64 (m, H-α, -β, -γ), 2.23 (s, CH_3/tolyl), 7.26 (m, C_6H_4, C_6H_5)	[21, 25]
		^{119}Sn-γ (77 K): δ = 1.31, Δ = 3.24	[25]
		IR (KBr): ν(C=O) 1550 (vs), ν(N-O) 942 (vs), 930 (s, sh), and 920 (m), ν(SnC) 595 (m, sh) and 535 (m), ν(SnO) 510 (s) and 435 (w), along with unassigned bands in the 1700 to 300 range UV (CH_3OH): λ_{max}(log ε) = 261(4.38) *trans*-octahedral structure like No. 7 suggested (angle C-Sn-C = 134.5° calculated)	
		cis-octahedral structure deduced	[21]

References on p. 345

Table 32 (continued)

No.	ON group method of preparation (yield in %)	properties and remarks	Ref.
*16	$ON(C_6H_4CH_3\text{-}4)COC_6H_5$ I [21, 25] II [24]	m.p. 98° [21, 25], 110° [24]	[21, 24, 25]
		^{1}H NMR ($CDCl_3$): 0.92 (t, H-δ), 1.60 and 1.80 (m's, H-α, -β, -γ), 2.28 (s, CH_3/tolyl), 7.15 (m, C_6H_4, C_6H_5)	[21, 25]
		^{119}Sn-γ (77 K): δ = 1.31, Δ = 3.24	[25]
		IR (KBr): ν(C=O) 1538 (vs), ν(N-O) 921 (m, sh), 915 (m), and 912 (m, sh), ν(SnC) 605 (m) and 508 (m), ν(SnO) 480 (m) and 400 (m), along with unassigned bands in the 1700 to 300 range	
		trans-octahedral structure like No. 7 suggested (angle C-Sn-C = 134.5° calculated)	
		cis-octahedral structure deduced	[21]
17	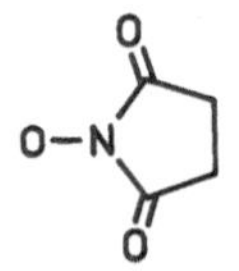 II (99 [22, 26])	m.p. 103°	[22, 26]
		IR (Nujol): ν(CH) 2940, 2900, and 2840, ν(C=O) 1700, 1690, and 1635, along with unassigned bands in the 1600 to 500 range	[22]
		catalyst in polyurethane synthesis	[26]
		with $(\text{-}(C_4H_9)_2SnO\text{-})_n$ → 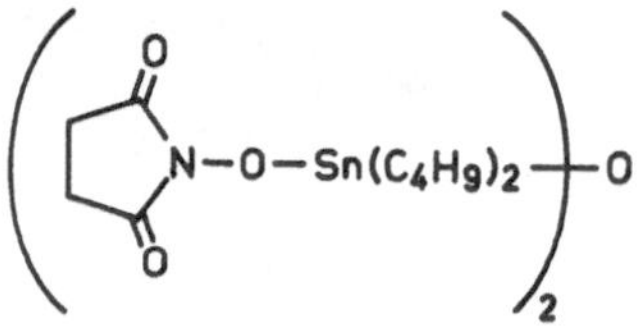	[7]
18	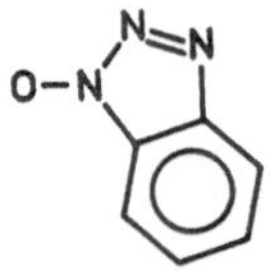 II (96 [22, 26])	m.p. 59 to 60°	[22, 26]
		IR (Nujol): ν(CH) 2940, 2910, and 2850, along with unassigned bands in the 1700 to 500 range	[22]
		catalyst in polyurethane synthesis	[26]
19	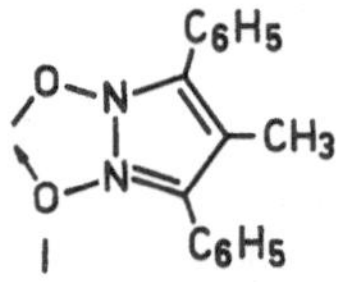 I	m.p. 109 to 112° ^{119}Sn-γ (80 K): δ = 1.06, Δ = 2.49; highly distorted six-coordinate structure	[16]
20	$ON{=}CHCH_3$ III (72)	b.p. 102°/0.2, n_D^{26} = 1.4845	[20]
		^{1}H NMR: 0.70 to 1.70 (m, C_4H_9), 1.78 (d, CH_3, ^3J(H,H) = 6), 6.50 to 7.00 (q, CH=N, anti), 7.10 to 7.60 (q, CH=N, syn)	
		IR and structure discussion	

References on p. 345

Table 32 (continued)

No.	ON group method of preparation (yield in %)	properties and remarks	Ref.
21	$ON{=}CHC_3H_7$ III	stabilizer for PVC	[1]
22	$ON{=}CHC_6H_5$ I [6] II [8, 9, 10] III (68 [20])	b.p. 169°/0.7, n_D^{26} = 1.5250 additive to organopolysiloxane-based compositions suitable for surface treatment, e.g., of paper	[20] [8, 9, 10, 26]
23	$ON{=}C(CH_3)_2$ I [6] II [8], (70 [20]) III (78 [20])	b.p. 100°/0.08 [6], 108°/0.2, n_D^{26} = 1.4738 [20] ^1H NMR: 0.67 to 1.82 (m, C_4H_9), 1.96 (s, CH_3) IR and structure discussion with $(C_4H_9)_2SnCl_2$ (C_6H_6, 110°/4 h) → $(C_4H_9)_2Sn(Cl)ON{=}C(CH_3)_2$ with $CH_3COCH_2COCH_3$ (neat, 80°/6 h) → $(C_4H_9)_2Sn(OC(CH_3){=}CHCOCH_3)ON{=}C(CH_3)_2$ with A=B (neat, room temperature) → $(C_4H_9)_2Sn(A\text{-}B\text{-}ON{=}C(CH_3)_2)_n(ON{=}C(CH_3)_2)_{2-n}$ (n = 1, 2; A=B: CCl_3CHO, C_6H_5NCO, t-$C_4H_9CH_2NCO$,C_6H_5NCS) used as an additive like No. 22	[6, 20] [20] [6, 8]
24	$ON{=}C(\overset{1}{C}H_3)\overset{2}{C}H_2\overset{3}{C}H_3$ III (72)	b.p. 148°/1.5, n_D^{26} = 1.4762 ^1H NMR: 0.30 to 1.73 (m, C_4H_9), 1.20 (t, H-3, $^3J(H,H)$ = 7), 1.80 (s, H-1), 1.95 to 2.50 (m, H-2)	[20]
25	$ON{=}C(CH_3)CH_2Cl$ II (58)	b.p. 110°/0.3	[20]
26	$ON{=}C(CH_3)CH_2CH_2CH_2$-$N(CH_3)_2$ III	n_D^{20} = 1.4902 tested as a fungicide and bactericide	[3]
27	$ON{=}C(CH_3)C_6H_5$ III (70)	b.p. 191°/2.5, n_D^{26} = 1.5220 ^1H NMR: 0.63 to 1.79 (m, C_4H_9), 2.22 (s, CH_3), 7.13 to 7.52 (m, H-3,4,5/C_6H_5), 7.52 to 7.88 (m, H-2,6/C_6H_5)	[20]
28	$ON{=}C(CH_3)C_6H_4Cl$ II (56)	ring position of Cl not reported b.p. 164°/0.3, n_D^{26} = 1.5478	[20]
29	$ON{=}C(CH_3)C_6H_4Br$ II (60)	ring position of Br not reported b.p. 182°/0.5, n_D^{26} = 1.5430	[20]
30	$ON{=}C(CN)_2$ I	light yellow oil IR: ν(C≡N) 2238 (neat), ν(C-C) 1245 (KBr); ν_{as}(CNO) 1435 and ν_s(CNO) 1140 (neat)	[11]
31	$ON{=}N(O)C_6H_5$ I (90)	m.p. 52°	[15]

References on p. 345

* Further information:

$(C_4H_9)_2Sn(ON(C_6H_5)COC_6H_5)_2$ (Table **32**, No. **8**). Some IR data of the compound are discussed in a comparative manner in [23]. The assigned IR and Raman data reported in [17] are listed in Table 33. Furthermore, from the IR frequencies appearing in the region 1602 to 340 cm^{-1} the following have been assigned: ν(C=O) 1544 (vs), ν(N-O) 922 (ms) and 915 (ms), ν(SnC) 592 (m) and 522 (m), ν(SnO) 475 (m) and 407 (m) [25].

From spectroscopic data the compound is suggested to possess a distorted *trans*-octahedral structure with bidentate, C=O→Sn coordinating ligands [17, 23] and a calculated C-Sn-C angle of 136° [25]. In [21] a *cis*-octahedral structure is deduced.

Formation constants and thermodynamic parameters for the formation of the title compound from $(C_4H_9)_2SnCl_2$ and $C_6H_5CON(C_6H_5)OH$ were determined in dioxane-water (75% v/v) by potentiometric titration: ΔH = 18.07 cal/mol and ΔS = 41.80 $cal \cdot mol^{-1} \cdot K^{-1}$ [23, 27].

Table 33
Assigned IR and Raman Frequencies of $(C_4H_9)_2Sn(ON(C_6H_5)COC_6H_5)_2$ [17].
Wave numbers in cm^{-1}.

IR (Nujol)	Raman (compacted powder)	assignment	IR (Nujol)	Raman (compacted powder)	assignment
3070 (ms)	3065 (vw)	ν(C-H)	1004 (w)	1001 (vvs)	ring
1930 (vvw, br)		combination frequency	978 (vvw)		γ(C-H)
			964 (vvw)		
1601 (m)	1602 (vs)	ν(C-C)	937 (s)		ν(N-O)
1586 (vvs)		ν(C-C)	928 (vs)		
1564 (vvs)		ν(C=O)	922 (ms, sh)		
1497 (s, sh)	1497 (mw)	ν(C-C)	882 (vw)		γ(C-H)
1459 (s)	1459 (s)	ν(C-C)	839 (vw)		
1422 (s)	1423 (m)	δ_{as}(C-H)	774 (s)		X-sensitive
1317 (vw)		ν(C-C)	723 (s)	725 (vw)	γ(C-H)
1308 (vw)		β(C-H)	619 (w)	615 (mw)	α(C-C-C)
1296 (vw)		δ_s(C-H)	610 (mw)		ν_{as}(Sn-C)
1291 (vw)	1285 (w)			592 (m)	ν_s(Sn-C)
1278 (vw)	1276 (mw)	X-sensitive	550 (w)		ν_{as}(Sn-O)
1172 (w, sh)		β(C-H)	498 (s)	493 (w)	
1156 (m, sh)	1157 (mw)			404 (vvw)	Φ(C-C)
1078 (m)					
1017 (vs)	1018 (vvw)				

$(C_4H_9)_2Sn(ON(C_6H_4CH_3\text{-}4)COC_6H_5)_2$ (Table **32**, No. **16**). Stepwise formation constants and thermodynamic parameters have been estimated for the formation of the compound from $(C_4H_9)_2SnCl_2$ and $C_6H_5CON(C_6H_4CH_3\text{-}4)OH$ in dioxane-water (75% v/v): ΔH = 2.97 cal/mol and ΔS = 44.62 $cal \cdot mol^{-1} \cdot K^{-1}$ [23, 27].

References:

[1] Mack, G. P., Parker, E., Advance Solvents and Chemical Corp. (U.S. 2727917 [1955]; C.A. **1956** 10761), Carlisle Chemical Works, Inc. (Ger. 953079 [1956]; C.A. **1959** 5197; Brit. 766875 [1957]; C.A. **1957** 8788).
[2] Kerr, K. B., Walde, A. W. (Experim. Parasitol. **5** [1956] 560/70).
[3] Farbenfabriken Bayer A.-G. (Brit. 945068 [1960/63]; C.A. **60** [1964] 12051).
[4] Gormley, J. J., Rees, R. G. (J. Organometal. Chem. **5** [1966] 291/2).
[5] Yasuda, K., Matsumoto, H., Okawara, R. (J. Organometal. Chem. **6** [1966] 528/34).
[6] Pande, K. C., Stauffer Chemical Co. (Fr. 1506186 [1965/67]; C.A. **69** [1968] No. 106878).
[7] Wagner, D., Gertner, D., Zilkha, A. (Can. J. Chem. **46** [1968] 3612/4).
[8] Gibbon, R. M., Imperial Chemical Industries Ltd. (Ger. Offen. 1942670 [1968/70]; C.A. **73** [1970] No. 132083).
[9] Gibbon, R. M., Pierpoint, E. K., Imperial Chemical Industries Ltd. (U.S. 3527728 [1968/70]).
[10] Gibbon, R. M., Pierpoint, E. K., Imperial Chemical Industries Ltd. (Brit. Amended 1186571 [1968/71]; C.A. **76** [1972] No. 35410).

[11] Köhler, H., Lange, U., Eichler, B. (J. Organometal. Chem. **35** [1972] C17/C19).
[12] Singh, A., Gupta, V. D., Srivastava, G., Mehrotra, R. C. (J. Organometal. Chem. **64** [1974] 145/69).
[13] Sharma, C. K., Gupta, V. D., Mehrotra, R. C. (Indian J. Chem. A **14** [1976] 64).
[14] Pradhan, B., Ghosh, A. K. (J. Organometal. Chem. **131** [1977] 23/30).
[15] Yandovskii, V. N., Traore, U., Zavgorodnii, V. S. (Zh. Obshch. Khim. **48** [1978] 708; J. Gen. Chem. [USSR] **48** [1978] 651/2).
[16] Petridis, D., Lockwood, T., O'Rourke, M., Naik, D. V., Mullins, F. P., Curran, C. (Inorg. Chim. Acta **33** [1979] 107/11).
[17] Harrison, P. G., Richards, J. A. (J. Organometal. Chem. **185** [1980] 9/51).
[18] Narula, C. K., Gupta, V. D. (Indian J. Chem. A **19** [1980] 491/3).
[19] Pelizzi, C., Pelizzi, G., Tarasconi, P. (Inorg. Chim. Acta **40** [1980] 183/6).
[20] Rupani, P., Singh, A., Rai, A. K., Mehrotra, R. C. (J. Organometal. Chem. **185** [1980] 209/17).

[21] Das, M. K., Nath, M. (Indian J. Chem. A **20** [1981] 1224/7).
[22] Gordetsov, A. S., Pereshein, V. V., Skobeleva, S. E., Pavlova, L. A., Tyutina, T. P., Karlik, V. M., Dergunov, Yu. I. (Zh. Obshch. Khim. **52** [1982] 2762/7; J. Gen. Chem. [USSR] **52** [1982] 2435/9).
[23] Singh, G., Narula, C. K., Gupta, V. D. (Indian J. Chem. A **21** [1982] 738/40).
[24] Singh, G., Gupta, V. D. (Natl. Acad. Sci. Letters [India] **5** [1982] 423/6).
[25] Das, M. K., Nath, M., Zuckerman, J. J. (Inorg. Chim. Acta **71** [1983] 49/59).
[26] Gordetsov, A. S., Noskov, N. M., Tasalova, M. E., Pavlova, L. A., Karlik, V. M., Kuzina, V. I., Dergunov, Yu. I. (Zh. Prikl. Khim. **56** [1983] 2635/8; J. Appl. Chem. [USSR] **56** [1983] 2453/60; C.A. **100** [1984] No. 210021).
[27] Singh, G., Singh, B., Gupta, V. D. (J. Indian Chem. Soc. **60** [1983] 987/9).
[28] Chaudhuri, S. K., Roy, P. S., Gosh, A. K. (Indian J. Chem. A **23** [1984] 533/4).
[29] Chaudhuri, S. K., Roy, P. S., Gosh, A. K. (Indian J. Chem. A **23** [1984] 917/9).

1.4.1.2.1.5.11 $(C_4H_9)_2SnOSO_2$, $(C_4H_9)_2SnOSO_3$, $(C_4H_9)_2\overline{SnO_3S(CH_2)_6SO_2O}$, and Dibutyltin Compounds of the $(C_4H_9)_2Sn(OSO_n(X,Y))_2$ Type with n = 1 or 2

$(C_4H_9)_2SnOSO_2$

Dibutyltin sulfite is formed in a yield of 15% in the reaction between $Sn(C_4H_9)_4$ and H_2O containing SO_2 at 60°C for 1d in a sealed tube [14, 16]. The change in the shape of the Mössbauer spectrum of $(C_4H_9)_2SnOSO_3$ after γ irradiation with a dose of 3×10^{22} eV/cm^3 corresponds to the formation of $SnSO_4$ and probably of the title compound [4]. The thermally stable, quite insoluble, polymeric compound melts at 204 to 205°C [14, 16].

^{119}Sn Mössbauer spectrum (78 K): $\delta = 1.3\pm0.2$, $\Delta = 4.0\pm0.2$ mm/s [3, 8, 10, 11].

IR spectrum (in KBr): $\nu(SO_3)$ 975 (sh) [14], 898 (s, br), 872 (vs, br) [14, 16], and 855 (sh), $\varrho(CH_2)$/Sn 691 (m) [14], $\delta_{as}(SO_3)$ 651 (m) and 638 (m), ν(SnC) 605 (w) [14, 16] and 554 (w)? [16], $\delta_s(SO_3)$ 445 (mw) cm^{-1} [14, 16]. The spectroscopic data indicate five-coordinate Sn in polymeric chain or network structures [16, 22].

$(C_4H_9)_2SnOSO_3$

Dibutyltin sulfate is obtained by the reaction of $Sn(C_4H_9)_4$ with H_2O containing SO_2 in a sealed tube at 90°C for 2d (40% yield) [16], or by the reaction of dry SO_2 and $(C_4H_9)_3SnSSn(C_4H_9)_3$ in a sealed tube at 60°C for 1 d (30% yield) [24]. Addition of 35% aqueous H_2O_2 to a solution of $Cl(C_4H_9)_2SnSSn(C_4H_9)_2Cl$ in C_6H_{14} leads to the title compound in a 65% yield [13]. It should be noted here that cryoscopic and conductivity measurements on solutions of $(C_4H_9)_3SnCl$ in $H_2S_2O_7$ [20] or of $(C_4H_9)_2Sn(OOCCH_3)_2$ in 100% H_2SO_4 [9] show that both compounds undergo solvolysis to yield $(C_4H_9)_2Sn(OSO_3H)_2$ which in turn is protonated to give $[(C_4H_9)_2Sn(OSO_3H)_2H]^+$ [20]. $(C_4H_9)_2SnOSO_3$ melts at temperatures above 300°C with some decomposition [16].

^{119}Sn Mössbauer spectrum: $\delta = -0.3$ (α-Sn) at 78 [10] or 80 K [8], -0.58 (α-Sn) at 78 K [10], 1.52 [12] or 1.8 mm/s [3] (SnO_2) at 78 K. $\Delta = 4.7$ [8], 4.75 [10, 12], or 4.8 mm/s [3, 10]. The LCAO approach has been used for the interpretation of the Mössbauer spectra of organotin compounds among them the title compound, and correlations between δ or Δ values and the substituent electronegativity have been found [11]. Studies on the decomposition of $(C_4H_9)_2SnOSO_3$ by γ irradiation during Mössbauer measurements are described [18] and the formation of $(C_4H_9)_2SnOSO_2$ has been established [4]. The field and concentration dependence of the ^{119}Sn Mössbauer spectra of solutions of $(C_4H_9)_3SnSn(C_4H_9)_3$ in 100% H_2SO_4 were studied [31].

IR spectrum: $\nu_{as}(OSO_3)$ 1106 (m) and 1072 (m)?, $\delta_{as}(OSO_3)$ 661 (w), and ν_s(SnC) 597 (s) cm^{-1} [16]. Unassigned bands in the 3000 to 400 cm^{-1} [13] or 1500 to 500 cm^{-1} [24] range were listed.

A polymeric chain structure with *trans*-C_4H_9 groups and four O atoms surrounding the Sn atom is consistent with the spectra [16].

$(C_4H_9)_2\overline{SnO_3S(CH_2)_6SO_2O}$

The compound, the preparation of which is not described, has been tested as a thermooxidative stabilizer for ester type synthetic lubricants [32].

References on p. 349

$(C_4H_9)_2(OS(O)_n(X,Y))_2$ Compounds (n = 1 or 2)

The compounds of this type are listed in Table 34 and the methods used for their preparation are presented below. The general formula in the heading stands for sulfinates, -OS(O)R (Nos. 1 and 2), sulfonamides, -OS(O)NR_2 (NO. 3), halogenosulfonates, -OSO_2X (No. 4), sulfamic acid derivatives, -OSO_2NR_2 (No. 5), and sulfonates, -OSO_2R (Nos. 6 to 14).

Method I: No. 4 or 9 has been obtained by the instantaneous elimination of HCl from $(C_4H_9)_2SnCl_2$ and a small excess over the twofold molar amount of $HOSO_2F$ or $HOSO_2CF_3$ in $CFCl_3$ at room temperature [17].

Method II: Azeotropic dehydration of 1:2 molar mixtures of $(-(C_4H_9)_2SnO-)_n$ and $HOSO_2NH_2$ (in C_6H_6), freshly prepared $HOSO_2CH{=}CH_2$ (in C_6H_{14}), or $HOSO_2C_6H_5$ (in C_6H_6) yields No. 5 [15], 8 [25], or 11 [23], respectively. The reaction between $(-(C_4H_9)_2SnO-)_n$ and the monohydrate of $HOSO_2C_6H_4CH_3$-4 yielding No. 12 has been conducted in refluxing CH_3OH until all of the oxide was dissolved [2]. Equimolar mixtures of $(-(C_4H_9)_2SnO-)_n$ and $C_4H_9SO_2Cl$ react in refluxing C_6H_6 in the presence of a small amount of H_2O (it initiates the hydrolysis of the acid chloride to give $HOSO_2C_4H_9$ and HCl) during 18 h with the formation of $(C_4H_9)_2SnCl_2$ and No. 7 [7]. Concentrating a solution of $(C_4H_9)_3SnOSn(C_4H_9)_3$ and $HOSO_2CH_3$ in 95% C_2H_5OH until only a little solvent is left, keeping this concentrate hot for 20 min and then cooling it, leads to No. 6 [6].

Method III: The insertion of SO_2 into Sn-C bonds of $Sn(C_4H_9)_4$ (H_2O containing SO_2, sealed tube, 60°C, 1 d) or of $(C_4H_9)_2Sn(SC_4H_3\text{-}2)_2$ (liquid SO_2, −20°C, dry conditions) affords No. 1 [16] or No. 2 [21], respectively. The reaction between equimolar amounts of $(C_4H_9)_2Sn(C_6H_5)_2$ or $(C_4H_9)_2Sn(C_6H_4CH_3\text{-}4)_2$ and $HOSO_2Cl$ in CCl_4 yielding $(C_4H_9)_2SnCl_2$, C_6H_6, or $C_6H_5CH_3$, and No. 11 or No. 12, respectively, can be considered formally as a SO_3 insertion [26].

Method IV: Heating 1:2 molar mixtures of $Sn(C_4H_9)_4$ and $HOSO_2CH_3$ without a solvent to 120°C leads to No. 6 by fission of two Sn-C bonds and evolution of C_4H_{10} [5].

Method V: Ligand exchange between 1:2 molar mixtures of $(-(C_4H_9)_2SnO-)_n$ and $(C_2H_5)_3SiO$-$SO_2CH{=}CH_2$, $(C_4H_9)_3SiOSO_2CF{=}CF_2$, $(CH_3)_3SiOSO_2C_6H_4CF{=}CFCl$-4, or $(CH_3)_3SiO$-$SO_2C_6H_4CF{=}CF_2$-4 in dry C_6H_{14} at 20 to 30°C for up to 12 h affords Nos. 8, 10 [25], 13, or 14 [19], respectively, always along with the appropriate hexaalkyldisiloxane.

Table 34
Dibutyltin Compounds of the $(C_4H_9)_2Sn(OSO_n(X,Y))_2$ Type with n = 1 or 2.
Explanations, abbreviations, and units on p. X.

No.	OSO_n(X,R) group method of preparation (yield in %)	properties and remarks	Ref.
1	OS(O)C_4H_9 III (3)	m.p. 190° (dec.) IR (solid/KBr): $\nu_{as,s}(SO_2)$ 945 (s, br), ν(SnC) 600 (mw) R (solid): ν(SnC) 592 (m) and 507 (m) polymeric chain structure with *trans*-C_4H_9 and bidentate $C_4H_9SO_2$ groups suggested	[16]

References on p. 349

Table 34 (continued)

No.	$OSO_n(X,R)$ group method of preparation (yield in %)	properties and remarks	Ref.
2	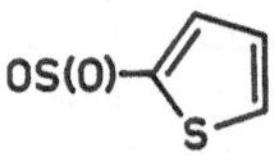 III	m.p. 148°	[21]
3	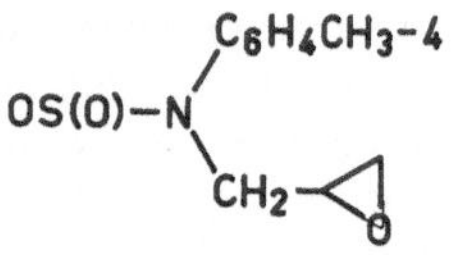	no preparation reported used as a stabilizer for resin compositions	 [1]
4	OSO_2F I	m.p. 198° (dec.) ^{119}Sn-γ (80 K): δ = 1.97, Δ = 5.42; structure with six-coordinate Sn, *trans*-C_4H_9 groups, and bidentate bridging anionic ligands coordinating through oxygen suggested	[17]
5	OSO_2NH_2 II	tested as a fungicide for wood and plastics and as a pharmaceutical and agricultural fungicide	[15]
6	OSO_2CH_3 II (91 [6]) IV (96 [5])	m.p. 306 to 309° [5], 312° [6] IR (KBr): $\nu_{as}(OSO_2)$ 1190, $\nu_s(OSO_2)$ 1059 and 1050 catalyst for urethane polymerization insecticide and fungicide stabilizer for plastics	[5, 6] [6] [5, 27] [5]
7	$OSO_2C_4H_9$ II (59)	white powder, m.p. 273 to 275° (dec.) IR (Nujol): $\nu(OSO_2)$ 1900, 1750, and 1415 crosslinking agent for chlorosulfonated polyethylene	[7]
8	$OSO_2CH{=}CH_2$ II (98) V	insoluble in organic solvents, e.g., petroleum ether, C_6H_{14}, $CHCl_3$, or $C_6H_5CH_3$	[25]
9	OSO_2CF_3 I	m.p. 250° (dec.) ^{119}Sn-γ (80 K): δ = 1.96, Δ = 5.40; suggested structure like No. 4	[17]
10	$OSO_2CF{=}CF_2$ V (65)	soluble in DMF and dioxane, insoluble in aliphatic and aromatic hydrocarbons	[25]

Table 34 (continued)

No.	$OSO_n(X,R)$ group method of preparation (yield in %)	properties and remarks	Ref.
11	$OSO_2C_6H_5$ II [23] III [26]	white amorphous powder; sparingly soluble in common organic solvents	[23]
		m.p. 318° [26], 320° [23]	[23, 26]
		^{119}Sn-γ (? K): δ = 1.76, Δ = 4.91	[23]
		IR: $\nu(OSO_2)$ 1250 (s), 1140 (vs), and 1035 (s), $\delta(OSO_2)$ 552 (s), 540 (s), and 519 (ms)	
		polymeric sheet structure suggested, with six-coordinate Sn and the sulfonate groups bridging linear $(C_4H_9)_2Sn$ moieties (cf. dimethyltin analogue)	[23]
		catalyst for polyurethane synthesis	[28, 30]
12	$OSO_2C_6H_4CH_3$-4 II [2] III [26]	m.p. >320° [2], >330° [26]	[2, 26]
		IR: $\nu(OSO_2)$ 1440 (w), 1250 (s), 1148 (vs), and 1040 (s), $\delta(OSO_2)$ 554 (s), 538 (s), and 521 (ms)	[26]
		catalyst for polyamine synthesis from polyisocyanates	[29]
13	$OSO_2C_6H_4CF{=}CFCl$-4 V (92)	m.p. >300°	[19]
14	$OSO_2C_6H_4CF{=}CF_2$-4 V (83)	m.p. >300°	[19]

References:

[1] Mack, G. P., M and T Chemicals, Inc. (U.S. 3147285 [1956/64]; C.A. **62** [1965] 11973).

[2] Seyferth, D., Stone, F. G. A. (J. Am. Chem. Soc. **79** [1957] 515/7).

[3] Aleksandrov, A. Yu., Delyagin, N. N., Mitrifanov, K. P., Polak, L. S., Shpinel, V. S. (Zh. Eksperim. Teor. Fiz. **43** [1962] 1242/7; Soviet Phys.-JETP **16** [1963] 879/82).

[4] Aleksandrov, A. Yu., Delyagin, N. N., Mitrifanov, K. P., Polak, L. S., Shpinel, V. S. (Zh. Eksperim. Teor. Fiz. **43** [1962] 2074/6; Soviet Phys.-JETP **16** [1963] 1467/8).

[5] Stamm, W. A., Breindel, A. W., Freiberg, A. H., Stauffer Chemical Co. (U.S. 3095434 [1961/63]; C.A. **59** [1963] 14023).

[6] Anderson, H. H. (Inorg. Chem. **3** [1964] 108/9).

[7] Nersasian, A., King, K. F., Johnson, P. R. (J. Appl. Polym. Sci. **8** [1964] 337/54).

[8] Cordey-Hayes, M. (Tech. Rept. Ser. Intern. At. Energy Agency No. **50** [1966] 156/63).

[9] Gillespie, R. J., Kapoor, R., Robinson, E. A. (Can. J. Chem. **44** [1966] 1197/202).

[10] Gol'danskii, V. I., Khrapov, V. V., Okhlobystin, O. Yu., Rochev, V. Ya. (in: Gol'danskii, V. I., Herber, R. H., Chemical Applications of Mössbauer Spectroscopy, New York **1968**, pp. 336/76).

[11] Kothekar, V., Shpinel, V. S. (Zh. Strukt. Khim. **10** [1969] 37/42; C.A. **70** [1969] No. 110326).

[12] Smith, P. J. (Organometal. Chem. Rev. A **5** [1970] 373/402).

[13] Stapfer, C. H., Dworkin, R. D. (Inorg. Chem. **9** [1970] 421/3).
[14] Lindner, E., Kunze, U. (Z. Naturforsch. **26b** [1971] 164/5).
[15] Nakanishi, M., Tsuda, A., Yoshitomi Pharmaceutical Industries, Ltd. (Japan. 71-30180 [1961/71]; C.A. **75** [1971] No. 151908).
[16] Kunze, U., Lindner, E., Koola, J. (J. Organometal. Chem. **38** [1972] 51/68).
[17] Tan, T. H., Dalziel, J. R., Yeats, P. A., Sam, J. R., Thompson, R. C., Aubke, F. (Can. J. Chem. **50** [1972] 1843/51).
[18] Llabador, Y., Friedt, J. M. (J. Inorg. Nucl. Chem. **35** [1973] 2351/9).
[19] Nikitina, A. A., Panov, E. M., Rybakova, L. F., Karandi, I. V., Kocheshkov, K. A. (Zh. Obshch. Khim. **43** [1973] 1319/21; J. Gen. Chem. [USSR] **43** [1973] 1311/2).
[20] Paul, R. C., Puri, J. K., Malhotra, K. C. (J. Inorg. Nucl. Chem. **35** [1973] 403/12).

[21] Gopinathan, S., Gopinathan, C., Gupta, J. (Indian J. Chem. **12** [1974] 623/5).
[22] Kunze, U., Völker, H. P. (Chem. Ber. **107** [1974] 3818/34).
[23] Harrison, P. G., Phillips, R. C., Richards, J. A. (J. Organometal. Chem. **114** [1976] 47/52).
[24] Kunze, U., Hengel, R. (Chem. Ber. **109** [1976] 2793/804).
[25] Rybakova, L. F., Panov, E. M., Nikitina, A. A., Fainshtein, I. Z., Khodkevich, O. M., Kocheshkov, K. A. (Khim. Elementoorg. Soedin. [Moscow] **1976** 68/71; C.A. **85** [1976] No. 177572).
[26] Bhattacharya, S. N., Raj, P., Husain, I. (Indian. J. Chem. A **16** [1978] 1108/10).
[27] Coe, C. G., Air Products and Chemicals, Inc. (U.S. 4286073 [1980/81]; C.A. **95** [1981] No. 170106).
[28] Robins, J., Edwards, B. H., Tokach, S. K. (Polym. Mater. Sci. Eng. **49** [1983] 331/5; C.A. **101** [1984] No. 6275).
[29] Rasshofer, W., Bayer A.-G. (Ger. Offen. 3244913 [1982/84]; C.A. **101** [1984] No. 111910).
[30] Robins, J., Edwards, B. H., Tokach, S. K. (Advan. Urethane Sci. Technol. **9** [1984] 65/76; C.A. **102** [1985] No. 7130).

[31] Birchall, T., Manivannan, V. (Can. J. Chem. **63** [1985] 2211/6).
[32] Hronec, M., Cvengrosova, Z., Kizlink, J., Stolcova, M., Malik, L., Ilavsky, J., Sitek, J. (Oxid. Commun. **8** [1985/86] 51/64).

1.4.1.2.1.5.12 Dibutyltin Borate and Dibutyltin Compounds Containing the $(C_4H_9)_2Sn(OB)_2$ Unit

Dibutyltin Borate

The reaction between $(-(C_4H_9)_2SnO-)_n$ and boric acid in refluxing $C_6H_5CH_3$ for 4 h yields a white powder [1]. Its constitution, as well as that of the so-called dibutyltin diborate [3] was not stated. Both products are used as stabilizers for halogen-containing plastics, as accelerators for polymerization reactions, as antioxidants, and as additives for lubricants [1, 3].

Dibutyltin Compounds Containing the $(C_4H_9)_2Sn(OB)_2$ Unit

The compounds in this section, listed in Table 35, have been prepared by azeotropic dehydration reactions in $C_6H_5CH_3$. Thus the reaction between $(-(C_4H_9)_2SnO-)_n$, $B(OH)_3$, and $C_4H_9CH(C_2H_5)CH_2OH$ (1:2:4 mole ratio) leads to the isolation of No. 2 [1]. Nos. 3 to 6 are obtained by condensation of $(-(C_4H_9)_2SnO-)_n$, $C_6H_5B(OH)_2$, and 4-RC_6H_4N=CHC_6H_4OH-2, or Nos. 7 to 10 by condensation of $(-(C_4H_9)_2SnO-)_n$, $C_6H_5B(OH)_2$, and 1,10-(4-RC_6H_4N=CH)$C_{10}H_6OH$ (1:2:2 mole ratio, 5 to 6 h, in each case), respectively [4].

Table 35
Dibutyltin Compounds Containing the $(C_4H_9)_2Sn(OB)_2$ Unit.
Explanations, abbreviations, and units on p. X.

No.	OB group	properties and remarks	Ref.
1	$OB(OH)OC_{12}H_{25}$	no preparation reported tested as a heat and color stabilizer for polypropylene	[2]
2	$OB(OCH_2CH(C_2H_5)C_4H_9)_2$	clear colorless liquid used as a stabilizer for halogen-containing resins, polymerization accelerator, antioxidant, and as an additive to lubricants	[1]
3	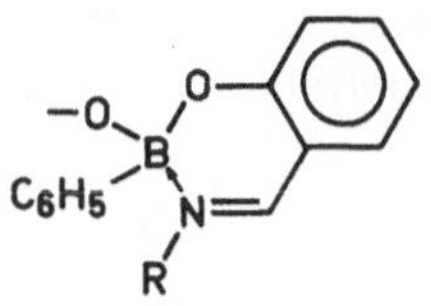 R = C_6H_4F-4	light yellow, m.p. 110° ^1H NMR ($CDCl_3$): 0.80 and 1.80 (m's C_4H_9), 6.80 to 7.20 (C_6H_4), 8.80 (CH=N) ^{11}B NMR (CD_3SOCD_3): 2.2 IR and structure discussion	[4]
4	like No. 3 R = C_6H_4Cl-4	light yellow, m.p. 80° ^1H NMR ($CDCl_3$): 0.80 and 1.80 (m's, C_4H_9), 6.85 to 7.30 (C_6H_4), 8.85 (CH=N)	[4]
5	like No. 3 R = C_6H_4Br-4	light yellow, m.p. 130°	[4]
6	like No. 3 R = $C_6H_4NO_2$-4	light brown, m.p. 180°	[4]
7	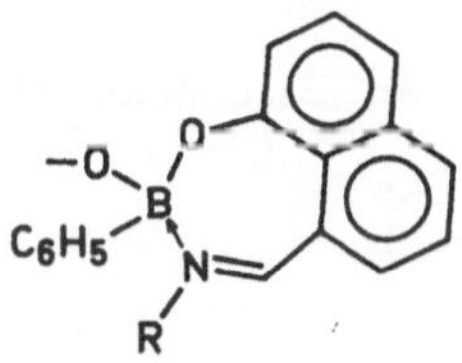 R = C_6H_4F-4	dark yellow, m.p. 246°	[4]
8	like No. 7 R = C_6H_4Cl-4	dark yellow, m.p. 135°	[4]
9	like No. 7 R = C_6H_4Br-4	light yellow, m.p. 210°	[4]
10	like No. 7 R = $C_6H_4NO_2$-4	red, m.p. 280°	[4]

References on p. 352

References:

[1] Metal and Thermit Corp. (Brit. 772646 [1957]; C.A. **1957** 15551; Ger. 1007328 [1957]; C.A. **1959** 14004; U.S. 2867641 [1959]; C.A. **1959** 6082).
[2] Kresta, J., Jadrnicek, B. (Chem. Prumysl **20** [1970] 222/3; C.A. **73** [1970] No. 99299).
[3] Shah, N. R., Shook, E. G., Van Cleve, R., Union Carbide Corp. (U.S. 4242476 [1978/80]; C.A. **94** [1981] No. 104364).
[4] Chaturvedi, V., Bhal, L., Tandon, J. P. (Indian J. Chem. A **24** [1985] 1039/41).

1.4.1.2.1.5.13 Dibutyltin Compounds with Sn-O-Si Bonds

$(C_4H_9)_2Sn(OSi(CH_3)_3)_2$

The compound has been prepared by the reaction of $(C_4H_9)_2Sn(OOCCH_3)_2$ with $(CH_3)_3SiN(CH_3)_2$ in a 1:2 mole ratio at 80°C for 2 h. It is used as a catalyst in the preparation of polyurethane foams and elastomers [5].

$(C_4H_9)_2Sn(OSi(C_6H_5)_3)_2$

The compound has been prepared by the 1:2 molar reactions of $(C_4H_9)_2SnCl_2$ and $(C_6H_5)_3SiONa$ (in C_6H_6, room temperature, 15 min, 70% yield) [1], $(\text{-}(C_4H_9)_2SnO\text{-})_n$ and $(C_6H_5)_3SiOH$ (azeotropic dehydration in refluxing C_6H_6) [2, 4], or $(C_4H_9)_2Sn(OOCCH_3)_2$ and $(C_6H_5)_3SiOC_2H_5$ (168°C, 23 h, 63% yield) [1]. It melts at 69 to 70°C [1], 70°C [4], or 71°C [2].

^{119}Sn NMR: $\delta = -36 \pm 2$ (saturated solution in CCl_4 or C_6H_6), or -45 ± 2 ppm (neat liquid at the melting point) [4]. ^{119}Sn-γ (78 K): $\delta = 1.19$, $\Delta = 2.40$ mm/s [3]. IR spectrum: $\nu_s(SnOSi)$ 961 cm^{-1} [2].

$(C_4H_9)_2Sn(OSi(OC_2H_5)_3)_2$

The synthesis of the title compound was not reported, but it is used as a component in the preparation of moisture-curable siloxane coating materials [6].

$(C_4H_9)_2\overline{Sn(\text{-}OSi(C_6H_5)_2\text{-})_2O}$ (Formula I)

The compound, which also may be polymeric, is obtained by the azeotropic dehydration of 1:2 molar mixtures of $(\text{-}(C_4H_9)_2SnO\text{-})_n$ and $(C_6H_5)_2Si(OH)_2$ in C_6H_6 or $C_6H_5CH_3$. It melts at 190 to 200°C and shows a $\nu_s(SnOSi)$ at 996 cm^{-1} [2].

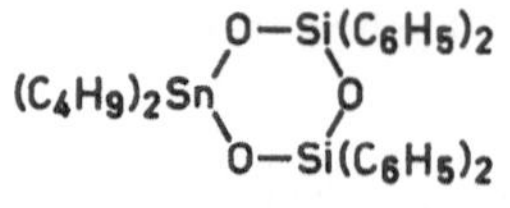

I

References:

[1] Thies, C., Kinsinger, J. B. (Inorg. Chem. **3** [1964] 551/4).
[2] Davies, A. G., Harrison, P. G., Silk, T. A. G. (Chem. Ind. [London] **1968** 949/50).
[3] Smith, P. J. (Organometal. Chem. Rev. A **5** [1970] 373/402).
[4] Smith, P. J., White, R. F. M., Smith, L. (J. Organometal. Chem. **40** [1972] 341/53).
[5] Baskent, F. O., Reedy, J. D., Union Carbide Corp. (Eur. Appl. 84183 [1981/83]; C.A. **100** [1984] No. 35616).
[6] Toshiba Silicone Co., Ltd. (Japan. Kokai Tokkyo Koho 58 57460 [1981/83]; C.A. **100** [1984] No. 8650).

1.4.1.2.1.5.14 Dibutyltin Compounds with Sn-O-P Bonds

1.4.1.2.1.5.14.1 Dibutyltin Compounds Containing the $(C_4H_9)_2Sn(OP)_2$ Unit

The compounds belonging to this section are listed in Table 36. They were prepared by the following methods.

Method I: Reaction of $(C_4H_9)_2SnCl_2$ with free acids or their sodium salts (1:2 mole ratio). The slow addition of an aqueous solution of $C_6H_5P(O)(OH)_2$ to a methanolic solution of $(C_4H_9)_2SnCl_2$ (vigorous stirring) yields No. 12 after 5 h at room temperature [19] and the addition of a small excess of $F_2P(O)OH$ to $(C_4H_9)_2SnCl_2$ under reduced pressure and without a solvent leads to No. 19 with instantaneous evolution of HCl [12].
The reaction of $(C_4H_9)_2SnCl_2$ with the sodium salt of the appropriate acid yields No. 2 (ether-H_2O, room temperature) [13]; No. 6 (CH_3COCH_3, reflux/3 h [4, 5] or C_6H_6, reflux/2 h [16]); No. 7 (conditions not clear) [7]; No. 13 (i-C_3H_7OH-CH_3OH, 45 to 50°C/1 h) [3]; or No. 18 (C_2H_5OH, reflux/1 h) [21].

Method II: Reaction of $(-(C_4H_9)_2SnO-)_n$ with free acids (1:2 mole ratio). Heating mixtures of $(-(C_4H_9)_2SnO-)_n$ and the appropriate acid in refluxing C_6H_6 or $C_6H_5CH_3$ and removal of the solvent-water azeotrope affords No. 10 [1], No. 16 [22], or No. 17 [18].

Method III: Reaction of $(C_4H_9)_2SnH_2$ with free acids (1:2 mole ratio). Nos. 3, 4, or 5 have been prepared by mixing C_6H_{12} solutions containing stoichiometric amounts of $(C_4H_9)_2SnH_2$ and $(C_5H_{11})_2P(O)OH$, $(C_6H_{13})_2P(O)OH$, or $C_6H_5CH_2(C_6H_{13})P(O)OH$ and refluxing the mixture ca. 12 h under a stream of high purity nitrogen. The reactions have the advantage to produce only H_2 as the byproduct and therefore give virtually pure products in quantitative yields [6].

Method IV: No. 1 has been obtained from $(C_4H_9)_2Sn(N(C_2H_5)_2)_2$ and $(C_6H_5)_2P(O)H$ without a solvent at room temperature for 2 h. $NH(C_2H_5)_2$ formed during the reaction is removed under vacuum [8].

Method V: No. 12 and No. 14 have been obtained by transesterification of $(C_4H_9)_2Sn(OOCCH_3)_2$ with $C_6H_5P(O)(OH)_2$ (in CH_3OH-H_2O, room temperature/5 h) [19], or with $C_6H_5(C_6H_5O)P(O)OH$ (in CH_3OH) [17]. Treatment of $(C_4H_9)_2Sn(OC_4H_9)_2$ with $(C_2H_5O)_2P(O)OH$ in refluxing $C_6H_5CH_3$ yields No. 15 along with C_4H_9OH [9, 14, 15].

Table 36
Dibutyltin Compounds Containing the $(C_4H_9)_2Sn(OP)_2$ Unit.
Further information on compounds preceded by an asterisk is given at the end of the table.
Explanations, abbreviations, and units on p. X.

No.	OP group method of preparation (yield in %)	properties and remarks	Ref.
1	$OP(C_6H_5)_2$ IV (48)	m.p. 109 to 113° with C_6H_5NCS (ether, 20°/5 h) → $(C_4H_9)_2Sn(N(C_6H_5)C(S)\text{-}P(O)(C_6H_5)_2)_2$	[8]
*2	$OP(O)H_2$ I	m.p. 183° ^{119}Sn-γ (80 K): $\delta = 1.42$, $\Delta = 4.47$ IR (Nujol): $\nu_s(PH_2)$ 2380 (sh), $\nu_{as}(PH_2)$ 2370 (s), $\delta(PH_2)$ 1160 (s), $\nu_{as}(PO_2)$ 1145 (s), $\delta(PH_2)$ 1082 (s), $\nu_s(PO_2)$ 1058 (s), $\varrho(PH_2)$ 812 (s), $\varrho(SnC)$ 690 (m), $\nu_{as}(SnC)$ 645 (m), $\delta(PO_2)$ 480 (m)	[13]
*3	$OP(O)(C_5H_{11})_2$ III (100)	m.p. 310° IR (Nujol): $\nu_{as}(PO_2)$ 1125 (vs), $\nu_s(PO_2)$ 1042 (s), $\nu(PC)$750 (mw), $\nu_{as}(SnC)$ 594 (w), $\nu_s(SnC)$ 518 (vw, sh), $\delta(PO_2)$ 483 (w)	[6]
*4	$OP(O)(C_6H_{13})_2$ III (100)	m.p. 250° ^{31}P NMR (saturated solution in C_6H_6, 27°): −52.0 and −41.8 ^{119}Sn-γ (78 K): $\delta = -0.15$ (Pd_3Sn), $\Delta = 4.00$ IR (Nujol): $\nu_{as}(PO_2)$ 1140 (vs), $\nu_s(PO_2)$ 1071 (vs), $\nu(PC)$ 742 (vw, sh), $\nu_{as}(SnC)$ ca. 590 (vw), $\nu_s(SnC)$ 513 (vw, sh), $\delta(PO_2)$ 483 (vw)	[6]
*5	$OP(O)(C_6H_{13})CH_2C_6H_5$ III (100)	m.p. 230° IR (Nujol): $\nu_{as}(PO_2)$ 1130 (vs), $\nu_s(PO_2)$ 1042 (vs), $\nu(PC)$ 750 (vw), $\nu_{as}(SnC)$ 582 (m), $\nu_s(SnC)$ 513 (w, sh), $\delta(PO_2)$ 490 (s)	[6]
*6	$OP(O)(C_6H_5)_2$ I [4, 5, 16] special [2]	dec. 230° with NaOH (C_2H_5OH, air, reflux/2 h) → $(\text{-}(C_4H_9)_2SnO\text{-})_n,(C_6H_5)_2P(O)OH$ stabilizer for polyolefins against heat and UV light	[2] [4, 5]

References on p. 357

Table 36 (continued)

No.	OP group method of preparation (yield in %)	properties and remarks	Ref.
7	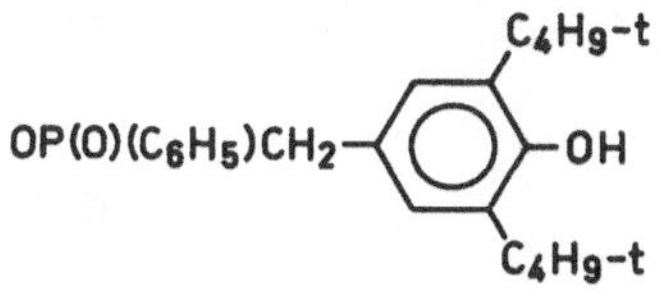 I	m.p. 275 to 283° stabilizer for polymers, oil, and paraffin waxes improves the polymer dyeability	[7]
8	$OP(O)(H)OC_4H_9$	no preparation reported stabilizer for polyolefins against heat and sunlight	[1]
9	$OP(OC_2H_5)_2$	no preparation reported catalyst for the preparation of colorless linear polyesters	[10]
10	$OP(OC_4H_9)_2$ II	stabilizer for polypropylene against heat and sunlight	[1]
11	$OP(OC_6H_5)_2$	no preparation reported catalyst for polyester preparation	[11]
12	$OP(O)(OH)C_6H_5$ I (50) V	^{119}Sn-γ (78 K): $\delta = 1.24$, $\Delta = 3.79$; chain structure with six-coordinate Sn and chain linking by strong hydrogen bonds suggested	[19]
13	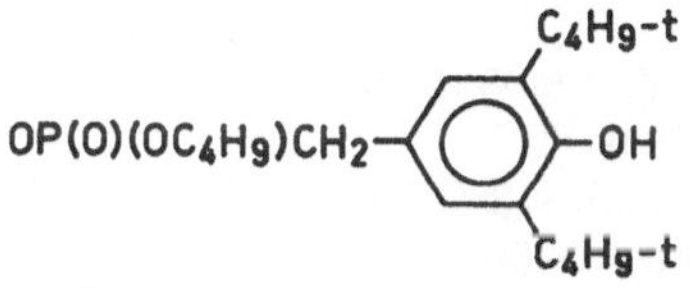 I	m.p. 201 to 207° stabilizer for polyolefins, lubricating oils, fatty acids, varnishes, and soaps	[3]
14	$OP(O)(OC_6H_5)C_6H_5$ V (100)	m.p. >230° ^{119}Sn-γ (77 K): $\delta = 1.47$, $\Delta = 4.61$ IR (Nujol): $\nu_{as}(OPO_2)$ 1070 (vs), $\nu_s(OPO_2)$ 1004 (m, sh); other bands in the 1600 to 500 range unassigned MS (70 eV): $[M - OC_6H_5]^+$ (4.6), $[M - OP(O)(OC_6H_5)C_6H_5]^+$ (21.0), $[SnOP(O)(OC_6H_5)C_6H_5]^+$ (73.1), $[SnOC_6H_5]^+$ (27.5), $[Sn]^+$ (3.3) polymeric *trans*-octahedral structure with intermolecular P=O → Sn coordination suggested	[17]

References on p. 357

Table 36 (continued)

No.	OP group method of preparation (yield in %)	properties and remarks	Ref.
15	$OP(O)(OC_2H_5)_2$ V	curing catalyst for polysiloxanes	[9, 14, 15]
16	$OP(O)(OC_4H_9)_2$ II (70 [22])	m.p. 269 to 272° ^1H NMR spectrum depicted ^{119}Sn NMR ($CHCl_3$): −469 pyrolysis at 170 to 190° studies of the catalytic effect in the polymerization of epoxides	[22] [20, 22]
17	$OP(O)(OC_6H_5)_2$ II (99)	m.p. >250° ^{119}Sn-γ (77 K): δ = 1.54, Δ = 4.58 IR (Nujol): $\nu_{as}(PO_4)$ 1115 (vs) MS (70 eV): $[M]^+$ not detectable, $[M - OP(O)(OC_6H_5)_2]^+$ is the highest mass fragment	[18]
*18	$OP(S)(OCH_3)_2$ I (55)	m.p. 178 to 188° ^{31}P NMR (C_6D_6): −20.37, −16.08, +1.57, and +3.45 ^{119}Sn-γ (77 K): δ = 1.41, Δ = 4.49 IR (Nujol): ν(PO) 1184 (s), 1160 (s), and 1080 (s), ν(PS) 690 (w), $\nu(POCH_3)$ 990 (w) and 970 (vw); other bands in the 3000 to 250 range unassigned	[21]
19	$OP(O)F_2$ I	m.p. 235° (dec.) ^{119}Sn-γ (80 K): δ = 1.67, Δ = 5.03 (5.10 from point charge calculations) IR (Nujol): $\nu_{as}(PO_2)$ 1260 (s), $\nu_s(PO_2)$ 1157 (s), $\nu_{as}(PF_2)$ 920 (s), $\nu_s(PF_2)$ 884 (s), $\nu_{as}(SnC)$ 644 (w), $\delta(PO_2)$ 533 (m), ϱ(POF) 501 (w) and 484 (s), together with unassigned bands in the 1425 to 480 range six-coordinate Sn with *trans*-octahedral C_4H_9 groups and bidentate ligands bridging through oxygen	[12]

* Further information:

$(C_4H_9)_2Sn(OP(O)H_2)_2$ (Table **36**, No. **2**). In accordance with the spectroscopic results No. 2 is suggested to have the sheet-like polymeric structure I (R = C_4H_9) with six-coordination for Sn, *trans*-C_4H_9 groups, and bidentate, bridging PO_2H_2 ligands [13].

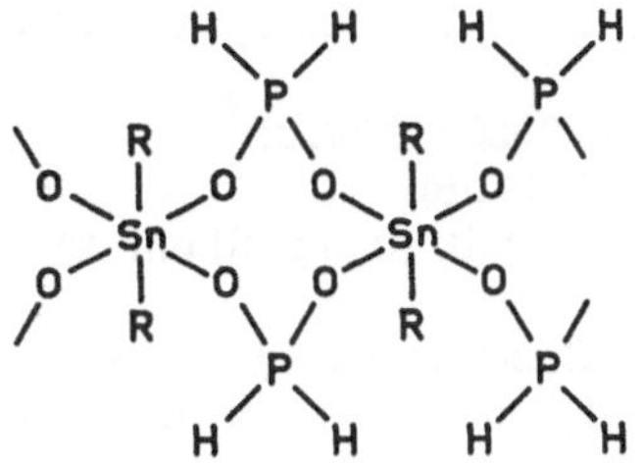

I

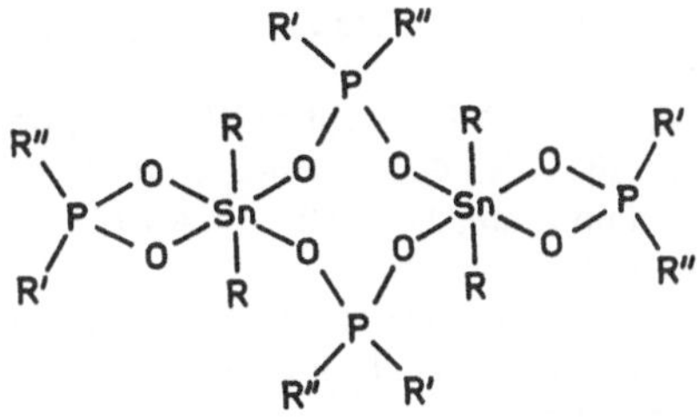

II

$(C_4H_9)_2Sn(OP(O)R'R'')_2$ (R′ = R″ = C_5H_{11}, C_6H_{13}; R′ = C_6H_{13}, R″ = $CH_2C_6H_5$, Table **36**, Nos. **3**, **4**, and **5**). Molecular weight data indicate a dimeric structure II (R = C_4H_9) consistent with the ^{119}Sn Mössbauer and ^{31}P NMR spectra [6].

$(C_4H_9)_2Sn(OP(O)(C_6H_5)_2)_2$ (Table **36**, No. **6**) is also formed in a quantitative yield by oxidation of $(C_4H_9)_2Sn(P(C_6H_5)_2)_2$ with 30% H_2O_2 in C_2H_5OH [2].

$(C_4H_9)_2(OP(S)(OCH_3)_2)_2$ (Table **36**, No. **18**) shows the following fragments in the 70 eV mass spectrum: $[(C_4H_9)_2SnOP(S)OCH_3][SP(O)H_5]^+$ (15.3), $[(C_4H_9)_2SnOP(S)OCH_3][OPH_5]^+$ (17.9), $[(C_4H_9)_2SnP(S)OCH_3\text{-}4H]^+$ (16.3), $[C_4H_9SnOP(S)(OCH_3)PH_3]^+$ (22.8), $[(C_4H_9)_2SnOP]^+$ (26.2), $[SnOP(S)(OCH_3)_2]^+$ (58.0), $[SnP(S)(OCH_3)_2]^+$ (22.8), $[SnOP(OCH_3)_2]^+$ (100), $[SnOP(S)OCH_3\text{-}H]^+$ (19.7), $[SnOP]^+$ (60.1), $[CH_3SnO]^+$ (26.7), and $[Sn]^{\bullet+}$ (32.7) [21].

References:

[1] Katsamura, T., Kawakami, Y. (Japan. 63-22483 [1961/63]; C.A. **60** [1964] 8199).
[2] Schumann, H., Köpf, H., Schmidt, M. (J. Organometal. Chem. **2** [1964] 159/65).
[3] Spivack, J. D., Geigy Chemical Corp. (U.S. 3310575 [1965/67]; C.A. **67** [1967] No. 22011).
[4] Walsh, E. N., Kopacki, A. F., Stauffer Chemical Co. (U.S. 3296193 [1962/67]; C.A. **66** [1967] No. 55981).
[5] Walsh, E. N., Kopacki, A. F., Stauffer Chemical Co. (U.S. 3358006 [1962/67]; C.A. **68** [1968] No. 30705).
[6] Ridenour, R. E., Flagg, E. E. (J. Organometal. Chem. **16** [1969] 393/404).
[7] Spivack, J. D., Geigy, J. R., A.-G. (Fr. 1558606 [1967/69]; C.A. **71** [1969] No. 81532).
[8] Issleib, K., Walther, B. (J. Organometal. Chem. **22** [1970] 375/86).
[9] Lengnick, G. F., Stauffer Chemical Co. (U.S. 3525778 [1968/70]; C.A. **73** [1970] No. 99839).
[10] Toda, T., Yoda, K., Kimoto, K., Toyo Spinning Co., Ltd. (Japan. 70-21118 [1965/70]; C.A. **73** [1970] No. 88374).

[11] Toda, S., Imai, M., Kuzumoto, E., Mizuguchi, K., Toyo Spinning Co., Ltd. (Japan. 71-25018 [1965/71]; C.A. **76** [1972] No. 4357).
[12] Tan, T. H., Dalziel, J. R., Yeats, P. A., Sams, J. R., Thompson R. C., Aubke, F. (Can. J. Chem. **50** [1972] 1843/51).
[13] Chivers, T., van Roode, J. H. G., Ruddick, J. N. R., Sams, J. R. (Can. J. Chem. **51** [1973] 3702/11).

[14] Lengnick, G. F., Stauffer-Wacker Silicone Corp. (Brit. 1326075 [1970/73]; C.A. **80** [1974] No. 38098).
[15] Lengnick, G. F., Stauffer-Wacker Silicone Corp. (Ger. 2028320 [1970/73]).
[16] Pandit, S. K., Gopinathan, C. (Indian J. Chem. A**15** [1977] 463/5).
[17] Cunningham, D., Kelly, L. A., Molloy, K. C., Zuckerman, J. J. (Inorg. Chem. **21** [1982] 1416/21).
[18] Molloy, K. C., Nasser, F. A. K., Zuckerman, J. J. (Inorg. Chem. **21** [1982] 1711/4).
[19] Cunningham, D., Firtear, P., Molloy, K. C., Zuckerman, J. J. (J. Chem. Soc. Dalton Trans. **1983** 1523/7).
[20] Nakata, T. (Polym. Sci. Technol. [Plenum] **19** [1983] 55/74).

[21] Nasser, F. A. K., Zuckerman, J. J. (J. Organometal. Chem. **244** [1983] 17/33).
[22] Otera, J., Yano, T., Kunimoto, E., Nakata, T. (Organometallics **3** [1984] 426/31).

1.4.1.2.1.5.14.2 Dibutyltin Compounds of the $(C_4H_9)_2\overline{SnOP(O,S)(R,OR)O}$ Type

The compounds of this type are summarized in Table 37. They have been prepared by the methods described below.

Method I: $(C_4H_9)_2SnCl_2$ and free acids or their alkali salts (1:1 mole ratio).
The addition of a solution of $(C_4H_9)_2SnCl_2$ in CH_3OH to a solution of 3,5-(t-$C_4H_9)_2$-4-$HOC_6H_2CH_2P(O)(OH)_2$ in i-C_3H_7OH and keeping the mixture at 45 to 50°C for 1 h with a slight vacuum to remove HCl and CH_3OH leads to the formation of No. 5 [2]. Addition of $(C_4H_9)_2SnCl_2$ in ether to an aqueous solution of $Na_2HPO_3 \cdot 5H_2O$ gives No. 1 which precipitates after standing at room temperature [5].

Method II: $(C_4H_9)_2SnH_2$ and free acids (1:1 mole ratio).
The reaction between $(C_4H_9)_2SnH_2$ and $RP(O)(OH)_2$ (R = C_6H_{13}, C_8H_{17}, or $CH_2C_6H_5$) in refluxing C_6H_{12} yields Nos. 2, 3, or 4, respectively, with liberation of H_2 [3].

Table 37
Dibutyltin Compounds of the $(C_4H_9)_2\overline{SnOP(O,S)(R,OR)O}$ Type.
Further information on compounds preceded by an asterisk is given at the end of the table.
Explanations, abbreviations, and units on p. X.

No.	OP(O,S)(R,OR)O group method of preparation (yield in %)	properties and remarks	Ref.
1	OP(O)(H)O I	m.p. 285° ^{119}Sn-γ (80 K): δ = 1.41, Δ = 3.87 IR (Nujol): ν(PH) 2360 (m), 2340 (w, sh), 1150 (m, sh), $\nu_{as}(PO_3)$ 1070 (s, br), $\nu_s(PO_3)$ 1005 (s), 950 (w, sh), ϱ(SnC) 672 (m), $\delta_s(PO_3)$ + ν_{as}(SnC) 588 (m) and 580 (m, br), $\delta_{as}(PO_3)$ + ν_s(SnC) 540 (w, sh) and 515 (w, br), 420 (w) polymeric structure with a non-linear C-Sn-C arrangement suggested	[5]

Table 37 (continued)

No.	OP(O,S)(R,OR)O group method of preparation (yield in %)	properties and remarks	Ref.
*2	$OP(O)(C_6H_{13})O$ II (100)	m.p. 312°; molecularity n = 37 (osmometric in THF), or n = 12 (ebullioscopic in C_6H_6) IR (Nujol): $\nu_{as}(PO_3)$ 1075 (vs), $\nu_s(PO_3)$ 990 (s), $\nu(PC)$ 750 (vw), 695 (vw), $\nu_{as}(SnC)$ 559 (w), $\nu_s(SnC)$ 509 (mw), $\delta_{as}(PO_3)$ ~ 490 (vw, sh)	[3]
*3	$OP(O)(C_8H_{17})O$ II [4], (100 [3])	m.p. 290°; molecularity n = 8 [3], 37 [4] (osmometric in $C_6H_5CH_3$), or n = 5 (ebullioscopic in C_6H_6) [3]	[3, 4]
		^{31}P NMR (THF, 27°): one singlet with J(Sn,P) = 144/172 ^{119}Sn-γ (78 K): δ = −0.37 (Pd_3Sn), Δ = 3.12 IR (Nujol): $\nu_{as}(PO_3)$ 1075 (vs, sh), $\nu_s(PO_3)$ ~995 (sh), $\nu(PC)$ 750 (vw), 702 (vw, sh), $\nu_{as}(SnC)$ 583 (m), $\nu_s(SnC)$ 523 (w)	[3]
		useful as a lubricant, stabilizer, fire retardant, and biocide	[4]
*4	$OP(O)(CH_2C_6H_5)O$ II (100)	m.p. 290°; molecularity n = 20.8 (osmometric in $C_6H_5CH_3$), or n = 10 (ebullioscopic in C_6H_6) IR (Nujol): $\nu_{as}(PO_3)$ 1075 (s, sh), $\nu_s(PO_3)$ 995 (s, sh), $\nu(PC)$ 749 (vw), $\nu_{as}(SnC)$ 582 (vw, sh), $\nu_s(SnC)$ 528 (ms), $\delta_{as}(PO_3)$ ~490 (w, sh)	[3]
5	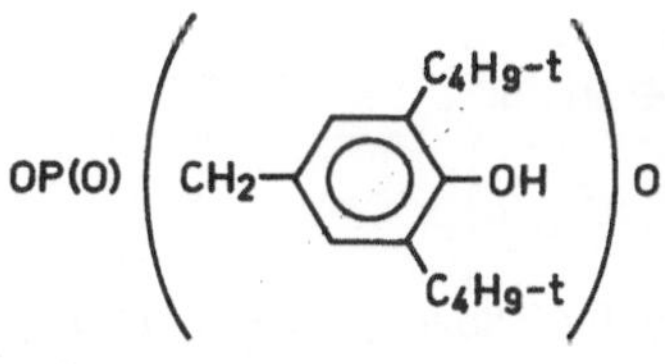 I	m.p. >300°; monomeric utilized as a stabilizer for polymers, oils, lubricants, fatty acids, varnishes, and soaps	[2]
6	OP(O)(OH)O	no preparation reported	
		tested as an anthelmintic against Raillietina cesticillus and Ascaridia galli from chicken	[1]
*7	$OP(S)(OC_6H_2Cl_3$-2,4,5)O special	revealed by chromatography	[6]

References on p. 360

* Further information:

$(C_4H_9)_2SnOP(O)(R)O$ (R = C_6H_{13}, C_8H_{17}, and $CH_2C_6H_5$, Table **37**, Nos. **2**, **3**, and **4**). A structure with trigonal bipyramidal Sn atoms, approximately planar $O \rightarrow SnC_2$ moieties, and $P{=}O \rightarrow Sn$ bridges (Formula I) is proposed [3].

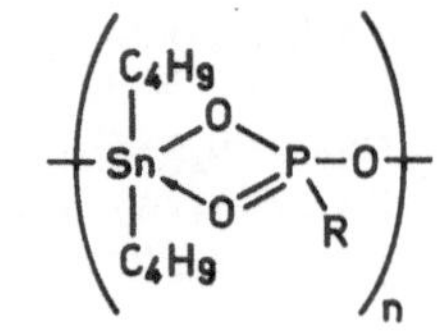

I

$(C_4H_9)_2SnOP(S)(OC_6H_2Cl_3\text{-}2,4,5)O$ (Table **37**, No. **7**) is formed by slow intramolecular rearrangement of $(C_4H_9)_3SnOP(S)(OC_6H_2Cl_3\text{-}2,4,5)OSn(C_4H_9)_3$ along with $Sn(C_4H_9)_4$ [6].

References:

[1] Kerr, K. B., Walde, A. W. (Experim. Parasitol. **5** [1956] 560/70).
[2] Spivack, J. D., Geigy Chemical Corp. (U.S. 3310575 [1965/67]; C.A. **67** [1967] No. 22011).
[3] Ridenour, R. E., Flagg, E. E. (J. Organometal. Chem. **16** [1969] 393/404).
[4] Ridenour, R. E., Flagg, E. E., Dow Chemical Co. (U.S. 3634479 [1968/72]; C.A. **76** [1972] No. 127770).
[5] Chivers, T., van Roode, J. H. G., Ruddick, J. N. R., Sams, J. R. (Can. J. Chem. **51** [1973] 3702/11).
[6] Lecat, J. L., Devaud, M. (Polyhedron **2** [1983] 1087/90).

1.4.1.2.1.5.14.3 Other Dibutyltin Compounds with Sn-O-P Bonds

$(C_4H_9)_2\overline{SnOP(S)(OC_2H_5)OC_6H_3(NHP(S)(OC_2H_5)_2\text{-}3)COO}\text{-}6$

The title compound is formed along with $Sn(C_4H_9)_4$ by the following slow reaction [1]:

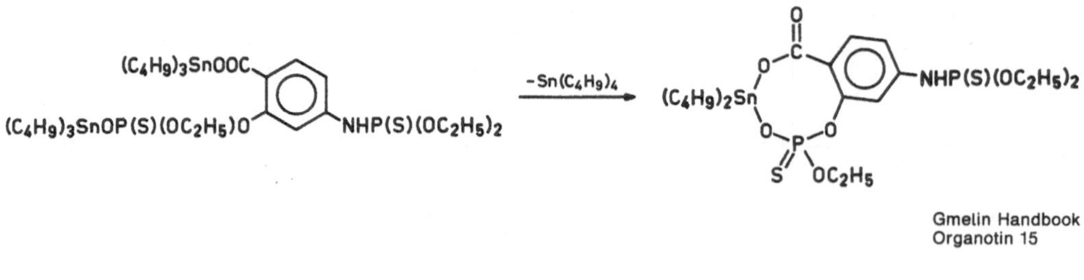

$(C_4H_9)_2\overline{SnOP(O)(OC_4H_9)OP(O)(OC_4H_9)O}$

Dibutyltin P,P′-dibutyl pyrophosphate is prepared in a good yield by the addition of $(C_4H_9)_2SnCl_2$ in CH_2Cl_2 to an aqueous solution of K_2CO_3 and $HO(C_4H_9O)(O)P\text{-}O\text{-}P(O)(OC_4H_9)OH$ and stirring the mixture at room temperature for 3 h.

^{119}Sn NMR spectrum (in $CHCl_3$): $\delta = -270$ ppm.

According to a ^{119}Sn NMR investigation, the pyrolysis at 170 to 190°C leads to a derivative of tetraphosphoric acid, with elimination of $C_4H_9OC_4H_9$:

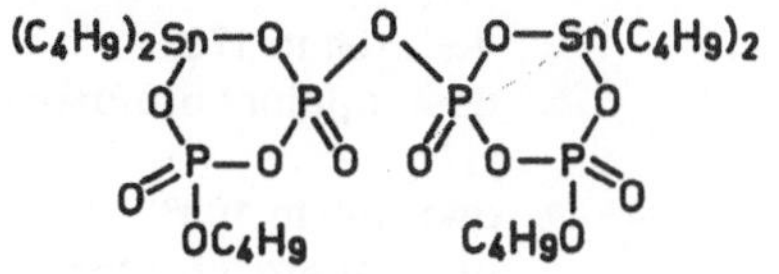

The catalytic activity of the title compound in the polymerization of epoxides has been investigated [2].

References:

[1] Lecat, J. L., Devaud, M. (Polyhedron **2** [1983] 1087/90).
[2] Otera, J., Yano, T., Kunimoto, E., Nakata, T. (Organometallics **3** [1984] 426/31).

1.4.1.2.1.5.15 Dibutyltin Compounds with Sn-O-As Bonds

1.4.1.2.1.5.15.1 Dibutyltin Compounds Containing the $(C_4H_9)_2Sn(OAs)_2$ Unit

The compounds belonging to this section are listed in Table 38. They have been prepared by the reaction between $(C_4H_9)_2Sn(OOCCH_3)_2$ and the appropriate arylarsonic acid (1:2 mole ratio) in CH_3OH at room temperature for No. 1 [3], or without a solvent at 100°C/1 h for Nos. 2, 3, and 4 [2].

Table 38
Dibutyltin Compounds Containing the $(C_4H_9)_2Sn(OAs)_2$ Unit.
Explanations, abbreviations, and units on p. X.

No.	OAs group (yield in %)	properties and remarks	Ref.
1	$OAs(O)(OH)C_6H_5$ (60)	m.p. >250°, insoluble in H_2O and organic solvents ^{119}Sn-γ (78 K): $\delta = 1.23$, $\Delta = 4.11$ polymeric chain structure suggested, with Sn in a *trans*-octahedral environment, intermolecular bridging ligands, and strong hydrogen bonds linking the chains	[3]

Table 38 (continued)

No.	OAs group (yield in %)	properties and remarks	Ref.
2	$OAs(O)(OH)C_6H_4NO_2$-4	dec. 165 to 172° used against protozoal infections	[2]
		tested as an anthelmintic against Raillietina cesticillus and Ascaridia galli from chicken	[1]
3	$OAs(O)(OH)C_6H_3(NO_2$-3)OH-4	dec. 164 to 170° used against protozoal infections	[2]
4	$OAs(O)(OH)C_6H_3(NO_2$-3)NH_2-4	dec. 189 to 193° used against protozoal infections	[2]

References:

[1] Kerr, K. B., Walde, A. W. (Experim. Parasitol. **5** [1956] 560/70).
[2] Walde, A. W., van Essen, H. E., Zbornik, T. W., Dr. Salsbury's Laboratories (U.S. 2762821 [1956]; C.A. **1957** 4424).
[3] Cunningham, D., Firtear, P., Molloy, K. C., Zuckerman, J. J. (J. Chem. Soc. Dalton Trans. **1983** 1523/7).

1.4.1.2.1.5.15.2 Dibutyltin Compounds of the $(C_4H_9)_2\overline{SnOAs(O)(R)O}$ Type

The compounds belonging to this section are listed in Table 39. They have been prepared by the following general methods.

Method I: The slow addition of $(C_4H_9)_2SnCl_2$ in CH_3OH to a vigorously stirred aqueous solution of $C_6H_5As(O)(OH)ONa$ or $C_6H_5As(O)(ONa)_2$ (1:1 mole ratio) yields, after further stirring for 3 h, No. 3 in its α modification or as a monohydrate, respectively [7].

Method II: $(-(C_4H_9)_2SnO-)_n$ and free acids, $As(O)(R)(OH)_2$ (1:1 mole ratio). Nos. 1 and 2 [4] and Nos. 18 to 20 [5] have been prepared from $(-(C_4H_9)_2SnO-)_n$ and the appropriate arsonic acid in refluxing C_2H_5OH-C_6H_6 within 1 to 2 h. Nos. 3 to 9, 11, and 15 [3], and Nos. 16 and 17 [6] precipitate quantitatively within 2 h from a refluxing mixture of $(-(C_4H_9)_2SnO-)_n$ and the appropriate arylarsonic acid in absolute C_2H_5OH.

Method III: $(C_4H_9)_2Sn(OOCCH_3)_2$ and $As(O)(R)(OH)_2$ (1:1 mole ratio).
The transesterification of $(C_4H_9)_2Sn(OOCCH_3)_2$ with the appropriate arylarsonic acid without a solvent at 100°C during 1 h yields Nos. 10 and 12 to 14, with elimination of CH_3COOH [2].

Table 39
Dibutyltin Compounds of the $(C_4H_9)_2\overline{SnOAs(O)(R)O}$ Type.
Further information on compounds preceded by an asterisk is given at the end of the table.
Explanations, abbreviations, and units on p. X.

No.	OAs(O)(R)O group method of preparation	properties and remarks	Ref.
*1	OAs(O)(CH_3)O II	m.p. 220°; soluble in C_2H_5OH, C_6H_6, and $CHCl_3$; dimeric IR in Table 40 on p. 365	[4]
*2	OAs(O)($CH_2C_6H_5$)O II	m.p. 260°, solubility like No. 1; dimeric IR in Table 40 on p. 365	[4]
*3	OAs(O)(C_6H_5)O I [7] II [3]	m.p. >250° [7], >280° [3], α and β modification and 1-H_2O solvate [7] ^{119}Sn-γ (78 K): δ = 1.08, Δ = 2.92 (α form), δ = 0.97, Δ = 2.60 (β form), or δ = 1.03, Δ = 2.98 (1-H_2O solvate) IR in Table 40 on p. 365	[3, 7] [7] [3]
*4	OAs(O)(C_6H_4Cl-4)O II	m.p. >280° IR in Table 40 on p. 365	[3]
*5	OAs(O)(C_6H_4Br-4)O II	m.p. >280° IR in Table 40 on p. 365	[3]
*6	OAs(O)(C_6H_4OH-4)O II	1-H_2O solvate m.p. >280° IR in Table 40 on p. 365	[3]
*7	OAs(O)($C_6H_4OCH_3$-2)O II	m.p. >280° IR in Table 40 on p. 365	[3]
*8	OAs(O)($C_6H_4OCH_3$-4)O II	m.p. >280° IR in Table 40 on p. 365	[3]
*9	OAs(O)($C_6H_4NH_2$-2)O II	light brown solid, m.p. >280° IR in Table 40 on p. 365	[3]
10	OAs(O)($C_6H_4NH_2$-4)O III	dec. 175 to 185° effective against protozoal infections	[2]
*11	OAs(O)($C_6H_4NO_2$-2)O II	yellow solid, m.p. >280° IR in Table 40 on p. 365	[3]
12	OAs(O)($C_6H_4NO_2$-4)O III [2]	dec. 249° effective against protozoal infections tested as an anthelmintic against Raillietina cesticillus and Ascaridia galli from chicken	[2] [1]

References on p. 367

Table 39 (continued)

No.	OAs(O)(R)O group method of preparation	properties and remarks	Ref.
13	$OAs(O)(C_6H_3(NO_2\text{-}3)OH\text{-}4)O$ III [2]	dec. 191° use and test like No. 12	[2] [1, 2]
14	$OAs(O)(C_6H_3(NO_2\text{-}3)NH_2\text{-}4)O$ III [2]	dec. 262° use and test like No. 12	[2] [1, 2]
*15	$OAs(O)(C_6H_4CH_3\text{-}4)O$ II	m.p. >280° IR in Table 40 on p. 365	[3]
*16	$OAs(O)(C_6H_4COOH\text{-}2)O$ II	m.p. 225 to 227°; dimeric ^{119}Sn-γ (80 K): $\delta = 1.34$, $\Delta = 3.46$ IR in Table 40 on p. 365	[6]
*17	$OAs(O)(C_6H_4COOH\text{-}4)O$ II	m.p. 290° (dec.); polymeric ^{119}Sn-γ (80 K): $\delta = 1.29$, $\Delta = 3.16$ IR in Table 40 on p. 365	[6]
*18	$OAs(O)(C_6H_4R'\text{-}2)O$ $R' = \text{-}N{=}CHC_6H_4Cl\text{-}2$ II	light green solid; m.p. 245 to 250°; polymeric IR in Table 40 on p. 365	[5]
*19	like No. 18 $R' = \text{-}N{=}CHC_6H_4NO_2\text{-}2$ II	red brown solid; m.p. 230°; monomeric ^{1}H NMR ($CDCl_3$): 0.6 to 0.875 (m, C_4H_9), 7.00 to 8.25 (m, C_6H_4) IR in Table 40 on p. 365	[5]
*20	like No. 18 $R' = \text{-}N{=}CHC_6H_4NO_2\text{-}3$ II	yellow solid; m.p. 198°; monomeric ^{1}H NMR ($CDCl_3$): 0.2 to 1.50 (m, C_4H_9), 7.525 to 8.75 (m, C_6H_4) IR in Table 40 on p. 365	[5]

* Further information:

$(C_4H_9)_2\overline{SnOAs(O)(C_6H_5)O}$ (Table **39**, No. **3**) is isolated as its α modification from the reaction of $(C_4H_9)_2SnCl_2$ with the monosodium salt of phenylarsonic acid, $C_6H_5As(O)(OH)ONa$. The β modification is obtained by removal of H_2O from $(C_4H_9)_2SnOAs(O)(C_6H_5)O \cdot H_2O$, which in turn is the product of the reaction between $(C_4H_9)_2SnCl_2$ and $C_6H_5As(O)(ONa)_2$. The spectroscopic data suggest that the β modification consists of infinite chains, and that chains with a similar backbone structure occur in the solid hydrate precursor. The α modification appears to have a two-dimensional sheet structure in which Sn achieves five-coordination, but the data do not preclude a network structure with Sn in highly distorted sites [7].

$(C_4H_9)_2\overline{SnOAs(O)(R)O}$ (Table **39**, Nos. **1** to **9**, **11**, and **15** to **20**). The reported selected IR frequencies of the compounds are listed in Table 40 (all spectra recorded in Nujol) [3 to 6].

References on p. 367

Table 40
Selected IR Frequencies of $(C_4H_9)_2\overline{SnOAs(O)(R)O}$ Compounds.
Wave numbers in cm^{-1}. Compound numbers as in Table 39.

No.	R	ν(As=O)	$ν(AsO_3)$	ν(As-O)	ν(AsC)	ν(SnO)	ν(SnC)	R vibrations	Ref.
1	CH_3	870 830		765 745	675	630 575 425	525 480		[4]
2	$CH_2C_6H_5$	830		750 720	685	620 570 410	530 470		[4]
3	C_6H_5		885, 860, 825, 730		1095				[3]
4	C_6H_4Cl-4		885, 860, 830, 730		1090			1090 (C-Cl)	[3]
5	C_6H_4Br-4		860, 830, 810, 740		1085			1065 (C-Br)	[3]
6	C_6H_4OH-4		895, 850, 825, 750		1080			3300 (OH)	[3]
7	$C_6H_4OCH_3$-2		910, 880, 870, 795, 780, 755		1070			1020 ($C-OCH_3$)	[3]
8	$C_6H_4OCH_3$-4		885, 860, 820, 735		1090			1025 ($C-OCH_3$)	[3]
9	$C_6H_4NH_2$-2		900, 880, 850, 820, 800, 790, 740		1100			3340, 3100, 1595, 1570 } (NH_2) 1320 (C-N)	[3]
11	$C_6H_4NO_2$-2		900 to 845, 785		1065			1580, 1350 (NO_2)	[3]
15	$C_6H_4CH_3$-4		890, 855, 825, 800		1090				[3]

References on p. 367

Table 40 (continued)

No.	R	ν(As=O)	ν(AsO_3)	ν(As-O)	ν(AsC)	ν(SnO)	ν(SnC)	R vibrations	Ref.
16	C_6H_4COOH-2	850			1100	360	500	1590 (C=O)	[6]
17	C_6H_4COOH-4	830			1085	350	470	1612 (C=O) 2770 (COOH)	[6]
18	$C_6H_4(N{=}CHC_6H_4Cl$-2)-2	840			1175	425		1620 (C=N) 1275 (C-N) 1040 (C-Cl) 468 (Sn ← N)	[5]
19	$C_6H_4(N{=}CHC_6H_4NO_2$-2)-2	820			1155			1630 (C=N) 1340 (C-NO_2) 1300 (C-N)	[5]
20	$C_6H_4(N{=}CHC_6H_4NO_2$-3)-2	820			1145			1635 (C=N) 1350 (C-NO_2) 1305 (C-N)	[5]

References:

[1] Kerr, K. B., Walde, A. W. (Experim. Parasitol. **5** [1956] 560/70).
[2] Walde, A. W., van Essen, H. E., Zbornik, T. W., Dr. Salsbury's Laboratories (U.S. 2762821 [1956]; C.A. **1957** 4434).
[3] Sandhu, S. S., Kaur, J., Sandhu, G. K. (Syn. React. Inorg. Metal-Org. Chem. **4** [1974] 437/45).
[4] Sandhu, S. S., Sandhu, G. K., Pushkarna, S. K. (Syn. React. Inorg. Metal-Org. Chem. **11** [1981] 197/210).
[5] Sandhu, G. K., Sandhu, S. S. (Syn. React. Inorg. Metal-Org. Chem. **12** [1982] 215/28).
[6] Sandhu, S. S., Sandhu Jr., S. S., Sandhu, G. K., Parish, R. V., Parry, O. (Inorg. Chim. Acta **58** [1982] 251/6).
[7] Cunningham, D., Firtear, P., Molloy, K. C., Zuckerman, J. J. (J. Chem. Soc. Dalton Trans. **1983** 1523/7).

1.4.1.2.1.5.15.3 Other Dibutyltin Compounds with Sn-O-As Bonds

$(C_4H_9)_2\overline{SnOAs(O)(C_6H_5)CH_2COO}$

The compound has been prepared by refluxing a 1:1 molar mixture of $(-(C_4H_9)_2SnO-)_n$ and $HOAs(O)(C_6H_5)CH_2COOH$ in C_2H_5OH for 2 h. It melts at 180°C and is monomeric.

IR spectrum (in Nujol): ν(C=O) 1570, ν(C-O) 1180, ν(As-C) 1110 and 1090, ν(As=O) 860 and 830, ν(As-O) 730 and 690, ν(SnO) 630 and 570, ν(SnC) 480, ν(SnO) 415 and 370 cm^{-1}.

The title compound is suggested to have an octahedral structure with *trans*-C_4H_9 groups and C=O → Sn and As=O → Sn coordination.

$(C_4H_9)_2\overline{SnOAs(O)(C_6H_5)\text{-}CH_2CH_2\text{-}(C_6H_5)(O)AsO}$

The compound has been obtained from $(-(C_4H_9)_2SnO-)_n$ and the free acid, $HOAs(O)(C_6H_5)$-CH_2CH_2-$(C_6H_5)(O)AsOH$, after refluxing C_2H_5OH for 3 h. It melts at 253°C and is polymeric.

IR spectrum (in Nujol): ν(AsC) 1090, ν(As=O) 860 and 830, ν(As-O) 730 and 710, ν(SnO) 620 and 575, ν(SnC) 475, ν(SnO) 380.

Reference:

Sandhu, S. S., Sandhu, G. K., Pushkarna, S. K. (Syn. React. Inorg. Metal-Org. Chem. **11** [1981] 197/210).

1.4.1.2.1.5.16 Dibutyltin Compounds Containing Sn-O-Se Bonds

$(C_4H_9)_2Sn(OSe(O)CH_3)_2$

$(C_4H_9)_2Sn(OSe(O)CH_3)_2$ is the product of the reaction between $(C_4H_9)_2SnCl_2$ and $CH_3Se(O)ONa$ in THF at room temperature for 20 h (88% yield). The compound decomposes at 179 °C.

IR spectrum (in Nujol): ν_s(OSeO) 798 (s), ν_{as}(SeO) 756 (s), ν(SeC) 585 (m), ν_{as}(SnC) 620 (m), ν_{as}(SnO) 440 (m) cm^{-1}.

Raman spectrum (solid): ν_s(OSeO) 811 (s), ν(SeC) 585 (s), ν_s(SnC) 585 (s), ν_{as}(SnO) 434 (w) cm^{-1} [1].

$(C_4H_9)_2Sn(OSe(O)C_6H_5)_2$

The compound is synthesized from $(C_4H_9)_2SnCl_2$ and $C_6H_5Se(O)ONa$ like the previous product (70% yield). It decomposes at 201 °C.

IR spectrum (in Nujol): ν_s(OSeO) 808 (s), γ(CH) 750 (s), ν_{as}(SeO) 689 (s), ν(SeC) 670 (mw), ν_{as}(SnC) 622 (m), ν_{as}(SnO) 415 (mw) cm^{-1}.

Raman spectrum (solid): ν_s(OSeO) 824 (s), ν(SeC) 674 (m), ν_s(SnC) 596 (s), ν_{as}(SnO) 404 (w) cm^{-1}.

Reference:

Lindner, E., Ansorge, U. (Z. Anorg. Allgem. Chem. **442** [1978] 189/94).

1.4.1.2.1.5.17 Dibutyltin Compounds Containing Sn-O-Ti Bonds

$(C_4H_9)_2SnOTiO_2$

The compound is described as a stabilizer for silicones [1].

$(C_4H_9)_2Sn\overline{SnOTi(OR)_2O}$, (OR = $C_7H_5O_2$, $C_8H_7O_3$, and $C_{15}H_{11}O_2$) **and**
$(C_4H_9)_2Sn(OTi(O)OC_6H_4CHO\text{-}6)_2$

Three compounds of the type $(C_4H_9)_2\overline{SnOTi(OR)_2O}$ with R = C_6H_4CHO-2, $C_6H_4COOCH_3$-2, and $C(C_6H_5)$=$CHCOC_6H_5$, and a dibutyltin derivative with the formula $(C_4H_9)_2Sn$-$(OTi(O)OC_6H_4CHO\text{-}6)_2$ are reported to be the products of the reaction of $(\text{-}(C_4H_9)_2SnO\text{-})_n$ with bis-salicylaldehydo-di-n-butoxy titanium, bis-methylsalicylato-di-n-butoxy titanium, bis(dibenzoyl)methano-di-n-butoxy titanium, and salicylaldehydo-n-propoxy oxotitanium. Decomposition points are given as 250, 240, 78, and 250 °C, respectively. The first three compounds should be monomeric in benzene [2].

References:

[1] Sumitomo Bakelite Co., Ltd. (Japan. Kokai Tokkyo Koho 58 225103 [1982/83]; C.A. **100** [1984] No. 157683).
[2] Gopinathan, S., Gopinathan, C., Gupta, J. (Indian J. Chem. **8** [1970] 303/4).

1.4.1.2.1.6 Diisobutyltin-Oxygen Compounds, $(i\text{-}C_4H_9)_2Sn(OR)_2$ and $(i\text{-}C_4H_9)_2\overline{SnO\text{-}R\text{-}O}$

The compounds belonging to this section are listed in Table 41. They are prepared by the following methods.

Method I: Reaction of $(i\text{-}C_4H_9)_2SnCl_2$ with ROH, RONa, or RCOOH (1:2 mole ratio).
The reactions of $(i\text{-}C_4H_9)_2SnCl_2$ with 8-hydroxiquinoline in C_2H_5OH yields No. 4 [8]; with $NaOCH_3$ in CH_3OH, No. 1 [13]; and with CH_3COOH, No. 10 [2]; $(i\text{-}C_4H_9)_2SnCl_2(NH_3)_2$ and lauric acid yield No. 11 [7, 9, 10].

Method II: Condensation of $(\text{-}(i\text{-}C_4H_9)_2SnO\text{-})_n$ with allyl 2-O-benzoyl-6-O-benzyl-α-D-galactopyranoside in CH_3OH has been used for the synthesis of No. 5 [19], with $CH_3(i\text{-}C_4H_9)(C_6H_5)SiOH$ for the synthesis of No. 15 [5], and with the appropriate acids in H_2O for the synthesis of Nos. 12 [1, 3, 12, 15] and 13 [4].

Method III: Transalkoxylation reactions (1:2 mole ratio).
The transalkoxylation of $(i\text{-}C_4H_9)_2Sn(OCH_3)_2$ with the appropriate ROH derivative leads to No. 6 [17] and Nos. 7, 8, 9 [18].

Method IV: $(i\text{-}C_4H_9)_2SnH_2$ reacts with $(CH_3)_2CHCHO$ with formation of No. 2 and with salicylic aldehyde with formation of No. 3 [11].

Table 41
Diisobutyltin-Oxygen Compounds, $(i\text{-}C_4H_9)_2Sn(OR)_2$ and $(i\text{-}C_4H_9)_2\overline{SnO\text{-}R\text{-}O}$.
Further information on compounds preceded by an asterisk is given at the end of the table.
Explanations, abbreviations, and units on p. X.

No.	$(OR)_2$ or O-R-O group method of preparation (yield in %)	properties and remarks	Ref.
*1	$(OCH_3)_2$ I (48)	m.p. 36 to 38°	[13]
2	$(OC_4H_9\text{-}i)_2$ IV (47)	m.p. 20°; b.p. 96°/0.002	[11]
3	$(OC_6H_4CH_2OH\text{-}2)_2$ IV (94)	m.p. 146°	[11]
4	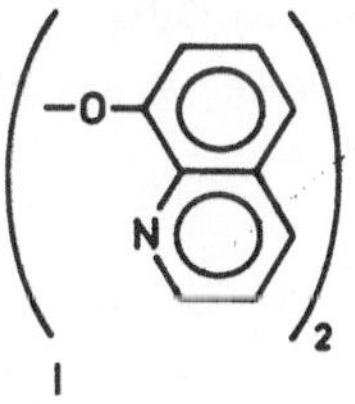 I	m.p. 188 to 189° $^{119}Sn\text{-}\gamma$ (80 K): $\delta = 0.85$, $\Delta = 1.82$	[8] [14]
5	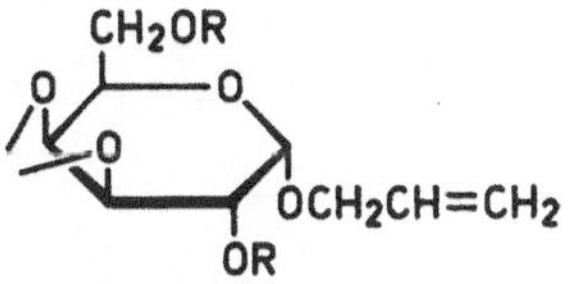 $R = CH_2C_6H_5$ II	colorless syrup with $C_6H_5CH_2Br \rightarrow$ 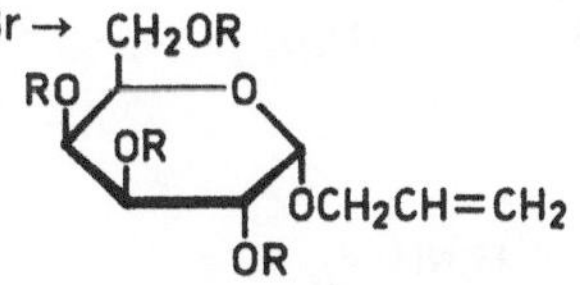	[19]

References on p. 371

Table 41 (continued)

No.	(OR)$_2$ or O-R-O group method of preparation (yield in %)	properties and remarks	Ref.
6	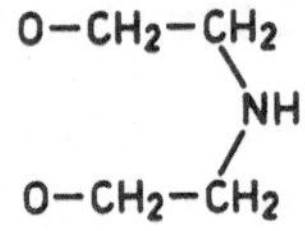 O—CH$_2$—CH$_2$ NH O—CH$_2$—CH$_2$ II	bactericide, fungicide	[17]
7	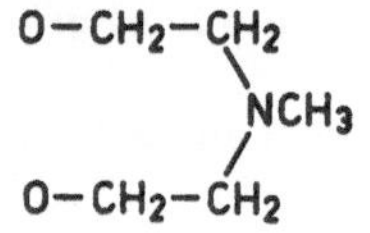 O—CH$_2$—CH$_2$ NCH$_3$ O—CH$_2$—CH$_2$ II	bactericide	[18]
8	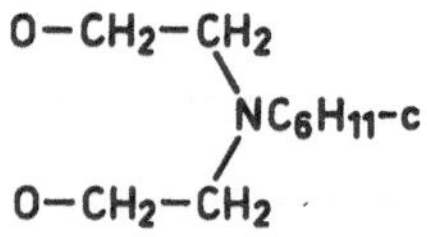 O—CH$_2$—CH$_2$ NC$_6$H$_{11}$-c O—CH$_2$—CH$_2$ II	bactericide	[18]
9	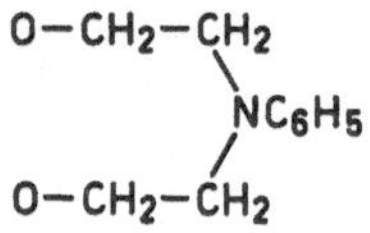 O—CH$_2$—CH$_2$ NC$_6$H$_5$ O—CH$_2$—CH$_2$ II	bactericide	[18]

10	(OOCCH$_3$)$_2$
	I
		b.p. 140 to 141°/10
		with Sn(OC$_2$H$_5$)$_4 \rightarrow$
		  CH$_3$COO(-Sn(C$_4$H$_9$-i)$_2$OSn(C$_2$H$_5$)$_2$O-)$_n$C$_2$H$_5$
		with HOOC(CH$_2$)$_8$COOH $\rightarrow$
		  (-(i-C$_4$H$_9$)$_2$SnOOC(CH$_2$)$_8$COO-)$_n$
			[2]
			
			[6]

11	(OOCC$_{11}$H$_{23}$)$_2$ I	n_D^{20} = 1.4689 stabilizer for PVC	[7, 9, 10]
12	(OOCC(CH$_3$)=CH$_2$)$_2$ II	m.p. 36°	[1, 3, 12, 15]
13	(OOCC$_6$H$_4$NH$_2$-4)$_2$ II	m.p. 181 to 182°	[4]

Table 41 (continued)

No.	$(OR)_2$ or O-R-O group method of preparation (yield in %)	properties and remarks	Ref.
14	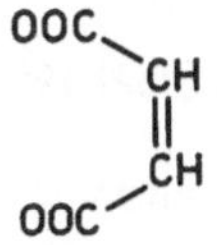	used for antiseptic treatment of Kozhmatol	[16]
15	$(OSi(CH_3)(C_4H_9\text{-}i)C_6H_5)_2$ II (34)	b.p. 210 to 212°/2.5 D^{20} = 1.0912; n_D^{20} = 1.5112 IR spectrum mentioned	[5]

* Further information:

$(i\text{-}C_4H_9)_2Sn(OCH_3)_2$ (Table **41**, No. **1**). The IR spectrum [13] is given in Table 42.

Table 42
IR Spectrum of $(i\text{-}C_4H_9)_2Sn(OCH_3)_2$.
Wave numbers in cm^{-1}.

	assignment		assignment
1462 (s)	$\delta_{as}(CH_3)$	965 (w)	
1447 (s)		949 (m)	
1400 (m)		918 (w)	
1380 (s)	$\delta_s(CH_3)$	890 (w)	
1363 (s)		815 (w)	
1325 (m)		800 (w)	
1310 (m)		727 (s)	$\varrho(CH_2C)$
1280 (w)		710 (w)	
1268 (w)		605 (s)	$\nu_{as}(SnC) + \nu_{as}(SnO)$
1199 (m)		590 (m)	
1159 (s)		540 (m)	
1065 (s)	$\nu(CO)$	511 (s)	$\nu_s(SnC)$
1035 (s)		466 (s)	$\nu_s(SnO)$
987 (w)			

References:

[1] Kochkin, D. A., Kotrelev, V. N., Kalinina, S. P., Kuznetsova, G. I., Laine, L. V., Chernova, L. V., Borisova, A. I., Borisenko, V. V. (Vysokomol. Soedin. **1** [1959] 1507/12; Polym. Sci. [USSR] **1** [1959] 30/8).

[2] Koton, M. M., Kiseleva, T. M. (Dokl. Akad. Nauk SSSR **130** [1960] 86/7; Proc. Acad. Sci. USSR Chem. Sect. **130/135** [1960] 13/4).

[3] Shostakovskii, M. F., Kalinina, S. P., Kotrelev, V. N., Kochkin, D. A., Kuznetsova, G. I., Laine, L. V., Borisova, A. I., Borisenko, V. V. (Mezhdunar. Simp. Makromol. Khim. Dokl. Aftoveraty, Moscow 1960, Vol. 1, pp. 160/6; C.A. **1961** 7273).
[4] Kochkin, D. A., Verenikina, S. G., Chekmareva, J. B. (Dokl. Akad. Nauk SSSR **139** [1961] 1375/8; Proc. Acad. Sci. USSR Chem. Sect. **136/141** [1961] 855/8).
[5] Kochkin, D. A., Chirzadze, Yu. N. (Zh. Obshch. Khim. **32** [1962] 4007/12; J. Gen. Chem. [USSR] **32** [1962] 3932/7).
[6] Korshak, V. V., Rogozhin, S. V., Makarova, T. A. (Vysokomol. Soedin. **4** [1962] 1297/302; C.A. **58** [1963] 14112).
[7] Koninklijke Industrielle Maatschappij voorheen Noury and van der Lande N.V. (Fr. 1320473 [1961/63]; C.A. **59** [1963] 8788/9).
[8] Gerrard, W., Mooney, E. F., Rees, R. G. (J. Chem. Soc. **1964** 740/5).
[9] Koninklijke Industrielle Maatschappij voorheen Noury and van der Lande N.V. (Neth. 109491 [1962/64]; C.A. **62** [1965] 9173).
[10] Viveen, W. J. C., Schröder, A., Koninklijke Industrielle Maatschappij voorheen Noury and van der Lande N.V. (Ger. Offen. 1167836 [1962/64]).

[11] Neumann, W. P., Heymann, E. (Liebigs Ann. Chem. **683** [1965] 11/23).
[12] Kochkin, D. A., Azerbaev, I. N. (Vestn. Akad. Nauk Kaz. SSR **22** No. 12 [1966] 53/61; C.A. **67** [1967] No. 64467).
[13] Maire, J. C., Ouaki, R. (Helv. Chim. Acta **51** [1968] 1150/4).
[14] Ali, K. M., Cunningham, D., Frazer, M. J., Donaldson, J. D., Senior, B. J. (J. Chem. Soc. [1969] 2836/40).
[15] Kochkin, D. A., Azerbaev, I. N. (Dokl. 4th Vses. Konf. Khim. Atsetilena, Alma-Ata 1972, Vol. 3, pp. 209/16; C.A. **79** [1973] No. 126823).
[16] Petrova, V. I. (Kozh. Obuvn. Prom. **1973** No. 12, pp. 22/4; C.A. **81** [1974] No. 38976).
[17] Plum, H., Buschhoff, M., Cejka, A., Schering A.-G. (Ger. Offen. 2526711 [1975/76]; C.A. **86** [1977] No. 116058).
[18] Plum, H., Buschhoff, M., Cejka, A., Schering A.-G. (Ger. Offen. 2610931 [1976/77]; C.A. **88** [1978] No. 37968).
[19] El-Shenawy, H. A., Schuerch, C. (Carbohydr. Res. **131** [1984] 227/38).

1.4.1.2.1.7 Di-sec-butyltin Oxygen Compounds $(sec\text{-}C_4H_9)_2Sn(OR)_2$

$(sec\text{-}C_4H_9)_2Sn(OOCC_{11}H_{23})_2$

The compound is described in a patent to be effective as a light stabilizer for vinyl chloride resins, Motohashi, A., Kogo, Y., Tsukahara, Y., Sankyo Organic Chemicals Co., Ltd. (Japan. Kokai 77-80349 [1975/77]; C.A. **87** [1977] No. 185540).

1.4.1.2.1.8 Di-tert-butyltin Oxygen Compounds, $(t\text{-}C_4H_9)_2Sn(OR)_2$ and $(t\text{-}C_4H_9)_2\overline{SnO\text{-}R\text{-}O}$

The compounds belonging to this section are listed in Table 43. They are prepared by the following methods listed according to the starting material.

Method I: From $(t\text{-}C_4H_9)_2SnX_2$ (X = Cl, Br).
No. 1 is formed when $(t\text{-}C_4H_9)_2SnCl_2$ is hydrolyzed in warm $C_6H_5CH_3$ in the presence of Na_2CO_3 [5], and in the reaction of $(t\text{-}C_4H_9)_2SnBr_2$ with KOH in ether [1]. The addition of tropolone, 8-hydroxiquinoline, 2-methyl-8-hydroxiquinoline, or acetylacetone to $(t\text{-}C_4H_9)_2SnCl_2$ in CH_3OH and stirring for 2 h at room temperature yields Nos. 3, 4, 5, or 16, respectively [13]. Di-t-butyltin dihalides react with thallium acetylacetonate with formation of No. 16 [7].

Method II: From $(\text{-}(t\text{-}C_4H_9)_2SnO\text{-})_n$.
Heating equimolar amounts of $(\text{-}(t\text{-}C_4H_9)_2SnO\text{-})_n$ and the appropriate HO-R-OH in xylene and removal of H_2O, usually by azeotropic distillation, yields Nos. 7 [14], 8, and 9 [8].

Method III: From $(t\text{-}C_4H_9)_2Sn(OCH_3)_2$.
Nos. 6 and 7 have been obtained by refluxing equimolar amounts of $(t\text{-}C_4H_9)_2Sn(OCH_3)_2$ and the appropriate HO-R-OH in $C_6H_5CH_3$ [8], Nos. 14 and 15 in C_6H_6 [15], and No. 18 in C_6H_6 in the presence of $(CH_3)_2NCHO$ [10].

Table 43
Di-tert-butyltin-Oxygen Compounds, $(t\text{-}C_4H_9)_2Sn(OR)_2$ and $(t\text{-}C_4H_9)_2\overline{SnO\text{-}R\text{-}O}$.
Further information on compounds preceded by an asterisk is given at the end of the table.
Explanations, abbreviations, and units on p. X.

No.	$(OR)_2$ or O-R-O group method of preparation (yield in %)	properties and remarks	Ref.
*1	$(OH)_2$ I [1, 5]	dec. 180 to 190° [2] IR: 2840 (s), 2270 (m), 1650 (s), 1150 (m), 790 (s)	[1, 2, 5] [5]
2	$(OCH_3)_2$	preparation not reported with HO-R-OH → Nos. 6, 7, 10 to 15, 18, 19	 [8, 10, 15]
3	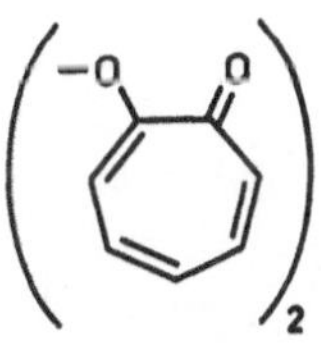 I (25 to 55)	m.p. 166 to 168° ^{1}H NMR: 1.18 (s, CH_3C, $^3J(Sn,H)$ = 111.4) ^{119}Sn NMR: −326	[13]
4	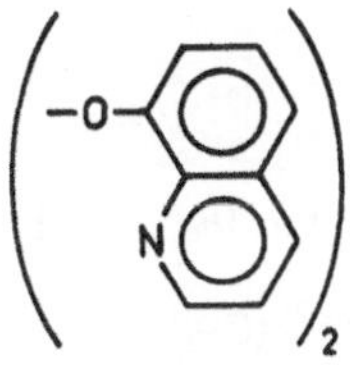 I (25 to 55)	m.p. 199 to 202° ^{1}H NMR: 1.13 (s, CH_3C, $^3J(Sn,H)$ = 107.2) ^{119}Sn NMR: −338 additive to improve the moldability of polypropylenes	[13] [4]

References on p. 379

Table 43 (continued)

No.	$(OR)_2$ or O-R-O group method of preparation (yield in %)	properties and remarks	Ref.
5	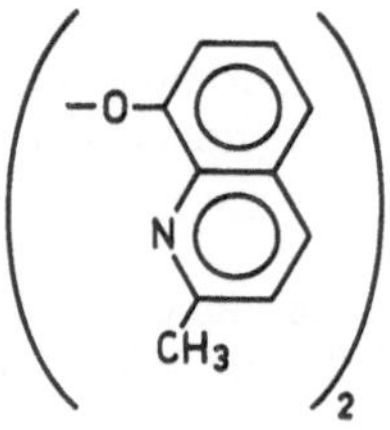 I (25 to 55)	m.p. 207 to 208° ^{1}H NMR: 1.03 (s, CH_3C, $^3J(Sn,H)$ = 109.4) ^{119}Sn NMR: −311	[13]
*6	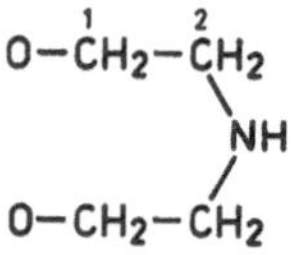 III (90 [8])	m.p. 76 to 112°; b.p. 115°/0.15 ^{13}C NMR: 30.4 (C-β), 37.5 (C-α), 51.7 (C-2), 60.3 (C-1) ^{119}Sn NMR: −209.5 in CH_2Cl_2; −210.5 in CD_3COCD_3	[8] [9] [9]
*7	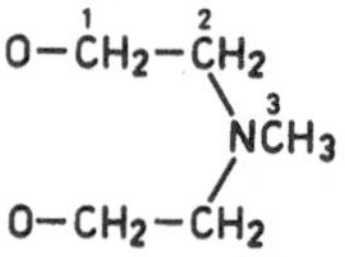 II [14] III (90 [8])	m.p. 68 to 70° [8], 74 to 75° [14]; b.p. 130 to 131°/0.15 [8] ^{13}C NMR: 30.4 (C-β), 37.7 (C-α), 39.5 (C-3), 57.5 (C-2), 62.0 (C-1) ^{119}Sn NMR: −205 in CH_2Cl_2; −204 in CD_3COCD_3 ^{119}Sn-γ (77 K): δ = 1.42, Δ = 2.54	[8, 14] [9] [9] [11]
*8	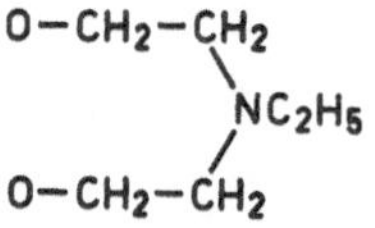 II (72 [8])	m.p. 47 to 48°; b.p. 119 to 121°/0.13	[8]
*9	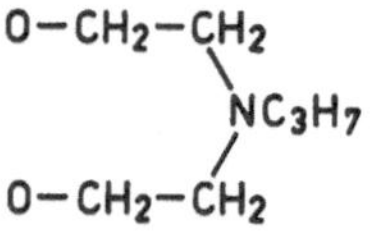 II (72)	m.p. 49 to 51°; b.p. 110 to 112°/0.03	[8]
10	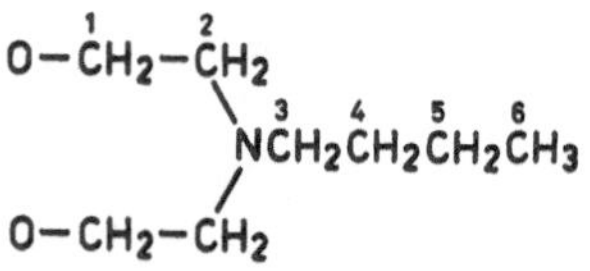	prepared by Methods II and III used for analogous compounds in [8] ^{13}C NMR: 13.7 (C-6), 20.6 (C-5), 22.5 (C-4), 30.6 (C-β), 39.0 (C-α), 46.4 (C-3), 56.1 (C-2), 57.5 (C-1)	[9]

References on p. 379

Table 43 (continued)

No.	$(OR)_2$ or O-R-O group method of preparation (yield in %)	properties and remarks	Ref.
*11	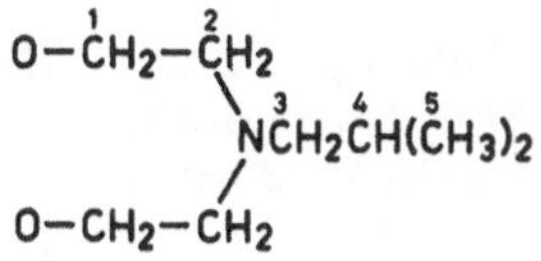	prepared by Methods II and III used for analogous compounds in [8] ^{13}C NMR: 20.6 (C-5), 23.3 (C-4), 30.5 (C-β), 38.2 (C-α), 54.2 (C-3), 56.4 (C-2), 57.5 (C-1) ^{119}Sn-γ (77 K): δ = 1.36, Δ = 1.85	[9] [11]
*12	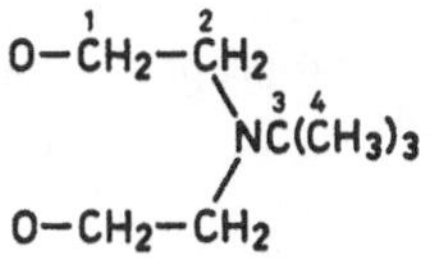	prepared by Methods II and III used for analogous compounds in [8] ^{13}C NMR: 26.8 (C-3), 30.3 (C-β), 38.2 (C-α), 53.5 (C-2), 55.3 (C-3), 66.7 (C-1) ^{119}Sn-γ (77 K): δ = 1.37, Δ = 1.87	[9] [11]
*13	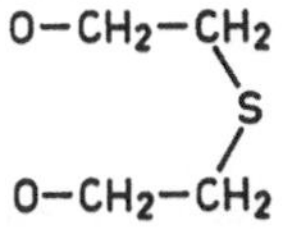	prepared by Method II used for analogous compounds in [8] ^{119}Sn-γ (77 K): δ = 1.64, Δ = 1.96	[11]
14	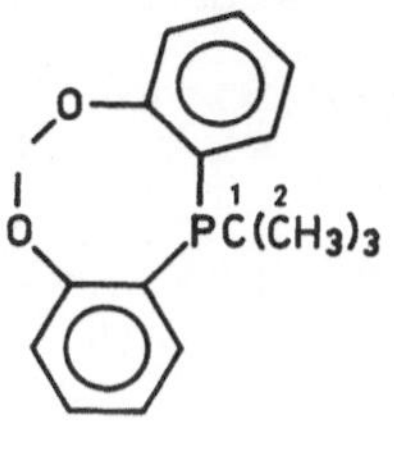 III (56)	m.p. 119 to 129° ^{1}H NMR (CH_2Cl_2): 1.14 (s, CH_3CSn, $^3J(Sn,H)$ = 104), 1.88 (s, CH_3CSn, $^3J(Sn,H)$ = 100), 1.50 (d, CH_3CP, $^3J(P,H)$ = 18) ^{13}C NMR (C_6D_6): 25.95 (C-1), 30.45 and 31.76 (C-α), 34.18 and 34.79 (C-β), 36.71 (C-2) ^{31}P NMR (C_6D_6): −63.43 ($^1J(Sn,P)$ = 466.3/488.1) ^{119}Sn NMR: −226.40 ($^1J(Sn,P)$ = 488.7) in C_6D_6; −224.33 ($^1J(Sn,P)$ = 494.3) in C_6D_6-C_5H_5N	[15]
15	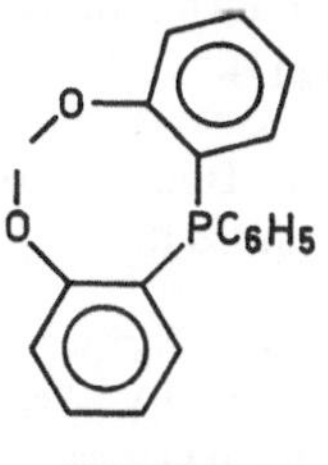 III (60)	m.p. 161 to 162° ^{1}H NMR (C_6D_6): 1.16 (s, CH_3CSn, $^3J(Sn,H)$ = 107), 1.22 (s, CH_3CSn, $^3J(Sn,H)$ = 96) ^{13}C NMR (C_6D_6): 30.52 and 30.82 (C-β), 37.61 and 37.80 (C-α) ^{31}P NMR (C_6D_6): −78.09 ($^1J(Sn,P)$ = 518.85/542.36) ^{119}Sn NMR (C_6D_6): −236.36 ($^1J(Sn,P)$ = 545.85)	[15]

References on p. 379

Table 43 (continued)

No.	$(OR)_2$ or O-R-O group method of preparation (yield in %)	properties and remarks	Ref.
16	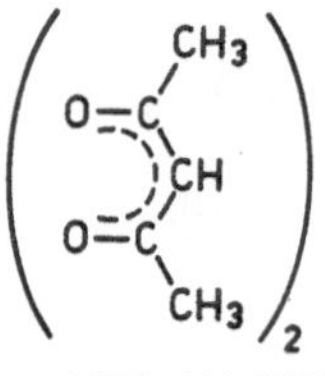 I [7], (41 [13])	m.p. 101 to 103°	[13]
		^{1}H NMR: 1.19 (s, CH_3CSn, $^3J(Sn,H)$ = 115.6/121.2), 1.98 (s, CH_3), 5.37 (s, CH) in $CDCl_3$ [7]; 1.01 (s, CH_3CSn, $^3J(Sn,H)$ = 130.6) in CCl_4 [13]	[7, 13]
		^{119}Sn NMR (CCl_4): −486	[13]
		IR: ν(CO) 1575 (vs), 1510 (vs)	[7]
17	$(OOCCH_3)_2$	proposed to be formed during polarography of $(t\text{-}C_4H_9)_2SnCl_2$ in CH_3COOH	[6]
*18	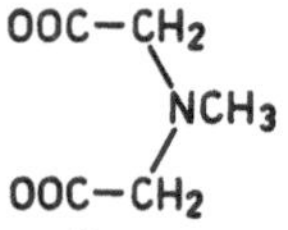 III	m.p. 269°	[10]
		^{119}Sn-γ (77 K): δ = 1.54, Δ = 3.88	[11]
		IR: ν(CO) 1682	[10]
*19	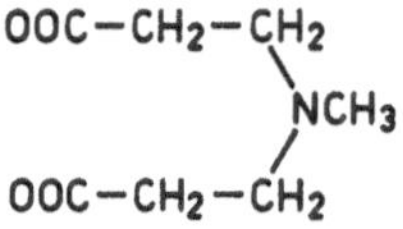	prepared by Method II used for analogous compounds in [8]	[11]
		^{119}Sn-γ (77 K): δ = 1.76, Δ = 3.24	

* Further information:

$(t\text{-}C_4H_9)_2Sn(OH)_2$ (Table **43**, No. **1**). The compound is proposed to be formed during polarography of $(t\text{-}C_4H_9)_2SnCl_2$ [6].

The compound is found to be trimeric in C_6H_6 [5]. It is soluble in THF and $CHCl_3$, slightly soluble in hot ether and acetone, and fairly soluble in water, when freshly prepared [2].

DTA shows an initial decomposition at 180°C, TGA at 190°C, whereas visual decomposition starts at 250°C [2]. $(t\text{-}C_4H_9)_2Sn(OH)_2$ reacts with HBr to give $(t\text{-}C_4H_9)_2SnBr_2$ [1].

The compound is used as a bactericide against Staphylococcus aureus, Bacillus subtilis, Aerobacter aerogenes, Pseudomonas aeruginosa, Escherichia coli and others [3].

$(t\text{-}C_4H_9)_2SnO_2C_4H_8X$ (Table **43**, No. **6**, X = NH; No. **7**, X = NCH_3; No. **8**, X = NC_2H_5; No. **9**, X = NC_3H_7; No. **11**, X = NC_4H_9-iso; No. **12**, X = NC_4H_9-t; No. **13**, X = S), **$(t\text{-}C_4H_9)_2SnO_4C_4H_4$-$NCH_3$** (Table **43**, No. **18**), and **$(t\text{-}C_4H_9)_2SnO_4C_6H_8NCH_3$** (Table **43**, No. **19**). The ^{1}H NMR spectra of Nos. 6 to 9 (Table 44) show two different signals for the t-C_4H_9 groups at −40°C which

References on p. 379

coalesce at room temperature. A mechanism of dissociation – inversion of the five-coordinate species was discussed [8]:

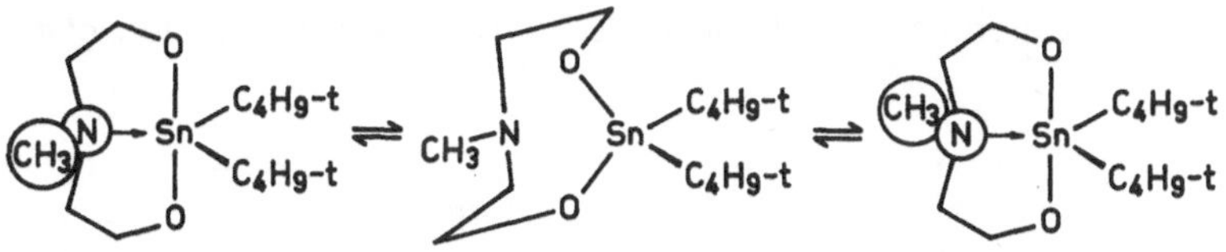

Table 44
^{1}H NMR Spectra of (t-$C_4H_9)_2SnO_2C_4H_8$X Compounds [8].
δ in ppm, J in Hz.

compound	No. 6 in $CDCl_3$		No. 7 in CH_2Cl_2		No. 8 in CH_2Cl_2		No. 9 in $CDCl_3$
δ(CH_3C) (−40°C)	1.17	1.24	1.07	1.16	1.04	1.11	
^{3}J(Sn,H)	93/95	90/93	91/95	92/96	90/94	91/95	
δ(CH_3C) (32°C)	1.25		1.20		1.18		1.32
^{3}J(Sn,H)	91.5/95.5		90.5/95.5		90/94		92/96
δ(CH_2O)	3.61		3.81		3.76		3.83
δ(CH_2N)	2.70		2.69		2.71		2.59
δ(NH, CH_3N, CH_2N)	3.19		2.08		2.85		

The dependence of the ^{3}J(Sn,H) coupling and the activation parameters of the inversion on the donor strength of solvents indicate the presence of six-coordinate species; the following data are given for No. 7 [8]:

	in $CHCl_3$	in $CDCl_3$	in CH_2Cl_2	in CH_3COCH_3
^{3}J(Sn,H) in Hz	93.0/97.5		92/96	89.5/93.5
	92.5/97.5		91/95	88.0/92.5
E_a in kcal/mol		15 ± 3		9 ± 3
$\Delta H^{\neq}$ in kcal/mol		14 ± 3		7 ± 2
$\Delta G_c^{\neq}$		15 ± 3		15 ± 3

The same dissociation-inversion mechanism is derived from the temperature and solvent dependent ^{1}H and ^{13}C NMR spectra of No. 18 [10]. The activation parameters of Nos. 6 and 7 have also been obtained from the ^{13}C NMR spectra (E_a, $\Delta H^{\neq}$, and $\Delta G_c^{\neq}$ in kcal/mol) [9]:

compound	solvent	T_c (K)	E_a	$\Delta H^{\neq}$	$\Delta G_c^{\neq}$
(t-$C_4H_9)_2SnO_2C_4H_8$NH	$CDCl_3$	284	12.2	11.6	15.2
	CD_3COCD_3	297	23.2	22.6	15.9
(t-$C_4H_9)_2SnO_2C_4H_8NCH_3$	$CDCl_3$	283	21.6	20.9	14.9
	CD_3COCD_3	286	22.5	21.9	15.0

References on p. 379

On the basis of the partial quadrupole splitting concept, Δ values were calculated for 35 compounds, among them Nos. 7, 11, 12, 13, 18, and 19, assuming different coordination numbers and, in the case of five-coordination, different configurations of the compounds. Comparison of the experimental and calculated Δ values gives the structure of the compounds [11]:

compound	Δ_{exp}	Δ_{calc} for structures I	II	III	suggested structure
No. 7	2.54	+2.20	+1.626	1.87	I
No. 11	1.85	+2.20	+1.625	1.87	II
No. 12	1.87	+2.20	+1.625	1.87	II
No. 13	1.96	+2.33	+1.52	1.87	II
No. 18	2.98	+3.29	+2.73	2.81	I or III
No. 19	3.24	+3.29	+2.73	2.81	I

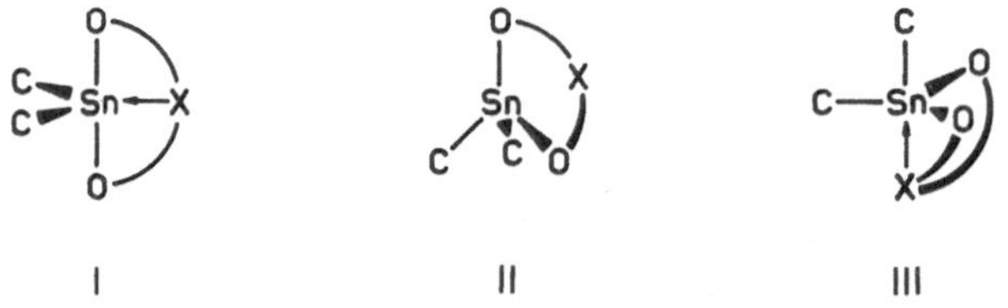

Compound No. 7 crystallizes in the monoclinic space group $P2_1/c-C^5_{2h}$ with a = 22.655(5), b = 25.990(4), c = 9.036(2) Å, and β = 106.97(2)°; Z = 12. The refined structure (R = 0.074) showed three independent molecules present in the unit cell; one of these molecules with its structural parameters is given in **Fig. 5**. The individual molecules essentially form distorted trigonal bipyramids with N→Sn dative bonds. The ring oxygen atoms are located in apical sites and the nitrogen atom in an equatorial position, with Sn lying in the equatorial plane [12, 14].

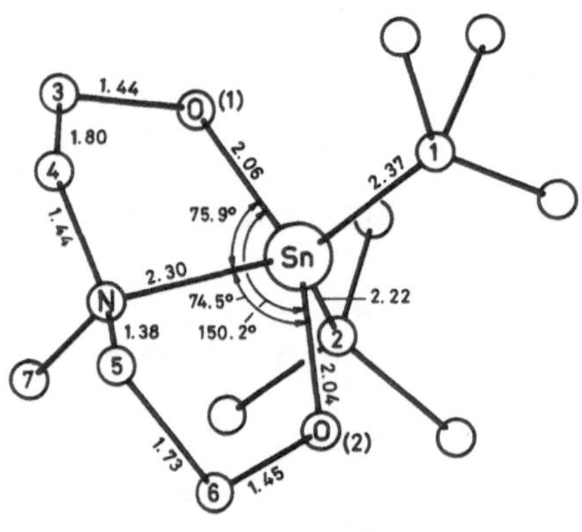

Fig. 5. Molecular structure of $(t\text{-}C_4H_9)_2Sn(OCH_2CH_2)_2NCH_3$ [12, 13].

Other selected bond angles (°):

O(1)-Sn-C(1)	103.6(7)	O(1)-C(3)-C(4)	109.3(18)
O(1)-Sn-C(2)	95.1(9)	C(3)-C(4)-N	92.4(19)
O(2)-Sn-C(1)	88.7(9)	C(4)-N-Sn	108.9(15)
O(2)-Sn-C(2)	96.7(8)	C(4)-N-C(5)	100.4(21)
N-Sn-C(1)	107.8(11)	O(2)-C(6)-C(5)	105.8(18)
N-Sn-C(2)	120.2(8)	C(6)-C(5)-N	101.2(21)
C(1)-Sn-C(2)	131.4(12)	C(5)-N-Sn	113.3(16)
Sn-O(1)-C(3)	116.6(13)	C(4)-N-C(7)	121.9(21)
Sn-O(2)-C(6)	121.8(11)	C(5)-N-C(7)	102.2(20)
Sn-N-C(7)	109.7(15)		

References:

[1] Krause, E., Weinberg, K. (Ber. Deut. Chem. Ges. **63** [1930] 3481/5).
[2] Baum, G. A., Considine, W. J. (J. Polym. Sci. B **1** [1963] 517/8).
[3] Baum, G. A., Considine, W. J., M and T Chemicals Inc. (Fr. 1396634 [1963/65]; C.A. **64** [1966] 8678).
[4] Kubota, Y., Koda, Y., Toyo Rayon Co., Ltd. (Japan. 69-27465 [1966/69]; C.A. **72** [1970] No. 101460).
[5] Chu, C. K., Murray, J. D. (J. Chem. Soc. A **1971** 360/7).
[6] Leroux, P., Devaud, M. (Bull. Soc. Chim. France **1973** 2254/8).
[7] Gielen, M., Topart, J. (Bull. Soc. Chim. Belges **84** [1975] 13/9).
[8] Zschunke, A., Tzschach, A., Jurkschat, K. (J. Organometal. Chem. **112** [1976] 273/8).
[9] Jurkschat, K., Mügge, C., Tzschach, A., Zschunke, A., Larin, M. F., Pestunovich, V. A., Voronkov, M. G. (J. Organometal. Chem. **139** [1977] 279/82).
[10] Tzschach, A. Jurkschat, K., Zschunke, A., Mügge, C. (J. Organometal. Chem. **193** [1980] 299/305).

[11] Korecz, L., Saghier, A. A., Burger, K., Tzschach, A. Jurkschat, K. (Inorg. Chim. Acta **58** [1982] 243/9).
[12] Swisher, R. G. (Diss. Univ. Massachusetts 1982, pp. 1/285; Diss. Abstr. Intern. B **42** [1982] 4786/7).
[13] Otera, J., Yano, T., Kusakabe, K. (Bull. Chem. Soc. Japan **56** [1983] 1057/9).
[14] Swisher, R. G., Holmes, R. (Organometallics **3** [1984] 365/9).
[15] Tzschach, A. Nietzschmann, E., Mügge, C. (Z. Anorg. Allgem. Chem. **532** [1985] 21/4).

Empirical Formula Index

In the following index the compounds are listed by their empirical formulas in the order of increasing carbon content (first column). The second column contains the substance formulas wherein cyclic ligands are partly also written as empirical formula. Formulas of ionic compounds are given in brackets; ions as well as components of solvates and adducts are separated by a period.

In the third column, page references are printed in ordinary type, table numbers in bold face and compounds numbers within the table in italics.

$\mathbf{C}_8H_{18}B_2O_4Sn$	$(C_4H_9)_2Sn(OBO)_2$	350
$C_8H_{18}Cl_2O_8Sn$	$(C_4H_9)_2Sn(OClO_3)_2$	336
$C_8H_{18}F_2O_6S_2Sn$	$(C_4H_9)_2Sn(OSO_2F)_2$	348, **34**, *4*
$C_8H_{18}F_4O_4P_2Sn$	$(C_4H_9)_2Sn(OP(O)F_2)_2$	356, **36**, *19*
$C_8H_{18}N_2O_6Sn$	$(C_4H_9)_2Sn(ONO_2)_2$	337
$C_8H_{18}O_3SSn$	$(C_4H_9)_2SnOSO_2$	346
$C_8H_{18}O_3SnTi$	$(C_4H_9)_2SnOTiO_2$	368
$C_8H_{18}O_4SSn$	$(C_4H_9)_2SnOSO_3$	346
$C_8H_{19}O_3PSn$	$(C_4H_9)_2SnOP(O)(H)O$	358, **37**, *1*
$C_8H_{19}O_4PSn$	$(C_4H_9)_2SnOP(O)(OH)O$	359, **37**, *6*
$C_8H_{20}O_2Sn$	$(C_4H_9)_2Sn(OH)_2$	1
	$((CH_3)_3C)_2Sn(OH)_2$	373, **43**, *1*
$C_8H_{22}N_2O_6S_2Sn$	$(C_4H_9)_2Sn(OSO_2NH_2)_2$	348, **34**, *5*
$C_8H_{22}O_4P_2Sn$	$(C_4H_9)_2Sn(OP(O)H_2)_2$	354, **36**, *2*
$\mathbf{C}_9H_{21}AsO_3Sn$	$(C_4H_9)_2SnOAs(O)(CH_3)O$	363, **39**, *1*
$\mathbf{C}_{10}H_{18}F_6O_6S_2Sn$	$(C_4H_9)_2Sn(OSO_2CF_3)_2$	348, **34**, *9*
$C_{10}H_{18}O_4Sn$	$(C_4H_9)_2SnOOC\text{-}COO$	301
$C_{10}H_{20}O_4Sn$	$(C_4H_9)_2Sn(OOCH)_2$	103
$C_{10}H_{22}O_2Sn$	$(C_4H_9)_2SnOCH_2CH_2O$	48, **14**, *1*
$C_{10}H_{22}O_3Sn$	$(C_4H_9)_2SnOOCH(CH_3)O$	334, **31**, *3*
$C_{10}H_{24}O_2Sn$	$(C_4H_9)_2Sn(OCH_3)_2$	3, **1**, *1*
	$((CH_3)_2CHCH_2)_2Sn(OCH_3)_2$	369, **41**, *1*
	$((CH_3)_3C)_2Sn(OCH_3)_2$	373, **43**, *2*
$C_{10}H_{24}O_4Se_2Sn$	$(C_4H_9)_2Sn(OSe(O)CH_3)_2$	368
$C_{10}H_{24}O_6S_2Sn$	$(C_4H_9)_2Sn(OSO_2CH_3)_2$	348, **34**, *6*
$\mathbf{C}_{11}H_{20}O_4Sn$	$(C_4H_9)_2SnOOCCH_2COO$	322, **30**, *1*
$C_{11}H_{22}O_3Sn$	$(C_4H_9)_2SnOOCCH(CH_3)O$	331
$C_{11}H_{23}ClO_2Sn$	$(C_4H_9)_2SnOCH(CH_2Cl)CH_2O$	49, **14**, *4*
$C_{11}H_{24}O_2Sn$	$(C_4H_9)_2SnOCH(CH_3)CH_2O$	49, **14**, *2*

	$(C_4H_9)_2SnOCH_2CH_2CH_2O$	53, **14**, *20*
$C_{11}H_{24}O_3Sn$	$(C_4H_9)_2SnOOCH(C_2H_5)O$	334, **31**, *4*
$C_{12}H_{18}Br_2O_4Sn$	$(C_4H_9)_2SnOOCCBr{=}CBrCOO$	325, **30**, *24*
$C_{12}H_{18}Cl_2O_4Sn$	$(C_4H_9)_2SnOOCCCl{=}CClCOO$	325, **30**, *23*
$C_{12}H_{18}Cl_6O_4Sn$	$(C_4H_9)_2Sn(OOCCCl_3)_2$	256, **23**, *6*
$C_{12}H_{18}F_4O_4Sn$	$(C_4H_9)_2SnOOCCF_2CF_2COO$	322, **30**, *4*
$C_{12}H_{18}F_6O_4Sn$	$(C_4H_9)_2Sn(OOCCF_3)_2$	255, **23**, *1*
$C_{12}H_{18}F_6O_6S_2Sn$	$(C_4H_9)_2Sn(OSO_2CF{=}CF_2)_2$	348, **34**, *10*
$C_{12}H_{18}O_4Sn$	$(C_4H_9)_2SnOOCC{\equiv}CCOO$	326, **30**, *30*
$C_{12}H_{20}Cl_4O_4Sn$	$(C_4H_9)_2Sn(OOCCHCl_2)_2$	255, **23**, *5*
$C_{12}H_{20}O_4Sn$	$(C_4H_9)_2SnOOCCH{=}CHCOO$	301/4
	$((CH_3)_2CHCH_2)_2SnOOCCH{=}CHCOO$	371, **41**, *14*
$C_{12}H_{22}Cl_2O_4Sn$	$(C_4H_9)_2Sn(OOCCH_2Cl)_2$	255, **23**, *4*
$C_{12}H_{22}F_6O_2Sn$	$(C_4H_9)_2Sn(OCH_2CF_3)_2$	5, **1**, *17*
$C_{12}H_{22}O_4Sn$	$(C_4H_9)_2SnOOCCH_2CH_2COO$	322, **30**, *2*
$C_{12}H_{22}O_5Sn$	$(C_4H_9)_2SnOCH_2COOCOCH_2O$	70, **14**, *102*
$C_{12}H_{24}O_2Sn$	$(C_4H_9)_2SnOC(CH_3){=}C(CH_3)O$	57, **14**, *42*
	$(C_4H_9)_2SnOC_4H_6O$	52, **14**, *15*
	$(C_4H_9)_2SnOCH_2CH{=}CHCH_2O$	58, **14**, *45*
$C_{12}H_{24}O_4Sn$	$(C_4H_9)_2Sn(OOCCH_3)_2$	104/26
	$((CH_3)_2CHCH_2)_2Sn(OOCCH_3)_2$	370, **41**, *10*
	$((CH_3)_3C)_2Sn(OOCCH_3)_2$	376, **43**, *17*
$C_{12}H_{24}O_6S_2Sn$	$(C_4H_9)_2Sn(OSO_2CH{=}CH_2)_2$	348, **34**, *8*
$C_{12}H_{24}O_6Sn$	$(C_4H_9)_2Sn(OOCCH_2OH)_2$	256, **23**, *14*
	$(C_4H_9)_2Sn(OOCOCH_3)_2$	332
$C_{12}H_{25}NO_4Sn$	$(C_4H_9)_2SnOCH_2C(CH_3)(NO_2)CH_2O$	56, **14**, *31*
$C_{12}H_{26}N_2O_2Sn$	$(C_4H_9)_2Sn(ON{=}CHCH_3)_2$	342, **32**, *20*
$C_{12}H_{26}N_2O_4Sn$	$(C_4H_9)_2Sn(ONHCOCH_3)_2$	339, **32**, *2*
$C_{12}H_{26}O_2SSn$	$(C_4H_9)_2SnOCH_2CH_2SCH_2CH_2O$	73, **14**, *116*
	$((CH_3)_3C)_2SnOCH_2CH_2SCH_2CH_2O$	375, **43**, *13*
$C_{12}H_{26}O_2Sn$	$(C_4H_9)_2SnOC(CH_3)_2CH_2O$	49, **14**, *3*
	$(C_4H_9)_2SnOCH(CH_3)CH(CH_3)O$	50, **14**, *8*
	$(C_4H_9)_2SnOCH(CH_3)CH_2CH_2O$	54, **14**, *21*
	$(C_4H_9)_2SnOCH_2CH_2CH_2CH_2O$	57, **14**, *39*
$C_{12}H_{26}O_3Sn$	$(C_4H_9)_2SnOCH_2C(CH_3)(OH)CH_2O$	55, **14**, *28*
	$(C_4H_9)_2SnOCH_2CH_2OCH_2CH_2O$	71, **14**, *103*
	$(C_4H_9)_2SnOOC(CH_3)(C_2H_5)O$	335, **31**, *12*
$C_{12}H_{27}NO_2Sn$	$(C_4H_9)_2SnOCH_2C(CH_3)(NH_2)CH_2O$	55, **14**, *30*
	$(C_4H_9)_2SnOCH_2CH_2NHCH_2CH_2O$	71, **14**, *107*
	$((CH_3)_2CHCH_2)_2SnOCH_2CH_2NHCH_2CH_2O$	370, **41**, *6*
	$((CH_3)_3C)_2SnOCH_2CH_2NHCH_2CH_2O$	374, **43**, *6*

$C_{12}H_{28}O_2Sn$	$(C_4H_9)_2Sn(OC_2H_5)_2$	3, **1**, *2*
$C_{12}H_{30}O_6P_2S_2Sn$	$(C_4H_9)_2Sn(OP(S)(OCH_3)_2)_2$	356, **36**, *18*
$C_{13}H_{18}F_6O_4Sn$	$(C_4H_9)_2SnOOC(CF_2)_3COO$	322, **30**, *6*
$C_{13}H_{21}NO_2Sn$	$(C_4H_9)_2SnOC_5H_3NO$	60, **14**, *55*
$C_{13}H_{22}O_4S_3Sn$	$(C_4H_9)_2SnOOCCH_2SC(=S)SCH_2COO$	324, **30**, *14*
$C_{13}H_{22}O_4Sn$	$(C_4H_9)_2SnOOCC(=CH_2)CH_2COO$	322, **30**, *3*
$C_{13}H_{24}O_4S_2Sn$	$(C_4H_9)_2SnOOCCH_2SCH_2SCH_2COO$	323, **30**, *11*
$C_{13}H_{25}NO_4Sn$	$(C_4H_9)_2SnOOCCH_2N(CH_3)CH_2COO$	325, **30**, *22*
	$((CH_3)_3C)_2SnOOCCH_2N(CH_3)CH_2COO$	376, **43**, *18*
$C_{13}H_{26}O_2SSn$	$(C_4H_9)_2SnOCH_2C(CH_2SCH_2)CH_2O$	56, **14**, *35*
$C_{13}H_{26}O_2Sn$	$(C_4H_9)_2SnOC_5H_8O$	52, **14**, *16*
$C_{13}H_{26}O_3Sn$	$(C_4H_9)_2SnOCH_2C(CH_2OCH_2)CH_2O$	56, **14**, *34*
$C_{13}H_{27}NO_4Sn$	$(C_4H_9)_2SnOCH_2C(C_2H_5)(NO_2)CH_2O$	56, **14**, *32*
$C_{13}H_{28}O_2Sn$	$(C_4H_9)_2SnO(CH_2)_5O$	57, **14**, *40*
	$(C_4H_9)_2SnOCH(CH_3)CH_2CH(CH_3)O$	54, **14**, *22*
	$(C_4H_9)_2SnOCH_2C(CH_3)_2CH_2O$	54, **14**, *23*
$C_{13}H_{28}O_3Sn$	$(C_4H_9)_2SnOCH_2C(CH_3)(CH_2OH)CH_2O$	55, **14**, *29*
$C_{13}H_{29}NO_2Sn$	$(C_4H_9)_2SnOCH_2CH_2N(CH_3)CH_2CH_2O$	72, **14**, *108*
	$((CH_3)_2CHCH_2)_2SnOCH_2CH_2NCH_3CH_2CH_2O$	370, **41**, *7*
	$((CH_3)_3C)_2SnOCH_2CH_2NCH_3CH_2CH_2O$	374, **43**, *7*
$C_{14}H_{18}Cl_2O_4Sn$	$(C_4H_9)_2SnO\text{-}C_6Cl_2(=O)_2\text{-}O$	59, **14**, *51*
$C_{14}H_{18}N_6O_2Sn$	$(C_4H_9)_2Sn(ON=C(CN)_2)_2$	343, **32**, *30*
$C_{14}H_{20}Cl_3O_3PSSn$	$(C_4H_9)_2SnOP(S)(OC_6H_2Cl_3\text{-}2,4,5)O$	359, **37**, *7*
$C_{14}H_{20}F_{12}O_2Sn$	$(C_4H_9)_2Sn(OCH(CF_3)_2)_2$	5, **1**, *18*
$C_{14}H_{22}AsBrO_3Sn$	$(C_4H_9)_2SnOAs(O)(C_6H_4Br\text{-}4)O$	363, **39**, *5*
$C_{14}H_{22}AsClO_3Sn$	$(C_4H_9)_2SnOAs(O)(C_6H_4Cl\text{-}4)O$	363, **39**, *4*
$C_{14}H_{22}AsNO_5Sn$	$(C_4H_9)_2SnOAs(O)(C_6H_4NO_2\text{-}2)O$	363, **39**, *11*
	$(C_4H_9)_2SnOAs(O)(C_6H_4NO_2\text{-}4)O$	363, **39**, *12*
$C_{14}H_{22}AsNO_6Sn$	$(C_4H_9)_2SnOAs(O)(C_6H_3(NO_2\text{-}3)OH\text{-}4)O$	364, **39**, *13*
$C_{14}H_{22}Cl_2O_4Sn$	$(C_4H_9)_2Sn(OOCCH=CHCl)_2$	269, **24**, *4*
$C_{14}H_{22}O_2Sn$	$(C_4H_9)_2SnOC_6H_4O\text{-}1,2$	59, **14**, *48*
	$(C_4H_9)_2SnOC_6H_4O\text{-}1,4$	59, **14**, *49*
$C_{14}H_{23}AsN_2O_5Sn$	$(C_4H_9)_2SnOAs(O)(C_6H_3(NO_2\text{-}3)NH_2\text{-}4)O$	364, **39**, *14*
$C_{14}H_{23}AsO_3Sn$	$(C_4H_9)_2SnOAs(O)(C_6H_5)O$	363, **39**, *3*
$C_{14}H_{23}AsO_4Sn$	$(C_4H_9)_2SnOAs(O)(C_6H_4OH\text{-}4)O$	363, **39**, *6*
$C_{14}H_{23}NO_3Sn$	$(C_4H_9)_2SnOOCH(C_5H_4N)O$	335, **31**, *11*
$C_{14}H_{24}AsNO_3Sn$	$(C_4H_9)_2SnOAs(O)(C_6H_4NH_2\text{-}2)O$	363, **39**, *9*
	$(C_4H_9)_2SnOAs(O)(C_6H_4NH_2\text{-}4)O$	363, **39**, *10*
$C_{14}H_{24}O_4Sn$	$(C_4H_9)_2Sn(OOCCH=CH_2)_2$	268, **24**, *1*
$C_{14}H_{26}Cl_2O_4Sn$	$(C_4H_9)_2Sn(OOCCHClCH_3)_2$	256, **23**, *7*

$C_{14}H_{26}Cl_2O_4Sn$	$(C_4H_9)_2Sn(OOCCH_2CH_2Cl)_2$	256, **23**, *8*
$C_{14}H_{26}Cl_6O_4Sn$	$(C_4H_9)_2Sn(OCH(CCl_3)OCH_3)_2$	5, **1**, *20*
$C_{14}H_{26}O_4SSn$	$(C_4H_9)_2SnOOCCH_2CH_2SCH_2CH_2COO$	324, **30**, *17*
$C_{14}H_{26}O_4Sn$	$(C_4H_9)_2SnOOC(CH_2)_4COO$	323, **30**, *7*
$C_{14}H_{26}O_5Sn$	$(C_4H_9)_2SnOC_5H_5O(OH)(OCH_2)O$	68, **14**, *94*
$C_{14}H_{26}O_6Sn$	$(C_4H_9)_2SnOCH(COOCH_3)CH(COOCH_3)O$	51, **14**, *13*
$C_{14}H_{28}Cl_2N_2O_2Sn$	$(C_4H_9)_2Sn(ON{=}C(CH_3)CH_2Cl)_2$	343, **32**, *25*
$C_{14}H_{28}O_2Sn$	$(C_4H_9)_2Sn(OCH_2CH{=}CH_2)_2$	29, **11**, *1*
	$(C_4H_9)_2SnOC_6H_{10}O$	52, **14**, *17*
	$(C_4H_9)_2SnOCH_2C(C_3H_6)CH_2O$	56, **14**, *33*
$C_{14}H_{28}O_3Sn$	$(C_4H_9)_2SnOOC_6H_{10}O$	335, **31**, *13*
$C_{14}H_{28}O_4S_2Sn$	$(C_4H_9)_2Sn(OOCCH_2SCH_3)_2$	258, **23**, *34*
$C_{14}H_{28}O_4S_2Sn$	$(C_4H_9)_2Sn(OOCCH_2CH_2SH)_2$	259, **23**, *40*
$C_{14}H_{28}O_4Sn$	$(C_4H_9)_2SnOC_5H_7O(OCH_3)O$	63, **14**, *73*
	$(C_4H_9)_2Sn(OOCC_2H_5)_2$	127
$C_{14}H_{28}O_5Sn$	$(C_4H_9)_2SnOC_4H_4O(CH_2OH)(OCH_3)O$	60, **14**, *56*
	$(C_4H_9)_2SnOC_5H_6O(OCH_3)(OH)O$	64, **14**, *75*
$C_{14}H_{28}O_6Sn$	$(C_4H_9)_2Sn(OOCCH(OH)CH_3)_2$	256, **23**, *15*
$C_{14}H_{30}N_2O_2Sn$	$(C_4H_9)_2Sn(ON{=}C(CH_3)_2)_2$	343, **32**, *23*
$C_{14}H_{30}O_2Sn$	$(C_4H_9)_2SnOC(CH_3)_2C(CH_3)_2O$	50, **14**, *9*
	$(C_4H_9)_2SnOCH(CH_3)CH_2C(CH_3)_2O$	54, **14**, *24*
	$(C_4H_9)_2SnO(CH_2)_6O$	57, **14**, *41*
$C_{14}H_{30}O_4Sn$	$(C_4H_9)_2SnOCH(CH_2OCH_3)CH(CH_2OCH_3)O$	50, **14**, *10*
$C_{14}H_{30}O_6S_2Sn$	$(C_4H_9)_2SnO_3S(CH_2)_6SO_2O$	346
$C_{14}H_{31}NO_2Sn$	$(C_4H_9)_2SnOCH_2CH_2N(C_2H_5)CH_2CH_2O$	72, **14**, *109*
	$((CH_3)_3C)_2SnOCH_2CH_2N(C_2H_5)CH_2CH_2O$	374, **43**, *8*
$C_{14}H_{31}NO_3Sn$	$(C_4H_9)_2SnOCH_2CH_2N(CH_2CH_2OH)CH_2CH_2O$	72, **14**, *111*
$C_{14}H_{31}O_3PSn$	$(C_4H_9)_2SnOP(O)(C_6H_{13})O$	359, **37**, *2*
$C_{14}H_{32}O_2Sn$	$(C_4H_9)_2Sn(OC_3H_7)_2$	3, **1**, *3*
	$(C_4H_9)_2Sn(OCH(CH_3)_2)_2$	3, **1**, *4*
$C_{14}H_{32}O_4Sn$	$(C_4H_9)_2Sn(OCH(CH_3)OCH_3)_2$	5, **1**, *19*
$C_{14}H_{34}N_2O_2Sn$	$(C_4H_9)_2Sn(OCH_2CH_2NHCH_3)_2$	6, **1**, *34*
	$(C_4H_9)_2Sn(OCH(CH_3)CH_2NH_2)_2$	7, **1**, *38*
	$(C_4H_9)_2Sn(OCH_2CH_2CH_2NH_2)_2$	7, **1**, *40*
$C_{14}H_{36}O_2Si_2Sn$	$(C_4H_9)_2Sn(OSi(CH_3)_3)_2$	352
$C_{14}H_{40}B_{20}O_4Sn$	$(C_4H_9)_2Sn(OOCB_{10}H_9C_2H_2\text{-}1,7)_2$	299, **28**, *6*
	$(C_4H_9)_2Sn(OOCC_2HB_{10}H_{10}\text{-}1,2)_2$	299, **28**, *7*
$C_{15}H_{21}NO_4Sn$	$(C_4H_9)_2SnOOC\text{-}C_5H_3N\text{-}COO$	328, **30**, *41*
$C_{15}H_{22}O_3Sn$	$(C_4H_9)_2SnOOCC_6H_4O$	331
$C_{15}H_{23}AsO_5Sn$	$(C_4H_9)_2SnOAs(O)(C_6H_4COOH\text{-}2)O$	364, **39**, *16*
	$(C_4H_9)_2SnOAs(O)(C_6H_4COOH\text{-}4)O$	364, **39**, *17*

$C_{15}H_{23}ClO_3Sn$	$(C_4H_9)_2SnOOCH(C_6H_4Cl\text{-}4)O$	335, **31**, *7*
$C_{15}H_{23}NO_3Sn$	$(C_4H_9)_2SnOOCC_6H_3(NH_2\text{-}4)O$	332
$C_{15}H_{23}NO_5Sn$	$(C_4H_9)_2SnOOCH(C_6H_4NO_2\text{-}4)O$	335, **31**, *8*
$C_{15}H_{24}O_2Sn$	$(C_4H_9)_2SnOCH_2C_6H_4O$	60, **14**, *53*
$C_{15}H_{24}O_3Sn$	$(C_4H_9)_2SnOOCH(C_6H_5)O$	335, **31**, *5*
$C_{15}H_{25}AsO_3Sn$	$(C_4H_9)_2SnOAs(O)(CH_2C_6H_5)O$	363, **39**, *2*
	$(C_4H_9)_2SnOAs(O)(C_6H_4CH_3\text{-}4)O$	364, **39**, *15*
$C_{15}H_{25}AsO_4Sn$	$(C_4H_9)_2SnOAs(O)(C_6H_4OCH_3\text{-}2)O$	363, **39**, *7*
	$(C_4H_9)_2SnOAs(O)(C_6H_4OCH_3\text{-}4)O$	363, **39**, *8*
$C_{15}H_{25}O_3PSn$	$(C_4H_9)_2SnOP(O)(CH_2C_6H_5)O$	359, **37**, *4*
$C_{15}H_{26}O_3Sn$	$(C_4H_9)_2SnOC(CH_3){=}CHCOCH{=}C(CH_3)O$	58, **14**, *46*
$C_{15}H_{28}O_4S_2Sn$	$(C_4H_9)_2SnOOCCH_2SC(CH_3)_2SCH_2COO$	323, **30**, *12*
$C_{15}H_{28}O_4Sn$	$(C_4H_9)_2SnOOCCH_2CH(CH_3)CH_2CH_2COO$	323, **30**, *8*
$C_{15}H_{29}NO_4Sn$	$((CH_3)_3C)_2SnOOCCH_2CH_2N(CH_3)CH_2CH_2COO$	376, **43**, *19*
$C_{15}H_{30}O_2Sn$	$(C_4H_9)_2SnOC_7H_{12}O$	53, **14**, *18*
$C_{15}H_{30}O_4Sn$	$(C_4H_9)_2SnOC_5H_6O(CH_3)(OCH_3)O$	64, **14**, *77*
$C_{15}H_{30}O_5Sn$	$(C_4H_9)_2SnOC_5H_5O(OCH_3)(CH_3)(OH)O$	65, **14**, *78*
$C_{15}H_{30}O_6Sn$	$(C_4H_9)_2SnOC_5H_5O(CH_2OH)(OCH_3)(OH)O$	65, **14**, *81*
	$(C_4H_9)_2SnOC_5H_5O(OCH_3)(CH_2OH)(OH)O$	65, **14**, *82*
		66, **14**, *83*
$C_{15}H_{32}O_2Sn$	$(C_4H_9)_2SnOC(CH_3)_2CH_2C(CH_3)_2O$	55, **14**, *25*
	$(C_4H_9)_2SnOCH_2C(CH_3)(C_3H_7)CH_2O$	55, **14**, *26*
$C_{15}H_{33}NO_2Sn$	$(C_4H_9)_2SnOCH_2CH_2N(C_3H_7)CH_2CH_2O$	72, **14**, *110*
	$(C_4H_9)_2SnOCH(CH_3)CH_2N(CH_3)CH_2CH(CH_3)O$	72, **14**, *114*
	$((CH_3)_3C)_2SnOCH_2CH_2N(C_3H_7)CH_2CH_2O$	374, **43**, *9*
$C_{16}H_{18}Cl_4O_4Sn$	$(C_4H_9)_2SnOOCC_6Cl_4COO$	327, **30**, *33*
$C_{16}H_{18}F_{14}O_4Sn$	$(C_4H_9)_2Sn(OOCC_3F_7)_2$	255, **23**, *2*
$C_{16}H_{21}NO_6Sn$	$(C_4H_9)_2SnOOCC_6H_3(NO_2\text{-}2)COO$	327, **30**, *34*
$C_{16}H_{22}O_4Sn$	$(C_4H_9)_2SnOOCC_6H_4COO\text{-}2$	326, **30**, *31*
	$(C_4H_9)_2SnOOCC_6H_4COO\text{-}4$	326, **30**, *32*
$C_{16}H_{24}O_3Sn$	$(C_4H_9)_2SnOOCCH(C_6H_5)O$	331
$C_{16}H_{24}O_4S_4Sn$	$(C_4H_9)_2Sn(OS(O)C_4H_3S)_2$	348, **34**, *2*
$C_{16}H_{24}O_5Sn$	$(C_4H_9)_2SnOOCH(C_6H_4COOH\text{-}4)O$	335, **31**, *10*
$C_{16}H_{24}O_8Sn$	$(C_4H_9)_2Sn(OOCCH{=}CHCOOH)_2$	273, **25**, *1*
$C_{16}H_{25}AsO_4Sn$	$(C_4H_9)_2SnOAs(O)(C_6H_5)CH_2COO$	367
$C_{16}H_{25}NO_2Sn$	$(C_4H_9)_2SnOC_6H_4N{=}C(CH_3)O$	74, **14**, *120*
$C_{16}H_{25}N_3O_2Sn$	$(C_4H_9)_2SnOC_6H_4CH{=}NN{=}C(NH_2)O\text{-}2$	73, **14**, *118*
$C_{16}H_{26}N_2O_6Sn$	$(C_4H_9)_2Sn(ONC_4H_4O_2)_2$	342, **32**, *17*
$C_{16}H_{26}O_2Sn$	$(C_4H_9)_2SnOCH(C_6H_5)CH_2O$	49, **14**, *6*
$C_{16}H_{26}O_3Sn$	$(C_4H_9)_2SnOOCH(C_6H_4CH_3\text{-}4)O$	335, **31**, *9*
$C_{16}H_{26}O_4Sn$	$(C_4H_9)_2SnOOCC_6H_8COO$	325, **30**, *26*

$C_{16}H_{26}O_4Sn$	$(C_4H_9)_2SnOOCH(C_6H_4OCH_3\text{-}4)O$	335, **31**, *6*
$C_{16}H_{28}O_4Sn$	$(C_4H_9)_2Sn(OOCC(CH_3)=CH_2)_2$	268, **24**, *2*
	$(C_4H_9)_2Sn(OOCCH=CHCH_3)_2$	269, **24**, *3*
	$((CH_3)_2CHCH_2)_2Sn(OOCC(CH_3)=CH_2)_2$	370, **41**, *12*
$C_{16}H_{28}O_8Sn$	$(C_4H_9)_2Sn(OOCCH_2COOCH_3)_2$	262, **23**, *66*
$C_{16}H_{30}Cl_2O_4Sn$	$(C_4H_9)_2Sn(OOCCH_2CH_2CH_2Cl)_2$	256, **23**, *9*
$C_{16}H_{30}O_6Sn$	$(C_4H_9)_2SnOCH(COOC_2H_5)CH(COOC_2H_5)O$	52, **14**, *14*
$C_{16}H_{32}O_2Sn$	$(C_4H_9)_2Sn(OCH_2C_3H_5)_2$	5, **1**, *16*
	$(C_4H_9)_2Sn(OCH_2CH_2CH=CH_2)_2$	29, **11**, *2*
	$(C_4H_9)_2Sn(OCH_2CH=CHCH_3)_2$	29, **11**, *3*
	$(C_4H_9)_2Sn(OCH_2C(CH_3)=CH_2)_2$	29, **11**, *4*
$C_{16}H_{32}O_4Sn$	$(C_4H_9)_2SnOCH(CH_3)CH(OC(CH_3)_2OCH_2)CHO$	56, **14**, *37*
	$(C_4H_9)_2Sn(OOCC_3H_7)_2$	127
$C_{16}H_{32}O_5Sn$	$(C_4H_9)_2SnOC_5H_5O(CH_3)(OCH_3)_2O$	65, **14**, *79*
$C_{16}H_{32}O_6Sn$	$(C_4H_9)_2Sn(OOCCH_2CH_2OCH_3)_2$	257, **23**, *21*
$C_{16}H_{34}N_2O_2Sn$	$(C_4H_9)_2Sn(ON=CHC_3H_7)_2$	343, **32**, *21*
	$(C_4H_9)_2Sn(ON=C(CH_3)CH_2CH_3)_2$	343, **32**, *24*
$C_{16}H_{34}N_2O_4Sn$	$(C_4H_9)_2Sn(ONHCOCH(CH_3)_2)_2$	339, **32**, *3*
$C_{16}H_{35}NO_2Sn$	$((CH_3)_3C)_2SnOCH_2CH_2N(CH_2CH_2CH_2CH_3)CH_2CH_2O$	374, **43**, *10*
	$((CH_3)_3C)_2SnOCH_2CH_2N(CHCH(CH_3)_2)CH_2CH_2O$	375, **43**, *11*
	$((CH_3)_3C)_2SnOCH_2CH_2N(C(CH_3)_3)CH_2CH_2O$	375, **43**, *12*
$C_{16}H_{35}O_3PSn$	$(C_4H_9)_2SnOP(O)(C_8H_{17})O$	359, **37**, *3*
$C_{16}H_{36}O_2Sn$	$(C_4H_9)_2Sn(OC_4H_9)_2$	4, **1**, *5*
	$(C_4H_9)_2Sn(OCH(CH_3)C_2H_5)_2$	4, **1**, *6*
	$(C_4H_9)_2Sn(OCH_2CH(CH_3)_2)_2$	4, **1**, *7*
	$(C_4H_9)_2Sn(OC(CH_3)_3)_2$	4, **1**, *8*
	$((CH_3)_2CHCH_2)_2Sn(OCH_2CH(CH_3)_2)_2$	369, **41**, *2*
$C_{16}H_{36}O_4S_2Sn$	$(C_4H_9)_2Sn(OS(O)C_4H_9)_2$	347, **34**, *1*
$C_{16}H_{36}O_4Sn$	$(C_4H_9)_2Sn(OOC(CH_3)_3)_2$	334, **31**, *1*
$C_{16}H_{36}O_6S_2Sn$	$(C_4H_9)_2Sn(OSO_2C_4H_9)_2$	348, **34**, *7*
$C_{16}H_{36}O_7P_2Sn$	$(C_4H_9)_2SnOP(O)(OC_4H_9)OP(O)(OC_4H_9)O$	361
$C_{16}H_{38}N_2O_2Sn$	$(C_4H_9)_2Sn(OCH_2CH_2NHC_2H_5)_2$	6, **1**, *35*
	$(C_4H_9)_2Sn(OCH_2CH_2N(CH_3)_2)_2$	6, **1**, *36*
	$(C_4H_9)_2Sn(ON(C_2H_5)_2)_2$	339, **32**, *1*
$C_{16}H_{38}O_6P_2Sn$	$(C_4H_9)_2Sn(OP(O)(H)OC_4H_9)_2$	355, **36**, *8*
	$(C_4H_9)_2Sn(OP(OC_2H_5)_2)_2$	355, **36**, *9*
$C_{16}H_{38}O_8P_2Sn$	$(C_4H_9)_2Sn(OP(O)(OC_2H_5)_2)_2$	356, **36**, *15*
$C_{17}H_{20}Cl_6O_4Sn$	$(C_4H_9)_2SnOOCC_7H_2Cl_6COO$	326, **30**, *29*
$C_{17}H_{22}O_6Sn$	$(C_4H_9)_2SnOOCC_6H_3(COOH\text{-}3)COO$	327, **30**, *35*
$C_{17}H_{26}O_4Sn$	$(C_4H_9)_2SnOOCC_7H_8COO$	326, **30**, *28*
$C_{17}H_{27}FN_2O_6Sn$	$(C_4H_9)_2SnOC_4H_4O(CH_2OH)(NCONHCOCF=CH)O$	61, **14**, *59*

$C_{17}H_{27}NO_2Sn$	$(C_4H_9)_2SnOC_6H_4CH{=}NCH_2CH_2O$	73, **14**, *117*
$C_{17}H_{27}N_3O_2Sn$	$(C_4H_9)_2SnOC_6H_4C(CH_3){=}NN{=}C(NH_2)O\text{-}2$	73, **14**, *119*
$C_{17}H_{28}N_2O_6Sn$	$(C_4H_9)_2SnOC_4H_4O(CH_2OH)(NCONHCOCH{=}CH)O$	60, **14**, *58*
$C_{17}H_{29}N_3O_5Sn$	$(C_4H_9)_2SnOC_4H_4O(CH_2OH)(NCON{=}C(NH_2)CH{=}CH)O$	61, **14**, *60*
$C_{17}H_{32}O_4S_2Sn$	$(C_4H_9)_2SnOOCCH_2CH_2SC(CH_3)_2SCH_2CH_2COO$	324, **30**, *19*
$C_{17}H_{32}O_4Sn$	$(C_4H_9)_2SnOOC(CH_2)_7COO$	323, **30**, *9*
$C_{18}H_{20}F_{12}O_4Sn$	$(C_4H_9)_2Sn(OC(CF_3){=}CHCOCF_3)_2$	98, **19**, *9*
$C_{18}H_{22}Br_2O_6Sn$	$(C_4H_9)_2Sn(OOCC_4H_2BrO)_2$	299, **28**, *2*
$[C_{18}H_{22}O_8S_2Sn]^{2-} \cdot 2Na^+$	$[(C_4H_9)_2SnOC_{10}H_4(SO_3)_2O]^{2-} \cdot 2Na^+$	59, **14**, *52*
$C_{18}H_{24}N_4O_2Sn$	$(C_4H_9)_2Sn(OC(CH_3){=}C(CN)_2)_2$	30, **11**, *11*
$C_{18}H_{24}O_6Sn$	$(C_4H_9)_2Sn(OOCC_4H_3O)_2$	298, **28**, *1*
$C_{18}H_{26}N_2O_2Sn$	$(C_4H_9)_2Sn(OC_5H_4N)_2$	40, **13**, *4*
$C_{18}H_{26}N_2O_4Sn$	$(C_4H_9)_2Sn(OC_5H_4NO)_2$	40, **13**, *5*
	$(C_4H_9)_2Sn(OOCC_4H_3NH)_2$	299, **28**, *3*
$C_{18}H_{26}O_8Sn$	$(C_4H_9)_2SnOOCCH{=}CHCOOCH_2CH_2OOCCH{=}CHCOO$	325, **30**, *25*
$C_{18}H_{28}BrN_5O_4Sn$	$(C_4H_9)_2SnOC_4H_4O(CH_2OH)(C_5H_3N_5Br)O$	62, **14**, *68*
$C_{18}H_{28}N_4O_5Sn$	$(C_4H_9)_2SnOC_4H_4O(CH_2OH)(C_5H_3N_4O)O$	61, **14**, *61*
$C_{18}H_{28}O_4Sn$	$(C_4H_9)_2Sn(OCH_2C_4H_3O)_2$	7, **1**, *44*
$C_{18}H_{28}O_8Sn$	$(C_4H_9)_2Sn(OOCCH{=}CHCOOCH_3)_2$	273, **25**, *2*
$C_{18}H_{29}BrN_6O_4Sn$	$(C_4H_9)_2SnOC_4H_4O(CH_2OH)(C_5H_4N_6Br)O$	63, **14**, *69*
$C_{18}H_{29}N_5O_4Sn$	$(C_4H_9)_2SnOC_4H_4O(CH_2OH)(C_5H_4N_5)O$	61, **14**, *62*
$C_{18}H_{29}N_5O_5Sn$	$(C_4H_9)_2SnOC_4H_4O(CH_2OH)(C_5H_4N_5O)O$	62, **14**, *67*
$C_{18}H_{30}O_2Sn$	$(C_4H_9)_2SnOC_6H_3(C(CH_3)_3\text{-}4)O$	59, **14**, *50*
$C_{18}H_{30}O_3Sn$	$(C_4H_9)_2SnOCH(CH_2OCH_2C_6H_5)CH_2O$	49, **14**, *5*
	$(C_4H_9)_2SnOCH_2CH(OCH_2C_6H_5)CH_2O$	55, **14**, *27*
$C_{18}H_{31}NO_2Sn$	$(C_4H_9)_2SnOCH_2CH_2N(C_6H_5)CH_2CH_2O$	72, **14**, *112*
	$((CH_3)_2CHCH_2)_2SnOCH_2CH_2N(C_6H_5)CH_2CH_2O$	370, **41**, *9*
$C_{18}H_{32}Cl_6N_2O_4Sn$	$(C_4H_9)_2Sn(OCH(CCl_3)ON{=}C(CH_3)_2)_2$	5, **1**, *21*
$C_{18}H_{32}O_4Sn$	$(C_4H_9)_2Sn(OC(CH_3){=}CHCOCH_3)_2$	97, **19**, *1*
	$((CH_3)_3C)_2Sn(OC(CH_3){=}CHCOCH_3)_2$	376, **43**, *16*
$C_{18}H_{32}O_6Sn$	$(C_4H_9)_2Sn(OCH_2COCH_2COCH_3)_2$	6, **1**, *30*
	$(C_4H_9)_2Sn(OC(CH_3){=}CHCOOCH_3)_2$	99, **19**, *18*
$C_{18}H_{32}O_8Sn$	$(C_4H_9)_2Sn(OOCCH_2COOC_2H_5)_2$	262, **23**, *67*
$C_{18}H_{34}Cl_2O_4Sn$	$(C_4H_9)_2Sn(OOC(CH_2)_3CH_2Cl)_2$	256, **23**, *10*
$C_{18}H_{34}N_2O_2Sn$	$(C_4H_9)_2Sn(OC(CH_3){=}CHC(CH_3){=}NH)_2$	100, **19**, *29*
$C_{18}H_{34}N_2O_6Sn$	$(C_4H_9)_2Sn(OOCCH(CH_3)NHCOCH_3)_2$	260, **23**, *52*
$C_{18}H_{34}O_4Sn$	$(C_4H_9)_2SnOOC(CH_2)_8COO$	323, **30**, *10*
$C_{18}H_{34}O_6Sn$	$(C_4H_9)_2SnOCH_2CH(C_4H_4O(OCH_3)O_2C(CH_3)_2)O$	50, **14**, *7*
$C_{18}H_{36}O_2Sn$	$(C_4H_9)_2Sn(OCH_2CH_2CH_2CH{=}CH_2)_2$	29, **11**, *5*

$C_{20}H_{32}O_8Sn$	$(C_4H_9)_2Sn(OOCCH{=}CHCOOC_2H_5)_2$	273, **25**, *3*
$C_{20}H_{32}O_{10}Sn$	$(C_4H_9)_2Sn(OOCCH{=}CHCOOCH_2CH_2OH)_2$	276, **25**, *30*
$C_{20}H_{33}N_5O_4Sn$	$(C_4H_9)_2SnOC_4H_4O(CH_2OH)(C_7H_8N_5)O$	61, **14**, *63*
$C_{20}H_{36}O_4Sn$	$(C_4H_9)_2Sn(OC(CH_3){=}CHCOC_2H_5)_2$	98, **19**, *2*
	$(C_4H_9)_2Sn(OOCC_5H_9\text{-c})_2$	253
$C_{20}H_{36}O_6Sn$	$(C_4H_9)_2Sn(OC(CH_3){=}CHCO(OC_2H_5))_2$	100, **19**, *19*
$C_{20}H_{36}O_8Sn$	$(C_4H_9)_2Sn(OOCCH_2CH_2COOC_2H_5)_2$	262, **23**, *68*
	$(C_4H_9)_2Sn(OOC(CH_2)_4COOH)_2$	263, **23**, *74*
$C_{20}H_{40}O_2Sn$	$(C_4H_9)_2Sn(OC_6H_{11})_2$	5, **1**, *15*
	$(C_4H_9)_2Sn(OCH(C_2H_5)CH{=}CHCH_3)_2$	30, **11**, *9*
	$(C_4H_9)_2Sn(OC(CH_3)(C_2H_5)CH{=}CH_2)_2$	30, **11**, *10*
	$(C_4H_9)_2SnOC_{12}H_{22}O$	53, **14**, *19*
$C_{20}H_{40}O_4Sn$	$(C_4H_9)_2Sn(OOCC_5H_{11})_2$	128
$C_{20}H_{40}O_6Sn$	$(C_4H_9)_2Sn(OCH(C_2H_5)COOC_2H_5)_2$	6, **1**, *28*
	$(C_4H_9)_2Sn(OC(CH_3)_2COOC_2H_5)_2$	6, **1**, *29*
	$(C_4H_9)_2Sn(OOCOC_5H_{11})_2$	336, **31**, *14*
$C_{20}H_{40}O_{14}Sn$	$(C_4H_9)_2Sn(OOC(CH(OH))_4CH_2OH)_2$	257, **23**, *18*
$C_{20}H_{41}NO_2Sn$	$(C_4H_9)_2SnOCH(CH_3)CH_2N(C_6H_{11})CH_2CH(CH_3)O$	73, **14**, *115*
$C_{20}H_{46}N_2O_2Sn$	$(C_4H_9)_2Sn(OCH_2CH_2N(C_2H_5)_2)_2$	7, **1**, *37*
$C_{20}H_{48}O_8Si_2Sn$	$(C_4H_9)_2Sn(OSi(OC_2H_5)_3)_2$	352
$C_{21}H_{27}AsClNO_3Sn$	$(C_4H_9)_2SnOAs(O)(C_6H_4(N{=}CHC_6H_4Cl\text{-}2)\text{-}2)O$	364, **39**, *18*
$C_{21}H_{27}AsN_2O_5Sn$	$(C_4H_9)_2SnOAs(O)(C_6H_4(N{=}CHC_6H_4NO_2\text{-}2)\text{-}2)O$	364, **39**, *19*
	$(C_4H_9)_2SnOAs(O)(C_6H_4(N{=}CHC_6H_4NO_2\text{-}3)\text{-}2)O$	364, **39**, *20*
$C_{21}H_{31}NO_6Sn$	$(C_4H_9)_2SnOOCCH(NHCOOCH_2C_6H_5)CH_2CH_2COO$	322, **30**, *5*
$C_{21}H_{32}O_4S_2Sn$	$(C_4H_9)_2SnOOCCH_2CH_2SCH(C_6H_5)SCH_2CH_2COO$	324, **30**, *20*
$C_{21}H_{32}O_5S_2Sn$	$(C_4H_9)_2SnOOCCH_2CH_2SCH(C_6H_4OH\text{-}2)SCH_2CH_2COO$	324, **30**, *21*
$C_{21}H_{34}O_5Sn$	$(C_4H_9)_2SnOC_5H_6O(OCH_3)(OCH_2C_6H_5)O$	64, **14**, *76*
$C_{21}H_{37}NO_7P_2S_2Sn$	$(C_4H_9)_2SnOP(S)(OC_2H_5)OC_6H_3(NHP(S)(OC_2H_5)_2\text{-}3)COO\text{-}6$	360
$C_{22}H_{24}Cl_2N_2O_8Sn$	$(C_4H_9)_2Sn(OOCC_6H_3(Cl\text{-}2)NO_2\text{-}4)_2$	291, **27**, *28*
$C_{22}H_{24}Cl_4O_4Sn$	$(C_4H_9)_2Sn(OOCC_6H_3Cl_2\text{-}2,4)_2$	288, **27**, *6*
$C_{22}H_{24}N_4O_{12}Sn$	$(C_4H_9)_2Sn(OOCC_6H_3(NO_2)_2\text{-}3,5)_2$	291, **27**, *30*
$C_{22}H_{26}Br_2O_4Sn$	$(C_4H_9)_2Sn(OOCC_6H_4Br\text{-}3)_2$	288, **27**, *7*
	$(C_4H_9)_2Sn(OOCC_6H_4Br\text{-}4)_2$	288, **27**, *8*
$C_{22}H_{26}Cl_2O_4Sn$	$(C_4H_9)_2Sn(OOCC_6H_4Cl\text{-}2)_2$	288, **27**, *3*
	$(C_4H_9)_2Sn(OOCC_6H_4Cl\text{-}3)_2$	288, **27**, *4*
	$(C_4H_9)_2Sn(OOCC_6H_4Cl\text{-}4)_2$	288, **27**, *5*
$C_{22}H_{26}F_2O_4Sn$	$(C_4H_9)_2Sn(OOCC_6H_4F\text{-}4)_2$	287, **27**, *2*
$C_{22}H_{26}N_2O_8Sn$	$(C_4H_9)_2Sn(OOCC_6H_4NO_2\text{-}2)_2$	290, **27**, *25*
	$(C_4H_9)_2Sn(OOCC_6H_4NO_2\text{-}3)_2$	291, **27**, *26*
	$(C_4H_9)_2Sn(OOCC_6H_4NO_2\text{-}4)_2$	291, **27**, *27*

$C_{22}H_{26}O_4S_2Sn$	$(C_4H_9)_2SnOC_6H_3(CHO)SSC_6H_3(CHO)O$	71, **14**, *105*
$C_{22}H_{28}Cl_2N_2O_4Sn$	$(C_4H_9)_2Sn(OOCC_6H_3(NH_2\text{-}2)Cl\text{-}5)_2$	290, **27**, *24*
	$(C_4H_9)_2Sn(ONHCOC_6H_4Cl\text{-}4)_2$	339, **32**, *5*
$C_{22}H_{28}O_2Sn$	$(C_4H_9)_2SnOC(C_6H_5)=C(C_6H_5)O$	58, **14**, *43*
$C_{22}H_{28}O_4S_2Sn$	$(C_4H_9)_2Sn(OOCC_6H_4SH\text{-}2)_2$	289, **27**, *14*
$C_{22}H_{28}O_4Sn$	$(C_4H_9)_2Sn(OC_7H_5O)_2$	31, **11**, *16*
	$(C_4H_9)_2Sn(OC_6H_4CHO\text{-}2)_2$	34, **12**, *17*
	$(C_4H_9)_2Sn(OOCC_6H_5)_2$	287, **27**, *1*
	$((CH_3)_3C)_2Sn(OC_7H_5O)_2$	373, **43**, *3*
$C_{22}H_{28}O_6Sn$	$(C_4H_9)_2Sn(OOCCH=CHC_4H_3O)_2$	269, **24**, *10*
	$(C_4H_9)_2Sn(OOCC_6H_4OH\text{-}2)_2$	288, **27**, *9*
	$(C_4H_9)_2Sn(OOCC_6H_4OH\text{-}4)_2$	288, **27**, *10*
$C_{22}H_{28}O_6SnTi$	$(C_4H_9)_2SnOTi(OC_6H_4CHO\text{-}2)_2O$	368
$C_{22}H_{28}O_8SnTi_2$	$(C_4H_9)_2Sn(OTi(O)OC_6H_4CHO\text{-}2)_2$	368
$C_{22}H_{30}N_2O_2Sn$	$(C_4H_9)_2Sn(ON=CHC_6H_5)_2$	343, **32**, *22*
$C_{22}H_{30}N_2O_4Sn$	$(C_4H_9)_2Sn(OOCC_6H_4NH_2\text{-}2)_2$	289, **27**, *17*
	$(C_4H_9)_2Sn(OOCC_6H_4NH_2\text{-}4)_2$	289, **27**, *18*
	$(C_4H_9)_2Sn(ONHCOC_6H_5)_2$	339, **32**, *4*
	$((CH_3)_2CHCH_2)_2Sn(OOCC_6H_4NH_2\text{-}4)_2$	370, **41**, *13*
$C_{22}H_{30}N_2O_6Sn$	$(C_4H_9)_2Sn(OC_6H_3(NO_2\text{-}2)CH_3\text{-}4)_2$	34, **12**, *14*
$C_{22}H_{30}N_2O_8S_2Sn$	$(C_4H_9)_2Sn(OOCC_6H_4SO_2NH_2\text{-}4)_2$	289, **27**, *16*
$C_{22}H_{30}O_2Sn$	$(C_4H_9)_2SnOCH(C_6H_5)CH(C_6H_5)O$	51, **14**, *12*
$C_{22}H_{32}As_2O_4Sn$	$(C_4H_9)_2SnOAs(O)(C_6H_5)\text{-}CH_2CH_2\text{-}(C_6H_5)(O)AsO$	367
$C_{22}H_{32}N_4O_2Sn$	$(C_4H_9)_2Sn(OC_6H_4CH=N(NH_2))_2$	35, **12**, *25*
$C_{22}H_{32}O_2Sn$	$(C_4H_9)_2Sn(OCH_2C_6H_5)_2$	7, **1**, *42*
	$(C_4H_9)_2Sn(OC_6H_4CH_3\text{-}4)_2$	34, **12**, *6*
$C_{22}H_{32}O_4Sn$	$((CH_3)_2CHCH_2)_2Sn(OC_6H_4CH_2OH\text{-}2)_2$	369, **41**, *3*
$C_{22}H_{32}O_6S_2Sn$	$(C_4H_9)_2Sn(OSO_2C_6H_4CH_3\text{-}4)_2$	349, **34**, *12*
$C_{22}H_{32}O_8Sn$	$(C_4H_9)_2Sn(OOCCH=CHCOOCH_2CH=CH_2)_2$	275, **25**, *25*
$C_{22}H_{33}BrO_6Sn$	$(C_4H_9)_2SnOC_5H_5O(OCH_3)(CH_2Br)(OOCC_6H_5)O$	65, **14**, *80*
$C_{22}H_{34}O_4Sn$	$(C_4H_9)_2SnOOCC_6H_2((CH_3)_2\text{-}3,4)(CH_2CH(CH_3)_2\text{-}6)COO$	327, **30**, *36*
$C_{22}H_{34}O_6Sn$	$(C_4H_9)_2SnOC_5H_5O(OCH(C_6H_5)OCH_2)(OCH_3)O$	68, **14**, *95*
	$(C_4H_9)_2SnOC_5H_5O(OCH(C_6H_5)OCH_2)(OCH_3)O$	69, **14**, *96*
	$(C_4H_9)_2SnOC_5H_5O(OCH(C_6H_5)OCH_2)(OCH_3)O$	69, **14**, *97*
	$(C_4H_9)_2SnOC_5H_5O(OCH(C_6H_5)OCH_2)(OCH_3)O$	70, **14**, *98*
$C_{22}H_{36}O_4Sn$	$(C_4H_9)_2SnOOCC_6H_5[(CH_3)_2\text{-}3,4](CH=C(CH_3)_2\text{-}6)COO$	325, **30**, *27*
$C_{22}H_{36}O_6Sn$	$(C_4H_9)_2Sn(OC(CH_3)=CHCOCH_2COCH_3)_2$	99, **19**, *17*
$C_{22}H_{36}O_8S_2Sn$	$(C_4H_9)_2Sn(OOCCH_2SCH_2COOCH_2CH=CH_2)_2$	258, **23**, *36*
$C_{22}H_{36}O_8Sn$	$(C_4H_9)_2Sn(OOCCH=CHCOOC_3H_7)_2$	273, **25**, *4*
	$(C_4H_9)_2Sn(OOCCH=CHCOOCH(CH_3)_2)_2$	273, **25**, *5*
$C_{22}H_{36}O_{10}Sn$	$(C_4H_9)_2Sn(OOCCH=CHCOOCH_2CH(OH)CH_3)_2$	276, **25**, *31*
	$(C_4H_9)_2Sn(OOCCH=CHCOOCH_2CH_2OCH_3)_2$	276, **25**, *32*

$C_{22}H_{38}Cl_2O_4Sn$	$(C_4H_9)_2Sn(OOCC_6H_{10}Cl)_2$	263, **23**, *78*
$C_{22}H_{40}O_4Sn$	$(C_4H_9)_2Sn(OOCC_6H_{11}\text{-c})_2$	253
$C_{22}H_{40}O_6Sn$	$(C_4H_9)_2Sn(OC(CH_3)=C(OC_2H_5)COCH_3)_2$	99, **19**, *14*
$C_{22}H_{40}O_8Sn$	$(C_4H_9)_2Sn(OC(OC_2H_5)=CHCO(OC_2H_5))_2$	100, **19**, *21*
$C_{22}H_{41}O_6S_3SbSn$	$(C_4H_9)_2SnOOCCH_2SSb(SCH_2COOC_8H_{17})SCH_2COO$	324, **30**, *15*
$C_{22}H_{42}Cl_2O_4Sn$	$(C_4H_9)_2Sn(OOC(CH_2)_5CH_2Cl)_2$	256, **23**, *11*
$C_{22}H_{44}O_4Sn$	$(C_4H_9)_2Sn(OOCC_6H_{13})_2$	128
	$(C_4H_9)_2Sn(OOCCH_2CH(CH_3)CH_2CH_2CH_3)_2$	128
$C_{22}H_{44}O_6Sn$	$(C_4H_9)_2Sn(OOCCH_2CH_2OC_4H_9)_2$	257, **23**, *22*
$C_{22}H_{48}N_4O_2Sn$	$(C_4H_9)_2Sn(ON=C(CH_3)CH_2CH_2CH_2N(CH_3)_2)_2$	343, **32**, *26*
$C_{22}H_{50}N_2O_2Sn$	$(C_4H_9)_2Sn(OCH_2CH_2CH_2N(C_2H_5)_2)_2$	7, **1**, *41*
$C_{23}H_{30}N_2O_2Sn$	$(C_4H_9)_2SnOC_6H_4CH=NN=C(CH_3)C_6H_4O$	74, **14**, *122*
$C_{23}H_{30}N_4O_2SSn$	$(C_4H_9)_2SnOC_6H_4CH=NNHCSNHN=CHC_6H_4O$	75, **14**, *127*
$C_{23}H_{32}O_2Sn$	$(C_4H_9)_2SnOC_6H_4C(CH_3)_2C_6H_4O$	60, **14**, *54*
$C_{23}H_{37}N_5O_4Sn$	$(C_4H_9)_2SnOC_4H_4O(CH_2OH)(C_5H_2N_4NHCH_2CH=C(CH_3)_2)O$	62, **14**, *64*
$C_{23}H_{37}N_5O_5Sn$	$(C_4H_9)_2SnOC_4H_4O(CH_2OH)(C_5H_2N_4NHCH_2CH=C(CH_3)\text{-}$ $CH_2OH)O$	62, **14**, *65*
$C_{23}H_{40}O_6Sn$	$(C_4H_9)_2SnOC_5H_5O(CH_2OCH_2CH=CH_2)(OCH_2CH=CH_2)_2O$	66, **14**, *84*
$C_{23}H_{40}O_8Sn$	$(C_4H_9)_2SnOC(OC_2H_5)=C(COOC_2H_5)CH_2\text{-}$ $C(COOC_2H_5)=C(OC_2H_5)O$	58, **14**, *47*
$C_{23}H_{41}O_4PSn$	$(C_4H_9)_2SnOP(O)(CH_2C_6H_2(C(CH_3)_3)_2\text{-}3,5\text{-}OH\text{-}4)O$	359, **37**, *5*
$C_{24}H_{18}F_{30}O_4Sn$	$(C_4H_9)_2Sn(OOCC_7F_{15})_2$	255, **23**, *3*
$C_{24}H_{22}F_{10}O_6Sn$	$(C_4H_9)_2Sn(OOCCH_2OC_6F_5)_2$	257, **23**, *25*
$C_{24}H_{26}Cl_2F_4O_6S_2Sn$	$(C_4H_9)_2Sn(OSO_2C_6H_4CF=CFCl\text{-}4)_2$	349, **34**, *13*
$C_{24}H_{26}Cl_6O_6Sn$	$(C_4H_9)_2Sn(OOCCH_2OC_6H_2Cl_3\text{-}2,4,5)_2$	258, **23**, *29*
$C_{24}H_{26}F_6O_4S_2Sn$	$(C_4H_9)_2Sn(OC(CF_3)=CHCO(c\text{-}C_4H_3S\text{-}2))_2$	99, **19**, *10*
$C_{24}H_{26}F_6O_6S_2Sn$	$(C_4H_9)_2Sn(OSO_2C_6H_4CF=CF_2\text{-}4)_2$	349, **34**, *14*
$C_{24}H_{28}Cl_4O_6Sn$	$(C_4H_9)_2Sn(OOCCH_2OC_6H_3Cl_2\text{-}2,4)_2$	258, **23**, *28*
$C_{24}H_{28}N_2O_4Sn$	$(C_4H_9)_2SnOOCC_6H_4N=CHCH=NC_6H_4COO$	328, **30**, *40*
$C_{24}H_{30}Cl_2O_6Sn$	$(C_4H_9)_2Sn(OOCCH_2OC_6H_4Cl\text{-}2)_2$	257, **23**, *26*
	$(C_4H_9)_2Sn(OOCCH_2OC_6H_4Cl\text{-}4)_2$	258, **23**, *27*
$C_{24}H_{30}N_2O_{10}Sn$	$(C_4H_9)_2Sn(OOCC_6H_3(NO_2\text{-}3)OCH_3\text{-}4)_2$	291, **27**, *29*
$C_{24}H_{30}O_4S_2Sn$	$(C_4H_9)_2SnOC_6H_3(COCH_3)SSC_6H_3(COCH_3)O$	71, **14**, *106*
$C_{24}H_{32}Br_2N_2O_2Sn$	$(C_4H_9)_2Sn(ON=C(CH_3)C_6H_4Br)_2$	343, **32**, *29*
$C_{24}H_{32}Cl_2N_2O_2Sn$	$(C_4H_9)_2Sn(ON=C(CH_3)C_6H_4Cl)_2$	343, **32**, *28*
$C_{24}H_{32}N_2O_2Sn$	$(C_4H_9)_2SnOC_6H_4CH=NCH_2CH_2N=CHC_6H_4O$	74, **14**, *123*
$C_{24}H_{32}N_2O_4Sn$	$(C_4H_9)_2SnON(C_6H_4CH_3\text{-}4)COCON(C_6H_4CH_3\text{-}4)O$	337,
$C_{24}H_{32}O_4S_2Sn$	$(C_4H_9)_2Sn(OOCCH_2SC_6H_5)_2$	259, **23**, *38*
$C_{24}H_{32}O_4Sn$	$(C_4H_9)_2SnOC(C_6H_4OCH_3\text{-}4)=C(C_6H_4OCH_3\text{-}4)O$	58, **14**, *44*
	$(C_4H_9)_2Sn(OOCCH_2C_6H_5)_2$	261, **23**, *59*

$C_{26}H_{26}Br_4N_2O_2Sn$	$(C_4H_9)_2Sn(OC_9H_4(Br_2\text{-}5,7)N)_2$	41, **13**, *9*
$C_{26}H_{26}Cl_4N_2O_2Sn$	$(C_4H_9)_2Sn(OC_9H_4(Cl_2\text{-}5,7)N)_2$	40, **13**, *8*
$C_{26}H_{26}I_4N_2O_2Sn$	$(C_4H_9)_2Sn(OC_9H_4(I_2\text{-}5,7)N)_2$	41, **13**, *10*
$C_{26}H_{26}N_6O_{10}Sn$	$(C_4H_9)_2Sn(OC_9H_4N(NO_2)_2\text{-}5,7)_2$	42, **13**, *15*
$C_{26}H_{28}N_4O_6Sn$	$(C_4H_9)_2Sn(OC_9H_5N(NO_2\text{-}7))_2$	41, **13**, *11*
	$(C_4H_9)_2Sn(OC_9H_5N(NO_2\text{-}5))_2$	41, **13**, *12*
$C_{26}H_{30}Cl_2O_4Sn$	$(C_4H_9)_2Sn(OOCCH{=}CHC_6H_4Cl\text{-}4)_2$	269, **24**, *6*
$C_{26}H_{30}N_2O_2Sn$	$(C_4H_9)_2Sn(OC_9H_6N)_2$	40, **13**, *7*
	$((CH_3)_2CHCH_2)_2Sn(OC_9H_6N)_2$	369, **41**, *4*
	$((CH_3)_3C)_2Sn(OC_9H_6N)_2$	373, **43**, *4*
$C_{26}H_{30}N_2O_8Sn$	$(C_4H_9)_2Sn(OOCCH{=}CHC_6H_4NO_2\text{-}4)_2$	269, **24**, *9*
$C_{26}H_{31}O_2PSn$	$((CH_3)_3C)_2SnOC_6H_4P(C_6H_5)C_6H_4O$	375, **43**, *15*
$C_{26}H_{32}O_4Sn$	$(C_4H_9)_2Sn(OOCCH{=}CHC_6H_5)_2$	269, **24**, *5*
	$(C_4H_9)_2Sn(OOCC_6H_4CH{=}CH_2\text{-}4)_2$	291, **27**, *35*
$C_{26}H_{32}O_6Sn$	$(C_4H_9)_2Sn(OOCC_6H_4COCH_3\text{-}4)_2$	292, **27**, *38*
$C_{26}H_{34}Cl_2O_6Sn$	$(C_4H_9)_2Sn(OOCCH_2OC_6H_3(CH_3\text{-}2)Cl\text{-}4)_2$	258, **23**, *31*
$C_{26}H_{36}Cl_2N_2O_4Sn$	$(C_4H_9)_2Sn(ON(C_2H_5)COC_6H_4Cl\text{-}4)_2$	339, **32**, *6*
$C_{26}H_{36}O_4S_2Sn$	$(C_4H_9)_2Sn(OOCCH_2SCH_2C_6H_5)_2$	258, **23**, *37*
$C_{26}H_{36}O_6S_2Sn$	$(C_4H_9)_2Sn(OOCC_6H_2(OH\text{-}2)(CH_3\text{-}4)SCH_3\text{-}5)_2$	292, **27**, *37*
$C_{26}H_{36}O_6Sn$	$(C_4H_9)_2Sn(OOCCH_2OC_6H_4CH_3\text{-}2)_2$	258, **23**, *30*
	$(C_4H_9)_2Sn(OOCC_6H_4OCH_2CH_3\text{-}4)_2$	289, **27**, *12*
$C_{26}H_{38}N_6O_4Sn$	$(C_4H_9)_2Sn(OC_6H_4C(CH_3){=}N(NHCONH_2)\text{-}2)_2$	36, **12**, *28*
$C_{26}H_{40}O_4Sn$	$(C_4H_9)_2Sn(OOC(CH_3)_2C_6H_5)_2$	334, **31**, *2*
$C_{26}H_{41}NO_6Sn$	$(C_4H_9)_2SnOCH_2C_5H_5O(OCH_2CH{=}CH_2)(NHCOCH_3)\text{-}$ $(OCH_2C_6H_5O)$	66, **14**, *85*
$C_{26}H_{42}N_2O_4Sn$	$(C_4H_9)_2Sn(OOCC(CN){=}C(CH_3)CH_2CH(CH_3)_2)_2$	270, **24**, *20*
$C_{26}H_{44}O_8Sn$	$(C_4H_9)_2Sn(OOCCH{=}CHCOOC_5H_{11})_2$	274, **25**, *8*
	$(C_4H_9)_2Sn(OOCCH{=}CHCOO(CH_2)_2CH(CH_3)_2)_2$	274, **25**, *9*
$C_{26}H_{44}O_{10}Sn$	$(C_4H_9)_2Sn(OOCCH{=}CHCOOCH_2CH_2OC_3H_7)_2$	276, **25**, *34*
$C_{26}H_{50}Cl_2O_4Sn$	$(C_4H_9)_2Sn(OOC(CH_2)_7CH_2Cl)_2$	256, **23**, *12*
$C_{26}H_{52}O_4Sn$	$(C_4H_9)_2Sn(OOCC_8H_{17})_2$	134
	$(C_4H_9)_2Sn(OOCC_8H_{17}\text{-}i)_2$	135
$C_{27}H_{38}O_5Sn$	$(C_4H_9)_2SnOC_4H_4O(CH_2OCH_2C_6H_5)(OCH_2C_6H_5)O$	60, **14**, *57*
$C_{28}H_{28}Cl_{12}O_8Sn$	$(C_4H_9)_2Sn(OOCC_7H_2Cl_6COOCH_3)_2$	284, **26**, *5*
$C_{28}H_{28}N_4O_2Sn$	$(C_4H_9)_2Sn(OC(C_6H_5){=}C(CN)_2)_2$	30, **11**, *12*
$C_{28}H_{30}Cl_2O_6Sn$	$(C_4H_9)_2Sn(OOCCH{=}CHCOC_6H_4Cl\text{-}4)_2$	270, **24**, *14*
$C_{28}H_{30}N_2O_4Sn$	$(C_4H_9)_2Sn(OC_{10}H_6NO)_2$	36, **12**, *30*
$C_{28}H_{30}N_2O_6Sn$	$(C_4H_9)_2Sn(OOCC(CN){=}CHC_6H_4OH\text{-}4)_2$	270, **24**, *21*
$C_{28}H_{32}N_2O_2Sn$	$(C_4H_9)_2SnOC_6H_4CH{=}NC_6H_4N{=}CHC_6H_4O$	75, **14**, *126*
$C_{28}H_{32}O_6Sn$	$(C_4H_9)_2Sn(OOCCH{=}CHCOC_6H_5)_2$	270, **24**, *13*

$C_{28}H_{34}Br_2O_4Sn$	$(C_4H_9)_2Sn(OC(CH_3)=CHCOC_6H_4Br\text{-}4)_2$	98, **19**, *7*
$C_{28}H_{34}Cl_2O_4Sn$	$(C_4H_9)_2Sn(OC(CH_3)=CHCOC_6H_4Cl\text{-}4)_2$	98, **19**, *6*
$C_{28}H_{34}F_2O_4Sn$	$(C_4H_9)_2Sn(OC(CH_3)=CHCOC_6H_4F\text{-}4)_2$	98, **19**, *5*
$C_{28}H_{34}N_2O_2Sn$	$(C_4H_9)_2Sn(OC_9H_5N(CH_3\text{-}2))_2$	42, **13**, *19*
	$((CH_3)_3C)_2Sn(OC_9H_5(CH_3\text{-}2)N)_2$	374, **43**, *5*
$C_{28}H_{34}O_4S_2Sn$	$(C_4H_9)_2Sn(OC(CH_3)=CHCOC_6H_4S\text{-})_2$	98, **19**, *8*
$C_{28}H_{36}O_4Sn$	$(C_4H_9)_2Sn(OC(CH_3)=CHCOC_6H_5)_2$	98, **19**, *4*
$C_{28}H_{36}O_6Sn$	$(C_4H_9)_2Sn(OOCCH=CHC_6H_4OCH_3\text{-}4)_2$	269, **24**, *7*
$C_{28}H_{36}O_8Sn$	$(C_4H_9)_2Sn(OOCC_6H_4COOC_2H_5\text{-}2)_2$	292, **27**, *40*
$C_{28}H_{38}Cl_2O_8Sn$	$(C_4H_9)_2Sn(OC_6H_4COOCH_2CH(OH)CH_2Cl\text{-}2)_2$	35, **12**, *23*
	$(C_4H_9)_2Sn(OOCC_6H_4OCH_2CH(OH)CH_2Cl\text{-}2)_2$	289, **27**, *13*
$C_{28}H_{38}N_2O_4Sn$	$(C_4H_9)_2Sn(OC(CH_3)=CHCONHC_6H_5)_2$	100, **19**, *24*
$C_{28}H_{38}N_2O_6Sn$	$(C_4H_9)_2Sn(OOCCH_2ON=C(CH_3)C_6H_5)_2$	258, **23**, *32*
$C_{28}H_{38}N_2O_8Sn$	$(C_4H_9)_2Sn(OOCCH_2ON=CHC_6H_4OCH_3\text{-}4)_2$	258, **23**, *33*
$C_{28}H_{38}O_6Sn$	$(C_4H_9)_2SnOC_5H_5O(OCH(C_6H_5)OCH_2)(OCH_2C_6H_5)O$	70, **14**, *99*
$C_{28}H_{40}Cl_4O_8Sn$	$(C_4H_9)_2Sn(OOCCCl=CClCOOC_6H_{11}\text{-}c)_2$	277, **25**, *49*
$C_{28}H_{40}N_2O_2Sn$	$(C_4H_9)_2SnOC_6H_3(CH_3)C(CH_3)=NCH_2CH_2N=C(CH_3)\text{-}C_6H_3(CH_3)O$	75, **14**, *124*
$C_{28}H_{40}N_4O_{10}Sn$	$(C_4H_9)_2Sn(OC_6H_2((NO_2)_2\text{-}2,4)C(CH_3)_3\text{-}6)_2$	34, **12**, *15*
$C_{28}H_{40}O_8Sn$	$(C_4H_9)_2Sn(OOCCH=CHCOOC\equiv CC_4H_9)_2$	276, **25**, *29*
	$(C_4H_9)_2Sn(OOCC_7H_8COOCH_3)_2$	284, **26**, *4*
$C_{28}H_{40}O_{12}Sn$	$(C_4H_9)_2Sn(OOCCH=CHCOOCH_2CH_2OOCC(CH_3)=CH_2)_2$	276, **25**, *41*
	$(C_4H_9)_2Sn(OOCCH=CHCOOCH(CH_3)CH_2OOCCH=CH_2)_2$	276, **25**, *42*
$C_{28}H_{42}N_2O_6S_2Sn$	$(C_4H_9)_2Sn(OS(O)N(C_6H_4CH_3\text{-}4)CH_2C_2H_3O)_2$	348, **34**, *3*
$C_{28}H_{42}N_6O_6Sn$	$(C_4H_9)_2Sn(OOCCH_2CH_2CON(CH_2CH_2CN)_2)_2$	263, **23**, *73*
$C_{28}H_{44}O_2Sn$	$(C_4H_9)_2Sn(OC_6H_4C(CH_3)_3\text{-}4)_2$	34, **12**, *7*
$C_{28}H_{44}O_6Sn$	$(C_4H_9)_2SnOC_5H_5O(CH_2OCH_2C_6H_5)(OC_3H_5)\text{-}(OCH_2CH=CHCH_3O)$	66, **14**, *86*
$C_{28}H_{44}O_8Sn$	$(C_4H_9)_2Sn(OOCCH=CHCOOC_6H_{11}\text{-}c)_2$	275, **25**, *20*
$C_{28}H_{48}O_{10}Sn$	$(C_4H_9)_2Sn(OOCCH=CHCOOCH_2CH_2OC_4H_9)_2$	276, **25**, *35*
$C_{28}H_{48}O_{12}Sn$	$(C_4H_9)_2Sn(O(CH_2)_3OOCCH=CHCOO(CH_2)_3OH)_2$	5, **1**, *22*
	$(C_4H_9)_2Sn(OOCCH=CHCOOCH(CH_3)CH_2O\text{-}CH_2CH(OH)CH_3)_2$	276, **25**, *40*
$C_{28}H_{52}O_4Sn$	$(C_4H_9)_2Sn(OC(CH_3)=C(C_4H_9)COC_2H_5)_2$	99, **19**, *13*
$C_{28}H_{52}O_8S_2Sn$	$(C_4H_9)_2Sn(OOCCH_2CH_2SCH_2CH_2COOC_4H_9)_2$	259, **23**, *42*
$C_{28}H_{52}O_8Sn$	$(C_4H_9)_2Sn(OOC(CH_2)_4COOC_4H_9)_2$	263, **23**, *75*
$C_{28}H_{56}O_4S_2Sn$	$(C_4H_9)_2Sn(OOCCH_2SC_8H_{17}\text{-}i)_2$	258, **23**, *35*
$C_{28}H_{56}O_4Sn$	$(C_4H_9)_2Sn(OOCC_9H_{19})_2$	135
	$(C_4H_9)_2Sn(OOCCH_2CH_2C(C_2H_5)_3)_2$	135
$C_{28}H_{56}O_6Sn$	$(C_4H_9)_2Sn(OCH_2COOCH_2CH(C_2H_5)C_4H_9)_2$	6, **1**, *26*
$C_{28}H_{60}O_2Sn$	$(C_4H_9)_2Sn(OC_{10}H_{21})_2$	5, **1**, *13*
$C_{28}H_{60}O_4Si_2Sn$	$(C_4H_9)_2Sn(OOCCH(CH_3)CH_2Si(C_2H_5)_3)_2$	263, **23**, *77*

$C_{28}H_{62}O_4P_2Sn$	$(C_4H_9)_2Sn(OP(O)(C_5H_{11})_2)_2$	354, **36**, *3*
$C_{29}H_{32}O_5S_2Sn$	$(C_4H_9)_2SnOOCC_6H_4SCH(C_6H_4OH\text{-}2)SC_6H_4COO$	328, **30**, *39*
$C_{30}H_{30}N_2O_8Sn$	$(C_4H_9)_2Sn(OOCC_6H_4N(CO)_2C_2H_2\text{-}3)_2$	290, **27**, *21*
	$(C_4H_9)_2Sn(OOCC_6H_4N(CO)_2C_2H_2\text{-}4)_2$	290, **27**, *22*
$C_{30}H_{34}Cl_2O_6Sn$	$(C_4H_9)_2Sn(OOCCH{=}CHCOC_6H_3(CH_3\text{-}5)Cl\text{-}2)_2$	270, **24**, *18*
$C_{30}H_{34}N_2O_4Sn$	$(C_4H_9)_2Sn(OC_9H_5N(COCH_3\text{-}5))_2$	43, **13**, *20*
$C_{30}H_{34}N_2O_8Sn$	$(C_4H_9)_2Sn(OOCCH(CH_3)NC_8H_4O_2)_2$	260, **23**, *55*
$C_{30}H_{36}Fe_2O_4Sn$	$(C_4H_9)_2Sn(OOCC_5H_4FeC_5H_5)_2$	299, **28**, *5*
$C_{30}H_{36}O_6Sn$	$(C_4H_9)_2Sn(OOCCH{=}CHCOC_6H_4CH_3\text{-}4)_2$	270, **24**, *16*
$C_{30}H_{36}O_8Sn$	$(C_4H_9)_2Sn(OOCCH{=}CHCOC_6H_4OCH_3\text{-}4)_2$	270, **24**, *15*
	$(C_4H_9)_2Sn(OOCCH{=}CHCOOCH_2C_6H_5)_2$	275, **25**, *23*
$C_{30}H_{40}O_6S_2Sn$	$(C_4H_9)_2Sn(OOCCH_2CH(C_6H_5)SCOCH_3)_2$	260, **23**, *51*
$C_{30}H_{40}O_6Sn$	$(C_4H_9)_2Sn(OC(C_6H_5){=}CHCOOC_2H_5)_2$	100, **19**, *20*
$C_{30}H_{42}N_2O_4Sn$	$(C_4H_9)_2Sn(OOCCH{=}CHC_6H_4N(CH_3)_2\text{-}4)_2$	269, **24**, *8*
$C_{30}H_{42}N_2O_6Sn$	$(C_4H_9)_2Sn(OOCCH(CH_2C_6H_5)NHCOCH_3)_2$	260, **23**, *54*
$C_{30}H_{44}O_4Sn$	$(C_4H_9)_2Sn(OOCC_6H_4C(CH_3)_3\text{-}4)_2$	291, **27**, *34*
$C_{30}H_{46}N_6O_8Sn$	$(C_4H_9)_2Sn(OCH_2CH_2N_3(CO)_3(CH_2CH{=}CH_2)_2$	7, **1**, *46*
$C_{30}H_{48}O_8Sn$	$(C_4H_9)_2Sn(OOCCH{=}CHCOOC_6H_{10}\text{-c-}CH_3\text{-}2)_2$	275, **25**, *21*
	$(C_4H_9)_2Sn(OOCC_6H_6(CH_3)_2COOCH_3)_2$	284, **26**, *1*
$C_{30}H_{52}O_2Si_2Sn$	$((CH_3)_2CHCH_2)_2Sn(OSi(CH_3)(CH_2CH(CH_3)_2)C_6H_5)_2$	371, **41**, *15*
$C_{30}H_{56}O_4Sn$	$(C_4H_9)_2Sn(OOC(CH_2)_8CH{=}CH_2)_2$	270, **24**, *12*
$C_{30}H_{56}O_8Sn$	$(C_4H_9)_2Sn(OC(OC_4H_9){=}CHCOOC_4H_9)_2$	100, **19**, *22*
$C_{31}H_{42}O_7Sn$	$(C_4H_9)_2SnOC_5H_5O(CH_2OCH_2C_6H_5)(OC_3H_5)(OCOC_6H_5)O$	66, **14**, *86*
$C_{31}H_{44}O_6Sn$	$(C_4H_9)_2SnOC_5H_5O(CH_2OCH_2C_6H_5)(OC_3H_5)(OCH_2C_6H_5)O$	67, **14**, *87*
	$(C_4H_9)_2SnOC_5H_5O(CH_2OC_3H_5)(OCH_2C_6H_5)_2O$	67, **14**, *88*
	$(C_4H_9)_2SnOC_5H_5O(CH_2OCH_2C_6H_5)(OCH_2C_6H_5)\text{-}(OC_3H_5)O$	67, **14**, *89*
	$((CH_3)_2CHCH_2)_2SnOC_5H_5O(CH_2OCH_2C_6H_5)\text{-}(OC_3H_5)(OCH_2C_6H_5)O$	369, **41**, *5*
$C_{32}H_{34}N_6O_2Sn$	$(C_4H_9)_2Sn(OC_6H_4N_3C_6H_4)_2$	36, **12**, *29*
$C_{32}H_{38}O_2P_2Sn$	$(C_4H_9)_2Sn(OP(C_6H_5)_2)_2$	354, **36**, *1*
$C_{32}H_{38}O_3Si_2Sn$	$(C_4H_9)_2Sn(\text{-}OSi(C_6H_5)_2\text{-})_2O$	352
$C_{32}H_{38}O_4P_2Sn$	$(C_4H_9)_2Sn(OP(O)(C_6H_5)_2)_2$	354, **36**, *6*
$C_{32}H_{38}O_6P_2Sn$	$(C_4H_9)_2Sn(OP(OC_6H_5)_2)_2$	355, **36**, *11*
	$(C_4H_9)_2Sn(OP(O)(OC_6H_5)C_6H_5)_2$	355, **36**, *14*
$C_{32}H_{38}O_8P_2Sn$	$(C_4H_9)_2Sn(OP(O)(OC_6H_5)_2)_2$	356, **36**, *17*
$C_{32}H_{40}N_4O_4Sn$	$(C_4H_9)_2Sn(OC(CH_3){=}CC(CH_3){=}NN(C_6H_5)CO)_2$	100, **19**, *25*
$C_{32}H_{40}O_{10}Sn$	$(C_4H_9)_2Sn(OOCCH{=}CHCOOCH_2CH_2OC_6H_5)_2$	276, **25**, *38*

$C_{32}H_{44}O_4Sn$	$(C_4H_9)_2Sn(OC(CH_3)=C(CH_2C_6H_5)COCH_3)_2$	99, **19**, *15*
$C_{32}H_{44}O_8Sn$	$(C_4H_9)_2Sn(OOCC_6H_4COOC_4H_9\text{-}2)_2$	292, **27**, *41*
$C_{32}H_{46}N_6O_{10}Sn$	$(C_4H_9)_2Sn(OOCCH_2CH_2N_3(CO)_3(CH_2CH=CH_2)_2)_2$	261, **23**, *58*
$C_{32}H_{48}O_{10}Sn$	$(C_4H_9)_2Sn(OOCCH=CHCOOC_7H_{10}CH_2OH)_2$	275, **25**, *22*
$C_{32}H_{52}Br_4O_8Sn$	$(C_4H_9)_2Sn(OOCCBr=CBrCOOCH_2CH(C_2H_5)C_4H_9)_2$	277, **25**, *52*
$C_{32}H_{52}Cl_4O_8Sn$	$(C_4H_9)_2Sn(OOCCCl=CClCOOCH_2CH(C_2H_5)C_4H_9)_2$	277, **25**, *50*
$C_{32}H_{52}O_{10}Sn$	$(C_4H_9)_2Sn(OOCCH=CHCOOCH_2CH_2OC_6H_{11}\text{-c})_2$	276, **25**, *37*
$C_{32}H_{56}N_4O_4Sn$	$(C_4H_9)_2Sn(OC(CH_3)=CHC(CH_3)=NCH_2CH_2N=C(CH_3)\text{-}$ $CH_2COCH_3)_2$	31, **11**, *15*
$C_{32}H_{56}O_8Sn$	$(C_4H_9)_2Sn(OOCCH=CHCOOC_8H_{17})_2$	274, **25**, *10*
	$(C_4H_9)_2Sn(OOCCH=CHCOOCH_2CH(C_2H_5)C_4H_9)_2$	274, **25**, *11*
	$(C_4H_9)_2Sn(OOCCH=CHCOO(CH_2)_5CH(CH_3)_2)_2$	274, **25**, *12*
	$(C_4H_9)_2Sn(OOCCH=CHCOOC_8H_{17}\text{-i})_2$	274, **25**, *13*
$C_{32}H_{56}O_{14}Sn$	$(C_4H_9)_2Sn(OC(COOC_2H_5)(CH_2COOC_2H_5)_2)_2$	6, **1**, *32*
$C_{32}H_{60}O_8Sn$	$(C_4H_9)_2Sn(OOCCH_2CH_2COOC_8H_{17})_2$	263, **23**, *69*
$C_{32}H_{60}O_{12}Sn$	$(C_4H_9)_2Sn(OCH(COOC_4H_9)CH(OH)COOC_4H_9)_2$	6, **1**, *31*
$C_{32}H_{64}O_4Sn$	$(C_4H_9)_2Sn(OOCC_{11}H_{23})_2$	135/249
	$(C_4H_9)_2Sn(OOC(CH_2)_7C(CH_3)_3)_2$	250
	$((CH_3)_2CHCH_2)_2Sn(OOCC_{11}H_{23})_2$	370, **41**, *11*
	$(C_2H_5(CH_3)CH)_2Sn(OOCC_{11}H_{23})_2$	372
$C_{32}H_{68}O_2Sn$	$(C_4H_9)_2Sn(OC_{12}H_{25})_2$	5, **1**, *14*
$C_{32}H_{70}B_2O_6Sn$	$(C_4H_9)_2Sn(OB(OH)OC_{12}H_{25})_2$	351, **35**, *1*
$C_{32}H_{70}O_4P_2Sn$	$(C_4H_9)_2Sn(OP(O)(C_6H_{13})_2)_2$	354, **36**, *4*
$C_{34}H_{34}N_6O_{12}Sn$	$(C_4H_9)_2Sn(ON(C_6H_5)COC_6H_3(NO_2)_2\text{-}3,5)_2$	340, **32**, *12*
$C_{34}H_{36}Cl_2N_2O_4Sn$	$(C_4H_9)_2Sn(ON(C_6H_5)COC_6H_4Cl\text{-}4)_2$	340, **32**, *9*
	$(C_4H_9)_2Sn(ON(C_6H_4Cl\text{-}4)COC_6H_5)_2$	341, **32**, *13*
$C_{34}H_{36}F_2N_2O_2Sn$	$(C_4H_9)_2Sn(OC_6H_4(CH=N(C_6H_4F\text{-}4)\text{-}2)_2$	35, **12**, *24*
$C_{34}H_{36}I_2N_2O_4Sn$	$(C_4H_9)_2Sn(ON(C_6H_5)COC_6H_4I\text{-}2)_2$	340, **32**, *10*
$C_{34}H_{36}N_4O_8Sn$	$(C_4H_9)_2Sn(ON(C_6H_5)COC_6H_4NO_2\text{-}4)_2$	340, **32**, *11*
$C_{34}H_{36}O_4Sn$	$(C_4H_9)_2Sn(OC_6H_4COC_6H_5\text{-}2)_2$	35, **12**, *18*
$C_{34}H_{38}N_2O_4Sn$	$(C_4H_9)_2Sn(OOCC_6H_4NHC_6H_5\text{-}2)_2$	290, **27**, *20*
	$(C_4H_9)_2Sn(ON(C_6H_5)COC_6H_5)_2$	340, **32**, *8*
$C_{34}H_{40}O_8Sn$	$(C_4H_9)_2Sn(OOCCH=CHCOOCH_2CH=CHC_6H_5)_2$	276, **25**, *28*
$C_{34}H_{44}N_4O_4Sn$	$(C_4H_9)_2Sn(OC(C_2H_5)=CC(CH_3)=NN(C_6H_5)CO)_2$	100, **19**, *26*
$C_{34}H_{44}O_6Sn$	$(C_4H_9)_2SnOC_5H_5O(CH_2OC(C_6H_5)_3)(OCH_3)(OH)O$	68, **14**, *92*
	$(C_4H_9)_2Sn(OOCCH=CHCOC_6H_2(CH_3)_3\text{-}2,4,6)_2$	270, **24**, *17*
$C_{34}H_{48}O_6Sn$	$(C_4H_9)_2SnOC_6H_6(OCH_2C_6H_5)_2(OCH_2CH=CH_2)_2O$	70, **14**, *101*
$C_{34}H_{52}O_8S_2Sn$	$(C_4H_9)_2Sn(OOCC_7H_8COOCH_2CH_2SC_2H_5)_2$	285, **26**, *11*
$C_{34}H_{52}O_8Sn$	$(C_4H_9)_2Sn(OOCC_7H_8COOC_4H_9)_2$	285, **26**, *6*
$C_{34}H_{58}O_4P_2Sn$	$(C_4H_9)_2Sn(OP(O)(C_6H_{13})CH_2C_6H_5)_2$	354, **36**, *5*
$C_{34}H_{60}O_8Sn$	$(C_4H_9)_2Sn(OOCCH=CHCOOC_9H_{19})_2$	274, **25**, *14*

$C_{34}H_{64}O_4Sn$	$(C_4H_9)_2Sn(OC(CH_3)=C(CH_2CH(C_2H_5)C_4H_9)COCH_3)_2$	99, **19**, *16*
$C_{34}H_{66}Cl_2O_4Sn$	$(C_4H_9)_2Sn(OOC(CH_2)_{11}CH_2Cl)_2$	256, **23**, *13*
$C_{34}H_{68}O_6Sn$	$(C_4H_9)_2Sn(OOCCH_2CH_2OC_{10}H_{21})_2$	257, **23**, *23*
$C_{35}H_{46}O_6Sn$	$(C_4H_9)_2SnOC_5H_5O(CH_2OCH_2C_6H_5)(OCH_2C_6H_5)_2O$	67, **14**, *90*
	$(C_4H_9)_2SnOC_5H_5O(CH_2OC(C_6H_5)_3)(OCH_3)_2O$	68, **14**, *91*
	$(C_4H_9)_2SnOC_5H_5O(CH_2OCH_2C_6H_5)(OCH_2C_6H_5)_2O$	68, **14**, *93*
$C_{36}H_{36}O_6Sn$	$(C_4H_9)_2Sn(OOCC_6H_4COC_6H_5\text{-}2)_2$	292, **27**, *39*
$C_{36}H_{38}N_2O_4Sn$	$(C_4H_9)_2Sn(OOCC_6H_4N=CHC_6H_5\text{-}2)_2$	290, **27**, *23*
$C_{36}H_{40}O_6Sn$	$(C_4H_9)_2Sn(OC_6H_3(OCH_3\text{-}5)COC_6H_5\text{-}2)_2$	35, **12**, *19*
	$(C_4H_9)_2Sn(OOCC(C_6H_5)_2OH)_2$	262, **23**, *63*
$C_{36}H_{40}O_8Sn$	$(C_4H_9)_2Sn(OC_6H_3(OCH_3\text{-}5)(COC_6H_4OH\text{-}2)\text{-}2)_2$	35, **12**, *20*
$C_{36}H_{42}N_2O_4Sn$	$(C_4H_9)_2Sn(ON(C_6H_4CH_3\text{-}2)COC_6H_5)_2$	341, **32**, *14*
	$(C_4H_9)_2Sn(ON(C_6H_4CH_3\text{-}3)COC_6H_5)_2$	341, **32**, *15*
	$(C_4H_9)_2Sn(ON(C_6H_4CH_3\text{-}4)COC_6H_5)_2$	342, **32**, *16*
$C_{36}H_{42}N_6O_6Sn$	$(C_4H_9)_2Sn(OOCC_6H_4CON(CH_2CH_2CN)_2\text{-}2)_2$	292, **27**, *45*
$C_{36}H_{44}Br_8O_{10}Sn$	$(C_4H_9)_2Sn(OOCC_6(COOCH_2CH_2OC_4H_9\text{-}2)Br_4)_2$	292, **27**, *46*
$C_{36}H_{46}N_2O_8Sn$	$(C_4H_9)_2Sn(OOCCH(CH_2CH(CH_3)_2)NC_8H_4O_2)_2$	261, **23**, *56*
$C_{36}H_{48}O_8Sn$	$(C_4H_9)_2Sn(OOCCH=CHCOOCH(CH(CH_3)_2)C_6H_5)_2$	275, **25**, *24*
$C_{36}H_{50}N_6O_6Sn$	$(C_4H_9)_2Sn(OOCC_6H_8CON(CH_2CH_2CN)_2)_2$	284, **26**, *2*
$C_{36}H_{52}O_8Sn$	$(C_4H_9)_2Sn(OC(OC_2H_5)=C(CH_2C_6H_5)COOC_2H_5)_2$	100, **19**, *23*
$C_{36}H_{60}O_2Sn$	$(C_4H_9)_2Sn(OC_6H_4C_8H_{17}\text{-i-}4)_2$	34, **12**, *8*
	$(C_4H_9)_2Sn(OC_6H_4C_8H_{17}\text{-t-}4)_2$	34, **12**, *9*
$C_{36}H_{60}O_{10}S_2Sn$	$(C_4H_9)_2Sn(OOCCH=CHCOSCH_2COOC_8H_{17})_2$	277, **25**, *54*
$C_{36}H_{64}O_{10}Sn$	$(C_4H_9)_2Sn(OOCCH=CHCOOCH_2CH_2OCH_2CH(C_2H_5)C_4H_9)_2$	276, **25**, *36*
$C_{36}H_{68}O_8S_2Sn$	$(C_4H_9)_2Sn(OOCCH_2CH_2SCH_2CH_2COOC_8H_{17})_2$	259, **23**, *43*
$C_{36}H_{68}O_8Sn$	$(C_4H_9)_2Sn(OOC(CH_2)_8COOC_4H_9)_2$	263, **23**, *76*
$C_{36}H_{72}O_4Sn$	$(C_4H_9)_2Sn(OOCC_{13}H_{27})_2$	250
$C_{36}H_{72}O_6Sn$	$(C_4H_9)_2Sn(O(CH_2)_4OOCC_9H_{19})_2$	6, **1**, *24*
$C_{38}H_{36}O_6Sn$	$(C_4H_9)_2Sn(OC_9H_4O(C_6H_5)(=O)_2$	40, **13**, *3*
$C_{38}H_{36}O_8Sn$	$(C_4H_9)_2SnOOCC_6H_4COOC(C_6H_5)=C(C_6H_5)OOCC_6H_4COO$	327, **30**, *38*
$C_{38}H_{38}N_6O_2Sn$	$(C_4H_9)_2Sn(OC_9H_5N(N=NC_6H_5\text{-}5))_2$	42, **13**, *16*
$C_{38}H_{38}O_4S_2Sn$	$(C_4H_9)_2Sn(OC(C_6H_5)=CHCOC_6H_4S\text{-})_2$	99, **19**, *12*
$C_{38}H_{40}O_4Sn$	$(C_4H_9)_2Sn(OC(C_6H_5)=CHCOC_6H_5)_2$	99, **19**, *11*
$C_{38}H_{40}O_6SnTi$	$(C_4H_9)_2SnOTi(OC(C_6H_5)=CHCOC_6H_5)_2O$	368
$C_{38}H_{44}N_2O_2Sn$	$(C_4H_9)_2SnOC_6H_3(CH_3)C(C_6H_5)=NCH_2CH_2N=C(C_6H_5)\text{-}$ $C_6H_3(CH_3)O$	75, **14**, *125*
$C_{38}H_{48}O_2Sn$	$(C_4H_9)_2Sn(OC_6H_4(C_6H_4CH(CH_3)_2\text{-}2)\text{-}4)_2$	34, **12**, *16*
$C_{38}H_{56}O_8Sn$	$(C_4H_9)_2Sn(OOCC_7H_8COOC_6H_{11}\text{-c})_2$	285, **26**, *10*
$C_{38}H_{60}O_4S_2Sn$	$(C_4H_9)_2Sn(OOCC_6H_4SC_8H_{17}\text{-}2)_2$	289, **27**, *15*

$C_{38}H_{60}O_6Sn$	$(C_4H_9)_2Sn(OOCC_6H_2(OH-4)(C(CH_3)_3)_2-3,5)_2$	291, **27**, *36*
$C_{38}H_{64}O_2Sn$	$(C_4H_9)_2Sn(OC_6H_4C_9H_{19}-i-4)_2$	34, **12**, *10*
$C_{38}H_{70}N_6O_2Sn$	$(C_4H_9)_2Sn(OC_6H_2(CH_2N(CH_3)_2)_3-2,4,6)_2$	34, **12**, *13*
C_{40} $H_{38}N_2O_4Sn$	$(C_4H_9)_2Sn(OC_9H_5N(COC_6H_5-5))_2$	43, **13**, *21*
$C_{40}H_{38}N_6O_6Sn$	$(C_4H_9)_2Sn(OC_9H_5N(N=NC_6H_4COOH-2)-5)_2$	42, **13**, *17*
$C_{40}H_{38}O_4Sn$	$(C_4H_9)_2SnOOCC_6(C_6H_5)_4COO$	327, **30**, *37*
$C_{40}H_{44}N_4O_4Sn$	$(C_4H_9)_2Sn(ON(C(C_6H_5)C(CH_3)C(C_6H_5)=)NO)_2$	342, **32**, *19*
$C_{40}H_{48}O_8Sn$	$(C_4H_9)_2Sn(OCH(CH_2OC_6H_5)CH_2OOCC_6H_5)_2$	6, **1**, *23*
$C_{40}H_{60}O_8Sn$	$(C_4H_9)_2Sn(OOCC_6H_4COOC_8H_{17}-2)_2$	292, **27**, *42*
	$(C_4H_9)_2Sn(OOCC_6H_4COOCH_2CH(C_2H_5)C_4H_9-4)_2$	292, **27**, *43*
$C_{40}H_{64}O_6Sn$	$(C_4H_9)_2Sn(OOCCH_2C_6H_2(OH-4)(C_4H_9)_2-3,5)_2$	261, **23**, *60*
$C_{40}H_{64}O_{14}S_2Sn$	$(C_4H_9)_2Sn(OOCCH=CHCOSCH(COOC_4H_9)CH_2COOC_4H_9)_2$	277, **25**, *53*
$C_{40}H_{68}O_2Sn$	$(C_4H_9)_2Sn(OC_6H_4C_{10}H_{21}-4)_2$	34, **12**, *11*
$C_{40}H_{72}O_8Sn$	$(C_4H_9)_2Sn(OOCCH=CHCOOC_{12}H_{25})_2$	274, **25**, *15*
$C_{40}H_{76}O_8Sn$	$(C_4H_9)_2Sn(OOCCH_2CH_2COOC_{12}H_{25})_2$	263, **23**, *70*
$C_{40}H_{78}N_2O_6Sn$	$(C_4H_9)_2Sn(OOCCH_2CH_2CONHC_{12}H_{25})_2$	263, **23**, *72*
$C_{40}H_{80}O_4S_2Sn$	$(C_4H_9)_2Sn(OOCCH(CH_3)CH_2SC_{12}H_{25})_2$	259, **23**, *50*
$C_{40}H_{80}O_4Sn$	$(C_4H_9)_2Sn(OOCC_{15}H_{31})_2$	250
$C_{40}H_{80}O_{10}Sn$	$(C_4H_9)_2Sn(OOC(CH_2)_7CH(OH)CH(OH)(CH_2)_5CH_2OH)_2$	256, **23**, *17*
$C_{40}H_{86}B_2O_6Sn$	$(C_4H_9)_2Sn(OB(OCH_2CH(C_2H_5)C_4H_9)_2)_2$	351, **35**, *2*
C_{42} $H_{40}F_2N_2O_2Sn$	$(C_4H_9)_2Sn(OC_{10}H_6CH=NC_6H_4F-4)_2$	36, **12**, *31*
$C_{42}H_{40}O_8Sn$	$(C_4H_9)_2Sn(OOCCH=C(COC_6H_5)_2)_2$	270, **24**, *19*
$C_{42}H_{42}Cl_2N_4O_4Sn$	$(C_4H_9)_2Sn(OC(C_6H_4Cl-4)=CC(CH_3)=NN(C_6H_5)CO)_2$	100, **19**, *28*
$C_{42}H_{42}N_2O_8Sn$	$(C_4H_9)_2Sn(OOCCH(CH_2C_6H_5)NC_8H_4O_2)_2$	261, **23**, *57*
$C_{42}H_{42}N_6O_6Sn$	$(C_4H_9)_2Sn(OC_9H_5N(N=NC_6H_4COOCH_3-2)-5)_2$	42, **13**, *18*
$C_{42}H_{44}N_4O_4Sn$	$(C_4H_9)_2Sn(OC(C_6H_5)=CC(CH_3)=NN(C_6H_5)CO)_2$	100, **19**, *27*
$C_{42}H_{46}Fe_2N_6O_2Sn$	$(C_4H_9)_2Sn(OC(C_5H_5N)=N-N=CHC_5H_4FeC_5H_5)_2$	332
$C_{42}H_{52}O_6Sn$	$(C_4H_9)_2SnOC_6H_6(OCH_2C_6H_5)_4O$	70, **14**, *100*
$C_{42}H_{68}O_6Sn$	$(C_4H_9)_2Sn(OOCCH_2CH_2C_6H_2((C(CH_3)_3-3,5)_2(OH-4))_2$	262, **23**, *61*
$C_{42}H_{68}O_8Sn$	$(C_4H_9)_2Sn(OOCC_7H_8COOC_8H_{17})_2$	285, **26**, *7*
	$(C_4H_9)_2Sn(OOCC_7H_8COOCH_2CH(C_2H_5)C_4H_9)_2$	285, **26**, *8*
C_{44} $H_{46}Cl_2Fe_2N_4O_2Sn$	$(C_4H_9)_2Sn(OC(C_6H_4Cl-4)=N-N=CHC_5H_4FeC_5H_5)_2$	332
$C_{44}H_{46}Fe_2N_6O_6Sn$	$(C_4H_9)_2Sn(OC(C_6H_4NO_2-2)=N-N=CHC_5H_4FeC_5H_5)_2$	332
$C_{44}H_{48}Fe_2N_4O_2Sn$	$(C_4H_9)_2Sn(OC(C_6H_5)=N-N=CHC_5H_4FeC_5H_5)_2$	332
$C_{44}H_{48}Fe_2N_4O_4Sn$	$(C_4H_9)_2Sn(OC(C_6H_4OH-2)=N-N=CHC_5H_4FeC_5H_5)_2$	332
$C_{44}H_{48}O_2Si_2Sn$	$(C_4H_9)_2Sn(OSi(C_6H_5)_3)_2$	352
$C_{44}H_{62}N_2O_6Sn$	$(C_4H_9)_2Sn(OOCC(CN)=CH-C_6H_2(OH-4)(C(CH_3)_3)_2-3,5)_2$	270, **24**, *22*
$C_{44}H_{80}O_{14}Sn$	$(C_4H_9)_2Sn(OC(COOC_4H_9)(CH_2COOC_4H_9)_2)_2$	6, **1**, *33*
$C_{44}H_{84}O_4Sn$	$(C_4H_9)_2Sn(OOC(CH_2)_7CH=CH(CH_2)_7CH_3)_2$	269, **24**, *11*

$C_{44}H_{84}O_6Sn$	$(C_4H_9)_2Sn(OOC(CH_2)_7(C_2H_2O)(CH_2)_7CH_3)_2$	257, **23**, *19*
$C_{44}H_{84}O_8S_2Sn$	$(C_4H_9)_2Sn(OOCCH_2CH_2SCH_2CH_2COOC_{12}H_{25})_2$	259, **23**, *44*
$C_{44}H_{84}O_8Sn$	$(C_4H_9)_2Sn(OOC(CH_2)_7(C_2H_2O)CH_2CH(OH)(CH_2)_5CH_3)_2$	257, **23**, *20*
$C_{44}H_{84}O_{12}Sn$	$(C_4H_9)_2Sn(OCHCH_2OCH(CH_2OOCC_{11}H_{23})CH(OH)CH(OH))_2$	39, **13**, *1*
$C_{44}H_{88}O_4Sn$	$(C_4H_9)_2Sn(OOCC_{17}H_{35})_2$	250
	$(C_4H_9)_2Sn(OOCC_{17}H_{35}\text{-i})_2$	251
$C_{44}H_{88}O_6Sn$	$(C_4H_9)_2Sn(OOC(CH_2)_{10}CH(OH)C_6H_{13})_2$	256, **23**, *16*
$C_{46}H_{46}As_2O_4Sn$	$(C_4H_9)_2Sn(OOCC_6H_4As(C_6H_5)_2\text{-}2)_2$	291, **27**, *31*
$C_{46}H_{46}B_2Br_2N_2O_4Sn$	$(C_4H_9)_2Sn(OB(C_6H_5)OC_6H_4(CH{=}NC_6H_4Br\text{-}4)\text{-}2)_2$	351, **35**, *5*
$C_{46}H_{46}B_2Cl_2N_2O_4Sn$	$(C_4H_9)_2Sn(OB(C_6H_5)OC_6H_4(CH{=}NC_6H_4Cl\text{-}4)\text{-}2)_2$	351, **35**, *4*
$C_{46}H_{46}B_2F_2N_2O_4Sn$	$(C_4H_9)_2Sn(OB(C_6H_5)OC_6H_4(CH{=}NC_6H_4F\text{-}4)\text{-}2)_2$	351, **35**, *3*
$C_{46}H_{46}B_2N_4O_8Sn$	$(C_4H_9)_2Sn(OB(C_6H_5)OC_6H_4(CH{=}NC_6H_4NO_2\text{-}4)\text{-}2)_2$	351, **35**, *6*
$C_{46}H_{82}O_8P_2Sn$	$(C_4H_9)_2Sn(OP(O)(OC_4H_9)CH_2C_6H_2((C(CH_3)_3)_2\text{-}3,5)OH\text{-}4)_2$	355, **36**, *13*
$C_{46}H_{88}O_6Sn$	$(C_4H_9)_2Sn(OCH(C_6H_{13})CH_2CH{=}CH(CH_2)_7COOCH_3)_2$	30, **11**, *13*
$C_{48}H_{88}O_8Sn$	$(C_4H_9)_2Sn(OOCCH{=}CHCOOC_{16}H_{33})_2$	274, **25**, *16*
	$(C_4H_9)_2Sn(OOCCH{=}CHCOOCH_2CH(C_6H_{13})C_8H_{17})_2$	274, **25**, *17*
$C_{48}H_{90}O_7Sn$	$(C_4H_9)_2SnOCH(C_6H_{13})CH_2CH{=}CH(CH_2)_7COO(CH_2)_2O\text{-}$ $(CH_2)_2OOC(CH_2)_7CH{=}CHCH_2CH(C_6H_{13})O$	71, **14**, *104*
$C_{50}H_{54}As_2O_4Sn$	$(C_4H_9)_2Sn(OOCC_6H_4As(C_6H_4CH_3\text{-}4)_2\text{-}2)_2$	291, **27**, *32*
$C_{50}H_{68}O_6Sn$	$(C_4H_9)_2Sn(OC_6H_3(OC_8H_{17}\text{-}5)COC_6H_5\text{-}2)_2$	35, **12**, *21*
$C_{50}H_{74}O_6P_2Sn$	$(C_4H_9)_2Sn(OP(O)(C_6H_5)CH_2C_6H_2((C(CH_3)_3)_2\text{-}3,5)OH\text{-}4)_2$	355, **36**, *7*
$C_{50}H_{80}O_8Sn$	$(C_4H_9)_2Sn(OOCC_7H_8COOC_{12}H_{23})_2$	285, **26**, *9*
$C_{50}H_{80}O_{10}S_4Sn$	$(C_4H_9)_2Sn(OOCCH_2CH_2SCH(C_6H_4OH\text{-}2)S(CH_2)_2COO\text{-}$ $C_8H_{17}\text{-i})_2$	259, **23**, *45*
$C_{50}H_{88}O_2Sn$	$(C_4H_9)_2Sn(OC_6H_4C_{15}H_{31}\text{-}3)_2$	34, **12**, *12*
$C_{50}H_{100}O_4Sn$	$(C_4H_9)_2Sn(OOCCH(CH_3)C_{18}H_{37})_2$	251
$C_{52}H_{92}O_8Sn$	$(C_4H_9)_2Sn(OOCCH{=}CHCOO(CH_2)_8CH{=}CHC_8H_{17})_2$	275, **25**, *27*
$C_{52}H_{92}O_{10}Sn$	$(C_4H_9)_2Sn(OOCCH{=}CHCOO(CH_2)_8CH{=}CHCH_2CH(OH)\text{-}$ $C_6H_{13})_2$	277, **25**, *44*
$C_{52}H_{96}O_8Sn$	$(C_4H_9)_2Sn(OOCCH{=}CHCOOC_{18}H_{37})_2$	275, **25**, *18*
$C_{52}H_{98}O_{12}S_6Sb_2Sn$	$(C_4H_9)_2Sn(OOCCH_2S\text{-}Sb(SCH_2COOC_8H_{17})_2)_2$	259, **23**, *39*
$C_{52}H_{104}O_4Sn$	$(C_4H_9)_2Sn(OOCC_{21}H_{43})_2$	251
$C_{54}H_{50}B_2Br_2N_2O_4Sn$	$(C_4H_9)_2Sn(OB(C_6H_5)OC_{10}H_6CH{=}NC_6H_4Br\text{-}4)_2$	351, **35**, *9*
$C_{54}H_{50}B_2Cl_2N_2O_4Sn$	$(C_4H_9)_2Sn(OB(C_6H_5)OC_{10}H_6CH{=}NC_6H_4Cl\text{-}4)_2$	351, **35**, *8*
$C_{54}H_{50}B_2F_2N_2O_4Sn$	$(C_4H_9)_2Sn(OB(C_6H_5)OC_{10}H_6CH{=}NC_6H_4F\text{-}4)_2$	351, **35**, *7*

Ligand Formula Index

The ligands containing carbon atoms can be used to locate a compound in this volume. These ligands are listed in the Ligand Formula Index by number of carbon atoms. The number of identical ligands in a compound and the nature of bonding are not taken into consideration. Thus several compounds may be listed at one position. Compounds having two or more different carbon-containing ligands occur at more than one position. The variable organic ligands are placed in the first two columns, while nonorganic ligands appear in the third column.

Page references are printed in ordinary type, table numbers in bold face, and compound numbers within the tables in italics.

CF_3O_3S	C_4H_9	–	348, **34**, *9*
CHO_2	C_4H_9	–	103
CH_3AsO_3	C_4H_9	–	363, **39**, *1*
CH_3O	C_4H_9	–	3, **1**, *1*
			369, **41**, *1*
			373, **43**, *2*
CH_3O_2Se	C_4H_9	–	368
CH_3O_3S	C_4H_9	–	348, **34**, *6*
C$_2Cl_3O_2$	C_4H_9	–	256, **23**, *6*
$C_2F_3O_2$	C_4H_9	–	255, **23**, *1*
$C_2F_3O_3S$	C_4H_9	–	348, **34**, *10*
$C_2HCl_2O_2$	C_4H_9	–	255, **23**, *5*
$C_2H_2ClO_2$	C_4H_9	–	255, **23**, *4*
$C_2H_2F_3O$	C_4H_9	–	5, **1**, *17*
$C_2H_3O_2$	C_4H_9	–	104/26
			370, **41**, *10*
			376, **43**, *17*
$C_2H_3O_3$			
$OOCCH_2OH$	C_4H_9	–	256, **23**, *14*
$OOCCOCH_3$	C_4H_9	–	332
$C_2H_3O_3S$	C_4H_9	–	348, **34**, *8*
C_2H_4NO	C_4H_9	–	342, **32**, *20*
$C_2H_4NO_2$	C_4H_9	–	339, **32**, *2*
$C_2H_4O_2$	C_4H_9	–	48, **14**, *1*
$C_2H_4O_3$	C_4H_9	–	334, **31**, *3*
C_2H_5O	C_4H_9	–	3, **1**, *2*
$C_2H_6O_3PS$	C_4H_9	–	356, **36**, *18*
C_2O_4	C_4H_9	–	301
C$_3HF_6O$	C_4H_9	–	5, **1**, *18*

$C_3H_2ClO_2$	C_4H_9	–	269, **24**, *4*
$C_3H_2O_4$	C_4H_9	–	322, **30**, *1*
$C_3H_3O_2$	C_4H_9	–	268, **24**, *1*
$C_3H_4ClO_2$			
$OOCCHClCH_3$	C_4H_9	–	256, **23**, *7*
$OOCCH_2CH_2Cl$	C_4H_9	–	256, **23**, *8*
$C_3H_4Cl_3O_2$	C_4H_9	–	5, **1**, *20*
$C_3H_4O_3$	C_4H_9	–	331
C_3H_5ClNO	C_4H_9	–	343, **32**, *25*
$C_3H_5ClO_2$	C_4H_9	–	49, **14**, *4*
C_3H_5O	C_4H_9	–	29, **11**, *1*
$C_3H_5O_2$	C_4H_9	–	127
$C_3H_5O_2S$			
$OOCCH_2SCH_3$	C_4H_9	–	258, **23**, *34*
$OOCCH_2CH_2SH$	C_4H_9	–	259, **23**, *40*
$C_3H_5O_3$	C_4H_9	–	256, **23**, *15*
C_3H_6NO	C_4H_9	–	343, **32**, *23*
$C_3H_6O_2$			
$OC(CH_3)CH_2O$	C_4H_9	–	49, **14**, *2*
$O(CH_2)_3O$	C_4H_9	–	53, **14**, *20*
$C_3H_6O_3$	C_4H_9	–	334, **31**, *4*
C_3H_7O			
OC_3H_7	C_4H_9	–	3, **1**, *3*
$OCH(CH_3)_2$	C_4H_9	–	3, **1**, *4*
$C_3H_7O_2$	C_4H_9	–	5, **1**, *19*
C_3H_8NO			
$OCH_2CH_2NHCH_3$	C_4H_9	–	6, **1**, *34*
$OCH(CH_3)CH_2NH_2$	C_4H_9	–	7, **1**, *38*
$O(CH_2)_3NH_2$	C_4H_9	–	7, **1**, *40*
C_3H_9OSi	C_4H_9	–	352
$C_3H_{11}B_{10}O_2$			
$CH(OOC)B_1OH_9CH$-1,2	C_4H_9	–	299, **28**, *6*
$CH(OOC)B_1OH_9CH$-1,7	C_4H_9	–	299, **28**, *7*
C_3N_3O	C_4H_9	–	343, **32**, *30*
C$_4Br_2O_4$	C_4H_9	–	325, **30**, *24*
$C_4Cl_2O_4$	C_4H_9	–	325, **30**, *23*
$C_4F_4O_4$	C_4H_9	–	322, **30**, *4*
$C_4F_7O_2$	C_4H_9	–	255, **23**, *2*
$C_4H_2O_4$	C_4H_9	–	301/4
			371, **41**, *14*
$C_4H_3O_2S_2$	C_4H_9	–	348, **34**, *2*

$C_4H_3O_4$	C_4H_9	–	273, **25**, *1*
$C_4H_4NO_3$	C_4H_9	–	342, **32**, *17*
$C_4H_4O_4$	C_4H_9	–	322, **30**, *2*
$C_4H_4O_5$	C_4H_9	–	70, **14**, *102*
$C_4H_5O_2$			
$OOCC(CH_3)=CH_2$	C_4H_9	–	268, **24**, *2*
			370, **41**, *12*
$OOCCH=CHCH_3$	C_4H_9	–	269, **24**, *3*
$C_4H_5O_4$	C_4H_9	–	262, **23**, *66*
$C_4H_6ClO_2$	C_4H_9	–	256, **23**, *9*
$C_4H_6O_2$			
1,2-Cyclobutandiyl-dioxy	C_4H_9	–	52, **14**, *15*
$OC(CH_3)=C(CH_3)O$	C_4H_9	–	57, **14**, *42*
$OCH_2CH=CHCH_2O$	C_4H_9	–	58, **14**, *45*
$C_4H_7NO_4$	C_4H_9	–	56, **14**, *31*
C_4H_7O			
Cyclopropylmethoxy	C_4H_9	–	5, **1**, *16*
$O(CH_2)_2CH=CH_2$	C_4H_9	–	29, **11**, *2*
$OCH_2CH=CHCH_3$	C_4H_9	–	29, **11**, *3*
$OCH_2C(CH_3)=CH_2$	C_4H_9	–	29, **11**, *4*
$C_4H_7O_2$	C_4H_9	–	127
$C_4H_7O_3$	C_4H_9	–	257, **23**, *21*
C_4H_8NO			
$ON=CHC_3H_7$	C_4H_9	–	343, **32**, *21*
$ON=C(CH_3)CH_2CH_3$	C_4H_9	–	343, **32**, *24*
$C_4H_8NO_2$	C_4H_9	–	339, **32**, *3*
$C_4H_8O_2$			
$OC(CH_3)_2CH_2O$	C_4H_9	–	49, **14**, *3*
$OCH(CH_3)CH(CH_3)O$	C_4H_9	–	50, **14**, *8*
$OCH(CH_3)(CH_2)_2O$	C_4H_9	–	54, **14**, *21*
$O(CH_2)_4O$	C_4H_9	–	57, **14**, *39*
$C_4H_8O_2S$	C_4H_9	–	73, **14**, *116*
			375, **43**, *13*
$C_4H_8O_3$			
$OCH_2C(CH_3)(OH)CH_2O$	C_4H_9	–	55, **14**, *28*
$O(CH_2)_2O(CH_2)_2O$	C_4H_9	–	71, **14**, *103*
$OC(CH_3)(C_2H_5)O_2$	C_4H_9	–	335, **31**, *12*
C_4H_9			
C_4H_9	–	BO_2	350
C_4H_9	–	ClO_4	336/7
C_4H_9	–	FO_3S	348, **34**, *4*

C_4H_9	—	F_2O_2P	356, **36**, *19*
C_4H_9	—	HO	1
C_4H_9	—	HO_3P	358, **37**, *1*
C_4H_9	—	HO_4P	359, **37**, *6*
C_4H_9	—	348, **34**, *5*	
	H_2NO_3S		
C_4H_9	—	H_2O_2P	354, **36**, *2*
C_4H_9	—	NO_3	337
C_4H_9	—	O_3S	346
C_4H_9	—	O_3Ti	368
C_4H_9	—	O_4S	346
C_4H_9	CF_3O_3S	—	348, **34**, *9*
C_4H_9	CHO_2	—	103
C_4H_9	CH_3AsO_3	—	363, **39**, *1*
C_4H_9	CH_3O	—	3, **1**, *1*
C_4H_9	CH_3O_2Se	—	368
C_4H_9	CH_3O_3S	—	348, **34**, *6*
C_4H_9	$C_2Cl_3O_2$	—	256, **23**, *6*
C_4H_9	$C_2F_3O_2$	—	255, **23**, *1*
C_4H_9	$C_2F_3O_3S$	—	348, **34**, *10*
C_4H_9	$C_2HCl_2O_2$	—	255, **23**, *5*
C_4H_9	$C_2H_2ClO_2$	—	255, **23**, *4*
C_4H_9	$C_2H_2F_3O$	—	5, **1**, *17*
C_4H_9	$C_2H_3O_2$	—	104/26
C_4H_9	$C_2H_3O_3$	—	256, **23**, *14*
			332
C_4H_9	$C_2H_3O_3S$	—	348, **34**, *8*
C_4H_9	C_2H_4NO	—	342, **32**, *20*
C_4H_9	$C_2H_4NO_2$	—	339, **32**, *2*
C_4H_9	$C_2H_4O_2$	—	48, **14**, *1*
C_4H_9	$C_2H_4O_3$	—	334, **31**, *3*
C_4H_9	C_2H_5O	—	3, **1**, *2*
C_4H_9	$C_2H_6O_3PS$	—	356, **36**, *18*
C_4H_9		—	301
	C_2O_4		
C_4H_9	C_3HF_6O	—	5, **1**, *18*
C_4H_9	$C_3H_2ClO_2$	—	269, **24**, *4*
C_4H_9	$C_3H_2O_4$	—	322, **30**, *1*
C_4H_9	$C_3H_3O_2$	—	268, **24**, *1*
C_4H_9	$C_3H_4ClO_2$	—	256, **23**, *7*
		—	256, **23**, *8*
C_4H_9	$C_3H_4Cl_3O_2$	—	5, **1**, *20*

C_4H_9	$C_3H_4O_3$	—	331
C_4H_9	C_3H_5ClNO	—	343, **32**, *25*
C_4H_9	$C_3H_5ClO_2$	—	49, **14**, *4*
C_4H_9	C_3H_5O	—	29, **11**, *1*
C_4H_9	$C_3H_5O_2$	—	127
C_4H_9	$C_3H_5O_2S$	—	258, **23**, *34*
			259, **23**, *40*
C_4H_9	$C_3H_5O_3$	—	256, **23**, *15*
C_4H_9	C_3H_6NO	—	343, **32**, *23*
C_4H_9	$C_3H_6O_2$	—	49, **14**, *2*
			53, **14**, *20*
C_4H_9	$C_3H_6O_3$	—	334, **31**, *4*
C_4H_9	C_3H_7O	—	3, **1**, *3*
			3, **1**, *4*
C_4H_9	$C_3H_7O_2$	—	5, **1**, *19*
C_4H_9	C_3H_8NO	—	6, **1**, *34*
			7, **1**, *38*
			7, **1**, *40*
C_4H_9	C_3H_9OSi	—	352
C_4H_9	$C_3H_{11}B_{10}O_2$	—	299, **28**, *6*
			299, **28**, *7*
C_4H_9	C_3N_3O	—	343, **32**, *30*
C_4H_9	$C_4Br_2O_4$	—	325, **30**, *24*
C_4H_9	$C_4Cl_2O_4$	—	325, **30**, *23*
C_4H_9	$C_4F_4O_4$	—	322, **30**, *4*
C_4H_9	$C_4F_7O_2$	—	255, **23**, *2*
C_4H_9	$C_4H_2O_4$	—	301/4
C_4H_9	$C_4H_3O_2S_2$	—	348, **34**, *2*
C_4H_9	$C_4H_3O_4$	—	273, **25**, *1*
C_4H_9	$C_4H_4NO_3$	—	342, **32**, *17*
C_4H_9	$C_4H_4O_4$	—	322, **30**, *2*
C_4H_9	$C_4H_4O_5$	—	70, **14**, *102*
C_4H_9	$C_4H_5O_2$	—	268, **24**, *2*
			269, **24**, *3*
C_4H_9	$C_4H_5O_4$	—	262, **23**, *66*
C_4H_9	$C_4H_6ClO_2$	—	256, **23**, *9*
C_4H_9	$C_4H_6O_2$	—	52, **14**, *15*
			57, **14**, *42*
			58, **14**, *45*
C_4H_9	$C_4H_7NO_4$	—	56, **14**, *31*
C_4H_9	C_4H_7O	—	5, **1**, *16*
			29, **11**, *2*

C_4H_9	$C_5H_5O_2$	—	7, **1**, *44*
C_4H_9	$C_5H_5O_4$	—	273, **25**, *2*
C_4H_9	$C_5H_6O_4S_2$	—	323, **30**, *11*
C_4H_9	$C_5H_7Cl_3NO_2$	—	5, **1**, *21*
C_4H_9	$C_5H_7NO_4$	—	325, **30**, *22*
C_4H_9	$C_5H_7O_2$	—	97, **19**, *1*
C_4H_9	$C_5H_7O_3$	—	6, **1**, *30*
			99, **19**, *18*
C_4H_9	$C_5H_7O_4$	—	262, **23**, *67*
C_4H_9	$C_5H_8ClO_2$	—	256, **23**, *10*
C_4H_9	C_5H_8NO	—	100, **19**, *29*
C_4H_9	$C_5H_8NO_3$	—	260, **23**, *52*
C_4H_9	$C_5H_8O_2$	—	52, **14**, *16*
C_4H_9	$C_5H_8O_2S$	—	56, **14**, *35*
C_4H_9	$C_5H_8O_3$	—	56, **14**, *34*
C_4H_9	$C_5H_9NO_4$	—	56, **14**, *32*
C_4H_9	C_5H_9O	—	29, **11**, *5*
			29, **11**, *6*
			30, **11**, *7*
			30, **11**, *8*
C_4H_9	$C_5H_9O_2$	—	127/8
			128
C_4H_9	$C_5H_9O_3$	—	6, **1**, *27*
			7, **1**, *45*
C_4H_9	$C_5H_{10}O_2$	—	54, **14**, *22*
			54, **14**, *23*
			57, **14**, *40*
C_4H_9	$C_5H_{10}O_3$	—	55, **14**, *29*
C_4H_9	$C_5H_{11}NO_2$	—	72, **14**, *108*
C_4H_9	$C_5H_{11}O$	—	4, **1**, *9*
			4, **1**, *10*
C_4H_9	$C_5H_{12}NO$	—	7, **1**, *39*
C_4H_9	$C_6Cl_2O_4$	—	59, **14**, *51*
C_4H_9	C_6Cl_5O	—	33, **12**, *3*
C_4H_9	$C_6H_2Br_2NO_3$	—	33, **12**, *5*
C_4H_9	$C_6H_2Cl_3O_3PS$	—	359, **37**, *7*
C_4H_9	$C_6H_4AsBrO_3$	—	363, **39**, *5*
C_4H_9	$C_6H_4AsClO_3$	—	363, **39**, *4*
C_4H_9	$C_6H_4AsNO_5$	—	363, **39**, *11*
			363, **39**, *12*
C_4H_9	$C_6H_4AsNO_6$	—	364, **39**, *13*
C_4H_9	C_6H_4ClO	—	33, **12**, *2*

C_4H_9	$C_6H_4NO_2$	—	299, **28**, *4*
C_4H_9	$C_6H_4N_3O$	—	342, **32**, *18*
C_4H_9	$C_6H_4O_2$	—	59, **14**, *48*
			59, **14**, *49*
C_4H_9	$C_6H_5AsNO_5$	—	362, **38**, *2*
C_4H_9	$C_6H_5AsNO_6$	—	362, **38**, *3*
C_4H_9	$C_6H_5AsN_2O_5$	—	364, **39**, *14*
C_4H_9	$C_6H_5AsO_3$	—	363, **39**, *3*
C_4H_9	$C_6H_5AsO_4$	—	363, **39**, *6*
C_4H_9	$C_6H_5NO_3$	—	335, **31**, *11*
C_4H_9	$C_6H_5N_2O_2$	—	343, **32**, *31*
C_4H_9	C_6H_5O	—	33, **12**, *1*
C_4H_9	$C_6H_5O_2Se$	—	368
C_4H_9	$C_6H_5O_3S$	—	349, **34**, *11*
C_4H_9	$C_6H_5O_4$	—	39, **13**, *2*
C_4H_9	$C_6H_6AsNO_3$	—	363, **39**, *9*
			363, **39**, *10*
C_4H_9	$C_6H_6AsN_2O_5$	—	362, **38**, *4*
C_4H_9	$C_6H_6AsO_3$	—	361, **38**, *1*
C_4H_9	C_6H_6NO	—	33, **12**, *4*
C_4H_9	$C_6H_6O_3P$	—	355, **36**, *12*
C_4H_9	$C_6H_7O_4$	—	273, **25**, *3*
C_4H_9	$C_6H_7O_5$	—	276, **25**, *30*
C_4H_9	$C_6H_8O_4$	—	323, **30**, *7*
C_4H_9	$C_6H_8O_4S$	—	324, **30**, *17*
C_4H_9	$C_6H_8O_5$	—	68, **14**, *94*
C_4H_9	$C_6H_8O_6$	—	51, **14**, *13*
C_4H_9	$C_6H_9O_2$	—	98, **19**, *2*
			253
C_4H_9	$C_6H_9O_3$	—	100, **19**, *19*
C_4H_9	$C_6H_9O_4$	—	262, **23**, *68*
			263, **23**, *74*
C_4H_9	$C_6H_{10}O_2$	—	52, **14**, *17*
			56, **14**, *33*
C_4H_9	$C_6H_{10}O_3$	—	335, **31**, *13*
C_4H_9	$C_6H_{10}O_4$	—	63, **14**, *73*
C_4H_9	$C_6H_{10}O_5$	—	60, **14**, *56*
			64, **14**, *75*
C_4H_9	$C_6H_{11}O$	—	5, **1**, *15*
			30, **11**, *9*
			30, **11**, *10*
C_4H_9	$C_6H_{11}O_2$	—	128

C_4H_9	$C_7H_5O_4Ti$	–	368
C_4H_9	C_7H_6NO	–	343, **32**, *22*
C_4H_9	$C_7H_6NO_2$	–	289, **27**, *17*
			289, **27**, *18*
			339, **32**, *4*
C_4H_9	$C_7H_6NO_3$	–	34, **12**, *14*
C_4H_9	$C_7H_6NO_4S$	–	289, **27**, *16*
C_4H_9	$C_7H_6O_2$	–	60, **14**, *53*
C_4H_9	$C_7H_6O_3$	–	335, **31**, *5*
C_4H_9	$C_7H_7AsO_3$	–	363, **39**, *2*
			364, **39**, *15*
C_4H_9	$C_7H_7AsO_4$	–	363, **39**, *7*
			363, **39**, *8*
C_4H_9	$C_7H_7N_2O$	–	35, **12**, *25*
C_4H_9	C_7H_7O	–	7, **1**, *42*
			34, **12**, *6*
C_4H_9	$C_7H_7O_3P$	–	359, **37**, *4*
C_4H_9	$C_7H_7O_3S$	–	349, **34**, *12*
C_4H_9	$C_7H_7O_4$	–	275, **25**, *25*
C_4H_9	$C_7H_8ClO_2$	–	263, **23**, *78*
C_4H_9	$C_7H_8O_3$	–	58, **14**, *46*
C_4H_9	$C_7H_9O_3$	–	99, **19**, *17*
C_4H_9	$C_7H_9O_4$	–	273, **25**, *4*
			273, **25**, *5*
C_4H_9	$C_7H_9O_4S$	–	258, **23**, *36*
C_4H_9	$C_7H_9O_5$	–	276, **25**, *31*
			276, **25**, *32*
C_4H_9	$C_7H_{10}O_4$	–	323, **30**, *8*
C_4H_9	$C_7H_{10}O_4S_2$	–	323, **30**, *12*
C_4H_9	$C_7H_{11}O_2$	–	253
C_4H_9	$C_7H_{11}O_3$	–	99, **19**, *14*
C_4H_9	$C_7H_{11}O_4$	–	100, **19**, *21*
C_4H_9	$C_7H_{12}ClO_2$	–	256, **23**, *11*
C_4H_9	$C_7H_{12}O_2$	–	53, **14**, *18*
C_4H_9	$C_7H_{12}O_4$	–	64, **14**, *77*
C_4H_9	$C_7H_{12}O_5$	–	65, **14**, *78*
C_4H_9	$C_7H_{12}O_6$	–	65, **14**, *81*
			65, **14**, *82*
			66, **14**, *83*
C_4H_9	$C_7H_{13}O_2$	–	128
C_4H_9	$C_7H_{13}O_3$	–	257, **23**, *22*
C_4H_9	$C_7H_{14}O_2$	–	55, **14**, *25*

		–	55, **14**, *26*
C_4H_9	$C_7H_{15}NO_2$	–	72, **14**, *110*
			72, **14**, *114*
C_4H_9	$C_7H_{15}N_2O$	–	343, **32**, *26*
C_4H_9	$C_7H_{16}NO$	–	7, **1**, *41*
C_4H_9	$C_8Cl_4O_4$	–	327, **30**, *33*
C_4H_9	$C_8F_{15}O_2$	–	255, **23**, *3*
C_4H_9	$C_8H_2F_5O_3$	–	257, **23**, *25*
C_4H_9	$C_8H_3NO_6$	–	327, **30**, *34*
C_4H_9	$C_8H_4ClF_2O_3S$	–	349, **34**, *13*
C_4H_9	$C_8H_4Cl_3O_3$	–	258, **23**, *29*
C_4H_9	$C_8H_4F_3O_2S$	–	99, **19**, *10*
C_4H_9	$C_8H_4F_3O_3S$	–	349, **34**, *14*
C_4H_9	$C_8H_4O_4$	–	326, **30**, *31*
			326, **30**, *32*
C_4H_9	$C_8H_5Cl_2O_3$	–	258, **23**, *28*
C_4H_9	$C_8H_6ClO_3$	–	257, **23**, *26*
			258, **23**, *27*
C_4H_9	$C_8H_6NO_5$	–	291, **27**, *29*
C_4H_9	$C_8H_6O_3$	–	331
C_4H_9	$C_8H_6O_5$	–	335, **31**, *10*
C_4H_9	$C_8H_7AsO_4$	–	367
C_4H_9	C_8H_7BrNO	–	343, **32**, *29*
C_4H_9	C_8H_7ClNO	–	343, **32**, *28*
C_4H_9	$C_8H_7NO_2$	–	74, **14**, *120*
C_4H_9	$C_8H_7N_3O_2$	–	73, **14**, *118*
C_4H_9	$C_8H_7O_2$	–	261, **23**, *59*
			291, **27**, *33*
C_4H_9	$C_8H_7O_2S$	–	259, **23**, *38*
C_4H_9	$C_8H_7O_3$	–	35, **12**, *22*
			257, **23**, *24*
			262, **23**, *62*
			289, **27**, *11*
C_4H_9	C_8H_8NO	–	343, **32**, *27*
C_4H_9	$C_8H_8NO_2$	–	290, **27**, *19*
			339, **32**, *7*
C_4H_9	$C_8H_8N_3O_2$	–	36, **12**, *27*
C_4H_9	$C_8H_8O_2$	–	49, **14**, *6*
C_4H_9	$C_8H_8O_3$	–	335, **31**, *9*
C_4H_9	$C_8H_8O_4$	–	325, **30**, *26*
			335, **31**, *6*
C_4H_9	$C_8H_9Br_2O_4$	–	277, **25**, *51*

C_4H_9	$C_8H_9Cl_2O_4$	—	277, **25**, *48*
C_4H_9	$C_8H_9N_2O$	—	36, **12**, *26*
C_4H_9	C_8H_9O	—	7, **1**, *43*
C_4H_9	$C_8H_9O_2$	—	284, **26**, *3*
C_4H_9	$C_8H_9O_4$	—	275, **25**, *26*
C_4H_9	$C_8H_{11}O_4$	—	273, **25**, *6*
			274, **25**, *7*
C_4H_9	$C_8H_{11}O_4S$	—	277, **25**, *43*
C_4H_9	$C_8H_{11}O_5$	—	276, **25**, *33*
C_4H_9	$C_8H_{11}O_6$	—	276, **25**, *39*
C_4H_9	$C_8H_{12}O_6$	—	52, **14**, *14*
C_4H_9	$C_8H_{13}O_2$	—	98, **19**, *3*
C_4H_9	$C_8H_{13}O_4S$	—	259, **23**, *41*
C_4H_9	$C_8H_{14}NO_3$	—	260, **23**, *53*
C_4H_9	$C_8H_{14}NO_4$	—	263, **23**, *71*
C_4H_9	$C_8H_{14}O_4$	—	56, **14**, *37*
C_4H_9	$C_8H_{14}O_5$	—	65, **14**, *79*
C_4H_9	$C_8H_{15}O_2$	—	129/30
			130/1
C_4H_9	$C_8H_{17}O$	—	4, **1**, *11*
			5, **1**, *12*
C_4H_9	$C_8H_{17}O_3P$	—	359, **37**, *3*
C_4H_9	$C_8H_{18}O_3P$	—	355, **36**, *10*
C_4H_9	$C_8H_{18}O_4P$	—	356, **36**, *16*
C_4H_9	$C_8H_{18}O_7P_2$	—	361
C_4H_9	$C_9H_2Cl_6O_4$	—	326, **30**, *29*
C_4H_9	$C_9H_3O_5$	—	292, **27**, *44*
C_4H_9	$C_9H_4BrN_2O_3$	—	41, **13**, *13*
			42, **13**, *14*
C_4H_9	$C_9H_4Br_2NO$	—	41, **13**, *9*
C_4H_9	$C_9H_4Cl_2NO$	—	40, **13**, *8*
C_4H_9	$C_9H_4I_2NO$	—	41, **13**, *10*
C_4H_9	$C_9H_4N_3O_5$	—	42, **13**, *15*
C_4H_9	$C_9H_4O_6$	—	327, **30**, *35*
C_4H_9	$C_9H_5N_2O_3$	—	41, **13**, *11*
			41, **13**, *12*
C_4H_9	$C_9H_6ClO_2$	—	269, **24**, *6*
C_4H_9	C_9H_6NO	—	40, **13**, *7*
C_4H_9	$C_9H_6NO_4$	—	269, **24**, *9*
C_4H_9	$C_9H_7O_2$	—	269, **24**, *5*
			291, **27**, *35*
C_4H_9	$C_9H_7O_3$	—	292, **27**, *38*

C_4H_9	$C_9H_8ClO_3$	—	258, **23**, *31*
C_4H_9	$C_9H_8O_4$	—	326, **30**, *28*
C_4H_9	$C_9H_9ClNO_2$	—	339, **32**, *6*
C_4H_9	$C_9H_9FN_2O_6$	—	61, **14**, *59*
C_4H_9	$C_9H_9NO_2$	—	73, **14**, *117*
C_4H_9	$C_9H_9N_3O_2$	—	73, **14**, *119*
C_4H_9	$C_9H_9O_2S$	—	258, **23**, *37*
C_4H_9	$C_9H_9O_3$	—	258, **23**, *30*
			289, **27**, *12*
C_4H_9	$C_9H_9O_3S$	—	292, **27**, *37*
C_4H_9	$C_9H_{10}N_2O_6$	—	60, **14**, *58*
C_4H_9	$C_9H_{10}N_3O_2$	—	36, **12**, *28*
C_4H_9	$C_9H_{11}N_3O_5$	—	61, **14**, *60*
C_4H_9	$C_9H_{11}O_2$	—	334, **31**, *2*
C_4H_9	$C_9H_{12}NO_2$	—	270, **24**, *20*
C_4H_9	$C_9H_{13}O_4$	—	274, **25**, *8*
			274, **25**, *9*
C_4H_9	$C_9H_{13}O_5$	—	276, **25**, *34*
C_4H_9	$C_9H_{14}O_4$	—	323, **30**, *9*
C_4H_9	$C_9H_{14}O_4S_2$	—	324, **30**, *19*
C_4H_9	$C_9H_{16}ClO_2$	—	256, **23**, *12*
C_4H_9	$C_9H_{17}O_2$	—	134
			135
C_4H_9	$C_{10}H_4O_8S_2$	—	59, **14**, *52*
C_4H_9	$C_{10}H_5Cl_6O_4$	—	284, **26**, *5*
C_4H_9	$C_{10}H_5N_2O$	—	30, **11**, *12*
C_4H_9	$C_{10}H_6ClO_3$	—	270, **24**, *14*
C_4H_9	$C_{10}H_6NO_2$	—	36, **12**, *30*
C_4H_9	$C_{10}H_6NO_3$	—	270, **24**, *21*
C_4H_9	$C_{10}H_7O_3$	—	270, **24**, *13*
C_4H_9	$C_{10}H_8BrO_2$	—	98, **19**, *7*
C_4H_9	$C_{10}H_8ClO_2$	—	98, **19**, *6*
C_4H_9	$C_{10}H_8FO_2$	—	98, **19**, *5*
C_4H_9	$C_{10}H_8NO$	—	42, **13**, *19*
C_4H_9	$C_{10}H_8O_2S$	—	98, **19**, *8*
C_4H_9	$C_{10}H_8O_8$	—	325, **30**, *25*
C_4H_9	$C_{10}H_9O_2$	—	98, **19**, *4*
C_4H_9	$C_{10}H_9O_3$	—	269, **24**, *7*
C_4H_9	$C_{10}H_9O_4$	—	292, **27**, *40*
C_4H_9	$C_{10}H_{10}BrN_5O_4$	—	62, **14**, *68*
C_4H_9	$C_{10}H_{10}ClO_4$	—	35, **12**, *23*
			289, **27**, *13*

C_4H_9	$C_{11}H_9FeO_2$	–	299, **28**, *5*
C_4H_9	$C_{11}H_9O_3$	–	270, **24**, *16*
C_4H_9	$C_{11}H_9O_4$	–	270, **24**, *15*
			275, **25**, *23*
C_4H_9	$C_{11}H_{10}O_4S_2$	–	323, **30**, *13*
C_4H_9	$C_{11}H_{11}O_3$	–	100, **19**, *20*
C_4H_9	$C_{11}H_{11}O_3S$	–	260, **23**, *51*
C_4H_9	$C_{11}H_{12}NO_2$	–	269, **24**, *8*
C_4H_9	$C_{11}H_{12}NO_3$	–	260, **23**, *54*
C_4H_9	$C_{11}H_{13}O_2$	–	291, **27**, *34*
C_4H_9	$C_{11}H_{14}N_3O_4$	–	7, **1**, *46*
C_4H_9	$C_{11}H_{14}O_8S_2$	–	324, **30**, *16*
C_4H_9	$C_{11}H_{15}NO_2$	–	72, **14**, *113*
C_4H_9	$C_{11}H_{15}O_4$	–	275, **25**, *21*
			284, **26**, *1*
C_4H_9	$C_{11}H_{19}O_2$	–	270, **24**, *12*
C_4H_9	$C_{11}H_{19}O_4$	–	100, **19**, *22*
C_4H_9	$C_{12}H_8CoO_4$	–	328, **30**, *43*
C_4H_9	$C_{12}H_8FeO_4$	–	328, **30**, *42*
C_4H_9	$C_{12}H_8N_3O$	–	36, **12**, *29*
C_4H_9	$C_{12}H_{10}BrN_5O_4$	–	63, **14**, *70*
C_4H_9	$C_{12}H_{10}OP$	–	354, **36**, *1*
C_4H_9	$C_{12}H_{10}O_2P$	–	354, **36**, *6*
C_4H_9	$C_{12}H_{10}O_3P$	–	355, **36**, *11*
			355, **36**, *14*
C_4H_9	$C_{12}H_{10}O_4P$	–	356, **36**, *17*
C_4H_9	$C_{12}H_{11}N_2O_2$	–	100, **19**, *25*
C_4H_9	$C_{12}H_{11}O_5$	–	276, **25**, *38*
C_4H_9	$C_{12}H_{12}BrN_5O_5$	–	63, **14**, *71*
C_4H_9	$C_{12}H_{13}BrN_6O_4$	–	63, **14**, *72*
C_4H_9	$C_{12}H_{13}O_2$	–	99, **19**, *15*
C_4H_9	$C_{12}H_{13}O_4$	–	292, **27**, *41*
C_4H_9	$C_{12}H_{14}N_3O_5$	–	261, **23**, *58*
C_4H_9	$C_{12}H_{14}O_4$	–	56, **14**, *38*
		–	64, **14**, *74*
C_4H_9	$C_{12}H_{15}N_5O_4$	–	61, **14**, *63*
C_4H_9	$C_{12}H_{15}O_5$	–	275, **25**, *22*
C_4H_9	$C_{12}H_{17}Br_2O_4$	–	277, **25**, *52*
C_4H_9	$C_{12}H_{17}Cl_2O_4$	–	277, **25**, *50*
C_4H_9	$C_{12}H_{17}O_5$	–	276, **25**, *37*
C_4H_9	$C_{12}H_{19}N_2O_2$	–	31, **11**, *15*
C_4H_9	$C_{12}H_{19}O_4$	–	274, **25**, *10*

			274, **25**, *11*
			274, **25**, *12*
			274, **25**, *13*
C_4H_9	$C_{12}H_{19}O_7$	—	6, **1**, *32*
C_4H_9	$C_{12}H_{21}O_4$	—	263, **23**, *69*
C_4H_9	$C_{12}H_{21}O_6$	—	6, **1**, *31*
C_4H_9	$C_{12}H_{22}O_2$	—	53, **14**, *19*
C_4H_9	$C_{12}H_{23}NO_2$	—	73, **14**, *115*
C_4H_9	$C_{12}H_{23}O_2$	—	135/249
			250
C_4H_9	$C_{12}H_{25}O$	—	5, **1**, *14*
C_4H_9	$C_{12}H_{26}BO_2$	—	351, **35**, *1*
C_4H_9	$C_{12}H_{26}O_2P$	—	354, **36**, *4*
C_4H_9	$C_{13}H_8N_3O_6$	—	340, **32**, *12*
C_4H_9	$C_{13}H_9AsClNO_3$	—	364, **39**, *18*
C_4H_9	$C_{13}H_9AsN_2O_5$	—	364, **39**, *19*
			364, **39**, *20*
C_4H_9	$C_{13}H_9ClNO_2$	—	340, **32**, *9*
			341, **32**, *13*
C_4H_9	$C_{13}H_9FNO$	—	35, **12**, *24*
C_4H_9	$C_{13}H_9INO_2$	—	340, **32**, *10*
C_4H_9	$C_{13}H_9N_2O_4$	—	340, **32**, *11*
C_4H_9	$C_{13}H_9O_2$	—	35, **12**, *18*
C_4H_9	$C_{13}H_{10}NO_2$	—	290, **27**, *20*
			340, **32**, *8*
C_4H_9	$C_{13}H_{11}O_4$	—	276, **25**, *28*
C_4H_9	$C_{13}H_{13}NO_6$	—	322, **30**, *5*
C_4H_9	$C_{13}H_{13}N_2O_2$	—	100, **19**, *26*
C_4H_9	$C_{13}H_{13}O_3$	—	270, **24**, *17*
C_4H_9	$C_{13}H_{14}O_4S_2$	—	324, **30**, *20*
C_4H_9	$C_{13}H_{14}O_5S_2$	—	324, **30**, *21*
C_4H_9	$C_{13}H_{16}O_5$	—	64, **14**, *76*
C_4H_9	$C_{13}H_{17}O_4$	—	285, **26**, *6*
C_4H_9	$C_{13}H_{17}O_4S$	—	285, **26**, *11*
C_4H_9	$C_{13}H_{19}NO_7P_2S_2$	—	360
C_4H_9	$C_{13}H_{20}O_2P$	—	354, **36**, *5*
C_4H_9	$C_{13}H_{21}O_4$	—	274, **25**, *14*
C_4H_9	$C_{13}H_{23}O_2$	—	99, **19**, *16*
C_4H_9	$C_{13}H_{24}ClO_2$	—	256, **23**, *13*
C_4H_9	$C_{13}H_{25}O_3$	—	257, **23**, *23*
C_4H_9	$C_{14}H_8O_4S_2$	—	71, **14**, *105*
C_4H_9	$C_{14}H_9O_3$	—	292, **27**, *39*

C_4H_9	$C_{14}H_{10}NO_2$	—	290, **27**, *23*
C_4H_9	$C_{14}H_{10}O_2$	—	58, **14**, *43*
C_4H_9	$C_{14}H_{10}O_6Ti$	—	368
C_4H_9	$C_{14}H_{11}O_3$	—	35, **12**, *19*
C_4H_9	$C_{14}H_{11}O_4$	—	35, **12**, *20*
			262, **23**, *63*
C_4H_9	$C_{14}H_{12}NO_2$	—	341, **32**, *14*
			341, **32**, *15*
			342, **32**, *16*
C_4H_9	$C_{14}H_{12}N_3O_3$	—	292, **27**, *45*
C_4H_9	$C_{14}H_{12}O_2$	—	51, **14**, *12*
C_4H_9	$C_{14}H_{13}Br_4O_5$	—	292, **27**, *46*
C_4H_9	$C_{14}H_{14}As_2O_4$	—	367
C_4H_9	$C_{14}H_{14}NO_4$	—	261, **23**, *56*
C_4H_9	$C_{14}H_{15}BrO_6$	—	65, **14**, *80*
C_4H_9	$C_{14}H_{15}O_4$	—	275, **25**, *24*
C_4H_9	$C_{14}H_{16}N_3O_3$	—	284, **26**, *2*
C_4H_9	$C_{14}H_{16}O_4$	—	327, **30**, *36*
C_4H_9	$C_{14}H_{16}O_6$	—	68, **14**, *95*
			69, **14**, *96*
			69, **14**, *97*
			70, **14**, *98*
C_4H_9	$C_{14}H_{17}O_4$	—	100, **19**, *23*
C_4H_9	$C_{14}H_{18}O_4$	—	325, **30**, *27*
C_4H_9	$C_{14}H_{21}O$	—	34, **12**, *8*
			34, **12**, *9*
C_4H_9	$C_{14}H_{21}O_5S$	—	277, **25**, *54*
C_4H_9	$C_{14}H_{23}O_5$	—	276, **25**, *36*
C_4H_9	$C_{14}H_{23}O_6S_3Sb$	—	324, **30**, *15*
C_4H_9	$C_{14}H_{25}O_4$	—	263, **23**, *76*
C_4H_9	$C_{14}H_{25}O_4S$	—	259, **23**, *43*
C_4H_9	$C_{14}H_{27}O_2$	—	250
C_4H_9	$C_{14}H_{27}O_3$	—	6, **1**, *24*
C_4H_9	$C_{15}H_9O_3$	—	40, **13**, *3*
C_4H_9	$C_{15}H_{10}N_3O$	—	42, **13**, *16*
C_4H_9	$C_{15}H_{10}O_2S$	—	99, **19**, *12*
C_4H_9	$C_{15}H_{11}O_2$	—	99, **19**, *11*
C_4H_9	$C_{15}H_{12}N_2O_2$	—	74, **14**, *122*
C_4H_9	$C_{15}H_{12}N_4O_2S$	—	75, **14**, *127*
C_4H_9	$C_{15}H_{14}O_2$	—	60, **14**, *54*
C_4H_9	$C_{15}H_{15}O$	—	34, **12**, *16*
C_4H_9	$C_{15}H_{19}N_5O_4$	—	62, **14**, *64*

C_4H_9	$C_{15}H_{19}N_5O_5$	—	62, **14**, *65*
C_4H_9	$C_{15}H_{19}O_4$	—	285, **26**, *10*
C_4H_9	$C_{15}H_{21}O_2S$	—	289, **27**, *15*
C_4H_9	$C_{15}H_{21}O_3$	—	291, **27**, *36*
C_4H_9	$C_{15}H_{22}O_6$	—	66, **14**, *84*
C_4H_9	$C_{15}H_{22}O_8$	—	58, **14**, *47*
C_4H_9	$C_{15}H_{23}O$	—	34, **12**, *10*
C_4H_9	$C_{15}H_{23}O_4P$	—	359, **37**, *5*
C_4H_9	$C_{15}H_{26}N_3O$	—	34, **12**, *13*
C_4H_9	$C_{16}H_{10}NO_2$	—	43, **13**, *21*
C_4H_9	$C_{16}H_{10}N_2O_4$	—	328, **30**, *40*
C_4H_9	$C_{16}H_{10}N_3O_3$	—	42, **13**, *17*
C_4H_9	$C_{16}H_{12}O_4S_2$	—	71, **14**, *106*
C_4H_9	$C_{16}H_{13}N_2O_2$	—	342, **32**, *19*
			4, **14**, *123*
C_4H_9	$C_{16}H_{14}N_2O_4$	—	337/8
C_4H_9	$C_{16}H_{14}O_4$	—	58, **14**, *44*
C_4H_9	$C_{16}H_{14}O_8Ti$	—	368
C_4H_9	$C_{16}H_{15}O_4$	—	6, **1**, *23*
C_4H_9	$C_{16}H_{21}O_4$	—	292, **27**, *42*
			292, **27**, *43*
C_4H_9	$C_{16}H_{23}O_3$	—	261, **23**, *60*
C_4H_9	$C_{16}H_{23}O_7S$	—	277, **25**, *53*
C_4H_9	$C_{16}H_{25}O$	—	34, **12**, *11*
C_4H_9	$C_{16}H_{27}O_4$	—	274, **25**, *15*
C_4H_9	$C_{16}H_{28}O_4S_2$	—	324, **30**, *18*
C_4H_9	$C_{16}H_{29}O_4$	—	263, **23**, *70*
C_4H_9	$C_{16}H_{30}NO_3$	—	263, **23**, *72*
C_4H_9	$C_{16}H_{31}O_2$	—	250
C_4H_9	$C_{16}H_{31}O_2S$	—	259, **23**, *50*
C_4H_9	$C_{16}H_{31}O_5$	—	256, **23**, *17*
C_4H_9	$C_{16}H_{34}BO_3$	—	351, **35**, *2*
C_4H_9	$C_{17}H_{11}FNO$	—	36, **12**, *31*
C_4H_9	$C_{17}H_{11}O_4$	—	270, **24**, *19*
C_4H_9	$C_{17}H_{12}ClN_2O_2$	—	100, **19**, *28*
C_4H_9	$C_{17}H_{12}NO_4$	—	261, **23**, *57*
C_4H_9	$C_{17}H_{12}N_3O_3$	—	42, **13**, *18*
C_4H_9	$C_{17}H_{13}N_2O_2$	—	100, **19**, *27*
C_4H_9	$C_{17}H_{14}FeN_3O$	—	332
C_4H_9	$C_{17}H_{14}N_2O_2$	—	74, **14**, *121*
C_4H_9	$C_{17}H_{15}N_5O_5$	—	62, **14**, *66*
C_4H_9	$C_{17}H_{16}N_4O_2S$	—	76, **14**, *128*

C_4H_9	$C_{17}H_{18}O_2$	—	50, **14**, *11*
C_4H_9	$C_{17}H_{25}O_3$	—	262, **23**, *61*
C_4H_9	$C_{17}H_{25}O_4$	—	285, **26**, *7*
			285, **26**, *8*
C_4H_9	$C_{18}H_{14}ClFeN_2O$	—	332
C_4H_9	$C_{18}H_{14}FeN_3O_3$	—	332
C_4H_9	$C_{18}H_{15}FeN_2O$	—	332
C_4H_9	$C_{18}H_{15}FeN_2O_2$	—	332
C_4H_9	$C_{18}H_{15}OSi$	—	352
C_4H_9	$C_{18}H_{22}NO_3$	—	270, **24**, *22*
C_4H_9	$C_{18}H_{23}NO_6$	—	66, **14**, *85*
C_4H_9	$C_{18}H_{31}O_7$	—	6, **1**, *33*
C_4H_9	$C_{18}H_{33}O_2$	—	269, **24**, *11*
C_4H_9	$C_{18}H_{33}O_3$	—	257, **23**, *19*
C_4H_9	$C_{18}H_{33}O_4$	—	257, **23**, *20*
C_4H_9	$C_{18}H_{33}O_4S$	—	259, **23**, *44*
C_4H_9	$C_{18}H_{33}O_6$	—	39, **13**, *1*
C_4H_9	$C_{18}H_{35}O_2$	—	250
C_4H_9	$C_{18}H_{35}O_2$	—	251
C_4H_9	$C_{18}H_{35}O_3$	—	256, **23**, *16*
C_4H_9	$C_{19}H_{14}AsO_2$	—	291, **27**, *31*
C_4H_9	$C_{19}H_{14}BBrNO_2$	—	351, **35**, *5*
C_4H_9	$C_{19}H_{14}BClNO_2$	—	351, **35**, *4*
C_4H_9	$C_{19}H_{14}BFNO_2$	—	351, **35**, *3*
C_4H_9	$C_{19}H_{14}BN_2O_4$	—	351, **35**, *6*
C_4H_9	$C_{19}H_{20}O_5$	—	60, **14**, *57*
C_4H_9	$C_{19}H_{32}O_4P$	—	355, **36**, *13*
C_4H_9	$C_{19}H_{35}O_3$	—	30, **11**, *13*
C_4H_9	$C_{20}H_{14}N_2O_2$	—	75, **14**, *126*
C_4H_9	$C_{20}H_{20}O_6$	—	70, **14**, *99*
C_4H_9	$C_{20}H_{22}N_2O_2$	—	75, **14**, *124*
C_4H_9	$C_{20}H_{26}O_6$	—	66, **14**, *86*
C_4H_9	$C_{20}H_{35}O_4$	—	274, **25**, *16*
			274, **25**, *17*
C_4H_9	$C_{21}H_{14}O_5S_2$	—	328, **30**, *39*
C_4H_9	$C_{21}H_{18}AsO_2$	—	291, **27**, *32*
C_4H_9	$C_{21}H_{25}O_3$	—	35, **12**, *21*
C_4H_9	$C_{21}H_{28}O_3P$	—	355, **36**, *7*
C_4H_9	$C_{21}H_{31}O_4$	—	285, **26**, *9*
C_4H_9	$C_{21}H_{31}O_5S_2$	—	259, **23**, *45*
C_4H_9	$C_{21}H_{35}O$	—	34, **12**, *12*
C_4H_9	$C_{21}H_{41}O_2$	—	251

C_4H_9	$C_{22}H_{37}O_4$	—	275, **25**, *27*
C_4H_9	$C_{22}H_{37}O_5$	—	277, **25**, *44*
C_4H_9	$C_{22}H_{39}O_4$	—	275, **25**, *18*
C_4H_9	$C_{22}H_{40}O_6S_3Sb$	—	259, **23**, *39*
C_4H_9	$C_{22}H_{43}O_2$	—	251
C_4H_9	$C_{23}H_{16}BBrNO_2$	—	351, **35**, *9*
C_4H_9	$C_{23}H_{16}BClNO_2$	—	351, **35**, *8*
C_4H_9	$C_{23}H_{16}BFNO_2$	—	351, **35**, *7*
C_4H_9	$C_{23}H_{16}BN_2O_4$	—	351, **35**, *10*
C_4H_9	$C_{23}H_{24}O_7$	—	66, **14**, *86*
C_4H_9	$C_{23}H_{26}O_6$	—	67, **14**, *87*
			67, **14**, *88*
			67, **14**, *89*
C_4H_9	$C_{23}H_{41}O_6S_2$	—	259, **23**, *47*
C_4H_9	$C_{24}H_{20}O_3Si_2$	—	352
C_4H_9	$C_{24}H_{43}O_4$	—	275, **25**, *19*
C_4H_9	$C_{25}H_{47}O_4S_2$	—	259, **23**, *48*
C_4H_9	$C_{26}H_{26}O_6$	—	68, **14**, *92*
C_4H_9	$C_{26}H_{30}O_6$	—	70, **14**, *101*
C_4H_9	$C_{27}H_{28}O_6$	—	67, **14**, *90*
			68, **14**, *91*
			68, **14**, *93*
C_4H_9	$C_{29}H_{55}O_4S_2$	—	259, **23**, *49*
C_4H_9	$C_{30}H_{18}O_8$	—	327, **30**, *38*
C_4H_9	$C_{30}H_{22}O_6Ti$	—	368
C_4H_9	$C_{30}H_{26}N_2O_2$	—	75, **14**, *125*
C_4H_9	$C_{32}H_{20}O_4$	—	327, **30**, *37*
C_4H_9	$C_{32}H_{47}O_4$	—	262, **23**, *64*
C_4H_9	$C_{33}H_{49}O_4$	—	262, **23**, *65*
C_4H_9	$C_{33}H_{49}O_{12}$	—	277, **25**, *46*
C_4H_9	$C_{33}H_{57}O_8$	—	277, **25**, *45*
C_4H_9	$C_{33}H_{61}O_7S_2$	—	259, **23**, *46*
C_4H_9	$C_{34}H_{34}O_6$	—	70, **14**, *100*
C_4H_9	$C_{40}H_{72}O_7$	—	71, **14**, *104*
C_4H_9	$C_{40}H_{73}O_7$	—	31, **11**, *14*
C_4H_9	$C_{68}H_{135}O_{27}$	—	6, **1**, *25*
C_4H_9	$C_{136}H_{265}O_{45}$	—	277, **25**, *47*
$CH_2CH(CH_3)_2$	CH_3O	—	369, **41**, *1*
$CH_2CH(CH_3)_2$	$C_2H_3O_2$	—	370, **41**, *10*
$CH_2CH(CH_3)_2$	$C_4H_2O_4$	—	371, **41**, *14*
$CH_2CH(CH_3)_2$	$C_4H_5O_2$	—	370, **41**, *12*
$CH_2CH(CH_3)_2$	$C_4H_9NO_2$	—	370, **41**, *6*

$CH_2CH(CH_3)_2$	C_4H_9O	—	369, **41**, *2*
$CH_2CH(CH_3)_2$	$C_5H_{11}NO_2$	—	370, **41**, *7*
$CH_2CH(CH_3)_2$	$C_7H_6NO_2$	—	370, **41**, *13*
$CH_2CH(CH_3)_2$	$C_7H_7O_2$	—	369, **41**, *3*
$CH_2CH(CH_3)_2$	C_9H_6NO	—	369, **41**, *4*
$CH_2CH(CH_3)_2$	$C_{10}H_{13}NO_2$	—	370, **41**, *9*
$CH_2CH(CH_3)_2$	$C_{10}H_{19}NO_2$	—	370, **41**, *8*
$CH_2CH(CH_3)_2$	$C_{11}H_{17}OSi$	—	371, **41**, *15*
$CH_2CH(CH_3)_2$	$C_{12}H_{23}O_2$	—	370, **41**, *11*
$CH_2CH(CH_3)_2$	$C_{23}H_{26}O_6$	—	369, **41**, *5*
$CH(CH_3)C_2H_5$	$C_{12}H_{23}O_2$	—	372
$C(CH_3)_3$	—	HO	373, **43**, *1*
$C(CH_3)_3$	CH_3O	—	373, **43**, *2*
$C(CH_3)_3$	$C_2H_3O_2$	—	376, **43**, *17*
$C(CH_3)_3$	$C_4H_8O_2S$	—	375, **43**, *13*
$C(CH_3)_3$	$C_4H_9NO_2$	—	374, **43**, *6*
$C(CH_3)_3$	$C_5H_7NO_4$	—	376, **43**, *18*
$C(CH_3)_3$	$C_5H_7O_2$	—	376, **43**, *16*
$C(CH_3)_3$	$C_5H_{11}NO_2$	—	374, **43**, *7*
$C(CH_3)_3$	$C_6H_{13}NO_2$	—	374, **43**, *8*
$C(CH_3)_3$	$C_7H_5O_2$	—	373, **43**, *3*
$C(CH_3)_3$	$C_7H_{11}NO_4$	—	376, **43**, *19*
$C(CH_3)_3$	$C_7H_{15}NO_2$	—	374, **43**, *9*
$C(CH_3)_3$	$C_8H_{17}NO_2$	—	374, **43**, *10*
			375, **43**, *11*
			375, **43**, *12*
$C(CH_3)_3$	C_9H_6NO	—	373, **43**, *4*
$C(CH_3)_3$	$C_{10}H_8NO$	—	374, **43**, *5*
$C(CH_3)_3$	$C_{16}H_{17}O_2P$	—	375, **43**, *14*
$C(CH_3)_3$	$C_{18}H_{13}O_2P$	—	375, **43**, *15*
$C_4H_9NO_2$			
$OCH_2C(CH_3)(NH_2)CH_2O$	C_4H_9	—	55, **14**, *30*
$O(CH_2)_2NH(CH_2)_2O$	C_4H_9	—	71, **14**, *107*
			370, **41**, *6*
			374, **43**, *6*
C_4H_9O			
OC_4H_9	C_4H_9	—	4, **1**, *5*
$OCH(CH_3)C_2H_5$	C_4H_9	—	4, **1**, *6*
$OCH_2CH(CH_3)_2$	C_4H_9	—	4, **1**, *7*
			369, **41**, *2*
$OC(CH_3)_3$	C_4H_9	—	4, **1**, *8*
$C_4H_9O_2$	C_4H_9	—	334, **31**, *1*

$C_4H_9O_2S$	C_4H_9	—	347, **34**, *1*
$C_4H_9O_3S$	C_4H_9	—	348, **34**, *7*
$C_4H_{10}NO$			
$OCH_2CH_2NHC_2H_5$	C_4H_9	—	6, **1**, *35*
$OCH_2CH_2N(CH_3)_2$	C_4H_9	—	6, **1**, *36*
$ON(C_2H_5)_2$	C_4H_9	—	339, **32**, *1*
$C_4H_{10}O_3P$			
$OP(O)(H)OC_4H_9$	C_4H_9	—	355, **36**, *8*
$OP(OC_2H_5)_2$	C_4H_9	—	355, **36**, *9*
$C_4H_{10}O_4P$	C_4H_9	—	356, **36**, *15*
C_4O_4	C_4H_9	—	326, **30**, *30*
$C_5F_6O_4$	C_4H_9	—	322, **30**, *6*
$C_5HF_6O_2$	C_4H_9	—	98, **19**, *9*
$C_5H_2BrO_3$	C_4H_9	—	299, **28**, *2*
$C_5H_3NO_2$	C_4H_9	—	60, **14**, *55*
$C_5H_3N_2O$	C_4H_9	—	30, **11**, *11*
$C_5H_3O_3$	C_4H_9	—	298, **28**, *1*
C_5H_4NO	C_4H_9	—	40, **13**, *4*
$C_5H_4NO_2$			
2-Oxypyridinyl-1-oxide	C_4H_9	—	40, **13**, *5*
[(2-Pyrrolyl)carbonyl]oxy	C_4H_9	—	299, **28**, *3*
$C_5H_4O_4$	C_4H_9	—	322, **30**, *3*
$C_5H_4O_4S_3$	C_4H_9	—	324, **30**, *14*
$C_5H_5O_2$	C_4H_9	—	7, **1**, *44*
$C_5H_5O_4$	C_4H_9	—	273, **25**, *2*
$C_5H_6O_4S_2$	C_4H_9	—	323, **30**, *11*
$C_5H_7Cl_3NO_2$	C_4H_9	—	5, **1**, *21*
$C_5H_7NO_4$	C_4H_9	—	325, **30**, *22*
			376, **43**, *18*
$C_5H_7O_2$	C_4H_9	—	97, **19**, *1*
			376, **43**, *16*
$C_5H_7O_3$			
$OCH_2COCH_2COCH_3$	C_4H_9	—	6, **1**, *30*
$OCC(CH_3)CHC(OCH_3)CO$	C_4H_9	—	99, **19**, *18*
$C_5H_7O_4$	C_4H_9	—	262, **23**, *67*
$C_5H_8ClO_2$	C_4H_9	—	256, **23**, *10*
C_5H_8NO	C_4H_9	—	100, **19**, *29*
$C_5H_8NO_3$	C_4H_9	—	260, **23**, *52*
$C_5H_8O_2$	C_4H_9	—	52, **14**, *16*
$C_5H_8O_2S$	C_4H_9	—	56, **14**, *35*
$C_5H_8O_3$	C_4H_9	—	56, **14**, *34*

$C_5H_9NO_4$	C_4H_9	–	56, **14**, *32*
C_5H_9O			
$O(CH_2)_3CH{=}CH_2$	C_4H_9	–	29, **11**, *5*
$OCH(CH_3)CH_2CH{=}CH_2$	C_4H_9	–	29, **11**, *6*
$OCH(CH_3)CH{=}CHCH_3$	C_4H_9	–	30, **11**, *7*
$OC(CH_3)_2CH{=}CH_2$	C_4H_9	–	30, **11**, *8*
$C_5H_9O_2$			
$OOCC_4H_9$	C_4H_9	–	127/8
$OOCC(CH_3)_3$	C_4H_9	–	128
$C_5H_9O_3$			
$OCH(CH_3)COOC_2H_5$	C_4H_9	–	6, **1**, *27*
(2-Methyl-1,3-dioxolan-4-yl)methoxy	C_4H_9	–	7, **1**, *45*
$C_5H_{10}O_2$			
$OCH(CH_3)CH_2CH(CH_3)O$	C_4H_9	–	54, **14**, *22*
$OCH_2C(CH_3)_2CH_2O$	C_4H_9	–	54, **14**, *23*
$O(CH_2)_5O$	C_4H_9	–	57, **14**, *40*
$C_5H_{10}O_3$	C_4H_9	–	55, **14**, *29*
$C_5H_{11}NO_2$	C_4H_9	–	72, **14**, *108*
			370, **41**, *7*
			374, **43**, *7*
$C_5H_{11}O$			
OC_5H_{11}	C_4H_9	–	4, **1**, *9*
$OCH(CH_3)C_3H_7$	C_4H_9	–	4, **1**, *10*
$C_5H_{12}NO$	C_4H_9	–	7, **1**, *39*
$\mathbf{C_6}Cl_2O_4$	C_4H_9	–	59, **14**, *51*
C_6Cl_5O	C_4H_9	–	33, **12**, *3*
$C_6H_2Br_2NO_3$	C_4H_9	–	33, **12**, *5*
$C_6H_2Cl_3O_3PS$	C_4H_9	–	359, **37**, *7*
$C_6H_4AsBrO_3$	C_4H_9	–	363, **39**, *5*
$C_6H_4AsClO_3$	C_4H_9	–	363, **39**, *4*
$C_6H_4AsNO_5$			
$OAs(O)(C_6H_4NO_2\text{-}2)O$	C_4H_9	–	363, **39**, *11*
$OAs(O)(C_6H_4NO_2\text{-}4)O$	C_4H_9	–	363, **39**, *12*
$C_6H_4AsNO_6$	C_4H_9	–	364, **39**, *13*
C_6H_4ClO	C_4H_9	–	33, **12**, *2*
$C_6H_4NO_2$	C_4H_9	–	299, **28**, *4*
$C_6H_4N_3O$	C_4H_9	–	342, **32**, *18*
$C_6H_4O_2$			
$OC_6H_4O\text{-}2$	C_4H_9	–	59, **14**, *48*
$OC_6H_4O\text{-}4$	C_4H_9	–	59, **14**, *49*
$C_6H_5AsNO_5$	C_4H_9	–	362, **38**, *2*

$C_6H_5AsNO_6$	C_4H_9	–	362, **38**, *3*
$C_6H_5AsN_2O_5$	C_4H_9	–	364, **39**, *14*
$C_6H_5AsO_3$	C_4H_9	–	363, **39**, *3*
$C_6H_5AsO_4$	C_4H_9	–	363, **39**, *6*
$C_6H_5NO_3$	C_4H_9	–	335, **31**, *11*
$C_6H_5N_2O_2$	C_4H_9	–	343, **32**, *31*
C_6H_5O	C_4H_9	–	33, **12**, *1*
$C_6H_5O_2Se$	C_4H_9	–	368
$C_6H_5O_3S$	C_4H_9	–	349, **34**, *11*
$C_6H_5O_4$	C_4H_9	–	39, **13**, *2*
$C_6H_6AsNO_3$			
$OAs(O)(C_6H_4NH_2\text{-}2)O$	C_4H_9	–	363, **39**, *9*
$OAs(O)(C_6H_4NH_2\text{-}4)O$	C_4H_9	–	363, **39**, *10*
$C_6H_6AsN_2O_5$	C_4H_9	–	362, **38**, *4*
$C_6H_6AsO_3$	C_4H_9	–	361, **38**, *1*
C_6H_6NO	C_4H_9	–	33, **12**, *4*
$C_6H_6O_3P$	C_4H_9	–	355, **36**, *12*
$C_6H_7O_4$	C_4H_9	–	273, **25**, *3*
$C_6H_7O_5$	C_4H_9	–	276, **25**, *30*
$C_6H_8O_4$	C_4H_9	–	323, **30**, *7*
$C_6H_8O_5$	C_4H_9	–	68, **14**, *94*
$C_6H_8O_4S$	C_4H_9	–	324, **30**, *17*
$C_6H_8O_6$	C_4H_9	–	51, **14**, *13*
$C_6H_9O_2$			
$OCC(CH_3)CHC(C_2H_5)CO$	C_4H_9	–	98, **19**, *2*
(Cyclopentylcarbony)oxy	C_4H_9	–	253
$C_6H_9O_3$	C_4H_9	–	100, **19**, *19*
$C_6H_9O_4$			
$OOC(CH_2)_2COOC_2H_5$	C_4H_9	–	262, **23**, *68*
$OOC(CH_2)_4COOH$	C_4H_9	–	263, **23**, *74*
$C_6H_{10}O_2$			
Cyclohex-1,2-diyldioxy	C_4H_9	–	52, **14**, *17*
Cyclobutylidene-1,1-bis(methoxy)	C_4H_9	–	56, **14**, *33*
$C_6H_{10}O_3$	C_4H_9	–	335, **31**, *13*
$C_6H_{10}O_4$	C_4H_9	–	63, **14**, *73*
$C_6H_{10}O_5$			
[1-*O*-Methyl-D-ribofuranosyl-3,4-diyl]dioxy	C_4H_9	–	60, **14**, *56*
[1-*O*-Methyl-ß-L-arabopyranosyl-3,4-diyl]dioxy	C_4H_9	–	64, **14**, *75*
$C_6H_{11}O$			
Cyclohexyloxy	C_4H_9	–	5, **1**, *15*
$OCH(C_2H_5)CH{=}CHCH_3$	C_4H_9	–	30, **11**, *9*
$OC(CH_3)(C_2H_5)CH{=}CH_2$	C_4H_9	–	30, **11**, *10*

$C_6H_{11}O_2$	C_4H_9	–	128
$C_6H_{11}O_3$			
$OCH(C_2H_5)COOC_2H_5$	C_4H_9	–	6, **1**, *28*
$OC(CH_3)_2COOC_2H_5$	C_4H_9	–	6, **1**, *29*
$OO\text{-}COC_5H_{11}$	C_4H_9	–	336, **31**, *14*
$C_6H_{11}O_7$	C_4H_9	–	257, **23**, *18*
$C_6H_{12}O_2$			
$OC(CH_3)_2C(CH_3)_2O$	C_4H_9	–	50, **14**, *9*
$OCH(CH_3)CH_2C(CH_3)_2O$	C_4H_9	–	54, **14**, *24*
$O(CH_2)_6O$	C_4H_9	–	57, **14**, *41*
$C_6H_{12}O_4$	C_4H_9	–	50, **14**, *10*
$C_6H_{12}O_6S_2$	C_4H_9	–	346
$C_6H_{13}NO_2$	C_4H_9	–	72, **14**, *109*
			374, **43**, *8*
$C_6H_{13}NO_3$	C_4H_9	–	72, **14**, *111*
$C_6H_{13}O_3P$	C_4H_9	–	359, **37**, *2*
$C_6H_{14}NO$	C_4H_9	–	7, **1**, *37*
$C_6H_{15}O_4Si$	C_4H_9	–	352
$C_7H_3ClNO_4$	C_4H_9	–	291, **27**, *28*
$C_7H_3Cl_2O_2$	C_4H_9	–	288, **27**, *6*
$C_7H_3NO_4$	C_4H_9	–	328, **30**, *41*
$C_7H_3N_2O_6$	C_4H_9	–	291, **27**, *30*
$C_7H_4BrO_2$			
$OOCC_6H_4Br\text{-}3$	C_4H_9	–	288, **27**, *7*
$OOCC_6H_4Br\text{-}4$	C_4H_9	–	288, **27**, *8*
$C_7H_4ClO_2$			
$OOCC_6H_4Cl\text{-}2$	C_4H_9	–	288, **27**, *3*
$OOCC_6H_4Cl\text{-}3$	C_4H_9	–	288, **27**, *4*
$OOCC_6H_4Cl\text{-}4$	C_4H_9	–	288, **27**, *5*
$C_7H_4FO_2$	C_4H_9	–	287, **27**, *2*
$C_7H_4NO_4$			
$OOCC_6H_4NO_2\text{-}2$	C_4H_9	–	290, **27**, *25*
$OOCC_6H_4NO_2\text{-}3$	C_4H_9	–	291, **27**, *26*
$OOCC_6H_4NO_2\text{-}4$	C_4H_9	–	291, **27**, *27*
$C_7H_4O_3$	C_4H_9	–	331
$C_7H_5AsO_5$			
$OAs(O)(C_6H_4COOH\text{-}2)O$	C_4H_9	–	364, **39**, *16*
$OAs(O)(C_6H_4COOH\text{-}4)O$	C_4H_9	–	364, **39**, *17*
$C_7H_4ClNO_2$			
$OOCC_6H_3(NH_2\text{-}2)Cl\text{-}5$	C_4H_9	–	290, **27**, *24*
$ONHCOC_6H_4Cl\text{-}4$	C_4H_9	–	339, **32**, *5*

$C_7H_5ClO_3$	C_4H_9	–	335, **31**, *7*
$C_7H_5NO_3$	C_4H_9	–	332
$C_7H_5NO_5$	C_4H_9	–	335, **31**, *8*
$C_7H_5O_2$			
(7-Oxo-1,3,5-cycloheptatrienyl)oxy	C_4H_9	–	31, **11**, *16*
			373, **43**, *3*
OC_6H_4CHO-2	C_4H_9	–	34, **12**, *17*
$OOCC_6H_5$	C_4H_9	–	287, **27**, *1*
$C_7H_5O_2S$	C_4H_9	–	289, **27**, *14*
$C_7H_5O_3$			
[2-(Furan-2-yl)ethenylcarbonyl]oxy	C_4H_9	–	269, **24**, *10*
$OOCC_6H_4OH$-2	C_4H_9	–	288, **27**, *9*
$OOCC_6H_4OH$-4	C_4H_9	–	288, **27**, *10*
$C_7H_5O_4Ti$	C_4H_9	–	368
C_7H_6NO	C_4H_9	–	343, **32**, *22*
$C_7H_6NO_2$			
$OOCC_6H_4NH_2$-2	C_4H_9	–	289, **27**, *17*
$OOCC_6H_4NH_2$-4	C_4H_9	–	289, **27**, *18*
			370, **41**, *13*
$ONHCOC_6H_5$	C_4H_9	–	339, **32**, *4*
$C_7H_6NO_3$	C_4H_9	–	34, **12**, *14*
$C_7H_6NO_4S$	C_4H_9	–	289, **27**, *16*
$C_7H_6O_2$	C_4H_9	–	60, **14**, *53*
$C_7H_6O_3$	C_4H_9	–	335, **31**, *5*
$C_7H_7AsO_3$			
$OAs(O)(CH_2C_6H_5)O$	C_4H_9	–	363, **39**, *2*
$OAs(O)(C_6H_4CH_3$-4$)O$	C_4H_9	–	364, **39**, *15*
$C_7H_7AsO_4$			
$OAs(O)(C_6H_4OCH_3$-2$)O$	C_4H_9	–	363, **39**, *7*
$OAs(O)(C_6H_4OCH_3$-4$)O$	C_4H_9	–	363, **39**, *8*
$C_7H_7N_2O$	C_4H_9	–	35, **12**, *25*
C_7H_7O			
$OCH_2C_6H_5$	C_4H_9	–	7, **1**, *42*
$OC_6H_4CH_3$-4	C_4H_9	–	34, **12**, *6*
$C_7H_7O_2$	C_4H_9	–	369, **41**, *3*
$C_7H_7O_3P$	C_4H_9	–	359, **37**, *4*
$C_7H_7O_3S$	C_4H_9	–	349, **34**, *12*
$C_7H_7O_4$	C_4H_9	–	275, **25**, *25*
$C_7H_8ClO_2$	C_4H_9	–	263, **23**, *78*
$C_7H_8O_3$	C_4H_9	–	58, **14**, *46*
$C_7H_9O_3$	C_4H_9	–	99, **19**, *17*
$C_7H_9O_4$			

$OOCCH{=}CHCOOC_3H_7$	C_4H_9	–	273, **25**, *4*
$OOCCH{=}CHCOOCH(CH_3)_2$	C_4H_9	–	273, **25**, *5*
$C_7H_9O_4S$	C_4H_9	–	258, **23**, *36*
$C_7H_9O_5$			
$OOCCH{=}CHCOOCH_2CH(OH)CH_3$	C_4H_9	–	276, **25**, *31*
$OOCCH{=}CHCOOCH_2CH_2OCH_3$	C_4H_9	–	276, **25**, *32*
$C_7H_{10}O_4$	C_4H_9	–	323, **30**, *8*
$C_7H_{10}O_4S_2$	C_4H_9	–	323, **30**, *12*
$C_7H_{11}NO_4$	C_4H_9	–	376, **43**, *19*
$C_7H_{11}O_2$	C_4H_9	–	253
$C_7H_{11}O_3$	C_4H_9	–	99, **19**, *14*
$C_7H_{11}O_4$	C_4H_9	–	100, **19**, *21*
$C_7H_{12}ClO_2$	C_4H_9	–	256, **23**, *11*
$C_7H_{12}O_2$	C_4H_9	–	53, **14**, *18*
$C_7H_{12}O_4$	C_4H_9	–	64, **14**, *77*
$C_7H_{12}O_5$	C_4H_9	–	65, **14**, *78*
$C_7H_{12}O_6$			
[1-*O*-Methyl-D-galacto-pyranosyl-3,4-diyl]dioxy	C_4H_9	–	65, **14**, *81*
[1-*O*-Methyl-D-galactopyranosyl-2,3-diyl]dioxy	C_4H_9	–	65, **14**, *82*
[1-*O*-Methyl-α-D-glucopyranosyl-2,3-diyl]-dioxy	C_4H_9	–	66, **14**, *83*
$C_7H_{13}O_2$	C_4H_9	–	128
$C_7H_{13}O_3$	C_4H_9	–	257, **23**, *22*
$C_7H_{14}O_2$			
$OC(CH_3)_2CH_2C(CH_3)_2O$	C_4H_9	–	55, **14**, *25*
$OCH_2C(CH_3)(C_3H_7)CH_2O$	C_4H_9	–	55, **14**, *26*
$C_7H_{15}NO_2$			
$O(CH_2)_2N(C_3H_7)(CH_2)_2O$	C_4H_9	–	72, **14**, *110*
			374, **43**, *9*
$OCH(CH_3)CH_2N(CH_3)CH_2CH$-$(CH_3)O$	C_4H_9	–	72, **14**, *114*
$C_7H_{15}N_2O$	C_4H_9	–	343, **32**, *26*
$C_7H_{16}NO$	C_4H_9	–	7, **1**, *41*
C_8 Cl_4O_4	C_4H_9	–	327, **30**, *33*
$C_8F_{15}O_2$	C_4H_9	–	255, **23**, *3*
$C_8H_2F_5O_3$	C_4H_9	–	257, **23**, *25*
$C_8H_3NO_6$	C_4H_9	–	327, **30**, *34*
$C_8H_4ClF_2O_3S$	C_4H_9	–	349, **34**, *13*
$C_8H_4Cl_3O_3$	C_4H_9	–	258, **23**, *29*
$C_8H_4F_3O_2S$	C_4H_9	–	99, **19**, *10*
$C_8H_4F_3O_3S$	C_4H_9	–	349, **34**, *14*
$C_8H_4O_4$			

$OOCC_6H_4COO$-2	C_4H_9	–	326, **30**, *31*
$OOCC_6H_4COO$-4	C_4H_9	–	326, **30**, *32*
$C_8H_5Cl_2O_3$	C_4H_9	–	258, **23**, *28*
$C_8H_6ClO_3$			
$OOCCH_2OC_6H_4Cl$-2	C_4H_9	–	257, **23**, *26*
$OOCCH_2OC_6H_4Cl$-4	C_4H_9	–	258, **23**, *27*
$C_8H_6NO_5$	C_4H_9	–	291, **27**, *29*
$C_8H_6O_3$	C_4H_9	–	331
$C_8H_6O_5$	C_4H_9	–	335, **31**, *10*
$C_8H_7AsO_4$	C_4H_9	–	367
C_8H_7BrNO	C_4H_9	–	343, **32**, *29*
C_8H_7ClNO	C_4H_9	–	343, **32**, *28*
$C_8H_7NO_2$	C_4H_9	–	74, **14**, *120*
$C_8H_7N_3O_2$	C_4H_9	–	73, **14**, *118*
$C_8H_7O_2$			
$OOCCH_2C_6H_5$	C_4H_9	–	261, **23**, *59*
$OOCC_6H_4CH_3$-4	C_4H_9	–	291, **27**, *33*
$C_8H_7O_2S$	C_4H_9	–	259, **23**, *38*
$C_8H_7O_3$			
$OC_6H_4COOCH_3$-2	C_4H_9	–	35, **12**, *22*
$OOCCH_2OC_6H_5$	C_4H_9	–	257, **23**, *24*
$OOCCH(C_6H_5)OH$	C_4H_9	–	262, **23**, *62*
$OOCC_6H_4OCH_3$-4	C_4H_9	–	289, **27**, *11*
C_8H_8NO	C_4H_9	–	343, **32**, *27*
$C_8H_8NO_2$			
$OOCC_6H_4NHCH_3$-2	C_4H_9	–	290, **27**, *19*
$ON(C_6H_5)COCH_3$	C_4H_9	–	339, **32**, *7*
$C_8H_8N_3O_2$	C_4H_9	–	36, **12**, *27*
$C_8H_8O_2$	C_4H_9	–	49, **14**, *6*
$C_8H_8O_3$	C_4H_9	–	335, **31**, *9*
$C_8H_8O_4$			
Cyclohex-4-en-1,2-diyl-bis(carbonyloxy)	C_4H_9	–	325, **30**, *26*
$OCH(C_6H_4OCH_3$-4)O-O	C_4H_9	–	335, **31**, *6*
$C_8H_9Br_2O_4$	C_4H_9	–	277, **25**, *51*
$C_8H_9Cl_2O_4$	C_4H_9	–	277, **25**, *48*
$C_8H_9N_2O$	C_4H_9	–	36, **12**, *26*
C_8H_9O	C_4H_9	–	7, **1**, *43*
$C_8H_9O_2$	C_4H_9	–	284, **26**, *3*
$C_8H_9O_4$	C_4H_9	–	275, **25**, *26*
$C_8H_{11}O_4$			
$OOCCH{=}CHCOOC_4H_9$	C_4H_9	–	273, **25**, *6*
$OOCCH{=}CHCOOCHCH_2(CH_3)_2$	C_4H_9	–	274, **25**, *7*

$C_8H_{11}O_4S$	C_4H_9	–	277, **25**, *43*
$C_8H_{11}O_5$	C_4H_9	–	276, **25**, *33*
$C_8H_{11}O_6$	C_4H_9	–	276, **25**, *39*
$C_8H_{12}O_6$	C_4H_9	–	52, **14**, *14*
$C_8H_{13}O_2$	C_4H_9	–	98, **19**, *3*
$C_8H_{13}O_4S$	C_4H_9	–	259, **23**, *41*
$C_8H_{14}NO_3$	C_4H_9	–	260, **23**, *53*
$C_8H_{14}NO_4$	C_4H_9	–	263, **23**, *71*
$C_8H_{14}O_4$	C_4H_9	–	56, **14**, *37*
$C_8H_{14}O_5$	C_4H_9	–	65, **14**, *79*
$C_8H_{15}O_2$			
$OOCC_7H_{15}$	C_4H_9	–	129/30
$OOCCH(C_2H_5)C_5H_9$	C_4H_9	–	130/1
$C_8H_{17}NO_2$			
$O(CH_2)_2N(C_4H_9)(CH_2)_2O$	C_4H_9	–	374, **43**, *10*
$O(CH_2)_2NCHCH_2(CH_3)_2$-$(CH_2)_2O$	C_4H_9	–	375, **43**, *11*
$O(CH_2)_2NC(CH_3)_3(CH_2)_2O$	C_4H_9	–	375, **43**, *12*
$C_8H_{17}O$			
OC_8H_{17}	C_4H_9	–	4, **1**, *11*
$OCH_2CH(C_2H_5)C_4H_9$	C_4H_9	–	5, **1**, *12*
$C_8H_{17}O_3P$	C_4H_9	–	359, **37**, *3*
$C_8H_{18}O_3P$	C_4H_9	–	355, **36**, *10*
$C_8H_{18}O_4P$	C_4H_9	–	356, **36**, *16*
$C_8H_{18}O_7P_2$	C_4H_9	–	361
$C_9H_2Cl_6O_4$	C_4H_9	–	326, **30**, *29*
$C_9H_3O_5$	C_4H_9	–	292, **27**, *44*
$C_9H_4BrN_2O_3$			
(7-Bromo-5-nitroquinolin-8-yl)oxy	C_4H_9	–	41, **13**, *13*
(5-Bromo-7-nitroquinolin-8-yl)oxy	C_4H_9	–	42, **13**, *14*
$C_9H_4Br_2NO$	C_4H_9	–	41, **13**, *9*
$C_9H_4Cl_2NO$	C_4H_9	–	40, **13**, *8*
$C_9H_4I_2NO$	C_4H_9	–	41, **13**, *10*
$C_9H_4N_3O_5$	C_4H_9	–	42, **13**, *15*
$C_9H_4O_6$	C_4H_9	–	327, **30**, *35*
$C_9H_5N_2O_3$			
(7-Nitroquinolin-8-yl)oxy	C_4H_9	–	41, **13**, *11*
(5-Nitroquinolin-8-yl)oxy	C_4H_9	–	41, **13**, *12*
$C_9H_6ClO_2$	C_4H_9	–	269, **24**, *6*
C_9H_6NO	C_4H_9	–	40, **13**, *7*
			369, **41**, *4*

			373, **43**, *4*
$C_9H_6NO_4$	C_4H_9	—	269, **24**, *9*
$C_9H_7O_2$			
$OOCCH{=}CHC_6H_5$	C_4H_9	—	269, **24**, *5*
$OOCC_6H_4CH{=}CH_2$-4	C_4H_9	—	291, **27**, *35*
$C_9H_7O_3$	C_4H_9	—	292, **27**, *38*
$C_9H_8ClO_3$	C_4H_9	—	258, **23**, *31*
$C_9H_8O_4$	C_4H_9	—	326, **30**, *28*
$C_9H_9ClNO_2$	C_4H_9	—	339, **32**, *6*
$C_9H_9FN_2O_6$	C_4H_9	—	61, **14**, *59*
$C_9H_9NO_2$	C_4H_9	—	73, **14**, *117*
$C_9H_9N_3O_2$	C_4H_9	—	73, **14**, *119*
$C_9H_9O_2S$	C_4H_9	—	258, **23**, *37*
$C_9H_9O_3$			
$OOCCH_2C_6H_4CH_3$-2	C_4H_9	—	258, **23**, *30*
$OOCC_6H_4C_2H_5$-4	C_4H_9	—	289, **27**, *12*
$C_9H_9O_3S$	C_4H_9	—	292, **27**, *37*
$C_9H_{10}N_2O_6$	C_4H_9	—	60, **14**, *58*
$C_9H_{10}N_3O_2$	C_4H_9	—	36, **12**, *28*
$C_9H_{11}N_3O_5$	C_4H_9	—	61, **14**, *60*
$C_9H_{11}O_2$	C_4H_9	—	334, **31**, *2*
$C_9H_{12}NO_2$	C_4H_9	—	270, **24**, *20*
$C_9H_{13}O_4$			
$OOCCH{=}CHCOOC_5H_{11}$	C_4H_9	—	274, **25**, *8*
$OOCCH{=}CHCOO(CH_2)_2$- $CH(CH_3)_2$	C_4H_9	—	274, **25**, *9*
$C_9H_{13}O_5$	C_4H_9	—	276, **25**, *34*
$C_9H_{14}O_4$	C_4H_9	—	323, **30**, *9*
$C_9H_{14}O_4S_2$	C_4H_9	—	324, **30**, *19*
$C_9H_{16}ClO_2$	C_4H_9	—	256, **23**, *12*
$C_9H_{17}O_2$			
$OOCC_8H_{17}$	C_4H_9	—	134
$CH_2C(CH_3)_2CH_2CH(CH_3)_2$	C_4H_9	—	135
C$_{10}H_4O_8S_2$	C_4H_9	—	59, **14**, *52*
$C_{10}H_5Cl_6O_4$	C_4H_9	—	284, **26**, *5*
$C_{10}H_5N_2O$	C_4H_9	—	30, **11**, *12*
$C_{10}H_6ClO_3$	C_4H_9	—	270, **24**, *14*
$C_{10}H_6NO_2$	C_4H_9	—	36, **12**, *30*
$C_{10}H_6NO_3$	C_4H_9	—	270, **24**, *21*
$C_{10}H_7O_3$	C_4H_9	—	270, **24**, *13*
$C_{10}H_8BrO_2$	C_4H_9	—	98, **19**, *7*

$C_{10}H_8ClO_2$	C_4H_9	–	98, **19**, *6*
$C_{10}H_8FO_2$	C_4H_9	–	98, **19**, *5*
$C_{10}H_8NO$	C_4H_9	–	42, **13**, *19*
			374, **43**, *5*
$C_{10}H_8O_2S$	C_4H_9	–	98, **19**, *8*
$C_{10}H_8O_8$	C_4H_9	–	325, **30**, *25*
$C_{10}H_9O_2$	C_4H_9	–	98, **19**, *4*
$C_{10}H_9O_3$	C_4H_9	–	269, **24**, *7*
$C_{10}H_9O_4$	C_4H_9	–	292, **27**, *40*
$C_{10}H_{10}BrN_5O_4$	C_4H_9	–	62, **14**, *68*
$C_{10}H_{10}ClO_4$			
$OC_6H_4COOCH_2CH(OH)$-			
CH_2Cl-2	C_4H_9	–	35, **12**, *23*
$OOCC_6H_4OCH_2CH(OH)$-			
CH_2Cl-2	C_4H_9	–	289, **27**, *13*
$C_{10}H_{10}NO_2$	C_4H_9	–	100, **19**, *24*
$C_{10}H_{10}NO_3$	C_4H_9	–	258, **23**, *32*
$C_{10}H_{10}NO_4$	C_4H_9	–	258, **23**, *33*
$C_{10}H_{10}N_4O_5$	C_4H_9	–	61, **14**, *61*
$C_{10}H_{11}BrN_5O_4$	C_4H_9	–	63, **14**, *69*
$C_{10}H_{11}Cl_2O_4$	C_4H_9	–	277, **25**, *49*
$C_{10}H_{11}N_2O_5$	C_4H_9	–	34, **12**, *15*
$C_{10}H_{11}N_5O_4$	C_4H_9	–	61, **14**, *62*
$C_{10}H_{11}N_5O_5$	C_4H_9	–	62, **14**, *67*
$C_{10}H_{11}O_4$			
$OOCCH=CHCOOC\equiv CC_4H_9$	C_4H_9	–	276, **25**, *29*
[(Bicyclo[2.2.1]hept-5-en-2-yl)carbonyl]oxy	C_4H_9	–	284, **26**, *4*
$C_{10}H_{11}O_6$			
$OOCCH=CHCOOCH_2CH_2$-			
$C(CH_3)=CH_2$	C_4H_9	–	276, **25**, *41*
$OOCCH=CHCOOCH(CH_3)CH_2$-			
$OOCCH=CH_2$	C_4H_9	–	276, **25**, *42*
$C_{10}H_{12}NO_3S$	C_4H_9	–	348, **34**, *3*
$C_{10}H_{12}N_3O_3$	C_4H_9	–	263, **23**, *73*
$C_{10}H_{12}O_2$	C_4H_9	–	59, **14**, *50*
$C_{10}H_{12}O_3$			
$OCH(CH_2OCH_2C_6H_5)CH_2O$	C_4H_9	–	49, **14**, *5*
$OCH_2CH(OCH_2C_6H_5)CH_2O$	C_4H_9	–	55, **14**, *27*
$C_{10}H_{13}NO_2$	C_4H_9	–	72, **14**, *112*
			370, **41**, *9*
$C_{10}H_{13}O$	C_4H_9	–	34, **12**, *7*
$C_{10}H_{13}O_4$	C_4H_9	–	275, **25**, *20*

$C_{10}H_{15}O_5$	C_4H_9	—	276, **25**, *35*
$C_{10}H_{15}O_6$			
$O(CH_2)_3OOCCH{=}CHCOO-(CH_2)_3OH$	C_4H_9	—	5, **1**, *22*
$OOCCH{=}CHCOOCH(CH_3)CH_2O-CH_2CH(OH)CH_3$	C_4H_9	—	276, **25**, *40*
$C_{10}H_{16}O_4$	C_4H_9	—	323, **30**, *10*
$C_{10}H_{16}O_6$	C_4H_9	—	50, **14**, *7*
$C_{10}H_{17}O_2$	C_4H_9	—	99, **19**, *13*
$C_{10}H_{17}O_4$	C_4H_9	—	263, **23**, *75*
$C_{10}H_{17}O_4S$	C_4H_9	—	259, **23**, *42*
$C_{10}H_{18}O_4S_4$	C_4H_9	—	56, **14**, *36*
$C_{10}H_{19}NO_2$	C_4H_9	—	370, **41**, *8*
$C_{10}H_{19}O_2$			
$OOCC_9H_{19}$	C_4H_9	—	135
$OOCCH_2CH_2C(C_2H_5)_3$	C_4H_9	—	135
$C_{10}H_{19}O_2S$	C_4H_9	—	258, **23**, *35*
$C_{10}H_{19}O_3$	C_4H_9	—	6, **1**, *26*
$C_{10}H_{21}O$	C_4H_9	—	5, **1**, *13*
$C_{10}H_{21}O_2Si$	C_4H_9	—	263, **23**, *77*
$C_{10}H_{22}O_2P$	C_4H_9	—	354, **36**, *3*
C$_{11}H_6NO_4$			
[3-(2,5-Dihydro-2,5-dioxopyrol-1-yl)phenylcar-bonyl]oxy	C_4H_9	—	290, **27**, *21*
[4-(2,5-Dihydro-2,5-dioxopyrol-1-yl)phenylcar-bonyl]oxy	C_4H_9	—	290, **27**, *22*
$C_{11}H_8ClO_3$	C_4H_9	—	270, **24**, *18*
$C_{11}H_8NO_2$	C_4H_9	—	43, **13**, *20*
$C_{11}H_8NO_4$	C_4H_9	—	260, **23**, *55*
$C_{11}H_9FeO_2$	C_4H_9	—	299, **28**, *5*
$C_{11}H_9O_3$	C_4H_9	—	270, **24**, *16*
$C_{11}H_9O_4$			
$OOCCH{=}CHCOC_6H_4OCH_3$-4	C_4H_9	—	270, **24**, *15*
$OOCCH{=}CHCOOCH_2C_6H_5$	C_4H_9	—	275, **25**, *23*
$C_{11}H_{10}BrN_6O_4$	C_4H_9	—	63, **14**, *70*
$C_{11}H_{10}O_4S_2$	C_4H_9	—	323, **30**, *13*
$C_{11}H_{11}O_3$	C_4H_9	—	100, **19**, *20*
$C_{11}H_{11}O_3S$	C_4H_9	—	260, **23**, *51*
$C_{11}H_{12}NO_2$	C_4H_9	—	269, **24**, *8*
$C_{11}H_{12}NO_3$	C_4H_9	—	260, **23**, *54*
$C_{11}H_{13}O_2$	C_4H_9	—	291, **27**, *34*

$C_{11}H_{14}N_3O_4$	C_4H_9	—	7, **1**, *46*
$C_{11}H_{14}O_8S_2$	C_4H_9	—	324, **30**, *16*
$C_{11}H_{15}NO_2$	C_4H_9	—	72, **14**, *113*
$C_{11}H_{15}O_4$			
[2-[(2-Methylcyclohexyloxy)carbonoyl]ethenyl-carbonyl]oxy	C_4H_9	—	275, **25**, *21*
[[2-(Methoxycarbonyl)-4,5-dimethylcyclohex-4-enyl]carbonyl]oxy	C_4H_9	—	284, **26**, *1*
$C_{11}H_{17}OSi$	C_4H_9	—	371, **41**, *15*
$C_{11}H_{19}O_2$	C_4H_9	—	270, **24**, *12*
$C_{11}H_{19}O_4$	C_4H_9	—	100, **19**, *22*
$C_{12}H_8CoO_4$	C_4H_9	—	328, **30**, *43*
$C_{12}H_8FeO_4$	C_4H_9	—	328, **30**, *42*
$C_{12}H_8N_3O$	C_4H_9	—	36, **12**, *29*
$C_{12}H_{10}OP$	C_4H_9	—	354, **36**, *1*
$C_{12}H_{10}O_2P$	C_4H_9	—	354, **36**, *6*
$C_{12}H_{10}O_3P$			
$OP(OC_6H_5)_2$	C_4H_9	—	355, **36**, *11*
$OP(O)(OC_6H_5)C_6H_5$	C_4H_9	—	355, **36**, *14*
$C_{12}H_{10}O_4P$	C_4H_9	—	356, **36**, *17*
$C_{12}H_{11}N_2O_2$	C_4H_9	—	100, **19**, *25*
$C_{12}H_{11}O_5$	C_4H_9	—	276, **25**, *38*
$C_{12}H_{12}BrN_5O_5$	C_4H_9	—	63, **14**, *71*
$C_{12}H_{13}BrN_6O_4$	C_4H_9	—	63, **14**, *72*
$C_{12}H_{13}O_2$	C_4H_9	—	99, **19**, *15*
$C_{12}H_{13}O_4$	C_4H_9	—	292, **27**, *41*
$C_{12}H_{14}N_3O_5$	C_4H_9	—	261, **23**, *58*
$C_{12}H_{14}O_4$	C_4H_9	—	56, **14**, *38*
	C_4H_9	—	64, **14**, *74*
$C_{12}H_{15}N_5O_4$	C_4H_9	—	61, **14**, *63*
$C_{12}H_{15}O_5$	C_4H_9	—	275, **25**, *22*
$C_{12}H_{17}Br_2O_4$	C_4H_9	—	277, **25**, *52*
$C_{12}H_{17}Cl_2O_4$	C_4H_9	—	277, **25**, *50*
$C_{12}H_{17}O_5$	C_4H_9	—	276, **25**, *37*
$C_{12}H_{19}N_2O_2$	C_4H_9	—	31, **11**, *15*
$C_{12}H_{19}O_4$			
$OOCCH{=}CHCOOC_8H_{17}$	C_4H_9	—	274, **25**, *10*
$OOCCH{=}CHCOOCH_2CH(C_2H_5)$-$C_4H_9$	C_4H_9	—	274, **25**, *11*
$OOCCH{=}CHCOO(CH_2)_5$-$CH(CH_3)_2$	C_4H_9	—	274, **25**, *12*

$OOCCH{=}CHCOOCH_2C(CH_3)_2$- $CH_2CH(CH_3)_2$	C_4H_9	—	274, **25**, *13*
$C_{12}H_{19}O_7$	C_4H_9	—	6, **1**, *32*
$C_{12}H_{21}O_4$	C_4H_9	—	263, **23**, *69*
$C_{12}H_{21}O_6$	C_4H_9	—	6, **1**, *31*
$C_{12}H_{22}O_2$	C_4H_9	—	53, **14**, *19*
$C_{12}H_{23}NO_2$	C_4H_9	—	73, **14**, *115*
$C_{12}H_{23}O_2$			
$OOCC_{11}H_{23}$	C_4H_9	—	135/249 370, **41**, *11* 372
$OOC(CH_2)_7C(CH_3)_3$	C_4H_9	—	250
$C_{12}H_{25}O$	C_4H_9	—	5, **1**, *14*
$C_{12}H_{26}BO_2$	C_4H_9	—	351, **35**, *1*
$C_{12}H_{26}O_2P$	C_4H_9	—	354, **36**, *4*
$C_{13}H_8N_3O_6$	C_4H_9	—	340, **32**, *12*
$C_{13}H_9AsClNO_3$	C_4H_9	—	364, **39**, *18*
$C_{13}H_9AsN_2O_5$			
$OAs(O)(C_6H_4NO_2$-2)O	C_4H_9	—	364, **39**, *19*
$OAs(O)(C_6H_4NO_2$-3)O	C_4H_9	—	364, **39**, *20*
$C_{13}H_9ClNO_2$			
$ON(C_6H_5)COC_6H_4Cl$-4	C_4H_9	—	340, **32**, *9*
$ON(C_6H_4Cl$-4)COC_6H_5	C_4H_9	—	341, **32**, *13*
$C_{13}H_9FNO$	C_4H_9	—	35, **12**, *24*
$C_{13}H_9INO_2$	C_4H_9	—	340, **32**, *10*
$C_{13}H_9N_2O_4$	C_4H_9	—	340, **32**, *11*
$C_{13}H_9O_2$	C_4H_9	—	35, **12**, *18*
$C_{13}H_{10}NO_2$			
$OOCC_6H_4NHC_6H_5$-2	C_4H_9	—	290, **27**, *20*
$ON(C_6H_5)COC_6H_5$	C_4H_9	—	340, **32**, *8*
$C_{13}H_{11}O_4$	C_4H_9	—	276, **25**, *28*
$C_{13}H_{13}NO_6$	C_4H_9	—	322, **30**, *5*
$C_{13}H_{13}N_2O_2$	C_4H_9	—	100, **19**, *26*
$C_{13}H_{13}O_3$	C_4H_9	—	270, **24**, *17*
$C_{13}H_{14}O_4S_2$	C_4H_9	—	324, **30**, *20*
$C_{13}H_{14}O_5S_2$	C_4H_9	—	324, **30**, *21*
$C_{13}H_{16}O_5$	C_4H_9	—	64, **14**, *76*
$C_{13}H_{17}O_4$	C_4H_9	—	285, **26**, *6*
$C_{13}H_{17}O_4S$	C_4H_9	—	285, **26**, *11*
$C_{13}H_{19}NO_7P_2S_2$	C_4H_9	—	360
$C_{13}H_{20}O_2P$	C_4H_9	—	354, **36**, *5*

$C_{14}H_{23}O_5$	C_4H_9	–	276, **25**, *36*
$C_{14}H_{23}O_6S_3Sb$	C_4H_9	–	324, **30**, *15*
$C_{14}H_{25}O_4$	C_4H_9	–	263, **23**, *76*
$C_{14}H_{25}O_4S$	C_4H_9	–	259, **23**, *43*
$C_{14}H_{27}O_2$	C_4H_9	–	250
$C_{14}H_{27}O_3$	C_4H_9	–	6, **1**, *24*
$C_{15}H_9O_3$	C_4H_9	–	40, **13**, *3*
$C_{15}H_{10}N_3O$	C_4H_9	–	42, **13**, *16*
$C_{15}H_{10}O_2S$	C_4H_9	–	99, **19**, *12*
$C_{15}H_{11}O_2$	C_4H_9	–	99, **19**, *11*
$C_{15}H_{12}N_2O_2$	C_4H_9	–	74, **14**, *122*
$C_{15}H_{12}N_4O_2S$	C_4H_9	–	75, **14**, *127*
$C_{15}H_{14}O_2$	C_4H_9	–	60, **14**, *54*
$C_{15}H_{15}O$	C_4H_9	–	34, **12**, *16*
$C_{15}H_{19}N_5O_4$	C_4H_9	–	62, **14**, *64*
$C_{15}H_{19}N_5O_5$	C_4H_9	–	62, **14**, *65*
$C_{15}H_{19}O_4$	C_4H_9	–	285, **26**, *10*
$C_{15}H_{21}O_2S$	C_4H_9	–	289, **27**, *15*
$C_{15}H_{21}O_3$	C_4H_9	–	291, **27**, *36*
$C_{15}H_{22}O_6$	C_4H_9	–	66, **14**, *84*
$C_{15}H_{22}O_8$	C_4H_9	–	58, **14**, *47*
$C_{15}H_{23}O$	C_4H_9	–	34, **12**, *10*
$C_{15}H_{23}O_4P$	C_4H_9	–	359, **37**, *5*
$C_{15}H_{26}N_3O$	C_4H_9	–	34, **12**, *13*
$C_{16}H_{10}NO_2$	C_4H_9	–	43, **13**, *21*
$C_{16}H_{10}N_2O_4$	C_4H_9	–	328, **30**, *40*
$C_{16}H_{10}N_3O_3$	C_4H_9	–	42, **13**, *17*
$C_{16}H_{12}O_4S_2$	C_4H_9	–	71, **14**, *106*
$C_{16}H_{13}N_2O_2$	C_4H_9	–	342, **32**, *19*
$C_{16}H_{14}N_2O_2$	C_4H_9	–	74, **14**, *123*
$C_{16}H_{14}N_2O_4$	C_4H_9	–	337/8
$C_{16}H_{14}O_4$	C_4H_9	–	58, **14**, *44*
$C_{16}H_{14}O_8Ti$	C_4H_9	–	368
$C_{16}H_{15}O_4$	C_4H_9	–	6, **1**, *23*
$C_{16}H_{17}O_2P$	C_4H_9	–	375, **43**, *14*
$C_{16}H_{21}O_4$			
$OOCC_6H_4COOC_8H_{17}$-2	C_4H_9	–	292, **27**, *42*
$OOCC_6H_4COOCH_2CH(C_2H_5)$-$C_4H_9$-4	C_4H_9	–	292, **27**, *43*
$C_{16}H_{23}O_3$	C_4H_9	–	261, **23**, *60*

$C_{16}H_{23}O_7S$	C_4H_9	–	277, **25**, *53*
$C_{16}H_{25}O$	C_4H_9	–	34, **12**, *11*
$C_{16}H_{27}O_4$	C_4H_9	–	274, **25**, *15*
$C_{16}H_{28}O_4S_2$	C_4H_9	–	324, **30**, *18*
$C_{16}H_{29}O_4$	C_4H_9	–	263, **23**, *70*
$C_{16}H_{30}NO_3$	C_4H_9	–	263, **23**, *72*
$C_{16}H_{31}O_2$	C_4H_9	–	250
$C_{16}H_{31}O_2S$	C_4H_9	–	259, **23**, *50*
$C_{16}H_{31}O_5$	C_4H_9	–	256, **23**, *17*
$C_{16}H_{34}BO_3$	C_4H_9	–	351, **35**, *2*
$C_{17}H_{11}FNO$	C_4H_9	–	36, **12**, *31*
$C_{17}H_{11}O_4$	C_4H_9	–	270, **24**, *19*
$C_{17}H_{12}ClN_2O_2$	C_4H_9	–	100, **19**, *28*
$C_{17}H_{12}NO_4$	C_4H_9	–	261, **23**, *57*
$C_{17}H_{12}N_3O_3$	C_4H_9	–	42, **13**, *18*
$C_{17}H_{13}N_2O_2$	C_4H_9	–	100, **19**, *27*
$C_{17}H_{14}FeN_3O$	C_4H_9	–	332
$C_{17}H_{14}N_2O_2$	C_4H_9	–	74, **14**, *121*
$C_{17}H_{15}N_5O_5$	C_4H_9	–	62, **14**, *66*
$C_{17}H_{16}N_4O_2S$	C_4H_9	–	76, **14**, *128*
$C_{17}H_{18}O_2$	C_4H_9	–	50, **14**, *11*
$C_{17}H_{25}O_3$	C_4H_9	–	262, **23**, *61*
$C_{17}H_{25}O_4$			
[[3-(Octyloxycarbonyl)-bicyclo[2.2.1]hept-5-en-2-yl]carbonyl]oxy	C_4H_9	–	285, **26**, *7*
[[3-[(2-Ethylhexyl)-oxycarbonyl]bicyclo-[2.2.1]hept-5-en-2-yl]-carbonyl]oxy	C_4H_9	–	285, **26**, *8*
$C_{18}H_{13}O_2P$	C_4H_9	–	375, **43**, *15*
$C_{18}H_{14}ClFeN_2O$	C_4H_9	–	332
$C_{18}H_{14}FeN_3O_3$	C_4H_9	–	332
$C_{18}H_{15}FeN_2O$	C_4H_9	–	332
$C_{18}H_{15}FeN_2O_2$	C_4H_9	–	332
$C_{18}H_{15}OSi$	C_4H_9	–	352
$C_{18}H_{22}NO_3$	C_4H_9	–	270, **24**, *22*
$C_{18}H_{23}NO_6$	C_4H_9	–	66, **14**, *85*
$C_{18}H_{31}O_7$	C_4H_9	–	6, **1**, *33*
$C_{18}H_{33}O_2$	C_4H_9	–	269, **24**, *11*

$C_{18}H_{33}O_3$	C_4H_9	–	257, **23**, *19*
$C_{18}H_{33}O_4$	C_4H_9	–	257, **23**, *20*
$C_{18}H_{33}O_4S$	C_4H_9	–	259, **23**, *44*
$C_{18}H_{33}O_6$	C_4H_9	–	39, **13**, *1*
$C_{18}H_{35}O_2$			
$OOCC_{17}H_{35}$	C_4H_9	–	250
$OOC(CH_2)_{14}CH(CH_3)_2$	C_4H_9	–	251
$C_{18}H_{35}O_3$	C_4H_9	–	256, **23**, *16*
$C_{19}H_{14}AsO_2$	C_4H_9	–	291, **27**, *31*
$C_{19}H_{14}BBrNO_2$	C_4H_9	–	351, **35**, *5*
$C_{19}H_{14}BClNO_2$	C_4H_9	–	351, **35**, *4*
$C_{19}H_{14}BFNO_2$	C_4H_9	–	351, **35**, *3*
$C_{19}H_{14}BN_2O_4$	C_4H_9	–	351, **35**, *6*
$C_{19}H_{20}O_5$	C_4H_9	–	60, **14**, *57*
$C_{19}H_{32}O_4P$	C_4H_9	–	355, **36**, *13*
$C_{19}H_{35}O_3$	C_4H_9	–	30, **11**, *13*
$C_{20}H_{14}N_2O_2$	C_4H_9	–	75, **14**, *126*
$C_{20}H_{20}O_6$	C_4H_9	–	70, **14**, *99*
$C_{20}H_{22}N_2O_2$	C_4H_9	–	75, **14**, *124*
$C_{20}H_{26}O_6$	C_4H_9	–	66, **14**, *86*
$C_{20}H_{35}O_4$			
$OOCCH{=}CHCOOC_{16}H_{33}$	C_4H_9	–	274, **25**, *16*
$OOCCH{=}CHCOOCH_2CH(C_6H_{13})$-$C_8H_{17}$	C_4H_9	–	274, **25**, *17*
$C_{21}H_{14}O_5S_2$	C_4H_9	–	328, **30**, *39*
$C_{21}H_{18}AsO_2$	C_4H_9	–	291, **27**, *32*
$C_{21}H_{25}O_3$	C_4H_9	–	35, **12**, *21*
$C_{21}H_{28}O_3P$	C_4H_9	–	355, **36**, *7*
$C_{21}H_{31}O_4$	C_4H_9	–	285, **26**, *9*
$C_{21}H_{31}O_5S_2$	C_4H_9	–	259, **23**, *45*
$C_{21}H_{35}O$	C_4H_9	–	34, **12**, *12*
$C_{21}H_{41}O_2$	C_4H_9	–	251
$C_{22}H_{37}O_4$	C_4H_9	–	275, **25**, *27*
$C_{22}H_{37}O_5$	C_4H_9	–	277, **25**, *44*
$C_{22}H_{39}O_4$	C_4H_9	–	275, **25**, *18*
$C_{22}H_{40}O_6S_3Sb$	C_4H_9	–	259, **23**, *39*
$C_{22}H_{43}O_2$	C_4H_9	–	251

$C_{23}H_{16}BBrNO_2$	C_4H_9	–	351, **35**, *9*
$C_{23}H_{16}BClNO_2$	C_4H_9	–	351, **35**, *8*
$C_{23}H_{16}BFNO_2$	C_4H_9	–	351, **35**, *7*
$C_{23}H_{16}BN_2O_4$	C_4H_9	–	351, **35**, *10*
$C_{23}H_{24}O_7$	C_4H_9	–	66, **14**, *86*
$C_{23}H_{26}O_6$			
[1-*O*-(2-Propenyl)-2,6-bis-*O*-(phenylmethyl)-α-D-galactopyranyosyl-3,4-diyl]dioxy	C_4H_9	–	67, **14**, *87*
			369, **41**, *5*
[6-*O*-(2-Propenyl)-1,2-bis-*O*-(phenylmethyl)-α-D-galactopyranosyl-3,4-diyl]dioxy	C_4H_9	–	67, **14**, *88*
[2-*O*-(2-Propenyl)-1,6-bis-*O*-(phenylmethyl)-α-D-galactopyranosyl-3,4-diyl]dioxy	C_4H_9	–	67, **14**, *89*
$C_{23}H_{41}O_6S_2$	C_4H_9	–	259, **23**, *47*
$C_{24}H_{20}O_3Si_2$	C_4H_9	–	352
$C_{24}H_{43}O_4$	C_4H_9	–	275, **25**, *19*
$C_{25}H_{47}O_4S_2$	C_4H_9	–	259, **23**, *48*
$C_{26}H_{26}O_6$	C_4H_9	–	68, **14**, *92*
$C_{26}H_{30}O_6$	C_4H_9	–	70, **14**, *101*
$C_{27}H_{28}O_6$			
[1,2,3-Tris-*O*-(phenyl-methyl)-α-D-galactopyranosyl-3,4-diyl]dioxy	C_4H_9	–	67, **14**, *90*
[1-*O*-Methyl-6-*O*-[(triphenyl)methyl]-α-D-galactopyranosyl-3,4-diyl]-dioxy	C_4H_9	–	68, **14**, *91*
[2,4,6-Tris-*O*-(phenylmethyl)-β-D-mannopyranosyl-1,2-diyl]dioxy	C_4H_9	–	68, **14**, *93*
$C_{29}H_{55}O_4S_2$	C_4H_9	–	259, **23**, *49*
$C_{30}H_{18}O_8$	C_4H_9	–	327, **30**, *38*
$C_{30}H_{22}O_6Ti$	C_4H_9	–	368
$C_{30}H_{26}N_2O_2$	C_4H_9	–	75, **14**, *125*
$C_{32}H_{20}O_4$	C_4H_9	–	327, **30**, *37*
$C_{32}H_{47}O_4$	C_4H_9	–	262, **23**, *64*
$C_{33}H_{49}O_4$	C_4H_9	–	262, **23**, *65*
$C_{33}H_{49}O_{12}$	C_4H_9	–	277, **25**, *46*

Table of Conversion Factors

Following the notation in Landolt-Börnstein [7], values that have been fixed by convention are indicated by a bold-face last digit. The conversion factor between calorie and Joule that is given here is based on the thermochemical calorie, cal_{thch}, and is defined as 4.1840 J/cal. However, for the conversion of the "Internationale Tafelkalorie", cal_{IT}, into Joule, the factor 4.1868 J/cal is to be used [1, p. 147]. For the conversion factor for the British thermal unit, the Steam Table Btu, BTU_{ST}, is used [1, p. 95].

Force	N	dyn	kp
1 N (Newton)	1	10^5	0.1019716
1 dyn	10^{-5}	1	1.019716×10^{-6}
1 kp	9.80665	9.80665×10^5	1

Pressure	Pa	bar	kp/m^2	at	atm	Torr	lb/in^2
1 Pa (Pascal) = 1 N/m^2	1	10^{-5}	1.019716×10^{-1}	1.019716×10^{-5}	0.986923×10^{-5}	0.750062×10^{-2}	145.0378×10^{-6}
1 bar = 10^6 dyn/cm^2	10^5	1	10.19716×10^3	1.019716	0.986923	750.062	14.50378
1 kp/m^2 = 1 mm H_2O	9.80665	0.980665×10^{-4}	1	10^{-4}	0.967841×10^{-4}	0.735559×10^{-1}	1.422335×10^{-3}
1 at = 1 kp/cm^2	0.980665×10^5	0.980665	10^4	1	0.967841	735.559	14.22335
1 atm = 760 Torr	1.01325×10^5	1.01325	1.033227×10^4	1.033227	1	760	14.69595
1 Torr = 1 mm Hg	133.3224	1.333224×10^{-3}	13.59510	1.359510×10^{-3}	1.315789×10^{-3}	1	19.33678×10^{-3}
1 lb/in^2 = 1 psi	6.89476×10^3	68.9476×10^{-3}	703.069	70.3069×10^{-3}	68.0460×10^{-3}	51.7149	1

Work, Energy, Heat	J	kWh	kcal	Btu	MeV
1 J (Joule) = 1 Ws = 1 Nm = 10^7 erg	1	2.778×10^{-7}	2.39006×10^{-4}	9.4781×10^{-4}	6.242×10^{12}
1 kWh	3.6×10^{6}	1	860.4	3412.14	2.247×10^{19}
1 kcal	4184.0	1.1622×10^{-3}	1	3.96566	2.6117×10^{16}
1 Btu (British thermal unit)	1055.06	2.93071×10^{-4}	0.25164	1	6.5858×10^{15}
1 MeV	1.602×10^{-13}	4.450×10^{-20}	3.8289×10^{-17}	1.51840×10^{-16}	1

1 eV ≙ 23.0578 kcal/mol = 96.473 kJ/mol

Power	kW	PS	kp m/s	kcal/s
1 kW = 10^{10} erg/s	1	1.35962	101.972	0.239006
1 PS	0.73550	1	75	0.17579
1 kp m/s	9.80665×10^{-3}	0.01333	1	2.34384×10^{-3}
1 kcal/s	4.1840	5.6886	426.650	1

References:

[1] A. Sacklowski, Die neuen SI-Einheiten, Goldmann, München 1979. (Conversion tables in an appendix.)

[2] International Union of Pure and Applied Chemistry, Manual of Symbols and Terminology for Physicochemical Quantities and Units, Pergamon, London 1979; Pure Appl. Chem. **51** [1979] 1/41.

[3] The International System of Units (SI), National Bureau of Standards Spec. Publ. 330 [1972].

[4] H. Ebert, Physikalisches Taschenbuch, 5th Ed., Vieweg, Wiesbaden 1976.

[5] Kraftwerk Union Information, Technical and Economic Data on Power Engineering, Mülheim/Ruhr 1978.

[6] E. Padelt, H. Laporte, Einheiten und Größenarten der Naturwissenschaften, 3rd Ed., VEB Fachbuchverlag, Leipzig 1976.

[7] Landolt-Börnstein, 6th Ed., Vol. II, Pt. 1, 1971, pp. 1/14.

[8] ISO Standards Handbook 2, Units of Measurement, 2nd Ed., Geneva 1982.